WATER RESOURCES PERSPECTIVES: EVALUATION, MANAGEMENT AND POLICY

DEVELOPMENTS IN WATER SCIENCE, 50

VOLS. 1 - 40, 42, 45 ARE OUT OF PRINT

OTHER TITLES IN THIS SERIES

41 D. STEPHENSON AND M.S. PETERSON
WATER RESOURCES DEVELOPMENT IN DEVELOPING COUNTRIES
43 J. ZÁRUBA
WATER HAMMER IN PIPE-LINE SYSTEMS
44 W.E. KELLY AND S.MARES (EDITORS)
APPLIED GEOPHYSICS IN HYDROGEOLOGICAL AND ENGINEERING PRACTICE
46 G.A. BRUGGEMAN
ANALYTICAL SOLUTIONS OF GEOHYDROLOGICAL PROBLEMS
47 S.M. HASSANIZADEH, R.J. SCHOTTING, W.G. GRAY AND G.F. PINDER
COMPUTATIONAL METHODS IN WATER RESOURCES
49 P. WILDERER AND S. WUERTZ
MODERN SCIENTIFIC TOOLS IN BIOPROCESSING

WATER RESOURCES PERSPECTIVES: EVALUATION, MANAGEMENT AND POLICY

Abdulrahman S. Alsharhan
Professor of Geology and Dean
Faculty of Science
United Arab Emirates University
P.O. Box: 17551, Al-Ain
United Arab Emirates

Warren W. Wood
Research Hydrologist
MS 430, National Center
U.S. Geological Survey
12201 Sunrise Valley Drive
Reston, VA 20192, U.S.A.

2003
ELSEVIER
Amsterdam - Boston - Heidelberg - London - New York - Oxford
Paris - San Diego - San Francisco - Singapore - Sydney - Tokyo

ELSEVIER B.V.
Sara Burgerhartstraat 25
P.O. Box 211, 1000 AE
Amsterdam, The Netherlands

ELSEVIER Inc.
525 B Street, Suite 1900
San Diego, CA 92101-4495
USA

ELSEVIER Ltd
The Boulevard, Langford Lane
Kidlington, Oxford OX5 1GB
UK

ELSEVIER Ltd
84 Theobalds Road
London WC1X 8RR
UK

First edition 2003
Second impression 2004

Library of Congress Cataloging in Publication Data
A catalog record is available from the Library of Congress.

British Library Cataloguing in Publication Data
A catalogue record is available from the British Library.

ISBN: 0-444-51508-9
ISSN: 0167-5648

∞ The paper used in this publication meets the requirements of ANSI/NISO Z39.48-1992 (Permanence of Paper).
Printed in The Netherlands.

Preface

Water resources and management activities played a central role in the development of the earliest known civilizations, and to this day they remain factors of the highest importance in the sustainable development of nations. Although fresh water exists both in the form of surface water and ground water, reliance in arid regions is mainly on ground water. It is therefore essential that the water held in these regions be studied, protected, and developed with the greatest care. The imbalance between availability and demand is expected to grow as population increases, pollution intensifies, and global climate change impacts desertification.

All plans, action programs, and measures should start with a careful assessment of sources and availability of fresh water. Efficient water resources management coupled with appropriate physical planning, can contribute to the development of dry lands in a sustainable manner, reducing the over-exploitation of water resources to unsustainable levels. Development and implementation of integrated policies and strategies to resolve the complex and interrelated problems must be an integral part of conservation and management.

The objectives of this publication is to assemble current knowledge on water resources management strategies; identify approaches for sustainable water resources development and prioritize issues and challenges; and evaluate the future development of conjunctive use required for the appropriate exploitation of available water resources. The papers selected for this volume illustrate some of these generalizations and are organized under three themes: Water resources management and policy, arid and semi-arid regional water resources management and policy, and science and engineering advances in support of water resources management.

The Editors
A.S. ALSHARHAN (UAE) & W.W. WOOD (USA)
May 2003

Acknowledgement

We would like to express our sincere gratitude and thanks to the General **H.H. Sheikh Mohamed Bin Rashid Al Maktoum**, Crown Prince of Dubai, UAE Defense Minister, Patron of the Zayed International Prize for the Environment, for his unceasing support and inspiration that made this event a great success.

We would like to thank **H.H. Sheikh Nahyan Mubarak Al Nahyan**, Minister of Higher Education and Scientific Research and Chancellor of the United Arab Emirates University for his inspiration, encouragement and support.

Special thanks go to the United Arab Emirates University and Zayed International Prize for the Environment for helping in the preparation and follow up. We also thank the United States Geological Survey for their scientific supports.

Our sincere appreciation goes to all of the authors of this volume for their efforts in writing and revising their manuscripts to meet our deadlines and the publisher's standard.

The authors and editors of this book would like to thank Dr. Mohamed Ahmed Bin Fahad, Chairman of the Higher Committee of the Zayed International Prize for the Environment, and Dr. Eissa M. Abdellatif for their inspiration and encouragement. Without their support, this publication would not have been possible.

This volume fully supported by **Zayed International Prize for the Environment**, Dubai, UAE. (For more information, please visit the website: **www.zayedprize.org**). This Prize is worth US$ 1 million, and considered the largest and most valuable environment prize in the world. Each winner will also receive a Trophy and a Zayed Prize Diploma from the Crown Prince of Dubai at a special award ceremony to be held in Dubai. The aim of the Prize is to recognize and promote pioneering contributions in the field of the environment in accordance with the philosophy and vision of **H.H. Sheikh Zayed Bin Sultan Al Nahyan**, President of the United Arab Emirates.

We greatly appreciate the effort of Mr. Mohamed Shahid who assisted us in more ways than could be imagined. He processed the chapters for this volume from inception to final completion, incorporated the authors' changes, and handled all correspondences with the authors. We would also like to express our thanks to Mr. Hamdi Kandil for drafting all the figures, arranging them in proper position in this book, and producing the final camera-ready copy of this volume.

We thank Elsevier Science Publishers for their patience and encouragement from the inception of this book to its completion.

As editors, it is our privilege to present this volume to all of you. It is a worthy publication for geoscientists and hydrologists working in comparable areas around the world.

We dedicate this publication to General **H.H. Sheikh Mohamed Bin Rashid Al Maktoum**, Crown Prince of Dubai, UAE Minister of Defense, and Patron of the Zayed International Prize for the Environment.

Table of Contents

SCIENCE AND ENGINEERING ADVANCES IN SUPPORT OF WATER RESOURCES MANAGEMENT

INTRODUCTION

Managing and Allocating Water Resources: Adopting the Integrated Water Resource Management Approach

STEPHEN J. TOOPE
Faculty of Law and Institute of Comparative Law, McGill University, Canada. E-mail: stephen.toope@mcgill.ca

KEN RAINWATER
Texas Tech University, Water Resources Center, Box 41022, Lubbock, Texas 79409-1022, U.S.A.
E-mail: kenrainwater@ttu.edu

TONY ALLAN
SOAS/King's College London, Water Research Group, SOAS, Thornhaugh Street, London WC1H 0XG, U.K.
Email: tony.allan@soas.ac.uk

Water is elemental. It is integral to everything that is living and has made a contribution to everything to which the living world relates. These universal features of water can, however, be susceptible to de-constructive scientific analysis. Water can also be territorialized environmentally, politically and legally. It can also be privatized and subjected to separate legal regimes. The disciplines and professions concerned with water are legion and address separate features of water. The users of water include every individual and institution in the world throughout history. Such universal relevance and significance has not escaped the concern and sometimes the pre-occupation of political interests and the institutions that formalize those interests.

The hydrologic cycle is dynamic spatially and temporally. Water flows in response to gravity, evaporation and transpiration. Water presents itself to different potential users at different moments in its environmental cycling. The environmental outcomes of the competition for water in Nature and in the human economy create a very complex set of dependent relationships. In Nature the competition is substantially defined by gravity, transpiration and evapotranspiration and the ecological relations of the biosphere. In the human economy the competition is determined largely by power relations, but can be influenced by normative considerations as they affect the construction of identities and interests (Brunnée and Toope, 2002).

Environmental scientists and engineers have been putting integration back in for almost four decades. The water management literature of the 1970s is rich in material on river basin management. The international agencies also adopted the hydrologic logic of the river basin (Brunnée and Toope, 1997). Environmental holism had rational appeal and provided a useful framework for environmental scientists. The professionals who intervene in the environment to enable economic systems to withdraw water also accommodated to the integrative principles of river basin management.

Any integrative process is intrinsically challenging. The participating professionals use different languages and perhaps more importantly work at different levels of precision. Many are the accounts of interdisciplinary endeavor that led to cross-sterilization. The river basin was a conceptual ideal but incorporating it into scientific and professional practice demonstrated that even technical integration is associated with problems of social relations and human frailty.

Awareness that a more comprehensive integration was necessary came also during the 1970s. Questions were raised about the evaluation of natural resources for what and for whom (Food and Agriculture Organization, 1974). To answer these questions it was necessary to involve economists and other social scientists, who could throw light on the perceptions and expectations of communities that used and managed natural resources.

The evolving integrative approaches to allocating and managing water up to and including the 1970s had been shaped by the professionals and scientists within the water sector. It was, however, an external force that shaped the way water was managed in the North during the last two decades of the twentieth century. The green social movement was especially effective in the water sector (Carter, 1982; Reisner, 1984; Worster, 1985). It had gained momentum in the 1960s, had become persuasive in the 1970s and was delivering significant policy change in the 1980s. The US government cut the dam construction program and the environment benefited by the lowering of demand for water for

Water Resources Perspectives: Evaluation, Management and Policy. Edited by A.S. Alsharhan and W.W. Wood.
Published in 2003 by Elsevier Science, Amsterdam, The Netherlands, p. 1-8.

agriculture. Environmental services were given priority.

The response of the water policy-making community has been to discover a new and more radical approach to integration. Environmentalists have given the environment a voice. This voice has contended successfully and water has been re-allocated. The process was one of discursive politics. The process was not comfortable for the water professionals. They have assumed for decades that they were involved in technical activities that were politically and ethically neutral. They had in practice been serving a political project driven by the notion that the environment could be controlled and it was the task of scientists and engineers to progress that hydraulic mission.

Integrated water resource management (IWRM) in its 1990s manifestation requires that scientists and engineers recognize the changed political circumstances – at least in the North. The word "management" when used in relation to water resources always has the contentious process of "allocation" embedded in it. It is unfortunate that the word "management" is used rather than the word "allocation." The term "management" sounds responsible, wise and even benign. The term "allocation" clarifies the nature of the process in which water users and policy makers are engaged. At the same time, it is important to note that allocation is not a synonym for management, for allocation is only a part of the processes of management. The point is that allocation is politically charged, and is implicit in all management regimes.

Allocation requires awareness of politics of allocation and re-allocation and some willingness to engage in them. Farmers have long experience in making their presence felt in the contentious circumstances of water allocation. Environmentalists have learned fast and if anything have deployed the power of knowledge and constructed knowledge to greater effect than any other stakeholder – at least in the North.

The impact of politics on observation, planning and water policy making

One of the perverse features of politics is that those who are most practiced in its arts know that information can be as readily de-emphasized as it can be emphasized. Emphasis will depend on whose interests the knowledge put into currency would advance. Science – whether awkward and de-constructed or exquisitely integrated via models that speak to each other – contends for political attention. The other knowledge that new science knowledge might contradict can be secure through being in place for millennia and deeply integrated in cultural and social norms. Water used in irrigated agriculture has symbolic power as well as underpinning respected and familiar livelihoods. Science has little leverage over symbols, yet symbols have extraordinary social and political significance.

Those involved in the *rapport de forces* of water policy-making try to avoid contention by de-emphasizing conflict heightening information. The notion of discursive politics will be introduced to contextualize such processes and explain why policy outcomes are not much influenced by the recommendations of water scientists and professionals. If science based information is likely to heighten conflict over water within an economy the information will almost certainly be de-emphasized.

It is unlikely that a few chapters in a collection of scientific papers on *Water in the next millennium* will steer the unwieldy vessel of water science and the water professions. The concept that integration is a political process because it is unavoidably concerned with allocation is unwelcome and intuitively resistible by environmental scientists and professionals. What is argued here can be seen only as a preliminary step in the discursive process of gaining wide recognition for the political and normative nature of integrated water management.

Integration [IWRM] at the national level

Within national societies, integration of water resource management has proven to be hard, both technically and politically. To understand that difficulty, one must recognize that the two categories – the technical and the political – are not in fact separate and parallel. They are intertwined. One of the reasons that technical integration is such a challenge is that in many countries the management of water resources is divided amongst a range of political and social actors. Historically, the key bureaucratic player was the ministry of agriculture, because of the disproportionate influence of farm lobbies in northern states and the realities of economic life in the south. More recently, planning ministries have become involved as have, of course, water ministries. In many countries, coordinating efforts has been hugely problematic. This problem is exacerbated by the influence of local and regional authorities whose policies, rightly or wrongly, may not be aligned with national goals. To complicate the picture still further, non-governmental actors may also play a significant policy role. Village-level committees well versed in traditional methods of water management, meaning allocation, may not be open to any form of centralized direction, however well-based scientifically. Farmers' associations will work hard to promote the perceived interests of their membership, even when it conflicts with comprehensive policy initiatives.

These complexities cannot be understood purely in terms of "conflicting interests," for interests themselves are constructed in political and social interaction (Risse, 2000; Wendt, 1994). For example, in traditional societies where water possesses a sacred quality, plans that focus upon markets in water, where the good is viewed purely as a commodity, are likely to be strongly resisted. The resistance is not purely interest-based, but is generated in some measure by considerations of identity: we are not people who treat the sacred as something to be bought and sold. Similarly, for families who have worked the land for generations, even if only at the level of subsistence agriculture, telling them that it is irrational to farm in their location because climate change is producing extended drought conditions, or that it would be wiser for the government to import virtual water in the form of grain, rather than supporting irrigation projects, is not likely to be well received. The examples are not limited to the developing world. Consider the worldwide industry that is golf. Why is it so hard to convince people that planting golf courses in Arizona and Dubai is not rational or environmentally sustainable? The wealthy are by no means immune to constructed identities: we are people of leisure who have worked hard and deserve both sun and golf. The purely rational often succumbs to the powerfully normative or the radically political.

For scientists, the lesson is obvious. Telling people "the truth" is not always enough to prompt rational decisions. This is not to say that scientists must become spin doctors. The rigorous standards of scientific inquiry must be maintained, but successful integrated water management at the national level must come to grips with normative constraints, and with politically and socially constructed interests. Scientific proof is not self-evidently convincing in political and normative processes. If there is any lesson to draw from the environmental movements of the 1970s onward, it is that "norm entrepreneurs" can have an enormous influence on public policy and legal development (Finnemore and Sikkink, 1998). Drawing on scientific knowledge (but sometimes misrepresenting it), actors in civil society and political life argued convincingly of the need to reorient public policy to address environmental degradation. This was an active process of political and legal discourse. Strategies of advocacy were every bit as important as the substantive message. In many countries, this advocacy led not only to revised public policy, but to the implantation of legal structures such as environmental impact assessment that have profoundly altered political equations. To offer but one example, building dams has become a far more politicized process today than it was thirty years ago, for reasons that relate to legal requirements for impact assessment and to the mobilization of oppositional constituencies, including environmentalists and indigenous peoples.

The dam example, and the reference to "norm entrepreneurs," both point to another factor that has altered the processes of national integrated water management, the increasing penetration of international norms and international politics into domestic systems. The politicization of dam construction is fostered in part by requirements of the International Financial Institutions (IFIs), and was recognized in the consultative processes promoted by The World Commission on Dams. The IFIs and The World Commission were influenced by norm entrepreneurs who worked internationally to foster new understandings of the impact of dam construction. The separate opinion of Justice Weeramantry of Sri Lanka in the famous *Gabcikovo Case* before the International Court of Justice (1999 GET CITE apparently a citation hasn't been found yet?) reflects these discursive social and political processes. In these processes, a new norm that is not always very comfortable (or even understandable) to scientists has been employed to re-balance the relative weights of economic development and environmental protection. That vague but powerful norm is sustainable development (Lowe, 2000). Sustainable development and other related principles, such as intergenerational equity and precaution, are employed in discursive processes at the international and national levels that allow for public and interest group participation in regime building. Scientists can ill afford to absent themselves from these processes if they wish to affect the course of public policy. Because of the influence of international norms and international politics within state borders, it is important to explore, albeit briefly, the integration of water resource management at the international level.

Integration [IWRM] at the international level

In international society, efforts at integration have been undertaken largely through the construction and amendment of treaty-based regimes. Most early freshwater regimes were sectoral in one or more respects. They displayed almost no integrative tendencies. Many dealt only with individual issues that seemed to demand coordination amongst states sharing a river or lake. Aquifers seemed not to spark much attention or interest, despite their importance to overall supply. The earliest legal agreements sought to govern specific activities such as navigation or fishing. Over time, apportionment and related issues such as flood control and irrigation became the primary concerns of treaty making. Once environmental problems were recognized as transboundary concerns, legal regimes continued to track the selective pattern, and focused upon individual pollutants or specific types of activities generating pollution. Moreover, agreements were typically selective of the environmental medium protected. Freshwater treaties were also generally quite narrow in geographic scope, treating a particular watercourse, or even a portion of a river, lake or basin. The interaction between water and other environmental media, such as land and air, was ignored (Brunnée and Toope, 1997).

Many freshwater legal regimes enshrined what can only be described as competitive and use-oriented approaches, establishing rigid, and supposed immutable, formulae for the allocation of water uses amongst riparians. The 1959 Nile Treaty is a striking example of this approach. To the extent that treaties included any substantive principles, aside from use allocation, these were limited to the balancing of competing sovereign interests, rather than the recognition of any common environmental interests (Brunnée and Toope, 1994). In particular, the principle limiting water use by one riparian only where it causes significant transboundary harm and the principle of equitable utilization remained, explicitly or implicitly, the foundation of almost all freshwater agreements. Cooperation amongst states was limited to specific procedures for information exchange, prior notification and consultation regarding planned activities, and early warning of potential harm. Even where international commissions existed, they tended to be granted only a narrow, technical jurisdiction.

Over time, and especially with the effective norm entrepreneurship of environmentalists, the early freshwater regimes were subjected to harsher and harsher criticism, largely on the ground that they failed utterly in encouraging, much less promoting, joint or cooperative transboundary management. The criticisms were legion. Sectoral agreements were inevitably embedded in and affected by a wider ecosystem, so the narrow goals of the agreement could often be impossible to achieve. Effectiveness could also be compromised by the spillover of conflicts relating to issues not treated in the agreement. Overly narrow procedural obligations often made it difficult to adapt agreements to changing circumstances. The treaties tended to polarize the parties along the lines of "acquired rights," resulting in normative stagnation, even desuetude (Asafuddowlah, 1995). Most importantly, the regimes simply failed to integrate water within its broader ecosystem context.

In recent times, freshwater regimes have evolved that adopt a far more integrative and imaginative approach to the management of the shared resource, in part by adopting what has been called an "ecosystem orientation" (Brunnée and Toope, 1997). A relatively early example was the set of principles contained in Chapter 18 of Agenda 21 (1992), including the need to adopt "holistic management of freshwater as a finite and vulnerable resource." Further, Chapter 18 called for "integrated water resources management, including the integration of land- and water-related aspects,...at the level of the catchment basin or sub-basin." Such principles contained in Agenda 21 are by now classic examples of how principles can be employed strategically to shape a political, and ultimately a legal, discourse that comes to re-define the governing regime. An even more robust version of the same process can be found in the ECE Convention on the Protection and Uses of Transboundary Watercourses and International Lakes (Helsinki Convention, 1992). Through a broad definition of "transboundary impact," the Convention stresses the interaction between water and other components of the ecosystem. The parties are to practice and promote "sustainable water-resources management, including...the ecosystems approach." Already the Convention has had a discernable impact upon individual basin agreements within Europe (Brunnée and Toope, 1997). Even without substantive ecosystem orientation, other regimes have fostered strong management integration through flexible procedural devices that foster continuous interaction amongst scientists, technical experts and policy makers. A striking and successful example can be found in the Canada-US Great Lakes Water Quality Board and Great Lakes Science Advisory Board, created in the 1970s (GLWQA, 1972 and GLWQA, 1978).

These international developments, tending towards the promotion of integrated water management, intersect with the domestic processes identified above. Through what one leading scholar has called "transnational legal process" (Koh, 1997), discursive understandings are generated internationally that then grow roots in the political and legal systems of individual states. The inverse can also happen. The precautionary principle emerged first as a construction of German public law, but fed into and now fundamentally shapes the attitudes of many participants in international environmental debates. Scientists hoping to affect the course of public policy cannot rely merely on the production of new knowledge. They must work with norm entrepreneurs (sometimes even becoming norm entrepreneurs), to ensure that new insights are fed into the discursive processes that condition public decision making nationally and internationally.

Why has the IWRM approach gained currency recently?

The issue of integrated water resource management is high amongst the concerns of the water science community and of water management professionals at the millennium because of a number of forces at work in the hydro-political complex. The increasing demand for freshwater caused by the continuing increase in the world's population is the underlying challenge where communities and political economies are poor. Secondly, many remedies have already been recommended and even tried to ameliorate the water resource stress for water short communities as well as for water short environments. Supply management structures and systems have been installed – dams, reservoirs, water transfer schemes and well-fields – to achieve technical efficiencies – more crop per drop. Demand management via technical innovation as well as institutional innovation – water pricing – has been recommended and in a few cases adopted. Principles of allocative efficiency – more jobs per drop – have been adopted with spectacular effect. Although they have been fiercely resisted because of the re-allocative political prices in water scarce regions in the South where they would be most effective. Thirdly, the environment has been privileged in the water sector.

Technical measures, the pursuit of economic efficiency and of environmental consideration have not been enough to get communities and political economies to adapt to the worsening water resource situation. By the late 1990s it was widely recognized amongst a group of water international professionals that the pace of adaptation could only be accelerated by new modes of interaction (World Water Council, 1990; Global Water Partnership, 2000). A major international conference took place in the Hague in March 2000 and a landmark consultative process took place between 1998 and 2001 – The World Commission on Dams process. The Commission was innovative and emblematic in that it brought together high profile representatives from international and national social movements-NGOs-representing civil society priorities and their own agendas as well as stakeholders from governments and the private sector (World Commission on Dams, 2001). Both the WCD process and its after-math have been extremely contentious. The level of contention is a measure of its significance. Many water professionals and government officials in the South have indicated that they wish the project had never been conceived. The different perspectives of Northern and Southern water users and officials will be examined in the chapter by Allan which refers to the WCD process as a Fifth Water Management Paradigm. He also identifies the other four paradigms as the first pre-modern approach, the second hydraulic mission paradigm until about 1980 in the North, the environmental paradigm from about 1980, and the economic paradigm in the 1990s.

Integrated water management requires a fifth paradigm approach involving inclusive and transparent policy-making. The fifth paradigm principles – inclusiveness, consultation, legitimacy and transparency – have been adopted by national agencies such as the Corps of Engineers and the US Bureau of Reclamation in the United States. They have also been projected into the South since the late 1990s by international agencies such as the World Bank and UNDP and by the donor agencies of OECD countries. There is a predictable serious discursive distance between the agencies and governments of the rich one-fifth of the world's population in the North and the governments of the Southern economies. This contentious distance is entirely explainable via political and social analysis and can be illustrated via the comparative water sector narratives of the North and the South. It is not suggested that water scientists and professionals immerse themselves in political and social analysis. Rather it is recommended that they gain confidence in such explanation and adopt approaches that anticipate that political and social circumstances will determine what is optional and feasible in managing and allocating water. Above all they should shape their research and professional expertise so that it takes into account the politically feasible. It is not enough to know that the strength of the profile beneath the dam is weak; the capacity of the water managing institutions and the expectations of the water users need to be accounted for in recommending policy options based upon scientific inquiry.

Adopting IWRM in dry [agriculturally marginal] regions: some water politics

Application of integrated water resource management in arid regions is obviously desirable for the noble purpose of stewardship of limited water supplies. Unfortunately, the definition of arid is quite debatable, leading to the use of even more confusing terminology such as "semi-arid," which is somehow superior to a desert. Politically, loyal residents of an arid locale often take pride maintaining acceptable lifestyles while overcoming the challenge of water scarcity through their own cleverness and/or government aid programs. Advances in cultivation and irrigation have bolstered agriculture in drier regions in the United States and Australia, and government price supports have encouraged production, and thus water use, beyond the needs of world markets.

The well-established strength of agriculture in national and regional politics, based on the sanctity of the traditional family farm, makes it difficult to set policies that control or curtail irrigation diversion and withdrawals. For younger regimes in the Middle East, the dream of purchasing more independence through irrigated agriculture is further complicated by severe problems in climate, water quality, and soil productivity. As societies become more urban, partly through migration from farming communities, the competition between urban and rural water demands increases. Development of industries that consume water, whether for energy production, processing of agricultural products, or manufacturing, also causes competing water demands. IWRM serves as a means to negotiate allocations between the competitors, but it is easier said than done.

The current politically correct model is known as "grass-roots" or "bottom-up" regional water planning, which utilizes discussion by designated stakeholder groups. This model was adapted from the environmental impact analysis procedure (Jain et al., 2002). The approach flies in the face of the previous tradition, in which scientists and engineers were trusted to set priorities based on logical decisions that would maximize the greater good for the residents of the region (Allan's hydraulic mission). The necessity of government funding of most sizable water projects in arid, and often relatively low in population, regions quickly made the government decision-makers more important than the technical analysts and designers. Opportunities for pork barrel projects abounded, and even the noblest projects suffered criticism. The stakeholder planning approach, such as that now employed for regional forecasting in Texas, was intended to remove the setting of priorities from the potentially corrupt government officials and place in the hands of the local water users. This lofty goal is appreciated, but the major shortcoming of this approach is that the stakeholders typically do just that, hold onto their individual stakes. The stakeholders are selected to represent agriculture, industry, municipalities, water utilities, environmentalists, general public (whatever that means) and government regulatory agencies. The diversity insures that no single stakeholder can empathize with all members of the planning group. Application of consensus building rather that democratic voting leads to few bold or innovative solutions. The scientists, engineers, and economists serving as consultants to these planning groups often find that their limited communication skills, especially with laymen, prevent implementation of their most prized proposals.

Our societies find it difficult to admit that agriculturally marginal regions may not be worth farming. The amount of local water supply is a major component of the agricultural viability of a region. Irrigating such marginal lands by mining groundwater can make money in the short run, but IWRM is necessary to consider longer term issues. In the Southern High Plains of Texas, for example, groundwater levels have been declining since the 1940's, with cotton as the leading irrigated crop. Texas groundwater rights were based solely on the English common law doctrine of "right of capture" until 2001. Improvements in irrigation efficiency and mechanization, as well as the removal of some land from irrigation through the conservation reserve program, have slowed the declines since the 1970's. However, in the 1990's, many farmers in that area began planting peanuts, which require even more irrigation than cotton. How did this change come about?

The farmer's business planning period is a single year. Last year's profit or this year's loan is used to buy seed, crop acreage is planned based on predictions of commodity prices at harvest, and inputs of water, pesticides, and fertilizers are applied to protect the season's yield. As it turns out, mechanized center pivot systems have made it easier to apply more water, simply by flipping switches, than the old furrow irrigation process that necessitated manual placement of moveable pipelines. As the water level declines beneath the farm, the local groundwater conservation district helps the farmer calculate a groundwater depletion allowance for deduction from that year's taxes. Finally, the farmers look to each year's federal farm legislation for price supports and other subsidies.

Irrigated agriculture uses 95 percent of all groundwater pumped in the Southern High Plains of Texas, and as such contributed in 2001 several billion dollars to the economy of the region annually. However, cotton is still cultivated after the market price dropped by 50 percent. Recent predictive groundwater modeling for the region forecasts that future irrigation demands will not be met in some parts of the region within 20 years. Municipalities that depend on local groundwater will be impacted even sooner. After two years of discussion by a sizeable stakeholder group, negative predictions about agriculture by the consultants were buried in the appendices of the planning report. Farmers are still left with all the risk as they do their best to make a living from the land.

In regions where independent property ownership is highly valued, increased government regulation is resisted. Overlapping jurisdictions between different agencies that regulate and/or promote agriculture, freshwater quality, water resources management, and other environmental concerns complicate the issues that must be considered to truly integrate the water planning and operational processes. Arid regions have the most risk if residents choose to stretch their water supplies to meet unrealistic demands. Negative impacts are not only financial, as changes in the hydrologic cycle can also affect valuable ecosystems through changes in streamflows and aquatic life diversity. As the perception of water as a commodity, whether actual or virtual, continues to grow, our stewardship of that commodity must improve. Hopefully, that stewardship will include realistic evaluation of good and bad investment.

CONCLUSION

Global, regional and local water resources are subject to increasing demands. Implicit in the management of all water is the task of allocation. Allocation between uses that have become competing because of the rising overall demand will be contentious. The purpose of the chapter has been to show that the information that underpins such contention is not just based on well-observed hydrology and rational economics. It has been shown that it is normal for identities, beliefs and long established water using practices to have more weight in political discourses that mediate allocative outcomes than science. Public knowledge is constructed through discursive politics where scientists hold only marginal sway, unless they strategize like other interested actors. Hydrology, hydraulics and economics will always have an important potential place in the way systems of integrated water resource management are developed, but science-based information can easily be overwhelmed by the arguments of water using interests who have a life and death stake in "who gets what, when and how" (Lasswell, 1936). Such interests are accustomed to contention and can quickly find new tactics which bury negative predictions about the future by the consultants in the appendices of the construction interests and politicians to make 'water planning reports. Coalitions of the users of big volumes of water can easily build coalitions with flow up-hill to money and power' (Reisner, 1984). In addition to doing good science, scientists must learn how to communicate their analysis and especially their wisdom in ways that are as tough and persuasive, and where necessary as subtle, as the highly motivated and very tough stakeholders who use the water.

REFERENCES

Agenda Twenty one, 1992. Chapter 18 – Protection of the Quality and Supply of Freshwater Resources: Application of Integrated Approaches to the Development, management and Use of Water Resources, UN Documentation, A/CONF.151/26 (Vol. 2).

Allan, J.A., 2001. The Middle East water question: hydro-politics and the global economy. London: I. B. Tauris Academic Publication.

Asafuddowlah, M., 1995. *Sharing of Transboundary Rivers: The Ganges Tragedy*. In: The Peaceful Management of Transboundary Resources 209 (G.H. Blake, et al., eds).

Brunnée, Jutta and Toope, Stephen J., 1994. *Environmental Security and Freshwater Resources: A Case for International Ecosystem Law*. 5 Y.B. International Environmental Law, 41p.

Brunnée, Jutta and Toope, Stephen J., 1997. *Environmental Security and Freshwater Resources: Ecosystem Regime Building*. 91 American Journal of International Law, 26p.

Brunnée, Jutta and Toope, Stephen J., 2002. *The Changing Nile Basin Regime: Does Law Matter?* 43 Harvard, International Law Journal, 105p.

Carter, Jimmy, 1982. *Keeping faith: memories of a president.* New York: Bantam Books, ISBN 0 553 05023 0, 622p.

Finnemore, Martha and Sikkink, Kathryn, 1998. 52 INT'L ORG. 887.

Food and Agriculture Organization, 1974. *A framework for land evaluation.* Food and Agriculture Organization Soils Bulletin No 32: Food and Agricultural Organization, Rome.

Gabćikovo-Nagymaros Project, 1998. Case Concerning the (Hungary vs. Slovakia), 37 I.L.M. 162 (I.C.J.).

Great Lakes Water Quality Agreement (GLWQA), 1972. 15 April 1972, 1972 Can. T.S. No. 12.

Great Lakes Water Quality Agreement (GLWQA), 1978. 22 November 1978, 1978 Can. T.S. No. 20.

Global Water Partnership, 2000. *Towards water security: a framework for action.* Report of the Global Water Partnership to the Hague Water Forum, March 2000. Wallingford (UK): HR Wallingford.

Helsinki Convention, 1992. (ECE Convention on the protection and Uses of Transboundary Watercourses and International Lakes), 17 March 1992, reprinted in 31 ILM 1312.

Jain, R., Urban, L.V., Stacey, G.S., Balbach, H., and Webb, M.D., 2002. Environmental Assessment. 2nd ed. McGraw-Hill, New York, 655 p.

Koh, H., 1997. Why do Nations Obey International Law? 106 Yale Law Journal, 2599.

Lasswell, H.D., 1936. *Who gets what, when, and how?* New York, 264p.

Lowe, Vaughan, 2000. *The Politics of Law-Making: Are the method and Character of Norm Creation Changing?* In: The Role of Law in International Politics: Essays in International Relations and International Law 207 (Michael Byers, ed.), Oxford: Oxford U.P.

Reisner, Marc, 1984. *Cadillac desert.* New York: Penguin Books.

Risse, Thomas, 2000. *"Let's Argue: Communicative Action in World Politics*, 52 INT'L ORG. 1.

Wendt, Alexander, 1994. *Collective Identity Formation and the International State.* 88 American Political Science Review, 384p.

World Commission on Dams, 2001. *Dams and development a new framework for decision-making.* London: Earthscan.

World Water Council, 2000. *A Vision for World Water at the Millennium.* Prepared for the World Water Forum March 2000, The Hague: World Water Council.

Worster, Donald, 1985. *Rivers of Empire,* Oxford: Oxford University Press.

WATER RESOURCES MANAGEMENT AND POLICY

Integrated Water Resources Management is more a Political than a Technical Challenge

JOHN ANTHONY ALLAN
SOAS/King's College London Water Research Group, SOAS, Thornhaugh Street, London WC1H 0XG, UK.
Email tony.allan@soas.ac.uk

ABSTRACT: The Integrated Water Resource Management (IWRM) approach was developed by environmental scientists, water resource engineers and economists in the late 1980's and the 1990's. Integrated Water Resource Management was a response by water resource planners to the negative outcomes of past water resource policies. Past practice did not for example properly recognise that sectoral demands could be competitive. The conflict between irrigated agriculture and the environmental services provided by water is a classic case. Upstream and downstream uses can also be incompatible. Integrated Water Resource Management was supposed to address these potential contests by planning. The study will show first, that Integrated Water Resource Management is a discursive political process and not just a planning process. Secondly it will show how social and political theory can be used to explain how and why attitudes to water resource allocation and management have changed through the last half of the twentieth century. Thirdly, such theory will also be used to show why the assumptions of Northern water policy makers and politicians differ from those in the South on how to achieve 'sustainability'. The analysis will focus on the last three decades of the twentieth century when green social movements had a decisive impact on the way environmental services were perceived. Finally, the study will argue that the pace of adoption of water policy reform has been very much slowed in the Middle East in particular through the import of 'virtual water' to ameliorate the region's serious water deficit. Virtual water is the water embedded in water intensive commodities such as wheat. Political leaders have been able to solve their serious watershed constraints by resorting to the irresistible virtual water solution in problem sheds outside the region.

INTRODUCTION

The purposes of the paper are *first* to draw attention the political nature of the processes associated with the construction and re-construction of our perceptions of water and water environments. It is argued that accepting that awareness of the social role and political significance of water are preliminaries to introducing technical and institutional reforms. Such innovation is envisaged in recent initiatives to raise awareness of the need to take environmentally considerate and economically efficient approaches to the allocation and management of water (WWC, 2000; Global Water Partnership, 2000). *Secondly,* it will be emphasized that the water allocating and managing communities worldwide are diverse. They are variously endowed and differentially developed socially, politically and economically. The economies of the plural North are relatively rich and deeply integrated into global information and trading systems. Those of the plural South are diverse, are frequently poor and not globalize and where global systems have impinged they have often been damaging. *Thirdly*, it will be argued that it follows that different approaches will be necessary to address the information rich and risk aware North from those which could effectively impact poor isolated rural communities of the South. *Fourthly*, it will be shown that those concerned to communicate new knowledge and to promote innovatory water policy reform must be clear about the phase of innovation with which they are concerned. The assumption that the unrevealed benefits of innovation will prevail over the valuation of traditional practice has no substance. Water users and communities must *want* as well as *know about* new methods before new systems can be introduced and operated effectively. *Fifthly*, it will be shown that there are important global economic processes that severely impact the place at which water policy reform takes place, including Integrated Water Resource Management initiatives. The Middle East and North African (MENA) region in particular is subject to the distortions of importing 'virtual water'. Because vast volumes of water can be imported in virtual form; that is embedded in for example food imports the urgency of the local water deficits endured by the Middle East and North African political economies can be de-emphasized.

Water Resources Perspectives: Evaluation, Management and Policy. Edited by A.S. Alsharhan and W.W. Wood.
Published in 2003 by Elsevier Science, Amsterdam, The Netherlands, p. 9-23.

Beliefs, knowledge and action

'We are in the knowledge construction business and the belief making game'

Identifying the trajectories to which regional water policy reform are subordinate

Those involved in the advocating Integrated Water Resource Management as well as the billions who are targeted by the knowledge that is being constructed are involved in political processes. These political processes have determined the directions that water-managing policies have taken in the past. Attempting to set new directions in an area so fundamental to existence, traditional livelihoods and social relations as water is to struggle with core politics.

The Integrated Water Resource Management approach can be situated historically. However, this aspect has not been given enough attention. The process of constructing NEW ideas [knowledge] about the economic and environmental values of water gained currency among water specialists only a short time ago; in the mid 1970s. This newly constructed knowledge is being disseminated throughout the world, and whilst this is a relatively simple process in the plural North it is more difficult in the equally plural South. In the North the new knowledge may be better known but there is still a need to disseminate the sound principles comprehensively. In the South long evolved beliefs are entrenched. Such beliefs contradict the new knowledge. Ownership of the new water knowledge has scarcely been taken on by water professionals in the South. New water knowledge has had no impact on most water using communities world-wide.

Making the Integrated Water Resource Management message politically credible to the South is an unavoidable preliminary challenge; the acquisition of ownership of the new knowledge highlighted in for example the Vision (WWC, 2000) is a political process. It is the crux of the challenge for those attempting to identify facilitating Actions (Global Water Partnership, 2000) Achieving widespread awareness, at all levels of the water using and water policy-making communities, that water's role in livelihood futures is as important as its current role in securing existing livelihoods, is essential. Such a shift in perception is politically counter-intuitive. Those who have a stake in the status quo will effectively resist the economically and environmentally inspired innovations of water use efficiency and the newest concepts of environmental sustainability. Fortunately it is coming to be realized that *sustainability* is no longer just an ecological concept. Economic and social dimensions have been recognized to be integral to the definition of sustainability as are the environmental priorities that were constructed by the green movement from the 1950s and 1960s (Carson, 1965).

In the Vision/Framework for Action (WWC, 2000 and GFWP, 2000) processes are trying to export Northern perceptions of the risk of a global water deficit to peoples who have overwhelming local anxieties resulting from extreme poverty and to politicians fearful of social instability. Water poverty is just one of their deficits.

'New knowledge' about approaches to water and water management

The following section provides an account of water relevant events in the North. These events in the North are recounted not because they are more important than events in the South. The narrative is provided to help understand where and when in the North its recently adopted economic and environmental wisdom on water emerged.

The political economies of the industrialized countries have been inspired for just over a century by the belief that nature, including water resources, could be controlled. Since the late nineteenth century the entrepreneurs and state agencies involved in delivering water for economic and social purposes believed that Nature, including water, could, and should, be subject to the mastery of science and industry. This high phase of *industrial modernity* was possible because of the revolutions in science and industry in the early nineteenth century and the achievements of capitalist organization in marshaling the resources of labour, the environment and capital.

Unsatisfactory outcomes of this unprecedented synergy had become evident by the mid-nineteenth century. In the 1840s socialist writers drew attention to the dangerous tendency of this capitalist inspired system to ignore the interests of the other contributors to the capitalist mission and especially the contribution of labour. In the event, capitalism and the polities in which it was embedded in the second half of the nineteenth century addressed this 'first failure of capitalism'. The message of the philosophers and critics was heard. The extreme risks to political stability of the grotesque and rapidly expanding urban poverty of the late nineteenth century were avoided. Over the next century various forms of redistributive social democracy emerged to reshape the capitalist mode and confound the predictions of Marx. Capitalist interests were made aware of the social necessity of addressing the concerns of labour. Post-modernists might see a version of 'reflexiveness' in the way the ideologies – liberal, social democrat and especially conservative – adjusted in the industrializing modern Northern polities.

The second fundamental problem, some would say the second failure of capitalism, resulting from the drive for progress, whether in the economies of Europe and North America or of the Former Soviet Union, became evident just over a century later. The negative pressures on environmental resources of progressive industrial modernity with its assumption that Nature could be controlled began to be evident by the 1950s. Classic analyses by environmentalists such as Rachel Carson (1965) drew attention to the carelessness of what others (Beck, 1992 and 1995; Giddens, 1990) have identified as a century of industrial modernity which damaged rather than controlled Nature. Two decades of discourse later certainties had been replaced by uncertainty; environmentalist principles had entered Washington politics in the United States via President Jimmy Carter's presidency – 1976-1979. (See Carter, 1982; Allan, 1999) Carter became a champion for water and the environment challenging the institutions and political networks put in place by his predecessors to dam and control the wild waters of the United States. He was unsuccessful but he did accelerate widespread recognition that the approaches of supposedly progressive industrial modernity, "harnessing the forces of nature for the benefit of mankind", were full of risks and no longer viable. He emphasized also that they were not cost effective. By the mid-1970's - the progressive Nature controlling - ends 'had ceased to charm' (after J. S. Mill) at least in the North. But note the mid-1970's was a very short time ago.

The recognition that the mismanagement of water resources requires new attitudes amongst those at the commanding heights of the rich economies of the North is just one of many signals that there has been a significant shift in approach to the use and husbanding of environmental resources. The lesson from the North, however, is that the two or three decade long environmental discourse and especially the water discourse did not start in the corridors of power in Washington, nor in those of the multi-national corporations, nor in those of the Corps of Engineers or the United States Bureau of Reclamation (USBR). Water gained a place on the agenda of those allocating national budgets relating to water only after the argument had been made by individuals and activists, mainly ecologists and scientists (and the hippies of the sixties who have come into power). Economists remind us that Ricardo valued the environment at the beginning of the nineteenth century, but they cannot explain their century and a half of neglect of the subject thereafter. Their recent attempts to 'value' the environment was a response to the questioning by the green community of what the latter regards as an environmentally charmless outcome of industrial modernity. The economist's role was a reflexive one as has been that of engineers and of almost all environmental scientists.

By 1992 a suite of environmental issues had become global concerns; global warming, species diversity, desertification and water. Each had attracted activists and champions. In the event the environmentalists gained what for many was intuitively impossible. At the Rio UNCED (UN Conference on Environment and Development) meeting priority was achieved for the issues of global warming and of biodiversity. Global warming is scientifically controversial, especially when it is used to suggest that there are associated more general climatic trends for example in rainfall levels. Biodiversity is a complex topic hardly understandable by non-specialists. Despite the unsteady scientific foundations of the climate change and biodiversity arguments they were the major issues at Rio. Even the unstable concept of desertification was constructed to more political effect than water at Rio. These unlikely causes squeezed out the predicament of the hydrosphere. Water was given relatively little attention at Rio, but the commendable preparations at the meeting in Dublin in January 1992, resulted in a detailed chapter 18 on freshwater, the longest in Agenda 21. The lessons from the UN Conference on Environment and Development meeting in Rio appear to be that abstract issues can gain the attention of policy makers and their influential political leaders if the projects are cleverly and effectively constructed. Such a tendency fits the risk society theory of Beck (1999) on how individuals and communities in the North respond to risk after such emblematic shocks as Chernobyl. People stopped trusting progress and it became possible to persuade opinion formers and communities that they should become precautionary in dealing with the environment.

The discourse of risk begins where the unbroken trust in safety ('progress') ends and applies so long as the catastrophe has not (yet) occurred. The perception of threatening risks determines thought and action (Beck, 1999:75).

The disaster of Aids and then of BSE and other food scares has further broken the trust that communities in the North had in science and the industries that could impact on public health. They became especially fearful that science and industry could neither control Nature nor be trusted to understand its potential. These events accelerated the process of mind changing which environmentalists had labored for three decades to achieve. With an awful lot of help from the globalising media Northern societies have forced their governments to become more environmentally aware and industry to adopt precautionary principles. The risk theorists call the condition in which the North finds itself 'reflexive modernity'. Responses are being made to more and more risks as they are identified and awareness is quickly and widely diffused by the globalising media. The susceptibility of the risk-aware North to 'new knowledge' about water resource poverty and water resource stress is thus high. The globalising media instruments of awareness-raising are to hand. However, evidence shows that this awareness alone does not lead to changes in attitude - pressures and incentives are needed.

The Integrated Water Resource Management message of the Vision and Action processes (WWC, 2000 and Global Water Partnership, 2000) is a major orchestrated attempt to construct knowledge about global and local water to remedy the outcome of the Rio UN Conference on Environment and Development event in 1992 where climate change and biodiversity claimed priority. The process is the most serious attempt yet to include all interested water stakeholders in a worldwide consultation process. The goal is to provide the pressure that leads to changed attitudes and changed and most important - funded - water policy priorities.

Getting the Integrated Water Resource Management concept across

The North

In the Northern milieu it should not be a major challenge to 'construct knowledge' about the risk of water shortages, impaired water quality and to project such knowledge deeply into the North. Nor should it be difficult to make the case for water as an investment priority for the multi-lateral and bi-lateral donors. They have already got their minds around the notion that water is an economic and social good and a vulnerable natural resource. All that is needed is some good copy and some skilled public relations consultancy. Nevertheless we should not forget the dreadful water politics of the Jimmy Carter years (1976-1980). He was unable to persuade the water professionals in the Corps of Engineers nor those in the US Bureau of Reclamation, nor the elected legislators on Capitol Hill that they should adopt the a green or an economically efficient agenda. (Carter, 1982) This unholy coalition prevailed against his sound analysis until his successor, President Reagan, was able to put through the necessary but unwelcome reforms.

The South

Getting the message across to the communities, professionals and governments in the South is a different challenge. The condition of community-wide 'reflexiveness' has not emerged. Responsiveness to risk awareness in the sense that whole populations can be alerted almost overnight to a newly perceived threat - such as the recent example of genetically modified food in Europe - is not in place. The local media in the South do not entertain, inform and educate 'risk aware societies'. Meanwhile the North's global media with their global reach and their risk-aware material have little impact on two billion people living in poverty. Yet it is this massive population that must be one of the major targets of the Integrated Water Resource Management initiatives. Perceptions of the very numerous poor communities need to be influenced if their political leaders are to be able to adopt the 'new water knowledge' being recommended by Integrated Water Resource Management outsider advocates.

The poor

For many reasons it will be very difficult to get those living in poverty and their political leaders to be responsive to the Integrated Water Resource Management concept. First, the poor everywhere have deeply held beliefs and expectations about water that contradict the Integrated Water Resource Management message. Secondly, the poor are not susceptible to the notions of environmental risk in which the message from the North is couched. Thirdly, they do not have the political, social, economic and technical capacities to respond to the message. Fourthly, officials and government ministers in the policy-making institutions are aware of the economic and technical impediments to making an effective response to the message. More important they are even more aware of the political prices that would have to be paid if the principles of water use efficiency and the recognition of the environmental value of water were to be imposed via tough water policy reform.

The activists

The Integrated Water Resource Management message is prepared by concerned professionals: scientists, technical and economist professionals, policy makers and policy implementers. For the message to be legitimate it must be seen to have been formed by a plural political process. Such a plural process requires that inputs be made by water professionals and policy makers in public authorities; by those capable of providing water services formally and informally in local markets and through private sector bodies; by activists in NGOs and social movements. It is this last group, the NGOs, who most readily access the concerns of civil society. They can also represent, not always perfectly, civil society in the contentious process of creating acceptable new knowledge in the area of water policy.

Links exist between the activist outsider professionals and insiders in the technical agencies for national and regional water management. These links are the result of common education and professional experience with the international epistemic groups to which the enthusiasts for Integrated Water Resource Management belong. Activist scientists and the committed, neo-liberal and green-aware, engineers and economist proponents of Integrated Water Resource Management in the donor agencies are also part of the wide professional community who have embraced what Giddens (1984) calls 'mutual knowledge'. In this case the 'mutual knowledge' is the suite of principles being advocated by the Northern, and some Southern, water professionals – namely water use efficiency and environmental sustainability. Such 'mutual knowledge' is not readily assimilated, however, by the poor billions in the South. The local engineers and economists who have embraced the new outsider knowledge face the same difficulties as the outsiders in relating to the local political imperatives of water allocation and management..

How should those advocating Integrated Water Resource Management approach the challenge of communicating with the numerous poor? There is no space here to provide detail. Key activists and champions have to be found who have an essential appreciation of the unavoidable political nature of innovation. Here are the major generic local circumstances that have to be known about if new knowledge is to have an impact:

1. Knowledge of the deeply held beliefs and expectations about water that may contradict the message of the concept of Integrated Water Resource Management must be identified
2. Notions of 'environmental [water resource] risk' must be expressed in a language which is locally relevant.
3. The 'social adaptive capacity'-socio-political, economic and technical capacities to respond to the challenges must be evaluated (Ohlsson, 1998; Turton, 1999).
4. The political prices that would have to be paid if the principles of Integrated Water Resource Management, of water use efficiency and the recognition of the environmental value of water were to be imposed via insensitive water policy reform would have to be evaluated.

Water management policy – where have we been in the past century, where are we going: some paradigms?

The problems facing arid and semi-arid regions of the world with respect to the availability of sufficient good quality freshwater have been highlighted during the last three decades of the twentieth century. The hydrological capacity of such regions to meet regional needs and of the global hydrological system to meet global needs has been debated. The debate has been inconclusive.

Shifts in perception reflecting awareness of water resource scarcity have influenced the discourse on water. As discussed above, it was the green movement which proved to be the main agent of innovation with respect to paradigmatic shifts in water management policy after 1980. Figure 1 uses trends in levels of use of freshwater for agriculture as an indicator of the inspirations which influenced water allocation and management policy since the late nineteenth century. During the twentieth century there was a progressive rise in the diversion of high quality freshwater water from the environment to the agricultural sector. The rise in irrigation uses was primarily to meet the burgeoning demand for food to meet the needs of the rising global population.

Five water management paradigms can be identified. First, the paradigm associated with of *pre-modern communities* with limited technical or organizational capacity. The second paradigm is that of *industrial modernity*. In the water sector the onset of industrial modernity came in the second half of the nineteenth century. The ideas of the Enlightenment, engineering capacity, science and the investment initiatives of the state and the private sector characterized industrial modernity. Industrial modernity was manifest as the hydraulic mission of the mid-twentieth century. This project seized both

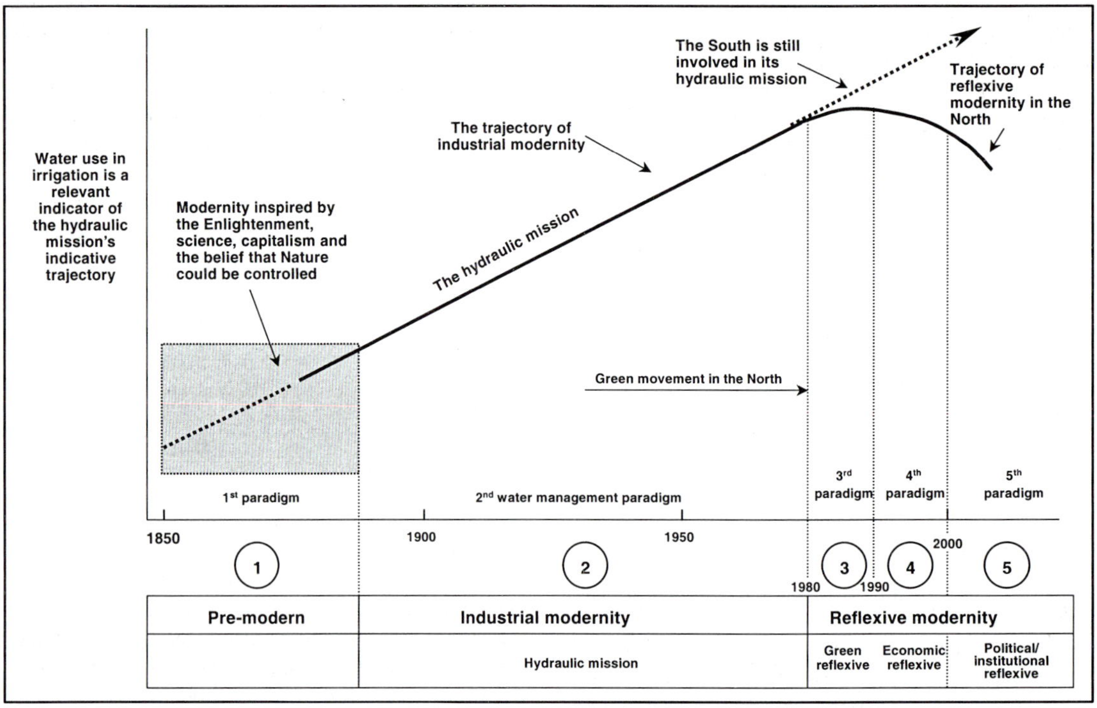

Figure 1. Five water management paradigms 1850-2000 (Allan, 2000).

liberal western economies, especially the United States federal government, as well as the centrally planned economies of the Soviet Union. The started in hydraulic mission also proved to be readily exportable to the South during the imperial era and especially in the second half of the twentieth century.

According to social theory the ideas underpinning industrial modernity shifted during the 1960's and the 1970's. The questioning led to reflexive responses and a phase which has come to be known as 'reflexive modernity'. Some theorists refer to the period as late modernity or post-modernity. In the North in the water sector the reflexive response (Beck, 1990) is evident in three water management paradigms. This phase witnessed a reduction of water use in agriculture in a number of semi-arid industrialized economies - Australia, California, Arizona and Israel. (Allan, 2001; p. 148) Water was in effect put back into the environment to rehabilitate environmental services.

This reflexive phase can be shown to have three sub-phases. The third paradigm is the change of water allocation and management priorities inspired by *the environmental awareness* of the green movement. These activists succeeded in persuading regions to allocate water to the environment and reduce allocations to agriculture. Their campaigns the 1960's but it was not until the 1980s that evidence of the green influence on policy became evident in water use figures (Allan, 2001; p. 148).

The fourth paradigm was inspired by economists who began to draw the attention of water users in the North to the *economic value of water* and its importance as a scarce economic input. These ideas gained currency in the early 1990's. There has been an attempt to export them to the South via such agencies as the World Bank and through the energies of institutions as UNCED, the WWC and the Global Water Partnership and the associated Global Water Forum in The Hague in March 2000. Those who have lived through the water discourses of the 1990's could not escape the economic elements of the Integrated Water Resource Management project. Water values, water pricing, water markets and privatization were constant and endlessly repeated themes.

The environmental and economic phases of Integrated Water Resource Management are still in train. It is argued here that they are being supplemented by a new fifth paradigm which is based on the notion that water allocation and management are political processes. The arguments in this analysis are converging on this big idea. Integrated Water Resource Management fundamentals such as the hydrological logic of the river basin as well as economic fundamentals relating to the value of water are central to the concept of Integrated. Water Resource Management. But Integrated Water Resource Management demands much more than the mere recognition of the environmental and economic value of water and planning, engineering and economic interventions. Integrated Water Resource Management is an intensely political process because water users have interests and they do not want them to be diminished by such interventions. Prioritizing water allocation with an eye on the economy and prioritizing investment to reduce environmental impacts will conflict with the interests of current water using practice. The fifth paradigm is bringing forward approaches which include participation, consultation and inclusive political institutions to enable the mediation of the conflicting interests of water users and the agencies which manage water. The inclusive political process of the fifth paradigm requires that the interests of civil society, hierarchy (government), social movements (NGOs) and the private sector are included in the policy making discourse (Thompson et al., 1990).

Evolving perspectives on sustainability and their special relevance to water management

Integrated Water Resource Management is also based on the concept of sustainability. This concept has been useful in emphasizing the role of environmental services. It was one of the ideas used by green activists to raise the profile of the importance of the environment. But the notion proved to be impossible to operationalise in a narrow environmental sense. This was because the environment was conceptualized as not being integrated with the social and economic users of environmental resources.

Figure 2 shows the way in which sustainability has been viewed in the second half of the 1990s. In this framework sustainability has three dimensions not just one - the environment. The three dimensions of sustainability relate to society, the economy and the environment. Voices reflecting the priorities associated with these three dimensions play a role in the political processes, which mediate the water using and water policy outcomes in a particular political economy. The strength of the voices changes according to the capacity of the different interests to construct and articulate their concerns. It will be shown in subsequent sections that social theory throws light on the reasons why the paradigms have been sequenced as they have.

For water to be managed to sustain the environment water management policies have to prioritize interventions and resource allocation so that society and the economy, as well as the environment, are also sustainable. This 'balance' is achieved, or not, via political processes. Managing water is a political process and 'integrated water resource management' – Integrated Water Resource Management, is particularly political.

Conceiving sustainability as a triangle makes it possible to locate the hydropolitical contest of the three concerns in the centre of the triangle. For every location or river basin the strengths of the voices will be differently expressed and more or less contentious. Such political contention is normal. The outcome of the discourse, whether in a political economy with parliamentary democracy or in apparently coercive authoritarian regimes, identifies what is politically feasible. Another way of describing at this complex political process is via the aphorism which captures a deal of sometimes challenging French philosophy (Foucault, 1971)-a politician stands firm when pressed from all sides.

An engineer would identify such a convergence of forces and the resultant, in this case an allocative water policy, as an equilibrium position reflecting the interacting forces.

It is possible to map the five water management paradigms identified in Figure 1 on to the conceptualised space of the sustainability triangle with its hydropolitical core-Figure 2. The first pre-modern paradigm expresses the concerns of *society* – water for drinking for local food production and local livelihoods. The second paradigm of industrial modernity was associated with the use of water for the development of the economy. Water was and remains an economic input strengthening the diverse elements of the economy. The third paradigm addressed the concerns expressed by environmentalists who emphasize the *environmental* services roles of freshwater. The fourth paradigm, strong during the 1990s in the North, was again in the *economy*. Value, pricing, markets and privatization were the driving ideas. Finally the fifth paradigm is located in the centre of the diagram. In this case Integrated Water Resource Management is seen to be part of politics. Integrated Water Resource Management policy options are honed in the contentious discourse resulting from the respective claims of society, the economy and that constructed for the environment.

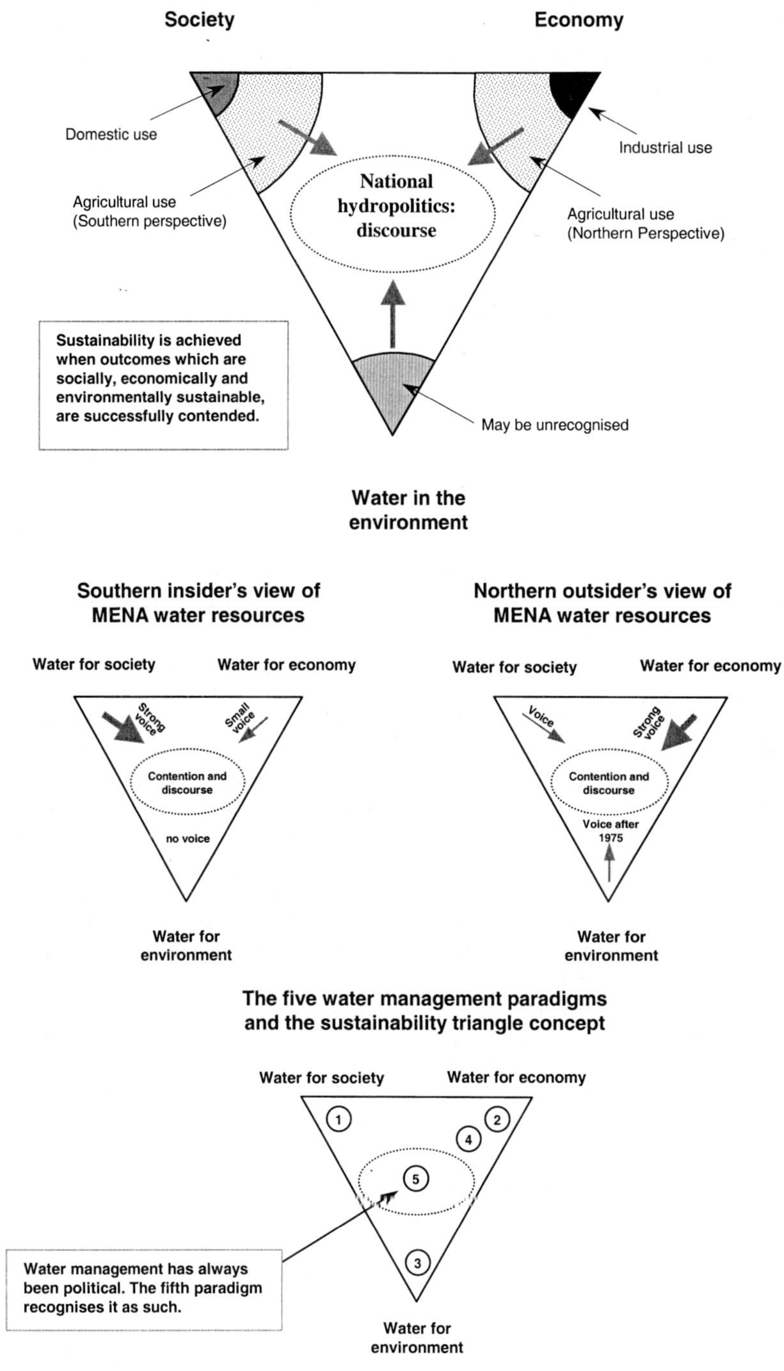

Figure 2. The concept of sustainability and the water sector; water management as a political process and in determining perceptions of the diverse values of water in the North and the South.

Contrasting water policy paradigms in the North and the South: contending trajectories

Figure 1 is also useful in conveying the difference between the priorities of the South and the North. Figure 1 relates to water using economies in semi-arid political economies. The semi-arid North can be shown to have partially adopted all five water management paradigms. The professional community associated with the water sector can easily recognize the first four paradigms. The fifth paradigm is gaining currency – albeit slowly. Water users and politicians in the North have been much slower to change their ways of perceiving water.

In the South, by contrast, the professional community generally, and all water users and politicians have resisted the adoption of the last three reflexive paradigms. Exceptions exist in the South at the local level where small communities manage their water via transparent institutions tested over time.

The South is generally deeply committed to its hydraulic mission. It has much economic development ground to make up. Socio-economic development priorities are urgent; environmental priorities are recognised but for the moment their voice is less powerful in the policy discourse than those of society and the economy.

The water policy discourses in the North and the South are different. The distance between water using trajectory in the South and that of the reflexive North reflects the discursive tension found everywhere that Northern outsider professionals and scientists share the logic of the third-environmental, and the fourth-economic, paradigms with Southern water professionals and policy makers. Those 'outsiders' from the North who insist on preaching the environmental and economic values of water have little impact on the 'insider' Southern water management discourses. Nor should they expect to. With rare exceptions the universities of the North did not have the third, environmental, and the fourth, economic, water management paradigms in their water management curricula until after the mid-1980's. Integrated Water Resource Management only gained currency, and even then without its essential political personality, until after the Dublin and Rio meetings of 1992. Figure 3 conceptualizes for the water challenged Middle East and North Africa the progressive utilization of water as perceived by insider professionals and politicians on the left of the diagram, and by outsider professionals on the right of the diagram. It is the perception of insiders that determines water allocation and management policies.

Some important exceptions

The social theory used in the preceding sections to underpin the notion of water management paradigms cannot be used beyond the semi-arid realm. The experience of France in managing its water sector shows that as early as the mid-1960s it was possible to install an inclusive and a decentralized and democratic political and management structure (Roche, 1999; Seine-Normandy Water Agency, 1999 and 2000). France legislated into existence in 1966 regional Water Parliaments based on the geographical river basins of the country. These parliaments enabled the diverse interests to be taken into account through their representation in the regional, basin determined, water parliaments. These structures reflect all the virtues of the fifth paradigm. The concerns of water users in agriculture and industry/services, of municipal authorities responsible for providing water services and related engineering, social and public health services, and more recently of those responsible for environmental services, were all represented. These institutional developments anticipated those in the world more generally by over forty years. That the utility of these apparently model institutions was being questioned at the millennium by the central government of France, anxious to bring back to the centre the control of the expanding budgets of France's decentralized Water Agencies, is a predictable reflection of the constant tension between the political centre and decentralized political institutions. In this case the tension is in the water sector (Water Academy, 1999). Water being a resource which is generally mobilized, treated, used and reused locally decentralized administrative services are essential.

Water policy reform: a classic case of the political challenge of innovation

Water policy reform: unavoidable phases and vital conditions

Water policy reforms are shaped by the discourses that precede their formulation. The outcomes of such discourses reflect the interests of the participants and the absence of the interests of any that might have been excluded. Some participants in the discourse might introduce new knowledge; others might contend old beliefs. Current politically correct procedures urge that no stakeholder be excluded if the consultative process is to be socially and politically safe and politically sustainable. The consultation process provides a potential for local approaches to be understood by

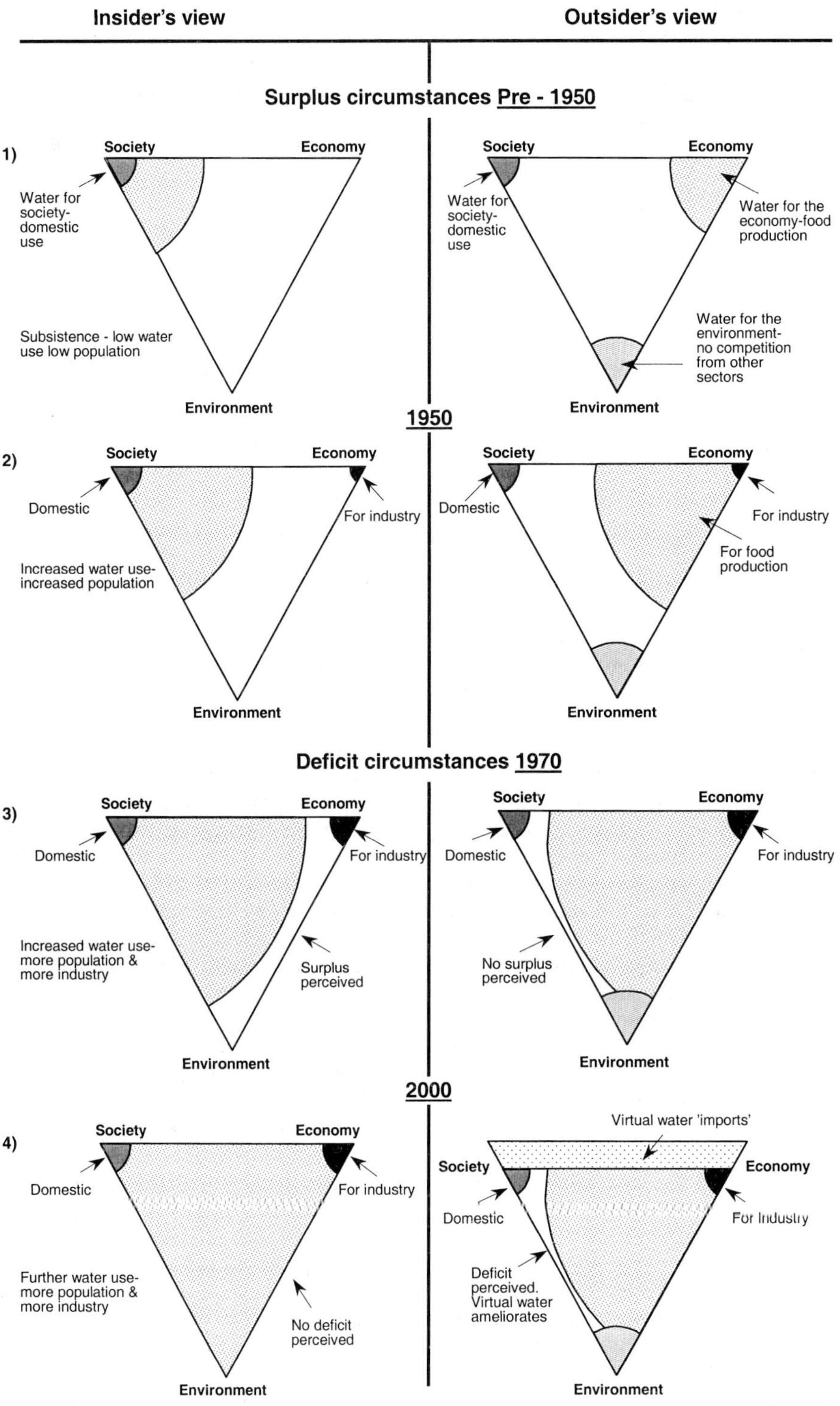

Figure 3. Sectoral growth in water demand and the perception of the consequences in the North and the South pre 1950 to 2000.

Figure 4a

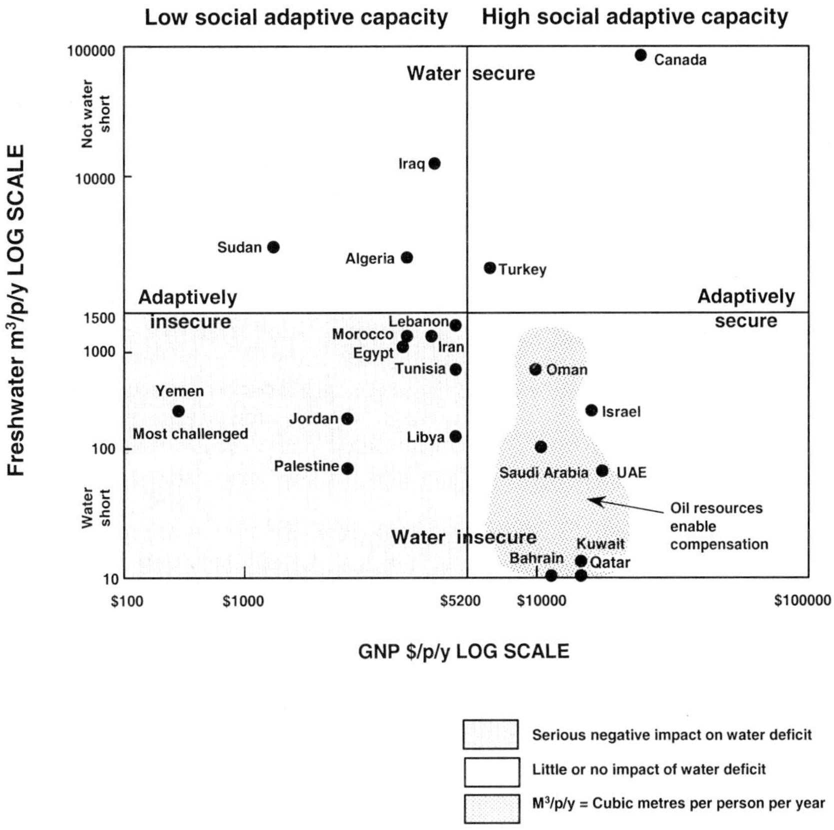

Figure 4b Water security trajectories of some MENA economies - 1950-2000

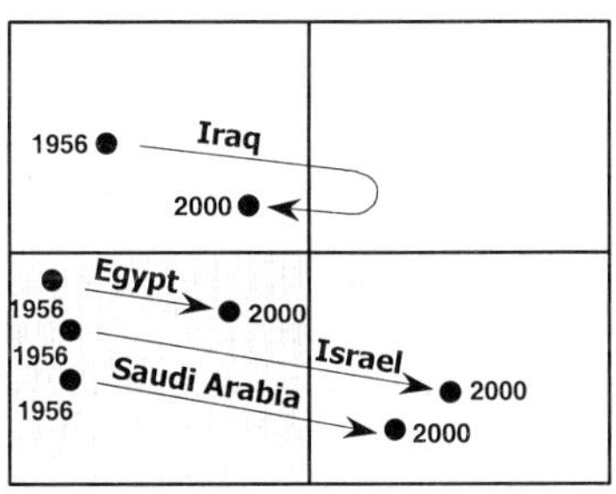

Figure 4c Description of the water/adaptive capacity nexus

Water rich and non-adaptive	Water rich and adaptive
Water poor and non-adaptive or at an early adaptive stage	Oil rich (can substitute) or adaptive and water poor

Figure 4. An analysis of the water security of some MENA economies in relation to water deficits and shortages of social capacity. Note: GDPs for Iraq, Kuwait, Palestine & Qatar are estimates. Concepts authors and Ohlsson and Turton, 1999. Data World Bank Atlas, 2000.

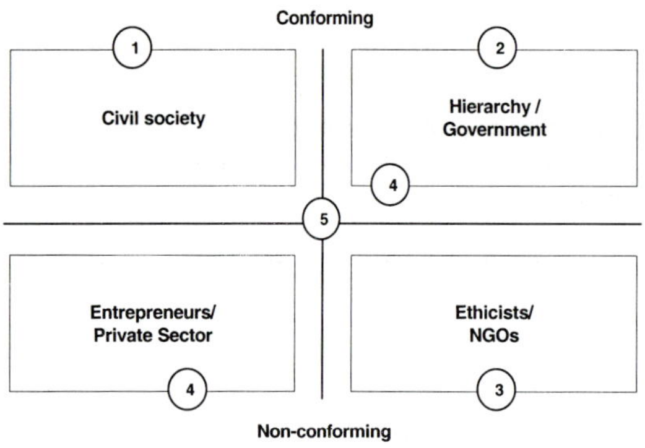

Figure 5. The Douglas / Thompson theory of ways of life, showing the sequencing of the pardigms of pre-modernity, industrial modernity and late modernity.

those wanting to introduce innovations based on newly constructed outsider knowledge.

Water policy reform, like all innovation is a political process. It has recognizable phases. Those attempting to understand, and intervene in innovative processes must be aware of the cycle if they are to make any analytical progress or have a substantive impact. In addition the frustration of those recommending radical water policy reforms will be reduced if they recognize that the benefits of the adoption of new socio-political approaches and new economic instruments are always retrospective rather than prospective for those stakeholders who have to adapt to an externally inspired innovation.

Water policy reform reduces transactional costs

Recent analysis of water policy reform by economists (Dinar et al., 1997; Dinar 2000) has rediscovered the powerful idea of Coase (1937) that institutions are transaction cost reducing. New institutional economists have been developing the idea for over a decade with reference to both the North and the South (North, 1981). 'Institutions are the rules of the game in a society, or more formally, are the humanly devised constraints that shape human interaction' (North, 1990).

Institutions and reformed institutions reduce transaction costs but they may not be perceived in this way by either the policy makers or those whom a policy reform might impact. This is unfortunate for the innovator because it takes little observation and analysis to detect the transaction cost savings in legally defined water management systems. For example the domestic water reticulation systems in cities across the world massively reduce the transaction costs of accessing safe water. The distribution of water to agriculture throughout the world in irrigation systems is also very successful indeed in reducing transaction costs. Would-be reformers at the millennium want to introduce legally defined water allocating and managing instruments at local, national and international/river basin levels which will reduce even further the transactional costs of making water available to user. They want to introduce environmental consideration and principles of economic efficiency. All this is intellectually worthy but not politically feasible.

Understanding the notion of reducing transaction costs is best achieved by considering the role of the familiar instrument of money (Coase, 1937 and 1960). The adoption of money to reflect mutually agreed values for diverse commodities and services required unprecedented levels of trust. Trust was essential in the socio-political domain. The progressive formal legitimizing of the instrument was also essential and was achieved by legal frameworks with regulatory capacities in the socio-economic domain. *Money is a spectacular example of how legally conceptualized instruments can reduce transaction costs.* Imagine gaining access to commodities and services as diverse as food, transport services and information without trusted and legal monetary instruments.

In the water sector *new instruments* based on principles of water-use efficiency and the need to protect the water resource will also be transaction cost reducing, but the nature and scale of the benefits of cost reduction are impossible to define and quantify at the level of the community and the nation. *This inability to define and quantify future benefits is one of the most serious impediments to water policy reform and the implementation of such reform.* It is not enough to cite the impressive benefits of the current water systems and institutions because the benefits are intangible and difficult to value in quantitative terms. Potential beneficiaries will resist any attempt to argue that the current systems should be reformed to reflect economic and environmental values if they would lead to reductions in the quantity of water used or the increase in price of water to users. The resistance and the discursive outcome is explained by the politics of the contention of existing interests and not by awareness of the benefits of potential innovation.

The unavoidable phases of innovation – KWHOE

If water policy reform innovators are to have any impact on water using stakeholders they have to recognize that such stakeholders have to *know about* the proposed innovation. They next have to *want* the proposed innovation. When the innovation is wanted it will be possible for them to *have* the new system or institution. With the innovation in place it will then be possible to *operate* it, or *comply* with it if it is a regulatory instrument. Operation of the new system may not at first be effective. The final stage, therefore is to achieve *effectively operation of* the system or achieve *effective compliance.* Knowing about, wanting, having, operating and effectively operating *[KWHOE]* water reforming policy and practice can be conceptualized in a sentence but the process can take decades.

The ***KWHOE*** process is relevant to the professionals involved in water policy reform in the North and the South. It is particularly relevant to the knowledge construction, information dissemination and general innovation central to developing the Integrated Water Resource Management:

- **Knowing** about the benefits of new [water reform] instruments which will reflect the environmental and economic values of water
- **Wanting** the new instruments which will reflect the environmental and economic values of water
- **Having** the new instruments which will reflect the environmental and economic values of water
- **Operating/complying with** the new instruments which will reflect the environmental and economic values of water
- **Effectively operating/complying with** the new instruments which will reflect the environmental and economic values of water

Virtual water makes resisting reform especially easy

'The innovator makes enemies of all those who prospered under the old order, and only lukewarm support is forthcoming from those who would prosper under the new'. Machiavelli, N, 1513, *The Prince.*

'Virtual water plays an economically invisible and a politically silent role in the solving the water deficit problems of peoples and political economies living in watersheds or parts of watersheds. '

The second half of the twentieth century has been unique in the problems faced by those allocating and managing water. This has been especially the case in the semi-arid South. In the South populations doubled in two or three decades and the demand for water for social uses and for economic uses increased proportionately. As did the pressure on the water which provided environmental services.

The Middle East and North Africa (MENA) region for example began to run out of water in the 1950's in the Gulf countries and in arid North Africa. Other economies such as Jordan ran out in the 1960s. The whole region entered water deficit in the early 1970's (Allan, 2001). The seriousness of this strategic insecurity has been hidden by the ready availability of an effective and very affordable solution – virtual water. Virtual water is the water required to produce commodities. The production of one tonne of wheat, for example, requires one thousand tonnes of water. The water used in the production may be water delivered in engineered irrigation systems or it may be soil moisture accessible to crops in the soil profile after periods of rainfall. The regions which receive abundant, well-distributed, rainfall have a significant comparative advantage. Temperate North America and Europe are favoured with relatively abundant soil water.

In the second half of the twentieth century international trade in grain has very effectively solved the rapidly increasing water deficit of the Middle East and North Africa region by moving grain from North America and Europe and other water surplus regions. By 2000 the region was importing fifty billion tonnes of water in virtual form. This volume is equivalent to the flow of the Nile into Egypt and to at least a quarter of the region's accessible freshwater. If an engineer were asked to find and move the water she/he would indicate that engineering solutions cannot handle such volumes of water. The farmer and trader manage the task without a thought.

The vast volumes of virtual water entering the Middle East and North Africa region are, however, seriously distort water policy making. They are especially influential with respective to initiatives to introduce the Integrated Water Resource Management principles of water use efficiency. Virtual water has two features, which make it very welcome to politicians. Virtual water plays an economically invisible and a politically silent role in the solving the water deficit problems of peoples and political economies living in watersheds or parts of watersheds. Politicians have no difficulty in turning to politically feasible solutions to their serious water problems outside their deficit watersheds. The virtual water solution has been additionally welcome because the world price of grain has been falling for the past century. Improvements in technical efficiency have been only been part of the story. Since the mid-twentieth century there has been an increasing tendency for the Northern producers to put the grain on the world market at half its production cost.

Outsiders advocating Integrated Water Resource Management can take in their theoretical stride the economic rationale of virtual water in balancing regional deficits with global water. They see the local freshwater deficit as subordinate to the international trade in virtual water. This trade they see as subordinate to global hydrology (Allan, 2001). Middle East and North Africa insider water professionals and water policy makers on the other hand are merely free to adopt what appears to be a very convenient socially and politically feasible option. This political optimization, albeit short-term, is however sub-optimizing. The outcome is what economists would call a 'second best solution' (Lipsey and Lancaster, 1956). The economists' theory of the second best runs that where in an economic system there are a number of interacting conditions that lead to an outcome then if one of the conditions is sub-optimum then all the interacting conditions will be sub-optimized to bring about a sub-optimal outcome. Virtual water brings about a classic example of second best policy making in the water sector in the Middle East and North Africa region. The seriousness of the Middle East and North Africa water deficits are obscured by the availability of virtual water. An optimum Integrated Water Resource Management solution would involve measures to increase returns to water by technical and allocative techniques. There is much evidence that the doubling of returns that can be achieved by investment in pipes and water control systems are introduced where financial capital is plentiful. The reallocation of water from one sector to another which can bring many thousand-fold improvements in returns to water are rarely adopted, however, because such reallocation is always politically stressful.

CONCLUSION

Water allocation and management in general and Integrated Water Resource Management in particular has been shown to be a political process subject to constantly changing social priorities. Approaches to water management in semi-arid economies differ in the plural North from those in the plural South. Cultural, social, political and economic circumstances determine such differences. Hydro-political processes mediate the interests of the social and economic users of water and the interests of the environment have been given voice by environmental activists at least in the North. Ideally sustainability is the outcome.

A major purpose of this study is to show that political contention associated with water policy making is dynamic and that in industrialized political economies in the semi-arid North five water management paradigms can be identified from pre-modernity to the present.

The plural South is very much engaged in its industrializing mode, which involves the control of water resources to increase agricultural output and to generate power. The environment is not such a high priority in the economical and institutional circumstances in the South. The contention between recently reformed Northern outsiders arguing for the inclusion of Integrated Water Resource Management principles, namely environmental and economic priorities, into water policy anticipates by some decades the politically feasible circumstances which will facilitate the adoption of the new approaches.

REFERENCES

Allan, J. A., 1996. The political economy of water. In: Allan, J. A. (ed.), *Water and Peace in the Middle East: the Negotiation of Resources in the Jordan Basin,* London: Tauris Academic Publications.

Allan, J.A., 1999. Productive efficiency and allocative efficiency: why better water management may not solve the problem. *Agricultural Water Management* 40: 71-75, ISSN 0378-3774, see also http://www.soas.ac.uk/geography/waterissues/OccasionalPapers/ *look at Paper 10*)

Allan, J.A., 2001. *The Middle East water question: hydropolitics and the global economy*. London: IB Tauris Academic Publications.

Arlosoroff, S., 1996. Managing scarce water: recent Israeli experience. In: Allan, J. A. (ed.), *Water, Peace and the Middle East.* London: Tauris Academic Studies, p. 21-48.

Arlosoroff, S, 1997. *Report to the Water Commissioner of Israel on the Israeli water sector.* Tel Aviv: Office of the Water Commissioner.

Beck, U., 1995. *Ecological politics in an age of risk.* Cambridge: Polity Press.

Beck, U., 1992. From industrial to risk society, *Theory, Culture and Society* 9: 97-123.

Beck, U., 1999. What is a 'risk (society)'?, *Prometheus,* 1.1: 73-79

Coase, R., 1937. As referred to by Coase when receiving the Nobel Prize for Economics in 1992.

Coase, R., 1960. The problem of social cost. *Journal of Law and Economics.* October 1960, 1-44.

Carson, Rachel, 1965. *Silent Spring*. London: Penguin Books.

Carter, J., 1982. *Keeping Faith: Memories of a President.* New York: Bantam Books, ISBN 0 553 05023 0, 622p.

Dinar, A., Rosegrant, M. W. and Meinzen-Dick, R., 1997.

Water Allocative Mechanisms: Principles and Examples. Policy Research Working Paper, Note 1779, Washington DC: World Bank.

Dinar, A. (editor), 2000. *The Political Economy of Water Pricing.* New York: Oxford University Press.

EXACT, 1998. *Overview of Middle East Water Resources,* Executive Action Team (EXACT), Middle East Water Data Banks Project, a Multilateral Working Group of the Middle East Peace Process. Published in Washington: United States Geological Survey.

Foucault, M., 1971. Orders of Discourse. *Social Science Information,* v.10.2 .

Giddens, A., 1990. *The Consequences of Modernity.* Cambridge: Polity Press.

Global Water Partnership (GWP), 2000. *Towards water security: a framework for action.* Stockholm: GWP.

Hajer, M., 1996. *The Politics of Environmental Discourse: Ecological Modernization and the Policy Process.* Oxford: Clarendon Press.

Lipsey, R. G. and Lancaster, K., 1956. The general theory of second best. *The Review of Economic Studies* 24(1): 11-32.

North, D, 1981. *Structure and Change in Economic History.* New York: Norton.

North, D, 1990. *Institutional Change and Economic Performance.* Cambridge: Cambridge University Press.

Ohlsson, L., 1998. *Environment, Scarcity and Conflict: A Study of Malthusian Concerns.* Göteborg: Göteborg University, Department of Peace and Development Research. ISBN 91-87380-43-49.

Reinhold, Niebuhr, C., 1940. *The Serenity Prayer,* New York: Commission on Social Justice of the Federal Council of Churches. http://www.wlbstuttgart.de/~www/referate/theologie/oetgeb00.html

Roche, P-A., 1999. Toward a basin integrated water resource management, Paper prepared by Pierre-Alain Roche - Executive Director of the Seine Normandie Water Agency, 51 Rue Salvador Allende, 92027 Nanterre Cedex, France.

Seine-Normandy Water Agency, 2000. Thirty five years of the Seine-Normandy Water Agency, Internet site: www@ aesn.fr

Seine-Normandy Water Agency, 1999. *Annual report of the Seine-Normandy Water Agency.* Nanterre: Seine-Normandy Water Agency.

Swyngedouw, E., 1999a. Modernity and hybridity - The production of nature: water and modernisation in Spain. Paper presented to the SOAS Water Issues Study Group, University of London, 25 January 1999.

Swyngedouw, E., 1999b. Hybrid waters: on water, nature and society. In: Swyngedouw, E. (ed.), *Sustainability, Risk and Nature: the Political Ecology of Water in Advanced Countries.* Proceedings of a workshop convened in the University of Oxford, 15-17 April 1999. Available from the Geography Department, University of Oxford.

Thompson, M., Ellis, R. and Wildavsky, A. (eds.), 1990. *Cultural Theory.* Boulder: Westview Press.

Turton, A.R., 1999. *Water scarcity and social adaptive capacity: towards an understanding of the social dynamics of managing water scarcity in developing countries.* MEWREW Occasional Paper 9, London: SOAS Water Issues Group, March 1999.

Water Academy, 1999. *The water Agencies facing the ecological fiscality.* Report of the congress of the Water Academy. Paris: Water Academy.

World Bank, 1998, *Middle East and North Africa water initiative.* Report on the proceedings of the May 1998 Cairo workshop convened by the World Bank, the EC and the European Investment Bank, Washington: World Bank.

World Bank, 2000. *The World Bank Atlas.* Washington DC: The World Bank.

World Commission on Dams, 2000. *Dams and development.* Cape Town: World Commission on Dams and London: Earthscan Publications.

World Water Commission, 2000. A report of the World Commission on Water for the 21st century. *Water International* 25(2): 194-201.

World Water Council, 2000. *Making water everybody's business.* Report for the Second World Water Forum in The Hague, March 2000, Marseilles: World Water Council.

Scenarios as a Tool in Water Management: Considerations of Scale and Application

CINDY WARWICK
Environmental Change Institute, University of Oxford, 1a Mansfield Road, Oxford, OX1 3SZ, U.K.
E-mail: cindy.warwick@eci.ox.ac.uk

KAREN B. BAKKER
Department of Geography, University of British Columbia, Room 217, 1984 West Mall, Vancouver, B.C., Canada, V6T 1Z2.

THOMAS E. DOWNING and KATE LONSDALE
Stockholm Environment Institute, Oxford Office, 10b Littlegate Street, Oxford, OX1 1QT, U.K.

ABSTRACT: In recent years, the use of scenarios for creative visioning of potential futures for planning and strategy testing has become increasingly popular. Water management has not been excluded in this trend and many case studies of scenario application can be identified. Three scenario exercises are considered in this paper – the regional Societal and Institutional Responses to Climate Change and Climatic Hazards project in southern England, the UK Environment Agency Water Resources Strategy and the World Water Council's World Water Vision. The paper discusses the role of scale in scenario approaches with regard to impacts, decision-making, participation and scalability. Scenarios are also compared with respect to their approaches and outcomes relative to issues in water resources planning, specifically demand management, extreme events and climate change. Conclusions point to the potential for innovative scenario approaches that consider adaptation, participatory interaction and local clustering.

1. INTRODUCTION

In recent years, the use of scenarios for creative visioning of potential futures for planning and strategy testing has become increasingly popular. Water resources management is a field where a variety of scenario approaches have been undertaken for diverse purposes. This paper outlines three case studies of water scenarios to examine how scale affects scenario development, outcomes and applications: the Societal and Institutional Response to Climate Change and Climatic Hazards (SIRCH) project funded by the European Union to understand process of drought and flood risk management in southern England; the UK Environment Agency's (EA) Water Resources Strategy scenarios developed at national and regional levels; and the World Water Council's World Water Vision global scenarios. The Thames Region in southeast England is used as a basis for comparing the case study scenarios.

The paper, in Section 2, introduces scenarios in general, the specific case of the Thames Region and a framework for developing and applying scenarios in the region. Section 3 sets out the three case studies individually. Section 4 compares the case studies considering issues related to scale including: impacts, decision-making, participation and the transfer of knowledge between scales. Section 5 discusses how the case studies relate to water resources issues of concern in the Thames Region, specifically demand management, extreme events and climate change. Finally, the paper proposes innovative approaches for future approaches scenarios in water resources management.

2. SCENARIOS AS A PLANNING TOOL IN THE THAMES REGION

2.1. What are scenarios?

A scenario may be defined as a coherent and systematic description of a 'possible future', expressed as a narrative about future environmental and socio-economic conditions. Scenarios usually have qualitative 'storylines' associated with quantitative indicators (for instance see Nakicenovic and Swart, 2000).

Scenarios can be used as a tool for delineating possible futures, which differ in crucial ways from the present. Rather than taking a predictive approach, scenarios are used to picture the future in an exploratory way, with the aim of identifying factors that will need to be taken into account in planning. More than one scenario is usually developed in a scenario-building exercise because the range of possible outcomes is of interest, rather

Water Resources Perspectives: Evaluation, Management and Policy. Edited by A.S. Alsharhan and W.W. Wood.
Published in 2003 by Elsevier Science, Amsterdam, The Netherlands, p. 25-43.

than a 'best guess' trend projection. The scenario methodology is particularly applicable in cases in which new factors may prove to be significant, or in which there is much uncertainty or disagreement about the future. Scenarios are also useful when adaptive time scales are longer than those used in conventional planning, or when the "worst case" needs to be planned for as well as more probable, expected, or central outcomes.

Scenarios can develop considering the current situation and drivers that propel the present into the future or by designing a snapshot of a future time and backcasting (working backwards, to determine how that state can be achieved) (Robinson, 1990). The consideration of 'branch points' (Gallopin et al., 1997), are predominantly associated with the former approach. Branch points represent critical decision points, or periods of instability in the system, where conventional beliefs in 'structural invariance' do not necessarily apply and constraints on the system are relaxed. At these points individual and collective decisions regarding the forward directions have more power to affect great transitions (Gallopin et al., 1997).

To frame our discussion of the application of scenarios to different issues and situations we make reference to the work of Ringland (1998). Ringland classified scenario exercises into three main categories: scenarios for learning (by those involved in the scenario building process), scenarios for building of a common vision to influence public attitudes outside of the business environment, and scenarios for strategy evaluation or development. As a result of his analysis of various scenario building case studies, Ringland concludes with what he sees as the two main advantages of scenario building: the unfreezing of the intellect to plan in an atmosphere of great uncertainty; and the creation of a scenario 'wind tunnel' to test strategic plans and ensure that they will 'stand-up' to different potential futures. In examining the water resources scenario case studies presented, Ringland's scenario classifications (which are not necessarily mutually exclusive) and the key advantages he has identified are helpful in clarifying the purpose, strengths and weakness of the different approaches undertaken. The goal of this discussion, as Ringland suggests, is to make useful observations and recommendations for scenario building and its application in water resources without detailing a prescriptive approach that would stifle the creative and innovative aspects of scenario development and use.

In relation to water resources management, planning for resource development and supply management is often predicated on assumptions about the stability of social and ecological systems which are actually highly variable short time scales and the rigidity of management paradigms that are open to evolution or change. Given the unreliability of projections over the medium-term (decades) and long-term (generations), and our limited understanding of the interrelationship between human and ecological processes, conventional trend projection methodologies seem particularly unsuited to the task of strategic adaptation.

As a process in water resources management, scenario building can facilitate creative engagement between different actors in the hydro-social cycle. A scenario-building exercise can also provide a forum for discussion, and can contribute to the creation of a shared framework within which alternative planning responses can be identified. On the other hand, scenarios may be taken as forecasts, or used for purposes for which they are not appropriate.

2.2. Water resources in the Thames Region

The Thames Region (shown in Figure 1) is the catchment area of the River Thames and its tributaries. There are 5,330 km of main river in the region 12 900 km^2 in area (EA, 1999). The region is located in the south east of England with the city of London at the down-stream end of the catchment. The region represents less than ten percent of the land area of England and Wales but contains twenty-three percent of the population, generates just over twenty-seven percent of Gross Domestic Product, and has a similar percentage of all construction work (Environment Agency, 1998). The region continues to attract high levels of growth and further housing development will place heavy demands on existing infrastructure networks (CPRE, 1999).

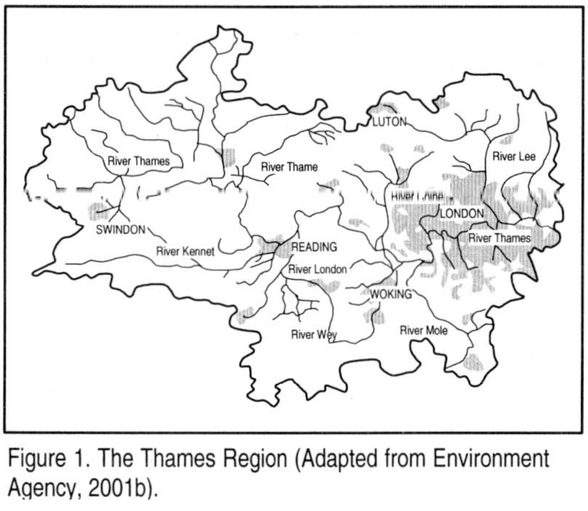

Figure 1. The Thames Region (Adapted from Environment Agency, 2001b).

The Thames Region is amongst the driest in the UK; annual effective rainfall is 235 mm (338 mm in west and 116 mm in east) with the northeast part of the region most at risk of drought (EA, 2000). Approximately 55% of the effective rainfall within the catchment every year is abstracted annually with 86% of this water going to public water supply (EA, 2001b). The majority of public water supply is abstracted to provide domestic water supply to the 12 million people living in the region mainly concentrated in the major population centres -- such as London and surrounding satellite cities. In recent years, rapidly growing urban centres in the upper Thames, have placed a strain on local water supply. Overall, the rate of water demand in the Thames Region has risen by over 50% since the early 1960's but until recently, water resources were not seen as a constraint on development in the region. Recent groundwater droughts and an increase in peak demands have raised awareness regarding the seriousness of the situation. Future developments in drought prone areas will have to be staged to coincide with new water resources. Potential shortfalls in the water supply-demand balance are already foreseen under the growth levels predicted for certain areas (EA, 2001b).

Development in the Thames Region is also constricted by land use restrictions. Over 40% of the Thames region is classified as 'Areas of Outstanding Natural Beauty' (26%), 'Green Belt' or has some other form of protection. The headwaters of the Thames' smaller groundwater-supported (or 'chalk') tributaries dry progressively through the summer as the water table falls, sometimes exacerbated by water abstractions; such persistent 'low flow' rivers in the region are one of the prime concerns of the environmental regulator (EA, 2001b). Additional pressure is placed upon resources through pollution of both ground and surface water, and through increasingly stringent environmental restrictions (which address, for example, low flows in rivers and water quality).

A remarkable variability in hydrological conditions has been witnessed in the Thames Region in the last decades (IOH, 1995; Marsh, 1995). Throughout the 1990s several hydrological systems set new records (e.g. flow and water levels), extending the average reporting range for system levels (Arnell et al., 1994). By the mid-1990s, meteorological events began to show a broad consistency with some climate change scenarios (Arnell et al., 1994; DOE, 1996) prompting the water industry to question the resilience of its water resource management strategies (Marsh, 1996).

The UK, as a whole, has experienced four major dry periods in the last twenty years (Brown, 1992; IOH, 1995; CEH, 1998). However, as water resources are groundwater-supported, single-season meteoro-logical droughts affect water resources in the Thames region less than in other regions in the UK. The most recent groundwater drought, running over consecutive seasons from 1995 to 1997, is the most severe since records began; many boreholes were at or below historic minima in 1997 and the early months of 1998 (CEH, 1998; 1999).

Fourteen counties, fifty eight district councils, and thirty three local planning authorities in London lie wholly or partly within the region, which is a planning and regulatory unit only for the Environment Agency (NRA, 1995). Privatised water service companies operate to boundaries that do not match any other geographical or political boundaries. The Thames Region is a planning and regulatory unit only for the Environment Agency Thames Region, one of eight regions in England and Wales (NRA, 1995).

The key stakeholders for water resource management in the Thames Region can be listed as:

- Water service companies - commercial enterprises with a natural monopoly (at present) to provide water to domestic, corporate and municipal users. There are five water service companies operating in the Thames Region.
- Environment Agency - the environmental regulator of the water industry in England and Wales is responsible for water resources planning, the issuing and management of abstraction licences as well as water quality and some aspects of nature conservation.
- Office of Water Services (Ofwat) - the economic regulator, set up at privatisation of the water industry in 1989 to stimulate competition and set prices for water and allowed capital expenditure.
- Department of the Environment, Food and Rural Affairs (DEFRA) - oversees both the EA and Ofwat has political representation in setting policy.
- Local authorities - planning and service level of government, with little direct representation in water issues other than planning approval for new development and major infrastructure.

2.3. A context for scenario development in the Thames Region

Though scenarios are receiving increasing attention as a tool for water resources planning

(Gallopin and Rijsberman, 2000; EA, 2001a, 2001b), they are building upon a more established framework of climate and socio-economic scenarios being used in environmental policy and research forums. Figure 2 shows the case study scenarios in the context of other scenario approaches used in the UK.

In 1996, the Intergovernmental Panel on Climate Change (IPCC) convened a Special Report on Emissions Scenarios (SRES), as a follow up to their earlier greenhouse gas emissions scenarios (called IS92) (Nakicenovic and Swart, 2000; see www.sres.ciesin.org). The four narrative storylines and families were labelled A1, A2, B1 and B2 (corresponding to quadrants along governance/sustainability axes), with 40 scenarios as realisations of the narratives. Six modelling teams implemented the scenarios (sometimes with alternative variants). Four scenarios were harmonised across the modelling teams to create marker scenarios that are now widely used in climate change studies. The panel involved a writing team of some 40 experts, widespread review and submission of storylines (the 'open process') and review and adoption by government representatives to the Panel on Climate Change. The scenarios were expressly oriented to provide a benchmark for greenhouse gas emissions and related climate forcing. While they cover a range of driving forces (population, economy, technology), they did not focus on water, vulnerability or climate impacts. For instance, disasters and surprises were excluded, and all regions experience significant economic growth (a 10-fold increase in global economic welfare in the 'poorest' scenario).

The Panel on Climate Change process overlapped with an independent panel, convened by the Stockholm Environment Institute in 1995. This "Global Scenario Group" produced four scenarios, each with two variants (Raskin et al., 2002); see www.gsg.org:

- Conventional Worlds—Market ('don't worry, be happy') and Policy Reform ('grown, environment, equity through better technology & management')
- Barbarization—Breakdown ('the end is coming') and Fortress World ('order through strong leaders')
- Great Transitions—Eco-communism ('small is beautiful') and New Sustainability Paradigm ('human solidarity, new values, the art of living')

The scenarios are different visions of the future represented by the first variant. The second variant illustrates a response to the scenario trend—a branch point for social, economic and political policy. These scenarios cover a wider range than most global projections-including bleak scenarios of increased inequitable development and fragmentation. Although water is included, the scenarios are holistic and do not provide great detail into water issues. The Global Scenarios Group is comprised of 14 experts drawn from around the world.

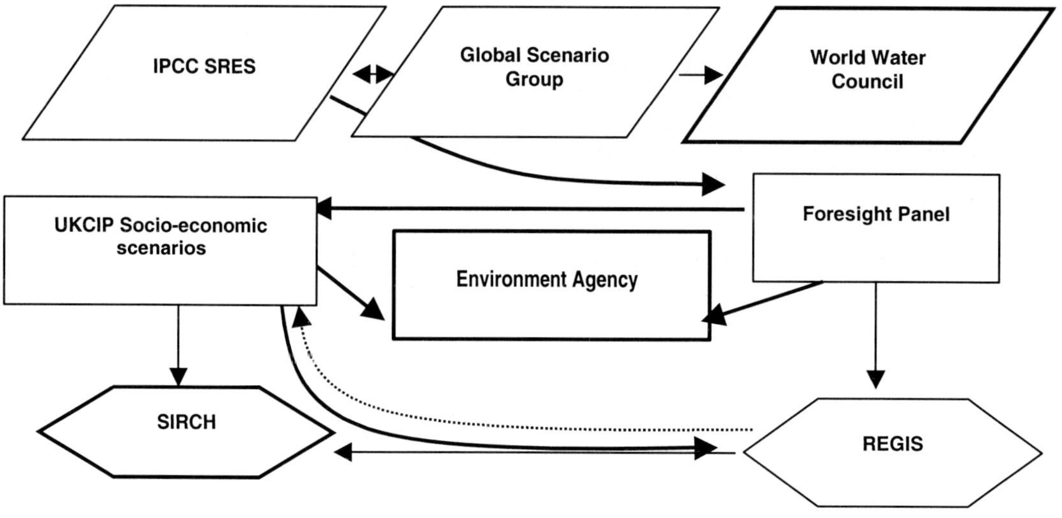

Figure 2. Socio-economic scenarios in the UK. Top row are international scenarios, middle row are national scenarios and bottom row are regional scenarios. Boxes in bold are compared in this paper. Arrows indicate downscaling processes and inter-relationships, with the width of the arrow reflecting the strength of the relationship.

The other global effort, by the World Water Council, is described below in Section 3.3.

In the UK, the Department of Transport and Industry (DTI) Foresight Programme on Environmental Futures scenarios (www.foresight.gov.uk) provided a key link between the Intergovernmental Panel on Climate Change (IPCC) Special Report on Emissions Scenarios (SRES) (see www.rivm.ni/env/int/ipcc) and other scenario programmes at the national and regional levels.

The Foresight Programme was begun in 1994 to identify future challenges for the UK, bring together diverse expertise to meet these challenges and encourage public debate about the future (http://www.foresight.gov.uk). The Environmental Futures programme, in particular, aims 'to inform and stimulate debate among businesses, regulators and Government departments about the environment and to encourage them to develop strategies and policies that will prove robust to a range of environmental futures' (DTI, 1999). The Foresight process involved a round-table to brainstorm ideas concerning possible futures, the sorting of ideas into scenario groupings and the creation of storylines and indicators for each of the scenarios. Four scenarios for a future United Kingdom were the result. The scenarios, Provincial Enterprise, World Markets, Global Sustainability and Local Stewardship, were developed in the quadrants of two axes representing changes in social values (consumerism to community) and governance (globalisation to regionalisation) (Figure 3). The scenarios describe possible UK futures between 2010 and 2040.

The UK Climate Impacts Programme initiated a process to create socio-economic scenarios, following on from the Foresight Programme (see www.ukcip.org.uk). The UK Climate Impacts Programme scenarios were developed by the same team as used in the Foresight Panel (this team had also been involved in the IPCC-SRES process) and the scenarios are similar to the Foresight scenarios, shown in Figure 3. After extensive review (particularly among government departments), quantified indicators were developed. One of the UK Climate Impacts Programme projects, Regional Integrated Assessment of Climate Change Impacts in the North West and East Anglia (REGIS), implemented the UK Climate Impacts Programme scenarios - indeed the project was instrumental in developing regional versions and involving extensive local stakeholder involvement (see Holman and Loveland, 2001).

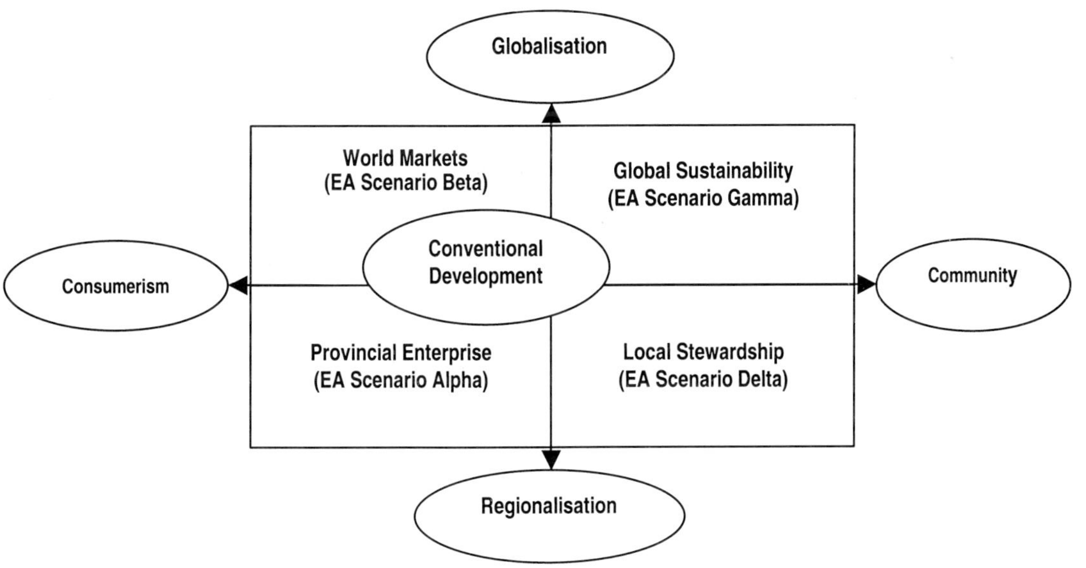

Figure 3. Classification of Foresight Socio-Economic Scenarios (Adapted from DTI, 1999).

3. SCENARIO CASE STUDIES

3.1. Coping with drought in the Thames region - the social and institutional responses to climatic change and climatic hazards scenarios

Building on the recent interest within the water industry in future droughts, the Social and Institutional Responses to Climatic Change and Climatic Hazards project, European Commission's Environment and Climate Programme (DGXII) Ref ENV4CT97 0447) initiated a participatory scenario development process in early 1998. This project included a case study, conducted by authors of this paper, on vulnerability and adaptation to drought hazard in the Thames Region of the United Kingdom. The case study used a scenario building process to identify possible futures and explore adaptation mechanisms under different institutional developments (Downing et al., 2001).

The Social and Institutional Responses to Climatic Change and Climatic Hazards scenarios were developed and assessed via a participatory methodology over the course of two workshops. Prior to the first workshop, in-depth interviews were conducted with key water managers and regulators in the region. The interviews provided information on current drought management strategies, and 'lessons learnt' from the most recent drought in the Thames Region (1984-1998 groundwater drought). The first scenario workshop involved sixteen water resource management experts from the region. The goal of the workshop was to identify factors that should be considered in evaluating adaptive capacity. Participants developed three plausible scenarios that described the possible development of social, economic and environmental drivers over the course of the 50 year time horizon.

One storyline involved Trend Projection and Conventional Wisdom - the experience of the last ten years projected forward, with increased wealth and consumer demand in a mixed regulatory regime. The other two scenarios were consciously extreme: Economic Growth and Market Enterprise -- high economic growth increases wealth, raises expectations, elevates demand for water, increases willingness-to-pay for water-related amenities, and also creates conflicts between water users; and Environmental Stewardship and Regulation -- a high general awareness of environmental issues, increased wealth and a more egalitarian society leads to a balance between water use and the environment. The three scenarios are related to the Foresight Panel scenarios (discussed in Section 2.3), to the extent of having similar orientations. However, two of the SIRCH scenarios (that is not the trend projection) were constructed be more extreme interpretations, and did not rely on the formal indicators developed by the Foresight Panel.

After the first workshop, storylines (the qualitative, overarching narrative element) were further developed, by the research team, into detailed scenarios with quantitative indicators. This process was aided by a four-part typology considering resources (i.e. hydrological indicators), institutions (the rules and norms which govern water management and planning), stakeholder influence, and key roles and issues for stakeholders. These socio-economic scenarios were mapped against four climate scenarios developed by the UK Climate Impacts Programme (Figure 4).

In the second Social and Institutional Responses to Climatic Change and Climatic Hazards workshop, participants were asked to explore the capacity of drought management institutions to adapt to the changes described by the socio-economic scenarios and the pattern of extreme events generated in the

		Climate scenario *increasing climatic risk* →			
Qualitative differences ↕	Socio-economic scenario	Low	Medium low	Medium high	High
	Market enterprise	Increased demand and increasing climate risk, but likely to be met through economic adjustments			High demand and climate impact could exceed economic management
	Trend projection	Moderate demand, risk management required		Increasing risk of supply/demand crises	
	Environmental stewardship	Lowest demand, little risk of adverse impacts		Potential climatic risk to environmental management	

Figure 4. Qualitative socio-economic scenarios related to increasing risk of climate change.

context of the climate change scenarios. Results of three models (showing dynamic behaviour, agent-based household consumption behaviour and game theoretic economic behaviour) developed during the course of the project were also presented to assist the participants.

The scenarios used in the Social and Institutional Responses to Climatic Change and Climatic Hazards project were developed to focus on a specific issue-- institutional capacity to manage future drought risk. That is, they are bottom-up analytical devices rather than global, alternative futures. The scenarios used were chosen on the basis of prior research within the project and for their capacity to capture and illustrate institutional and stakeholder capacity to adapt to plausible, but extreme, changing circumstances. The scenarios were not intended to span a range of all possible futures nor to select a preferred future for drought management and determine how that future could be achieved.

The planning horizon for the scenarios was the 50 years to 2050. This compares with the water company planning horizon of 2016 and the planning horizon of 2025 for the environmental regulator of the water industry (although both include longer views at some levels).

The project found that the "Economic Growth and Market Enterprise" scenario was unsustainable under mid-to-extreme climate change pointing to the need for a reconsideration of current policy goals of increased liberalisation in the water sector. The scenario exploration also provided possible support for a move away from 'predict-and-build' to risk-based water resources planning

Participants in the Social and Institutional Responses to Climatic Change and Climatic Hazards project were an expert group of stakeholders (including senior water company managers, representatives of regulatory agencies and a small number of academics and NGO representatives), specifically selected for the task of scenario building. The project was not conducted as an open, public exercise. The workshops drew upon an established community of water resource managers and regulators who are accustomed to working with each other at various scales. The process was participatory in nature. The project facilitators provided a general outline of several scenarios to facilitate discussion, while the storylines and full scenarios were developed in small working groups. This approach was undertaken to promote learning in the individuals involved and potentially influence drought management policy and approaches to implementation.

The Social and Institutional Responses to Climatic Change and Climatic Hazards project was undertaken almost immediately subsequent to an unprecedented drought in the Thames Region when stakeholders, primarily in the national government, were making decisions regarding the future management of drought in England and Wales. Indeed, the process was intentionally placed between two regulatory reviews, to explore research methods and long-term perspectives rather than to directly intervene in water management in the UK. In many cases, the recommendations of the project have been in line with recent developments (e.g. the increasing focus on demand management as a sustainable strategy for water management (EA, 2001b)) but no work has been undertaken to determine what linkages exist between the project and the generation of these outcomes. The most likely impact, if any, was indirect benefit to stakeholders involved in the 'unfreezing of the intellect' associated with scenario building in an atmosphere of uncertainty.

3.2. Water Resources for The Future - Environment Agency Scenarios

The scenarios that are playing a major role in water resources management in England and Wales are those generated through the Water Resources Strategy process undertaken by the Environment Agency (EA, 2001a). The scenario building process was undertaken as part of an integrated water resources planning process to secure the proper use of water resources in England and Wales (EA 2001a). The scenarios use a national framework for the development of regional scenarios and are used by the environmental regulator to develop water resources strategies for the future. The Thames Regional strategy is developed for roughly the same area as the Social and Institutional Responses to Climatic Change and Climatic Hazards scenarios (see Figure 1).

The scenarios developed by the Environment Agency examine the uncertainties related to the future demand for water. As a basis for this examination, the Environment Agency adopted the Foresight Programme Environmental Futures scenarios developed by the UK Department of Trade and Industry (DTI, 1999) as the framework for socio-economic change over the planning horizon. Scenario work undertaken by the Environmental Agency focussed on developing additional narrative and indicators for water demand that are consistent with the economic and social changes described in the Foresight scenarios. The augmented scenarios were relabelled as Alpha (Provincial Enterprise), Beta (World Markets), Gamma (Global Sustainability) and Delta (Local Stewardship) (see Figure 3) to distinguish the EA narrative and indicators from the base Foresight scenarios.

In the Environment Agency scenarios, water demand was broken into 4 key components: household demand; industrial and commercial demand; spray irrigation demand and leakage. For each component, key water policy, technology and sector-specific drivers of demand were identified and discussed relative to the socio-economic factors detailed in the Environmental Futures Scenarios. Different methods of calculating scenario demand were then used for each of the demand components.

Household demand under scenario conditions was calculated using a micro-component analysis. This method involved breaking aggregate household demand into its component uses and forecasting the Ownership, Frequency and Volume (OFV) of use for each component. Industrial and commercial demand was classified by selected Standard Industrial Classification codes and company size and examined relative to average water use per unit production or employee and uptake of water saving technology. Spray irrigation demand was forecast considering the relationship between optimum demand (determined by the crop and soil type, local climate and recommended irrigation plans) and economic demand (determined by cost, price and constraints on irrigation). Leakage was calculated for each scenario using scenario-specific assumptions concerning the regulatory, political and economic aspects of leakage. Full details of demand scenario development are available in Environment Agency (EA, 2001c). In general, assumptions regarding the components of demand were made nationally and applied to the unique conditions (climate, population projections, scenario descriptions of growth etc.) found in each water resource area to determine incremental demand over the elapsed scenario period.

The published result of the scenario exercise was the calculation of changes in water demand levels for the four scenarios at two snapshot dates in the future (2010 and 2025) for each of the eight Environment Agency Regions in England and Wales (see Table 1 for scenario results for the Thames Region). Regional results were then broken down for additional reporting by water company resource zones.

At the Environment Agency, the development of national and regional water demand scenarios was considered a first step in developing water resources strategies. Strategies were seen to consist of different options, or groups of options, for managing the supply-demand balance. In developing the strategies the goal was to 'identify mechanisms and management measures for achieving society's present aims within the constraints that the scenario presents' (EA, 2001a). Scenario constraints were seen, first of all, to consist of the quantitative levels of demand that would have to be balanced with supply through the implementation of various supply and demand management options. The options available to the Environment Agency were then subjected to a sustainability analysis based on present sustainability criteria to determine the recommended strategy. The robustness of the final strategy, its appropriateness and projected impact, under all scenarios was examined on a regional basis.

Strategy options for the Thames Region included demand management and water efficiency options, surface water yield improvements, groundwater developments, abstraction recovery to protect the environment and the potential for new reservoir storage or bulk transfers, contingent upon further investigation and developments. The full list of options is presented in Environment Agency (EA, 2001b).

Scenarios were not used to examine the impact that potential environmental, economic and social aspects of the scenario futures could have on the importance or definition of the sustainability criteria used. Neither was the resilience of the Agency itself (and therefore its ability to fully implement the strategies) examined as part of the strategy process. Therefore, scenarios were not used, in this instance, to envision potential futures for water resources management in England and Wales but to examine how options developed under current water resources management systems and values can potentially cope with future water demand.

Work on the Environment Agency demand scenarios was conducted in-house with reference to literature, expert opinion and widespread, but ad hoc, consultation with members of the water industry (EA, 2001c). The Environment Agency water demand scenario process contributed to learning within the Environment Agency resulting in the development of many key indicators useful for considering water demand levels under future conditions. However, learning was generally restricted to a small group of Environment Agency staff. Therefore, it remains to be seen whether or not the scenarios will result in a shared vision between the Agency and other key stakeholders in the water industry or environmental protection even though the details of scenario development and strategy findings have been published and widely distributed. The approach taken assumes that sufficient learning and creation of shared vision is already an intrinsic part of the Foresight scenarios, leaving the Agency to focus strictly on the extension and application of these scenarios to suit their strategic needs. However, the common vision and learning process generated through Foresight was not necessarily specific-to nor shared-within the water resources community.

With respect to strategic application, the scenario approach used created an effective 'wind tunnel' for strategy development and evaluation. The scenarios were used in a nationally consistent way to explore demand. The resilience of options to scenario conditions could then be explored in a regional manner. However, the use of demographic forecasts that were not a part of the Foresight scenario process and the informal nature of resilience testing can lead to some internal inconsistencies within the process and the results.

An additional point to note regarding the water resources strategies developed is that the scenarios were developed considering the demand on an annual basis. Annual variations in demand were accounted for by considering annual demand in a 'dry year'. The impact of climate on annual household water consumption was considered by adjusting micro-component levels to reflect dry-year demand and the impact of climate on spray irrigation was considered using a more complex analysis of optimum demand, economic demand and irrigation constraints. Seasonality in water supply was discussed relative to some of the strategy options discussed and the current differences between winter and summer surface water availability were shown in maps of the region (EA, 2001b). The heightened impact of drought on resource availability and demand was not considered in any detail.

To address the potential impact of climate change on water supply, the strategy document states that current research supports the conclusion that within the planning horizon of the strategy, 'public water supply systems will retain their existing yields' (EA, 2001a) though individual systems will respond in different ways. With respect to demand, climate change was considered in impacts to garden water use on a sample scenario and a preliminary assessment of climate change impact on spray irrigation demand was undertaken. The strategy also considered drought and the potential for climate change impacts through the sustainability appraisal of strategy options. Preference is given to strategy options that provide maximum security of supply during drought and are resilient to climate change. Drought is addressed in the Drought Management Plans (see EA, 2001d) that are produced for every Environment Agency Region separate from the scenario building and strategy process.

3.3. Global Water Scenarios - World Water Council

The scenarios developed under the auspices of the World Water Council will be an important reference for the global water industry for some years to come (Gallopin and Rijsberman, 2000). The World Water Council scenarios were developed as part of the World Water Vision programme.

Table 1. Percentage change from 1998 baseline in components of demand under scenario conditions (from Environment Agency, 2001b).

Component	Scenario							
	Alpha		Beta		Gamma		Delta	
	2010	2025	2010	2025	2010	2025	2010	2025
Household	+14	+26	+13	+14	+4	-29	+7	-21
Industrial and Commercial	+10	+5	+18	+38	-2	-32	-12	-43
Spray Irrigation	+24	+47	+24	+35	+12	-6	-6	-12
Leakage	-32	+143	-32	-33	-62	-72	-32	-49
Total	0	+53	+1	+5	-16	-42	-9	-34

The ultimate aim of the World Water Vision exercise is 'to generate global awareness of the water crisis that women and men face and of the possible solutions for addressing it' (Cosgrove and Rijsberman, 2000). The exercise is predicated on the belief in a crisis in the current management of water resources around the world and the need for action at all levels to create a sustainable future. The World Water Council's World Water Vision is a call to action, or in the words of Cosgrove and Rijsberman (2000) 'this is not an academic exercise – it is the start of a movement'. The global scenarios developed are key both to the formation of the World Water Vision and the development of strategies to achieve the World Water Vision.

The World Water Council's global water scenarios are based on 'alternative evolutions of major forces driving the global water situation: economic, demographic, technological, social, environmental and governance' (Gallopin and Rijsberman, 2000). The process began with a broad set of global drivers and a list of 'critical uncertainties' (factors which have an important role in shaping scenarios but have future values or outcomes difficult to determine at present) at a global scale. Critical uncertainties to be addressed at a global scale can be listed as

- water productivity trends
- expansion of irrigated agriculture
- massive increases in food production from rainfed agriculture
- dematerialisation of the economies
- national food self sufficiency vs. global food security
- availability of cheap water-purifying technologies
- public acceptance of genetically modified crops in the south and the north
- public opposition to large dams
- fundamental scientific discoveries
- scientific changes in human values and lifestyles

Source: (Gallopin and Rijsberman, 2000)

Three World Water Vision scenarios were developed. The first scenario was entitled "Business-as Usual" and was based primarily upon the projection of current trends. The second and third scenarios were built around the consideration of 'branch points'. The "Technology, Economics and the Private Sector" scenario is optimistic about uncertainties related to market systems, new technologies and regulatory control, in a context of decentralization of decision-making. The final scenario, "Values and Lifestyles" describes a future in which concerns over sustainability motivate a new ethic of water conservation, enacted by governments as well as civil society. The scenarios are meant to represent a wide span of plausible futures and to highlight issues that will need to be addressed if global water management is to become sustainable. The "Values and Lifestyles" scenario is put forward as the path to the most preferred future and it is used to define both the World Water Vision for 2025 and the process for achieving the Vision (Cosgrove and Rijsberman, 1998). The "Technology, Economics and Private Sector" and "Values and Lifestyles" scenarios are used to show the magnitude of change required (in investment or social change, respectively) diverting from the "Business as Usual" outcomes.

The scenarios were initially developed by a Scenario Development Panel consisting largely of academics, consultants and employees of multi-lateral lending and development institutions concerned with water management. They were subjected to four iterations of development through consultation and discussion with others involved in the World Water Vision process (modellers, reviewers and people working on regional and sectoral World Water Vision scenarios). Computational models were used to ensure scenario consistency and fill gaps. The World Water Vision process as a whole has a broader participatory component which has involved 15,000 people at various scales of engagement in various regions around the world (Gallopin and Rijsberman, 2000).

Climate change is not considered in the World Water Vision scenarios but drought frequency is anticipated to change over the course of some scenarios as a result of other factors. Drought frequency is not seen to be a determining factor in the scenario outcomes except in the final phases of the "Business as Usual" scenario where severe drought could tip the scenario into breakdown.

The World Water Council scenario process focuses strongly on the building of a common vision to influence attitudes in water resources management. Learning has also been promoted through the breadth of the exercise and publications generated by the programme. The question becomes the translation of that learning and influence into policy and its implementation. This raises the further question: who is the audience for the global scenarios and what influence does they hold? Of course, every citizen can act in the water polity through their consumption activities but the ability of global scenario building published in academic and practitioner forums to influence changes in public behaviour is questionable.

However, as a framework engaging regional stakeholders and creating regional scenarios the global WW Vision may be more effective in prompting change.

4. A DISCUSSION OF SCALE

In the formulation of a scenario exercise, consideration must be given to the scale of impacts, the scale of decision-making required to act on outcomes and the scale of stakeholders to be engaged. Where these issues cannot all be addressed at one scale, consideration must also be made of how to enact the process or transfer knowledge between scales.

In the three case studies discussed, a variety of scales are used to examine water resources issues. The World Water Council takes a global approach, the Social and Institutional Response to Climate Change and Climatic Hazards project focuses on the Thames catchment area and the Environment Agency applies a national approach to scenario building to each of it's eight regions in England and Wales. In each of these cases, scale affects the questions being asked in scenario building, the stakeholders involved, the possible futures generated and the potential applications for the results.

Gallopin et al. (1997) discuss the need for examination at different scalar levels to fully understand the entire system of socio-ecological interactions and to address the problem of sustainability. They also propose a division of different topics between the different geographical scales: a global scale to illuminate world changes in economic, cultural, demographic and environmental phenomena. A regional perspective is required to analyze the problems of acid rain, water allocation, and certain migration patterns. A national focus sheds light on many policies, trade patterns and security issues. A local view often is appropriate for evaluating land-change patterns, biodiversity and ground level pollution. In this framework water allocation is listed as an issue to be addressed at regional scale, however, no other aspects of water resources management are specifically discussed. The question therefore becomes 'what are the global, regional, national and local aspects of water resources management?'

4.1. Environmental Impacts

Water is a local resource, given not only the constraints of the hydrological cycle, but also of the hydro-social cycle (in which water is cheap to store but heavy to transport, limiting the spatial extent of water distribution systems in most parts of the world).

When looking at the issue of scale in the case studies, it is interesting to note that none of the scenarios are set at a sub-catchment scale where local environmental problems can be acute. However, water transfers between communities and catchments often make the relationship between supply and demand an issue that is larger than the area that is suffering the physical consequences of low flows. The challenge therefore becomes an issue of identifying a scale that encompasses the catchment-based impacts and the demand serviced by the supply network.

The Environment Agency case study tries to overcome the offset of demand and supply by focusing on balances in water company supply zones. Strategy options are put forward to meet the supply demand balance considering additional resources needed to meet current or potential environmental constraints to water provision as well as resources required to meet the scenario demands.

The Social and Institutional Responses to Climatic Change and Climatic Hazards project was conducted at catchment scale but due to the size of the catchment and the focus on management issues occurring across the region, or even nationally, local variations in demand and water management were not addressed.

Unlike the climate change problem, it is difficult to identify the global environmental commons in relation to the water debate. However, Gallopin and Rijsberman (2000) address this issue explicitly in the World Water Vision case by listing environmental drivers as: water-related diseases; soil salination; groundwater; and ecosystem health. In the "Business as Usual Scenario" a situation of stress and vulnerability evolves. However, this vulnerability generally only tips into crisis in the developing world where stress is greatest and fewer adaptive mechanisms are in place. In the "Technology, Economic and Private Sector" scenario new technology and economics are applied to reduce or eliminate environmental impacts, where it is economic to do so, while incurring some social costs. Again in this case, developed countries experience the greatest environmental improvement at the lowest social cost. However, environmental degradation and social impacts are more unevenly distributed between countries in the north and south. In the "Values and Lifestyle Scenario" a crisis prompts social and technological change that tackles environmental and social challenges in a sustainable manner.

Table 2: Scales of Water and Land Management Activities for the Thames Region.

	National	Regional	Supply Zone	Local
Department of Environment Food and Rural Affairs (DEFRA)	Water and Land Policy			
Environment Agency	Policy development with DEFRA	Policy implementation - strategy development	Policy implementation and Operations catchment management	
Ofwat	Policy (with DEFRA) and implementation	Some differentiation of policy		
Water Companies	Trade associations and consultation		Operations	
Secretary of State	National (Development) Planning Policy	Regional (Development) Planning Guidance	Public enquiries and development control	
Local Authorities		Input to Regional Planning Guidance Process through planning fora	Local Development Plans	

4.2. Decision Making

If scenarios are designed to prompt changes in a management system, the scale of the scenarios must consider the relevant scale of decision-making. This is not always as straight forward as it may sound. Decision-making for different aspects of a single system can be led by a variety of actors spread across various scales. This is certainly the case for water resources management in the Thames Region. Scales of decision making for land and water resource management activities are given in Table 2.

Above the national level, the UK, and therefore the Thames Region, are also subject to European Legislation. Article 130r of the Treaty of Rome place water quality issues under the environmental policy competency of the European Union (allowing decisions to be taken by a majority in the Council of Ministers and co-decision between Parliament and Council). In contrast, measures concerning town and country planning, land use and water resources planning (among others) require unanimity within the Council of Ministers under Article 130s (Kraemer, 1998). This distinction reflects the views of many European Union Member States that strategies related to regional development and infrastructure policy are of a national importance such that a veto to European legislation is required in these areas (Kallis and Butler, 2001). But more than answering questions concerning how to address water quality issues, the decision-making powers dictated by the Treaty of Rome raise questions regarding the appropriate scale of decision making for water resources management within the UK. The physical relationship between quality and quantity means that this division of power is not always uncomplicated. The European Union Water Framework Directive (2000/60/EC) which came into force on 22nd December 2000 is primarily a water quality document but concerns water quantity to the extent that it impacts quality.

The clear demarcation of water quality as a supra-national issue fits well with the issues addressed by the three case study scenario programmes. The World Water Vision programme, set at a global scale, was the only programme designed to examine water quality issues to any depth.

In England and Wales, water resources, and related land planning policies, are developed nationally and implemented nationally, regionally and locally through various public and private bodies of various levels of accountability (see Table 2). Policymaking occurs at a national level, water companies operate to supply zone boundaries (unrelated to any other planning or management boundaries) and regulators. Ofwat and the EA bridge scales. In the case of Ofwat this is translation of national policy to water company level. In the case of the Environment Agency bridging occurs between the national policy level, regional strategic level and local environmental protection and between all of these levels and water companies.

The Environment Agency scenarios can be seen as an example of the Environment Agency's need to cross scales. The scenario process was established nationally for regional application with outputs at regional and water company scales for the consideration of strategic action at all levels. The scenarios were used by Environment Agency decision makers to develop strategies for resource management. The majority of actions proposed through the strategy are to be carried out by the Agency, either independently or in partnership with other organisations. Externally, the strategy programme was focussed on communicating the Agency's strategic plan to specific decision makers, usually related to the national, regional and water company scales used in scenario development.

Scenario development was conducted at a regional scale in the Social and Institutional Responses to Climatic Change and Climatic Hazards project. However, many of the issues raised related to decision making at various scales. Environmental regulation and water resources planning are conducted by the Environment Agency at a regional level. However, this is set within a national framework for environmental regulation and planning set out by Environment Agency Head Office and Central Government. Water companies operate within the regional scale according to their own water company boundaries, but again within a framework set nationally by Ofwat. Therefore, the institutional capacity to manage drought risk at a regional level is limited. Tackling issues of national importance at a regional level may be ineffective in influencing the national debate on drought management unless a case can be made for the representative nature of the case study or the inordinate severity of regional impacts.

The appropriateness of these divisions in decision-making powers within and across scales can, in one way, be assessed through the examination of what capacity exists for visioning, strategising and creating cohesive, adaptive and effective water management futures. The inclusion of private companies and new regulatory agencies in water management and the institutional scales and structures developed through the privatisation of the water industry in England and Wales have added increased complexity to the water management polity. However, none of the scenario processes discussed have fully undertaken the challenge of examining this capacity or even explicitly acknowledging this problem.

4.3. Participation

Scenario building is an essential process of many participatory methods (van Asselt et al., 2001). It is a key means to build a group identity, to evaluate potential conflicts and to foster 'out of the box' thinking about complex issues. Indeed, personal experience by scenario 'facilitators' (based on informal discussions with several scenario teams in Europe) indicates that this is the true benefit of the technique, rather than the final report and quantified set of indicators. The scale of a scenario process can be related to both the size of the scenario building group and the individuals or groups involved.

What is the relationship between group size and creative processes of visioning? Do larger (and multi-national?) groups tend to create less extreme scenarios, or more variants? The Foresight and UKCIP 'official' national scenarios are less diverse than the Social and Institutional Responses to Climatic Change and Climatic Hazards 'small group' regional scenarios. And, the IPCC-SRES process (Nakicenovic and Swart, 2000), with a large panel and governmental review, resulted in 40 scenarios from four narrative storylines. The broad participation programme undertaken by the World Water Vision programme resulted in three scenarios.

Scenarios of what for whom? Are there differences in the legitimate constructions of futures? The sense of ownership is a function of group size, as well as the structure of the process. The SIRCH workshops and World Water Vision programme were driven by stakeholder issues, whereas the Environment Agency process was 'imposed' on stakeholders, although there was widespread consultation. A regulator's scenarios may imply restrictions in how the future is portrayed, and certainly raises questions of how the scenarios are to be used. If scenarios are meant to infer wisdom leading to action, this issue is critical.

Who is best placed to develop these scenarios? Should a water regulator formulate scenarios regarding the future of water supply? Is their use of the scenario approach limited by their regulatory role of private enterprise? Who facilitates the process—an independent entity or one of the leading stakeholders? To what extent are citizens involved and is their involvement substantive? At what scale do citizens feel they can best engage in scenario building and work towards building any normative futures envisioned. If the process is intended to alter long-term futures, build political support for different regulatory regimes or raise awareness among consumers, for instance, then citizen participation (including the institutions that affect the public) might be considered essential. For instance, one conclusion of the SIRCH project is that consumers, advocacy groups and the media are poorly represented in the water planning process, although they are key to dynamic demand management.

4.4. Transferring knowledge between scales

In the three examples of scenario building and application, various scales of analysis have been encountered – from the global to the regional. Each scale has its strengths and weaknesses. What recommendations can we offer for future scenario building and its role in water resources policy and application in regulation, operation and management?

As illustrated in Figure 2, past exercises in transferring scenario outputs between scales has focussed on the down-scaling of results (although the REGIS project was used to test the UKCIP scenarios (indicated by the dotted arrow)). This pattern is reinforced by the case study scenarios.

The World Water Vision scenarios address possible futures at a global scale, and are intended to be 'scaled-down' as required. National climate change scenarios were downscaled to regional impacts in the analysis of the Social and Institutional Responses to Climatic Change and Climatic Hazards scenarios (Downing, 2002).

In their scenario exercise, the Environmental Agency works downwards from a national framework to a regional strategy. The Agency also sums its regional plans to produce a national strategy but this is predominantly an exercise in summation as opposed to a formulation of an independent national vision that takes regional inconsistencies into account.

Downscaling, however, is not always a straightforward adjustment of scenario results to another scale. The UKCIP socio-economic scenarios, themselves based in part on the Foresight Panel, were 'downscaled' to the regional level in the REGIS project (Regional Climate Change impact and Response Studies in East Anglia and North West England, DETR Project CC0337). So, both the Social and Institutional Responses to Climatic Change and Climatic Hazards and REGIS scenarios have the same orientation. However, the two market-led scenarios have diverging interpretations of the changes to the supply/demand balance. Both indicate that increasing wealth and commodification of water leads to higher demand. However, In the Social and Institutional Responses to Climatic Change and Climatic Hazards 'vision' this is met through a variety of market mechanisms—that is drought risk is reduced through the ability of water companies to price water at its marginal value and develop new resources, leading to a healthy supply/demand balance although with some environmental impacts. This comparison illustrates the point that the interpretation of broad storylines, in the final analysis, rests with the stakeholders and analysts involved in the process—there is not a 'necessary' relationship between global driving forces and their interpretation at the local, sectoral level.

The possibilities for up-scaling scenario results were demonstrated in the project Integrated Visions for a Sustainable Europe (see http://www.icis. unimaas. nl/visions/) where scenario results were integrated across several sectors and scales. The Visions process involved the development of detailed and diverse regional scenarios in each of the project's case study regions along with research into global issues and futures (Rotmans et al., 2001). Global information was then down-scaled and regional scenarios up-scaled to guide and enrich scenarios built with stakeholders at the European level. The final scenarios addressed similarities and tensions between the regions under a European Framework (that did not necessarily include the maintenance of the European Union). The process of scenario building across these scales resulted in multi-dimensional scenarios with different regional winners and losers under the same European scenario conditions. The process also allows for learning across the scalar divide (how the tensions and interactions between different regional visions can work together or apart to form a cohesive vision with positive and negative aspects). At both the regional and European levels stakeholders were incorporated into various stages of the scenario building process.

In the water management polity in England and Wales, upscaling of scenarios would be most productive in areas were the potential for diverse regional, local or water company impacts needs to be accommodated within a national management plan.

5. WATER RESOURCES ISSUES

This section discusses some of the water resources issues that have been explored by the different case studies to see if the scenarios progress our understanding of the issues or have other practical application.

5.1. Demand management

Domestic demand for water is one of the key concerns for future water management in the Thames region. As all of the case studies dealt with changes in water demand, to some extent, comparisons between the different scenarios are informative in a discussion of scenario outcomes and applications.

The Social and Institutional Responses to Climatic Change and Climatic Hazards and World Water Council processes independently began their scenario-building processes with similar qualitative storylines upon which to base scenario development. The rubrics of the three Social and Institutional Responses to Climatic Change and Climatic Hazards scenarios (Trend Projection and Conventional Wisdom, Economic Growth and Market Enterprise, and Environmental Stewardship and Regulation) are broadly comparable to those of the World Water Vision scenarios (Business-as Usual; Technology, Economics and the Private Sector; and Values and Lifestyles). Both projects initiated a participatory process of scenario-building on the basis of these scenarios. However, differences emerge upon closer comparison of demand scenarios related to each project; for example, the Social and Institutional Responses to Climatic Change and Climatic Hazards "Trend Projection" and the World Water Vision "Business as Usual" projections differ on the degree to which efficiency measures will be implemented (the World Water Vision assuming increased efficiency of water use, the Social and Institutional Responses to Climatic Change and Climatic Hazards scenario arguing that customers will continue to demand no restrictions in demand or reductions in use). The 'grounding' of the Social and Institutional Responses to Climatic Change and Climatic Hazards scenarios in a particular region, in consultation with water management stakeholders, thus leads to very different specifics of each scenario despite the similar overall rubric. This leads us to question the applicability of the World Water Council scenarios at regional scales, and thus to question their robustness when applied to the Thames region.

Extending comparisons to the Environment Agency scenarios for water demand, points to further differences between scenario outcomes that result from the scenario building process, the stakeholders involved and the scaling of outcomes. In the Social and Institutional Responses to Climatic Change and Climatic Hazards project the Thames stakeholders, working at a regional level, envisioned that an "Economic Growth and Market Enterprise" scenario would result in an increase in domestic consumption by 200% for indoor use and 300% for outdoor use as householders would have an increased willingness to pay for water-related amenity. The Environment Agency's national assumptions for the "Beta" scenario (based on the "World Markets" socio-economic scenario) offset increases in discretionary water uses (power showers, swimming pools) with improved water efficiency of appliances through technical innovation and increased turn-over of home appliances (due to enlarged personal wealth) to arrive at a more modest 13% increase in household water demand in the Thames region. The Social and Institutional Responses to Climatic Change and Climatic Hazards "Environmental Stewardship and Regulation" scenario is similar to the Environment Agency "Gamma" and "Delta" ("Global Sustainability" and "Local Stewardship", respectively) scenarios with respect to in house water consumption. However, the Social and Institutional Responses to Climatic Change and Climatic Hazards scenarios translate environmental stewardship into increased garden water use while the Environment Agency carries water saving beliefs and behaviours into the garden as well as the home with the increasing prevalence of rainwater collection for garden watering.

The differences between demand in the Social and Institutional Responses to Climatic Change and Climatic Hazards and Environment Agency scenario programmes could possibly be explained by a variety of factors including: the Social and Institutional Responses to Climatic Change and Climatic Hazards project explicitly set out to create extreme scenarios so extreme possibilities for consumption are reasonable; the scenario building process involved a varied group of stakeholders (including private water companies) and interpretations were neither controlled by one stakeholder nor scrutinized by the participants; consumptive demands in Thames region could differ from other regions in England and Wales and the application of some national demand assumptions may be inappropriate; and the more detailed micro-component approach undertaken by the Environment Agency to calculate scenario demand may have resulted in more realistic results.

Scenarios are different from predictions and there is no correct answer for domestic demand within a scenario process. A more appropriate way to evaluate the demand scenarios is to look at their fitness for purpose within their individual scenario processes. The Social and Institutional Responses to Climatic Change and Climatic Hazards project sought extreme scenarios to test institutional adaptability. The Environmental Agency makes more detailed, less extreme, scenarios in line with national socio-economic predictions as a basis for making strategic decision regarding water supply. In this case, an extreme variation in scenarios could have severe repercussion for the over design of supply systems.

Though the scenario results must be looked at within the context of the scenario building process and the proposed application, this does not mean that the different scenarios should always stand separately. Global views on the possibilities for technological development of water saving appliances could inform the narrower national and regional views on water saving possibilities. The willingness of households to take up new technologies, however, is better assessed on a regional level were stakeholders have a hands-on feel for potential impact. It is also informative to break any demand prediction into detailed micro-components, as was done in the Environment Agency case study and some of the modelling conducted for the Social and Institutional Responses to Climatic Change and Climatic Hazards project, to ensure that numbers generated in visioning exercises have a basis in the realities of water consumption.

5.2. Connecting the Present to the Future: Pathways and Extreme Events

Two key issues in the development of scenarios are the timeline over which the scenario is developed and the method of development. These two factors work together to make the present into a possible future or to map a possible future back to the present. Development methods can be described in part by the extent to which behavioural or dynamic pathways are considered in moving from the present to the future. For example, extreme events may initiate step changes in innovation and institutional norms.

Extreme hydrological events such as droughts and floods pose significant challenges to water resource managers, and threats to water resource and supply system security. There is increasing evidence supporting the hypothesis that climate change will result in an increasing frequency and amplitude of extreme meteorological and hydrological events (Hulme et al., 2002). There is thus an increasingly pressing need for an examination of adaptive capacity, and planning for adaptive responses to crisis events.

In looking at the time element of the scenarios used, the Foresight scenarios are seen to describe potential futures in the UK between 2010 and 2050. The Environment Agency adaptation of these scenarios are assumed to gradually unfold over time starting from the present, however, the differing consequences of the four scenarios do not become immediately apparent due to the lead-in time required for many water resources projects and behavioural changes.

The Environment Agency scenarios do not consider the potential power of extreme events to effect changes in water supply or consumption. A dry-year scenario is used to calculate scenario water-demand levels but no dynamic between wet and dry years and how this might impact behaviour is discussed. Drought is seen to be a controllable event outside of resource planning (through drought management plans – for example see (EA, 2001d) In the Water Resources Strategies, the Agency does point to on-going and future studies where the links between climate change and demand and will be further explored. The outcomes of these studies will be considered in annual reviews of the water resources strategies.

The Social and Institutional Responses to Climatic Change and Climatic Hazards scenarios were not time dependent in that they looked at future possibilities to 2050 but did not tie these events to a specific time line. The Social and Institutional Responses to Climatic Change and Climatic Hazards project uses hazards as natural laboratories for water management in the belief that crises can provide insights into normal operations and the assumptions and priorities that support them. The project report discusses how real events can promote, or even force, change as part of a systematic learning process, by working in the direction of macro societal trends, by providing new opportunities for leadership, or working as a chain of events - for example the floods of autumn 2000 after the Easter floods of 1998. However, as the report points out, events can also be harnessed to prevent change or shift blame. Use of crisis or extreme events as a starting point for analysis may also detract from comprehending the human ecology of everyday situations (Downing et al., 2001).

In the Social and Institutional Responses to Climatic Change and Climatic Hazards example, changes to the frequency and severity of drought conditions are predominantly a result of climate change. Changes to the frequency and severity of drought impacts are a result of adaptation in the management of supply and demand.

The World Water Vision scenarios consider the unfolding of three possible scenarios from a branch point in 2005, a date chosen to reflect system lags and inertia. The branch point is a result of different human actions and beliefs culminating to pull society in a single direction.

The World Water Vision scenarios do consider water scarcity but do not discuss the possibility that the frequency or severity of drought events could occur as a result of changing climate patterns. Climate is not seen as a driver or critical uncertainty related to change in these scenarios. Drought is seen to occur mainly as a combination of deteriorating environmental quality, depleted groundwater levels and the difficulty and cost of establishing new sources of supply. The potential for drought to act as a 'sideswipe' (Gallopin et al., 1997) to the system trajectory is only mentioned in the later stages of the "Business as Usual" scenario. Differences in drought response (wither anticipatory or reactive) are not seen as branch points.

To deal with the potential for hazards adequately in scenario building, key drivers and uncertainties related to hazards, and the scales at which they occur, must be identified. One of the main purposes of pursuing a scenario approach is to explore a variety of futures and therefore it seems natural to explore a variety of risk, hazard event and adaptation scenarios. The occurrence of different degrees of hazard and resulting impact and adaptation can be examined through various scenarios.

In its integrated scenarios for Europe, the Visions project considers the ability of events (including flood events) to have significant impact on population patterns and water resource management strategies in Europe. The scenario 'Living on the Edge' was used to explore a 'period of chaos, crisis and crisis management. At both the European and regional levels coping with chaos was the rule. The patterns of coping, however, differed among regions and sectors.' (Rotmans et al., 2001). The Visions programme shows the advantage of the scenario approach where diverse scenarios covering crises and non-crises situations can promote further learning and communication regarding strategy robustness and similarities and tensions between changes prompted by hazard occurrence and those that develop as a result of more gradual societal change.To relate the discussion of events to the use of scenarios being discussed by this paper, we can refer back to Ringland's ideas of using scenarios for learning, building a common vision and developing strategy. Though learning and promoting common visions regarding hazard events are important, strategy development and testing is the key area where the consideration of events is vital to water resources scenarios. A strategy that is not robust to event occurrence could be fatally flawed.

5.3. Toward Scenarios of Vulnerability and Adaptation to Climate Change

The volatility of environmental change poses a challenge to traditional demand projections and industry design standards.

The conventional approach to climate change impact assessment relies on climate change scenarios to drive impacts models, working through a chain of linkages ending with an assessment of autonomous and planned adaptation. This climate scenario-driven impacts approach of linked models results in a cascade of uncertainty to the point where insight is lost.

More recently, people have sought to integrate socio-economic scenarios and climate change scenarios, with both following similar lines of construction. Scenarios of future greenhouse gas emissions are driven from global assumptions of economic growth, population growth and consumption. This translates fairly readily into global emissions, atmospheric concentrations, global climate change and regional to local climate scenarios (of course this chain is grand and entails significant uncertainties). A parallel track downscales the regional and global driving forces to local sensitivities to climate impacts.

The main problem with this approach is that the coarse regional scale hides the interesting issues of regional and local vulnerability. Local institutions, lives and livelihoods at risk cannot be easily located in the driving forces at a continental scale. As indicated above, the global-local link is not a unidirectional, necessary hierarchical relationship. In fact, it is almost entirely indeterminate. One would expect large local diversities even in a more uniform global economy.

An alternative is to cluster scenarios and regions according to the dominant axes of concerns. This is similar to the approach taken in the Foresight Panel (Figure 3), but involves axes of immediate concern to the analysis (rather than the large scale driving forces behind the scenarios themselves).

As a means of characterising risk and adaptive capacity, a simple matrix may be insightful (Figure 5). The qualitative 'vision' of future socio-economic conditions indicate the nature of vulnerability to climate change impacts, and the range of adaptive capacity. Climate impacts over the time scale of climate change (i.e., several decades and more into the future) are not themselves predictable, and should be considered impacts scenarios.

Figure 5 indicates how such a matrix of socio-economic scenarios of adaptive capacity and impacts scenarios can differentiate between regions and populations at-risk. For example, semi-arid urban areas already short of water and without adequate infrastructure or economic bases (regional towns in the Sahel come to mind) are almost certain

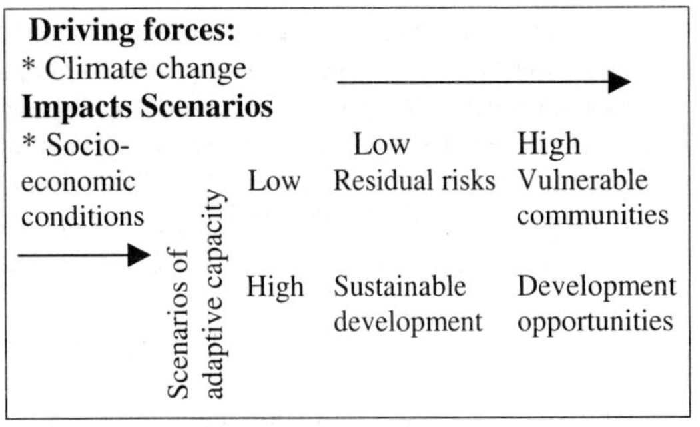

Figure 5. Matrix of scenarios of adaptive capacity and climate change impacts.

to fall in the upper right quadrant of potentially high climate impacts and low adaptive capacity (which might be labelled 'adapt to survive'). Where impacts could be high but adaptive capacity is also high - coastal tourism resorts in Antigua for example - development assistance could lead to new opportunities and a sustainable strategy to cope with increased risks (perhaps, 'adapt and thrive'). Where impacts are low, further information on the nature of climate change may be warranted, although a precautionary approach to situations of low capacity may be suitable to be prepared for surprises (such as a rapid increase in damaging extreme events) (corresponding to 'mainstreaming adaptation').

6. CONCLUSIONS

The three examples discussed in this paper can be seen to engage different decision-makers at different scales to differing degrees on different topics – all of which are productive and informative. However, the usefulness of these distinct scenario processes could potentially be improved if they were set into a framework where one set of scenarios could relate to and inform the development or interpretation of another scenario exercise or examination process. Explicitly discussing and comparing the differences in scale, engagement, learning, issues addressed and application of three examples of scenario building is a first step in this process. Though no two scenario processes or results will be the same (nor should they be), the discussion of these issues and the placement of the work within a larger common context will help to illustrate the key outcomes and findings that will frame work at smaller geographical scales or inform work at larger scales and higher levels of decision making.

Motivating effective adaptation to socio-institutional and environmental change requires participation across a broad spectrum of stakeholders and citizens. Scenario exercises tend to be static—initiated by a small group of experts (and stakeholders), developed in diverse ways with larger groups, then recorded, usually in reports and various visioning media (e.g., mock newspaper cuttings). More participatory interaction is necessary. For instance, a short-form of scenario development might be encouraged among user groups, leading to families of scenarios that take off from common themes. This would lead to more diversity and less control over the 'official' scenario, but that may be inevitable in any case (as concluded in the UKCIP process - (UK Climate Impacts Programme, 2000)).

Drawing upon our own background in integrated assessment, qualitative scenario building must blend the formalisms of models with the expert knowledge of institutional change. This is generally done off-line - by research teams that translate the qualitative storylines into quantitative indicators and test them in a range of models. More promising are ways to open up access to scenario models and exercises. Simple rule-based games available on the internet, for example, provide a new means to engage the public and key users over extended periods (for example, see FIRMA project water game at http://www.cpm.mmu.ac.uk/firma/).

A stakeholder-driven and public access exercise in scenario building is likely to lead to a large number of scenarios, particularly if regional and local variations are included. The International Panel on Climate Change resolved this issue by indicating marker scenarios that are typical of a family of scenarios. This is relatively tractable if the outcome of the family of scenarios has a singular identity-in the case of the IPCC greenhouse gas emissions leading to global climate change.

A more enterprising analysis would be to derive clusters, of regions and scenarios, by sampling across the population of plausible scenarios posed by diverse stakeholders. Of course, this requires consideration of the size of regional clusters (villages to nations?), the interpretation of clusters (e.g., common responses to globalisation?), and the link between local clusters and national and global processes. The analysis would at least show the diversity of plausible futures at different scales.

7. REFERENCES

Arnell, N. W., A. Jenkins, and D. G. George, 1994. The Implications of Climate Change for the National Rivers Authority. NRA Research and Development Report, 12, NRA, Bristol.

Brown, A. 1992. Inland water quality and pollution. Pages 87-105 *in* A. Brown, editor. The UK Environment. HMSO, London.

CEH, 1998. Hydrological Review of 1997. Centre for Ecology and Hydrology and British Geological Survey.

CEH, 1999. Hydrological Review of 1998. Centre for Ecology and Hydrology and British Geological Survey.

Cosgrove, W. J., and F. R. Rijsberman, 1998. Creating a vision for waterlife and the environment. *Water Policy* **1**:115-122.

Cosgrove, W. J., and F. R. Rijsberman, 2000. World Water Vision. Earthscan Publications Ltd, London.

CPRE, 1999. Hungry Housing. Council for the Protection of Rural England, London.

DOE, 1996. Review of the Potential Effects of Climate Change in the United Kingdom. Department of the Environment. HMSO, London.

Downing, T. E., 2002. Domestic water demand in southern England: Sensitivity to climate change and drought. *Water Policy* (submitted).

Downing, T. E., J. Handmer, K. Bakker, K. Lonsdale, and E. Crook, 2001. Hydrological Risk in the Thames Region: Prospects for Adapting to Climate Change. Environmental Change Institute, Oxford University, Oxford.

DTI, 1999. Environmental Futures. Department of Trade and Industry, Office of Science and Technology, London.

EA, 1999. Environment Agency Website - Thames statistics. *in*. http://www.environment-agency.gov.uk

EA, 2000. Tony Lawton, Hydrometry Information Group - pers. comm.

EA, 2001a. Water Resources for the Future: A Strategy for England and Wales. Environment Agency, Bristol.

EA, 2001b. Water Resources for the Future: A Strategy for Thames Region. Environment Agency, Thames Region, Reading.

EA, 2001c. A scenario approach to water demand forecasting. Environment Agency, National Water Demand Management Centre, Worthing.

EA, 2001d. Thames Region Drought Plan V.3.

Gallopin, G., A. Hammond, P. Raskin, and R. Swart, 1997. Branch Points: Global Scenarios and Human Choice. Stockholme Environment Institute, Stockholm.

Gallopin, G. C., and F. Rijsberman, 2000. Three global water scenarios. *International Journal of Water* **1**:16-40.

Hulme, M., J. Turnpenny, and J. Jenkins, 2002. Climate Change Scenarios for the United Kingdom: The UKCIP02 Briefing Report. Tyndall Centre for Climate Change Research, School of Environmental Sciences, University of East Anglia, Norwich.

IOH, 1995. Hydrological Data United Kingdom: 1994 Yearbook. Institute of Hydrology and British Geological Survey, Wallingford, Oxon, UK.

Kallis, G., and D. Butler, 2001. The EU water framework directive: measures and implications. *Water Policy* **3**:125-142.

Kraemer, R. A., 1998. Subsidiarity and water policy. *In:* F. N. Correia (editor). Selected issues in water resources management in Europe. A.A. Balkema, Rotterdam, p.387-417.

Marsh, T. J., 1995. Drought Returns to the United Kingdom. *Drought Network News* **7**:5-6.

Marsh, T. J., 1996. The 1995 UK Drought - a signal of climatic instability? Proceedings of the Institution of Civil Engineers: Water, Maritime & Energy 118:189 - 195.

Nakicenovic, N., and R. Swart (editors), 2000. Special Report on Emission Scenarios. Cambridge University Press, Cambridge.

NRA, 1995. Thames 21 - A Planning Perspective and a Sustainable Strategy for the Thames Region. National Rivers Authority, Reading, UK.

Raskin, P., T. Banuri, G. C. Gallopin, P. Gutman, A. Hammond, and R. Kates, 2002. Great Transitions: The Promose and Lure of the Times Ahead. Stockholme Environment Institute, Boston.

Ringland, G., 1998. Scenario Planning: Managing for the Future. John Wiley & Sons, Chichester, UK.

Robinson, J., 1990. Futures under glass: a recipe for people who hate to predict. Futures October 1990.

Rotmans, J., M. B. A. van Asselt, C. Anastasi, D. Rothman, S. Greeuw, and C. van Bers, 2001. Integrated Visions for a sustainable Europe: Summary of Project Results and Visions. ICIS, Maastricht.

UK Climate Impacts Programme, 2000. Socio-economic scenarios for climate change impact assessment: a guide to their use in the UK Climate Impacts Programme. UKCIP, Oxford.

van Asselt, M. B. A., J. Mellors, N. Rijkens-Klomp, S. C. H. Greeuw, K. G. P. Molendijk, P. J. Beers, and P. van Notten, 2001. Building Blocks for Participation in Integrated Assessment: A review of participatory methods. working paper I01-E003 ICIS, ICIS, Maastricht.

Water Sustainability: Science or Science Fiction? Perspective from One Scientist

WARREN W. WOOD
MS 430, U.S. Geological Survey, 12201 Sunrise Valley Drive, Reston, VA 20192, USA. E-mail: wwwood@usgs.gov and School of Geography and the Environment, University of Oxford, Mansfield Road, Oxford OX1 3TB, U.K.

ABSTRACT: A significant increase in the world's human population is predicted to occur in the next 100 years. This population will almost certainly require additional water resources for irrigated agriculture, domestic supply, and ecosystem support. Because most of the world's easily captured water is already identified and allocated, society must either change the present allocations or develop new sources to meet the expected demands. The philosophy of sustainability has been proposed as a management goal that will address water-resource-related issues. Sustainability is generally defined as the ability to meet the needs of the present generation without compromising the ability of future generations to meet their needs. "Meeting the needs" beyond basic survival requirements of a few liters of water per day requires many value judgments, and the assessment of these judgments lies outside the realm of science or engineering. Sustainability fails to address historical property rights or recognize that population increase is exponential although the resource is finite. Sustainability also implies stasis, yet uses and supply are temporally and spatially dynamic. Thus, the concept of sustainability is flawed not only as a goal, but also as a management objective and holistic sustainability (sustainability for all uses) is nearly impossible on a drainage-basin scale. Development of an alternative strategy based on water renewability that identifies and measures major sources, sinks, uses, and consumption in a system rather than "needs" can form the basis for informed policy decisions.

INTRODUCTION

A high probability exists for doubling the world's human population in the next 100 years, although the estimates vary widely (Lutz et al., 2001). This increase will almost certainly require significant additional water. Over a thousand million people have an inadequate diet (Latham, 1984); however, that appears to be improving slightly (United Nations, 1992). Improving nutrition of the existing population, a laudable goal, will require additional irrigated agriculture. It is also generally recognized that in the future, a larger share of the water will be demanded for fisheries restoration, biodiversity, and general ecosystem maintenance. Adding to these environmental pressures for additional water is a strong positive correlation between standard of living and water use. World population increased 42% from 3.8 thousand million to 5.4 thousand million during a twenty year period between the 1970's and 1990's, while water used increased 300% (Postel, 1999). If national economies continue to grow, there is likely to be increased water demand, even with a static population. What will be the source of this water? The water-resource perspectives expressed in this manuscript are global in nature, but it is recognized because of the limited-trading market that most water-resource problems and solutions are local in nature.

It is generally believed that there is no significant volume of water entering or leaving the earth from space; thus, water on earth is finite, but renewable. Approximately 97% of the world's water resources are saline and located in the oceans and seas. Of the 2.5% of the world's water resources that are fresh water, approximately 69% is stored in glaciers and ice caps that are not readily available, 30% is stored in ground water, and less than 1% resides in rivers and lakes at any instant (Shiklomanov, 1993). The distribution of water on earth is not uniform, nor is it correlated well with population. For example, the Amazon River accounts for 15% of the global runoff and 0.5% of the population, while China has 21% of the population and 7% of the global runoff (World Resources Report, 1994; Population Reference Bureau, 1998).

Approximately 110,000 km^3 of precipitation falls on the continents each year. Of this, approximately 50,000 km^3 are transpired from the native vegetation; 20,000 km^3 are transpired from irrigated crops and evaporated from surface reservoirs; 12,000 km^3 are utilized by man for industrial and domestic use; 20,000 km^3 are returned to the sea in unrecoverable floods; and 8,000 km^3 are unavailable for societal needs because of its remote locations (Postel et al., 1996). Of the water utilized by man, 69% is used for irrigation, 23% for industry, and 8% for domestic water supply (World Resources

Water Resources Perspectives: Evaluation, Management and Policy. Edited by A.S. Alsharhan and W.W. Wood.
Published in 2003 by Elsevier Science, Amsterdam, The Netherlands, p. 45-51.

Institute, 1992). This estimate for the percentage of irrigation use is approximately 10% larger than that given above by Postel et al. (1996). It is, however, within the uncertainty of the methods used in both approaches. Using the estimate from Postel et al. (1996) suggests that 32,000 km^3/yr is the amount of renewable water resources that society can utilize without removing additional native vegetation, building new dams, or transporting water from remote locations. Implicit in this estimate is the temporal stability of this resource. Yet, we know that there have been large shifts in climate in the last 20,000 years, and from historical records, it is clear that measurable climate changes can occur decadally.

The predicted increase in population will almost certainly require additional water for both domestic water supply and irrigated agriculture; however, the readily available water resources are already fully allocated. We are, in fact, borrowing from the future as unreplenished ground water from storage is used to produce approximately 8% of the total food crops (Postel, 2001). Worldwide, 12 to 15 million hectares of land, or about 1% of the total cropland, are lost each year to degradation due to salinization, erosion, desertification, and urbanization (Kendall and Pimentel et al., 1994). Additionally, nutrient depletion, over-cultivation, over-grazing, and acidification contribute to the reduction of yields. If other factors remain constant, the per capita food production will decline as the aquifers are de-watered and cropland is lost or its productivity is reduced. Irrigation of approximately 17% of the agricultural land produces 35 to 40% of the food crops (Postel, 1999). Thus, with increased food demands, there will almost certainly be a need for increased agricultural irrigation.

Associated with many grain crops is the loss of arable land due to salinization. Between 1.5 and 2 million hectares of arable land are lost each year to salinization, or approximately 0.1% of the total cropland (Kendall and Pimentel, 1994). High concentrations of solutes are detrimental to plant growth and to the tilth of the soil. In advanced cases of salinization, salts form on the land surface and completely inhibit plant growth.

Because of its relatively low value per unit mass and thus high transportation costs, water is not traded extensively like other commodities. Significant amounts of water, however, are used in the production of grains and meats, and this "virtual" or "embedded" water is traded worldwide (Allan, 1996). The lack of well-defined property rights to water is another limitation of trading. In the United States where some jurisdictions permit private ownership of water, an active market in trading water rights has recently developed (McCoy and Zachary, 1997), but these markets are local and constitute a minute portion of the total volume of annual water use. In some cases, water can be "wheeled" like electricity, thus reducing long transport for individual water molecules. That is, water moved to one area is used locally, and then new water from this area is passed on to the next user and so on until the transfer is complete. We are approaching, or in some areas have reached, a critical point in water resources where the total readily recoverable volume of water falling on a watershed has been allocated. The policy/management questions associated with these conditions are: 1) how does society reallocate the presently available water to accommodate changing needs, and/or 2) how does society acquire new water?

Water Uses

Water resources use can be classified into different categories: domestic water supply and waste treatment, livestock watering, irrigation, transportation, industrial (resource development/manufacturing), power generation, recreation, flood control, and ecosystem maintenance. These uses are not mutually exclusive as water is generally used in multiple ways as it moves through the basin, and any use typically impacts one or more of the other uses. One also must distinguish carefully between use and consumption. For example, in North America, power generation is the largest user of water, but its consumption is minuscule relative to irrigation. It appears that society's most critical, non-fungible water uses are domestic water supply, irrigation, livestock watering, and ecosystem support. Public recognition of the ecological importance of water-resource management has come to the forefront only in the last 40 years; thus, only small amounts of water have been formally allocated for ecosystem maintenance (Gleick, 2000). This late recognition, after most water already has been allocated, coupled with the difficulty of establishing the direct economic benefit of biodiversity and ecosystem maintenance in market-based economies, results in the smallest formal water allocation among the major uses. Improvement in freshwater fisheries and species diversity for ecosystem maintenance will almost certainly require significant additional water appropriation in the future.

In countries with democratic governments and property rights with environmental laws, and policies of pricing (including scaled water charges, subsidies, taxation) generally control water use. Because of the local nature of climate, topography, geology, historical development, infrastructure, and other factors, organizing water policy strategies around drainage basins might be a rational approach. Typically, however, there are many different and competing political jurisdictions in any given drainage basin, making decisive, coherent political action involving water-resources policy difficult. Additionally, thousands of private water and wastewater purveyors exist (National Research Council, 2001). This private ownership and a lack of universal property rights of water coupled with the fact that rivers are frequently used as boundaries between jurisdictions further inhibit the establishment of a coherent water policy.

Water policy as generated by legislative bodies tends to be descriptive, immediate, asks the question "how," seeks human-based assessment, and is rigid. That is, it is designed as a "one size fits all" solution. In contrast, scientists and engineers recognize the need for long-term studies, ask the question "why," seek peer-review assessment, and demand variability with each case, depending on the circumstances. Thus, in political environments, it is not surprising that there are few opportunities for scientists and engineers to make direct contributions to water policy. Water scientists and engineers can provide important answers to "what if" policy questions related to physical, biological, and chemical stresses to the environment. It is because of their understanding of the full dimension of the water cycle that hydrologists can frame discussion and provide insight into what the potential tradeoffs might be. I suspect that the greatest immediate contributions by scientists and engineers to the solution of societal water problems lie in improving the efficiency and productivity of present water uses. Although long-term water-resource allocation is clearly a policy issue, policy cannot create "new" water. It can, however, foster the intellectual and economic environment where scientists and engineers excel.

Potential Areas for Increased Efficiency and/or Conserving Water

Political or technological solutions to saving water must be distinguished from virtual and real forms of "saving" water. "Real" savings actually generate additional water supplies, but virtual savings are "paper" savings. Further, saving water is scale-dependent. On a global scale, the hydrologic cycle is closed: no water is lost; it only changes location and physical state. On the scale of river basins, however, it is extremely important to know where and how the major sources and sinks of the water occur. Postel (1999) uses an interesting example of irrigation in the ancient Nile valley to illustrate efficiency as a function of scale. This system was not efficient when viewed from the perspective of individual farms because much of the water spread over each field drained off without benefiting the crop. The runoff from upstream farmers became the supply for downstream farmers, however, and the system's efficiency was much greater when examined from the perspective of the entire Nile valley.

In general, there are five ways to "save" or make more productive use of the water available: 1) reduce evaporation and un-recycled seepage loss, such as those seepage losses to a closed basin or an ocean; 2) shift water from one use to another with higher food value and lower water demands; for example, growing wheat rather than cotton or raising chicken rather than beef; 3) develop more efficient crops, cultivation, and irrigation methods; 4) substitute capital for water, thereby increasing efficiency; and 5) capture of a greater percentage of the water in rivers before it is discharged to the oceans or closed basins. This last option is becoming less viable as the true costs of social disruption, inequitable benefits, loss of productive agricultural land, and loss of biodiversity and ecosystems are weighted against the societal benefits (McCully, 1996). In addition, the number of remaining topographically-suitable sites is limited. It also seems unlikely that a concerted world-wide effort will be mounted to harvest more water by removing native vegetation, although removal of invasive species is part of the Republic of South Africa's program of water conservation (Maitre et al., 1999). Deforestation continues to remove trees that use large quantities of water; thus, some gains in runoff are expected. In some cases, however, this deforestation also may lead to salinization (Cook et al., 1993), to increased sediment loading of streams, and to other harmful impacts. It appears that most of the increased productivity or savings must come from the other areas.

The "green revolution" (in the last half of the 20th century) that increased agricultural productivity so significantly through modification of crops and increased irrigation is expected to continue. Evaporative water losses from irrigated fields are difficult to measure because the loss by transpiration and evaporation is commonly lumped together as a measured or calculated parameter in water-balance studies. That is, both processes return water to the atmosphere in the vapor phase. Transpiration by crops is necessary and not considered a loss; so, it is necessary to separate transpiration and evaporation to get a measure of the true loss. If one assumes that 10% of the water applied to irrigation is lost to evaporation, this would amount to 250 x 10^9 m^3/yr worldwide. If this water could be saved and applied to grain production, it would generate approximately 250 million tons of grain, or 15% of the total annual grain harvest -- not an insignificant amount and about equal to the world's annual reserve amount. Note that approximately 1,000 tons of water are required to grow 1 ton of wheat (Ritschard and Tsao, 1978); 3,500 tons of water to grow one ton of chicken (Pimentel et al., 1997), 10,000 tons of water to grow one ton of beef (Falkenmark, 1994); and 17,000 tons of water to grow one ton of cotton (Kendall and Pimentel, 1994).

Approaches to reduce evapotranspiration can be grouped into four general areas: agronomic, technical, managerial, and policy/institutional. As stated above, scientists and engineers typically have little direct influence in policy/institutional aspects such as establishing water-use organizations, setting conservation-oriented pricing, or establishing a legal framework for property rights. They can, however, make important contributions in the other areas of reducing evaporation. Clearly, great strides are anticipated in plant physiology (whether by genetic engineering or traditional selection methods) so that plants will require less water for the same yields, or yields will be increased for the same water application (Somerville and Briscoe, 2001). Development of plant species that can produce high protein fodder from saltwater irrigation or plant growth in salt marshes is the long-term goal of genetic modification. Development of species that require fewer pesticides or lower concentration of nutrients will limit or minimize degradation of runoff water, making it available for other uses. Development and implementation of inexpensive micro-irrigation systems that deliver the correct amount of water to the roots at the precise time required by a plant is the ultimate goal of irrigation. For poor countries with subsistence agriculture, where the improvements are needed most, this must be accomplished on small, hectare-size plots with hand labor and a minimum of technically sophisticated hardware. Examples of this approach are Treadle pumps and portable, inexpensive drip-irrigation systems (Postel, 1999). Expansion of furrow irrigation using surge irrigation or low energy precision application (LEPA) on existing sprinkler systems result in significant increases in productivity and can quickly recover capital investment (Moseley, 1998); they are capital-intensive, however.

Minimizing degradation of water resources is a fruitful area for saving water. That is, if the properties are not changed or contaminants are either not introduced or successfully removed from water, the water is then available to downstream users in its original condition. Although technology permits relatively easy identification of point sources of contamination and has provided mechanisms for treating domestic sewerage and industrial discharge, agricultural runoff containing sediments, fertilizers, pesticides, and herbicides has resisted large-scale remediation in most areas. Technical remediation strategies for agricultural runoff range from land management and control of applications of agricultural chemicals to removal of these chemicals and sediments by buffer strips of riparian vegetation. Preliminary studies suggest that these buffer strips between crop fields and streams are a passive, economical method to significantly reduce sediment transport and some overland runoff of fertilizer, pesticides, and herbicides to streams (U. S. Environmental Protection Agency, 1995). Furthermore, riparian buffer strips have other environmental benefits to wildlife.

Although domestic water supply constitutes only about 8% of total water use, the cost per unit volume is the greatest of any water use. Thus, there are strong economic incentives for its conservation. Gleick (2001) uses a dramatic illustration of water loss by pointing out that a greater volume of water leaks from the Mexico City water distribution system than is used by the city of Rome! The use of treated sewerage for irrigation of ornamental plants (Gori and Lubello, 2000), Xeriscaping (Gregg and Curry, 1995), and treated wastewater for domestic supply (Haarhoff and Van der Merwe, 1996) are solutions that have proven effective in certain areas. The recent example of New York City saving 6 or 7 thousand million U. S. dollars by improving the environment of the catchment area rather than building filtration facilities (PCAST, 1998) illustrates the significant financial savings possible by alternative methods of water treatment. This approach not only improved the environment of the catchment area for many other uses, but also saved money that can be applied to repair leaking delivery pipes, install water meters, or supply low-flush toilets -- approaches that save water.

Policies of subsidies, taxes, and environmental regulations are the most common approaches to control of industrial uses of water in the market-driven economies of democratic societies. Because of the various uses in industry/manufacturing/

mining, it is impossible to make generalizations for savings and water reclamation. A recent example of the impact of environmental regulations on water use in the United States, however, is the requirement that heated water be cooled to the ambient temperature of the receiving body prior to its discharge. This requirement has reduced the amount of water withdrawn for thermal cooling (Solley et al., 1998). Seemingly, industry has found it more economical to cool and reuse the bulk of the previously used water rather than withdraw new water.

Sustainability

To address these water-resources problems, many popular press, radio, and television programs have proposed the concept of sustainability. Sustainability is on everyone's radar; it's the new "catchword." Sustainability is a relatively new term, less than 30 years old. According to the Oxford English Dictionary, it was first used in the early 1970's to explain an economy that was unchanging in time. In 1987, the Brundtland Commission of the United Nations popularized the word in the sense now used by managers of natural resources. This commission defined sustainability as "the ability to meet the needs of the present generation without compromising the ability of future generations to meet their needs." This philosophy appears to be a wealth distribution concept and does not address important water-management questions, such as sustainability at what level, for whom, in what time frame, using which economic model? Sustainability of any resource is considered a desirable goal by some analysts, although many resources are not renewable or recyclable on societally-relevant time scales, and thus, are not sustainable. Do we stop developing fossil fuels because they are not sustainable? The concept of sustainability does not provide the critical mechanisms for the distribution of the resource among competing factions, it is simply a philosophical goal without exact guidelines. Is this the correct way to cast a resource management problem? Because water has many different uses and functions in a drainage basin, sustainability requires a holistic approach; that is, what might be sustainable for one use is unsustainable for another, adding complexity to the concept of sustainability. Implicit in the concept of sustainability is the delay in use or consumption of a resource; yet the economic benefits from such a delay are not defined. A further concern is that sustainability implies stasis when, in fact, uses and consumption are dynamic both temporally and spatially. Clearly, the source is dynamic both in mass and space viewed on almost any relevant time scale. This awareness is of particular importance when predicting effects of long-term climate changes on water resources.

A more fundamental problem with the philosophy of sustainability is the flawed logic of comparing a finite resource with an exponential population growth - sustainability is ultimately impossible for critical, non-fungible uses. At some point, population demand will outstrip resource supply – it is the classic Malthusian argument. In the past, the availability of unallocated water and under-utilized, arable land as well as improvements in technology have saved humanity from the starvation predicted by Malthus (1798). We now have allocated most of the easily available water and arable land. Although we are adding cropland each year by deforestation, we are experiencing a net annual loss of arable land because of salinization, urbanization, desertification, and erosion (Barrow, 1991). Technical improvements continue to boost productivity but are ultimately constrained by the thermodynamics and kinetics of the systems, a concept not addressed by sustainability.

A strategy of water-resource management based on annual water renewability that identifies major uses, consumption, sources, and sinks in a system rather than "needs" is an approach that can be incorporated in policy decisions on how water should be allocated. In general, greater clarity of water-resources management issues can be gained by separating the different roles of policy, science, and engineering. It must be emphasized that scientists and engineering cannot make the value-based allocation decisions required for establishing policy, although they can help avoid unintended consequences by practicing a "what if" evaluation of suggested water policies that physically, chemically, or biologically stress a system.

CONCLUSIONS

It is my perception that there has been a failure by policy makers and the general public to recognize that utilization of water, land, and air isintimately related. What is needed in water-resources management is a more holistic conceptual framework that encompasses basin-scale hydrologic systems. It is clear that technological advances are to be the source of "new" water, as there remain opportunities for significant improvement in technology. The kinetic and thermodynamic limits of most processes are not presently constraining development. Specifically, it is desirable to identify methods and techniques that can economically save a significant portion of the water presently under-utilized or lost from the system. If, however, energy prices decline significantly or the rate of technological innovation for desalinization increases, then desalinization of seawater could provide some new water. Similarly, if one could grow grains or fodder with saline water, some of the world's food concerns would be assuaged. At this time, it appears that the greatest savings almost certainly will come from agriculture and reclamation of degraded water now discharged to the sea because these are the processes that constitute over 90% of the total human-related consumption.

The concept of sustainability has been proposed to resolve water-resource problems, but it is fundamentally flawed both as a goal and as an operational policy in several areas. It does not incorporate an economic model that justifies delayed use or non-use of a resource. It is based on the implied assumption that it is possible to satisfy all the demands on a finite resource in spite of an exponentially increasing population and increased water use. It does not address the important questions of distribution of the resource, and it implies a static use and supply as a function of time. The concept of water renewability and quantification of major water uses, sources, and sinks in a drainage basin can provide a viable base on which to evaluate water-resources policies for an increasingly thirsty and hungry world. This concept of additional water information for policy development is widely recognized by the scientific community (Food and Agriculture Organization of the United Nations, 2001).

If technologies are developed to address water-supply problems, they must become affordable and available in areas of subsistence agriculture where much of the population growth is expected to occur. There is no viable international market in water, so water savings generated in one area on earth cannot be readily transferred to another except by embedded water. That is, competitive advantages and trade may dictate the "water-rich" areas specializing in production of some goods, while "water poor" areas specialize in other goods. The goods are then traded, not the water. There is reluctance on the part of most consumers to pay for benefits that will not be delivered until sometime in the future. There is also significant inertia in customs (social behavior) and consumer choices. Furthermore, the long lifecycles of dams and other infrastructure investments are impediments to change in water use. Thus, overcoming economic, social, and political obstacles necessary to make use of technological changes is a significant challenge.

REFERENCES

Allan, J.A., 1996. Policy responses to the closure of water resources. In *Water policy: allocation and management in practice*; edited by P. Howsam and R. Cater. London: Chapman and Hall.

Barrow, C.J., 1991. *Land degradation – development and breakdown of terrestrial environments*. Cambridge, U.K.: Cambridge University Press.

Cook, P.G., A.L. Telfer, and G.R. Walker, 1993. Potential for salinization of groundwater beneath mallee areas of the Murray Basin. *Centre for Groundwater Studies Report No. 42/Engineering and Water Supply Department (S.A.) Report No. 93/6*. Adelaide, Australia.

Falkenmark, M., 1994. Landscape as life support provider, water-related limitations. In *Population: the complex reality*, ed. P. Graham-Smith, 103-116. London: The Royal Society.

Food and Agriculture organization of the United Nations, 2001, accessed January 2, 2001 at *http://www.fao.org/gtos/gtospub\pub26.hml.*

Gleick, P.H., 2000. The changing water paradigm - a look at twenty-first century water resources development. *Water International* 25: 127-138.

Gleick, P.H., 2001. Our water. *Scientific American* 284(2): 41-45.

Gregg, T., and J. Curry, 1995. Xeriscape: promises and pitfalls. *Proceedings of Conservation 96*, 165-168. Denver, CO: American Water Works Association.

Gori, R., and C. Lubello, 2000. Pilot plant for reclaimed wastewater reuse in nurseries. *Water Science and Technology* 42(1-2): 221-226.

Haarhoff, J., and B. Van der Merwe, 1996. Twenty-five years of wastewater reclamation in Windhoek Namibia. *Water Science and Technology* 33(10-11): 25-35.

Kendall, H.W., and D. Pimentel, 1994. Constraints on the expansion of the global food supply. *Ambio* 23(3): 198-205.

Latham, M.C, 1984. International nutrition and problems and policies. In *World food issues*. Ithaca, N.Y: Cornell University, Center for Analyses of World Food Issues, International Agriculture.

Lutz, W., W. Sanderson, and S. Scherbov, 2001. The end of world population growth. *Nature* 412: 543-545.

Malthus, T. R., 1798. An essay on the principle of population, as it affects the future improvement of society with remarks on the speculations of Mr. Godwin M., Condorcet, and other writers, Printed for J. Johnston in St. Pauls Church-yard, London.

Maitre, D.C.L., D.C. Scott, and C. Colvin, 1999. A review of information on interactions between vegetation and groundwater. *Water, South Africa* 25(2): 137-152.

McCoy, C., and G.P. Zachary, 1997. Bass play in water may presage big shift in its distribution. *Wall Street Journal*, New York, July 11.

McCully, P., 1997. *Silenced rivers*. London: Zed Books.

Moseley, L., 1998. Deaf Smith County producer pleased with drip irrigation. *The Cross Section,* 44, no. 3. Lubbock, TX: High Plains Underground Water Conservation District.

National Research Council, 2001. *Envisioning the agenda for water resources research in the twenty-first century*. Washington, D.C.: National Academy Press.

PCAST, 1998. *Teaming with life - investing in science to understand and use America's living capital.* Washington, D.C.: The President's Committee of Advisors on Science and Technology.

Pimentel, D., J. Houser, E. Preiss, O. White, H. Fang, L. Mesnick, T. Barsky, S. Tariche, J. Schreck, and S. Alpert, 1997. Water resources - agriculture, the environment, and society. *Bioscience* 47(2): 97-106.

Population Reference Bureau, 1998. *World population data sheet wall chart.* Washington, D.C.: Population Reference Bureau.

Postel, S.L., 1999. *Pillar of sand.* New York: W.W. Norton.

Postel, S.L., 2001. Growing more food with less water. *Scientific American* 284, no. 2: 46-51.

Postel, S.L., G.C. Daily, and P.R. Ehrlich, 1996. Human appropriation of renewable fresh water. *Science* 271: 785-788.

Ritschard, R. L., and K. Tsao, 1978. *Energy and water use in irrigated agriculture during drought conditions.* Lawrence Berkeley Laboratory Publication LBL-7866. University of California at Berkeley.

Shiklomanov, I.A., 1993. World fresh water resources. In *Water in crisis, a guide to the world's fresh water resources*, ed. P. H. Gleick. New York: Oxford University Press.

Somerville, C., and J. Briscoe, 2001. Genetic engineering and water. *Science* 292: 2217.

Solley, W.B., R.R. Pierce, and H.A. Perlman, 1998. Estimated use of water in the United States in 1995. *U. S. Geological Survey Circular 1200.* Reston, VA: U.S. Geological Survey.

United Nations, 1992. *World food supplies and prevalence of chronic undernutrition in developing regions as assessed in 1992.* New York: Statistical Analysis Service, Statistical Division, Economic and Social Policy Department, United Nations Food and Agriculture Organization.

U.S. Environmental Protection Agency, 1995. Water quality functions of riparian forest buffer systems in the Chesapeake Bay watershed. *U.S. Environmental Protection Agency Report EPA-R-95-004.* Annapolis, MD: Chesapeake Bay Program.

World Resources Institute, 1992. *World resources report 1992-1993.* New York: Oxford University Press.

World Resources Institute, 1994. *World resources report 1994-1995.* New York: Oxford University Press.

Improving Irrigation Efficiency through Technology Adoption: When will it Conserve Water?

NORMAN WHITTLESEY

Washington State University, Department of Agricultural Economics, P.O. Box 646210, Pullman, WA 99164-6210, U.S.A.
E-mail: whittles@wsu.edu

ABSTRACT: Irrigated agriculture accounts for as much as 90% of the water consumed in arid parts of the world. In these areas the availability of water for competing agriculture, domestic, industrial, and environmental uses is quite low. Since agriculture is normally the largest consumer of water, and the historical owner of water rights, it is frequently proposed that water could be made available for the other uses by increasing irrigation efficiency through investment in new irrigation technology. Unfortunately, this position is often advocated without full knowledge of how increased irrigation efficiency is achieved and the broader implications of this process.

This paper will explain the physical phenomena of irrigation management and plant growth involved in changing irrigation efficiency. The paper will then examine the conditions under which such changes might conserve water use in agriculture and make more available for other uses. The paper will also demonstrate conditions in which improvements in irrigation efficiency may actually reduce water availability for other uses, particularly in periods of drought or water shortage.

Irrigation return flows from inefficient irrigation systems are normally an important part of the water supply to subsequent users in irrigated river basins. In such settings the process of increasing irrigation efficiency can actually decrease instream water supplies during periods of drought when compared to use of older, less efficient technologies.

INTRODUCTION

A large portion of the world's population lives under conditions of water shortage, particularly in arid regions dependent upon irrigation for food production. Many different and innovative means are being sought to extend scarce water supplies in these regions to make them more productive and to satisfy increasing demands outside of agriculture. This competition for scarce water committed to agriculture causes proposed conservation measures to focus on water use in irrigated agriculture for two major reasons. First, irrigated agriculture accounts for more than 70% of total water withdrawals (Johnson et al., 2001), and 60-80% of total consumptive water use (Tiwari and Dinar, 2001), in producing about 40% of the world's food crops. These percentages are even larger in many arid regions of the world. Moreover, the demand for irrigation water is expected to grow with the needs of increasing populations. Second, water use efficiency in irrigated agriculture tends to be low worldwide, with national averages in the range of 25-50% (Tiwari and Dinar, 2001). One study recently noted that, "more than half the water entering irrigation distribution systems never makes it to the crops because of leakage and evaporation." (Johnson *et al.*, 2001, p. 1071). In many of these areas the on-farm irrigation efficiency is also quite low leading policy makers to seek means of inducing greater efficiency in the use of irrigation water.

A common policy prescription is to encourage irrigators to conserve water with economic incentives (Johnson *et al.*, 2001; Tiwari and Dinar, 2001; Willey and Diamant, 1995). A number of countries have instituted, or are considering instituting, higher agricultural water prices to discourage wasteful water use (Dinar and Subramanian, 1997). Scholars also have proposed subsidy programs to encourage the private adoption of improved on-farm irrigation technologies (Tiwari and Dinar, 2001). Regardless of the incentive for their achievement, proposals for improved irrigation efficiency through technology adoption and irrigation management are well intended. However, they do not always have the desired effect (Huffaker and Whittlesey, 2001; 1995 and Huffaker et al., 1998a; 1998b). This paper describes some of the basic concepts of irrigation management to illustrate the phenomena involved in changing irrigation efficiency and provides some guidance for expected water conservation under varying conditions.

Water Resources Perspectives: Evaluation, Management and Policy. Edited by A.S. Alsharhan and W.W. Wood.
Published in 2003 by Elsevier Science, Amsterdam, The Netherlands, p. 53-62.

Economics of Improved Irrigation Efficiency

Improving irrigation efficiency through adoption of new technology is not an end in itself. It is a means to other ends. Included in those ends are environmental improvement, increased crop yields and better quality crops, reduced labor requirements for irrigation, and possibly increased water supply for irrigation or other uses. There will always be some social tradeoffs involved in the adoption of new technology. While some individuals and segments of society will benefit from such change, there are likely to be others who will not benefit or will lose by changing the status quo conditions of the old technology. Changing irrigation technology can cause widespread effects in water allocation. Included are effects on upstream vs downstream users, point of diversion, groundwater vs surface flows, and timing of water diversions and return flows.

This paper will not attempt to formulate a benefit-cost framework for evaluating the process of technology adoption. Such a structure would have to focus on the level and incidence of investment costs, water reallocation issues, as well as formulate the various benefits that flow from a change in technology. However, this paper will focus on the single critical issue of whether and how changing irrigation efficiency through adoption of new technology will affect water conservation. It will be shown that, in most irrigated settings throughout the world, raising irrigation efficiency will not conserve water but will result in higher levels of water consumption by irrigated agriculture leaving less water for other uses. This is not to imply that such change is undesirable or should be discouraged in all cases. However, it should be considered for the right reasons and with full knowledge of all hydrologic and social effects that might flow from the proposed change.

Principles of Irrigation

To increase the productivity of scarce water and expensive energy resources used in irrigated agriculture, it is first necessary to understand the complex and integrated relationships among crops, soils, and irrigation systems that can affect the outcome of irrigation management decisions. The following discussion provides the reader with a basic understanding of irrigation to evaluate potential changes in irrigation efficiency and their effect on water conservation in agriculture (McNeal et al., 1986).

Irrigation Efficiency and Water Losses

Runoff and deep percolation water losses are a direct function of on-farm irrigation efficiency, which can be defined as the fraction of water applied to an irrigated field that is stored in the crop root zone and available for crop consumptive use. The remainder of the water is lost as evaporation, runoff, or deep percolation. In other words,

$$E = \frac{A - E - R - P}{A} = \frac{C}{A}$$

where:

E = on-farm irrigation efficiency

A = quantity of water applied to the field
E = quantity of water evaporated during application
R = quantity of water which runs off the field
P = quantity of water deep percolated below the root zone
C = quantity of water evapotranspired from the crop root

It should be noted that this form of efficiency function differs from the overall system efficiency, which includes seepage losses from delivery canals, laterals, and on-farm ditches. Such seepage losses, are operationally extremely important in terms of project-wide water requirements, localized groundwater accumulation, and salinity pickup. In general terms, the runoff losses of irrigation water are largely responsible for soil erosion from irrigated lands and associated phosphorous and nitrogen contained in eroded soil. Deep percolation losses of water carry most of the nitrogen fertilizer lost from cropland and are the cause of nitrate and salinity accumulation in subsurface and surface receiving waters.

Evaporative losses (E) can range from 3 to more than10 percent of applied water (A) for sprinkler-irrigated fields. Such losses during application are generally quite small for furrow-irrigated fields. Evaporative losses from the surfaces of recently-wetted (but empty) furrows are included in the C values for a given crop and area. Values of R are similarly assumed negligible for sprinkler-irrigated fields, though localized runoff can be substantial when water is applied at high rates to soils of low intake capacity and/or fields with steep slope. A common example can be observed near the periphery of center-pivot irrigation systems on fine-textured soils. Such runoff losses are generally assumed to remain primarily within-field. However, the within-field runoff does affect the uniformity of water application within the field. Runoff values are generally assumed to be more easily measured for furrow-irrigated fields than are values of deep percolation (P). This may not be true where well-designed subsurface drainage systems intercept virtually all deep percolation from an irrigated field without appreciable pickup of canal seepage or deep percolation from adjacent fields.

On-farm irrigation efficiency is a concept used widely throughout the irrigated world. Efficiency estimates are commonly available for the major crops and irrigation systems of a given physiographic region. Reliable values will depend upon crop, irrigation system, slope, soil depth, soil texture, set time, length of run, and irrigation stream size. Environmental factors such as temperature, humidity, and wind will also affect irrigation efficiency.

Rooting Zone, Depletable Moisture, and Number of Irrigations

The crop root zone can be defined as the depth of soil which is penetrated by crop roots. It may or may not include a limited depth of soil from which water is removed by upward flow into the actual crop root zone. Typical rooting depths for common irrigated crops range from 60 cm for peas, beans, and potatoes to 120 cm or more for irrigated alfalfa, (USDA, 1973). Typical values generally assume that no obstructions to root growth are present, such as hard pans, chemically indurated layers, or dry soil.

Deep percolation (P) is defined as that quantity of water which moves vertically through the soil below the crop root zone. Coarse-textured soils allow more rapid infiltration and percolation of water than do fine-textured (high-clay) soils. For this reason, coarse-textured soils have a larger potential for deep percolation, and fine-textured soils have a larger potential for surface runoff and erosion. Soils that are extremely coarse-textured may even result in such large percolation losses that furrow irrigation is not practical, and other irrigation methods must be used.

Fine-textured soils hold more water than do coarse-textured soils. Typical water holding capacities for major soil texture classes range from 25mm/30cm for sands to 100mm or more /30cm for clay loams (Whittlesey et al., 1986). These water holding capacities are derived from "field capacity" measurements, with field capacity defined as the amount of water remaining in a soil 24 to 48 hours after a heavy irrigation, under conditions of free drainage.

Depletable soil moisture can be defined as the quantity of moisture which may be withdrawn from the root zone without resultant crop moisture stress. At its maximum, it would be the water held between "field capacity" and the "permanent wilting point." At the permanent wilting point, however, soil water moves so slowly that it becomes insufficient to supply the daily crop water requirement. For sustained economic crop production under irrigated conditions, only about one-half of the total moisture range should be regarded as truly depletable. Unacceptable moisture stress would result if the wilting point were approached prior to each irrigation. Hence, the rooting depth of the crop, rate of water withdrawl by the crop, and water holding capacity of the soil combine to determine the frequency of irrigation required to meet crop needs for maximum production. Of course, the ability of the irrigation system to meet these needs will vary with the technology being employed.

For a given crop, coarse-textured soils have a smaller depletable moisture range than do fine-textured soils. Crops with deeper root systems also have larger depletable moisture values, and crops vary in their sensitivity to moisture stress. For example, although most crops appear able to repeatedly remove up to approximately 50 percent of the depletable soil moisture without resultant yield losses, potatoes are often assumed able to remove no more than 33 percent of the depletable soil moisture without a loss in yield or tuber quality. Irrigation frequency and number of irrigations per season are determined by the time required for depletable moisture to be extracted from the crop root zone.

Length of Set, Stream Size, and Length of Run

The time period that water is continuously applied to the field with furrow or sprinkler (not including center-pivot) irrigation can be defined as the length of set. Other factors being constant, the adequacy of irrigation should increase with length of set. Coarse-textured soils or shallow-rooted crops normally require a shorter length of set (but more frequent irrigations) than do fine-textured soils or deep-rooted crops.

In practice, the length of set is more commonly determined by convenience and water supply conditions rather than by exact field needs. The most common lengths of set are 12 or 24 hours, so that sprinkler irrigation lines, or siphon tubes, can be changed at specified times each day. A less common length of set, for fine-textured soils, deep-rooted crops, or sprinkler lines which apply water at uncommonly low rates, is 48 hours. For center-pivot systems, care is usually taken to ensure that a single revolution is completed in other than a 12, 24, 36, or 48 hour period. This is done to prevent recurring application of water to a given part of the field at the same time of day, for less water remains in the root zone following mid-day irrigation than following irrigation at other times. The same principles would apply to fixed systems such as drip or permanent sprinklers that require some rotation or sequencing in use to irrigate entire fields or farms.

When using gravity flow application systems, stream size normally refers to the size of irrigation water stream delivered to the head of irrigation furrows. Because of infiltration along the furrow, it differs markedly from the size of stream flowing from the tail-end of that same furrow. Stream size is an important factor affecting furrow advance rate (the velocity at which the initial water stream moves down a previously dry furrow). A larger stream size results in a faster furrow advance rate at a given site, other factors remaining constant.

Other factors also affect furrow advance rate. Fields with steeper slopes, finer-textured soils, or relatively smooth furrow surfaces all result in greater furrow advance rates. Conversely, recently tilled furrows result in slower furrow advance rates. A shorter period of furrow advance to the end of the field results in less deep percolation, since the head of the field is then being irrigated for a shorter time in relation to the total length of set. Hence, the uniformity of application is increased. A faster furrow advance rate may also result in greater runoff and erosion, however. The optimal furrow advance rate is one which minimizes the sum of deep percolation and runoff losses. This is commonly established as the advance rate for which the total advance time (for the stream to initially reach the end of the furrow) is approximately one-fourth the length of set (e.g., five to seven hours for a 24 hour set). An alternative approach is to use cut-back irrigation, where a larger stream size is used to push the stream through to the end of the furrow in minimal time, and then the stream size is cut back to minimize runoff and erosion.

The length of an irrigation furrow from top to bottom of the field is termed the length of run. The shorter the length of run, the shorter the furrow advance time, other factors remaining constant. Use of a shorter length of run also allows the use of a smaller stream size, which implies a slower furrow advance rate and reduced runoff losses. Hence, the use of a shorter length of run can reduce both deep percolation and runoff, and increase irrigation efficiency. But the labor and management costs of irrigation can be increased by reducing length of run.

Coefficient of Uniformity and Adequacy of Irrigation

Any system of irrigation, whether it is furrow or sprinkler, results in nonuniform application of water over the field surface. In practice, for example, considerable more deep percolation occurs immediately beneath irrigation furrows than beneath adjacent crop rows, particularly on coarse-textured soils. In addition, the top portions of furrow-irrigated fields generally receive more water than do the bottom portions, producing greater deep percolation. Under sprinkler irrigation, overlapping sprinkler patterns result in double applications to some parts of the field, while other portions receive less than a full irrigation. Variations in irrigation application uniformity are often described by the "coefficient of uniformity," with a value of 100% denoting completely uniform water application. Values for existing sprinkler systems are commonly in the range 60 to more than 80%. Uniformity coefficients for furrow irrigated fields may be 50 to 60% or less. Improving the uniformity of water application will normally result in a reduction in runoff and deep percolation losses, and increased irrigation efficiency.

The concept of "adequacy of irrigation" can be defined as the percentage of the root zone throughout a field which is restored to field capacity during each irrigation. Obtaining an adequacy level of 100% is generally not possible with most irrigation systems without incurring substantial percolation and/or runoff losses. In order to meet crop water needs, most fields are irrigated to an adequacy level of 75 to 87.5%. The latter level is commonly sought by U.S. Bureau of Reclamation design criteria, but is rarely met in fields for which variations in water application are statistically normally distributed. The value of the crop, the importance of uniform field appearance to the grower, and the availability of water and nutrients all interact to determine the appropriate adequacy level for given irrigated settings.

Modern irrigation technologies such as drip, low energy precision application (LEPA), and laser leveled border systems can consistently achieve higher coefficients of uniformity and levels of adequacy than most of the older systems. However, the capital costs and energy requirements can be quite large for the modern systems and such investments are only possible for high value crops.

Irrigation Timing

The timing of irrigations on growing crops was introduced in the previous discussion of field capacity and available moisture that can be withdrawn from the soil without affecting the yield potential of the crop. This topic is revisited to emphasize that the ability of irrigation management to precisely meet crop water demands is normally enhanced with improved irrigation technology. Older irrigation systems are limited in their ability to apply water to all parts of a field or farm in a timely manner. Older gravity flow systems generally require longer set times and result in less frequent irrigations than many of the modern irrigation technologies. Crops are required to extract a greater portion of the field capacity water supply than would be desired to achieve maximum potential crop yield. That is, some crop water stress is normally imposed on all or parts of the field due to the limitations of the irrigation system. One of the major contributions of modern irrigation technology is to increase the ability of management to meet crop water needs in a more timely manner. More timely applications will obviously increase crop yields through reduced water stress while also increasing the potential water consumption by crops, a matter frequently forgotten by the advocates of improved irrigation efficiency.

Uniformity, Adequacy and Irrigation Efficiency

Combining the concepts of uniformity and adequacy will demonstrate their relationship to irrigation efficiency. However, this relationship can only be described in a general way because it will depend upon many site specific conditions such as crop root depth, crop cover, soil type, soil slope, irrigation method (furrow or sprinkler), temperature, humidity, wind, etc. The general relationships of uniformity, adequacy and irrigation efficiency will be graphically demonstrated to show how changes in the uniformity coefficient or adequacy level will have predictable effects on irrigation efficiency.

Recall that the adequacy level measures the percent of a field that is irrigated to field capacity. The complement of the adequacy level is the percent of the field that is under irrigated or remains in a water deficit condition immediately after irrigation. The uniformity coefficient measures the amount of over irrigation and under irrigation throughout the field relative to a complete uniform application of water over the entire field area.

Figure 1 illustrates a field irrigated to a 75% level of adequacy with a uniformity coefficient of 60 percent and an assumed irrigation efficiency of 70%. The sloped line shows the relative amounts of water put into and through the root zone of the crop during irrigation. In this case 75% of the received more than sufficient water to reach field capacity. The area above the horizontal line showing field capacity is considered to be represents excess water application. In contrast 25% of the field is left with less than the field capacity quantity of water in the root zone. Of course, both over and under irrigation quantities will vary in amount throughout the field. Some small portions of the field will receive only small quantities of water exceeding field capacity, with a reminder of the 75% that is over irrigated ranging between these levels. The same kind of distribution will apply to that portion of the field that is under irrigated. The uniformity coefficient for applied water in Figure 1 is assumed to 60%, a measure of distortion from completely uniform application throughout the field. This is a relative term measuring the deviations of water application across the field relative to a complete uniform application.

It must be emphasized that Figure 1 illustrates the amount of water that is percolated below the crop root zone, shown by the horizontal dashed area, and the amount of additional water that would be required to bring the entire cropped area to field capacity, shown by the vertical dashed area. However, it does not show the amount of applied

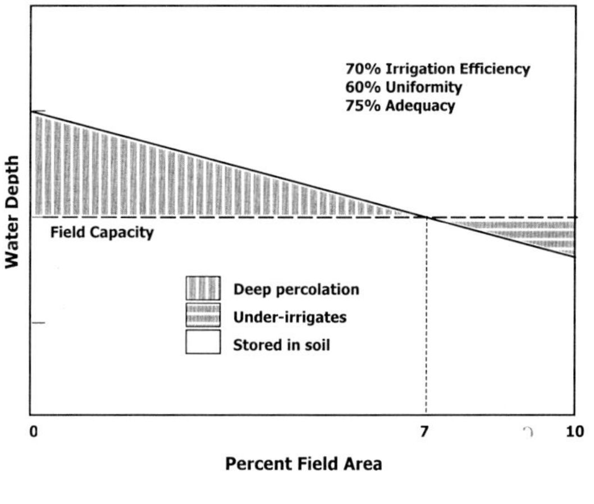

Figure 1. Irrigated field with 75 percent adequacy and 60 percent uniformity.

water that is lost to evaporation during application or the amount of water that leaves the field through surface runoff. With sprinkler application systems the evaporation during application can be 10% or more of total applied water but the expected field runoff is expected to be small. In contrast, gravity flow application systems (furrow or flood) will present negligible amounts of evaporative loss during an irrigation but will almost always provide some field runoff, the amount depending upon factors such as soils, length of run, and the technical sophistication of the irrigation system. Runoff with gravity flow systems can normally be avoided only with laser leveled fields with borders or severely under irrigating a major portion of the field.

The underlying irrigation technology is not indicated in Figure 1 and, hence, it is not possible to provide a precise measure of the resulting irrigation efficiency. Since both sprinkler and gravity flow irrigation systems can have losses not illustrated in the figure, the indicated irrigation efficiency of 70% is only an approximation for purposes of this discussion. However, regardless of the assumed irrigation technology, it is possible to predict the direction of change in the irrigation efficiency that will result from a change in either irrigation uniformity or adequacy.

Figure 2 holds the uniformity coefficient constant at 60% while observing three different assumed levels of adequacy. The slope of the line representing applied water is the same in each panel indicating a similar irrigation technology in each case. The first panel shows a case of 50% adequacy, leaving one-half the field under irrigated and a relative small amount of deep percolation on the remainder of the field. In this situation the irrigation efficiency will be relatively high, shown to be 85%. This scenario could be indicative of a drought situation where there is insufficient water to provide a greater level of adequacy. Of course, average crop yield will be well below full potential yield because much of the field receives insufficient water to meet full crop needs. It is shown, however, that under conditions of water shortage irrigation efficiency will normally increase even without changing irrigation technology.

The center panel of Figure 2 shows a more typical condition in which the field is irrigated to an adequacy level of 75%. In this case only 25% of the field is under irrigated providing a much lower level of crop water stress, but with more water devoted to deep percolation. Crop yields will increase over the condition of the first panel but irrigation efficiency decreases to 70%. The third panel of Figure 2 represents a situation of water abundance wherein the field is irrigated to an adequacy level of 90%. In order to push the field further toward maximum potential crop yield it is necessary to commit greater quantities of water to deep percolation. It is probable that other water losses such as runoff or evaporation will be also increased. The indicated irrigation efficiency for this situation is only 55 percent.

The process of changing irrigation efficiency through technology change that results in improvements in irrigation uniformity and adequacy is illustrated in Figure 3. The first panel of this figure reflects the same conditions as the center panel of Figure 2. This scenario includes a coefficient of uniformity of 60% and an adequacy level of 75 percent. These conditions are assumed to result in an irrigation efficiency of 70%.

The center panel of Figure 3 changes the coefficient of uniformity to 70% while maintaining the adequacy level at 75%. The amount of land that remains over irrigated and under irrigated does not change but the amount of water devoted to percolation is reduced while improving irrigation efficiency to 75%. Also, since the degree of under irrigation is reduced, water consumption by crops will rise and crop yields would be expected to increase.

The third panel of Figure (3) reflects a more modern technology that achieves an adequacy level of 90% while also increasing the coefficient of uniformity to 80%. In this case the irrigation efficiency is increased to 85%. Only 10% of land is under irrigated and to a small degree while the amount of over irrigation is simultaneously reduced.

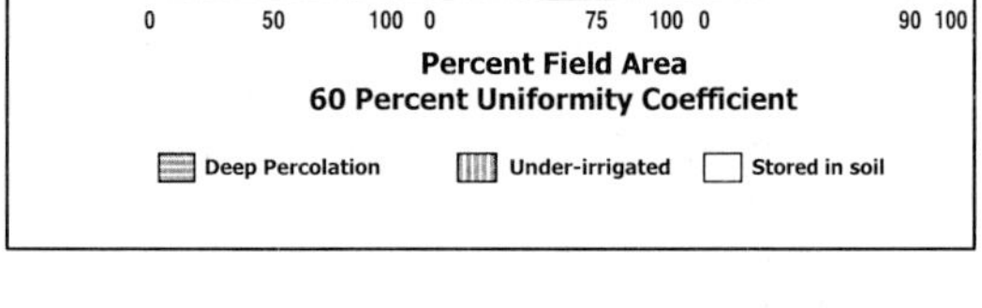

Figure 2. Changing quantities of applied water with the same irrigation technology.

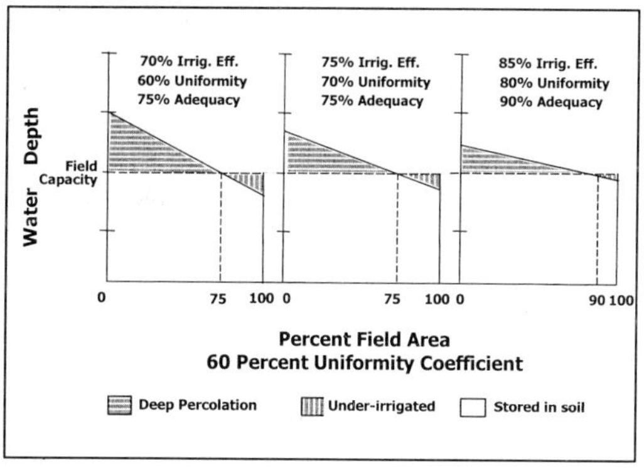

Figure 3.Changing irrigation efficiency through improvements in irrigation uniformity and adequacy.

Note that the amount of land as well as the degree of under irrigation is reduced as irrigation efficiency is increased through technology adoption. That is, the process of increasing irrigation efficiency in this manner provides more water for crop consumption. It is probable that the timing of umption. It is probable that the timing of water applications will also improve with the changing irrigation technology, increasing the actual water consumption by crops even more than indicated by Figure (3). While crop yields are normally increased with better irrigation technology, crop water consumption increases as well.

Figure (4) is illustrates the typical relationship of crop yield, crop water consumptive use, and applied irrigation water. For most irrigated field crops the response of yield to water consumption by the crop is linear to the point of maximum attainable yield (y^m). However, as illustrated by Figure (2) the process of increasing the portion of a field that is fully irrigated with a given technology will normally result in decreasing irrigation efficiency. This is shown in Figure (4) by the increasing horizontal distance between the applied water curve and the crop yield curve in approaching the level of maximum attainable crop yield. This distance, representing water losses to deep percolation, evaporation, and runoff, can be reduced by moving to modern, more efficient irrigation technologies. In fact, older irrigation technologies are generally incapable of providing water adequacy and timing to achieve maximum attainable crop yield for a given climatic condition without providing a very low level of irrigation efficiency. Only the modern irrigation technologies are generally capable of meeting crop water need throughout the growing season without significant losses of water to percolation, evaporation (in the case of sprinklers), and runoff.

Irrigation Efficiency vs Water Conservation

We now return to the question of whether water conservation is achieved by increasing irrigation efficiency through technology adoption. It has been shown that increasing irrigation efficiency will result in increased water consumption by crops. Hence, the answer to the posed question will depend upon other site-specific conditions including the nature and source of water supply and the fate of the existing water losses through percolation and runoff.

In most irrigated regions the water losses by percolation beyond the crop root zone or field runoff are termed irrigation return flows that either return to the aquifer or the surface water supplies of a river basin hydrology and are available for reuse. Hence, irrigated regions with low irrigation efficiency on individual farms can have relatively high levels of irrigation efficiency over the entire region through the reuse of irrigation return flows, though crops will not reach maximum attainable yield in most of the region. However, when the percolation and runoff portions of applied irrigation water escape the local hydrologic system and are totally lost to subsequent use it is more likely that changing irrigation technology (and irrigation efficiency) will actually conserve water. The extreme cases are examined in following figures.

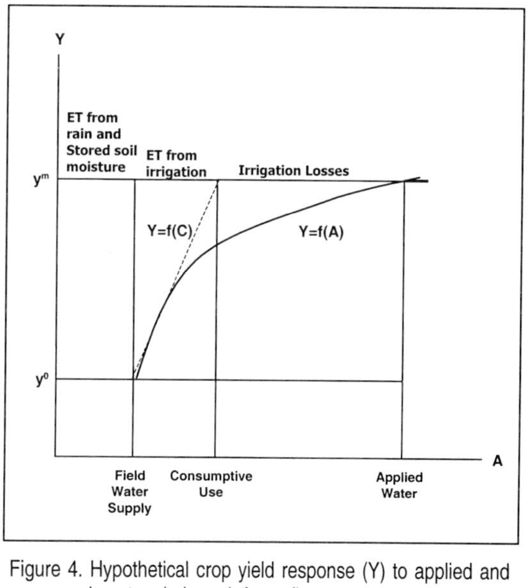

Figure 4. Hypothetical crop yield response (Y) to applied and consumed water (adapted from figure 1 in Bernardo and Whittlesey, 1989).

Before proceeding, however, it must be noted that moving from existing gravity flow irrigation systems to most forms of sprinkler irrigation will increase the evaporation losses of water during application. Since evaporated water, like crop consumptive use, is lost to the local hydrologic system, this process of increasing irrigation efficiency must offset both forms of water loss through savings of previous permanent losses of irrigation return flows if water savings are to occur.

Figure (5) illustrates a case of increasing irrigation efficiency where the field losses of water are not captured for subsequent reuse. The setting of this example is assumed to be a river with 10 units of water flow prior to the point of irrigation diversion. In panel (a) the irrigated agriculture of this example diverts eight units of water to provide four units of crop consumptive use. Irrigation efficiency is 50% in this case. The other half of irrigation diversions (four units of water) is presumed to be completely lost to the local hydrologic system and is unavailable for subsequent use. Water remaining in the river or water source is two units after irrigation diversion occurs.

Panel (b) of Figure (5) shows the same setting after raising irrigation efficiency to 83%. However, the increase in irrigation efficiency also raises crop consumptive use in the same land area from four to five units of water. Now irrigation diversions are six units of water leaving four units in the water source. Unrecoverable water losses from irrigation are now reduced to one unit of water. Though crop consumptive use has increased, more water remains in the river (water supply) and less water is lost from the hydrologic system. In this case, water conservation has occurred as a result of improved irrigation efficiency.

Throughout the irrigated world it is most common that the irrigation return flows are recaptured and reused in the local hydrologic system. However, there are concrete examples of where efficiency improvements might conserve water. If the water supply is an aquifer that cannot be recharged with irrigation return flows and the "lost" water enters a sink making it unrecoverable a sink making it unrecoverable for subsequent use, the reduction in irrigation diversions does extend the life of the aquifer and represents a form of water conservation. Other examples might be irrigation that returns surface runoff directly to saline lakes such as the Dead Sea, Great Salt Lake, or Aral Sea.

Figure (6) represents a case wherein the irrigation return flows are recaptured in the local hydrologic system and are available for reuse. Panel (a) is similar to panel (a) of Figure (5) in terms of irrigation diversion, irrigation efficiency, and crop consumptive use. Now, however, the four units of irrigation return flows are returned to the water source to provide six units of water available for further use.

Panel (b) shows this setting after increasing irrigation efficiency to 83%. Again, water diversions are reduced to six units of water while crop consumptive use increases to five units.

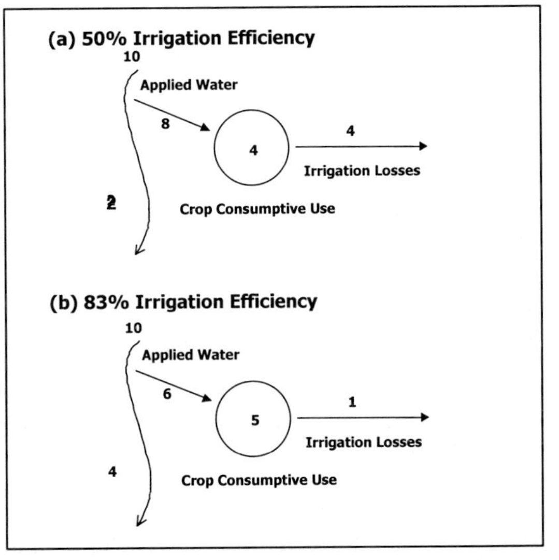

Figure 5. Improved irrigation efficiency without captured irrigation return flows.

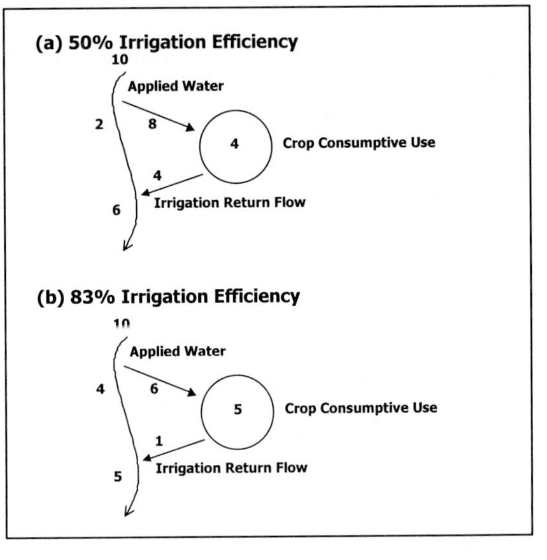

Figure 6. Improving irrigation efficiency with captured irrigation return flows.

Importantly, the water remaining in stream after accounting for one unit of irrigation return flows is only five units, compared to the six units before increasing irrigation efficiency. While the efficiency of water use by irrigated agriculture is improved by changing irrigation technology, there remains less water for other uses including downstream irrigation. Water conservation has not occurred in this case, though crop production will have increased.

Irrigation return flows are an essential component of water systems in many basins. For example, one study estimated that 63% of the water diverted from the Snake River in the western United States for irrigation eventually returns to the river (Hydrosphere, 1991). Another study estimated that irrigation return flows constitute 47% of all water diverted for irrigation in the western United States (Pulver, 1988). A reduction in applied water is not a measure of conserved water when irrigation retur flows are recycled in the river basin to become a portion of the usable water supply (Huffaker and Whittlesey, 1995, 2001).

CONCLUSIONS

The impact of increasing irrigation efficiency through technology adoption on water conservation will depend upon the setting in which it occurs. In the uncommon condition that irrigation return flows cannot be captured and reused, it is possible that water conservation will occur through technology adoption. Also, agriculture will benefit from the increased consumptive use of water through higher crop yields.

When the water source for irrigation is an aquifer that is being mined and cannot be recharged, the process of increasing irrigation efficiency can reduce irrigation diversions and extend the life of the aquifer and/or provide more water for other uses. This occurs whether or not the irrigation return flows can be recaptured for subsequent use. Though more water may be consumed by agriculture in this case, some benefits to water supplies can be derived by increasing irrigation efficiency.

The more common condition, however, is that the return flows from irrigation are available for reuse locally and downstream. Field runoff is quickly recaptured on the same farm or nearby farms for reuse. Percolated water can be picked up by field drains for subsequent use or returned to the river system through seepage and spring flow, sometimes with a considerable time lag. In all such settings, the process of increasing irrigation efficiency without reducing irrigated land area will result in more water consumption by irrigated crops, leaving less water for other uses including downstream irrigation. While improving irrigation efficiency through technology adoption can provide benefits to water quality and crop yields, such changes in irrigation technology generally should not be advocated to conserve water.

This paper has focused primarily on the principles of irrigation as they apply to a single field or farm. Broad areas or river basins with extensive irrigation development can experience compounded effects of improved irrigation efficiency if new technology is widely adopted. In these settings increased irrigation efficiency can leave more water instream at the bottom of a river basin in average or above average water supply periods. However, in drought periods the widespread increase in irrigation efficiency will leave less water instream at the bottom of the basin. It is possible that low priority or downstream irrigation water rights will be diminished during drought to a greater extent than without the improved irrigation efficiency. Certainly there will be less water available for other uses during drought periods with the increased irrigation efficiency.

REFERENCES

Bernardo, D. and Whittlesey, N., 1989. Factor Demand in Irrigated Agriculture Under Conditions of Restricted Water Supplies. United States Department of Agriculture, Economic Research Service, Technical Bulletin Number 1765.

Dinar, A. and Subramanian, A., 1997. Water Pricing Experiences: An International Perspective, Technical Paper No. 386, World Bank, Washington, DC.

Huffaker, R. and Whittlesey, N., 1995. Agricultural Water Conservation Legislation: Will It Save Water?, *Choices*, Fourth Quarter, 24-28.

Huffaker, R. and Whittlesey, N., 2003. A Theoretical Analysis of Economic Incentive Policies Encouraging Agricultural Water Conservation. *International Journal of Water Resources Development* (Accepted)

Huffaker, R., Whittlesey, N., Michelsen, A., Taylor, R., and McGuckin, T., 1998a. Evaluating the Effectiveness of Conservation Water-Pricing Programs. *Journal of Agricultural and Resource Economics* 23: 12-19.

Huffaker, R., Whittlesey, N., Michelsen, A., Taylor, R., and McGuckin, T., 1998b. Evaluating the Effectiveness of Conservation Water-Pricing Programs: Reply. *Journal of Agricultural and Resource Economics* 23: 571-572.

Hydrosphere Resource Consultants, 1991. Water Supplies to Promote Juvenile Anadromous Fish Migration in the Snake River Basin: A Report to the National Marine Fisheries Service. Contract No: 50ABNF900105, Boulder, Colorado.

James, L., 1988. Principles of Farm Irrigation System Design. John Wiley and Sons, Chapter 2.

Johnson, N., Revenga, C. and Echeverria, J., 2001. Managing water for People and Nature. *Science* 292: 1071-72.

Pulver, R., 1988. Liability Rules as a Solution to the Problem of Waste in Western Water Law: An Economic Analysis, *California Letters Review*, 76, p. 671, citing U.S. Bureau of Reclamation (1973), *Shut Off the Water—The Root Zone is Full: A Study of Irrigation Water Use.*

Tiwari, D. and Dinar, A., 2001. Role and Use of Economic Incentives in Irrigated Agriculture. World Bank, Washington, DC. http://wbln0018.worldbank.org/essd/essd.nsf/All/B258214F738E38FB852569FB00596684

Whittlesey, N., McNeal, B., and Obersinner, V., 1986. Concepts Affecting Irrigation Management. Chapter 5 in *Energy and Water Management in Western Irrigated Agriculture,* edited by Norman K. Whittlesey, Westview Press, USA.

Willey, Z. and Diamant, A., 1995. *Using Water Prices to Encourage Conservation: The Design and Evaluation of Water Rates for Irrigation Districts in Washington State*. Study commissioned by State of Washington, prepared by Environmental Defense Fund, New York.

Lessons Learnt from the Impact of the Neglected Role of Groundwater in Spain's Water Policy

RAMÓN LLAMAS
Department of Geodinamics, Complutense University, 28040 Madrid, Spain. E-mail: mrllamas@geo.ucm.es

ABSTRACT: Spain is the most arid country in Europe. Issues and conflicts related to water management are more relevant to the country's development than that of in other European countries. In July of 2001 the Spanish Parliament enacted the Law of the National Water Plan. This Plan has induced considerable controversy among diverse social sectors and lobbies: political parties, farmers associations, large construction firms, conservation groups, scholars, and others. The real role of groundwater in Spain was practically ignored in the initial proposal for the National Water Plan (NWP). The approved bill included a series of provisions about groundwater which, if they are enforced, will make it difficult to maintain some old paradigms. One of these is the need to continue the large subsides to build large hydraulic structures. One such structure is the aqueduct of the Ebro River, designed for the transfer of water to several Mediterranean regions. Spain is the country -except four small countries- with the largest number of large dams per person. It is also the country of the European Union which uses the lowest proportion of groundwater for urban water supply. This may be the result of the central government's policy in the nineteenth Century which restricted the use of groundwater as the source of supply for Madrid. This policy later spread trough out the rest of the country.

Nevertheless, farmers do not usually follow the government's paradigms and during the last 30 to 40 years they have considerably increased the use of groundwater for irrigation. Today about one million hectares, out of a total of 3.5 million hectares, are irrigated with groundwater. The economic value of the crops and the employment generated by the use of groundwater irrigation is higher than that from surface water irrigation. It can be said that groundwater irrigation has produced significantly "more crops and jobs per drop". Most of this spectacular agricultural development has been made with scarce planning and very limited control by the government water authorities. This has induced legal and ecological chaos in a few regions. The National Water Plan is a blend of new and old paradigms. The lobbies of the followers of the old paradigms are still powerful and it is presently difficult to foresee which will be the practical results of this plan.

1. INTRODUCTION

1.1. Background

Spain is the most arid country in Europe. Groundwater development has significantly increased during the last half century in most semiarid or arid countries of the world. This development has been mainly undertaken by a large number of small (private or public) developers and often the scientific or technological control of this development by the responsible regulatory agencies has been minor, if any at all. In contrast, the surface water projects developed during the same period are usually of larger dimension and have been designed, financed, and constructed by government entities which normally manage or control the operation of such irrigation or urban public water supply systems. This historical situation has often produced two effects: 1) most regulators have limited understanding and poor data on the groundwater situation and value; 2) in some cases the lack of control on groundwater development has caused problems such as decline of the water level in wells, decrease of well yield, degradation of water quality, land subsidence or collapse, interference with streams and/or surface water bodies, and ecological impact on wetlands or riparian forests.

In Spain, and almost everywhere, these problems have been frequently magnified or exaggerated by groups with lack of hydrogeological know-how, professional bias or vested interests. For instance, the World Water Council (2000, page 13) states that: *"Aquifers are being mined at an unprecedent rate -10% of world's agricultural production depends on using mineral groundwater..."*. However, this 10% estimation is not based on any reliable data. In recent decades, the term groundwater overexploitation has become a kind of *hydromyth* that has flooded the water resources literature. A usual axiom derived from this pervasive *hydromyth* is that groundwater is an unreliable and fragile resource that should only be developed if the conventional large surface water

Water Resources Perspectives: Evaluation, Management and Policy. Edited by A.S. Alsharhan and W.W. Wood.
Published in 2003 by Elsevier Science, Amsterdam, The Netherlands, p. 63-81.

projects are not feasible. This groundwater resource fragility concept has been dominant in Spain during the last twenty years (Llamas, 1985; López-Gunn and Llamas, 2000). In the last decade a good number of authors have also voiced this fragility as a common issue (Postel, 1999; Seckler *et al.*, 1998).

Another usual *hydromyth* is to consider that groundwater mining, that is the development of non-renewable groundwater resources, is always overexploitation. The implication of this word is that groundwater mining goes against basic ecological and ethical principles. Some authors (Selborne, 2001; Llamas, 2001; Llamas and delli Priscoli, 2000) have shown that such *hydromyth* may always not be correct.

1.2. Purpose

This article describes the intense groundwater development in Spain during the last three or four decades and the positive and negative aspects of such development. During this period, Spain has become an industrialized country.. The analysis of the changing role of groundwater in Spain's water policy may be useful for other countries which are undergoing or will undergo this change. The Spanish Water Code of 1985 is one of the few in the world which has regulated the concept of *overexploited aquifer*. Based mainly in the Spanish experience, the main aim of this paper is to present and discuss: 1) the many meanings of the terms groundwater (or aquifer) overexploitation and sustainability; 2) the main factors to take into consideration in analyzing the pros and cons of intensive groundwater development; and 3) the strategies to prevent or correct the unwanted effects of intensive groundwater development.

What is an intensively used or stressed aquifer? During the last decade the expression *water stressed regions* has become pervasive in the water resources literature. Usually this means that those regions are prone to suffer now or in the near future serious social and economic problems because of water scarcity. Some authors insist on the probable outbreak of violent conflicts, that is, water wars among water stressed regions. The usual threshold to consider a region under water stress is 1000 m^3/person/year (United Nations, 1997), but some authors increase this figure to 1700 m^3/person/year. If this ratio is only 500 m^3/person/year the country is considered in a situation of absolute water stress or water scarcity (Seckler *et al.*, 1998; Postel, 1999). This simplistic approach of considering only the ratio between water resources and population has scarce practical application and may be misleading. Most water problems are related to its quality degradation and not its relative scarcity. As an example, a good number of Spanish regions with a ratio lower than 500 m^3/year/person, are regions with high economic and social standard of living.

In its last Assessment of Global Water Resources, the United Nations did a more realistic classification of countries according to their water stress. This assessment considered not only the ratio water/population but also the Gross National Product per capita (United Nations, 1997). Other experts are also beginning to use other more sophisticated indices or concepts in order to diagnose the current or future regions with water problems. The result of these analyses will probably show that a certain *water-stress* may be an incentive to promote the development of these regions. In this case, it could be defined as an *eu-stress*, that is, a good stress. For example, during the last few decades in a good number of semi-arid or arid regions tourism or the production of high value crops have significantly intensified. The scarcity of precipitation has been fully compensated by the great amount of sunny days and the high solar energy received. Examples of these developments are the sun *belt* in the USA and most of the European Mediterranean coast. The necessary water for these activities may be from different sources. Groundwater is probably the most frequently used, but it also may be imported, recycled or desalinated water. An example described later is a desalination plant (40 Mm^3/year) that will be completed in 2002 southeastern Spain, for use in greenhouse irrigation.

This article will show that groundwater development during the last decades has significantly contributed to Spanish agricultural development and to alleviate poverty as well as and to improve public health in developing countries. These improvements should be maintained and increased but not in the same way. The generally uncontrolled and unplanned groundwater development has to be rationalized and the externalities of groundwater extraction and the temporary or intrinsic uncertainties related to water management considered. The implementation of this groundwater sustainable use requires, as a *conditio sine qua non*, the participation of educated and informed groundwater users and other stakeholders in groundwater management decisions. This demands urgently the development of institutional arrangements for groundwater management where users can work jointly with the corresponding water authorities.

2. WHAT DOES SUSTAINABILITY MEAN?

During the last decade, the concept of sustainability has been proposed by many as a philosophy to solve most water problems or conflicts. In 1987, the United Nations Commission on Environment and Development defined sustainability as "*the ability to meet the needs of the present generations without compromising the ability of future generations to meet their needs*". The European Union Water Framework Directive, enacted in December 2000, states that it is necessary to promote or foster a sustainable water use. Probably, most people agree with this general principle, but its practical application in natural resources management is challenging. For instance, in terms of water resources management even the concept of minimum basic water needs, often estimated in 20 to 50 litters per day per person, is controversial.

Another terminological problem is related to the concept of future generations. Are we talking about the people that will live in this planet in the 22nd century, in the whole third millennium or only in two generations, that is, within the next fifty years? No scientist is able to predict the situation one thousand years from now, and very few dare to present plausible scenarios for the 22nd century. Most current predictions refer to the needs of humans in one or two generations, i.e. not more than fifty years from now. The U.S. Geological Survey (1999) has published an interesting circular on the topic of groundwater sustainability but, in it the issue is considered almost exclusively from a hydrological point of view. It is clear that environmental problems have a natural science foundation but also, and even mainly, a social science foundation. There is no doubt that the issue is complex. Some authors consider different types of sustainability: ecological, social, economic, and political. In some regions there is plenty of water and other natural resources and by all the ecological indicators the people in those regions are *rich*. Nevertheless, because of the economic, social and political situation their sustainable development is in jeopardy. Wood (2001) points out the complexity of the sustainability concept and introduces the concept of "renewability" based on water balance to separate science and engineering aspects from those of water management strategies.

For many years, this author has put forth the idea that in Spain and in many other countries water scarcity is not usually the problem (Llamas, 1992). The real issue is widespread water mismanagement which very frequently results in the very serious problem of water quality degradation

The way to solve the existing water problems, mainly the lack of potable water, is not to persist on gloom and doom unrealistic campaigns, trying to create *environmental scares* and predicting *water wars in the near future* (see Kessler, 1998; The Economist, 1998; Asmal, 2000; World Humanity Action Trust, 2000) but to improve its management.

3. THE MANIFOLD COMPLEX, AND THE PRACTICALLY USELESS CONCEPT OF OVEREXPLOITATION: OVERVIEW

The term overexploitation has been frequently used during the last three decades. Nevertheless, most authors agree in considering that the concept of aquifer overexploitation is one that resists a useful and practical definition (Llamas, 1992; Collin and Margat, 1993). Custodio (2000) and Sophocleous (2000) have most recently dealt with this topic in detail.

A number of terms related to overexploitation can be found in the water resources literature. Some examples are: safe yield, sustained yield, perennial yield, overdraft, groundwater mining, exploitation of fossil groundwater, optimal yield and others (see glossaries in Todd, 1980, Fetter, 1994, and Acreman, 1999). In general, these terms have in common the idea of avoiding *undesirable effects* as a result of groundwater development. However, this *undesirability* depends mainly on the social perception of the issue. This social perception is more related to the legal, cultural and economic background of the region than to hydrogeological facts.

For example, in a research study on groundwater fed catchments, called GRAPES (Acreman, 1999), three pilot catchments were analysed: the Pang in the UK, the Upper Guadiana in Spain and the Messara in Greece. The main social value in the Pang has been to preserve the amenity of the river, related to the conservation of its natural low flows. In the Messara, the development of irrigation is the main objective and the disappearance of relevant wetlands has not been a social issue. In the Upper Guadiana the degradation of some important wetlands caused by groundwater abstraction for irrigation has caused a serious conflict between farmers and conservationists which has not been settled yet (see Llamas *et al.*, 2001; Bromley et al., 2001).

The Spanish Water Code of 1985 does not mention specifically the concept of sustainability in water resources development but frequently indicates that this development has to be in balance with nature. It basically considers an aquifer *overexploited* when the pumpage is close or larger than the natural recharge.

The Regulation for the Public Water Domain that developed the 1985 Water Act, says that "*an aquifer is overexploited or in risk of being overexploited, when the continuation of existing uses is in immediate threat as a consequence of abstraction being greater or very close to the mean annual volume of renewable resources, or when it may produce a serious water quality deterioration*". According to the law, 14 aquifers have been declared either provisionally or definitively overexploited. Also according to the law, strict regulatory measures have been designed to deal with these situations of stress. However, to a large extent, these measures have not been successfully implemented and a situation of legal chaos still persists in many of these aquifers, as is implicitly recognized in the White Paper on Water In Spain (MIMAM, 2000).

The misconception of considering that *safeyield* is practically equal to natural recharge, already shown by Theiss in 1940, has been voiced by many other hydrogeologists (see Custodio, 2000; Sophocleous, 2000; Hernández-Mora *et al.*, 2001). Bredehoeft *et al.* (1982, pages 53, 54 and 56) describe the issue in the following way: "*Water withdrawn artificially from an aquifer is derived from a decrease in storage in the aquifer, a reduction of the previous discharge from the aquifer, an increases in the recharge, or a combination of these changes. The decrease in the discharge plus the increase in recharge is termed capture. Capture may occur in the form of decreases in the groundwater discharge into streams, lakes, and the ocean, or from decreases in that component of evapotranspiration derived from the saturated zone. After a new artificial withdrawal from the aquifer has begun, the head of the aquifer will continue to decline until the new withdrawal is balanced by capture*". "*In many circumstances the dynamics of the groundwater system are such that long periods of time are necessary before any kind of an equilibrium conditions can develop*".

As an example of the change in the social perception of water values it is interesting to remark that in 1940, according to Theiss, the water *was gained* by lowering the water table in areas of rejected recharge or where the recharge was *lost* through transpiration from *non-beneficial vegetation* (phreatophytes). During the period of time when Theiss addressed these issues *wetlands were wastelands*.

Bredehoeft *et al.* (1982) present some theoretical examples to show that the time necessary to reach a new equilibrium or steady state between groundwater extraction and capture may take decades or centuries. Custodio (2000) and Sophocleous (2000) have also presented illustrations to show the relationship between the size of the aquifer, its difussivity and the time necessary to reach a new steady state after the beginning of a groundwater withdrawal.

Several national and international conferences have been organized by Spanish hydrogeologists over the past two decades to discuss and help dispel the misconceptions related to aquifer overexploitation (cf. Simmers *et al.*, 1992; Custodio and Dijon, 1991). Nevertheless, up to now the success of these activities has been limited in Spain and abroad.

It was suggested that a possible definition is to consider an aquifer overexploited when the economic, social and environmental costs that derive from a certain level of groundwater abstraction are greater than its benefits (Llamas *et al.*, 1992). Given the multi-faceted character of water, this comparative analysis should include hydrologic, ecological, socioeconomic and institutional variables. While some of these variables may be difficult to measure and compare, they must be explicitly included in the analysis so they can inform decision-making processes. Following and expanding on to the ideas discussed in Hernández-Mora *et al.* (2001), the criteria that can be used to evaluate the benefits and costs related to groundwater abstraction will be presented, after describing the Spanish situation in water resources and uses. In this article the so-called Total Economic Value (TEV) of groundwater, as defined by the U. S. National Research Council, has not been explicitly considered. Nevertheless, the basic categories of extractive services and *in situ* services are taken into account in the description of costs and benefits of groundwater development. The National Research Council (1997) recognizes that the monetary value of groundwater's *in situ* services (avoiding subsidence, conservation of wetlands, or maintaining the base flow of rivers, among others), is a rather complex and difficult task for which there is only limited information.

4. WATER RESOURCES OF SPAIN

4.1. Climatic and Hydrologic Setting

Spain has a surface of approximately 500,000 km^2. The average precipitation is 700 mm/year.

However, this average has considerable spatial deviation. From 100 mm/year in some islands in the Canary Archipelago to more than 2,000 mm/year in the humid North. The average annual temperature is 14 °C. Average potential evapotranspiration is about 900 mm/year. In most of Spain evapotranspiration is higher than precipitation. The average streamflow is about 110 km^3/year. From this amount about 80 km^3/year is surface runoff and about 30 km^3/year is groundwater discharge. At least one third of Spain is endowed with good aquifers. These aquifers may be detrital, calcareous or volcanic (figure 1). The Water Administration has formally identified 411 aquifer systems or hydrogeological units (see Llamas *et al.*, 2001, table 4.1) which cover an area of 180,000 km^2, approximately.

The estimated *natural* recharge of these aquifers averages 30 km^3/year but varies with weather conditions between 20 and 40 km^3/year (figure 2).

4.2. Water uses

Spain's current total water uses are about 33 km^3/year or about one third of this total water resources (110 km^3/year).These uses are distributed as follows (in km^3/year):

Irrigation	24
Urban Water supply and connected industrial	5
Independent industrial uses and cooling	3
Total	36

Spain has about 40 million inhabitants. This means an average of almost 3,000 m^3/person/year, considering the whole country. Nevertheless, this average may be misleading because in some areas this indicator is smaller than 200 or 300 m^3/person and year.

Table 1 shows the range of groundwater volumes used in Spain in recent years.. The higher numbers correspond to dry periods in which groundwater use increases. The dramatic increase in the use of ground water during the last forty years is illustrated in figure 3.

This groundwater use growth has been the result of groundwater development by individuals, small municipalities and industries. It has not been promoted or planned by government agencies. As a matter of fact, Spain is a serious case of *hydroschizophrenia*; that is of a almost complete separation of surface and groundwater in the mind of water planners (Llamas, 1985). As a result of this *disease*, Spain is among the countries with the highest number of large dams per person: 30 large dams per million inhabitants. Only four countries have a higher ratio: Norway, Cyprus, Iceland and Albania (figure 4). The rhythm of large dams construction in Spain during the last fifty years has been almost 20 large dams per year (figure 5).

Table 1. Spain's groundwater use summary (estimated from several sources).

Activity	Volume applied (Mm^3/year)	Percentage of total water (surface + ground water)
Urban	1,000-1,500	~ 20%
Irrigation	4,000-5,000	~ 25%
Industrial and cooling	300-400	~ 5%
Total	5,500-6,500	15-20%

Source: Llamas et al. (2001, table 5.16).

Inside the European Union, Spain is the country with the least percentage of groundwater uses for urban water supply. It is about 25% (see figure 6). The explanation of this anomaly is not the lack of aquifers, like in Norway's case, but the previously mentioned *hydroschizophrenia* of the government water planners who were the persons making decisions on the use of public funds for the water supply to large cities and for large surface water irrigation schemes.

4.3. Groundwater ownership and markets

Until the 1985 Water Code took effect, groundwater in Spain was private domain. In contrast, surface water was almost always public domain, ruled by government agencies. Because of the real or imagined problems related to the uncontrolled development of groundwater the 1985 Water Code declared all groundwater in Spain of public domain. Every new groundwater abstraction requires a permit by the corresponding water authority, except those abstracting less than 7,000 m^3 per year.

The groundwater developments made before the 1st January 1986 may continue as private domain, using the same amount of groundwater that they were using previously. All these wells, galleries or springs have to be inventoried and registered with the corresponding water agencies. The main problem is that the legislators and the water authorities underestimated the relevance and number of groundwater abstractions and did not provided the economic means and the man-power to register all the grandfathered groundwater rights. Sixteen years after the enactment of the Water Code the number of private groundwater abstraction rights remains uncertain. Llamas *et al.* (2001, chapter 8) have estimated that probably 90% of the private groundwater developments have an illegal or a-legal status. And what is worse, it seems that the number of illegal wells drilled after the 1st January 1986 is very large. In other words, the 1985 Water Code groundwater provisions have been scarcely enforced. This has created a situation which may be considered as legal and administrative chaos. The 1985 Water Code was partly amended in 1999, mainly to introduce in some way the water markets. This was mainly done to allow greater *flexibility* to sell or buy water rights. In principle, this new *flexibility* is not relevant to groundwater markets because in Spain still most groundwater resources are private ownership and they can be sold or bought like any other commodity. The importance of these groundwater market varies according to the different Spanish regions. For example, in the Canary Islands organized water markets have been in operation for more than a century. A detailed description of the Spanish groundwater markets can be read in Hernández-Mora and Llamas (2001).

5. BENEFITS OF GROUNDWATER DEVELOPMENT

The concept of groundwater overexploitation or sustainability must necessarily take into account the numerous socioeconomic and even ecological benefits that can derive from groundwater use. Socioeconomic benefits range from water supply to economic development, as a result of agricultural growth in a region. With respect to potential ecological benefits, the use of groundwater resources can often eliminate the need for new large and expensive hydraulic infrastructures that might seriously damage the natural regime of a river or stream and/or create serious social problems (World Commission on Dams, 2000).

5.1. Potable Water Supply

Groundwater is a key source of drinking water, particularly in rural areas and in island environments. In Spain, for example, medium and small municipalities (of less than 20,000 inhabitants) obtain 70% of their water supply from groundwater sources (MIMAM, 2000). In some coastal areas and islands the dependence on groundwater as a source of drinking water is even higher. Nevertheless Spain is one of the European countries that less proportion of groundwater uses per public urban water supply to large cities. This situation has the historical roots explained in Llamas (1985). The failure in the 19th century in continuing the water supply to Madrid by groundwater (Khanat systems) induced one of the most serious cases known of *hydroschizofrenia*. This neglect of groundwater role is the basic cause of today's most current water conflicts in Spain (Llamas and Pérez Picazo, 2001).

5.2. Irrigation

In many arid and semiarid countries, the main groundwater use is for agriculture. Although few studies have looked at the role that groundwater plays in irrigation, those that do exist point to a higher socio-economic productivity of irrigated agriculture using groundwater than that using surface water. A 1999 study done for Andalucía, in Southern Spain shows that irrigated agriculture using groundwater is significantly more productive than agriculture using surface water, per volume of water used (Hernández-Mora *et al.*, 2001).Table 2 shows the main results of the Andalucia study. It is important to note that these results were based on the average water volumes applied in each irrigation unit (or group of fields). The water losses from the source to the fields were not estimated. Nevertheless, these losses in surface water irrigation are significant. Other studies have calculated the volumes used in surface water irrigation as the water actually taken from the reservoirs. For example, the White Paper of Water in Spain (MIMAM, 2000) estimated an average consumption of 6,700 m^3/ha/year and 6,500 m^3/ha/year for the two catchments that are the subject of the Andalucía study without differentiating between surface and groundwater irrigation. Using these new figures and the volumes given for irrigation with groundwater in the Andalucía study, a more realistic average volume used for irrigation with surface water of 7,400 m^3/ha/year can be estimated. Table 2 shows that productivity of groundwater irrigation is 5 times greater than irrigation using surface water and generates more than 3 times the employment per m^3 used. It could be argued that the greater socio-economic productivity of groundwater irrigation in Andalucía can be attributed to the excellent climatic conditions that occur in the coastal areas. While good climatic conditions may influence the results, the situation is similar in other continental regions of Spain (Hernández-Mora and Llamas, 2001).

A literature search did not reveal similar studies in other parts of Europe. Studies in India show similar results. For instance, Dains and Pawar (1987) estimated that groundwater constitutes 30%

of all water used for irrigated agriculture in India but was responsible for 70-80% of all agricultural production. According to Dhawan (1995), in India research indicates that yields in groundwater irrigated areas are higher by one third to one half more than in areas irrigated from surface waters. First Postel (1999) and later the World Water Council (2000) and Rijsberman and Cosgrove (2000), consider that intensive groundwater use in India has been decisive in order to feed that country's current population of one billion people. Nevertheless, this *miracle*, as Postel describes it, is not sustainable because it has been mainly done mining groundwater, that is, using about 100 km^3/year of non-renewable groundwater resources. This figure is based on an undated and unpublished report, and does not fit the data provided by the Indian Water Resources Society (2000), or the estimation of groundwater recharge by Rangarajan and Athavale (2000). India seems to be the country which has most intensively developed groundwater irrigation during the last three or four decades (more than 20 million hectares developed). A thorough analysis of the benefits and costs of this phenomenon would be useful for the whole humanity. When examining the data included in this section it is important to keep in mind the strong uncertainties that are attached to hydrologic data. However, the results obtained are indicative of the greater productivity of irrigation using groundwater. This should not be attributed to any intrinsic quality of groundwater. Rather, causes should be found in the greater control and supply guarantee that groundwater provides mainly during droughts (see Llamas, 2000), and the greater dynamism that has characterized the farmers who have sought their own sources of water and bear the full (direct) costs of drilling, pumping and distribution (Hernández-Mora and Llamas, 2001).

Table 2. Comparison of irrigation using surface and groundwater in Andalucía.

Indicator for irrigation	Origin of irrigation water: Ground-water	Surface water	Total	Relation groundwater surface water
Irrigated surface (10^3 ha)	210	600	810	0.35
Average use at origin (m^3/ha/year)	4000	7400	6500	0.54
Water productivity (€/m^3) [1]	2.16	0.42	0.72	5.1
Employment generated (EAJ/10^6 m^3) [2]	58	17	25	3.4

Source: Llamas et al. (2001).

[1] 1 € ≅ 0.9 US$

[2] EAJ stands for Equivalent Annual Job which is the work of one person working full-time for one year.

5.3. Hydrologic benefits

Other potential benefit of groundwater development is the increase in net recharge in those aquifers that, under natural conditions, have the water level close to the land surface. The drawdown of the water table can result in: a decrease in evapotranspiration, an increase in the recharge from precipitation that was rejected under natural conditions, and an increase in indirect recharge from surface water bodies. This process was already described by the American hydrogeologist Theis in 1940 and was later developed by Bredehoeft *et al.* (1982). Johnston (1997) analyzed eleven American regional aquifers. He shows that intensive groundwater development in nine of these aquifers has resulted in significantly increased recharge.

A clear example of this situation is the increase in available resources for consumptive uses that followed intensive groundwater pumping in the Upper Guadiana basin in central Spain (see Bromley *et al.*, 2001). Cruces and Martínez (2000) have estimated that average renewable resources may have increased between one third -and one half- under disturbed conditions. Figure 7 illustrates some results from this monograph. Prior to the 1970s, groundwater pumping in the Guadiana basin did not have significant impacts on the hydrologic cycle. Intensive pumping for irrigated agriculture started in the early 1970s and reached a peak in the late 1980s. As a result, wetlands that under semi-natural conditions had a total extension of about 25,000 ha, today only cover 7,000 ha. In addition, some rivers and streams that were naturally fed by the aquifers, now have become net losing rivers.

The results of the decline of the water table have been two-fold. On one hand, a significant decrease in evapotranspiration from wetlands and the water table, from about 175 hm^3/year under quasi-natural conditions to less than 50 hm^3/year today. At the same time, there has been a significant increase in induced recharge to the aquifers from rivers and other surface water bodies. Consequently, more water resources have become available for other uses, mainly irrigation. Clearly, it is important to keep in mind the associated negative impacts that the drawdown of the water table has had on dependant natural wetlands.

6. DISADVANTAGES OF GROUNDWATER USE

6.1. Groundwater level decline

The observation of a trend of continuous significant decline in groundwater levels is frequently considered an indicator of imbalance between abstraction and recharge. While this may be most frequently the case, the approach may be somewhat simplistic. Management decisions based on this simplistic approach may sometimes be misguided. Custodio (2000) and Sophocleous (2000) agree that every groundwater withdrawal causes an increasing piezometric depletion until a new equilibrium is achieved between the pumpage and the new recharge (or capture). This transient situation can be however long depending on aquifer characteristics such as storage capacity, transmissivity, degree of stratification and heterogeneity. For instance, in large unconfined aquifers, the time necessary to reach a new equilibrium state in water table levels can of decades or centuries when transmissivity is low. On the other hand, in large confined aquifers, water level declines do not necessarily imply a significant decline in storage but, rather, a change in the elastic conditions of the system.

With respect to the climatic sequence, in arid and semiarid countries significant recharge can occur only every 5 to 10 years. Therefore, continuous decline in the water table during a dry climatic sequence of a few years, when recharge is low and abstraction high, may not be representative of long-term trends. Declines in water levels should indicate the need for further analysis. Whether they indicate possible abstraction greater than recharge is something that needs to be studied on a case-by-case basis, always taking into consideration the hydrogeological characteristics and the size of the aquifer, as well as the climatic sequence. In any case, declines in the water table can result in a decrease in the production of wells as well as increases in pumping costs. This economic impact can be more or less significant depending on the value of the crops obtained. For instance, in some zones of Andalucía, the value of crops in greenhouses may reach 40,000 to 60,000 US$/ha/year. The water volume used is between 4,000 and 6,000 m^3/ha. The energy needed to pump one cubic meter 100 m high is about 0,3 kWh. This means that the increase on the costs of pumping in the event of a drawdown of 100 m is almost irrelevant for the agribusiness. As previously described, a significant fact is that a large sea water desalination plant (40 Mm^3/year) will be completed in Almería (South Spain) in 2002. The main use of this treated water will be green-house irrigation. On the other hand, if the value of the crops is only about 1,000 US$/ha/year, and the water needed is about 10,000 m^3/ha, the increase in the cost of energy of a drawdown of 100 m can render those crops uneconomic.

The socio-economic situation created by water table depletion in rural areas of poor developing countries may be more complex and prone to social inequity. For instance, according to Moench (1999), this may occur in some poor rural regions of India. A similar situation may occur when a fragmentation of well ownership exist, as in Tamal Nadu, India, such as it is described (Janakarajan,1999). In both cases, a possible solution would be to require that the person causing the water table depletion compensate in money or water the person whose well becomes dry or sees its yield reduced. Perhaps in such regions, the main problem may be that the officers of the corresponding water authority have neither the knowledge nor the necessary means to enforce groundwater regulations. Then, probably the impacted *farmers* will emigrate to some urban area. Can this frequent migration towards urban areas be stopped or significantly reduced by forbidding groundwater development in order to avoid such aquifer depletion?

It seems clear that, before trying to extrapolate Spanish solutions to developing countries situations, a thorough assessment of the hydrogeological and socio-economic situation of these countries is required. The adequate solutions in a country where the average income per person and year is smaller than US$ 500, should usually be very different from those adequate for countries industrialized.

6.2. Degradation of groundwater quality

Groundwater quality is perhaps the most significant challenge to the long-term sustainability of groundwater resources. Restoration of contaminated aquifers can be a very costly and

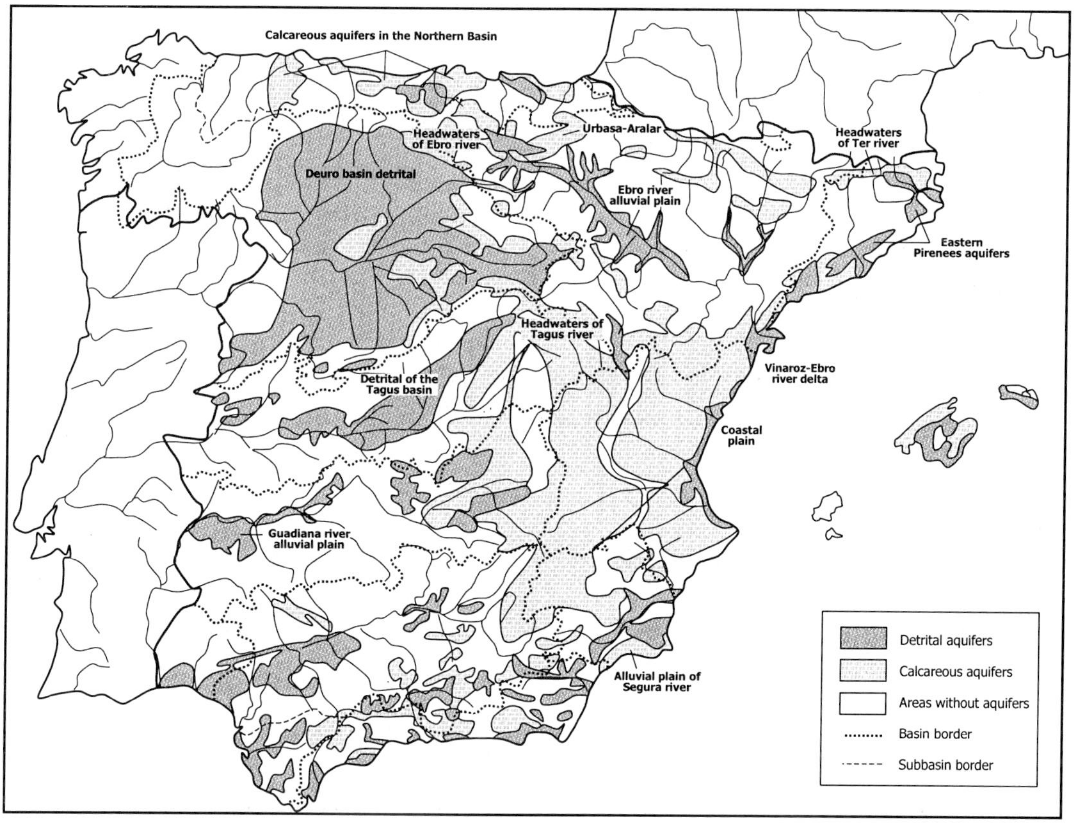

Figure 1. Spain's location and types of main aquifers (Source: MIMAM, 2000, figure 2).

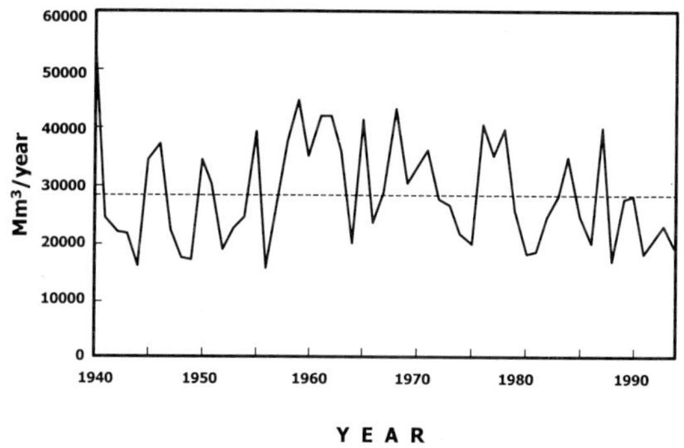

Figure 2. Estimated natural groundwater recharge in Spain from 1940 to 1995 (Source: MIMAM, 2000, figure. 124).

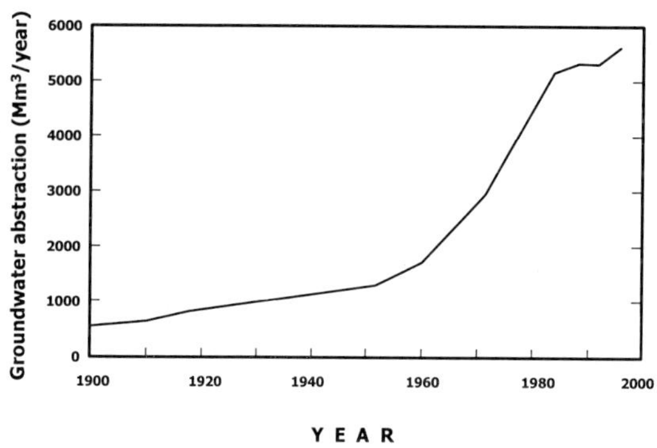

Figure 3. Groundwater consumption development of groundwater in Spain, showing its rapid increase from the 1960s to date (Source: MIMAM, 2000, figure 339).

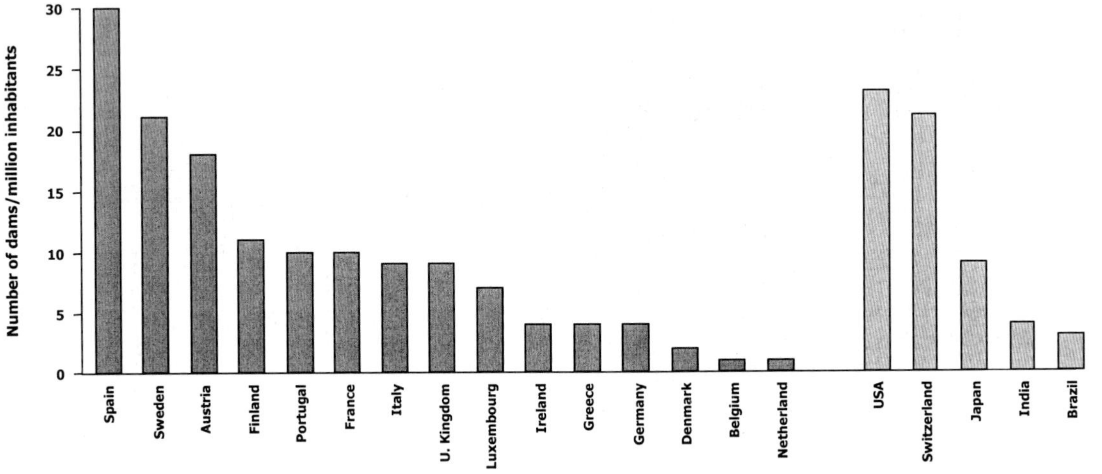

Figure 4. Number of dams per capita in different countries (Source: Llamas et al., 2001, figure 5.7).

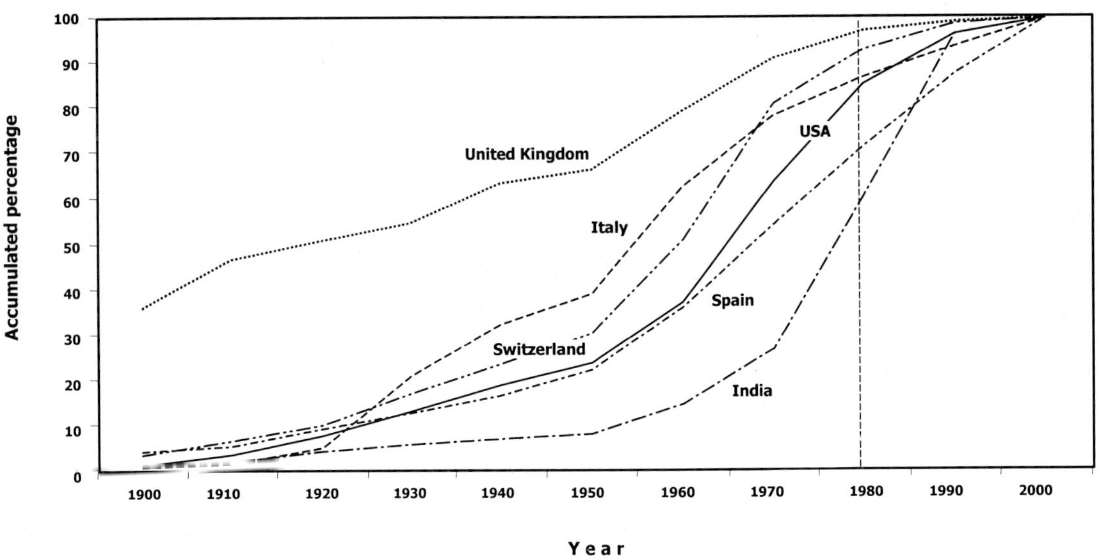

Figure 5. Temporal dam construction rhythm in several representative countries (Source: Llamas *et al.*, 2001, figure 5.8).

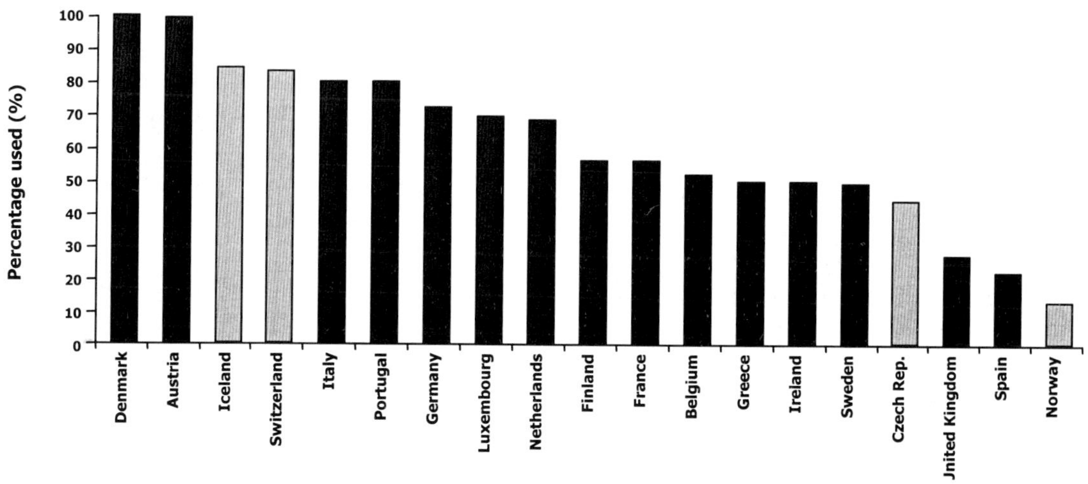

Figure 6. Percentage of groundwater used for urban supply in several European countries (Source: Llamas *et al.*, 2001, figure 5.3)

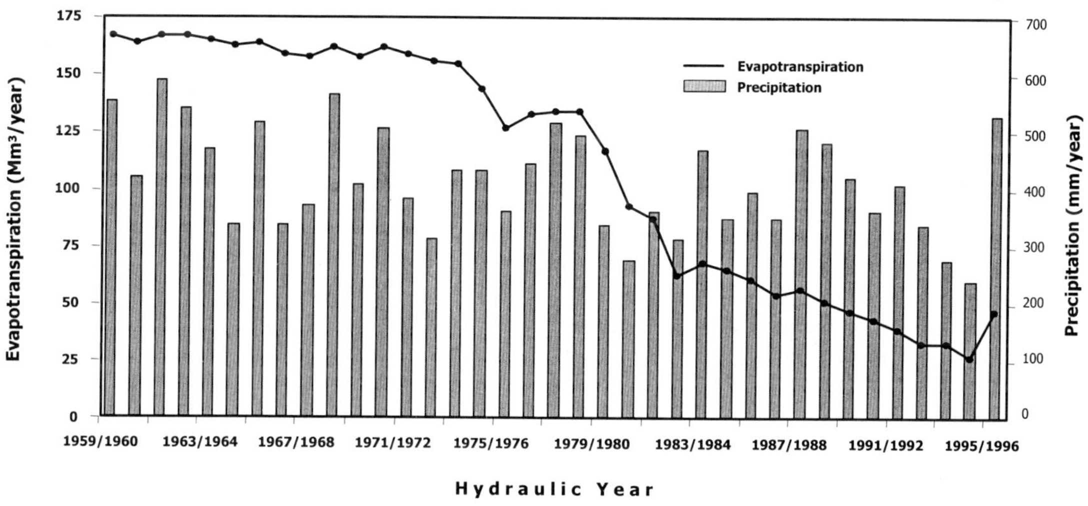

Figure 7. Temporal evolution of evapotranspiration from the water table in the Upper Guadiana basin, caused by water table depletion. (Source: Martínez Cortina as cited in Llamas *et al.*, 2001, figure 4.7).

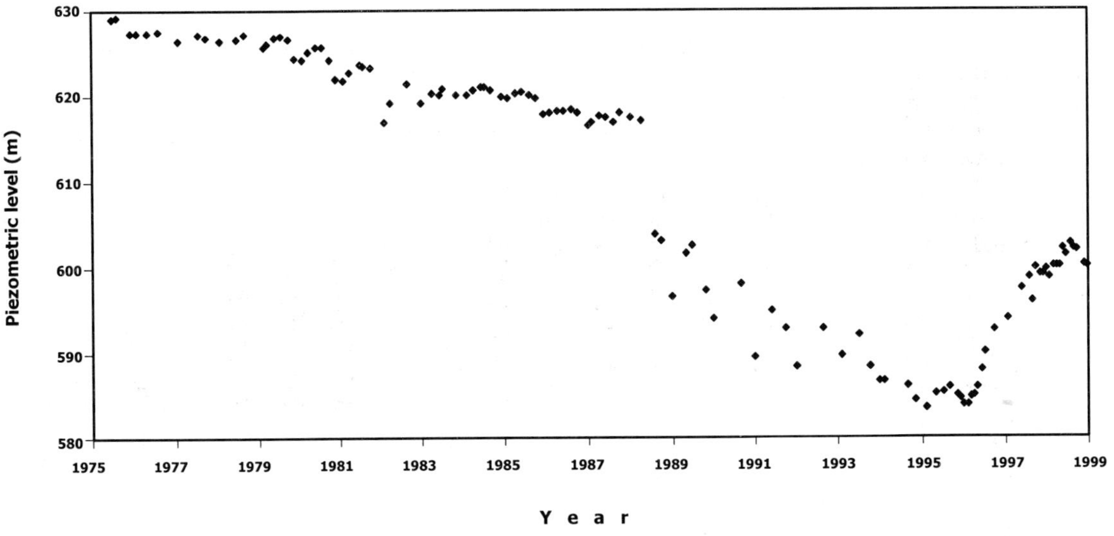

Figure 8. Water table evolution in Manzanares (Upper Guadiana catchment, Spain) (Source: Martínez Cortina 2001 as cited in Hernández-Mora et al., 2001)

difficult task. Most often, degradation of groundwater quality is not a result of excessive abstraction, but is related to other causes such as point or non-point source pollution from various sources such as return flows from irrigation, leakage from septic tanks and landfills or industrial liquid wastes. These problems are not exclusive of industrialized countries but also may be serious in developing countries (Janakarajan, 1999).

Groundwater abstraction can also cause changes in groundwater quality. Some indicators of the susceptibility of an aquifer to water quality degradation are the following (Custodio, 2000):

1. Proximity to saline water bodies: risk of saltwater intrusion which not only depends on the amount of abstraction relative to recharge, but also on the well field location and well design and on the geometry and hydrogeological parameters of the aquifer.
2. Hydraulic connection to low quality surface or groundwater bodies. Changes in the hydraulic gradient as a result of groundwater abstraction may result in the intrusion of poor quality water into the aquifer from adjacent water bodies.

In these cases, the problem is often related to inadequate well field location and not necessarily to the total volumes abstracted. Technical solutions to deal with problems of saline or lower quality water intrusion have been developed and applied successfully in some places such as California and Israel. Unfortunately, the public awareness in Spain about the groundwater pollution problems is still weak. The situation is improving but very slowly. Spain is not a region to imitate in this respect.

6.3. Susceptibility to subsidence and/or collapse of the land surface

Aquifers formed in young sedimentary formations are prone to compaction as a result of water abstraction and the resulting decrease in intergranular pore pressure. For example, this has been the case in the aquifers underlying Venice or Mexico City. More dramatic collapses are a common occurrence in karstic landscapes, where oscillations in water table as a result of groundwater abstractions can precipitate the occurrence of karstic collapses. In both cases, the amount of subsidence or the probability of collapses is related to the decrease in water pressure. This is a result not only of the amount of groundwater withdrawal, but also of well field location and design. Fortunately, these type of geotechnical problems are not relevant in Spain.

6.4. Interference with surface water bodies and streams

Decline in the water table as a result of groundwater withdrawals can affect the hydrologic regime of connected wetlands and streams. Loss of baseflow to streams, dissecation of wetlands, and transformation of previously gaining rivers to losing rivers, may all be potentially undesirable results of groundwater abstraction and serve as indicators of possible excessive abstraction. The already mentioned Upper Guadiana catchment in Spain is a typical example of this type of situation.

6.5. Ecological impacts on groundwater-dependant ecosystems

The ecological impacts of drawdowns of the water table on surface water bodies and streams are increasingly constraining new groundwater developments (Llamas, 1992). Drying up of wetlands, disappearance of riparian vegetation because of decreased soil moisture, or alteration of natural hydraulic river regimes, can all be used as indicators of overexploitation. Reliable data on the ecological consequences of these changes is not always available, and the social perception of such impacts varies in response to the cultural and economic situation of each region. The lack of adequate scientific data to evaluate the impacts of groundwater abstraction on the hydrologic regime of surface water bodies makes the design of adequate restoration plans difficult. For instance, wetland restoration programs often ignore the need to simulate the natural hydrologic regime of the wetlands, that is, not only restore its form but also its hydrological function. Similar problems result in trying to restore minimum low flows to rivers and streams. Oftentimes minimum streamflows are determined as a percentage of average flows, without emulating natural seasonal and year-to-year fluctuations to which native organisms are adapted.

The social perception of the ecological impacts of groundwater abstraction may differ from region to region and result in very different management responses. A European Union-funded project, previously mentioned, looked at the effects of intensive groundwater pumping in three different areas Greece, Great Britain and Spain (Acreman, 1999). In the Pang River in Britain, conservation groups and neighborhood associations with an interest in conserving the environmental and amenity values of the river that had been affected by groundwater abstraction, mainly drove management decisions. In the Upper Guadiana basin, dramatic drawdowns in the water table (30-40 m) caused jointly by groundwater abstraction and drought (see Figure 8) resulted in intense conflicts between nature conservation officials and environmental non-governmental organizations (NGOs), irrigation farmers and water authority officials. The conflicts have been ongoing for the past 20 years and have not yet been resolved. Management attempts to mitigate the impact of water level drops on the area's wetlands have so far had mixed results (Fornés and Llamas, 1999; Bromley *et al.*, 2001). On the other hand, in the Messara Valley in Greece, the wetland degradation caused by decline in the water table has not generated any social conflict. Conservation interests do not seem to be strong in Crete.

7. STAKEHOLDERS PARTICIPATION IN GROUNDWATER MANAGEMENT

Spain has a long tradition of collective management of common pool resources. Probably the *Tribunal de las Aguas de Valencia* (Water Court of Valencia) is the most famous example. This Court has been meeting at noon every Thursday for many centuries at the entrance of Cathedral of Valencia to solve all the claims among the water users of a surface irrigation system located close to Valencia. All the members of the Court are also farmers. The decisions or sentences are oral and can not be appealed to a higher court. The systems has worked and it is a clear proof that *The Tragedy of Commons* is not always true. Further evidence of social cooperation in Spain are the thousands of *Comunidades de Regantes* (Irrigation Communities or Water Users Associations). Some of them have been in operation for several centuries.

The 1985 Spanish Water Code preserved the traditional *Comunidades de Regantes* that existed before its enactment and recommended these institutions for surface water management. It is also extended these type of collective institution to groundwater management, and required the compulsory formation of *Comunidades de Usuarios de Aguas Subterráneas* (Groundwater Users Communities) when an aquifer system was legally declared overexploited. A description of these institutions is contained in Hernández-Mora and Llamas (2000). It has been difficult to enforce the formation of the groundwater users associations in overexploited aquifers and sixteen years after the enactment of the Water Code only a few are operating at this time (Llamas *et al.*, 2001, chapter 9; Hernández-Mora and Llamas, 2001). As recognized in the White Paper on Water in Spain (MIMAM, 2000), the main cause is that these new groundwater users communities were established top-down, that is the water authorities imposed their implementation without the agreement of the farmers who are the main stakeholders. Both, the 1999 amendments to the 1985 Water code and the

2001 Law of the National Water Plan, have provisions to overcome these difficulties and to foster the implementation of institutions for collective management of aquifers with ample participation of the stakeholders under a certain control of water authorities.

In Spain, independently of these officially created groundwater user communities, there are a large number of private collective institutions to manage groundwater. Some of them have been in operation for many years. There is a general consensus on the crucial need to implement these institutions but it will be necessary to wait a few years to know if the incentives (economic, fiscal, operational) offered to the stakeholders are the right ones to achieve this goal.

8. HYDROSOLIDARITY AND GROUNDWATER MANAGEMENT IN SPAIN

8.1. Overview

In Spain, like everywhere, ethical factors play a crucial role in water uses and in water management. Many recent publications address this topic (see Selborne, 2001; Llamas, 2001a; Llamas and Priscoli, 2000). An analysis of this scenario in Spain is described by Llamas (2001 b). Human solidarity is one of the ethical principles that underlies most water policy agreements or treaties. One of the meanings of the concept of solidarity, as it applies to the use of natural resources, is that a person's right to use those resources should be constrained or limited by the rights or needs of other or future human beings. These rights or needs should also include the water needed for care of the natural environment.

Nowadays, few people would dare to speak openly against hydrosolidarity (the need to share water resources), that is, the application of the principle of solidarity to water resources management. In practice, however, it might be difficult to find constructive ways to facilitate an equitable and fair share of water resources among concerned stakeholders, particularly in densely populated arid and semi-arid regions. Lack of knowledge, arrogance, vested interests, neglect, institutional inertia, corruption, are some of the obstacles frequently encountered to achieve hydrosolidarity (Llamas and Perez-Picazo, 2001). The noble and beautiful concept of hydrosolidarity may also be used in a corrupt or unethical way by some lobbies in order to withdraw public money, that is *perverse subsidies*, which are bad for the economy and the environment (Llamas and Delli Priscoli, 2000). An example of the improper use of hydrosolidarity is that of the Segura catchment area. It has influenced the first National Water Plan in Spain, which was approved as a Law in July 2001 by the Spanish Parliament.

8.2. The Segura Catchment

Hydrology

The Segura catchment is located in southeastern Spain. Its main features are: a) surface area 19,000 km^2; b) average annual precipitation of 400 mm, ranging from 800 mm in the headwater to 200 mm in the coastal plain.; c) annual potential evapotranspiration: 800-900 mm; d) average streamflow: 1000 Mm^3. The relief is abrupt with mountains that reach an altitude of 2000 m. The geology is complex with numerous faults and thrusts. Calcareous aquifers cover about 40% of the catchment's surface. Natural recharge is estimated at about 600 Mm^3/year (about 60% of the total streamflow). The climate is typical of Mediterranean regions: hot summers, frequent flash floods and long droughts.

Water Development Until the 1960s

Sixty percent of the Segura river basin is within the Murcia Autonomous Region with the reminder 40% divided between the Autonomous Regions of Valencia and Castilla-La Mancha. The mild climate and the important baseflow (typical of a karstic catchment) of the Segura river, encouraged the development of an important agricultural economy in the region on the basis of an irrigation network on the floodplains of the middle and lower part of the catchment area, that dates as far back as the Muslim occupation twelve hundred years ago. Vegetables, citrus and other fruits have been cultivated in the region for many centuries. Agro-industry (food processing) has also been significant at least since the beginning of the 20th century. Collective systems to manage surface irrigation were implemented several centuries ago.

Until recently agriculture was the main revenue generating activity in the Segura catchment area. Murcia was considered the orchard of Spain. Since the integration of Spain in the European Union (1986), the demand of its agricultural products increased significantly. The scarcity and/or variability in the availability of surface water resources has historically been considered a serious drawback for the development of this region. After the Spanish Civil War (1936-1939), an important program to control the Segura River was designed and implemented. The total capacity of the two dozen reservoirs built since that time is about 1,000 Mm^3. These dams have served to mitigate the destructive consequences of former catastrophic floods, but have not satisfied the farmers' water demands for irrigation. In Spain, like in most countries, the expectation is that surface water supplies for irrigation should be delivered to the farmers at an almost nominal price. Consequently, the corresponding dams and other hydraulic infrastructures were financed with public funds. According to these predominant paradigms (almost free water for irrigation), and in response to the active lobby of the Segura farmers, ever since the early twentieth century, politicians and engineers have advocated for the transfer of water resources from the "humid" Spain to the "dry" Spain It has been often argued, that it was necessary to correct the natural hydrologic imbalance of Spain. The first formal proposal to transfer water from the Tagus river headwaters (in central Spain) to the Segura River dates back to 1933. This project was dormant for more than thirty years. It was proposed again in the sixties, implemented, and completed in the seventies.

The Groundwater Abstraction Boom

In the 1950s and 1960s, the Spanish Ministry of Agriculture launched a significant effort to promote groundwater irrigation in Spain. The officers of this Ministry, together with the experts of the Spanish Geological Survey, elaborated the first regional hydrogeological surveys of Spain, introduced modern drilling technologies for deep waterwells, and promoted the use of submersible pumps, invented a few decades before. This initial activity, heavily subsidized with public funds, was soon a catalyst that promoted intensive waterwell drilling by many private farmers in many regions of Spain. The most active region in this respect was probable the Segura catchment area. There were several reasons for the special development of groundwater abstraction in this region: a) the area had a long tradition of irrigation with surface water and a traditional capacity to market high value crops in Spain and abroad; b) many farmers had the expectation that these groundwater irrigated areas would have some kind of preference in the allocation of surface water coming from the Segura reservoirs, from the Tagus rivers water transfer or, more recent, from the future Ebro River water transfer project(all these infrastructures built by the Government with public funds). In 1976 several years before the arrival of the first Tagus water, the new areas irrigated with groundwater required more water than the total theoretical water to be transferred to the Segura catchment in the eighties.

In Spain, according to the Water Law of 1879, groundwater was private ownership. The landowner could drill a waterwell in his/her land and pump as much groundwater as he/she wished, unless a third person was affected. Nevertheless, in the 1950's special regulations were enacted by the Government that theoretically made groundwater a part of the public domain in the "Vegas del Segura" (Segura floodplains). The lack of experts in hydrogeology inside the Segura Water Authority made this regulation difficult to enforce.

Even after the enactment of the 1985 Water Code the control of the old and new water wells in the Segura catchment area is rather scarce. The situation can accurately be described as one of administrative and legal "chaos" (see Llamas et al., 2001). For example, the official White Paper on Spain's Water (MIMAM, 2000, page 343) admits that in this region only about 2,500 waterwells -out of more than 20,000 drilled- are legally inventoried by the Segura Water Authority.

8.3. The Tagus River Water Transfer and the Future Ebro River Water Transfer

In 1979, almost fifty years after the first formal proposal, water from the Tagus River was transferred to the Segura catchment through a 300 km long aqueduct. The capacity of this aqueduct is about 33 m^3/second or 1,000 Mm^3/year, but the maximum volume approved for transfer during the first phase was only 600 Mm^3/year. The reality is that the average volume transferred during the first two decades of operation of the aqueduct has been about 300 Mm^3/year, that is, half of the maximum volume authorized and one third of the total capacity of the aqueduct. The theoretical 600 Mm^3 to be transferred were distributed in the following way: 110 Mm^3 for urban water supply, 400 Mm^3 for irrigation and 90 Mm^3 as estimated losses during transfer. It was also stipulated that when the water transferred is smaller that this theoretical amount, urban water supply had a clear priority. One interesting aspect of this project is that the beneficiaries of the transferred Tagus water pay a tariff for the water that is significantly higher than the tariff usually paid by surface water farmers in Spain (~0.005 €/m^3). In this case, they pay an average of about 0.1 €/m^3, although water for urban supply has a higher tariff than water for irrigation. In June of 2001 the Spanish Parliament enacted the Law of the National Water Plan. The most relevant aspect of the Law is that it approves a new water transfer of 1,050 Mm^3/year from the Ebro River in Northern Spain to several regions along the Mediterranean coast. Almost 50% of this volume is for delivering to the Segura catchment area. The planned aqueduct is almost 900 km long. Out of the total volume transferred, about 50% is for urban water supply and the rest to supply water to areas in which groundwater abstraction has been excessive and has impacted the storage and groundwater quality of the aquifers . Most of these *degraded* aquifers are located inside the Segura catchment region.

The Ebro water transfer has met strong opposition among many and different groups. The principal one is the Government of the Aragon autonomous region, located on the Ebro basin. Other interests opposed to the proposal include the people of the Ebro delta region, many environmental groups, a good number of scholars, and others that have shown their partial or total rejection to the transfer. Several massive demonstrations against the Ebro transfer have taken place. One of them in Zaragoza, the capital city of Aragon, with about 400,000 participants. Another in Brussels with more than 10,000 participants. According to the government, the real cost of the Ebro water transfer will be about 0,30 €/m^3. This figure has been contested by a good number of experts in economy, who consider that the real cost will be significantly higher. Some consider that sea or brackish water desalination is a better alternative both from an economic and an ecological point of view. In any case, the detailed technological and finantial design of the Ebro water transfer is going to take a certain number of years. The Socialist Party (now an opposition party), has announced that they will stop this water transfer if they win the next general elections (not later than May 2004).

8.4. The Conflict about the Ebro Water Transfer: Lack of Hydrosolidarity or False Paradigms?

In 2001 a poll was taken about the social perception of the Ebro water transfer. Fifty percent of those interviewed were in favor of the transfer, 30% were against and the reminder had no opinion on the issue. One could think that those who were against the Ebro transfer lacked solidarity with the Mediterranean regions because they denied water to *thirsty areas*, while the Ebro River has a surplus of water which is wasted uselessly into the Mediterranean sea. Most people, in every culture or religion think that it is a good action to give freshwater to the thirsty. In our western civilization this is an evangelical dogma. Nevertheless, are the people in the Segura catchment region really thirsty? Definitely, not. Almost 90% of the water used in this area is for irrigation of high value crops and not for urban water supply. Moreover, most of the people in Spain still think that the hydraulic infrastructures have to be financed with public funds, because the *poor farmers* cannot pay the full cost of water and they perform a task that is really essential and beneficial for the whole country. But irrigation in the modern Segura catchment area is essentially an economic activity.

Table 3 shows the evolution of irrigated lands in the Segura catchment region. It is shown that this surface has almost tripled since 1933, when the use of surface water reservoirs and groundwater was minimum. During the last twenty or thirty years, the increase in the irrigated surface has been done mainly through an excessive and uncontrolled (by the government) use of groundwater.

Table 3. Evolution of irrigated area in the Segura catchment area.

Year	Area (ha)
1933	90,000
1956	104,000
1963	115,000
1983	197,000
1993	235,000
2000	252,000

Source: Llamas and Pérez Picazo (2001).

The second old and current false paradigm is that farmers cannot (and should not) pay the *full cost* of the infrastructures to bring them water from the Ebro River. Most authors consider that if the *full cost* of the transfer was passed on to the farmers and urban users through water use fees, they would not support the Ebro water transfer or be willing to pay for it, since there are cheaper and faster solutions to meet their water needs. Nevertheless, considering the prevalent paradigm, farmers are sure that a great part of the cost of these infrastructures will not be paid by them but with public funds. As discussed in 5.2 detailed studies undertaken in Andalucía, Spain, have shown clearly that groundwater irrigation is much more efficient than surface water irrigation: it produces about 5 times more cash per m^3 used; and three times more jobs per cubic meter. With groundwater irrigation the goal "*more crops and jobs per drop*" is usually achieved. The analysis done for Andalucía (a sample of almost one million hectares), and the conclusions drawn from it, can be applied to most irrigated areas of Spain (3,5 million ha).

The need to provide cheap water to *thirsty* areas and the belief in the need to subsidize water infrastructures are still today the predominant paradigms in Spain. Llamas and Pérez Picazo (2001) consider that nowadays both paradigms are obsolete. However, some time will be necessary to change the mentality of the general public. These false paradigms are also frequent in other countries.

9. THE ROLE OF GROUNDWATER IN THE NATIONAL WATER PLAN

The main goal of the Law of the National Water Plan is the Ebro water transfer but it has also included several articles that, if enforced, will act as a time bomb against the Ebro water transfer, and dramatically improve groundwater management in Spain. For example:

1) Article 18 established that not a single drop of the Ebro River can go to an *overexploited aquifer* if detailed studies of the situation are not previously performed and approved by the central government. Groundwater user communities have to be formally established on these aquifers and none had been established in the Segura catchment area until October 2001, when suddenly several groundwater user communities have been approved by the Segura Water Authority.
2) Article 29 establishes the need to perform a thorough study of groundwater in Spain, and to foster, in general, the formation of groundwater user communities.
3) Article 34 establishes that a relevant education campaign on water resources addressed to the general public, has to be promptly carried out.

10. SUMMARY

In Spain like everywhere, complexity and variability characterize water management problems in general and even more so in the case of groundwater. Uncertainty is an integral part of water management. This uncertainty relates to scarcity of data, strong non-linearities in groundwater recharge values and changing social preferences. Honesty and prudence in recognizing current uncertainties is necessary. At the same time, there needs to be a concerted effort to obtain more and better hydrological data on which to base management decisions.

Intensive groundwater development is a recent situation in most arid and semiarid countries. Usually, it is less than 30-40 years old. Three technological advances have facilitated this: 1) turbine pumps, 2) cheap and efficient drilling methods, and 3) scientific hydrogeology advance.
Full cost (finantial, operation and maintenance) of groundwater abstraction is usually low in comparison to the direct benefits obtained.

Groundwater development has been carried out mainly by individual farmers, industries, or small municipalities. Finantial and technical assistance by conventional Water Authorities has been scarce.
The lack of planning and control of groundwater development has resulted in ecological or socio-economic impacts in a few regions.

Aquifer overexploitation is a complex concept that needs to be understood in terms of a comparison of the social, economic, and environmental benefits and costs that derive from a certain level of water abstraction. It is useless to define overexploitation in purely hydrogeological terms given uncertainties in recharge and abstraction values and the fact that the amount of available resources in a catchment area is variable and can be influenced by human actions and management decisions. The assumption that a long trend (10 years, for example) of decline in groundwater levels implies real overexploitation or overdraft may be too simplistic and misleading.

Increasing emphasis on cost-effective and environmentally sensitive management practices places a new thrust on broad public involvement in any water management decision-making process. But guaranteeing effective public participation in management processes requires informing and educating the public on increasingly complex scientific and technical issues. Effective information and education campaigns are therefore essential. The conflicts that are often a part of water management processes require the use of innovative conflict resolution mechanisms that will allow for the discovery of feasible solutions that are accepted by all and can be successfully implemented.

Because of the persistence of obsolete paradigms, the wonderful concept of Hydrosolidarity has been recently improperly-used in Spain to promote *perverse subsidies* mainly through the Ebro River water transfer to the Mediterranean regions.

The Law of the National Water Plan includes a certain number of articles that, if are actively enforced, would contribute efficiently to introduce a new water culture in Spain.

REFERENCES

Acreman, M.C. (compiler), 1999. Guidelines for the sustainable management of groundwater-fed catchments in Europe. Report of the Groundwater and River Resources Action Programme on a European Scale (GRAPES), *Institute of Hydrology*, Wallingford, UK., 82p.

Asmal, K., 2000. Water: from casus belli to catalyst for Peace. Address in the opening session in the Stockholm Water Symposium. 14 August 2000.

Bredehoeft, J.D., Papadopoulos, S.S. and Cooper, H.H., 1982. The water budget myth (Scientific basis of Water Management), *Studies in Geophysics, National Academy of Sciences,* 51-57.

Bromley; J., Cruces, J., Acreman, M., Martínez, L. and Llamas, M.R., 2001. Problems of Sustainable Management in an Area of Overexploitation: The Upper Guadiana Cathcment Central Spain. *Water Resources Development* 17: 379-396.

Collin, J.J. and Margat, J., 1993. Overexploitation of water resources: overreaction or an economic reality?. *Hydroplus*, nº 36, 26-37.

Cosgrove, W. J. and Rijsberman, F.R., 2000. World Water Vision. Earthscan Publications Ltd., London, 108p.

Cruces de Abia, J. y Martínez, L., 2000. La Mancha húmeda . Explotación intensiva de las aguas subterráneas en la cuenca alta del río Guadiana. Papeles del Proyecto Aguas Subterráneas, No. A3. Fundación Marcelino Botín, Santander, Spain, 66p.

Custodio, E., 2000. The Complex Concept of Groundwater Overexploitation. Papeles del Proyecto Aguas Subterráneas. No. A1. Fundación Marcelino Botín. Santander, Spain, 58p.

Custodio, E. and Dijon, R., 1991. *Groundwater overexploitation in developing countries. Report of an U. N. Interregional Workshop*, UN.INT/90/R43, 116p.

Dains, S.R. and Pawar, J.R., 1987. *Economic return to irrigation in India*, New Delhi. Report prepared by SDR Research Group Inc. For the U.S. Agency for International Development.

Dhawan, B.D., 1995. *Groundwater depletion, land degradation and irrigated agriculture in India.* Commonwealth Publisher. Delhi, India.

Fetter, C. W., 1994. Applied Hydrogeology. MacMillan, New York.

Fornés, J., and Llamas, M.R., 1999. Conflicts between Groundwater Abstraction for Irrigation and Wetlands Conservation: Achieving Sustainable Development in the La Mancha Húmeda Biosphere Reserve (Spain). In *Groundwater Ecology. A Tool for Management of Water Resources* (Griebler *et al.*, ed.). *European Commission. Environmental and Climate Programme*, 227-236.

Hernandez-Mora, N. and Llamas, M. R., 2000. "The Role of User Groups in Spain: Participation and Conflict in Grounwater Management". CD-ROM of the X World Water Congress (Melbourne, 12-16 March 2000), *International Association of Water Resources*, 9p.

Hernandez-Mora, N., Llamas, M.R. y Martinez, L., 2001. Misconceptions in Aquifer Overexploitation. Implications for Water Policy in Southern Europe. In: *Agricultural Use of Groundwater.* Towards Integration between Agricultural Policy and Water Resources Management (ed. C. Dosi), Kluwer Academic Publishers, 107-125.

Hernandez-Mora, N. y Llamas, M.R. (ed.), 2001. *La economía del agua subterránea y su gestión colectiva.* Fundación Marcelino Botín y Mundi Prensa, Madrid, 550p.

Indian Water Resources Society, 1999. Water: Vision 2050, New Delhi, 74p.

Janakarajan, S., 1999. Conflicts over the Invisible Resource in Tamil Nadu: Is there a way out?. In *Rethinking the Mosaic* (Moench *et al.*, ed.). The Institute for Social and Environmental Transition, Boulder, Colorado, p. 123-160.

Johnston, R.H., 1997. Sources of Water Supplying Pumpage from Regional Aquifer System of the United States. *Hydrogeology Journal*, 5(2): 54-63.

Kessler, E., 1998. Editorial, *Ambio* 27(6), p.427.

Llamas, M.R., 1985. Spanish Water Resources Policy: The Illogical Influence of Certain Physical and Administrative Factors. *Memoir of the 18th International Congress of the International Association of Hydrologists* XVIII (2): 160-168.

Llamas, M.R., 1992. Wetlands: An important issue in Hydrogeology. In *Selected Papers on Aquifer Overexploitation*, (Simmers *et al.*, de.), Heise, Hannover, 3: 69-86.

Llamas, M.R., 2000. Some Lessons Learnt during the Drought of 1991-1995 in Spain. In *Drought and Drought Mitigation in Europe* (Vogt and Somma, Ed.), Kluwer Academic Publishers, 253-264.

Llamas, M. R., 2001a. Considerations on Ethical Issues in Relation to Groundwater Development and/or mining. UNESCO Conference on International Aquifers Systems in Arid Zones - Managing non-renewable Resources. Tripoli, 20-24 November 1999. *Technical Documents in Hydrology*, V IHP, no. 42, UNESCO, París, 467-480.

Llamas, M.R., 2001b. "Cuestiones éticas en relación con la gestión del agua en España". Discurso de Ingreso en la Real Academia de Doctores, Madrid, 85p.

Llamas, M.R., Back, W. and Margat, J., 1992. Groundwater use: equilibrium between social benefits and potential environmental costs. *Applied Hydrogeology*, Heise Verlag, 1(2): 3-14.

Llamas, M.R. and Priscoli, J. Delli, 2000). Water and Ethics. In Papeles del Proyecto Aguas Subterráneas, no. A5, Fundación Marcelino Botín, Madrid, 56-99.

Llamas, M.R. and Pérez Picazo, M.T., 2001. The Segura Catchment Management and the Debate on Hydrosolidarity in Spain. 2001 Seminar of the Stockholm International Water Institute, Stockholm 18 Augusts 2001. 18 pages, *in press*.

López Gunn, E. and Llamas, M.R., 2000. New and Old Paradigms in Spain's Water Policy. In *Water Security in the Third Millenium: Mediterranean Countries Towards a Regional Vision*. UNESCO Science For Peace Services (9): 271-293.

Ministerio de Medio Ambiente (MIMAM), 2000. Libro Blanco del Agua en España, Secretaría de Estado para Aguas y Costas, Madrid, 640p.

Moench, M., 1999. Addressing constraints in Complex Systems. Meeting the Water Management Needs of South Asia in the 21st Century. In *Rethinking the Mosaic* (Moench et al, ed.). The Institute for Social and Environmental Transition. Boulder. Colorado, 1-56.

National Research Council, 1997. *Valuing Ground Water*. National Academy Press, Washington DC., 189p.

Postel, S. (1999), *The Pillar of Sand*, W.W. Norton and Co., New York, 313p.

Rangarajan, R. and Athavale, R.N., 2000. Annual Replenishable Ground Water Potential in India - An Estimate Based on Injected Tritium Studies. To be published in *Journal of Hydrology*, preprint, 30p.

Seckler, D., Amarashinge, U., Molden, D., de Silva, R. and Barker, R., 1998. World Water Demand and Supply, 1990 to 2025. *Scenarios and Issues*, Research Report 19. International Water Management Institute, Colombo, Sri Lanka, 42p.

Selborne, L., 2000. *The ethics of Freshwater Use: A Survey*. COMEST, UNESCO, Paris, 58p.

Simmers, I., Villarroya, F., and Rebollo, L.F. (editors), 1992. Selected papers on overexploitation. *Hydrogeology*, Selected Papers, Vol. 3, Heise, Hannover, 392p.

Sophocleous, M., 2000. From Safe Yield to Sustainable Development of Water Resources - The Kansas Experience . *Journal of Hydrology* 235: 27-43.

The Economist, 1998. *Environmental Scares*. December 20, 1998, 21-23

Tood, K.D. (1980). Ground Water Hydrology, 2nd Ed., McGraw Hill, New York.

United Nations, 1997. *Assessment of Global Water Resources*, New York.

U.S. Geological Survey, 1999. *Sustainability of Ground-Water Resources*. Circular 1186, 50p.

Wood, W.W., 2001. Water Sustainability-Science or Management. *Ground Water* 39(5): 641.

World Commissions on Dams, 2000. Dams and Developments. A New Frame for Decision-Making. *Earthscan*, 404p.

World Humanity Action Trust (WHAT), 2000. Governance for a Sustainable Future. Part IV. Working with Water, *Reports of WHAT*, London, 145-197.

World Water Council, 2000. A Water Secure World. *The Commission Report*.

Hydrologic Sciences and Water Resources Management Issues in a Changing World

SOROOSH SOROOSHIAN, BISHER IMAM, SHAYESTEH MAHANI, THOMAS PAGANO and MARTHA WHITAKER
University of Arizona, Department of Hydrology and Water Resources, Tucson, Arizona.
E-mail: soroosh@hwr.arizona.edu

ABSTRACT: The need for more effective management of water resources is greater than ever, particularly in arid and semi-arid regions of the world. Water resources managers must utilize more sophisticated hydrologic prediction tools. Depending on the problems, the hydrologic information needed may range from hourly forecasts (i.e., in the case of flash floods) to seasonal to interannual (i.e., in the case of reservoir operation), and to decadal to century (i.e., in the case of water supply structural design).

A further complication arises from the additional uncertainty resulting from global climate change and, hence, the intensification of the hydrologic cycle. Therefore, there is no doubt that climate variability and change will have a great impact on regional water resources and the way they should be managed. Further research and data-collection activities are needed in order to quantify the impact of the intensification of the hydrologic cycle on the magnitude, direction, and frequency of floods, droughts, and other weather hazards which impact water resources management. A number of international projects have been initiated for this purpose, including the Global Energy and Water cycle EXperiment (GEWEX) of the World Climate Research Program (WCRP) and the Earth Observing System (EOS) of NASA in the United States. These projects have a strong emphasis on state-of-the-art remote-sensing satellites which are capable of providing useful information for a variety of purposes related to hydrology and water resources. A review of these projects and their potential benefits to water resources management issues, particularly in the context of arid/semi-arid regions, will be presented.

INTRODUCTION

There is no doubt in the minds of many experts and scientists that issues related to worldwide water resources, both in terms of quantity and quality, will be in the forefront of challenges to be addressed in the 21st Century. The population of the Earth is expected to grow to about 8.3 billion by the year 2025 from the most recent estimate of 6.4 billion in 2000, with the largest potential increases in semi-arid regions, as shown in Figure 1. The availability of water resources will definitely be the most challenging in arid/semi-arid regions of the world. These regions constitute nearly one-third of the world's land mass and, as shown in Figure 2, they span all continents, with a dominant presence in Africa, the Middle East, central Asia, and western China. Several recent reports and discussions at international forums have brought much-needed attention to the various types of water resources stresses being experienced in major river basins worldwide. This challenge, of course, is not exclusive to the developing countries, but applies also to several countries in the developed world. The water problems of arid/semi-arid regions are caused primarily by rapidly growing populations in most countries located in these areas. For instance, the most rapidly growing U.S. states are located in the semi-arid Southwest, where the population is expected to increase from 45 million in 1995 to over 70 million by the year 2025. An example of the water resources stresses on the Colorado River basin in the southwest United States is given in Figure 3. This river basin is the primary source of water for a number of states in this area, including California, Arizona, New Mexico, Nevada, Utah, Colorado, and Wyoming. In addition, the discharge graph given in Figure 3 shows that, with the construction of the last major reservoir (Glen Canyon Dam) in 1960, all of the water resources of this river basin are fully utilized by the time the water reaches the Mexico border near the city of Yuma, Arizona. We may be facing a highly unsustainable situation with respect to the future availability of water, and all its intended purposes, to meet the needs of the added population.

Further complicating the situation is the issue of water quality. The head waters of the Colorado River are of excellent quality, where the Total Dissolved Solids (TDS) are about 55 ppm. By the time the flow reaches the border of Arizona and Mexico, the TDS concentration has increased to 800 ppm, primarily a result of the return flow from irrigated fields, as well as from some natural causes.

The status of ground-water resources worldwide is a major concern for the new millennium, particularly for the arid and semi-regions of the world. Many regions of the world are relying increasingly on

Water Resources Perspectives: Evaluation, Management and Policy. Edited by A.S. Alsharhan and W.W. Wood.
Published in 2003 by Elsevier Science, Amsterdam, The Netherlands, p. 83-92.

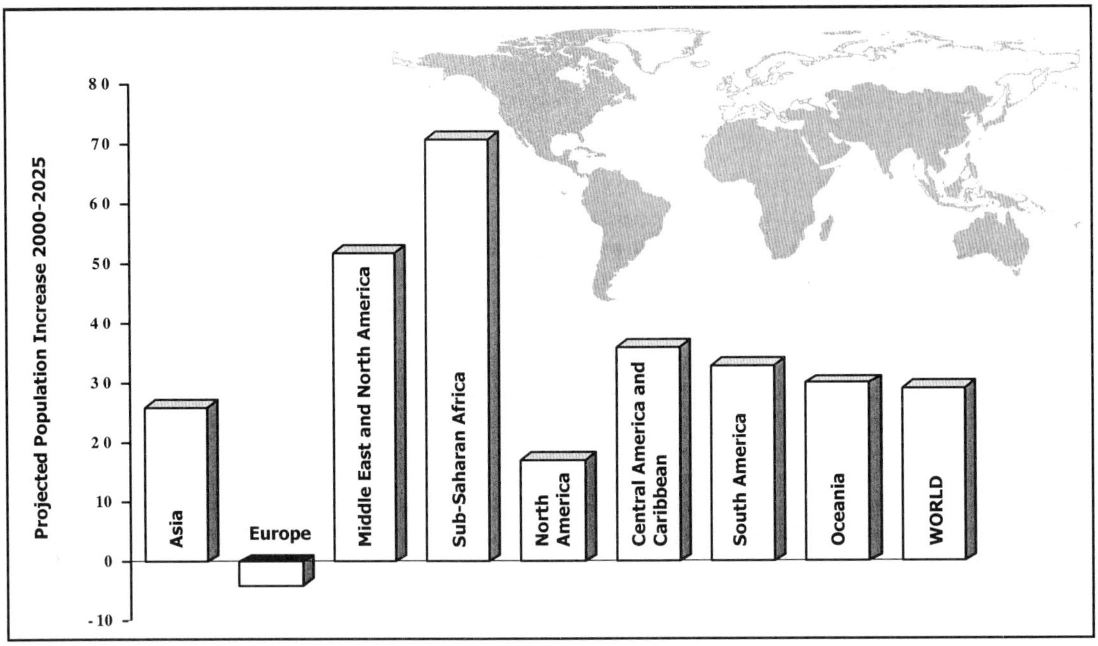

Figure 1. Projected population growth between 2000-2025 by region. Source: World Resources Institute.

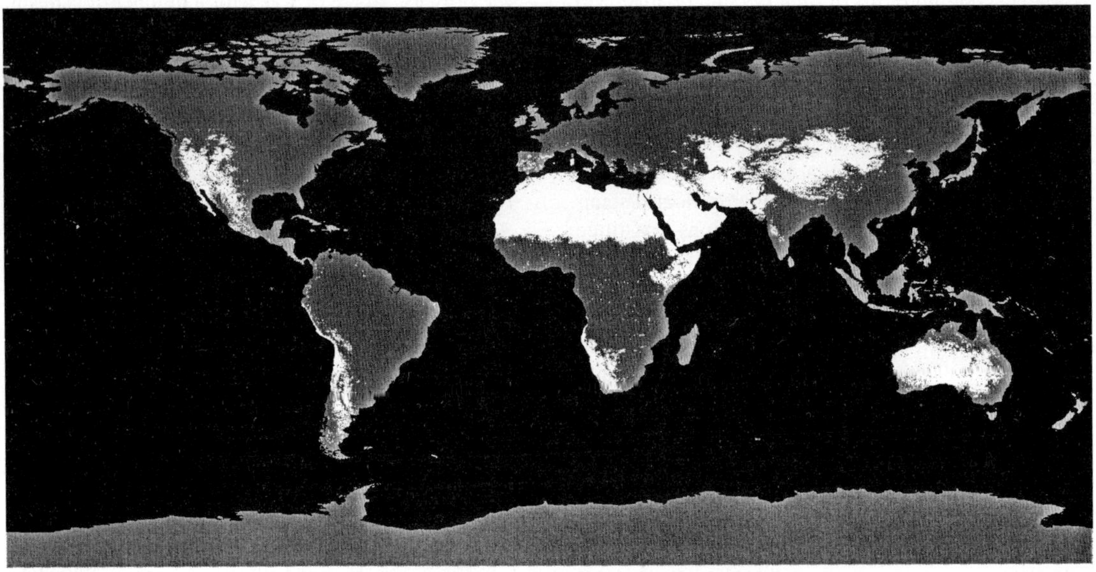

Figure 2. Global distribution of Arid/Semi-arid regions derived from USGS land cover data.

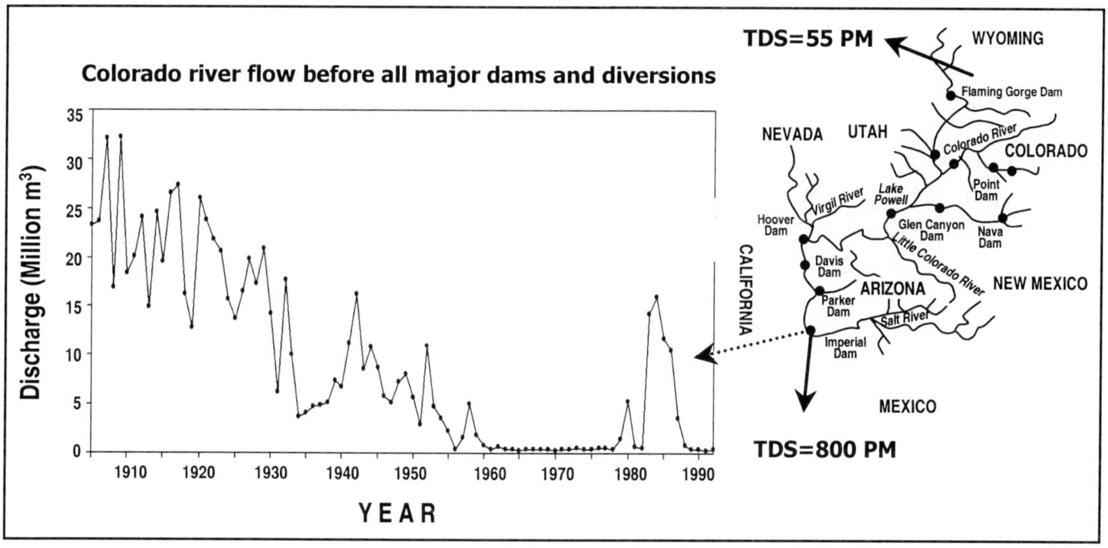

Figure 3. Colorado River Basin in the U.S. with flow hydrograph at the U.S.-Mexico border and measures of the amount of Total Dissolved Solids (TDS) at two different locations.

ground-water resources, not only for domestic consumption but also for irrigation and industrial purposes.

The invention of electrical and mechanical pumps has resulted, over the past 100 years, in a rapid utilization of ground-water resources in many regions of the world. Because this ground-water utilization has far exceeded the natural processes that recharge the ground-water aquifers, many areas are experiencing land subsidence and, hence, a major loss of aquifer storage capacity, which can never be replenished. Examples of localized and regional land subsidence due to depletion of water supplies are numerous worldwide (i.e., Beijing and southeastern Iran). The impact on land subsidence has varied from a uniform settling of ground levels to the occurrence of sinkholes, depending on the underlying geological stratiform. Examples from southeastern Iran, where ground-water depletion is now showing its impact through the occurrence of sinkholes, are given in Figure 4. The impacts of excessive ground-water pumping on water quality of the aquifers are of even more concern and are discussed extensively in the literature, not requiring further coverage in this paper.

It is clear that, from the perspective both of surface water and ground water, the potential for sustainable development will vary significantly from one place to another. Hence, uniform solution approaches, which use the philosophy of a "one size fits all" will not work. The emphasis must be placed on the regional strategies to address these problems, with special attention paid to sociocultural and geopolitical considerations.

An additional complicating dimension is the potential effect of future climate and global change on the hydrologic cycle, which may impact the development of long-term strategies for water resources management. This is a particularly critical issue in arid/semi-arid regions, because even minor changes in climate regimes are likely to have a significant incremental impact on the hydrologic cycle, both in terms of seasonal, interannual, and decadal variability and long-term trends in precipitation and drought patterns. Evidence in support of climatically induced changes in hydrologic regimes can be seen in analyses of historical records, as well as climate model scenarios. Analyses of observational records over the past 100 years by the United States National Climatic Data Center (Quayle and Karl, 1996) show an increase in annual mean in both global temperature and precipitation. In their latest assessment, the Intergovernmental Panel on Climate Change (IPCC, 2000) provided the scenario analyses by a number of climate models, in which the global mean temperature will increase over the next 100 years, ranging from 1.4-5.8°C. In the same report, climate models projected a trend in increased global mean precipitation, estimated from 7 to 15%, depending on the model. This trend suggests that the global hydrologic cycle is intensifying.

The above information supports the intensification hypothesis of the hydrologic cycle at the global scale. However, from a water resources point of view, knowledge of these changes will provide little guidance for developing long-term strategies in water resources management. Most water resources issues are regional in nature, and decisions are often made at river basin to catchment scales. Therefore, it is of paramount importance that future climate impacts are appropriately observed and studied at regional scales to make relevant and accurate information available for decision-making purposes. The remaining sections of this paper will focus on two aspects, namely our current understanding of the climate of the Middle Eastern region and some state-of-the-art global observational systems with potential benefit to the Middle Eastern region.

Hydroclimatology of the Middle Eastern Region

November 2001 marked the third consecutive year of drought in the Middle East. This drought has been the worst since 1964, inflicting at least \$3.5 billion (U.S. dollar equivalent) in damages, killing 800,000 head of livestock, and drying up major reservoirs (UN OCHA, 2000). By understanding the interannual variability of precipitation in the Middle East, it is possible to predict and prepare for such droughts, thereby mitigating the damages they incur. Indeed, the International Research Institute (IRI) produces seasonal climate forecasts for the entire globe and, over the past several winters, has predicted the development and intensification of the current drought over the eastern Arabian Peninsula, Pakistan and Afghanistan.

The Middle East, which is bisected by the Tropic of Cancer, is the nexus of several different synoptic weather features. Around the Persian Gulf, a semi-permanent trough of low pressure is capped at higher altitudes by anticyclone-inhibiting cloud growth and rainfall. The local seas somewhat influence the passage of cyclones, but the surrounding land masses diminish the effects which the seas may have. Strong orographic spatial heterogeneity in Iran and Turkey creates a complex mixture of orographic lifting, rain shadows, and small-scale local weather phenomena (such as intense localized seasonal winds). For further descriptions of the climatology of Iran and Turkey, see respectively, Ganji (1954) and Cullen and deMenocal (2000).

Generally during winter, the Icelandic Low, the Azores (Bermuda) High, and the Asiatic High modulate the passage of frontal cyclones over Europe and the Middle East (see Figure 5). Rainfall increases to the north, to the west, and with an increase in altitude, with Turkey, the eastern Mediterranean coast, and the Zagros Mountains receiving the greatest share of precipitation for the Middle East. Winter temperature variability is greatly affected by the Atlantic Ocean features (positively correlated with Icelandic Low intensity and negatively correlated with the Azores High [r = +0.4, -0.4, respectively, not shown]). The influence of the Atlantic Ocean features, especially on precipitation, diminishes to the east as more localized phenomena gain influence. Beyond Greece, individual depressions can take a mazy path before their eventual eastern exit from the region (Ganji, 1954). It is possible that the Pacific Ocean features (the Aleutian Low and the Hawaiian High) begin to gain influence as these features move eastward (i.e., Iran, Afghanistan).

Figure 4. Example of sinkholes occurring due to over pumping of the groundwater in southeast Iran.

During spring, the Asiatic High begins to weaken and, before the summer circulation is set into place, convective rainfall occurs over the snow-covered slopes of the northern Zagros Mountains. During the summer, strong subsiding outflow from the Indian monsoon makes the region oppressively hot, humid, and without cloud cover. Many locations, particularly central Saudi Arabia, Iraq, and Syria, have zero average rainfall during the summer months. The forested southern coast of the Caspian Sea experiences year-round rainfall associated with sea breezes and orographic lifting of local winds. Another exception during the wet summer is the portion of the southern Arabian Peninsula (i.e., Yemen) that comes under the influence of the Intertropical Convergence Zone and the rains of eastern Africa.

For years, scientists have attempted to detect linkages between seasonal precipitation and remote ocean temperatures (called "teleconnections"). For example, the most widely studied climate phenomenon is the El Niño/Southern Oscillation (ENSO). El Niño refers to an aperiodic warming of tropical Pacific Ocean surface temperatures, and the Southern Oscillation refers to the associated shifts in pressure and wind patterns in the tropics. La Niña is the opposite of El Niño, in the sense that it describes a cooling of the tropical Pacific Ocean. These events occur every 2-8 years and generally last from 1 to 3 years. For example, an extremely strong El Niño event occurred during 1997-1998, and a moderate-to-weak La Niña event followed during 1998-2001.

A variety of local studies have linked ENSO and Middle Eastern precipitation, although none specifically investigated the United Arab Emirates.

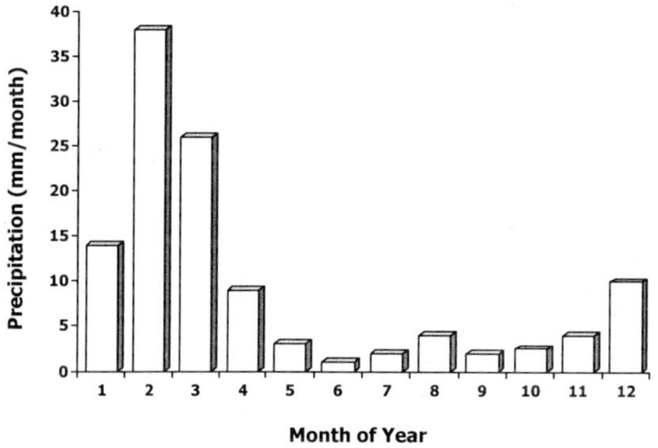

Figure 5. Seasonal cycle of United Arab Emirates monthly average precipitation (1981-1991) from a 19-station average based on data from UAE Ministry of Communications (1996).

Alpert and Reisin (1986) recognized an enhancement of snowfall in Jerusalem during El Niño events. Yakir et al. (1996) verified this precipitation enhancement during El Niño by using tree ring reconstructions. From October through April, Israeli rainfall is positively correlated with various El Niño indices (i.e., El Niño bringing wet conditions; Price et al., 1998).

Kadioglu et al. (1999) found a positive but small and spatially heterogeneous relationship between Turkish December rainfall and El Niño. Compositing revealed December precipitation enhancement in the northwest and a precipitation decrease in the southcentral and eastern parts of Turkey during El Niño. Cullen and deMenocal (2000) linked Turkish climate with various large-scale climate indices. They found insignificant correlations between an El Niño ocean temperature index and a composite index of 23 Turkish winter (DJFM) temperature stations and 27 winter (DJFM) precipitation stations. The correlations from 1930-1995 were 0.2 and 0.04 for temperature and precipitation, respectively (El Niño warms and wettens Turkey). Given that the correlation coefficients were so small, the authors dismissed this result as insignificant.

Felis et al. (2000) detected climate teleconnections in a 245-year coral oxygen isotope record from the northern tip of the Red Sea (south of El Tor, Egypt). The correlation varied from one decade to the next, but their results confirmed general impressions that El Niño brings wet conditions to northeastern Africa and the Arabian Peninsula.

Nazemosadat and Cordery (2000a) and Nazemosadat (1999) detected an enhancement in Iranian autumn precipitation during El Niño. The correlation reaches its maximum strength in the northwestern and central parts of Iran (northwest of Tehran) and then diminishes near the Caspian Sea. The signal of El Niño diminishes in Iran as one enters the heart of the winter. Seasonally, less (more) than usual winter precipitation on the west coast of the Caspian Sea tends to be weakly associated with El Niño (La Niña). In contrast, during the autumn, more (less) precipitation can be weakly associated with El Niño (La Niña) over the same area. Nazemosadat (1999) also studied the influence of ENSO on summer precipitation in Oman and found that precipitation is enhanced during La Niña and suppressed during El Niño.

While many of the cited authors studied specific parts of the Middle East, Barlow et al. (2002) were recently able to see the "big picture" with respect to El Niño and the Middle East. During the current drought in the Middle Eastern region, the area surrounding Indonesia has been exceptionally wet. This ocean region is often called the western Pacific "warm pool" because of its very warm surface temperatures. The warm pool creates intense rainfall over most of southeastern Asia. Barlow et al. (2002) correlated warm pool ocean temperatures with global winter precipitation and found a relatively strong horseshoe-shaped negative correlation from the Middle East to eastern Africa to the southern Indian Ocean (see Figure 6 for a similar analysis). By using a simplified linear dynamic model, they showed that heating in the warm pool produces Rossby waves, which disturb the jet stream over the Middle East. The authors attributed the recent extended drought in the Middle East to exceptionally warm pool temperatures.

This new work is particularly exciting because it documents a physical mechanism for the linkage between the tropical Pacific Ocean and the Middle East. It also explains why past authors found weak correlations when directly comparing precipitation and ENSO: the Middle East is influenced by warm pool ocean temperatures, which are influenced (but not entirely determined) by El Niño. The effect is generally strongest during the fringe of the Middle Eastern rainy season, namely September to November and March to May.

During the heart of winter, it is believed that the North Atlantic Oscillation (NAO) dominates. The NAO describes interactions between the ocean and atmosphere in the North Atlantic region, modulating the strength and position of the subtropical high and polar low air masses. NAO is often measured by the pressure difference between Iceland and Gibraltar, and this difference is strongest in winter. "Low" NAO years favor the passage of winter frontal cyclones over the Middle East, leading to wetter and warmer conditions. In "high" NAO years, these storm tracks are pushed farther to the north, causing the Middle East to dry out. Preliminary research has shown that this band of drying and cooling during high NAO years extends from Spain through northern Africa and Italy to Turkey and the Middle East (Figure 7). This signal for temperature in Iraq is exceptionally strong, with NAO explaining nearly 50% of the variance in wintertime temperatures (Figure 8). Such a strong signal is rare among climate studies, and it indicates a potential for accurate seasonal forecasts. NAO is also strongly linked to the flow of the Tigris-Euphrates River, and the recent upward trend in NAO has been responsible for unusually low streamflow in the Middle East. While the impacts of NAO on the Middle East appear to be relatively strong, the specific impacts have not been studied in detail.

Finally, there has been increasing interest in predicting climate on relatively short time scales. In the early 1970s, researchers discovered a distinct 40-50 day oscillation in tropical Pacific Ocean winds and rainfall, propagating from west to east. It wasn't until the mid- to late-1980s that this oscillation, the "Madden Julian Oscillation" (MJO), received much attention. The MJO explains a significant amount of short-term variability in precipitation in the tropics. However, fairly weak correlations with mid-latitude precipitation have been documented. If a forecaster recognized the development of a wave of convection over tropical Africa, s/he could predict subsidence and drier-than-normal conditions 10-15 days later over the Middle East as the wave propagates into the Arabian

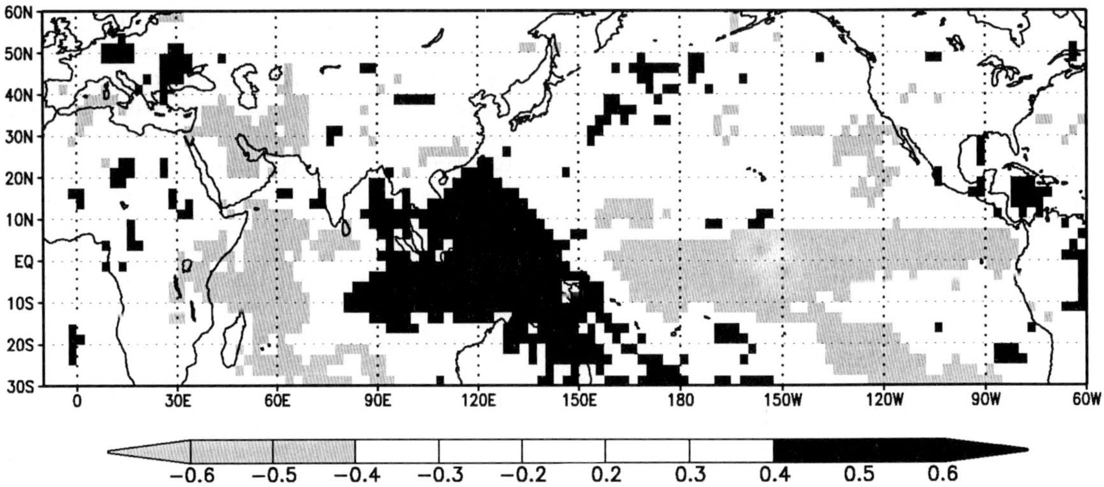

Figure 6. Correlation of a Warm Pool Sea Surface Temperature Index [105°N-100°S] with global precipitation from September-November. Dark hues (e.g., southeast Asia) indicate regions where warmer than usual Warm Pool SSTs are associated with increased precipitation. Light hues (e.g., the Middle East, west Indian Ocean and central Pacific) are regions were precipitation is decreased.

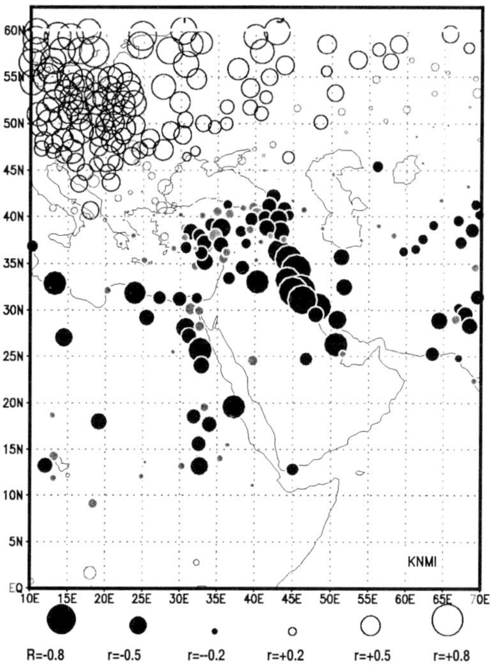

Figure 7. Correlation of a North Atlantic Oscillation index with monthly mean station temperatures for January-April. Filled circles indicate negative correlation, while hollow circles indicate positive correlation. The strength of the correlation is reflected in the size of the circle.

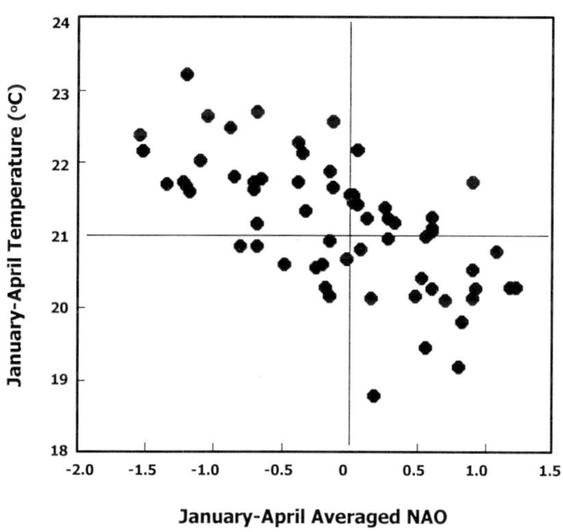

Figure 8. Correlation of a North Atlantic Oscillation index with monthly mean station temperatures at Bahrain International Airport for January-April. The overall strength of this correlation is -0.64, which is exceptionally strong among climate studies.

Sea. After the wave has passed, increased rainfall in the Middle East occurs approximately 30-45 days after the onset of oscillation. The impacts of MJO in the Middle East have not yet been specifically addressed in the literature, and this research topic has great potential.

Global Observation Technologies and their Potential Applications for Hydrology and Water Resources

Over the past several decades, utilization of space-based remote sensing of the various components of the Earth, oceans, and atmosphere has increased. NASA has been planning and launching sophisticated satellites with specific sensors designed to measure different fluxes and parameters of our environment, i.e., chemistry of the atmosphere, clouds, precipitation, vegetation, and ocean temperature. It is true that the application of remote sensing in fields such as meteorology and oceanography has progressed much further than we have experienced in hydrologic sciences. With the launch of NASA's comprehensive program, the Earth Observing System (EOS), in 1990, the scientific community began to develop and design special sensors that are becoming useful to hydrologists. The EOS satellites now allow for the study of clouds, water and energy cycles, land-surface features (such as areas of deforestation with potential for major erosion), and various aspects of the ecosystem processes. Several pictures of EOS-related satellites, which have either been recently launched or are in the final preparation launch phase, are given in Figure 9. The first primary EOS satellite, named TERRA, was launched in December 1999 and has numerous high-resolution instruments onboard its platform. Among these instruments are the Moderate Resolution Imaging SpectroRadiometer (MODIS), Clouds and the Earth's Radiant Energy System (CERES), and the Multi-angle Imaging SpectroRadiometer (MISR). The scientific community has been heavily engaged in the development of various algorithms for extraction of useful information from TERRA data. We can expect that, in a few years, TERRA data will become very helpful in areas such as parameterization of distributed hydrologic rainfall-runoff models and land-surface atmosphere models.

A Satellite-Based Rainfall Product

Traditional rainfall measurements from gauges are relatively sparse and are located primarily over the land. The use of ground-based radar now enables the measurement of rainfall over a larger area, but the coverage is essentially limited to land surface and coastal regions of developed countries. Recent advances in the development and use of satellite information has been showing great promise for monitoring the global distribution of rainfall over more remote regions and oceans.

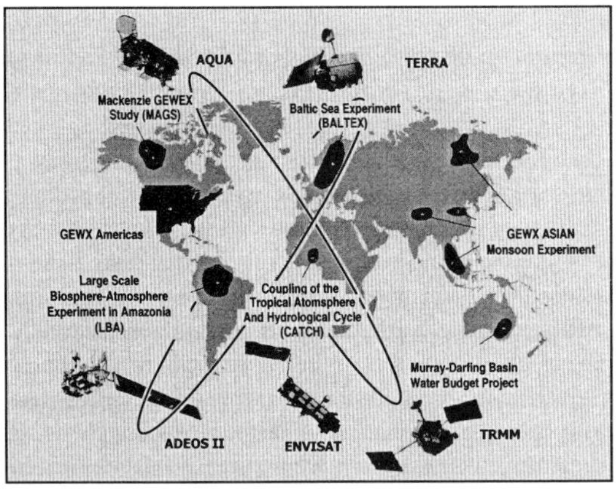

Figure 9. Satellite observations are integral to the Global Energy and Water Cycle Experiment (GEWEX) basin scale studies.

The Precipitation Estimation from Remotely Sensed Information using the Artificial Neural Network (PERSIANN) system for rainfall estimation has been under development at the University of Arizona. The fundamental algorithm is based on an Artificial Neural Network (ANN) and can therefore be easily adapted to incorporate relevant information as it becomes available. An adaptive training feature facilitates rapid updating of the network parameters whenever independent estimates of rainfall are available. The original system (Hsu et al., 1997) was based on geostationary infrared imagery and later extended (Hsu et al., 1999) to include the use of both infrared and daytime visible imagery. The latest version of the PERSIANN algorithm (Sorooshian et al., 2000) is based on the geostationary infrared imagery approach, but also uses the precipitation radar data provided by the low-inclination orbiting high-resolution Tropical Rainfall Measuring Mission (TRMM) satellite for regular updating of the network parameters. TRMM was launched as a U.S.-Japan research satellite over four years ago to measure rainfall over the tropical regions.

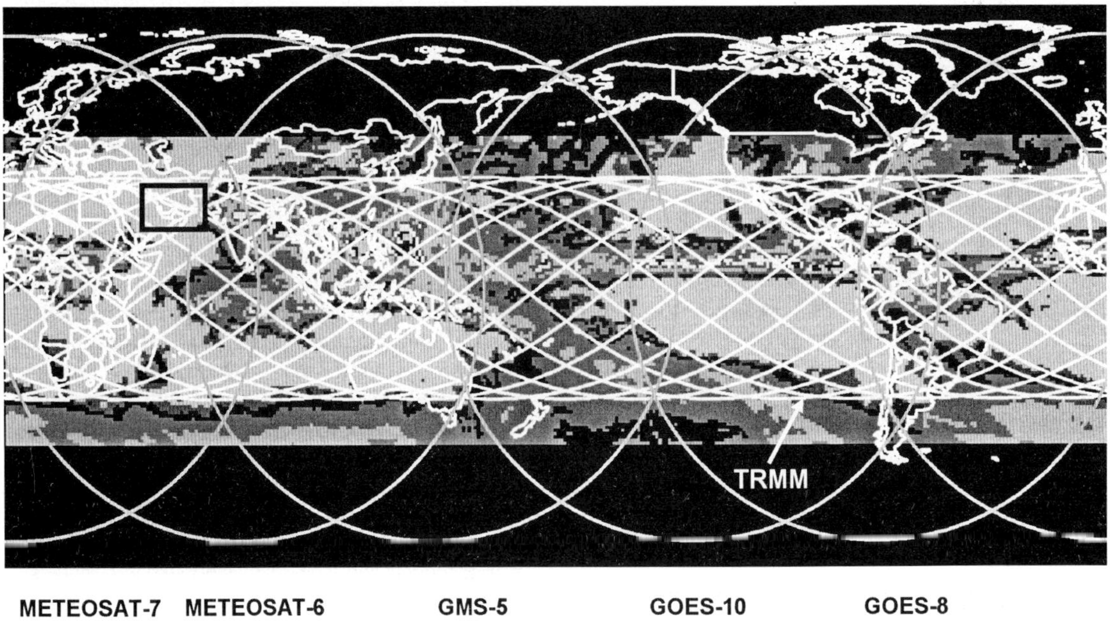

Figure 10. Coverage of PERSIANN rainfall products [50° S-50° N]. Circles indicate geostationary satellites used in generating the product, which include Meteosat-6 and-7 (Europe), GMS-5 (Japan), the GOES-8 and –10(U.S.A), spiral lines indicate daily TRMM swaths used in training the model. Operational PERSIANN rainfall data are made available over the United Arab Emirates by the Hydrologic Data and Information System (HyDIS) at http://hydis.hwr.arizona.edu.

Coverage of PERSIANN system rainfall products is shown in Figure 10. This area covers 50°S to 50°N globally. The system uses grid infrared images of global geosynchronous satellites (GOES-8, GOES-9/10, GMS-5, METEOSAT-6, and METEOSAT-7) provided by NCDC, NOAA, and TRMM TMI instantaneous rainfall products (2A12) of NASA. The "elemental" $0.25° \times 0.25°$ rain rates at 30-minute intervals are aggregated to various rainfall products of resolution $1° \times 1°$, 6-hour, daily, and monthly. The next phase in our development of the PERSIANN system is to move towards the finer spatial resolution of 4 km x 4 km.

Other Promising Areas of Remote-Sensing Applications in Hydrology

The Soil Moisture and Ocean Salinity (SMOS) mission is under development by the European Space Agency (Kerr et al., 2001). The baseline SMOS payload is an L band (1.4 GHz) 2D interferometric radiometer. SMOS aims at providing global maps of soil moisture over the land surfaces, with an accuracy better than 0.035 m^3/m^3 every 3 days, a space resolution better than 60 km, as well as vegetation water content with an accuracy of 0.2 kgm^2. Soil-moisture estimates provided by SMOS over the continents provide system continents provide system memory that can be exploited for seasonal forecasting and improving our ability to monitor water reservoirs, which are critical to the climate and economy.

The Gravity Recovery and Climate Experiment (GRACE) satellite mission, sponsored by NASA, provides promising opportunities for monitoring ground-water storages. GRACE will use a pair of satellites to map tiny variations in the Earth's gravitational field. Because variations in water storage on land affect the time-dependent component of the Earth's gravitational field, GRACE will accurately map the gravity field at 2- to 4-week intervals and may soon provide global data on temporal changes in continental water storage, particularly ground-water storage (Rodell and Famiglietti, 1999).

SUMMARY

In this paper, a brief review of some of the emerging water resources management issues that require a better understanding and characterization of the elements of the hydrologic cycle was presented. The potential impact of climate change and variability was discussed, with special emphasis on the Middle East region. In addition, the promising opportunities provided by state-of-the-art satellite remote sensing for observing and estimating a number of important hydrologic fluxes and variables were also discussed were also discussed.

As arid and semi-arid countries of the Middle East continue to plan for the future with various water resources issues, they may find that participation in international projects sponsored by a number of organizations, such as the World Climate Research Programme (WCRP), may be useful to their objectives. One such project is the Global Energy and Water cycle EXperiment (GEWEX), which was initiated by WCRP in 1988. The objectives of GEWEX are to observe (primarily through remote sensing) and model the hydrologic cycle and energy fluxes in the atmosphere and at the land and ocean surfaces. GEWEX is an integrated program of research, observations, and scientific activities ultimately leading to the prediction of variations in regional water resources. Given various scenarios of global climate change, arid and semi-arid countries may wish to investigate how the global hydrologic objectives of GEWEX can be of value, as we gain a greater understanding of the potential effects of hydrologic and climate variability on various parts of the globe.

At a smaller, more localized scale is the National Science Foundation (NSF)-funded Science and Technology Center at the University of Arizona, entitled "Sustainability of semi-Arid Hydrology and Riparian Areas" (SAHRA). The aim of SAHRA is to integrate stakeholder-driven research, education, and knowledge transfer efforts in semi-arid regions.

The research being addressed by SAHRA proceeds in five major trends, or thrust areas (TAs). The primary focus of Thrust Area 1 is measuring precipitation and snowpack above the treeline and improving predictions of runoff. This research is critical because snowpacks are often the main source for streamflow and ground-water recharge in many basins. Forecasts of spring snowmelt runoff are often critical for operating reservoirs and for planning water allocations that ultimately affect the economic and environmental health of a community. Modeling recharge and salt transport across basin floors and in ephemeral and intermittent water courses is the emphasis of Thrust Area 2. This work is of particular importance in arid and semi-arid areas, because recharge is a critical part of quantifying a basin's water budget, and knowledge of a basin's water budget is a basic facet of water resources planning. The emphasis of Thrust Area 3 is to improve our understanding of water, energy, and nutrient fluxes in riparian areas. In many arid and semi-arid regions, riparian areas support unique ecosystems and often serve as migratory flyways for a continent's bird populations.

Human communities often reside near riparian areas because of the accessibility to ground water and river water. Thrust Area 4 works in cooperation with researchers in all of the TAs to integrate physical models with policy tools, so that the models and/or the subsequent results will be of practical use for water policies and decision-makers. Finally, Thrust Area 5 uses decision support models and policy analyses to facilitate sustainable water resource management practices among policy and decision makers.

While the research that takes place within SAHRA spans both the physical and behavioral sciences at local and basin scales, researchers are also committed to producing information, analyses, curricula, expertise, and information that can ultimately increase hydrologic literacy and positively impact water policy and water resources management. It is an excellent, smaller-scale example of how one might consider more broad-reaching management of water resource issues in semi-arid regions.

The integrated research of SAHRA, combined with the insight and experience of global hydrologic research and research-coordinating programs (e.g., NASA's EOS program, GEWEX, etc.), provide excellent examples of how one might consider the imperative, broad-reaching management water resource issues facing water resource managers and policy makers, particularly in arid and semi-arid regions of the world.

ACKNOWLEDGEMENTS

This work was supported by NASA grants (NAG8502 and NAG8503), NOAA OGP grant (NA16GP1577), and NASA Raytheon award (300623). The authors also wish to thank Ms. Corrie Thies and Ms. Carly Davis for their help in proofreading and editing this manuscript.

REFERENCES

Alpert P. and T. Reisin, 1986. An early winter polar air mass penetration to the eastern Mediterranean. *Monthly Weather Review* 114:1411-1418.

Barlow, M., H. Cullen, and B. Lyon, 2002. Drought in central and southwest Asia: La Nina, the warm pool, and Indian Ocean precipitation. *Journal of Climate*, 15(7):697-700.

Cullen, H.M. and P. deMenocal, 2000. North Atlantic influence on Tigris-Euphrates streamflow. *International Journal of Climatology* 20(8):853-863.

Felis, T. et al., 2000. A coral oxygen isotope record from the northern Red Sea documenting NAO, ENSO, and North Pacific teleconnections on Middle East climate variability since the year 1750. *Paleoceanography* 15(6):679-694.

Ganji, M.H., 1954. A contribution to the climatology of Iran. Ph.D. Thesis, Clark University. Worcester, Massachusetts. 333 p.

Kadioglu, M., Y. Tulunay, and Y. Borhan, 1999. Variability of Turkish precipitation compared to El Niño events. *Geophysical Research Letters* 26(11):1597-1600.

Kerr, Y.H., P. Waldteufel, J.P. Wigneron, J.M. Martinuzzi, J. Font, and M. Berger, 2001. Soil moisture retrieval from space: the Soil Moisture and Ocean Salinity (SMOS) mission. *IEEE Transactions in Geosciences and Remote Sensing* 39(8):1729-1735.

Nazemosadat, M.J., 1999. ENSO's impact on the occurrence of autumnal drought in Iran. *Drought Network News* 11(2):15-16.

Nazemosadat, M.J. and I. Cordery, 2000a. On the relationships between ENSO and autumn rainfall in Iran. *International Journal of Climatology* 20:47-61.

Price, C. et al., 1998. A possible link between El Niño and precipitation in Israel. *Geophysical Research Letters* 25(21):3963-3966.

United Arab Emirates Ministry of Communications, Department of Meteorology, 1996. UAE Climate. Cultural Foundation Publications, Abu Dhabi, UAE. 217 p.

United Nations Office for the Coordination of Humanitarian Affairs (OCHA), 2000. Iran-Drought. OCHA Situation Report No. 5. OCHA/GVA-2000/0227. Geneva, Switzerland.

Yakir, D., S. Lev-Yadun, and A. Zangvil, 1996. El Niño and tree growth near Jerusalem over the last 20 years. *Global Change Biology* 2:101-105.

ARID AND SEMI-ARID REGIONAL WATER RESOURCES MANAGEMENT AND POLICY

The Nile Basin Regime: A Role For Law?

JUTTA BRUNNÉE
Faculty of Law, University of Toronto, Canada. E-mail: jutta.brunnee@utoronto.ca

STEPHEN J. TOOPE
Faculty of Law and Institute of Comparative Law, McGill University, Canada. E-mail: stephen.toope@mcgill.ca

ABSTRACT: Only a few years ago the regime of the Nile basin was one of unremitting and open conflict, or at least of incipient, and barely camouflaged, competition. The last few years have witnessed a remarkable shift in the tone and substance of state-to-state relationships along the Nile. The Nile Basin Initiative (NBI) is the most recent and far-reaching example of this new dynamic.

Our goal in this paper is to tease out factors that have contributed to nascent regime change in the Nile basin. We catalogue a variety of explanatory factors. Law has been accorded a subordinate role in academic and policy debates concerning the Nile. We argue, by contrast, that throughout the evolution of the Nile basin regime, legal norms have been influential and have historically hindered but now promote co-operation. We trace three distinct but interrelated facets of law's persuasiveness: how it shapes the individual and collective identities of states, enables or constrains convincing arguments, and promotes adherence through its specific form of legitimacy.

We suggest that the various historic Nile treaties have served not only to entrench the competitive attitudes of Nile riparians but, more fundamentally, have prompted separate and competitive identities amongst the Nile basin states. Law's control over what counts as persuasive argument has also played out in essentially negative terms within the Nile basin: The legal positions of various Nile riparians concerning international water law have echoed the historic patterns of conflict.

The unsettled and contradictory state of international water law over the last thirty years has also served to exacerbate competitive attitudes along the Nile. The long unresolved tension and continuing co-existence of two seemingly incompatible legal principles, "equitable use" and "no significant harm," allowed states to make entirely self-interested arguments that were often in bad faith. Recent changes in this "legal endowment," changes that begin to address the fundamental conflicts between upstream and downstream states, are one of the reasons for the new spirit in Nile basin relationships.

The recent informal processes of discussion on the Nile cannot appropriately be characterized as purely strategic bargaining. Rather, they have begun to involve genuine arguments, geared towards reaching common understandings and reasoned consensus. They are interactional processes of lawmaking, contributing simultaneously to shaping state identities, framing persuasive arguments and building legitimacy. Law's contribution to any treaty regime that may evolve is not merely in the delivery of "bindingness" at the end of the day. Law will already have helped to make agreement possible, and will have generated the sense of adherence to mutually constructed norms that serves as the basis of their persuasiveness.

Law has played a significant role in the Nile basin for generations, often in negative ways, by reinforcing competition and conflict. Law's persuasive influence has recently shifted, both through the elaboration of a less oppositional regime of international water law, and through the creation of various processes that have fostered regional collaboration. If law continues to play its current facilitative role in shaping interaction, prospects are good for the elaboration of a comprehensive approach to Nile management.

INTRODUCTION

Until very recently the regime of the Nile Basin was one of open conflict, or at least of barely camouflaged competition. Journalists, as well as political and legal analysts, delighted in quoting a succession of regional statesmen, especially Egyptians, who spoke of the threat of war over scarce Nile resources. In the 1970s, Egyptian President Sadat and Ethiopian strongman Mengistu Haile Miriam exchanged threats over the apportionment of Nile waters (Swain, 1997). President Sadat warned that "[t]ampering with the rights of a nation to water is tampering with its life, and a decision to go to war on this score is indisputable in the international community" (Krishna, 1988). As recently as 1991, media reports suggested that Sudan aimed missiles at the Aswan High Dam in Egypt during the Gulf War, rendering the Egyptian political hierarchy "apoplectic"

Water Resources Perspectives: Evaluation, Management and Policy. Edited by A.S. Alsharhan and W.W. Wood.
Published in 2003 by Elsevier Science, Amsterdam, The Netherlands, p. 93-117.

(Hundley, 1992). Such threats reflect the strongly competitive political and legal environment that dominated the Nile Basin for generations.

More recent public actions and pronouncements of regional leaders strike a markedly different note. In May of 1999, Ethiopian Foreign Minister Seyoum Mesfin told a press conference, after a meeting between Prime Minister Meles Zenawi and Egyptian President Hosni Mubarak, that the issues of the Nile "will never be the cause of a war" between Ethiopia and Egypt ("Ethiopia Rules", 1999). The Egyptian Minister of Public Works and Water Resources, Mahmoud Abu Zeid, had previously declared that "[t]here is no conflict or struggle between Egypt and any other Nile Basin country . . ." ("Egyptian", 1998).

Of course, mere declarations of peaceful intent do not in themselves signal a regime transition, but concrete evidence of a striking shift in attitudes and approaches is growing. Reports of tension between Ethiopia and Egypt over access to the Nile waters were denied at the highest levels in 1998. The Ethiopian Prime Minister admitted that the history of conflict was significant, but he suggested, in an interview with the Arabic-language London daily *Al-Hayat,* that Ethiopia's position was changing because Egypt had shown "more . . . understanding of the Ethiopian viewpoint" in the previous year ("Ethiopia Moves", 1998). After its annual meeting in May 1999, the Nile Council of Ministers, the high-level political body charged to pursue information sharing and to promote coordination, announced that it had passed resolutions "to facilitate the transition from the era of confrontation to cooperation among the Nile Basin countries" ("Nile Basin Council", 1999). Over the last three years, Ethiopia, Sudan, and Egypt have agreed to increase their cooperation with respect to the Blue Nile ("Ethiopia, Sudan", 1999), while nine Nile riparian (or water-supplying) states have agreed to the creation of a joint Nile Basin Initiative (NBI) and have established an NBI secretariat in Uganda ("Nile Basin Initiative", 1999).[1] The Initiative is designed to foster not only information-sharing and technical assistance, but also joint development initiatives. The NBI is cast as a "transitional arrangement," intended to facilitate basin-wide discourse until a "permanent legal framework" is in place (ICCON, 2001). At a meeting of water resources ministers from all ten Nile riparians, held in Khartoum, Sudan, in early August 2000, Osman el Tom, Vice-Chairman of Sudan's Water Resources Authority, reported "remarkable convergence toward future co-operation" ("Nile Basin Ministers", 2000). One year later, it appears that consensus on many parts of the framework is emerging. Indeed, at a June 2001 meeting of Nile Basin states, the international donor community, and NGOs, participants expressed hope that the NBI will emerge as "an example of how international waters can become catalysts for cooperation, development, and stability" ("Press Release", 2001).

This more cooperative spirit is new, and it is therefore hard to assess its long-term impact. Yet, anyone interested in regime transition can discern an emerging pattern that is worthy of notice.[2] Recent developments in the Nile Basin provide an opportunity to consider the processes that surround the emergence and evolution of regimes. In particular, these developments provide a unique opportunity for international lawyers to examine the role of legal norms in these processes and to apply the rich theoretical insights that have emerged from recent literature in the fields of international relations (IR) and international law. The goal of this essay is to tease out factors that have contributed to nascent regime change in the Nile Basin. The essay catalogues a variety of explanatory factors. Its most controversial argument is that evolving legal norms have influenced this change; but not through the creation of predictable rules and institutional structures, as international relations (IR) scholars often posit (Brunnée and Toope, 2000; Brunnée and Toope, 2001).

Throughout the evolution of the Nile Basin regime, legal norms have been influential and have both hindered and promoted cooperation. Building upon a previous description of "contextual regimes" (Brunnée and Toope, 1997) this essay will suggest that the evolving normative framework for shared freshwater has helped to redefine both the identities and interests of key state actors in the Nile Basin, moving them more recently towards more cooperative behavior. The essay flows directly from the overview provided by Professor Allan earlier in this volume, wherein he emphasises the social and political factors that influence water policy. Indeed, this essay is a concrete demonstration of the truth of the marvelous epigram offered up by Allan: "Constructed knowledge has just as much political purchase as probablistic science" (Allan, 2003). The approach adopted here draws on the interactional legal theory of Lon Fuller. Fuller understands law not as hierarchical ordering but as an ongoing generative activity, oriented towards the construction of relatively stable patterns of practices and normative expectations. Rules are persuasive and legal systems are seen as legitimate to the extent that they are consistent with this background of practices and expectations. Many of the central insights regarding the role of law in the Nile Basin are also gleaned from extending the analysis of IR constructivists such as Kratochwil, Onuf, Ruggie, and Wendt, into the discipline of international law.

The first part of the essay briefly reviews the linkages between constructivism and the "interactional theory of international law" (Brunnée and Toope, 2001). The second part provides an overview of the hydrological and geographic context of the Nile Basin. A third section details the social and political context, including the weak legal framework that has fostered the historical evolution of a purely competitive regime of freshwater in the Nile Basin. The fourth part of the essay extends the empirical examination to highlight recent moves toward greater cooperation, and canvasses various factors promoting regime change. The fifth and final part posits findings concerning the role of international legal norms in helping to explain significant political change. In so doing, the essay assesses the contributions of the historic treaties governing the Nile Basin, of international water law, including the new U.N. Convention on the Law of Non-Navigational Uses of International Water courses ("U.N. Convention", 1997), and of various informal institutions designed to promote cooperation among Nile riparians.

Within the last decade, the legal framework has been recast in ways that seem more conducive to cooperation in the Nile Basin. The hypothesis underlying this essay is that law's influence derives from three interrelated characteristics. First, law has the power to influence the individual and collective identities of states. Second, law both enables and constrains international discourse by establishing what counts as persuasive argument or rhetoric. Further, its inherent and specific form of legitimacy enhances the persuasiveness of law's postulates. This hypothesis is founded upon a particular theoretical construct of law and politics, which will be set out in summary form before turning to the fascinating case of the Nile Basin Regime.

Constructivism and Interactional Legal Theory

One of the greatest challenges to theories of international relations in the last century was how to explain major political change. The challenge is particularly pointed with respect to the shifting Nile Basin regime. Two broad streams of IR theory, neo-realism and neo-liberalism, have dominated the disciplinary debate. In a neo-realist framework, states are seen as homogenous actors that proceed on the basis of rationally assessed and pursued self-interest (Katzenstein et al, 1998; Morgenthau, 1954; Waltz). Participation in a regime or adherence to a norm occurs if the net benefits outweigh those of unilateral action. Regimes and norms are seen as reflections of underlying power and interest balances rather than as independent factors influencing behavior. While neo-liberal IR theory, particularly institutionalism, is also rooted in rationalist assumptions, states are not assumed to be homogenous actors. Rather, states must be understood, at least partly, in relation to the institutions in which they are engaged (Keohane and Nye, 1977; Keohane and Nye, 1987). Ultimately, however, both approaches treat actors and their interests as separate from processes of interaction, as largely 'given' (Ruggie, 1998; Wendt, 1994). Engagement in institutions may affect their behavior but not the 'givens' as such (Ratner, 2000).

Rationalist IR theory, almost by definition, attributes to law a highly limited role in influencing state behavior. Law, like other social norms, can provide predictable rules and stable institutional structures.[3] However, most IR theory devotes little time to distinguishing legal and non-legal norms nor to attributing distinctive effect to the former (Finnemore, 2000). This view has been supported by the predominantly positivistic stance of the discipline of law itself. This stance has significant implications for efforts, by IR theorists and legal theorists alike, to understand how norms function in international society. If law is understood as a hierarchical exercise in social control derived from a de facto sovereign (Austin, 1885), a grundnorm (Kelsen, 1952; Kelsen, 1961), or a rule of recognition (Hart, 1983; Hart, 1994) the horizontal nature of international law is problematic. International law either does not exist at all or is epiphenomenal, as realists would conclude, or it must be distorted to fit into the positivist framework, as much legal theory has done to justify international law as "real" law (Kratochwil, 1989).

One possible starting point for breaking out of this self-fulfilling program for law's ineffectiveness is constructivist IR theory (Onuf, 1989; Onuf, 1998), which emerged in part as a reaction to the dominant rationalist explanatory models (Katzenstein et al., 1998; Martin and Simmons, 1998). Like institutionalists, constructivists focus on interaction, communication, and discourse among actors.[4] The distinctive features of constructivist theory are its rejection of the assumption of interests as exogenous to interaction and its focus on the *identities* of the actors (including states) as generators of interests.[5] The main point is that identity formation is relational and *prior* to interest formation (Wendt, 1992). Identities are constructed through social interaction. The ends of the interaction are not predetermined but can be discovered and learned. Structures such as institutions, norms, and rules are not immutable but can be recast through changes in actor identity, which, in turn, are influenced by interaction and mutually created structures (Risse, 2000). Constructivists describe how structures foster "shared understandings" that can then shape both the identity of the actors and the further evolution of the structures themselves (Ruggie, 1998; Wendt, 1992). This emphasis on the shaping of identities contains a further important insight: ideas, shared understandings, or norms are seen not as direct causes of behavior but as structures that both constrain and enable choices (Ruggie, 1998). This conception opens new vistas on the role of international law as neither simply imposed social control nor as completely subordinate to the interests of states. Rather, law is seen as generated and molded through interaction and, in turn, as affecting actor behavior by influencing actor identity, thereby reconstructing interests.

Within legal theory, one of the richest repositories of insights to assist international lawyers in re-conceiving the role of law in international society is found in the writings of Lon Fuller (Brunnée and Toope, 2000). Fuller's work has much common ground with constructivism in that it articulates an interactional understanding of law. Law is seen as evolving through social practice -- a continuing challenge rather than a finished product (Fuller, 1969). Law is not by definition hierarchical, but is a construction dependent upon mutual generative activity (Fuller, 1969). Through interaction, relatively stable patterns of expectation ("shared understandings," in constructivist terms) must emerge to allow the application of norms in specific contexts. Ultimately, rules are persuasive and legal systems are perceived as legitimate when they are broadly congruent with the practices and shared understandings in society (Postema, 1994). These shared understandings create a framework for a rich conception of rhetorical activity whereby actors in a social system can be convinced of the need for legal norms (Brunnée and Toope, 2000). As well, actors learn to read the social context against which particular legal norms must be postulated and interpreted (Postema, 1994).

For Fuller, there are certain internal characteristics that distinguish law from other forms of social ordering. They may be summarized as requiring that rules be compatible with one another, that they ask reasonable things, that they are transparent and relatively predictable, and that known rules actually guide the discretion of officials.[6] It is these internal characteristics that account for the 'binding' quality of law, rather than hierarchical authority or pedigree. The greater the extent to which these characteristics are present the greater the legitimacy of the norms or legal system and the greater law's power to promote adherence.[7] The criteria, and the resulting legitimacy of law, are largely related to the "process" of law creation; the tests remain neutral as to substantive goals pursued by the system. However, it is important to stress that "means" (process) and "ends" (substance) are not radically distinct. In an interactional model of law, process is part of the ends that law serves.[8] The binding effect of law is achieved essentially through self-bindingness, created in processes of mutual construction, legitimacy gained by adherence to internal criteria, and congruence with existing social norms, practices, and aspirations.

The implications of such an interactional understanding of law for the role of law in the context of the shared waters of the Nile Basin are significant. It is the prevailing anemic conception of law that drives the conclusion of many commentators that international law has little to contribute to the relations among basin states with respect to Nile waters.[9] An understanding of law that would posit its direct impact upon state behavior in the absence of underlying congruence with shared expectations is misconceived. If one's expectation of law is that by its mere formal existence (as an enunciated "rule") it can directly cause behavior, then one would inevitably conclude that international water law has been ineffective in the Nile context. Conversely, if one attributes no independent effect to law, as realists would, or sees it mainly as a device to coordinate interstate relations and make them more efficient, as institutionalists might, then international law's role is confined to the mere formalizing of Nile agreements once reached. By contrast, an interactional view of law can lead to a better understanding of both the promise and the limitations of law in the Nile context. Specifically, this essay traces three distinct but interrelated facets of law's persuasiveness: how it shapes the individual and collective identities of states, how it enables or constrains convincing arguments, and how it promotes adherence through its specific form of legitimacy. The essay examines each of these facets in relation to the historic Nile treaties, international water law (in particular as manifested in the U.N. Convention 1997), and various informal institutions involving Nile riparians.

Hydrological and Geographical Context of the Nile Basin

The Nile is one of the world's great rivers, flowing for 6,825 kilometers through much of Northeastern Africa, draining approximately 2.9 million square kilometers of territory, and nourishing roughly 280 million people in ten riparian states. The geographic and climatic setting of the Basin is varied, ranging from tropical rain forest to mountains, plateaus, and deserts. Roughly 85% of the Nile's water originates in the highlands of Ethiopia from which three rivers flow into the main body of the Nile: the Blue Nile, draining Lake Tana, and the Atbara and Sobat Rivers (through Eritrea and Sudan, and Ethiopia and Sudan, respectively). The remainder of the water flows along the White Nile from Central and East Africa, originating in Lake Victoria (Kenya, Uganda, and Tanzania) and the mountains of Burundi, Rwanda, and the Democratic Republic of Congo. The White Nile flows through and feeds the great Sudanese swamps (The Sudd). The White Nile and Blue Nile meet at Khartoum, the former "clear and limpid" and the latter carrying "the fertilizing lime which in the past has been so helpful to Sudan and to Egypt" (Caponera, 1993, p.650). From Khartoum, the Nile is a single river, flowing through Egypt on its way to the Mediterranean Sea (Elhance, 1999; CIDA, 1994; Waterbury and Whittington, 1998; Kliot, 1994; Dellapenna, 1994; Marcus, 1997)

Although "flow" statistics for the Nile can be quite imprecise for a number of technical reasons, it is fair to conclude that annual flow of the Nile, measured at Aswan in Egypt, has diminished significantly over the last century (Swain, 1997; Kliot, 1994; CIDA, 1994). The river is also highly seasonal, with roughly 80% of its discharge occurring between August and October (Lowi, 1993; CIDA, 1994; Elhance, 1999). Many scientists believe that with the completion of the Aswan High Dam, and the creation of the huge reservoir known as Lake Nasser, controlled discharge has saved Egypt both from major floods and from major droughts, and has helped to markedly increase agricultural production (CIDA, 1994; Chesworth, 1994). However, as a result of the same project, only about 2% of the Nile's flow actually reaches the sea (Swain, 1997). Such low flow levels have contributed to ecological problems that have reached crisis proportions. In a number of Nile riparians, a further problem is presented by poor water distribution. For example, it is reported that in Ethiopia roughly forty million hectares of land are drought prone and 540,000 hectares are flood prone (El-Atawy, 1996).

The Nile is essentially Egypt's sole source of water, (Gleick, 1993a; Schwabach, 1998) sustaining a population of over sixty eight million people, estimated to be growing at a rate of one million every nine months (Theroux, 1997)[10] Over 95% of Egypt's population resides in the Nile Valley (Marcus, 1997). Population growth is even faster in Ethiopia, where it is expected that by 2025, there will be more people to feed than in Egypt (Swain, 1997). A World Bank study has predicted that by 2025, the amount of water available to each person in North Africa will have dropped by roughly 80% in a single lifetime (Berkoff, 1994; Vidal, 1995). Water planners describe a country as "water scarce" when annual renewable freshwater is less than 1,000 cubic meters per person per year. By this measure, as of 1990, Burundi, Kenya, and Rwanda were already subject to water scarcity. Egypt and Ethiopia are expected to fall into that category by 2025, and Tanzania and Uganda will join the group by 2050 (Laki, 1998).[11] Sudan, because of its limited population in relation to its size, its relatively accessible flood region, and its endowment of enough rain to sustain significant rain-based food production, is not expected to face the effects of water scarcity in the near future (Allan, 1994a; Stoner, 1994; Kliot, 1994).[12] However, development plans under consideration by upper riparians could have a significant negative impact on the Nile's flow into Sudan and Egypt, altering the current political equation enormously.[13] In addition, climate change could well contribute to a speeding up of the processes giving rise to water scarcity throughout the Nile Basin (Gleick, 1990).

A further troubling problem faces Egypt, the downstream riparian and the state most dependent on the Nile: pollution of the river and other negative environmental effects of water development choices made over the last few decades.[14] The high use of pesticides and fertilizers in the Nile valley, and particularly in Egypt, has caused pollution of the river through the leaching of chemicals from the soil (Sadik and Barghouti, 1994; Theroux, 1997; Kliot, 1994). The level of pollution has also been exacerbated by the inability to manage the industrial, domestic, and agricultural waste produced by a fast-burgeoning population (Mageed, 1994; Theroux, 1997). The building of the Aswan High Dam in 1963 had many positive results, but for almost forty years, the Dam "has kept the river from flooding and depositing renewing sediment at its mouth. The delta has instead been inundated with catastrophic superlatives: It is among the world's most densely cultivated lands, with one of the world's highest uses of fertilizers and highest levels of soil salinity" (Theroux, 1997). The Aswan reservoir, Lake Nasser, also permits high levels of water loss through seepage and evaporation, contributing to increased soil salinity along the course of the river (Laki, 1998; Kliot, 1994).

Water quality is especially important to Egypt because it is so dependent upon the Nile for agricultural production. Although Egypt has, over the last thirty years, substituted food imports (funded massively by the United States of America) for indigenous production (Allan, 1994a), it remains true that without a secure and clean water supply from the Nile, poorer Egyptians would starve. This is not currently true for any other Nile riparian. Even in Ethiopia, with its fast-growing population, "food supply ... is not much affected by Nile flow."(Allan, 1994b). Of course, Ethiopia has not previously exploited the Nile to support food production, depending instead upon rain-based agriculture. That trend is likely to change in the short term, so by expanding production and attempting to create more stable food supplies, Ethiopia may become more reliant on the Nile than is currently the case (Befekadu, 1999; Kendie, 1999a).

This brief sketch of geographical and hydrological factors suggests a classic case of international resource competition, especially given Egypt's high level of dependence upon the waters of the Nile. As will be shown, the competitive environment has been intensified by historical and legal factors, and by changing political dynamics affecting development options throughout the basin.

Historic Patterns of Competition and Control

To a significant degree, the political context in the Nile Basin is conditioned by the region's colonial history and the strategic concerns of its colonial powers. Control over, and competition for, the waters of the Nile were central preoccupations, pursued either through efforts to gain direct control over key areas, or through treaties designed to establish legal control over the Nile. The colonial patterns of competition and quest for control were subsequently replicated by the newly independent states in the region and the influence of a competitive legal environment continues to be felt.

Egypt's case exemplifies the tension between the desire for control and the reality of interdependence that underpins much of the geopolitics in the Nile Basin. An old saw runs "Egypt is the Nile and the Nile is Egypt" (Marcus, 1997).[15] The phrase is as misleading as it is accurate. While it encapsulates Egypt's dependence on the Nile and its aspirations for control over its waters, it fails to acknowledge the import of the Nile for other basin states. Throughout its long history, Egyptian civilization has fed on the lifeblood of the Nile. As noted above, contemporary Egypt remains strikingly dependent on Nile water for its very survival. Yet the same historical claim is made for Sudan, which despite being comparatively less dependent on the Nile than Egypt, nonetheless confers crucial importance on the River. Seventy-seven percent of Sudan's water originates outside its borders and most of that is carried into the country by the Nile system (Gleik, 1993a). The Nile and its sources also have profound effects on at least seven other countries. So while the Nile may, in a very real sense, *be* Egypt, the Nile is also *more than* Egypt. However, a major part of the historical story of resource competition is framed by hegemonic claims for the protection of Egyptian interests, claims offered first on behalf of Egypt by the British, and more recently by Egyptians themselves.

As colonial "protector," Britain undertook a series of initiatives to ensure the almost unimpeded flow of Nile waters into Egypt. Indeed, it was the British who imposed a basin-wide regime to the benefit of Egypt, a regime that is only now undergoing significant change (Ahmed, 1991; Waterbury, 1994; Babatunde, 1994).[16] Although the hegemonic aspect of the regime is deeply problematic, it had the virtue of treating the entire Nile Basin as a unit, a feature that the modern "ecosystem approach" to water management seeks to recreate, albeit for fundamentally different reasons (Brunnée and Toope, 1994). Believing initially that the main supply of Nile water came from the East African tropical lakes, Britain's first initiatives were concentrated in Central and Eastern Africa, particularly Uganda (Howell, 1994; Ntambirweki, 1996).

British efforts to protect the White Nile flow continued into the twentieth century with a 1906 British-Belgian agreement guaranteeing water flow from the Congo into the Nile Basin, and similar agreements with Italy and France regarding their colonial territories ("Agreement", 1906, article 3; Jacobs, 1993). The most far-reaching agreement was the 1929 Nile Waters Agreement, a colonial treaty involving Britain, as the governing power in Sudan, and Egypt (Jacobs, 1993). The Treaty specifically granted priority to Egyptian water needs and forbade any construction on the Nile in Sudan that would restrict Nile flow (Dellapenna, 1996).[17] Further, Britain granted Egypt the right to veto any future hydroelectric project in any of the British colonies along the Nile (El-Atawy, 1996).

Finally understanding the central importance of flow from the Ethiopian highlands to overall Nile water flow, Britain concluded a treaty with the Italian colonialists in 1891 to preclude the building of any structures that would impede the flow of the Atbara River into the Nile ("Protocols", 1891; El-Atawy, 1996).[18] When the Ethiopians succeeded in expelling the Italians and claiming the eastern banks of the White Nile, the British and Egyptians again seized control of Sudan in 1898, putting down the Mahdist rebellion and at the same time ensuring their control over the river (El-Atawy, 1996). Then, in 1902, King Edward VII sent an emissary to Addis Ababa to negotiate a treaty with Emperor Menelik II recognizing Sudan's borders and governing the use of the Blue Nile (Kendie, 1999a). Article III of the 1902 Anglo-Ethiopian Treaty contained an undertaking by the Ethiopian Emperor "not to construct or allow to be constructed, any works across the Blue Nile, Lake Tana or the Sobat, which could arrest the flow of their waters into the Nile" without the prior consent of the British and the Government of the Sudan ("Treaties", 1902, article 3; Hussein, 1997). This concession facilitated British recognition of Ethiopian independence.

Ethiopia later repudiated the Anglo-Ethiopian agreement of 1902. In an *Aide Memoire* of September 1957, the Ethiopian Government asserted that it "has the right and obligation to exploit its water resources for the benefit of present and future generations of its citizens" (Kendie, 1999a). Similarly, in 1958, the independent Sudan, prompted in large measure by a desire to undertake a dam project at Roseires, rejected the 1929 Treaty with Egypt concluded on Sudan's behalf by Britain, exacerbating border tensions between Egypt and Sudan (Jacobs, 1993). Major diplomatic efforts were undertaken to avoid war (Hussein, 1997; Okidi 1994).

In 1959, Egypt and Sudan concluded a treaty to replace the 1929 Treaty that Sudan had long found unpalatable ("Agreement", 1959). The 1959 version was not, however, markedly different from the 1929 Treaty (Mohamed, 1986; Teclaff, 1985), although it did implicitly repudiate the provision in the 1929 Treaty that granted Egypt complete control over the Nile and permitted Sudan to go ahead with the Roseires project (Tadros, 1996). The 1959 Treaty did not address broader issues such as water quality, flood control, or environmental protection (Goldenmann, 1990). Rather, it established fixed quotas on water distribution between Egypt and Sudan – quotas rooted in the two states' "acquired rights" under the 1929 Treaty and therefore highly favorable to Egypt ("Agreement", 1959, article 2(1)).[19] Moreover, the Treaty contained no provision for renegotiation or automatic alteration in the face of changing physical or political conditions.[20] Another problem with the 1959 Treaty, one that is the root of much regional tension, is that while it is purely bilateral, it seeks to apportion the entire flow of the Nile to Egypt and Sudan, excluding the interests of any other riparian, notably Ethiopia ("Agreement", 1959, article 1; LeRoy, 1995; Howell, 1994).

The conclusion of the 1959 Treaty provided Egypt with the security it needed to undertake the construction of the Aswan High Dam (Ahmed, 1991; Elhance, 1999) and established a political basis for the construction of the long-envisioned Jonglei Canal. The Canal would channel the waters of the White Nile past the South Sudanese swamps, preventing significant water loss.[21] In the 1959 Treaty, Egypt agreed to pay half of the cost of the project and negotiated for up to 50% of the water savings that would result ("Agreement", 1959, article 3). It is now more likely that, if the project ever moves forward, the Sudanese themselves will need to exploit any water gains (Waterbury and Whittington, 1998).

The Jonglei Canal Project was finally launched in 1978, but was suspended in 1984 (Hornstein, 1999; Kliot, 1994). It was controversial from the start, not only because of arguments over potential environmental damage, but also because the scheme has been considered a pawn in the hands of the Northern Sudanese who seek to dominate the South and extract the advantages of the project without bearing its costs (Deng, 1993; Jacobs, 1993, Kliot, 1994). It has even been argued that the project helped ignite a civil war in Sudan that has been fought since the 1970s (Lako, 1992). However, in recent years, the leader of the Southern rebellion, Dr. John Garang, has stated publicly that the Jonglei Canal Project should go forward as long as guarantees are provided that the local population and wildlife will be protected. As these statements were made on an "official visit" to Cairo, where Garang was soliciting Egyptian support for his cause ("Sudan", 1997),[22] it is hard to assess his sincerity. Whatever the real opinion of key actors, the war in Sudan has effectively halted the Jonglei Project, although it is rhetorically resurrected in ministerial pronouncements from time to time.[23]

The more immediate threat to the hegemonic aspirations of Egypt over the Nile is Ethiopia (Allan, 1994a; Krishna, 1988).[24] Ethiopia has never recognized the validity of the 1959 Sudanese-

Egyptian Apportionment Treaty (Allan, 1994c). Ethiopian distrust of Egypt has been strong almost since the time of Ethiopian independence, so much so that rumors of Egyptian policies of destabilization toward Ethiopia circulate widely.[25] In 1988, Ethiopia launched the first phase of an ambitious hydroelectric scheme called the Tana Beles Project. The goal was to double hydroelectric production in Ethiopia and to provide irrigation for new settlements with up to 200,000 farmers. Water was to be taken from Lake Tana to the Beles River, where five dams were to be built. Egypt was deeply concerned about the potential effect of the scheme upon the flow of the Blue Nile and in 1990 it blocked a loan provision for the Project that Ethiopia had requested from the African Development Bank (Dellapenna, 1997; Swain, 1997). This action prompted anger among other riparian states and led to complaints that Egyptian hegemonic goals had replaced the nineteenth-century British hegemony, and led to a view of Egypt as an indigenous colonialist (Tadros, 1996).[26]

In recent times, greater political stability and a modestly growing economy have allowed Ethiopia to plan two small hydroelectric projects and a series of micro-dam projects, largely for irrigation ("Ethiopian", 1998; Waterbury and Whittington, 1998; Swain, 1997); the micro-dam projects could reduce the flow of the Blue Nile and the Atbara, both of which flow into the Nile proper within Sudan. Creating new agricultural land will be increasingly important for the Ethiopians, as traditional farming in the highlands is adversely affected by environmental destruction, especially soil erosion due largely to deforestation (Whittington, 1995).

Meanwhile, Egypt has gone ahead with ambitious irrigation plans of its own. Work is said to be underway on the Salaam Canal, designed to carry 12.5 million cubic meters of water per day from Lake Nasser to the northern Sinai in support of a huge resettlement scheme that will eventually bring some three million new inhabitants to the region (Kendie, 1999b). The parallel New Valley Project will pump 4.94 billion cubic meters of water a year from Lake Nasser to irrigate up to 250,000 hectares in new settlement areas in the Western Desert outside the Nile Valley, the traditional home to almost all of Egypt's population (Huband, 1997; Marcus, 1997; Waterbury and Whittington, 1998).

There simply is not enough water in the Nile to complete the irrigation plans of both Ethiopia and Egypt, much less to satisfy the ambitions of all the Nile riparians (Marcus, 1997). Waterbury and Whittington suggest that even with zero use by upstream riparians, the Egyptians would have to "find" roughly five billion cubic meters of water a year to meet the requirements of the New Valley Project. Egyptian planners suggest that this water can come from more efficient use of existing supplies, but such projections are wildly optimistic. Moreover, Egyptian projections are based on the continuing availability of the portion of Nile flow accorded to Egypt under the 1959 Treaty with Sudan. Such an assumption is untenable, given the rising need and capacity of Ethiopia to exploit its own water resources (Waterbury and Whittington, 1998).

Given these social and political realities, combined with the physical limitations noted above, it is no surprise that the regime governing water apportionment and use in the Nile Basin has been purely competitive until recently. There exists a myriad of other factors inhibiting cooperation. Aside from concerns over water per se, there is a range of border disputes and political disagreements that have undermined cooperation among Nile riparians. Kenyan and Ugandan support for the rebel SPLA in southern Sudan has prompted tension, as has Sudanese support for the Lord's Day Army fighting Ugandan authorities in the northern parts of that country. Sudanese assistance to Islamic fundamentalists in Ethiopia has caused that government great concern, with similar issues clouding Egyptian-Sudanese relations (Tadros, 1996; Jacobs, 1993). Sudan was implicated in the assassination attempt on Egyptian President Mubarak during his 1995 visit to Ethiopia (Wright, 1996). In turn, Egypt is said to have fostered regional instability, such as the most recent Ethiopian-Eritrean border conflict, in order to maintain its position of strength among Nile Basin states ("Nile River", 2000; "Conflict", 1999).

Political pressures within the Nile states also discourage cooperative approaches. The regional distrust of Egypt and its hegemonic claims has already been mentioned. But the distrust is mutual. In one of the weekly opposition newspapers in Cairo, for instance, an article appeared in 1998 claiming that the United States and World Bank were about to loan $2.5 billion to Ethiopia "for the establishment of four giant dams on the Blue Nile… thus threatening the flow of water to Egypt and Sudan" (Salah'Azzazi, 1998). The article inaccurately reported Ethiopian plans and the support that might be derived from external agencies. Finally, at a structural level, it is important to note that until recently, states along the Nile displayed few integrative activities or interests. Their economies were largely disconnected, with trade patterns related to colonial connections and the desire to link with developed countries of the North, rather than with other poor regional economies. As is obvious from the listing of endemic conflicts in this paragraph, there are no strategic alliances centered on the Nile or its basin (Allan, 1994b).

Finally, a variety of additional reasons for non-cooperative behavior arise in the context of water distribution and use. First, interest in the Nile at a political level is highly disparate, with Egyptian reliance prompting intense strategic interest and all other states manifesting attitudes ranging from modest engagement to indifference (Allan, 1994b; Mallat, 1994). Second, given limited resources and generally scarce technical expertise within water ministries, national water plans tend to be designed in a vacuum, with little provision for basin-wide activities or even for information gathering (Allan, 1994b; Brichirieri-Colombi, 1996; Kliot, 1994). Third, until recently, the same resource and expertise problems ensured that many states along the Nile simply did not know what their water needs would be, and few had any clear understanding of available water resources within their countries (Tamrat, 1996).[27] Lack of knowledge on the ground also exacerbates latent distrust, especially *vis-à-vis* the Egyptians, who possess much more extensive technical capacity and can therefore dominate negotiations. Fourth, the historical pattern of bilateral agreements ignoring third party interests has bred legal confusion and has compounded the distrust among the Nile states (Mallat, 1994).

The unsettled and contradictory state of international water law over the last thirty years has also served to exacerbate competitive attitudes along the Nile. The evolution of the legal regime governing shared freshwater is discussed below, but it is important to note here that the long unresolved tension and continuing coexistence of two seemingly incompatible legal principles, "equitable use" and "no significant harm," have allowed states to make entirely self-interested arguments, often in bad faith (Waterbury and Whittington, 1998; Shapland, 1995; McCaffrey, 1991). Recent changes and improvements in this "legal endowment," changes that begin to address the fundamental conflicts between upstream and downstream states (Tamrat, 1996; Schwabach, 1998), are one of the reasons for a new, more cooperative spirit in Nile Basin relationships.

Moves Toward Cooperation Along the Nile

The last few years have witnessed a remarkable shift in the tone and substance of state-to-state relationships along the Nile. This shift can be demonstrated if one adopts the narrative approach advocated in Allan's discursive framework of water politics (Allan, 2003). From dire predictions of inevitable "water wars" to pronouncements of fraternal bonds and cooperative activity, from unilateralism to multilateral institutional development, the regime governing the waters of the Nile is evolving in surprising ways – that is, surprising if one relies only on self-interest and relative power as causal variables in explaining the behavior of states. The outlines of the new regime are emerging from patterns of interaction that have been fostered both by Nile states and external agents over the last two decades.

The story of change begins in the early 1980s, with a series of overlapping initiatives.[28] As an offshoot of the Egypt-Sudan Permanent Joint Technical Commission [PJTC] (Ahmed, 1994, p.358), and based on a 1967 agreement among Egypt, Sudan, Kenya, Tanzania, and Uganda, with support from the UNDP and the World Meteorological Organization (Okidi, 1994), a series of hydrometeorological studies (the Hydromet Project) was re-launched in the early 1980s. The goal was to provide a baseline set of measurements of water availability and future needs affecting the White Nile in and around lakes Victoria, Kyoga, and Albert (El-Atawy, 1996; Caponera, 1993; Elhance, 1999). Despite the political crises in the region, a treaty was concluded in 1977 ("Agreement", 1977), to which Uganda acceded in 1981, creating a basin-wide management regime for the Kagera River (Okidi, 1994).

The more ambitious UNDUGU – meaning "brotherhood" in Swahili (Ibrahim, 1994, p.334) – group was founded at Egypt's behest in 1983, drawing together all Nile riparians except, again, Ethiopia and Kenya. As has been typical of all but the most recent attempts at furthering basin-wide cooperation, these two states participated only as observers (Ibrahim, 1994). UNDUGU's unrealistic goal was to foster economic, social, cultural, and technical ties among the Nile riparians, leading to the foundation of a permanent sub-regional economic organization. The group brought together high level political actors, including Ministers of

Foreign Affairs (Ahmed, 1991). From an Egyptian perspective, the group was probably imagined as an exercise in hegemonic influence.[29] From a Pan-African perspective, UNDUGU was part of the failed drive toward "self-reliance and African inter-dependence. The creation of UNDUGU has not, in and of itself, led to significant concrete interaction among the riparian states, much less to coordinated economic initiatives. However, an UNDUGU-brokered call for a plan of action on economic development, including shared water, was to be addressed in other processes during the following decade.

In 1992, an intergovernmental Technical Cooperation Committee for the promotion of development and environmental protection on the Nile (TECCONILE) was established (Shady et al., 1994).[30] Even with the strong technical focus, Ethiopia and Kenya again refused to join as members because they deemed the framework for the organization to be inadequate as it did not address the fundamental equitable concerns of water apportionment. TECCONILE meetings in 1993 - 95 resulted in the creation of a Nile River Basin Action Plan. All Nile Basin states were involved in the development of this plan (El-Gamal, 1996) which the Council of Ministers for Water Affairs of all the Nile Basin states formally approved in February 1995 (TECCONILE, 1995).[31] It envisioned twenty-two projects, grouped into five categories of activities ("components"), with an estimated cost of $100 million.[32] Within component "D" on "Regional Cooperation," the Action Plan contemplated "the establishment of a basin-wide, multidisciplinary framework for legal and institutional arrangements" (Project D-3) (TECCONILE, 1995).[33] However, very little implementation occurred under the Action Plan as originally conceived,[34] in part because of resource constraints, and in part because of continuing competitive behavior among the basin states. The Action Plan was subsequently reviewed and, in 1999, it was superseded by a revised program, the Nile River Basin Strategic Action Program. In the same year, the Nile Basin Initiative (NBI) was launched to replace TECCONILE.[35] Before turning to the NBI and the most recent developments, it is worth describing the "Nile 2002 Conferences," a further process designed to promote basin-wide cooperation on shared freshwater. This process, when linked to an evolving normative framework, has played an important role in changing the political climate along the Nile. The first conference was held in 1993 in Aswan, Egypt, and yearly conferences have been held since in various basin states (Shady, et al., 1994).[36] The Conferences have been organized around the theme of comprehensive cooperation, and typically take the form of sessions devoted to individual papers canvassing issues of urgency, followed by the presentation of "country papers" sharing information on the challenges faced by each of the riparians. In addition, time is allotted for open discussion (Shady, et al., 1994). All basin states have sent participants to the conferences, though for some time Burundi, Eritrea, Ethiopia, and Kenya insisted that they were merely "observers" (Swain, 1997). One of the central advantages of the Nile 2002 series is its relative independence from immediate political calculations. In the closing session of the 1996 Conference in Kampala, one delegate noted that the process had not been "captured" by governments. The discussions were therefore more "direct and open" than would have been the case if the sessions were formally conceived of as negotiations (Comments from the Floor, 1996).[37] It should be noted that discussions at the first and second Nile 2002 Conferences fed into the TECCONILE process, helping to structure the Nile River Basin Action Plan.

Although the Nile 2002 series was conceived as a "technical" forum, its discussions assumed a decidedly normative turn. Papers often canvassed the legal status of riparian relationships (including the continued vitality of "colonial treaties"), and issues concerning the proper legal framework for apportionment (equitable utilization versus avoidance of harm) were raised in public debate. The prevalence of normative issues has not always been well received. At the 1996 conference in Kampala, an Egyptian delegate complained early in the sessions that Nile 2002 was supposed to be about technical issues, so why was there so much talk about law, even from non-lawyers? (Comments of an Egyptian, 1996). This concern was echoed, on the last day of the Conference, by the Sudanese delegation. Such reactions were predictable, as these discussions were taking place just as the international law framework for apportionment of shared freshwater was being actively debated in other fora and the stakes were therefore high.[38]

The Nile Basin Initiative

So far, this essay has traced various processes, none of which can be said to have led to a breakthrough from competition to cooperation. Until 1999, one could claim nothing more than a modest improvement in communications and a broader rhetorical commitment to cooperative behavior. Even the Nile 2002 Conferences had few tangible results. There was no joint management, or even coordinated planning and development, of Nile water resources. What may prove to be a breakthrough was the creation of the NBI in February 1999 and the official launching of its secretariat in Entebbe, Uganda in September 1999 ("Ethiopia, Sudan", 1999; "Nile Basin Initiative", 1999). The key development is active Ethiopian involvement in the Initiative. Long a careful bystander or active critic of efforts at coordination, Ethiopia, through its increasing engagement, signals real hope for joint water planning and management along the Nile.[39] Significantly, unlike any of the previous initiatives, the NBI has engaged all key state actors in the Nile Basin. Moreover, the Initiative is consciously designed to engage parallel technical and political processes, with regular communication between the two.[40]

As noted earlier, the NBI and its Nile River Basin Strategic Action Program have superseded TECCONILE and the Nile Basin Action Plan, respectively. The NBI is intended to promote "sustainable socio-economic development through the equitable utilization of, and benefit from, the common Nile Basin water resources" and hopes have been raised that a comprehensive Nile treaty, replacing the 1959 Egyptian-Sudan agreement, might emerge. Indeed, the NBI is viewed as a "transitional mechanism" to facilitate basin-wide discourse until a "permanent cooperative framework" is put in place. (ICCON, 2001). To this end, the NBI's Strategic Action Program comprises a "Shared Vision Program" which is designed to provide a basin-wide framework for action, and "Subsidiary Action Programs" which are to generate joint development projects at the sub-basin level ("Policy", 2000). The NBI places greater emphasis on sub-basin initiatives than the 1995 Action Plan, which acknowledged the need for sub-basin activities in specific cases but clearly emphasized a basin-wide process (TECCONILE, 1995).

The Strategic Action Program is guided by the belief that "real action on the ground" is key to the success of the NBI and that basin-wide cooperation is best promoted by "taking decisions at the lowest appropriate level" ("Policy", 2000; Saad, 2000). Notwithstanding the emphasis on subsidiarity, the NBI's Shared Vision Program does encompass, in its revised form, most of the basin-wide components of the earlier plan. The five broad action themes, designed to promote the shared vision, are: cooperative framework; confidence building and stakeholder involvement; socio-economic, environmental, and sectoral analyses; development and investment planning; and applied training ("Policy", 2000). The "cooperative framework" theme of the Shared Vision Program, supported by UNDP, is in fact the continuation of the earlier plan's "Project D-3" on legal and institutional matters.[41] Until that framework is established, sub-basin activities under the NBI are governed by a "common understanding" of the basin states regarding several implementation guidelines. These include guidelines on the need for involvement of all affected parties in the planning of sub-basin initiatives, and on the need to build on "principles of equitable utilization, no significant harm, and cooperation." The guidelines also call for "solutions both that have benefits for all involved and distribute benefits, costs, and risks equitably as well as use resources efficiently and protect the environment" ("Policy", 2000).

Work by a panel of experts on the highly sensitive issues arising in the context of the framework has been proceeding quietly, and secretly, over the last few years. There are indications that significant progress is being made. In early August 2000, the water resources ministers from all ten Nile Basin states, including Eritrea as an observer, met in Khartoum, Sudan, to review the panel of experts' final report on the framework. (UNDP, 2002; World Bank Group, 2002). At the conclusion of the meeting, Osman el Tom, vice-chairman of Sudan's water resources authority, reported "remarkable convergence toward future cooperation." Disagreements remained "mainly on the principle of prior notification of planned measures and the state of the existing agreements under the new cooperative framework" ("Nile Basin Ministers", 2000).

Notwithstanding the unresolved issues in the cooperative framework text, decisive progress was made with respect to a package of cooperative projects in the Nile Basin. In March 2001, the Nile Council of Ministers of Water Affairs endorsed several basin-wide Shared Vision Program projects and several sub-basin projects, the latter to be undertaken within the NBI's Eastern Nile (ENSAP) and Nile Equatorial Lakes (NELSAP) sub-programs.[42] According to the Ethiopian Water Affairs Minister, this development constitutes a "watershed agreement" and "reflects the new spirit of cooperation between the countries of the Nile" (ICCON, 2001). Building on the agreement reached on the cooperative projects, a new International Consortium for Cooperation on the Nile (ICCON) met for the first time in June 2001. ICCON brings together "the riparian countries of the Nile, the international donor community, public and private lenders, as well as other interested parties such as civil society, professional organizations, and NGOs" to work in support of the NBI ("NBI", 2001). A consultative group of interested donors, a sub-group of ICCON organized by the World Bank at the request of the Nile Council of Ministers, was also launched ("Press Release", 2001). The consultative group approved an initial grant of $140 million, while further financial assistance for the NBI's Shared Vision Program in the amount of $3 billion is contemplated ("Donors", 2001; Mburur, 2001).

The obvious question is: what prompted this promising set of developments, particularly the significant change of attitude on the part of previously reluctant participants, such as Ethiopia? Various explanations can be offered, not singular answers but rather a constellation of influences promoting changing patterns of behavior. The interest of this essay is in demonstrating the distinctive explanatory force of law without failing to acknowledge the importance of other factors with which legal norms co-exist and interact (Brunnée and Toope, 2002).

The Role of Legal Norms in the Nile Basin Regime

This essay began with the observation that international law's actual or potential contributions to promoting cooperation in the Nile Basin tend to be seen as limited at best. Specifically, the treaties that do apply to the Nile are perceived as having neither addressed water problems in the basin nor defused the potential for conflict (Carroll, 1999; Okaru-Bisant, 1998; Tadros, 1996; Kukk and Deese, 1996). Further, the United Nations Watercourses Convention ("U.N. Convention", 1997), adopted in 1997 to provide a global normative framework for shared freshwater, is viewed by many observers as unsuited to resolving the issues that arise in the Nile Basin (Carroll, 1999; Okaru-Bisant, 1998; Dellapenna, 2000). Finally, while policy-makers in the region have been contemplating the development of a basin-wide legal framework, its role tends to be seen largely as one of formalizing pre-existing agreements and enshrining states' entitlements (Interview, 2000; Carroll, 1999).

Legal norms have actually been influential in molding the evolving Nile Basin regime, but the influence until recently has been such that conflictual rather than cooperative behavior has been promoted. Within the last decade, legal norms have been recast and the contemporary pattern seems more oriented to the promotion of cooperation. This argument, drawing upon the interactional understanding of law outlined earlier in this paper, posits that law's influence derives from three interrelated characteristics. First, like all norms, law has the power to shape the individual and collective identities of states. Second, it both enables and constrains international discourse by establishing what counts as persuasive argument or rhetoric. Further, its inherent and specific form of legitimacy enhances the persuasiveness of legal conclusions.

The effect of the historic treaties is first studied by turning to the capacity of norms to shape identity. Wendt notes that norms and other intersubjective structures have the capacity either to "inhibit or facilitate the emergence of dynamics of collective identity formation" (Wendt, 1994). Wendt goes on to argue that:

... conflicts are also intersubjective phenomena, partly in virtue or [*sic*] rules shared by the parties ... but especially in virtue of shared *perceptions* of issues and threat ... The greater the degree of conflict in a system, the more the states will fear each other and defend egoistic identities by engaging in relative gains thinking and resisting the factors that might undermine it (Wendt, 1994).

Drawing on cognitive psychology, other constructivist IR theorists suggest that, even absent objective conflict and competition, language and the categories it employs can promote the formation of distinct identities (Kowert, 1998; Glenn, 2000). In this vein, to the extent that norms articulate "categories," they can contribute to the formation or hardening of separate, competitive identities. By the same token, norms can promote the construction of collective identities and collective definitions of interests by framing processes that foster mutual understanding. Through interaction, actors can learn "to see themselves as others do," and through cooperation, they can gradually change their beliefs about themselves and "internalize" their new, collective identity (Wendt, 1994). It is important to emphasize, however, that the formation of collective identities is not inevitable. Even where collective identities have emerged, actors will continue to be motivated by egoistic loyalties (U.N. 1983; U.N. 1987). Moreover, actors' social identities are typically multiple (Walzer, 1983; Sandel, 1998): varying "by issue, time, and place, and by whether they are bilateral, regional, or global" (Wendt, 1994).

Seen from this standpoint, the historic treaties regarding the Nile display a variety of features that limit their ability to foster cooperation in the basin. These treaties have not addressed the water problems in the region and are ill-suited to guide future relations among basin states. This conclusion is not surprising, given the strategic concern that prompted past treaty-making with respect to the Nile: securing control over or access to its waters. Moreover, the various treaties have served not only to entrench the competitive attitudes described above, but, more fundamentally, have reinforced, even encouraged, separate and competitive identities among the Nile Basin states.

A first salient feature of the existing treaty-based law is the pattern of bilateral treaty-making. This pattern has undercut the emergence of basin-wide shared understandings and the evolution of a collective identity of the riparian states as basin states. This is so in part because bilateral treaty-making did not provide opportunities for basin-wide interaction and trust-building. Second, the treaty terms, focused as they were upon allocation of states' shares in Nile waters, actually served to foster the distinct and egoistic identities of the Nile states. This certainly applies to the relations between those states that had concluded agreements and those that remained excluded. Since the treaty terms ultimately served to articulate separate or even competing interests, it would seem to be equally true for the relations between treaty parties. This reading applies with particular force to the 1959 agreement between Egypt and Sudan, which sought to enshrine acquired rights of the parties. Third, the lack of provision for adaptation of the various Nile treaties' terms to changing circumstances forestalled opportunities for the transformation of the shared competitive understandings of the parties. Similarly, the very limited procedural frameworks, such as technical cooperation committees, established by these treaties provided little opportunity for fostering mutual learning and understanding.

Similar to law's influence on identity formation, law's control over what counts as persuasive argument has also played out in essentially negative terms within the Nile Basin. The legal positions of various Nile riparians concerning international water law have echoed the conflictual patterns just described. The bilateral treaties have permitted the adoption of legal rhetoric that is entirely self-serving, fostering competition rather than cooperation. The central Egyptian claim to its current Nile water draw has been based upon reliance and historical usage. For generations the Egyptian government has stressed in various intergovernmental fora that states are entitled "to the quantities of water which they customarily took from that watercourse" and that the "extent of their dependence on the watercourse" should be the critical factor in apportioning draws among riparian states (U.N. 1983; U.N. 1986). A related argument is that the "sunk costs" already expended by Egypt in harnessing the Nile accord it a veritable "property right to the resource" (Waterbury and Whittington, 1998). These claims are categorically rejected by all other Nile riparians (U.N. 1993), with the exception of Sudan, linked as it is to Egypt in the symbiotic (if unequal) relationship defined by the terms of the 1959 bilateral treaty. As a downstream state like Egypt, Sudan has itself embraced claims of historical use and reliance to resist the limitation of its access that might result from water projects in Ethiopia ("Sudan Plans", 1998).

The failure of the historic treaties to foster cooperation is further explained by their lack of legitimacy, the third characteristic of persuasive law. Legal legitimacy depends upon congruence with underlying social practice and internal characteristics of fair process. Obviously, the mere fact that all of the treaties are bilateral, and in no way coordinated, means that they cannot legitimately be seen to govern all of the Nile waters. Moreover, an assessment of legitimacy within the framework of interactional international law reveals why this is so and points to other delegitimizing factors affecting the Nile Basin regime. First, to take the example of the Egyptian-Sudan Treaty of 1959, it asks the entirely unreasonable by purporting to divide all of the flow of the Nile between only two of the ten riparians. Also, because it contained no provision for reconsideration of the water allocation, at some point in time, the Treaty terms would come to ask the impossible, for the flow cannot match the assigned quotas. Furthermore, the fact that a shared resource crossing many boundaries is regulated by a series of bilateral treaties undercuts the notion of transparency, because there is no framework for multilateral information sharing. Even more pointedly, actors with clear interests in decisions on activities that affect Nile waters are simply excluded from the regime. For generations, claims have been made that the colonial era treaties are not binding upon riparian states. These claims are also rooted in legitimacy concerns. Inevitably, a treaty negotiated by a colonial power has trouble meeting the test of social congruence because the "real" parties are not present at the negotiations. To the extent that any "shared understandings" evolved, they are not shared by post-colonial governments, let alone by the citizenry of newly emerging states.

The framework of international water law has reinforced separate and competitive identities among Nile Basin states. It has also served to reinforce self-interested, and ultimately unconvincing, legal arguments (Brunnée and Toope,

1994; Brunnée and Toope, 1997). Until recently, the unresolved relationship between two core principles of international water law, 'equitable utilization' and 'no significant harm,' has allowed states to maintain irreconcilable positions. In brief terms, the basic approach of international water law has been rooted in these core rules and in the underlying idea of mutual limitation of sovereign rights. Under the principle of equitable utilization, riparian states are entitled to use international watercourses in a "reasonable" and "equitable" manner ("U.N. Convention", 1997, article V). What is reasonable and equitable must be determined in each individual case and depends upon various factors, none of which has inherent priority. These factors range from the geography, hydrology and climate of the basin, to economic, social and demographic factors, to existing uses or availability of alternative resources ("U.N. Convention", 1997, article VI). The mutual limitation approach also dictates the rule that a state's right to use its territory is limited by the duty not to cause significant harm to another state ("U.N. Convention", 1997, article VII). This rule not only is part of international water law but also constitutes a cornerstone of international environmental law (Declaration 1972; Rio Declaration 1992).

Although specific geographical, political, and economic contexts no doubt shape legal discourse, the equitable utilization principle is typically advanced by upper riparians, such as Ethiopia, looking to alter or increase the uses of an international watercourse. Lower riparians, such as Egypt, have tended to argue that the right to equitable utilization finds its limitation in the duty not to cause significant transboundary harm (Schwabach, 1998).[43] Upper riparians, in turn, countered that this argument would amount to a system of prior appropriation and effectively preclude their own development. Therefore, the argument goes, it is the principle of equitable utilization that ultimately takes priority, with downstream harm being merely one factor to be considered in the determination of what is equitable and reasonable (Brunnée and Toope, 1994).

The U.N. Convention 1997 includes the latest attempt to grapple with the relationship between the two principles. Its provisions on the equitable utilization (Articles 5, 6) and no harm (Article 7) rules were, inevitably, a delicate compromise. While Article 7(1) does require states, in utilizing an international watercourse in their territories, "to prevent the causing of significant harm to other watercourse States," this duty is merely one of due diligence ("U.N. Convention", 1997). Further, Article 7(2) suggests that, where significant harm is caused by diligent conduct, it is lawful so long as it results from activities that remain within a state's right to reasonable and equitable use under Article 5. This means that, in the vast majority of cases, issues of significant harm will ultimately be resolved through a balancing of interests under the equitable utilization rules. This result, in turn, is tempered by a requirement that the state causing the harm consult with the affected state "to eliminate or mitigate such harm, and where appropriate, discuss the question of compensation."[44]

At first glance, the U.N. Convention 1997 would seem to perpetuate the competitive paradigm. Its substantive core rules remain grounded in the mutual limitation of sovereign rights. The focus is not on identifying and promoting shared understandings or common interests but on demarcating individual, separate entitlements. As a result, the Convention's terms may promote the maintenance, or even the reinforcement, of the separate identities of riparian states. This is true to the extent that the Convention fails to offer a sufficiently developed alternative conceptual framework that could facilitate the formation of collective identities.[45]

Nonetheless, rather than perpetuate the old paradigm, the Convention may in fact accomplish the opposite. The Convention's terms effectively 'neutralize' the previously competing rules. Given the circular nature of the regime established by Articles 5 and 7, which ties the equitable utilization and no harm rules together without resolving the priority issue, neither rule can any longer be argued as overriding. The U.N. Convention 1997 deprives each side of convincing legal arguments for the priority of their claims. Upper and lower riparians are forced to re-examine their entrenched positions and to engage with one another to find fair solutions to their disagreements. Through its deliberately open-ended equitable utilization framework, the Convention provides little in the way of substantive guidance for the necessary balancing processes (Nollkaemper, 1996; Hey, 1998). However, while the provisions of the U.N. Convention 1997 cannot resolve the issues that arise in the Nile Basin, the Convention makes an important contribution toward cooperation in the basin. Ironically, this contribution is that the U.N. Convention 1997 undercuts the very riparian rights-based approach to water law that it was intended to codify and progressively develop and thus allows, though admittedly does not directly facilitate, process-based approaches to find common ground (McCaffrey and

Sinjela, 1998; Canelas de Castro, 2000; Benvenisti, 1996).[46]

The "coincidental" timing between the vote on the U.N. Convention, in May 1997, and the growing momentum in the work on the NBI is remarkable. In June 1997, the World Bank accepted the invitation to take the lead role among external agencies supporting Nile Basin cooperation, and in November of the same year the review of the Nile River Basin Action Plan was launched, culminating in the NBI in 1999 (ICCON, 2001). The language of the NBI appears to follow the lead of the U.N. Convention 1997 in drawing together the equitable utilization and no harm principles so as to neutralize both. The shared vision of the NBI states is the achievement of "sustainable socio-economic development through the equitable utilization of, and benefit from, the common Nile Basin water resources" (ICCON, 2001). At the same time, any subsidiary action programs in the basin are to "build on principles of equitable utilization, no significant harm and cooperation" ("Policy", 2000).

Before the U.N. Convention 1997, international water law allowed for the construction of legal arguments that undermined the legitimacy of its rules. Although it might have been possible to reconcile the rules of "no significant harm" and "equitable utilization," the normative framework contained no principles to guide states toward such reconciliation. This failure is manifest in the longstanding and inconclusive "priority" debate. Effectively, the two rules precluded the structuring of a legal framework that was internally consistent. Instead, states were permitted to offer up arguments that were incapable of coordination and reconciliation. The legal framework encouraged contradiction and invited abusive claims to rights. The contradiction in the two fundamental principles has been effectively erased through their melding in the U.N. Convention 1997. Of course, the influence of the traditional rules cannot be erased overnight.[47] Therefore, the new challenge to legitimacy will be to foster social practices along the Nile that have the capacity to generate adherence to emerging cooperative principles of water law, such as sustainable development, precaution and inter-generational equity.[48]

Informal Interaction Among Basin States

Finally, this Section evaluates the various informal processes that have taken place in the Nile Basin since the 1980s. Superficially, endeavors such as TECCONILE, the Nile 2002 Conferences, the Nile River Basin Action Plan, and the NBI might seem to suggest a move away from law. Conventional wisdom among many policy-makers in the region and academic observers has been that law has no constructive contribution to make at this stage. Rather, lawyers and legal arguments have been seen by many as hindering progress.[49] Law's moment, then, tends to be seen as coming at the end of a successful negotiation process when law is called upon to formalize the agreed outcome and render it "binding" (Interview, 2000). From an interactional perspective, this view is plainly wrong. The "final" legal framework is already emerging in the patterns of interaction among the riparians. These patterns are gradually re-shaping basin states' identities, shifting the parameters of persuasive argument and, crucially, promoting enhanced legitimacy for emerging legal norms. The point is not that the informal processes of interaction are themselves "legal" institutions, but rather that the cooperative frameworks they engender are part of what helps law emerge. At the same time, existing and crystallizing legal norms affect the interaction of states and other actors within informal processes. Law is both generative of changes in interaction and generated by further interaction. The informal Nile processes have aspired to promote the kind of interaction, trust-building, and mutual learning that are crucial to the emergence of shared understandings which in turn must underlie any effective normative evolution

The NBI process and the work on the "Cooperative Framework" do seem to mark a breakthrough on the path toward cooperation in the Nile Basin. Although promises of new development funding from the World Bank and other donors are relevant material inducements toward greater cooperation, dollars alone would not have been enough to prompt regime change. The developments culminating in the NBI evidence the emergence of modest shared understandings and norms that have helped to prompt concrete regime evolution. For example, the Nile 2002 and other related initiatives are all grounded in the notion of basin-wide cooperation. While the commitment to this approach was tentative initially, it is now the very underpinning of the NBI and its shared vision. This commitment is significant not just for hydrological reasons but also because it can help shape a collective identity of the Nile states as basin states and promote the identification of collective interests.

By the same token, the emergence of a basin identity and shared understandings as to basin-wide interests will serve to constrain Nile states from making entirely self-interested arguments. This process can be further assisted by norms that serve to recast perceived reasons for conflict or competition into reasons for cooperation. For example, under the long standing mutual limitation paradigm of international water law, water scarcity in the Nile Basin was a matter of competing sovereign interests, whereas it is now beginning to be recognized as requiring cooperation. A significant contribution to overcoming the competitive approach was made by the drawing together of the equitable utilization and no harm principles, as now reflected in the NBI. In addition, concepts such as sustainable development, benefit sharing, or environmental protection provide important normative guidance by directly focusing upon positive outcomes, best achieved through cooperation, not competition. These emerging norms, all reflected in the NBI, help guide the interaction of the Basin states by rendering purely competitive arguments, even in the context of water scarcity, untenable or, at least, unconvincing.

Inclusiveness and transparency of process are the hallmarks of recent Nile Basin initiatives. Indeed, the main goal of each of the processes, at least initially, was to involve all riparians in discourse. ICCON goes further by seeking to involve "civil society and NGOs" in basin-wide discussions. The subject matter was less important, although there was an explicit avoidance of discussions rooted in assertions of rights. An important thread in all gatherings was the principle of equal access and equal participation, furthering the goals of both inclusiveness and transparency. All states had access to information, ideas, and perspectives. This became "shared knowledge." On the basis of this shared knowledge, it became possible to discuss concepts such as sustainable development without viewing them as foreign impositions, bargaining chips, or ploys. The processes of discussion were imbued with legitimacy that enhanced the persuasiveness of the products of discussion. As a crucial further step, regular ministerial-level meetings complemented the technical discussions within the NBI. Special effort was expended to ensure that the technical discussions were shared with all relevant ministries and then fed into the inter-state negotiations. Thus, legitimacy generated at the technical level radiated into the political arena. In turn, as political leaders were convinced of the need for cooperation, their commitment reinforced the technical processes.

A further key development was the recognition, through the NBI process, that only a shared basin-wide set of normative principles would legitimize particular project activities by ensuring a perspective that was inclusive of the identities and interests of all basin states. At the same time, the states recognized that regional or sub-basin projects were more likely than basin-wide projects to generate concrete results that would, in turn, lend greater legitimacy to the governing principles by showing them to be effective. In sum, it was not possible to promote "action on the ground" before a shared framework of principles had been elaborated and inclusive processes of discussion had been created. But to insist on basin-wide cooperation on all issues and projects was unrealistic, if not impossible. So, perhaps ironically, it was a basin-wide, completely inclusive process of normative evolution that cut the Gordian knot to allow Nile states to move toward concrete cooperative projects, many of which will be undertaken at the sub-basin level.

The recent informal processes of discussion on the Nile cannot be appropriately characterized as either 'cheap talk' or purely strategic bargaining. Rather, they have begun to involve genuine arguments, geared toward reaching common understandings and reasoned consensus.[50] They are interactional processes of lawmaking, contributing simultaneously to shaping state identities and building legitimacy (Brunnée and Toope, 2000). Law's contribution is not simply formal, adding "legal varnish" to a pre-established strategic bargain, nor does it solely provide "bindingness" at the end of the day.[51] Law has already helped to make agreement possible and has generated the sense of adherence to mutually constructed norms that serves as the basis of their persuasiveness.

CONCLUSION

Is there a role for law in shaping the Nile Basin regime? The answer is emphatically yes. But law's effects are not adequately captured in the descriptions of positivist lawyers, most IR theorists, or even of some Nile negotiators. This essay's goal is to show how an interactional understanding of law illuminates both the promise and the limitations of law as a factor that influences the evolution of the Nile Basin regime.

This essay does not suggest that law is a fully autonomous, let alone controlling, influence in international society generally or on the Nile specifically. A variety of factors, including factors relating to a reassessment of material interests, interact to prompt regime change. The claim is simply that law is one of these factors, and that it possesses independent explanatory and normative force that should not be ignored. Therefore, law's persuasive influence is neither inevitable nor inevitably positive.

As illustrated through the analysis of the various strands of the evolving Nile Basin legal framework, law has exerted influence because of its ability to shape the identities of states and to guide international discourse by demarcating what counts as persuasive argument. Law's ability to generate its own adherence and to persuade Nile states depends upon the extent to which it meets specific, internal standards of legitimacy and is congruent with the shared understandings and practices in the Basin.[52] When these conditions are not present, law cannot fulfill its constructive role.

Finally, an assessment of the various informal processes in the Nile Basin illustrates that norms can generate adherence even if they are not formally binding. This is not to suggest that all attempts to create such formally binding rules, through the negotiation of explicit agreements, are misguided. In fact, in a setting such as the Nile, it ultimately may be necessary to craft formal rules of apportionment and other rights. Such formal rules must be rooted in interactional processes that frame persuasive argument, that help construct identities and legitimacy, and that involve all relevant actors.[53] Binding law can emerge only within settings where actors have agreed upon parameters for their interaction and have developed certain basic understandings about what they hope to achieve together. It follows that law cannot, by simply declaring a formal rule, ensure adherence.

The insights drawn together in this essay help explain the limitations of the older Nile treaties and the possible contributions of the law of international watercourses. Most significantly, however, they help explain why the slowly evolving processes in the Nile Basin, such as those embedded in Nile 2002 Conferences or the NBI, are crucial to the emergence and evolution of legal norms. Moreover, the interactional view of law that the essay sketches illuminates that law's role in the Nile Basin is confined neither to formalizing agreements previously reached through purely strategic bargaining, nor to establishing explicit entitlements. Rather, emerging legal norms, such as those related to sustainable, basin-wide management of the Nile, already serve to articulate common understandings and to influence - *not* control - the evolution of previously entrenched positions.

Along the Nile, law has played a significant role for generations, often in negative ways, by reinforcing competition and conflict. Law's persuasive influence has shifted recently, both through the elaboration of a less oppositional regime of international water law, and through the creation of various processes that have fostered regional collaboration. These processes have encouraged environmental learning, have begun to build a basin-wide identity among Nile riparians, and have re-oriented normative discourse. Normative frameworks to promote cooperation have greater legitimacy today than at any time in the Nile's recent history. If law continues to play its current facilitative role in shaping interaction, prospects are good for the elaboration of a comprehensive approach to Nile management.

ACKNOWLEDGEMENT

We thank the editors of the Harvard International Law Journal for their excellent editorial work, some of which finds its way into this condensed version of the earlier article. Our acknowledge with gratitude the research assistance of Sasha Nowicki, Sean Rehaag, Hiroko Sawai, and Richard Wodnicki.

REFERENCES

"Agreement Between Great Britain and His Majesty King Leopold II, Sovereign of the Independent State of the Congo modifying the Agreement signed at Brussels on May 12, 1894, May 9, 1906, 24" (1906). *Hertslet's Commercial Treaties* 24, 344.

"Agreement Between the Republic of the Sudan and the United Arab Republic for the Full Utilization of the Nile Waters, Nov. 8, 1959" (1959). *United Nations Treaty Series* 453, 51p.

"Agreement for the Establishment of the Organization for the Management and Development of the Kagera River Basin (The Rusumo Treaty), Aug. 24, 1977" (1977) [Rwanda - Burundi – Tanzania]. *United Nations Treaty Series* 1089, 165p.

Ahmed, M., 1991. Ma'arik al-Miyah al-Muqbillah fi al-Sharq-al-aswat: Ru'yah Mustaqbaliyah Hawla Ahammiyat al-Miyah Ka-'amil Silm aw Harb fi al-Sanawat al-Qadimah [Future Water Battles in the Middle East: A Futuristic Vision about the Importance of Water as a Factor in Peace or War in the Coming Years].

Ahmed, M., 1994. "Principles and Precedents in International Law Governing the Sharing of Nile Waters. In: *The Nile: Sharing a Scarce Resource*, Howell, P. and Allan, J.A. (eds.), Cambridge University Press, New York, 351p.

Allan, J.A., 1994a. Developing Policies for Harmonised Nile Waters Development and Management. In: *The Nile: Sharing a Scarce Resource*, Howell, P. and Allan, J.A. (eds.), Cambridge University Press, New York, 385p.

Allan, J.A., 1994b. Evolving Water Demands and National Development Options. In: *The Nile:*

Sharing a Scarce Resource, Howell, P. and Allan, J.A. (eds.), Cambridge University Press, New York, 301p.

Allan, J.A., 1994c. Nile Basin Water Management Strategies". In: *The Nile: Sharing a Scarce Resource*, Howell, P. and Allan, J.A. (eds.), Cambridge University Press, New York, 313p.

Allan, J.A., 2003. Integrated Water Resources Management is more a Political than a Technical Challenge. In: A.S. Alsharhan and W.W. Wood (eds.), *Water Resources Perspectives: Evaluation, Management and Policy*, Elsevier, 9-23.

Austin, J., 1885. *Lectures on Jurisprudence.* 5th edition (1885) (J. Murray, London).

Babatunde, N., 1994. Cooperation and Conflict in International Regimes: Water Resource Management in the Nile Drainage Basin. Ph.D. dissertation, University of South Carolina [unpublished] (on file with the authors).

Befekadu, S., 1999. Ethiopia: Taking a Lowly Option. *The Reporter*, Aug.3, 1999.

Benvenisti, E., 1996. Collective Action in Utilization of Shared Freshwater: The Challenges of International Water Resources Law. *American Journal of International Law* 90, 384p.

Berkoff, J., 1994. *A Strategy for Managing Water in the Middle East and North Africa.* World Bank, Washington, D.C., U.S.A.

Brichirieri-Colombi, J.S.A., 1996. Speech to the *Nile 2002 Conference,* February 26, 1996. [unpublished] (on file with the authors).

Brunnée, J. and Toope, S., 1994. Environmental Security and Freshwater Resources: A Case for International Ecosystem Law. *Yearbook of International Environmental Law* 5, 40p.

Brunnée, J. and Toope, S., 1997. Environmental Security and Freshwater Resources: Ecosystem Regime Building. *American Journal of International Law* 91, 26p.

Brunnée, J. and Toope, S., 2000. International Law and Constructivism: Elements of an Interactional Theory of International Law. *Columbia Journal of Transnational Law* 39, 19p.

Brunnée, J. and Toope, S., 2001. Interactional International Law. *International Legal Forum* 3, 186p.

Brunnée, J. and Toope, S., 2002. The Changing Nile Basin Regime: Does Law Matter? *Harvard International Law Journal* 43, 105p.

Canelas de Castro, P., 2000. The Future of International Water Law. *Shared Water Systems and Transboundary Issues with Special Emphasis on the Iberian Peninsula*, edited by Vlachos, E. (Luso American Development Foundation, Lisbon), 149p.

Caponera, D., 1993. Legal Aspects of Transboundary River Basins in the Middle East: The Al Asi (Orontes), The Jordan and The Nile. *Natural Resources Journal* 33, 629p.

Carroll, C., 1999. Past and Future Legal Framework of the Nile River Basin". In: *Georgetown International Environmental Law Review* 12, 269.

Chesworth, P., 1994. The History of Water Use in the Sudan and Egypt. *The Nile: Sharing a Scarce Resource*, Howell, P. and Allan, J.A. (eds.), Cambridge University Press, New York), 65p.

Canadian International Development Agency [CIDA],1994. *Water Resources Atlas of the Nile Basin.* Published in Canada by CIDA.

Collins, R., 1988. The Jonglei Canal: Illusion or Reality? *Water International* 13, 144p.

Comments from the Floor, 1996. *Nile 2002 Conference*, February 26 - 28, 1996 (on file with the authors).

Comments of an Egyptian and Sudanese Delegate, 1996. *Nile 2002 Conference*, February 26 - 28, 1996 (on file with the authors).

Conflict Spreads in Horn of Africa, 1999. AprIL 21, 1999 (available at: http://www.stratfor.com/services/Giu/042199.asp)

Cunningham, S., 2000. Do Brothers Divide Shares Forever: Obstacles to the Effective Use of International Law in Euphrates River Basin Issues. *University of Pennsylvania Law Review* 21, 131p.

Declaration, 1972. On the Human Environment, U.N. Conference on the Human Environment. 17th meeting at Principle 21, U.N. Doc. A/CONF.48/14 (1972).

Dellapenna, J., 1994. Treaties as Instruments for Managing Internationally Shared Water Resources: Restricted Sovereignty vs. Community of Property. *Case Western Reserve Journal of International Law* 26, 27p.

Dellapenna, J., 1996. Rivers as Legal Structures: The Examples of the Jordan and the Nile. *Natural Resources Journal* 36, 217p.

Dellapenna, J., 1997. The Nile as a Legal and Political Structure. In: *The Scarcity of Water: Emerging Legal and Policy Responses.* Edited by Brans, E. et al., Kluwer Law International, London, 121p.

Dellapenna, J., 2000. The Customary International Law of Internationally Shared Freshwaters. In: *Shared Water Systems and Transboundary Issues with Special Emphasis on the Iberian Peninsula.* Edited by Vlachos, E. Luso American Development Foundation, Lisbon, 79p.

Deng, F., 1993. Northern and Southern Sudan: The Nile. In: *Culture and Negotiation: The Resolution of Water Disputes.* Faure, G.O. and Rubin, J.Z. (eds.), Sage Publications, Newbury Park), 85p.

Donors Pledge US $ 140 Million for Nile Basin Projects, 2001. *United Nations Regional Information Network*, June 29, 2001 (available at: http://www.allafrica.com/stories/printable/200107010106.html

Downs, G., et al., 2000. The Transformational Model of International Regime Design: Triumph of Hope or Experience?" *Columbia Journal of Transnational Law* 38, 465p.

Egypt Rules Out Disputes Over Nile Water, 1999. *Xinhua News Agency,* March19, 1999.

Egyptian Minister Rules Out War Over Nile Waters, 1998. LEXIS, News Library, BBC Summary World Broadcast, April 27, 1998.

El-Atawy, M.H., 1996. Nilopolitics: A Hydrological Regime 1870-1990. *Cairo Papers in Social Science* 19:1 (American University in Cairo Press, Cairo).

El-Gamal, F., 1996. Water Resources Management Strategies and Action Programs in the Nile Basin: The Nile River Basin Action Plan. Paper presented to the *Nile 2002 Conference,* Feb. 26 - 28, 1996, [Unpublished] (on file with the authors).

Elhance, A.E., 1999. The Nile Basin. In: *Hyrdopolitics in the Third World,* Elhance, A.E. (ed.), U.S. Institute of Peace Press, Washington DC., 53p.

Ethiopia Moves Ahead on Nile Dams, 1988. *Middle East Economic Digest,* April 24, 1998.

Ethiopia Rules Out War with Egypt Over Nile Water, 1999. *Xinhua News Agency,* May 28, 1999.

Ethiopia, Sudan, Egypt to Join Hands in Water Project, 1999. *Xinhua News Agency,* September 16, 1999.

Ethiopian Prime Minister Calls for Changes in Nile Waters Agreement, 1998. *Al-Hayat* (London), April 10, 1998 [BBC Summary of World Broadcasts].

Falkenmark, M., 1986. Fresh Water – Time for a Modified Approach. *Ambio* 15: 192.

Finnemore, M, 2000. Are Legal Norms Distinctive? *New York University Journal of International Law and Politics* 32: 699.

Franck, T., 1990. *The Power of Legitimacy Among Nations.* Oxford University Press, New York.

Franck, T., 1995. *Fairness in International Law and Institutions.* Clarendon Press, Oxford.

Fuentes, X., 1998. Sustainable Development and the Equitable Utilization of International Watercourses. *British Yearbook of International Law* 69: 119.

Fuller, L., 1965. Irrigation and Tyranny. *Stanford Law Review* 17: 1021.

Fuller, L., 1969. *The Morality of Law,* revised edition. Yale University Press, New Haven.

Gleick, P., 1990. Climatic Changes, International Rivers, and International Security: The Nile and the Colorado. *Greenhouse Glasnost.* Edited by Minger, T., Ecco Press, New York, 147p.

Gleick, P., 1993a. Water and Conflict: Fresh Water Resources and International Security. *International Security* 18: 79.

Gleick, P., 1993b. Water Resources: A Long-Range Global Evaluation. *Ecology Law Quarterly* 20, 141.

Glenn, H.P., 2000. *Legal Traditions of the World: Sustainable Diversity in Law.* Oxford University Press, Oxford.

Goldenmann, G., 1990. Adapting to Climate Change: A Study of International Rivers and their Legal Arrangements. *Ecology Law Quarterly* 17: 41.

Handl, G., 1992. The International Law Commission's Draft Articles on the Law of International Watercourses (General Principles and Planned Measures): Progressive Development or Retrogressive Development of International Law? *Colorado Journal of International Environmental Law and Policy* 3:123.

Hart, H.L.A., 1983. *Essays in Jurisprudence and Philosophy.* Clarendon Press, Oxford.

Hart, H.L.A., 1994. *The Concept of Law,* revised edition. Clarendon Press, Oxford.

Hey, E., 1998. The Watercourses Convention: To What Extent Does it Provide a Basis for Regulating Uses of International Watercourses? *Review of European Community and International Environmental Law* 7: 291.

Hornstein, D., 1999. Environmental Sustainability and Environmental Justice at the International Level: Traces of Tension and Traces of Synergy. In: *Duke Environmental Law and Policy Forum* 9: 291.

Howell, P., 1994. East Africa's water requirements: the Equatorial Nile Project and the Nile Waters Agreement of 1929. *The Nile: Sharing a Scarce Resource.* Howell, P. and Allan, J.A. (eds.). Cambridge University Press, New York, 81p.

Huband, K.M., 1997. The Nile States Look to New Division of Waters. In: *Financial Times* (London Edition) February 28: 1997.

Hultin, J., 1992. Source of Life, Source of Conflict: Fear and Expectations Along the Nile. *Regional Case Studies of Water Conflicts,* Ohlsson, L. (ed.). Padrigu Papers, Göteborgs, 29p.

Hundley, T., 1992. Water: New Crisis is Waiting to Grip the Middle East. *The [Montreal] Gazette*, February 8, 1992, B1.

Hussein, F.A., 1997. *Al-miyah wa-awraq al-lu'bah al-siyasiyah fi al-sharq al-aswat [Water and the Political Game in the Middle East*] 79p.

Ibrahim, H.T., 1994. Mushkilat al-Miah fi Misr (The Problem of Water in Egypt), Mushkilat al-Miyah fi al-Sharq al-Awsat [The Problem of Water in the Middle East], p.293

International Consortium for Cooperation on the Nile (ICCON), 2001. Nile Basin Initiative, Strategic Action Program Overview Document (available at: http://www.nilebasin.org/iccon_overview.htm).

Interview with Aly Shady [Senior Policy Advisor, Canadian International Development Agency], 2000, May 10 & 17, 2000 (transcripts on file with authors).

Jacobs, L.M., 1993. Sharing the Gifts of the Nile: Establishment of a Legal Régime for Nile Waters Management. *Temple International and Comparative Law Journal* 7: 95.

Katzenstein, P.J., et al., 1998. International Organization and the Study of World Politics. *International Organization* 52: 645.

Kelsen, H., 1951. *Principles of International Law.* Rinehart, New York.

Kelsen, H., 1961. *General Theory of Law and State.* Russel & Russle, New York.

Kempf, H., 2000. La Planète est Menacée par des Graves Pénuries d'Eau au XXIe Siècle. In: *Le Monde*, March 17, 2000, 2.

Kendie, D., 1999a. Egypt and the Hydro-politics of the Blue Nile River: Part I. *Addis Tribune,* August 6, 1999.

Kendie, D.,1999b. Egypt and the Hydro-politics of the Blue Nile River: Part II. *Addis Tribune,* August 13, 1999.

Keohane, R. and Nye, J.S., 1977. *Power and Interdependence.* Little-Brown, Toronto.

Keohane, R. and Nye, J.S., 1987. Power and Interdependence Revisited. International Organization 4: 725.

Kliot, N., 1994. *Water Resources and Conflict in the Middle East.* Routledge, London.

Kowert, P., 1998. Agent versus Structure in the Construction of National Identity. *International Relations in a Constructed World,* Kubálková, V., et al. (eds.). M.E. Sharpe, Armonk, NY, 101.

Krasner, S.D., 1983. Structural Causes and Regime Consequences: Regimes as Intervening Variables. *International Regimes.* Krasner, S.D. (ed.). Cornell University Press, Ithica, 1.

Kratochwil, F., 1989. *Rules Norms, and Decisions: On the Conditions of Practical and Legal Reasoning in International Relations and Domestic Affairs.* Cambridge University Press, Cambridge.

Krishna, R., 1988. The Legal Regime of the Nile River Basin. *The Politics of Scarcity: Water in the Middle East.* Starr, J and Stoll, D. (eds.). Westview Press, Boulder, 23.

Kukk, C. and Deese, D., 1996. At the Water's Edge: Regional Conflict and Cooperation Over Fresh Water. *UCLA Journal of International Law and Foreign Affairs* 1: 21.

Lako, G.T., 1992. The Jonglei Canal Scheme as a Socio-Economic Factor in the Civil War in the Sudan. *African River Basins and Dryland Crises,* Darkoh, M.B.K. (ed.). Uppsala University Press, Uppsala, 45.

Laki, S.L., 1998. Management of Water Resources of the Nile Basin. *International Journal of Sustainable Development and World Ecology* 5: 288.

LeRoy, P., 1995. Troubled Waters: Population and Water Scarcity. *Colorado Journal of International Environmental Law and Policy* 6: 299.

Lowi, M., 1993. *Water and Power: The Politics of a Scarce Resource in the Jordan River Basin.* Cambridge University Press, Cambridge.

Macdonald, R.A., 1986. Pour la Reconnaissance d'une Normativité Juridique Implicite et 'Inférentielle'. *Sociologie et Société* 18: 47.

Mageed, Y.A., 1994. The Central Region: Problems and Perspectives. *Water in the Arab World. Perspectives and Prognoses.* Rogers, P. and Lydon, P. (eds.), *Harvard University Press, Cambridge,* 101.

Mallat, C., 1994. Law and the Nile River: emerging international rules and the Shari'a. In: *The Nile: Sharing a Scarce Resource,* Howell, P. and Allan, J.A. (eds.), Cambridge University Press, New York, 365p.

Marcus, A.D., 1997. Water Fight. *The Wall Street Journal,* August 25, 1997, p.1.

Martin, L.L. and Simmons, B., 1998. Theories and Empirical Studies of International Institutions. *International Organizations* 52: 729.

Mburu, C., 2001. West Pledges $ 140 Million to Kick Off Nile Basin Projects. *The East African* (Nairobi), July 23, 2001 (available at: http://www.allafrica.com/Stories/printable/200107270015.html).

McCaffrey, S., 1991. International Organizations and the Holistic Approach to Water Problems. *Natural Resources Journal* 31: 139.

McCaffrey, S. and Sinjela, M., 1998. The 1997 United Nations Convention on International Watercourses. *American Journal of International Law* 92: 97.

Mohamed, O.M.A., 1986. *Proposal for a Nile Waters Treaty.* Khartoum University Press, Khartoum.

Moorehead, A., 1962. *The Blue Nile.* Harper & Row, New York.

Morgenthau, H., 1954. *Politics Among Nations,* 2nd ed., Knopf, New York.

NBI, Nile Basin Ministers of Water Achieve Agreement on Cooperative Projects as Historic Meeting Wraps Up, 2001. (Press Release, March 29, 2001) http://nilebasin.org/pressreleases.htm.

Nile Basin Cooperative Projects Launched, 2001. *Tomric News Agency [Der Es Salaam],* April 4, 2001, (available at: http://www.allafrica.com/stories).

Nile Basin Council of Ministers' Meeting in Ethiopia Ends, 1999. *Radio Ethiopia External Service,* May 15, 1999.

Nile Basin Initiative Secretariat launched at ceremony in Uganda, 1999. *The New Vision,* September 6: 1999.

Nile Basin Ministers Discuss Water Development, 2000. *Panafrican News Agency (PANA),* August 7, 2000 (available at: http://allafrica.com/stories/200008070109.html)

Nile River Politics: Who Receives Water?, 2000, August 10, 2000 (available at: http://www.stratfor.com/MEAF/commentary).

Nollkaemper, A., 1996. The Contribution of the International Law Commission to International Water Law: Does it Reverse the Flight from Substance? *Netherlands Yearbook of International Law* XXVII: 39.

Ntambirweki, J., 1996. Colonial Treaties and the Legal Regime of the Nile Valley: Rethinking the Legal Framework into the 21st Century" (Paper presented to the *Nile 2002 Conference,* Feb. 26 - 29, 1996) [unpublished] (on file with the authors).

Okaru-Bisant, V., 1998. "Institutional and Legal Frameworks for Preventing and Resolving Disputes Concerning the Development and Management of Africa's Shared River Basins. *Colorado Journal of International Environmental Law and Policy* 9: 331.

Okidi, O., 1994. History of the Nile and Lake Victoria Basins through treaties. In: *The Nile: Sharing a Scarce Resource.* Howell, P. and Allan, J.A. (eds.), Cambridge University Press, New York, 321.

Onuf, N.G., 1998. Constructivism: A User's Manual. In: *International Relations in a Constructed World.*

Kubálková, V., et al. (eds.), M.E. Sharpe, Armonk, New York.
Onuf, N.G., 1989. *World of Our Making: Rules and Rule in Social Theory*. University of South Carolina Press, Columbia.
Policy Guidelines for the Nile River Basin Strategic Action Program, 2000. Council of Ministers of Water Affairs of the Nile Basin States. http://www.nilebasin.org/Documents/TACPolicy.html.
Postema, G.J., 1994. Implicit Law. *Law and Philosophy* 13: 361.
Press Release, NBI, Donor Community Supports Poverty Reduction, Prosperity and Peace Through Nile Basin Initiative, 2001. June 28, 2001 (available at: http://www.nilebasin.org/pressreleases.htm).
Protocols Between the Governments of Great Britain and Italy, for the Demarcation of their Respective Spheres of Influence in Eastern Africa, Protocol No. 2, April 15, 1891 (1891). *British and Foreign State Papers* 83, 19.
Ratner, S., 2000. Does International Law Matter in Preventing Ethnic Conflict? *New York University Journal of International Law and Politics* 32: 591.
Rio Declaration, 1992. On Environment and Development. U.N. Conference on Environment and Development, Agenda Item 9 at Principal 2, U.N. Doc.A/CONF.151/5/Rev. 1 (1992).
Risse, T., 2000. 'Let's Argue!': Communicative Action in World Politics. *International Organization* 54: 1.
Ruggie, J.G., 1998. What Makes the World Hang Together? Neo-utilitarianism and the Social Constructivist Challenge. *International Organization* 52: 855.
Saad, M.B.A., 2000. The Nile River Basin: A Vision for Development. *River Basin Management – Proceedings of the International Workshop* (The Hauge, 27-29 October, 1999), ed. Mostert, E. (UNESSCO, Paris), 231p.
Sadik, A. and Barghouti, S., 1994. The Water Problems of the Arab World: Management of Scarce Resources. *Water in the Arab World: Perspectives and Prognoses*. Rogers, P. and Lydon, P. (eds.). Harvard University Press, Cambridge, 1p.
Salah'Azzazi, 1998. $2.5 Billion from America to Ethiopia to Deprive Egypt and Sudan of 85 Percent of Nile Water. *Cairo Al-'Arabi*, March 2, 1998 [translated in FBIS-NES-98-064].
Sandel, M., 1998. *Liberalism and the Limits of Justice*. 2nd edition. Cambridge University Press, Cambridge.
Schwabach, A., 1998. The United Nations Convention on the Law of Non-navigational Uses of International Watercourses, Customary International Law, and the Interests of Developing Upper Riparians. *Texas International Law Journal* 33: 257.
Shady, A., et al., 1994. The Nile 2002: The Vision Toward Cooperation in the Nile Basin. *Water International* 19: 77.
Shapland, G., 1995. Policy Options for Downstream States in the Middle East. In: *Water in The Middle East: Legal, Political, and Commercial Implications*. Allan, J.A. and Mallat, C. (eds.). Tauris Academic Studies, London, 301p.
Stoner, R., 1994. Future Irrigation Planning in Egypt. In: *The Nile: Sharing a Scarce Resource*. Howell, P. and Allan, J.A. (eds.). Cambridge University Press, New York, 195.
Sudan: Garang on Nile Water, Sudan's Unity, 1997. *Cairo Mena*, November 30, 1997 [translated in FBIS-NES-97-334].
Sudan Plans to Build Three New Hydroelectric Dams across the Nile, 1998. *Africa News Service*, September 4, 1998.
Sutcliffe, J.V. and Parks, Y.P., 1999. *The Hydrology of the Nile*. International Associate of Hyrdological Sciences, Wallingford.
Swain, A., 1997. Ethiopia, the Sudan, and Egypt: The Nile River Dispute. *Modern African Studies* 35: 675.
Tadros, N., 1996. Shrinking Water Resources: The National Security Issue of this Century. *Nortwestern University Law International Law and Business* 17: 1091.
Tamrat, I., 1996. Problems and Prospects for Cooperation in the Nile Basin – Legal and Institutional Perspectives. Paper presented to the *Nile 2002 Conference*, February 26 - 29, 1996) [Unpublished] (on file with the authors).
Teclaff, L., 1985. *Water Law in Historical Perspective*. W.S. Hein, Buffalo, USA.
TECCONILE, Nile River Basin Action Plan, 1995.
Theroux, P., 1997. The Imperiled Nile Delta. *National Geographic* 191: 2.
Treaties Relative to the Frontiers Between the Sudan, Ethiopia, and Eritrea, May 15, 1902 (1902). *Hertslet's Commercial Treaties* 23: 1.
U.N. Convention on the Law of Non-Navigational Uses of International Watercourses, 1997. G.A. Res. 51/229, U.N. GAOR, 51st Session 99th plen. mtg., U.N. Document A/RES/51/229. [U.N. Convention, 1997].
U.N., 1983. GAOR 6th Comm., 38th Session, 48th meeting 1, U.N. Document A/C.6/58/SR.48 (1983) (statement of Mr. Ramadan of Egypt, discussing the ILC's work on international watercourses).
U.N., 1987. GAOR 6th Comm., 41st Session, 34th meeting 26, U.N. Document A/C.6/41/SR.34 (1986) (statement of Mr. El-Araby of Egypt, discussing the ILC's work on international watercourses).
U.N., 1993. GAOR 6th Comm., 48th Session, 28th meeting 40, U.N. Document A/C.6/48/SR.28 (1993) (statement of Mr. Nega of Ethiopia, articulating the Ethiopian position).
U.N.D.P. [United Nations Development Programme], 2002. The Nile River Basin Cooperative Framework Project (available at: http://www.undp.org/seed/Water/region/nile.html).
Vidal, J., 1995. Ready to fight to the last drop. *The Guardian Weekly,* August 20, 1995, p.13.

Waltz, K., 1979. *Theory of International Politics.* McGraw-Hill, New York.
Walzer, M., 1983. *Spheres of Justice: a Defense of Pluralism and Equality.* Basic Books, New York.
Waterbury, J., 1994. Transboundary Water and the Challenge of International Cooperation in the Middle East. In: *Water in the Arab World: Perspectives and Prognoses*, Rogers, P. and Lydon, P. (eds.), Harvard University Press, Cambridge, 39p.
Waterbury, J. and Whittington, D., 1998. Playing Chicken on the Nile? The Implications of Microdam Development in the Ethiopian Highlands and Egypt's New Valley Project. *Natural Resources Forum* 22: 155.
Wendt, A., 1992. Anarchy is What States Make of It: The Social Construction of Power Politics. *International Organization* 46: 391.
Wendt, A., 1994. Collective Identity Formation and the International State. *American Political Science Review* 88: 384.
Wendt, A., 1999. *Social Theory of International Politics.* Cambridge University Press, Cambridge.
Whittington, D., et al., 1995. Toward a New Nile Waters Agreement. In: *Water Quantity/Quality Management and Conflict Resolution*,. Dinar, A and Loehman, E. (eds.), Praeger, Westport, 167p.
World Bank Group, 2002. Nile Basin Initiative: Overview. (available at: http://www.worldbank.org/afr/nilebasin/overview.htm).
Wright, J., 1996. Egypt tells Sudan to learn lesson of U.N. vote, *REUTERS*, Feb. 1, 1996.

NOTES AND REMARKS

[1] The participating states are Burundi, Democratic Republic of Congo, Egypt, Ethiopia, Kenya, Rwanda, Sudan, Tanzania, and Uganda. Eritrea has not formally joined, but now participates as an observer. Note that the predecessor organization, TECCONILE, also had a secretariat. The NBI secretariat is housed in the old TECCONILE building and effectively replaces that organization's secretariat. On TECCONILE, *see infra*, page 100.

[2] What is happening in the Nile Basin is especially important, as the mood among water specialists and diplomats alike is undeniably gloomy when considering the state of freshwater resources around the world (c.f. Kempf, 2000).

[3] Institutionalist regime theory describes regimes as sets of "implicit or explicit principles, norms, rules, and decision-making procedures around which actors' expectations converge" in a given issue area (Krasner, 1983). Drawing upon game theory, Eyal Benvenisti has argued that international law's key role with respect to shared freshwater resources is to structure interactions as iterated games, thus maintaining indefinite interdependence amongst the players. In such circumstances, "cooperation may emerge despite the self interest of the players" (Benvenisti, 1996).

[4] Interaction includes bilateral and multilateral intercourse among diverse international actors, mostly states, but increasingly–at least in some issue areas such as human rights, environmental law, and trade–involving actors from civil society and private business (Risse, 2000).

[5] Although constructivists need to produce more empirical work demonstrating the role that identity plays in shaping actor behavior, Wendt has postulated a useful distinction between "type identities" (social categories of states that share characteristics such as regime type: liberal democracy, traditional monarchy, totalitarian dictatorship) and "role identities" (the product of relationships amongst states: friend, rival, enemy are the clearest categories) (Wendt, 1999).

[6] Fuller actually posited eight internal requirements: generality of rules; promulgation; limiting cases of retroactivity; clarity; avoidance of contradiction; not asking the impossible; consistency over time; and congruence of official action with the underlying rules. One of his most controversial theses was that law is recognizable by adhering to these eight requirements of "internal morality," and by subjecting its substantive conclusions to weak tests of "external morality." Because of his emphasis upon "internal morality," one can conclude that Fuller believed that the basis of legal obligation is found within the system of rules itself, and is not dependent upon an external validating principle. Adherence to an internal morality helps to render law more legitimate in the eyes of those to whom rules are directed. In addition, modest substantive commitments to external morality evidence an underlying congruence with commonly shared understandings in society, which also tends to support the legitimacy of rules (Fuller, 1969).

[7] "[T]he internal morality of the law is not something added to, or imposed on, the power of law, but is an essential condition of that power itself." (Fuller, 1969) These tests are not perfect and other formulations would be plausible, but the purpose here is to stress Fuller's interactional thesis, not his commitment to a specific formulation of the internal morality of the law (Macdonald, 1986).

[8] Fuller's internal tests of legality are to be distinguished from the conception of legitimacy developed by Thomas Franck. Franck argues that there are four variables affecting legitimacy: determinacy, symbolic validation, coherence, and adherence (Franck, 1990). With the exception, perhaps, of "symbolic validation," these variables map onto Fuller's conditions of internal morality. However, Franck treats his variables as elements of "process fairness," which is distinct, in his view, from "moral fairness," his term for distributive justice (Franck, 1995). For Fuller, the conditions of internal morality are genuinely moral, not mere desiderata of coherence or logical soundness, because they are both means and ends. In an interactional model of law, process is part of the ends that law serves because inclusive processes reinforce the commitments of

participants in the system to the substantive outcomes achieved by implicating participants in their generation.

[9] See *infra* page 105.

[10] Egypt's population has been growing at between 2.8 and 3.5% a year since the late 1980s. Unless this rate is slowed dramatically, the population could almost double to 118 million by 2050 (Huband, 1997).

[11] According to the definition of "water stress" proposed by Falkenmark (annual renewable supplies approaching 2,000 cubic meters per person per year), by 1992 Burundi, Egypt, Ethiopia, Kenya, Rwanda, and Tanzania were already experiencing such stress (Hultin, 1992; Falkenmark, 1986).

[12] It is important to note, however, that cotton, Sudan's principal agricultural export, is completely dependent upon Nile irrigation (Babatunde, 1994). Sudan is generally considered to have "great potential for increased irrigation" as well (Gleick, 1993b).

[13] See *infra* pages 96-97.

[14] Some Nile experts argue that pollution is not a major issue for the other Nile riparians. Indeed, the major pollution problems on the Nile arise downstream from Cairo (Whittington, et al., 1995). However, the negative environmental impact of upstream development could have effects throughout the Nile system as new dam projects, irrigation, and diversion schemes, and other uses come online in various riparian states. For a discussion of the prospects for upstream development, see *infra* pages 96-98.

[15] The original phrasing was more felicitous: "Egypt is an acquired country, the gift of the Nile" (Moorehead, 1962, p.3 [quoting Herodotus]).

[16] The style of British colonialism in Africa also allowed some forms of local hegemony to arise, for example, the so-called "Egyptian Equatorial African Empire" under Khedive Ismail in the 1860s-70s. This "Empire" was in part inspired by a desire to encourage total Egyptian control over the sources of the Nile (El-Atawy, 1996).

[17] The key clause regarding the waters of the Nile and its tributaries in the agreement reads: "Save with [sic] the previous agreement of the Egyptian Government no irrigation or power works or measures are to be constructed or taken which would, in such a manner as to entail prejudice to the interests of Egypt, either reduce the quantities of water arriving in Egypt, or modify the date of its arrival, or lower its level" (Jacobs, 1993).

[18] Jacobs notes that this agreement was concluded before the Blue Nile, which remained under Italian influence, was identified as the major source of Nile water (Jacobs, 1993, p.106). Okidi argues that this agreement became void with the termination of Italian and British colonial rule in the region (Okidi, 1994). Some authors suggest that international water law's emphasis on equitable apportionment would undermine a claim to validity for the colonial treaties (Kliot, 1994). Others, however, argue that all the colonial era treaties relating to territorial status are still in force and are binding on successor states (Ahmed, 1991; Ahmed, 1994). A full discussion of the many problems associated with state succession as to colonial treaties is beyond the scope of this paper.

[19] Under article 1(1) and (2), the quotas were established on the basis of "acquired rights" and set at "48 Milliards (billion) of cubic meters" per year for Egypt and four Milliards for Sudan. In addition, article 2(3) and (4) provides for division between Egypt and Sudan of the "net benefits" from water storage by the Aswan High Dam (Gleick, 1993a; Krishna, 1988; Okidi, 1994).

[20] While no provision is made for adjustments to the quotas, Article 2(5) of the agreement does provide that the allocation of net benefits from the Aswan High Dam "shall be the subject of revision by the two parties" ("Agreement", 1959, article 2(5)). However, since no amendment or adjustment process is set out in the agreement, any such revision would require a negotiated solution.

[21] It has been estimated that approximately half of the White Nile flow is lost through evaporation and transpiration in the Sudd (Sutcliffe and Parks, 1999; Collins, 1988; Kliot, 1994). The canal was to reduce evaporation in the Sudd swamps and to mitigate the effects of floods. It also was to generate hydroelectric power for Sudan and Egypt and provide irrigation for farming in Sudan (Elhance, 1999).

[22] But it may be important that Dr. Garang is himself an agricultural economist who received his Ph.D. in 1981 from Iowa State University, having written a dissertation on aspects of the Jonglei Project (Hornstein, 1999).

[23] Consider the hopeful comments of the Egyptian Minister of Public Works and Water Resources ("Egypt", 1999). The Egyptians are expected to continue to press for the completion of the Jonglei Canal, even though it would probably be more cost effective for the Sudanese to concentrate on small-scale projects on the Blue Nile (Kliot, 1994). However, the Sudanese may be forced into the right choices, despite otherwise powerful Egyptian pressure, because of significant resource limitations.

[24] As Waterbury states, Egypt is not a true hegemon because it cannot simply impose its desired solutions, but "it can coax and threaten its neighbors convincingly" (Waterbury, 1994).

[25] Consider the allegations of Egyptian malfeasance involving support for the Eritrean independence struggle and for other efforts promoting instability in Ethiopia (Kendie, 1999b).

[26] The benign view is that this regional hegemony has also caused stability, virtually eliminating the possibility of military confrontation (Jacobs, 1993).

[27] Contrast this with Allan's argument that the "political leaders of the Nile countries . . . have a very sophisticated knowledge" of scientific issues and the rate of water consumption (Allan, 1994b). Allan is probably correct as concerns the Egyptians, and increasingly the Sudanese and Ethiopians, but Tamrat's analysis is applicable to the other riparian state governments.

[28] Earlier initiatives did not succeed in fostering significant cooperation, despite modest attempts at coordination in committees such as the East African Nile Waters Coordination Committee (Howell, 1994). The Egypt-Sudan Permanent Joint Technical Commission was set up under the terms of the 1959 Treaty and its 1960

Protocol to observe activities on the Nile and to promote cooperation ("Agreement", 1959, article4). It was hoped that all Nile riparians might eventually participate, but Ethiopia and Kenya have participated only sporadically. The Commission has no formal authority, especially as to the eight riparians other than Egypt and Sudan, and little cooperative activity has resulted (Tadros, 1996; Allan, 1994b).

[29] Waterbury argues that UNDUGU and other related Egyptian initiatives were part of a strategy (rooted in Egyptian hegemonic aspirations) that sought to create "multi-good bargaining situations" more likely to result in agreement than negotiations purely devoted to water issues (Waterbury, 1994).

[30] TECCONILE stands for "Technical Co-operation Committee for the Promotion of the Development and Environmental Protection of the Nile Basin," an initiative agreed to by ministers responsible for water affairs in the Nile Basin countries at a meeting in Kampala, Uganda in December 1992. Its main long-term objectives were to "assist participating countries in the development, conservation, and use of the Nile Basin water resources in an integrated and sustainable manner, through basin-wide cooperation for the benefit of all" and to "assist participating countries in the determination of the equitable entitlement of each riparian country to the use of Nile waters" (TECCONILE, 1995).

[31] Egypt, Rwanda, Sudan, Tanzania, Uganda, and Zaire (now the Congo) participated in TECCONILE as members; Burundi, Eritrea, Ethiopia, and Kenya participated as observers.

[32] The five "components" were: A. Integrated Water Resources Planning and Management; B. Capacity Building; C. Training; D. Regional Cooperation; and E. Environmental Protection and Enhancement. (TECCONILE, 1995, p.vi-viii.).

[33] Work on Project D-3 was accorded "high priority" in the ranking of projects for action, and was to be "implemented under direct supervision of the Council of Ministers" ("TECCONILE, 1995). As conceived in the Action Plan, the project was to "include a comprehensive study and documentation of processes and institutional frameworks that have been used successfully in other international basins, and principles utilized in other international situations, as well as the Nile Basin" (TECCONILE, 1995). The project also was to "propose alternative institutional frameworks, principles and processes" (TECCONILE, 1995). The NRBAP observes that to achieve basin-wide cooperation, it "is advisable, in the long term, to have an international agreement on water management. An agreement would assure all countries that development of waters in any particular country would not result in a conflict over water management with other Nile Basin countries" (TECCONILE, 1995).

[34] But note that work on Project D-3, the "Co-operative Framework," has been continuing with UNDP funding (ICCON, 2001).

[35] See *infra* pages 101-102.

[36] Toope was privileged to participate as an academic observer at the Nile 2002 Conference in Kampala, Uganda (Feb. 26 - 28, 1996).

[37] It may be more accurate to say that the Conferences have not been "captured" by Ministries of Foreign Affairs and by Offices of Prime Ministers and Presidents. Most of the delegates are from "government." However, for the most part, they are from ministries and agencies devoted expressly to issues of water development and environmental protection. The shared background of most of the delegates goes some way to explaining the progress made in the Nile 2002 conference series.

[38] See *infra* pages 104-105.

[39] Egypt, Sudan, and Ethiopia are now collaborating under the auspices of the NBI's Eastern Nile Subsidiary Action Program (ENSAP), while Burundi, the Democratic Republic of Congo, Egypt, Kenya, Rwanda, Sudan, Tanzania, and Uganda are engaged through the Nile Equatorial Lakes Region Subsidiary Action Program (NELSAP) (ICCON, 2001).

[40] See the timeline of meetings (ICCON, 2001). The authors discussed communication between technical and political levels with participants who wish to remain anonymous.

[41] In the shorter term, the goal of the cooperative framework is to attain a regional framework "acceptable to all Basin countries;" in the longer term, the goal is to "pave the way for milestones which would determine net equitable entitlements for each riparian country for the use of the Nile waters" (UNDP, 2002).

[42] For the membership of ENSAP and NELSAP, see *supra* note xxix. The cooperative projects include confidence-building, stakeholder involvement, efficient water use for agriculture and hydropower interconnection, and watershed management ("NBI", 2001; "Nile Basin Cooperative", 2001).

[43] It has also been argued that environmental concerns deserve special consideration and that significant environmental harm should not be justifiable on the grounds of equitable utilization (Handl, 1992).

[44] Article 7(2) of the "U.N. Convention", 1997 reads in full:

Where significant harm is nevertheless caused to another watercourse State, the State whose use causes such harm shall, in the absence of agreement to such use, take all appropriate measures having due regard for the provisions of articles 5 and 6, in consultation with the affected State, to eliminate or mitigate such harm, and where appropriate, discuss the question of compensation ("U.N. Convention", 1997).

[45] The individualized effects of the Convention remain true despite the Convention's procedural provisions: Articles 8, 9, 11, 12, 15 - 18, 24. While designed to promote various forms of cooperation, the procedural provisions are limited by the substantive framework. Furthermore, this assessment holds irrespective of the fact that the Convention does contain a number of provisions that alert states to common concern: Articles 20 - 23 contain various environmental and ecosystem protection obligations. Albeit tentatively, these

provisions do move the Convention beyond the "mutual limitation" paradigm. The ecosystem protection duties in Articles 20 and 22 and the duty to protect the marine environment in Article 23 arise regardless of interferences with another riparian's sovereign rights. Nonetheless, given the predominance of the regime framed by Articles 5 to 7, it is not clear that the Convention's environmental protection provisions can significantly change the flavor of riparian relations that it otherwise models ("U.N. Convention", 1997; Brunnée and Toope, 1994).

[46] Some authors, however, wrongly argue that one of the primary deficiencies of the "U.N. Convention", 1997 is its lack of clarity and determinacy, which is manifested in the failure to resolve the priority issue and to clarify what constitutes significant harm or reasonable use and "contributes to the weakness of international law" (Cunningham, 2000).

[47] Indeed, in July 2001, unnamed Nile experts were quoted in the press as saying "Whether Egypt is legally willing to abandon its claims of an untouchable share of the Nile water ... and whether Ethiopia is willing to accept sharing the Nile on an equitable basis, regardless of its increasing needs, remains to be proven through a binding international agreement ... "(Mburu, 2001).

[48] This discussion of the U.N. Watercourses Convention focuses only upon the impact of the framework established in its Articles 5 and 7. This focus is not intended to deny that some of the Convention's more technical or procedural provisions could play a role as textual blueprints for similar provisions in the Nile Co-operation Framework. Evidently, to the extent that legal text is available that matches the shared expectations of the riparian states on a given issue, resort to models can obviate the need for detailed negotiation over precise wording.

[49] For example, one observer with long-standing involvement in the Nile Basin noted that the traditional legal approach had failed and that there was a need to "maximize the emotional side" and "minimize the legal side" (Interview, 2000).

[50] According to Risse, "rhetoric" should be understood in the limited sense of "strategic" (Risse, 2000). The need for participants to "argue," in Risse's terms, rather than bargain, resonates with Fuller's assertion that:

The good negotiator [in the negotiation of a complex agreement] must not only make a genuine effort to understand the declared aims of the opposing party, but must be capable of some sympathetic participation in those aims...To serve his principal well he must identify himself with the opposing party...In this reconciliation of altruism and self-interest there is revealed...a more mature conception of morality (Fuller, 1965).

[51] Fuller argues that:

Problems concerned with the sharing of water supplies and the joint utilization of river systems are inherently unsuited to adjudicative solution, involving as they do a complex interplay of diverse interests. Only those who know those interests intimately, who can feel their way to the best reciprocal adjustment of them, are competent to find a truly satisfactory solution. If this is not the easiest way to hydraulic peace, it is surely the best (Fuller, 1965).

[52] According to a recent piece by George Downs and colleagues, Brunnée and Toope are among a group of "Transformationalists," who are said to believe that participation in regimes that reflect certain design principles induces "a mutually reinforcing series of normative and cognitive shifts among member states because states in effect are socialized by the regime" (Downs, et al., 2000). This article lists the following four "transformational" design principles: universal or nearly universal regime membership, emphasis on non-binding or low-threshold commitments in early stages of regime formation, consensus-based decision-making, and promotion of compliance through managerial approaches (Downs, et al., 2000). The interactional conception of law outlined above provides a much fuller understanding of how and why actors may be "transformed" by regimes and norms than the design principles listed by the authors. It would also go a long way toward addressing the authors' concern that "Transformationalism" does not provide a coherent explanation of "how the Transformational process operates" (Downs, et al., 2000). Indeed, the elements of interactional theory help explain why certain "design principles" are particularly conducive to guiding regime participants toward cooperation.

[53] Some of Brunnée and Toope's previous work has been misunderstood in this respect. For example, Brunnée and Toope have argued, *inter alia*, for an approach to freshwater regime building that draws upon the framework-protocol model of regime building, which has been successful in international environmental law (Brunnée and Toope, 1997). The approach was not to be an exclusively "soft approach to obligations," as some commentators have concluded (Fuentes, 1998). Rather, the argument is that emphasis upon the elaboration of binding rules may not be appropriate or beneficial in the *early* stages of regime building. Indeed, an initial contextual regime, such as a framework agreement, including the soft principles that it articulates, is usually an essential precursor for the development of binding rules. Fuentes, therefore, misses the point when she critiques Brunnée and Toope's point by stating: "What would be Mexico's reaction if the obligation assumed by the United States to deliver 1,500,000 acre-feet of water from the Colorado were to be characterized as 'soft'?" (Fuentes, 1998). Binding apportionment rules may well be needed, but this need is not necessarily best addressed through the immediate enactment of formally binding rules.

Biosphere Reserves on North-western Egyptian Coast, a Site for Monitoring Biodiversity and Integrated Water Management

BOSHRA SALEM
Department of Environmental Sciences, Faculty of Science, University of Alexandria, P.O. Box 21511, Moharram Bey, Alexandria, Egypt. E-mail: Boshra.salem@dr.com

ABSTRACT: Biosphere Reserves (BRs) have an important role to play in the application of the ecosystem approach for the integrated management of land, water and biodiversity, using the World Network of Biosphere Reserves as pilot sites. Almost all terrestrial Biosphere Reserves in Arab Countries (14 in 6 countries) are drylands (fragile ecosystems) that could serve as reference sites for conducting research for evaluating the relationship between water and biodiversity. This paper presents a case study from El-Omayed Biosphere Reserves in Egypt that is now under stress due the introduction of fresh water through new irrigation canals with resultant environmental damage including habitat loss and fragmentation. This paper therefore has three objectives: (a) to present the main characteristics of Biosphere Reserves as possible indicators for water resource management, and sustainable development; (b) to use the Omayed Biosphere Reserve (OBR) as a case study of some of the problems encountered, emphasizing water related issues; and (c) to consider ways in which biosphere reserves and integrated ecological studies in general might be used more widely as part of improved environments, even in water-scarce parts of the world.

INTRODUCTION

Water is a decisive factor in areas with little or insufficient amounts of freshwater, in particular in drylands. While dryland ecosystems are intrinsically robust, under human pressure and misuse they rapidly become unstable and degrade, and can be regarded as "fragile". Human pressures can include over use of "ecological water", changes in disturbance regimes from the propagation of invasive alien species, overuse of some biodiversity elements, etc. Ecosystem management, and rehabilitation of degraded dryland ecosystems, are dual imperatives to achieve poverty eradication and the prescriptions of the key environmental conventions, such as the United Nation Convention to Combat Desertification (UNCCD) and the Convention on Biological Diversity (CBD). It is important in this context to define ways to characterize the water needs in any parts of the world ecosystems that have not been yet covered.

Biosphere Reserves (BRs) are multipurpose areas of terrestrial and coastal ecosystems, which are internationally designated by one of UNESCO's Science programmes, Man and the Biosphere (MAB) Programme. They serve as demonstration areas for cooperation in building harmonious relationships between human activities and conservation of ecosystems and biological diversity. Each Biosphere Reserve represents one of a characteristic ecosystem of one of the world's biogeographic regions, involving human communities as integral components and including resources that need to be sustainably managed. Biosphere Reserves are envisioned as areas for learning, in which monitoring, research education and training are encouraged and supported. Collectively, Biosphere Reserves constitute The World Network of Biosphere Reserves, within which exchange of information, experience and personnel are promoted. By 2000, the world network included 368 biosphere reserves in 91 countries, with a total area of over 260 million hectares (UNESCO, 2000). These Biosphere Reserves are further organized into regional Man and the Biosphere networks, among which are EuroMAB, AfriMAB, USMAB and the latest, ArabMAB network (www.arabmab.net). Biosphere Reserves could be used in particular as reference sites for evaluation of water and nutrients cycling conditions, biodiversity and potential consequences for sustainable development The relationship between species and water availability needs could also be studied at all levels of biodiversity. Biosphere reserves could then serve as model sites for quantifying ecological water needs.

Biosphere reserves are nominated by national governments and must meet a minimal set of criteria and adhere to a minimal set of conditions before being admitted into the World Network. Each Biosphere Reserve is intended to fulfill three basic functions: (a) a conservation function, contributing

Water Resources Perspectives: Evaluation, Management and Policy. Edited by A.S. Alsharhan and W.W. Wood.
Published in 2003 by Elsevier Science, Amsterdam, The Netherlands, p.119-128.

to the conservation of landscapes, ecosystems, species and genetic variation; (b) a development function, fostering economic development which is ecologically and culturally sustainable; and (c) a logistic support function, which supports research, monitoring, training and education related to local, regional, national, and global conservation and sustainable development issues (UNESCO, 2000). To carry out these functions, Biosphere Reserves are organized into three interrelated zones, to ensure a long-term protection of its ecosystems. (i) The core area(s), which may be used for very limited human activities such as research, education and monitoring of traditional practices that may contribute to the conservation of biodiversity e.g. grazing; (ii) The buffer zone, surrounding or contiguous to the core area(s), to provide more protection. The activities carried out in this zone are minimally disruptive and might include research on ecosystem management or rehabilitation; (iii) the transition area, which is the site of human activities e.g. cities and villages. This zone is the place where human activities are directed towards sustainable development. Activities that can be carried out in these three zones are shown in Figure 1. The above zonation can be applied in different ways in the real world to accommodate geographical conditions, socio-cultural settings, available legal protection measures and local constraints.

In the context of the present paper, it is proposed that Biosphere Reserves may be used as reference sites for evaluating the status of water bodies especially water quality. Disturbances to the surface water systems and water tables may have serious impacts to the ecosystem. However the demise of wetlands, by uncontrolled groundwater pumping or the diversion of surface water can also be early warning signs of non-sustainable use of water resources. Artificial solutions such as irrigation or water importation also bring separate problems to the habitat. The relationship between species and water availability needs could also be studied in biosphere reserves at all levels of biodiversity. The network of protected areas in Egypt includes two of Biosphere Reserves, one of, which is located at the Mediterranean Coastal Desert of Egypt, and is used as case study in this paper.

STUDY AREA

The Mediterranean coast of Egypt generally, forms the edge of the Western Coastal Desert Except for the narrow Mediterranean coastal belt, which is the wettest region of Egypt (mean annual rainfall about 200mm), the whole Western Desert is one of the arid parts of the world (Figure 2). Its aridity results from its distant position from the main oceans, coupled with the absence of high altitudes which may attract orographic rain (Zahran et al., 1992). This coastal region is considered to be important as a potential area for agriculture and tourism extension projects. It is characterized by having suitable soils for agriculture, groundwater resources and moderate climatic conditions. The mean elevation of 500m above sea level (Zahran et al., 1992), and the mean annual rainfall is about 18mm, ranging between 0 to 200mm/year. The Western Desert as a whole, though considered barren, supports certain types of plant, which occur in areas with enough water resources (fed either directly from rainfall and/or from groundwater). Physiographically, the region is characterized by having two units; the southern plateau and the northern coastal plain. (Figure 3). The plateau reaches a maximum elevation of about 200 m above

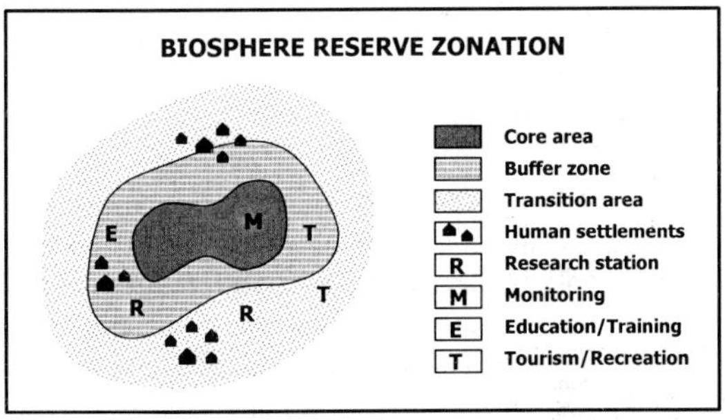

Figure 1. Structure and function of biosphere reserves.

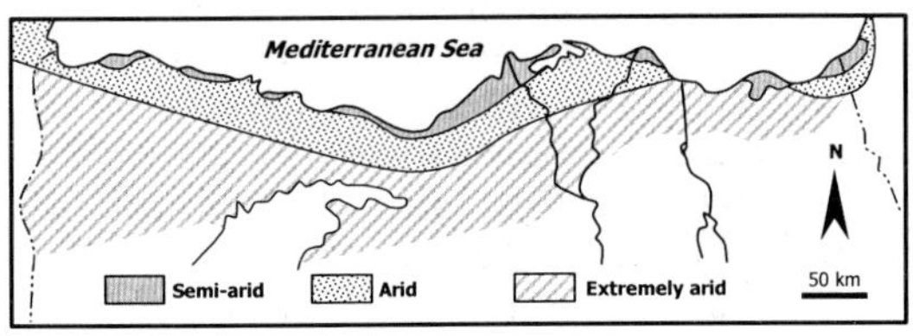

Figure 2. Aridity map of the coastal desert of Egypt.

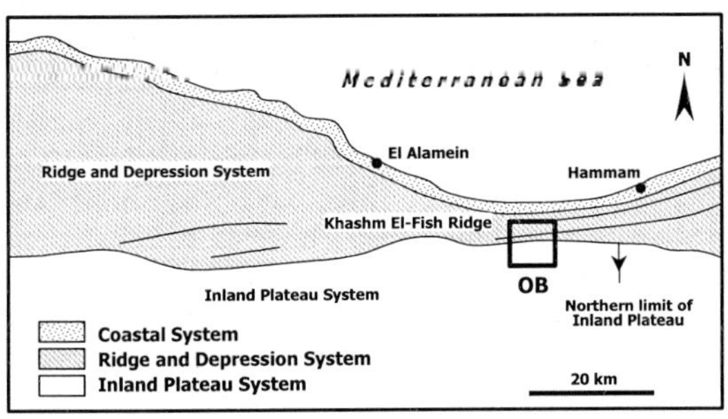

Figure 3. Habitats of the western coastal desert of Egypt

sea level at Salum (Libyan borders) and slopes gently towards the east, to a coastal plain west of Mersa Matruh (320 kms west of Alexandria) which varies from 10 to 20 m above sea level. Hydrologically, it comprises numerous shallow wadis of recent age that run northwards from the southern plateau towards the Mediterranean and crossing the coastal plain (Kassas and Gergis, 1964) (Figure 4). Thus the coastal plain is considered to be a basin for surface water collection.

Surface water occurrence in the region is controlled by a water divide line lying parallel to the coast and which varies in its location according to the points of high elevation (Figures 4 and 5) The surface water flows create a series of basins forming centers of activities and settlements. Figure 6 shows the coastal basins and classification of these basins according to the amount of water they receive. The first basin in which the Biosphere reserve is situated is the largest one and receives an additional amount of water due to seepage of fresh water from the newly irrigated canal (El Nasr Canal) that extends from the river Nile to Al Alamain in the west.

Groundwater in the region occurs under both artesian and non-artesian conditions, however, all the fresh groundwater available for agricultural and domestic uses occurs in relatively shallow (non-artesian) aquifers (about 3 to 15 m, according to the distance from the shore line). Large quantities of groundwater are found at the depths up to 200-1000 m in rocks ranging in age from Cretaceous to Miocene, although water is brackish or highly saline. Along the narrow belt near the Mediterranean Sea, wells and cisterns fed by local rainfall are found (Said, 1962).

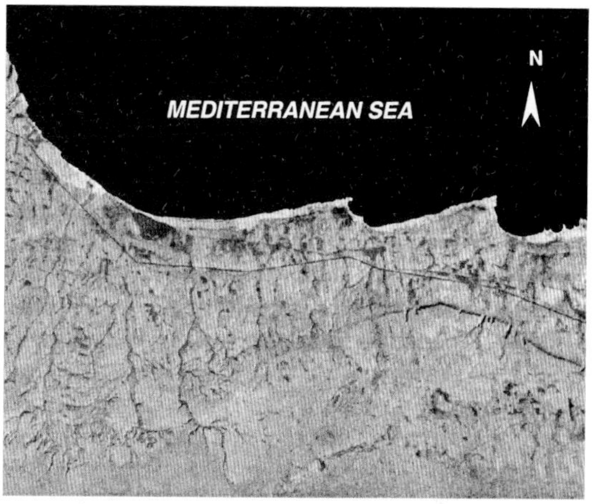

Figure 4. SPOT Satellite images showing shallow wadis run northwards from the southern plateau towards the Mediterranean and crossing the coastal plain . the water divide line is shown by the arrow.

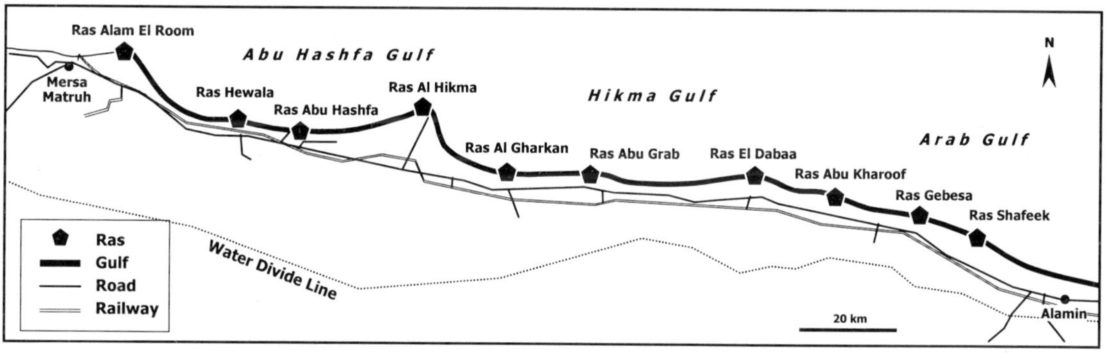

Figure 5. A Map showing the water divide line along the north coastal desert of Egypt.

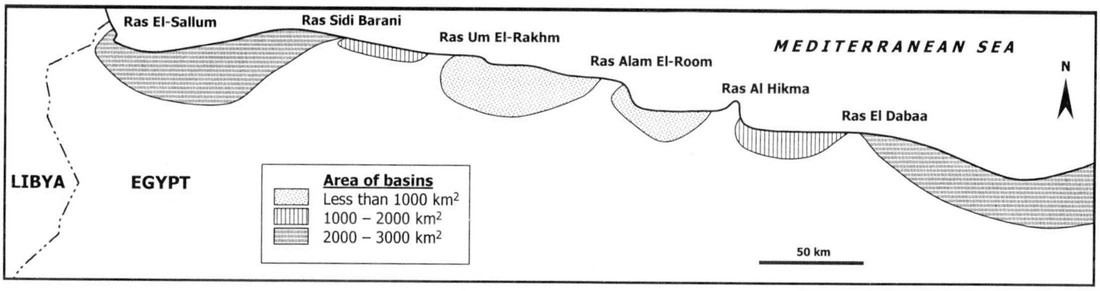

Figure 6. Coastal basins of the north coastal desert of Egypt.

Most of the species recorded from different habitats of wadis are Mediterranean species which do not occur in other parts of Egypt (E1-Hadidi and Ayyad, 1975). These Mediterranean species include perennials, mostly woody shrublets, which form the permanent framework of the vegetation; the rest are annuals which appear after winter rains. The wadi ecosystem may be physiographically differentiated into its upper, middle and lower slopes, and the wadi bed. The upper slopes are usually steep and almost completely devoid of soil cover; they support a typical cliff vegetation dominated by *Ephedra aphylla, Capparis orientalis, Umbilicus horizontahis, Periploca angustitloia, Phlomis fluccosa, Lycium europaeums* and *Asparagus stipularis* (El-Hadidi and Ayyad, 1975). The middle slopes are less steep and covered by shallow soil with mixed stones; the vegetation is dominated by shrubby species of chasmophytic nature as *Limonium sinuatum, Limonium tubiflorum, Gymnocarpos decandrum,* and *Artemisia inculta,* and some grasses as *Hyparrhenia hirta* and *Stipa capensis*. The lower slopes are gentle, where deep soil accumulates and supports meadow-like vegetation of annual species; the most common are *Picris sprengerianas, Aseragalus hamosus, Medicago truncatula, Hippocrepis bicontorta, Spergula fallax,* and *Medicago litoralis*. In the wadi bed, there is little chance for the fine sediments or soil to settle, due to high stream velocities during the rainy season. The wadis become filled mainly with large boulders and supports a sparse vegetation which is restricted more or less to shallow soil accumulations between rock fragments. Common perennials in the wadi bed are *Echium sericeum, Salvia lanigera, Euphorbia terracina, Cynara sibthorbiana, Allium erdelti* and *Atriplex halimus* dominates in locations with deep soil. Common annuals are *Astragalus boeticus, Pisum sativum, Erodim hirtum,* and *Erodium gruinum.*

Description of El Omayed Biophere Reserve

The Omayed Biophere Reserve (OBR), was the first Biosphere Reserve to be declared in 1981 in Egypt, as well as in Arab countries, in 1981, two years before the Egyptian Parliament promulgated a special law (Law 102/1983) for the establishment of nature reserves. OBR (Figure 7) was designed to fulfill the previously described three functions of a typical Biosphere Reserve as follows :

(1) Conservation, where the area includes samples of landscapes, ecosystems, species, and genetic variations representative of the Western Mediterranean coastal region of Egypt. It is up till now the only protected area in this region that is serving the purpose of conserving these biodiversity components, although, as discussed below, human activities threaten to obliterate them.
(2) Development, where the area is an ideal site for demonstration of sustainable development of the range resources in the region by rational management and organization of grazing activities to minimize grazing pressure and to make it compatible with the carrying capacity of the ecosystem.
(3) Logistic support, where the area supports demonstration projects of propagation of multi-purpose species of shrubs and trees. It is also an ideal site for education, training, continued research.

The Omayed Biophere Reserve is characterized by a distinct variety of habitats ranging from snow-white coastal sand dunes to clay/loam salt marshes and rocky ridges with skeletal shallow soils. Although it represents a large number of ecological systems but belongs to only one biogeographic region, it includes a remarkable gradation of human interventions, sometimes intercalating into a mosaic of tiny fragmented land use types. Omayed Biophere Reserve is distinguished into a variety of

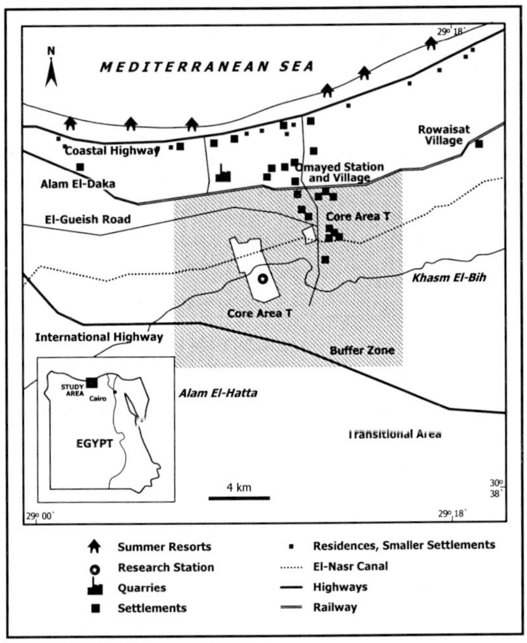

Figure 7. Zonation in omayed biosphere reserve (OBR) and location of El Nasr canal extending in the two core areas.

physiographic units, and is characterized by successive undulations running more or less parallel to the coast. These undulations are in the form of calcareous cemented ridges (ancient dunes) alternating with depressions. Several ridges start near lake Mariut and become gradually less obvious towards the west.

The life-form spectrum recorded in Omayed Biophere Reserve expresses a typical desert flora. The majority of species are annuals (ephemerals) and geophytes (perennial ephemeroid herbs and grasses), which are drought evaders, or chamaephytes (subshrubs and shrubs). The area has high floral diversity; over 183 plant species (belonging to 41 families of angiosperms) have been recorded and listed in a study of plant uses and value in the protected area (UNESCO, 1997). There are at least 26 plant species in the area that are considered to be nationally threatened, and a few are restricted-range and globally-threatened species.

Five natural habitat types are recognized at Omayed Biophere Reserve, moving inland: (1) coastal calcareous dunes (the First Ridge), (2) inland ridges with skeletal shallow soils (the other three ridges), (3) saline marshy depressions (between the second and third ridges, an extension of the Mariut Lake), (4) non-saline depressions (between the ridges, except the former two), and (5) the inland plateau. In the context of this paper reference will be made to only the first and the fourth habitats.

The first type of habitat extends all along the Mediterranean coast and is therefore of regional distribution. Its characteristic vegetation species are: *Ammophila arenaria, Thymelaea hirsuta, Crucianella maritima, Ononis vaginalis, Pancratium maritimum,* and *Echinops spinosissima,* among others (Ayyad, 1973). The important natural processes in this habitat are sand movement and wind erosion, as well as percolation of rain water into deeper layers. The main traditional uses are grazing, fig cultivation and catching of quail in autumn, but most of the strip has been transformed into lush resort villages for holiday makers from all over Egypt. The fourth habitat type, the non-saline depression, is also of regional distribution and has vegetation communities dominated by *Anabasis articulata* and *Zygophyllum album* on sandy soils with low calcium carbonate content, and *Asphodelus microcarpus* and *Thymelaea hirsuta* on fine textured soils (from deposition of erosion particles). Sand deposition is the main natural process. Dry farming, irrigated farming, grazing, settlements, and wood cutting are the main human activities here, because of a relatively wide expanse and good deep soils, together with run-off from the neighbouring ridges. Grazing, however is the most important occupation, and hence range management and improved dry farming should be the relevant management practice.

Main issues involving water stress

Deep drilled wells seem to be unknown in Omayed. There is only one well where water is of high salinity and was used only during the reconstruction of the Alexandria – Matruh highway, some twenty years ago. At El Omayed, groundwater is available for agricultural and domestic uses (Ayyad and Le Flock, 1983). Other water sources include i) about 13 cisterns which are used for drinking animals but only a few of which are in operation, ii) the pipeline extended from the nearest city to the east, which is used for supplying fresh drinking water for people and animals; and iii) surface runoff from Khashm El-Eish to the non-saline depression, which needs proper management to make full use of the amounts received.

The main water table aquifer is composed primarily of relatively fine-grained marly limestone, but near the coast, it is composed of alluvial silt and sand. Recharge to the main water table occurs through direct infiltration of precipitation and through infiltration of surface runoff, particularly near the coast. Discharge occurs through extraction from shallow wells, evapotranspiration where the water level is near the surface, and subsurface outflow to the sea.

The depth to the water table over the region varies from less than a meter to more than 50 meters depending on the season and topography. The highest water level occurs from February to April, depending on the length of rainy season, and the lowest water level occurs from October to November depending on the beginning of the rainy season. To the south of the Omayed Biophere Reserve area, wells with water deeper than 50m are utilized by bedouins. In the interdunal plain behind the coastal dunes, the water table is generally less than 5 m below the surface. In the coastal dunes, the water generally ranges from 5 to 10m below the surface depending on the height of the dunes.

The quality of water in the several aquifers found in the area varies widely. Depending on the hydrogeological setting, the content of dissolved solids may vary from a few hundred to several thousand parts per million. Water quality also varies with season, being best after winter rainfall and worst in late autumn before the beginning of the rainy season. In the area between Burg el Arab and Dabaa, where Omayed Biophere Reserve lies, ground water salinity is between 1000 and 6000 ppm, except in fresh water lenses, overlying the saline water in the north. Salinity in these lenses reaches a maximum of 3000ppm.

RESULTS AND ANALYSIS

Regular monitoring and research activities were carried out in the Omayed Biophere Reserve by Salem,1990, 1993, 1994, Salem *et al.* , 1994, 1995, 1997, and Le Floc'h *et al* , 1992, as well as recent studies conducted by Salem (2001a and b), to assess habitat loss and fragmentation and their impact on biodiversity, in the five main habitats in the Omayed Biophere Reserve. From these studies, it was observed that in the last decade, Omayed Biophere Reserve had been subjected to an intensive process of habitat fragmentation and loss due mainly to human interventions. Multidate SPOT satellite imagery of 20 x 20 m pixel resolution was used for the years 1987, 1993 and 1999 for monitoring and assessing the changes in two, of the five habitats, in the Omayed Biophere Reserve that are mostly affected by these human impacts. These are the coastal ridge habitat, which is encountering almost complete loss, and the non-saline depression habitat, which is encountering severe fragmentation process due to various development activities. The main image processing techniques used are change detection techniques using post land-use classification comparisons, and principal components analysis. The results were confirmed by ground-truthing surveys and classification accuracy methodologies (Salem, 2001a,b).

Results of the above work showed that throughout the 12-year period of study, about 33% of the area of coastal sand dunes habitat has been lost due to land transformation of the dunes ecosystem to an extended coastal strip of tourism resort. Almost all the natural resources sand dunes habitat has been removed and replaced by the sprawl of urbanization; buildings, utilities, and others. This natural habitat has been almost all lost due to the removal of and leveling of the sand dunes for the construction purposes. The result is an associated loss of the biodiversity of this habitat, in terms of its flora and fauna and mineral resources in Omayed Biophere Reserve. Figure 8 shows the habitat of sand dunes before and after the construction of summer resorts; Figure 9 shows the delineation of the coastal sand dune habitat on three sequential satellite image acquired from SPOT satellite in 1987, 1993 and 1999. It is assumed due to the above observations that this habitat loss would also affect the water table, and surface runoff system of this habitat because this habitat has been considered as an important fresh water reservoir in the area. This habitat also was supporting the best varieties of figs, palm trees and other rain-fed annual crops. It was also found that the existing fresh water lenses beneath the sand dunes are currently heavily depleted as a result of pumping out for domestic use by the inhabitants of the resorts as a supplementary water source (water from the extended water pipelines is always insufficient especially during summer time). This calls for further study to assess the water balance and water quality in Omayed Biosphere Reserves.

With regard to the non-saline depression, this habitat is considered as the most fertile habitat in the coastal desert in general, and in the Omayed Biophere Reserve specifically, supporting both rain fed and partially irrigated agricultural practices. The habitat also represents the main range-land used by local inhabitants for feeding their livestock. Recently, and after investigation using satellite imagery, it was found that the non-saline depression habitat has been intensively fragmented and its total area has been reduced by about 35%. The fragmentation is mainly due to the construction of a new irrigation canal network (Nasr canal) in the

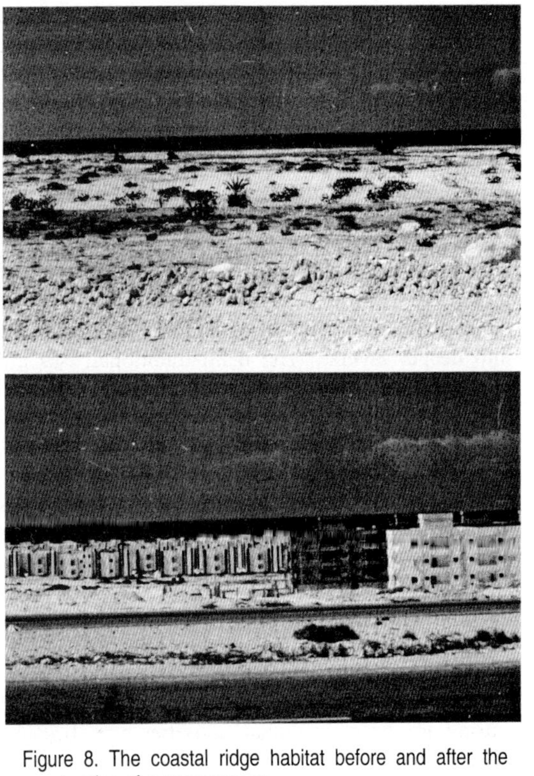

Figure 8. The coastal ridge habitat before and after the construction of summer resorts.

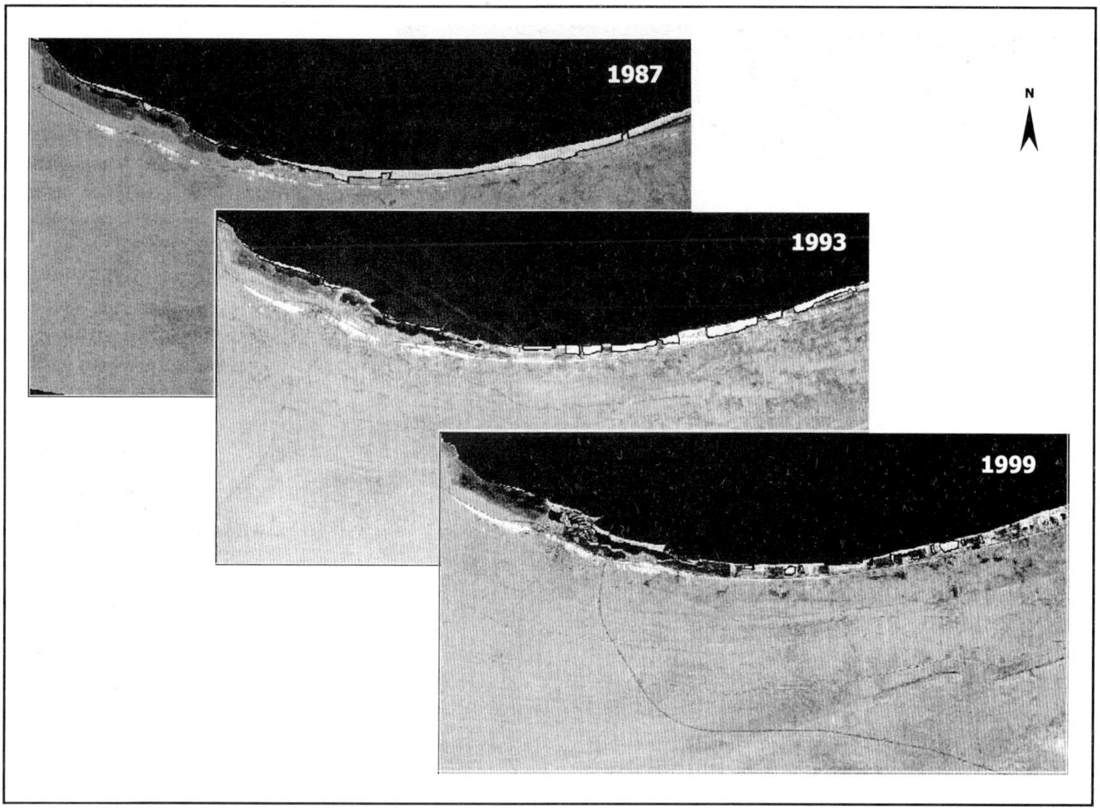

Figure 9. Habitat Fragmentation /Loss of the Coastal Dune Habitat as observed from SPOT Satellite Imagery for the three years 1987, 1993 and 1999.

Figure 10. The irrigation canal in two different locations of the coastal desert of Egypt.

area. The new canal is an extension (some 80 km) from the river Nile towards the Western Desert. This canal was designed for the supplementary irrigation of the annual cropping system after reclamation projects takes place. Figure 10 shows the irrigation canal in two areas in the Omayed Biophere Reserve. The direct environmental effect of the establishment of this irrigation canal is the fragmentation process that occurred in the non-saline habitat due to the constructions and disturbances that affected the homogeneity and caused spatial discontinuity of this habitat and turned it into separated and in some cases isolated fragments. From the biodiversity point of view, this fragmentation increases the risk of extinction in many ways. It may limit a potential for species dispersal and colonization, it also may cause population decline and extinction due to dividing an existing widespread population into two or more sub-populations in a restricted area. Figure 11 shows the non-saline depression habitat from satellites in the three acquired dates. The introduction of fresh water to this arid environment is also an important point to consider. This may introduce new water-borne species, that may replace

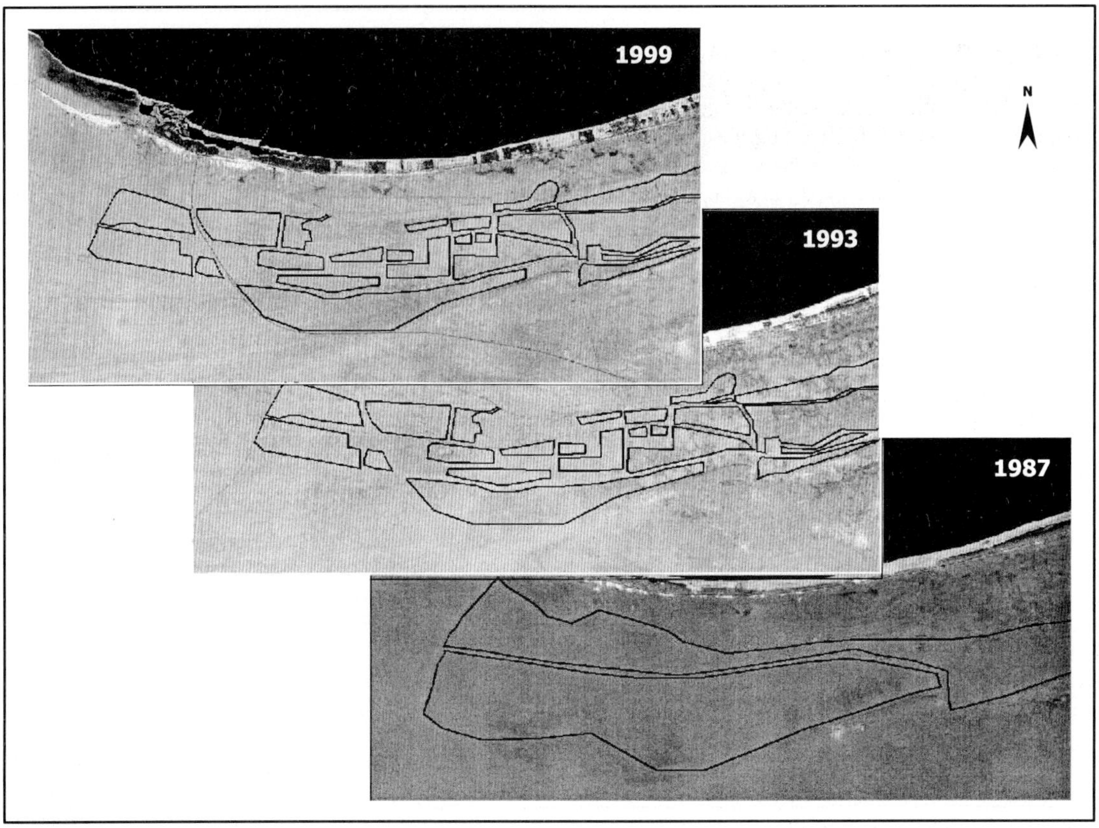

Figure 11. Habitat Fragmentation process as observed by SPOT Satellite Imagery for the three years 1987, 1993 and 1999.

the existing plant communities, change the ecological balance and consequently the structure of the ecosystem. The changes in the structure of the plant communities would definitely affect animal communities and their distribution.

CONCLUSIONS AND RECOMMENDATIONS

The waters of the Nile have always been regarded as the only water source available for irrigation requirements. Irrigation systems in the reclaimed lands are based on a succession of pumping stations from the main river. However, large irrigation projects which impound or divert the river water have the potential to cause major environmental changes, resulting from the change in hydrology and limnology of river basins (Food and Agriculture Organization, 1997).

Irrigation development may have both positive and negative impacts on the environment. To be sustainable, irrigation must avoid negative impacts. The negative impacts are more profound due to excessive pressure on arid and semiarid lands under rain-fed cultivation or grazing. Such lands are already undergoing a process of degradation. The transition of people, who have subsisted for generations on rain-fed lands to irrigated farming may be a difficult social change. It is also important to consider that potentially negative environmental impacts of irrigation development may occur off-site as well as on-site. Therefore it is important to explore the interdependence of land, water, and ecosystems for enhanced environmental resource management.

Existing information derived from the monitoring studies carried out in the Omayed Biophere Reserve have directed the attention to the possible impacts of developing irrigation schemes in the desert conditions. This information has created an improved picture of inland water biological diversity, its uses and its threats to Omayed Biophere Reserve. Although land reclamation and irrigation schemes have contributed significantly to poverty alleviation, food security, and improving the quality of life, the sustainability of these schemes is questioned environmentally. Inadequate attention to the ecology of deserts and their natural resources under reclamation schemes have frequently led to great difficulties. Decisions to embark these costly projects have often been made in the absence of sound assessments of their environmental and social implications. Deserts ecosystems are very fragile, and the expansion and intensification of irrigation schemes has the potential of causing increased erosion due to water logging and salinization, pollution of surface and ground water from agricultural biocides. Additionally, over-extraction of groundwater for irrigation can result in lowering the water table, land subsidence, a decrease water quality and salt-water intrusion in the coastal areas.

The above study of the Omayed Biophere Reserve shows the environmental deterioration and degradation due to mismanagement of its resources. Human impacts are expressed in terms of habitat loss and fragmentation, which are directly threatening the biodiversity of this area, characterized by its fragile ecosystems and where native species are highly sensitive to extinction. The habitats of coastal sand dunes and non-saline depression have been used to draw examples on habitat loss and habitat fragmentation due to establishment of summer resorts and extension of a network of irrigation canal respectively.

It is therefore recommended that for the Omayed biosphere to fulfill its objectives and mandate, efforts should be directed towards changing its zonation scheme, to include less fragmented areas. These results would be useful if considered by decision-makers in the regional planning and sustainable development of the coastal desert.

Additional studies particularly of key indicators are needed for the assessments of the effect of introducing fresh water into this arid environment, and the rational management of this water for the sustainable development of natural resources in Omayed Biophere Reserve and the coastal desert in general. The relationship between species and water availability needs to be studies at all levels of biodiversity. These studies should include investigations of the contributions of climate, site geomorphology, topography and vegetation to any observed effects in the Omayed Biophere Reserve. Howevre, it is proposed that careful monitoring of water levels and water quality be made since these changes detect the changes, both for better or worse in the ecosystem. The information derived from such biosphere reserve studies may also be transferred to aid development areas that have more impact, especially in projects involving irrigation, reclamation and water management that do not include biodiversity consideration.

It is therefore recommended that is any region where development is proposed a research project is conducted to assess the potential long-term impacts of the irrigation schemes on the ecological processes in the OBR, and more importantly the impact on the hydrologic processes in desert ecosystems. Such research would need to target five objectives:

1. The understanding of the relationships between ecosystem functions and processes on the overall water balance.
2. To use this information, along with water resource models, climate change scenarios and assumptions about the future intensities of existing irrigation project, and their effects on the future functioning of the ecosystem
3. To provide stakeholders with information on how the correct functioning of the ecosystem are likely to be affected, so that they can make informed decisions;
4. To develop a methodology that may be transferable to other large irrigation schemes in desert ecosystems, especially using hydrological and biodiversity indicators.
5. To assess the socioeconomic impacts of irrigation-scheme projects on the local inhabitants, and their reaction towards changing their traditional dry farming schemes.

REFERENCES

Ayyad, M.A., 1973. Vegetation and environment of the western Mediterranean coastal land of Egypt: the habitat of sand dunes. *Journal of Ecology* 61: 509-523.

Ayyad M.A. and Le Flockm, 1983. An ecological assessment of renewable resources for rural agricultural development in the western Mediterranean coastal region of Egypt: Case Study El Omayed Test area. C.N.R.S./ C.E.P.E.:l-Emberger Montpellier. p104.

Ayyad, M.A. and Ammar, M.Y., 1973. Relationship between local physiographic variations and the distribution of common Mediterranean desert species. *Vegetatio* 27: 163-176.

El-Hadidi, M.N. and Ayyad, M.A., 1975. Floristic and ecological features of Wadi Habis (Egypt). In: La flore du bassin mediterranean: Essaie de systematique synthetique. Colloques Internationaux du C.N.R.S., No. 235.

Food and Agriculture Organization, 1997. Small-scale irrigation for arid zones, Principles and options. Author: Daniel Hillel, www.fao.org/docrep/ w3094e00.htm

Food and Agriculture Organization, 1970. United Arab Republic: Pre-investment survey of North-Western coastal region; comprehensive account of the project. Technical report1, ESE: SF/UAR 49.

Le Flock, E., Kamal, S., Salem, B. and Ayyad, M., 1992. Utilization Des Donnees SPOT Pour La Cartographie Des Resources Renouvelables, Raport Final, ORSTOM, CNRS-France.

Salem, B., 2001a. Monitoring biodiversity of the Western Coastal Deserts of Egypt using multidate remote sensing imagery case study Omayed biosphere reserve. Paper presented at the International Conference on Detecting Environmental Change, Science and Society. Environmental Change Research Center (ECRC), University College, London, July 2001.

Salem, B., 2001b. Monitoring habitat fragmentation due to land use changes using Multidate Satellite Imagery. Case Study: Omayed Biosphere Reserve. Paper presented in the International Conference: Detecting Environmental Change: Science and Society, Environmental Change Research Center (ECRC), University College, London, July 2001.

Salem, B., 1994. Land evaluation of Western Coastal Desert of Egypt by Geographic Information Systems (Case Study: Omayed Transact). Proceeding of the Regional Environmental Symposium, UAE.

Salem, B., 1993. Detection of temporal environmental changes in arid land by Remote Sensing. Case study: North Coastal desert of Egypt. Proceeding of the 25th International Symposium, Remote Sensing and Global Environmental Change, Graz, Austria, 4-8 April, 1993.

Salem, B. 1990. Application of principal component analysis for image processing of the Mediterranean Desert of Egypt. Proceeding of the Regional Symposium on Environmental Studies (UNARC), Alexandria, Egypt.

Salem, B. and Ayyad, M., 1994. Monitoring of Human Impact on Land cover by Remote Sensing; Case Study: The Mediterranean Coastal Desert of Egypt. Proceedings of UNESCO/IUSSP/IGU conference on "Population and Environment in Arid Region" Amman-Jordan 24-27 October. IUSSP-UIESP Belgium.

Salem, B., El Cibahy, A. and El Raey, M., 1995. Detection of Land Cover classes in Agro-Ecosystems of Northern Egypt by Remote Sensing. *International Journal of Remote Sensing,* 16(14): 2581-2594.

Salem, B. and El Kenany, I., 1997. Monitoring Land Use/Cover Changes in El-Omayed Biosphere Reserve By Remote Sensing. Proceedings of the Earth Observation and Environmental Information Conference-Arab Academy for Science, Technology and Marine Transport, October 13-16, 1997, Egypt.

Said, R., 1962. The Geology of Egypt. Elsevier, Amsterdam.

UNESCO, 1997. Man and the Biosphere (MAB) Programme- Biosphere Nomination Form.

UNESCO, 2000. Poster of "The World Network of Biosphere Reserves".

Zahran , M.A. and Willis, A.J., 1992 . The Vegetation of Egypt. Chapman and Hall – London, 424p.

Contribution of UNESCO-International Hydrological Programme to Water Resources Management in the Arabian Gulf Countries

ABDIN SALIH
The International Hydrological Programme, Division of Water Sciences, UNESCO 1, rue Miollis 75732 Paris Cedex 15, France. E-mail : a.salih@unesco.org

ABSTRACT: Water availability in the countries of the Gulf Cooperation Council (GCC) has been critically reviewed. Though per capita annual renewable water resources in these countries is by far the lowest compared to all regions of the world, yet these countries have managed to secure considerable quantities of additional water supplies from non-conventional resources such as desalination, fossil groundwater and wastewater re-use. Outstanding problems related to the application of this policy have been outlined together with a vision for the future. Potential contributions of the International Hydrological Programme of UNESCO (IHP) and its initiatives for solving some of their water resources problems have been identified. The outlook is promising for the success of such initiatives.

INTRODUCTION

The Gulf Cooperation Council (GCC) countries are located in a prominent part of the extreme arid zones of the world, in which water scarcity is a permanent threat to socio-economic development. In this region, as underlined at the 2nd World Water Forum (The Hague, 2000), immediate action is needed to face the challenges imposed by the limited natural water resources and growing populations. Water security and sustainable management will increasingly depend on strategies and programmes that respond to water scarcity and other factors affecting the supply and the demand in water. Given the lack of adequate renewable surface and groundwater resources, water security and sound management will increasingly depend on advances in sea water desalinisation, re-use and recycling of waste water, water harvesting, non-renewable groundwater, imported freshwater and above all conservation practices and strict demand management.

Current experiences have proved that great adverse problems could be created if the current economic prosperity of these countries is directed mainly towards implementing actions that respond to the escalating demand for water by a policy of continuous increase of the supply, mainly from desalinisation sources or non-renewable groundwater. One of these problems is the widespread rising groundwater levels reported from most of the large cities of the Gulf Cooperation Council countries. This is in contrast to the escalating rate of pollution and lowering in the groundwater levels of both renewable and non-renewable aquifers heavily utilized for irrigation in many countries of the region. Therefore, at both the national and regional levels, wise policies, strategies and plans of action are needed to mitigate the looming water crisis and to secure a sustained future.

Full cooperation and partnership between public, private and NGO's in the region should be encouraged, in the interest of efficient water management and peaceful sharing of transboundary water resources. The necessity to exchange expertise and experience between countries of the region and abroad is to be emphasized, in order to use the comparative advantage of each country in resolving local and regional problems. The education system needs to be promoted, with curricula that include subjects relevant to water policy, technology and scientific research.

UNESCO, through its International Hydrological Decade (IHD, 1965-1974), the completed five phases of the International Hydrological Programme (IHP, 1975-2001) and the recently launched Sixth Phase of International Hydrological Programme (2002-2007) and its parallel new initiatives such as the World Water Assessment Programme (WWAP) undertaken with all relevant UN agencies, the International Groundwater Resources Assessment Centre (IGRAC), the Internationally Shared Aquifer Resources Management (ISARM), Hydrology for Environment, Life and Policy (HELP), the project "From Potential Conflict to Cooperation Potential" (PCCP), and the recently established UNESCO-IHE Institute for Water Education (IHE), have developed several techniques and approaches that could provide a solid framework for dealing with many of these problems in a scientific and rational way.

This paper broadly reviews the freshwater situation in the GCC countries, at present and in the

Water Resources Perspectives: Evaluation, Management and Policy. Edited by A.S. Alsharhan and W.W. Wood.
Published in 2003 by Elsevier Science, Amsterdam, The Netherlands, p. 129-139.

future, and comes out with concrete key examples that demonstrate problematic areas. Ways of dealing with the critical problems, whether partially or fully, through the framework and the substances provided by the forthcoming IHP-VI themes and related initiatives are highlighted.

OVERVIEW OF THE FRESHWATER SITUATION IN THE REGION

Global Freshwater Budget

One of the pioneering reliable efforts to quantify global freshwater resources is the work undertaken in the framework of the International Hydrological Programme of UNESCO by the Russian IHP National Committee (Shiklomanov, 1998). This work has indicated that there is a volume in the order of 35 million km³ of freshwater available in the globe, of which close to 70% is in the form of glaciers and permanent snow cover and close to 30% as groundwater which are both not readily available for direct human use. The main sources of renewable water resources (river systems, lakes, reservoirs, etc.) readily available for human livelihood do not exceed 90,000 km³ constituting only about 0,26% of the world freshwater budget. This could have been enough for a world-wide need in the order of 5,000 km³ per year, were it not for the very wide geographical variation in the distribution of this budget resulting in catastrophic floods in some regions while many vast areas suffer long periods of drought, water scarcity and extreme aridity. This situation is increasingly intensified by the fears of widespread quality deterioration and land degradation resulting from mismanagement of these resources as well as the highly publicized negative effects of climate change.

The result of Shiklomanov's work is compiled in Table 1 together with forecasted situations for the years 2010 and 2020 based on an average annual reduction rate estimated from the 1950 to 2000 reported values. The table demonstrates clearly the uneven distribution of the renewable freshwater resources between continents and regions as well as the rate of per capita reduction in each of these boundaries.

Among the obvious deductions from the table is that in spite of the wide variation between a maximum per capita annual availability of 179,000 m³ in North South America to a minimum of 2,300 m³ in North Africa, the year 1950 represented a period where none of the world regions had reached a per capita annual availability less than the 1,000 m³ minimum level described by many sources of the literature as a serious water poverty line. As demonstrated in the table, this minimum level has been violated only in the seventies within North Africa and is anticipated to be violated also in Western Asia beyond the year 2010. These two last regions have obviously higher rates of per capita annual reduction due mainly to their widely reported higher rates of population growth.

Table 1: Water Resources Availability in Different Regions of the World. *(Source: Shiklomanov, 1998)*

Continent	Area (10^6 km²)	Actual water availability (10^3 M³ per year per capita)							Average reduction per year (%)
		1950	1960	1970	1980	2000	2010[a]	2020[a]	
Europe	10.46	5.9	5.4	4.9	4.6	4.1	3.8	3.6	0.6
North & Central America	24.30	37.2	30.2	25.2	21.3	17.5	15.6	13.6	1.1
Africa	30.10	20.6	16.5	12.7	9.4	5.1	4.3	3.6	1.5
North Africa	8.78	2.3	1.6	1.1	0.69	0.21	0.17	0.13	1.8
Asia	43.50	9.6	7.9	6.1	5.1	3.3	2.9	2.4	1.3
West Asia	6.82	6.3	4.2	3.3	2.3	1.3	1.09	0.88	1.6
South America	17.90	105	80.2	61.7	48.8	28.3	24.0	19.8	1.5
North South America	2.55	179	128	94.8	72.9	97.4	81.8	66.2	1.6
Australia & Oceania	8.95	112	91.3	74.6	64.0	50.0	44.5	39.0	1.1

[a] : Figures in the last three columns are computed by the author.

Though Table 1 gives broad useful information on the quantitative distribution of renewable freshwater resources, it lacks important complementary information on the quality situation as well as potential contributions from non-renewable resources (groundwater, desalination, wastewater re-use, etc.). Another deficiency in these values is that they don't consider type of uses (irrigation, domestic, industrial, etc.) and climatic zones. For example, in countries where agriculture demands are small due to abundance of rain or to implementation of policies that give lesser priority to irrigation in favour of basic human needs such as domestic and industrial requirements, the assumed 1000 m³/year/capita water poverty limit will become unrealistic. Furthermore, the information on the table describes the situation in a broad regional framework that could disguise critical shortages in other smaller areas within the reported broad regions. These limitations would also apply to the currently growing concern on the subjectiveness of the concepts of "availability" and "scarcity limits" definition (Hoekstra, 1998; Salih, 1994; WMO, 1997).

Freshwater Situation in the Gulf Cooperation Council Countries

The freshwater situation in the countries of the Gulf Cooperation Council is broadly integrated with the numerical figures indicated in Table 1 for West Asia. Excluding North Africa, it is the region with the lowest per capita annual water availability in the world and one of the highest annual reduction rate due to its higher rate of population growth. Though Table 1 indicates a per capita annual availability of freshwater of less than one thousand cubic meter only beyond year 2010 for West Asia, yet Table 2, compiled from recent releases for year 2000, reveals a more detailed picture for the four sub-regions of the Arab States, including the Gulf Cooperation Council countries.

The critically low water availability in the whole Arab region is clearly demonstrated in this table with countries of the Arabian Peninsula (Gulf Cooperation Council countries and Yemen) sub-region having the lowest per capita renewable resources in the region (270 m³/capita in 1996). Noting that Salih (1998) has demonstrated that the Arab region has by far the lowest per capita annual freshwater availability in the world, it can thus be safely deducted that the countries of the Arabian Peninsula represent the world's poorest of the poor zone in renewable freshwater resources.

Table 2: Sub-regional Breakdown of Water Resources Availability for Year 2000 in Arab States.

Sub-region	Renewable water resources 10^9 M³	Population in Millions	Per capita renewable water resources 10^3 M³/year
Eastern (Syria, Lebanon, Palestine, Jordan and Iraq)	76.3	51	1.50
Arabian Peninsula (Gulf countries and Yemen)	13.1	49	0.27
Central (Egypt, Sudan, Somalia and Djibouti)	98.4	114	0.86
Western (Libya, Tunisia, Algeria, Morocco and Mauritania)	58.0	79	0.73
Grand total	245.8	29.3	0.84

A word of caution that must be repeated here is that, in spite of the voluminous reports and papers published on the freshwater situation in the Arab States, including the Gulf Cooperation Council countries, wide ranges of variations and contradicting values are continually published at the level of the region, the Gulf Cooperation Council countries and individual countries. Salih (1998) has given examples of these variations and emphasized the urgent need for a rational water resources assessment project at the national, sub-regional and regional levels. The same problem has been met by the author of this paper in compiling reliable numerical values for this document. For this reason, it is advisable to take the figures reported in this paper as guiding indicators in support of the suggested policy directions subsequently argued. Nonetheless, it must be stressed again that there is an urgent need for a water resources assessment programme in this region as a pre-requisite for a holistic and rational water resources management approach.

With the above warning and recommendation in mind, a more detailed picture of the conventional and non-conventional water resources of the Gulf Cooperation Council countries is given in Table 3. The table is adapted from a recent publication of the United Nations Economic and Social Commission for West Asia (Economic and Social Commission for Western Asia, 2001a) and supported with corresponding values from two relevant papers (Al-

Zubari, 1997; Bremere et al., 2001). Table 3 confirms the extreme scarcity of renewable freshwater resources in the six Gulf Cooperation Council countries with a maximum of 583 m³/year/capita in Oman, a minimum of only 74 m³/year/capita in Kuwait and an average of 277 m³/year/capita for the whole sub-region. The picture, however, looks different if one includes the non-conventional resources (groundwater, desalination and water re-use) in the available water resources. Through this last consideration water availability in Kuwait has substantially increased to over six fold, the availability in Oman has increased to 1265 m³/year/capita, which is well beyond the controversial water poverty limit of 1000 m³, and the regional average almost reached the 1000 m³ limit.

The main lesson learnt from this exercise is that the concept of water availability based on indicators driven from the so-called renewable water resources divided by the total population should be taken with great care in situation like the one of the Gulf Cooperation Council countries, as well as countries where irrigation water requirement is not considerable. The Gulf Cooperation Council countries have heavily utilized desalination, non-renewable groundwater resources and wastewater re-use to compensate for their renewable water scarcity. The instrumental factor in the success of this direction was the booming economy within the last three decades and the cheap cost of energy. Due to the importance of these three non-conventional water resources to the Gulf Cooperation Council countries, the paper will elaborate on the opportunities and constraints of these alternative water resources.

Table 3: Conventional and Non-Conventional Water Resources in the Gulf Cooperation Council Countries (in Million Cubic Meters per Year). *(Source: Adapted from Economic and Social Commission for Western Asia, 2001[a])*

Gulf Cooperation Council Countries	**Conventional water resources**			**Non-conventional water resources**		**Population in thousands (2000)**	**Water availability (m³/year/capita)**		**Total water availability as percentage of renewable water**
	Surface water	Ground-water recharge	Ground-water use	Desali-nation	Waste-water re-use		Renewable resources	Total water resources	
Bahrain	0.2	100	258	75	175 (56.2)[b]	613	163	829	508
Kuwait	0.1	160	405	388	30 (76.0)	2 165	74	454	614
Oman	918	550	1 644	51	23 (7.4)	2 518	583	1 265	217
Qatar	1.4	85	185	131 (206.2)[a]	28 (283)	599	144	719	499
Saudi Arabia	2 230	3 850	14 430	795 (1854.2)	131 (448.9)	21 930	277	977	353
United Arab Emirates	185	130	900	455.0	108 (719.8)	2 441	129	728	564
Total	3334.7	4 875	17 822	1 895	495	29 653	271	958	346

a. Figures in parentheses are reported by Bremere et al. (2001).

b. Figures in parentheses are reported by W.K. Al-Zubari (quoted from Economic and Social Commission for Western Asia, 2001[a]).

Water desalination

The growing technology of desalination is currently providing enormous quantities of water to meet the esclating needs for domestic and industrial sectors in many water scarce countries. It is reported that the desalination capacity currently exceeds 2-10 times the renewable groundwater resources in Qatar, Kuwait, Malta and Saudi Arabia (Bremere et al., 2001). Looking again to Table 3, this claim is only true for Qatar and Kuwait if one considers only the renewable resources. The figures shown for groundwater recharge in Saudi Arabia is much higher than the desalination contribution. However, it must be stressed that many sources in the literature believe that the renewable groundwater indicated for Saudi Arabia in Table 3 is an over-estimation of reality and it also represents double counting of the $2230X10^3$ m^3 reported under surface water (Salih, 2002). Though again one has to be careful in dealing with numerical figures, yet no doubt water desalination plays an important role in providing additional water resources to the GCC countries.

The future and sustainability of these resources have been emphasized in many publications that anticipated substantial reduction in the cost of both desalination sea and brackish water. In spite of this anticipation, Bushnak (2001) has forecasted that by the year 2020, an additional desalination water capacity of 1.5 million m^3/day will be needed in Saudi Arabia, which together with the foreseen wastewater treatment plants, requires 74 US$ billion dollars. The author indicated that this huge capital investment could only be met through public and private initiatives as well as international investors.

The intentions and the opportunities in desalination seem to be huge. However, it is high time to consider carefully and plan ahead of time the constraints and negative consequences resulting from the difficulties in integrating this source with the local natural hydrological balance. It is not the purpose of the paper to review comprehensively all of the negative consequences of desalinated water resources, yet, one opts to mention two areas as examples of existing and potential problems. These examples include the imbalances created in the local water budget of many urban areas resulting in catastrophic rising groundwater, and the vulnerability of desalination plans as a single or major sources of supplying domestic water to the bulk of the population of these countries.

Critical rising groundwater levels have been reported beneath and over the surface of many cities in the Gulf Cooperation Council countries (Riyadh, Jeddah, Kuwait, Doha,..etc..) as a result of inbalances in the local water budget created by the importation of huge quantities of desalinated water well beyond the storage/percolation capabilities of the geological formations of these urban areas. In Riyadh, for example, a comprehensive study undertaken by the High Commission for the Development of Riyadh has reported that the main sources of recharge are leakage from the freshwater distribution system, excess landscape irrigation and in-house leakage from both distribution/collection systems as well as storage tanks. The various environmental, health, structural and economic threats of this problem have been emphasized in many reports and published literature (Abu-Rizaiza 1999). Some of these problems are treatable, but in many cases only through strict managerial options that may not be politically easy to take and/or require expensive engineering solutions. Other problems such as the severe contamination of neighbouring shallow aquifers are very difficult to rectify. Many of these aquifers were historically the only source of potable water to these cities and could have been of great use, at this risky time, as strategic-storage reservoirs of artificially recharged freshwater for emergency situations.

Water scarcity, particularly in relation to domestic water supplies, is becoming an important element of national security. This issue is currently an important agenda of most countries of the world, including risks related to wars, accidental pollution and/or to terroristic actions. Nowadays, many cities in the Gulf Cooperation Council countries receive considerable part of their domestic water needs from desalination sources without substantial strategic reserves for emergency situations. This situation needs to be given serious consideration through the ultimate provision of alternative emergency sources. These sources could include well-protected naturally recharged aquifer systems, artificial recharge and recovery from suitable aquifers, as well as man-made storage facilities. The technology of aquifer recharge/recovery systems has substantially progressed within the last three decades utilising both freshwater and reclaimed wastewater. This technology could represent a good storage alternative for the Gulf Cooperation Council countries.

Wastewater re-uses

Wastewater re-use and recycling represent a great potential to help balance deficiencies in water availability in countries with scarce water situations. It is thus logical to observe from Table 3 that wastewater re-use represents an important component of the water resources of all of the six Gulf Cooperation Council countries whether considering Economic and Social Commission for Western Asia's values or the higher values reported by Al-Zubari (1997). According to Al-Zubari, the re-use of treated wastewater has reached about 90% in Qatar, over 60% in Kuwait and United Arab Emirates, while Saudi Arabia which produces the largest quantity utilizes only 22% of this valuable resource.

Wastewater re-use and recycling, however, require effective and feasible technical and economical measures to protect public health and the environment (Anderson, 2001). In this reference, and in Anderson and Dillons (2001) important tabular information were given on quantitative microbiological risk assessment, possible recycled water grades and treatment, and draft water recycling guidelines. Anderson and Dillons concluded that greater human needs can be met with less freshwater that could also have a potential to support three times as much activities before being thrown away. In both papers, the authors called for the creation of a single international framework for water recycling guidelines that links the high technology/ high cost/ low risk, and low technology/ low cost/ considerable risk approaches.

Both papers have recommended six grades of reclaimed wastewater with the top grade (5 stars) reaching a potable level while the sixth grade (1 star) is for restricted use. In between these two limits, the other four grades allow great varieties of applications such as surface water, groundwater, food and non-food crops, pasture and animal fodder, urban and residential, and commercial and industrial. The development of similar European guidelines for wastewater reclamation and use have been strongly called for by the 20 European countries (Angelakis and Bontoux, 2001). It must be emphasized, however, that most Northern European countries have abundant water resources but they view the importance of wastewater reclamation as a protection measure for the receiving environment rather than an additional water supply. The interaction with all of these experiences, however, would be of great benefit for a rational and safe integration of wastewater resources strategies of the Gulf Cooperation Council countries.

Non-renewable groundwater resources

Table 3 clearly indicates that for all six countries, groundwater use is many times higher than the natural groundwater recharge which suggests an increasing abstraction of non-renewable groundwater resources mainly for irrigation purpose. Economic and sustainability concerns on excessive abstraction from these aquifers have been voiced loudly. Economic and Social Commission for Western Asia /Food and Agriculture Organization of the United Nations (1993), for example, has reported that the rate of extraction from fossil water has increased in Saudi Arabia, from 2.36 billion cubic meters in 1980 to 13.5 billion cubic meters in 1990 resulting in water level drops in the order of 100 meters. Table 4, reported by Abdelrahman (2001), however, demonstrates a decrease trend in irrigation water use which reflects a decrease in the rate of abstraction from the non-renewable aquifers. The table also indicates a lowering trend in the percentage of agricultural water use beyond its peak of 94.1% in 1992. This new trend seems to be a result of a wise national policy welcomed by many experts.

Table 4: The Growth of Water Use in Saudi Arabia (MCM). *(Source: Abdelrahman, 2001)*

Year	Domestic and Industrial	Agricultural	Total
1980	502 (21.3)[a]	1850 (78.7)	2 352
1990	1650 (6.06)	25 589 (93.94)	27 239
1992	1870 (5.9)	29 826 (94.1)	31 696
1997	2063 (11.17)	16 406 (88.83)	18 469
2000	2900 (20.57)	11 200 (79.43)	14 100
2010	3600 (19.67)	14 700 (80.33)	18 300

[a] : Values in parentheses are percentage of total

Of course, there is no objection to the principle of utilization of fossil groundwater if realistic economic considerations have been adopted in addition to a clear policy that considers the needs of future generations. An important policy issue that needs to be considered is the question of "food security" at the national, regional and international levels. This issue is extremely important for the Gulf Cooperation Council countries where renewable freshwater is meagre but the current economic strength and low cost energy could advocate a self sufficiency concept that may not be economically and environmentally sustainable. It can be rationally resolved, however, if priorities set for the different uses (domestic, industrial, agricultural, ecosystem protections, ..etc.) are put right and equitable trade agreement reached through a win/win cooperation framework at the levels of the Gulf Cooperation Council countries, the Arab region, and the world at large. This approach can't be reached in the absence of a well-prepared water vision and sustainable plans of action.

THE VISION AHEAD

UNESCO had hosted between the years 1998 and 2000 a worldwide "Water Vision" project which reported that "there is a water crisis today. But the crisis is not about having too little water to satisfy our needs. It is a crisis of managing water so badly that billions of people (and the environment) suffer badly" (Cosgrove and Rijsberman, 2000). The results of the vision project represented the main material for the key agenda items of the second World Water Forum and Ministerial Conference, held in The Hague in March 2000. An important outcome of the Conference was the Ministerial Declaration of The Hague on Water Security in the 21st Century (World Water Council, 2000). The Declaration outlined seven challenges to achieve water security, namely:

- Meeting basic human needs to safe and sufficient water and sanitation. This particular need was given top priority and argued as a basic human right during the Conference deliberations. Water was previously raised and argued as a human right supported by international laws (Gleick, 1998).
- Securing food supply through more equitable allocation of water for food production.
- Protecting Ecosystem through sustainable water resources management.
- Sharing water resources through promoting peaceful cooperation.
- Managing risks from floods, droughts, pollution and other hazards.
- Valuing water to reflect its economic, social, environmental and cultural values.
- Governing water wisely by involving the public and all stakeholders in its management.

For meeting these challenges, the Declaration recognized the need for institutional, technological and financial innovation beyond the "business as usual" approach. The advocated actions are based on integrated water resources management that calls for drawing on all possible resources and depends on full partnership at all levels.

The Vision Project has drawn its conclusions from many sectorial and regional visions among which was the vision that tackled the Arab region including the Gulf Cooperation Council countries. This unpublished vision report has recognized the extreme aridity of this region and the indigenous water scarcity. The identified challenges were the same as those identified in The Hague Declaration.

To meet these challenges, the following issues have been considered in meetings organized by UNESCO, Economic and Social Commission for Western Asia and the World Water Council in Marseille (France, 1999), Beirut (1999), and The Hague (2000):

- Promote additional water supplies from desalination, wastewater re-use and recycling, non-renewable groundwater, water harvesting and imported freshwater.
- Strict conservation practices and strict management of demand to develop a water saving culture in the region.
- Develop at national and regional levels wise policies, management strategies and action plans that allow mitigation of the looming water crisis and secure sustainable future. These should give due considerations to the technical as well as the environmental, legal, institutional and cultural issues.
- Full cooperation and partnership at all levels to promote equitable sharing of transboundary freshwater resources, exchanges of expertise, natural and financial resources as well as trades.
- Enhancing research and education including public awareness.

COPING WITH FRESHWATER SCARCITY

In spite of the growing academic demand for finding globally accepted definitions and indicators for widely used terms such as "water scarcity", "water availability", "water demand management", it is practically unacceptable to stop using these terms in a situation described by the World Vision results as "a world crisis of today". This finding, which was endorsed by the second World Water Forum and meant to describe a worldwide picture, is certainly more applicable in the circumstances of the Gulf Cooperation Council countries. It is a fact that none of the Gulf Cooperation Council countries have enough water resources to secure all their needs. Nonetheless, water requirements for domestic, industrial and a limited portion of food security needs could be obtained through conventional and non-conventional water sources harnessed within the borders of these countries. However, for full food security, the only feasible option is reached through cooperation and trade partnership with neighbouring countries and the world at large.

The five vision directions mentioned previously could serve in a broad way as a route for meeting the challenges of water security in the sub-region. For these directions to be effective in making a real positive change, they need to be translated into policies, management strategies and plans of actions. In a recent study, Economic and Social Commission for Western Asia (2001b) has commented that traditional policies and management strategies in the Economic and Social Commission for Western Asia region (which includes the Gulf Cooperation Council countries) are not feasible if viewed from social, economic and environmental perspectives. The same author has iterated in the same paper that "despite previous policies aimed at achieving self-sufficiency in food, the Gulf Cooperation Council countries are now focusing on policies dealing with the augmentation of water supplies through sea-water desalination and groundwater to provide adequate potable water and improve desalination as a main goal." One could, however, conclude from this publication and experience in the region that there are policies and strategies, yet these policies need to be reviewed and updated to cater to the above five directions as well as the social, economic and environmental deficiencies reported by Economic and Social Commission for Western Asia. Full cooperation and partnership at all levels must particularly be stressed if a change is to be made towards achieving water security. Another deficiency which is noted in the Gulf Cooperation Council countries is the wide variations in the numerical values related to water availability and its various utilizations. This deficiency requires a comprehensive water resources assessment programme as a prerequisite for driving a rational water resources management strategy and for assisting in the review and up-date of the corresponding policies. The International Hydrological Programme of UNESCO, among other international bodies, could assist in the realization of actions directly related to the mentioned directions, particularly through a regional project on water resources assessment linked to the UN-wide World Water Resources Assessment Programme (WWAP) hosted by UNESCO.

A brief description of the mission of International Hydrological Programme, in general, and its current Sixth Phase (IHP-VI) in particular, will be included in this paper for easy reference. However, for more detailed information on IHP-VI, its cross-cutting components and new initiatives, the reader is referred to the UNESCO-IHP water portal: http://www.unesco.org/water.

UNESCO-IHP CONTRIBUTION TO WATER RESOURCES MANAGEMENT IN ARID ZONES

UNESCO's contribution to this subject started as far back as 1950 within the framework of its Arid Zones Programme. Building on that experience, water resources management of arid zones had continued to form a central part of the International Hydrological Decade (IHD, 1965-1974) as well as the International Hydrological Programme (IHP), all through its previous five consecutive phases (1975-2000).

The UNESCO intergovernmental scientific co-operative programmes in water resources (International Hydrological Decade, and its successor the International Hydrological Programme) were established because both the international scientific community and the Member States of UNESCO, realizing that water resources are often one of the primary limiting factors for harmonious development in many regions and countries of the world, saw the need for an internationally coordinated programme. The general objective of the International Hydrological Decade, and later of the International Hydrological Programme, was the improvement of the scientific and technological basis for the development of methods and the human resource base for the rational management of water resources, including the protection of the environment.

Direct considerations were given to arid zone problems in the Fourth Phase of the International Hydrological Programme (IHP-IV), through project H-5-3: "Hydrological research and water resources management strategies in arid and semi-arid zones". It was however, in the Fifth Phase of the International Hydrological Programme (IHP-V, 1996-2001): "Hydrology and Water Resources Development in a Vulnerable Environment", that this subject was well emphasized through a whole theme titled "Integrated water resources management in arid and semi-arid zones". This fifth theme was given the highest priority of all IHP-V's eight themes, in recognition of its importance. The subject has also been dealt with through other themes of IHP-V, namely: Theme 3 on "Groundwater resources at risk", Theme 4 on "Strategies for water resources management in emergency and conflicting situation", Theme 7 on "Integrated urban water management", and Theme 8 on "Transfer of knowledge, information and technology". Water resources management under arid conditions is still a priority of the current Sixth Phase of the International Hydrological Programme

(2002-2007), where the General Conference of UNESCO has explicitly called for special emphasis on arid and semi-arid zones.

The Sixth Phase of the International Hydrological Programme (IHP-VI)

In recognition of the shift in thinking about water from fragmented compartments of scientific inquiry to a more holistic, integrated approach, the general theme for IHP-VI has been defined as "Water Interactions: Systems at Risk and Social Challenges". In defining the critical themes for 2002-2007, it became clear that what had hitherto been missing is a close investigation of water science and policy "at the margins". What happens, for example, at the intersections of distinct components of water resources management?

Some of the interactions to be further investigated, or to be focused on during the Sixth Phase of IHP include those between:

- Surface water and ground water;
- Atmospheric and terrestrial part of the hydrological cycle;
- Fresh water and salt water;
- Global watershed and river reach scales;
- Quantity and quality;
- Water bodies and aquatic ecosystems;
- Science and policy; and
- Water and civilization

Bridging the gaps between these disparate components in an integrated fashion represented the major need to draw the IHP-VI Programme in line with the above-mentioned comprehension of water interactions, technological development of data acquisition and improved modelling of processes and interactions. The relevant IHP-VI topics on hydrological research, water resources management and education have been formulated in five themes; with the transition and interaction from the global scale to the watershed scale being the overall need for knowledge, information and technology transfer. The five themes are:

Theme 1. Global Changes and Water Resources
Theme 2. Integrated Watershed and Aquifer Dynamics
Theme 3. Land Habitat Hydrology
Theme 4. Water and Society
Theme 5. Water Education and Training

Complementary to these five themes are two cross-cutting programmes and four new initiatives. The two cross-cutting programme components, FRIEND (Flow Regimes for the International Experimental and Network Data) and HELP (Hydrology for Environment, Life and Policy), have been identified which, through their operational concept, interact with all of the five themes.

The New Initiatives

The new initiatives include:

- The UN World Water Assessment Programme ((WWAP): This programme is a UN-wide initiative hosted by UNESCO. Among its major product is the "World Water Development Report" (WWDR) which is a periodic review, continuously updated, designed to give a credible picture of the state of the world's water resources and their development. It will contain indicators and analysis of case studies that will identify and diagnose:
  - The needs for and uses of water;
  - The state of the world's freshwater resources;
  - The broad organizational and socio-economic context of their utilization;
  - Current problems and emerging threats to the resource base;
  - Approaches to sustainable development, management and use of freshwater resources.
- The project "From Potential Conflicts to Potential Cooperation" (PCCP): The main objective of this project is to foster cooperation between stakeholders in the management of shared water resources and mitigate the risk that potential conflicts turn into real ones.
- The Groundwater Initiatives: The two Groundwater initiatives include a project on "Internationally Shared Aquifer Resources Management (ISARM)". The International Hydrological Programme, in co-operation with many partners, have launched this project in recognition of the fact that there are many large regional aquifers, shared by more than one country, that represent a substantial and secure supply for many beneficial socio-economic uses and are thus critical for national and regional water security. The aim of this interagency initiative is to improve the regional co-ordination as well as the understanding of the scientific, socio-economical, legal, institutional and environmental issues of the management of shared aquifers.

The second initiative is the establishment of an "International Groundwater Resources Assessment Centre (IGRAC)" under the auspices of UNESCO and the World Meteorological Organization (WMO). The Center should act as a catalyst for stimulating and focussing national efforts in groundwater system monitoring and assessment.

- The UNESCO IHE Institute for Water Education and Training and Regional Centers: One area that received great emphasis and demand for immediate action from UNESCO was the strengthening of "Fresh Water Resources Education". As a response to this, the General Conference of UNESCO has approved a proposal from the Government of the Netherlands to establish the UNESCO-IHE for Water Education (IHE) in Delft as a UNESCO Center of Excellence in water resources education, starting January 2002. In the same session, the General Conference of UNESCO has also approved the establishment, under the auspices of UNESCO, of a "Regional Center for Training and Water Studies of Arid and Semi-arid Zones (RCTWS)" in Egypt and a "Regional Center for Urban Water Management (RCUWM)" in Iran.

Potential links

A detailed review to the five themes of IHP-VI, the two cross-cutting programmes and the four initiatives would reveal great potentials and opportunities for contributing significantly to the previously identified directions for meeting the challenges of achieving water security in the GCC countries. Table 5 identifies in a broad manner the IHP-VI's themes, cross-cutting components and new initiaitves where close linkages exist with the five major issues identified from regional and sub-regional vision exercises.

It is hoped that the current preliminary contacts with key governmental and non-governmental entities in the Gulf Cooperation Council countries may achieve positive results in realizing specific projects utilizing the International Hydrological Programme opportunities to respond to these issues and directions. A Gulf Water resources assessment programme could be a starting nucleus for this collaboration.

CONCLUSION

Per capita, annual renewable water resources in the Gulf Cooperation Council countries is by far the lowest compared to all regions of the world. Utilizing the economic prosperity of the last three decades and the low energy cost, these countries have managed to cope with their indigenous water scarcity through policies that lead to the provision of considerable quantities of additional water supplies from desalination plans, non-renewable groundwater and increasing utilization of treated wastewater. These policies have caused considerable concern among experts, particularly with regard to their sustainability from social, economic and environmental considerations as well as risks related to relying on desalination as a major source of water supply. Corrective directions have been suggested through a regional vision exercise conducted in the framework of the "World Water Vision Project". The reviewed features of UNESCO-IHP actions and its new initiatives could provide an opportunity for contributing to the corrective directions. Comprehensive water resources assessment is identified as an urgent action and a pre-requisite for reviewing the current policies and for drawing rational management strategies and plan of actions. Rational water management requires a holistic and integrated approach that could focus on specific goals, yet essentially considers all related parameters and their interacting elements utilizing the current development in trans-disciplinary knowledge and advanced tools.

Table 5: Potential Links between Major Water Issues in the Gulf Countries and Focal Areas of IHP-VI and its Initiatives.

Important Water Issues in Gulf Countries	IHP-VI Themes of Relevant Focal Areas (Theme number)	Cross-cutting Components and New Initiatives of IHP-VI
Promote additional water supplies	1, 2, 3, 4, 5	FRIEND, HELP, ISARM, IGRAC, IHE, RCTWS, RCUWM
Conservation and demand management	1, 3, 4	WWAP, IHE, RCTWS, RCUWM
Policies, management, strategies and action plans	1, 2 , 3, 4	WWAP, HELP, PCCP, ISARM, IGRAC, IHE, RCTWS, RCUWM
Cooperation and partnership	1, 2, 3, 4, 5	WWAP, PCCP
Research, education and public awareness	1, 2, 3, 4	WWAP, FRIEND, HELP, ISARM, IGRAC, IHE, RCTWS, RCUWM

REFERENCES

Abderrahaman, W.A., 2001. Energy and Water in Arid Development Countries: Saudi Arabia, a Case-study. *Arab Water World* XXV(4): 48-51.

Abu-Rizaiza, O.S., 1999. Threats from Groundwater Table Rise in Urban Areas in Developing Countries. *Water International* 24(1): 46-52.

Al-Zubari, W.K., 1997. Towards the Establishment of a Total Water Cycle Management and Re-use Programme in the GCC Countries. *The Third Gulf Water Conference, Muscat, Qatar*, 8-13 March 1997, p. 1-16.

Anderson, J.M. and Dillons, P.J., 2001. New Approaches to Water Recycling and Trading. 6th National Water Conference, Melbourne, 26 July 2001, Water Research Foundation of Australia, 1-12.

Anderson J., 2001. Prospects for International Guidelines for Water Recycling. Water 21, *Magazine of the International Water Association*, August 2001, 16-21.

Angelakis, A.N. and Bontoux, L., 2001. Wastewater Reclamation and Re-use in Eureau Countries. *Water Policy* 13(1): 47-59.

Bremere, I., Kennedy, M., Stikker, A., and Schippers, J., 2001. How Water Scarcity will Effect the Growth in the Desalination Market in the Coming 25 Years. *European Conference on Desalination and The Environmental Water Shortage, May 28-31 2001, Lemesos (Limassol), Cyprus, Proceedings I*, 138: 7-15.

Bushnak, A.A., 2001. Investment Needs and Privatization of Water Services in Saudi Arabia. *Arab Water World*, July/August, XXV(4): 45-47.

Economic and Social Commission for Western Asia (ESCWA), 2001a. Development of Freshwater Resources in the Rural Areas of the Economic and Social Commission for Western Asia Region Using Non-conventional Techniques. Document E/ESCWA/ENR/1999/16, 10 November 1999, UN/ESCWA, Beirut, Lebanon.

Economic and Social Commission for Western Asia (ESCWA), 2001b. Current Water Policies and Practices in Selected Economic and Social Commission for Western Asia Members Countries, Document E/ESCWA/ENR/1999/15, 28 October 1999, UN/ESCWA, Beirut, Lebanon.

Economic and Social Commission for Western Asia /Food and Agriculture Organization of the United Nations, 1993. Agriculture and Development in Western Asia. Joint Publication (Arabic and English), N° 15, Dec. 1993.

Cosgrove, W.J. and Rijsberman F.R., 2000. Making Water Everybody's Business. World Water Vision, World Water Council.

Gleick, P.H., 1998. The Human Right to Water. *Water Policy* 1(5): 487-503.

Hoekstra, A.Y., 1998. Appreciation of Water: Four Perspectives. *Water Policy* 1(6): 605-622.

Salih, A.M.A., 1994. Water Scarcity in the Arab Region: Options for Survival. *Proceedings of the Eighth Islamic Academy of Sciences' Conference, Water in the Islamic World: an Imminent Crisis*, 5-9 December 1994, Khartoum, Sudan.

Salih, A.M.A., 1998. Integrated Water Management in Arid and Semi-Arid Zones. *Proceedings of the International Conference on Coping with Water Scarcity, the African Division of the International Association of Hydraulic Research* (IAHR-AD), Hurghada, Egypt, 26-28 August 1998.

Salih, A.M.A. and Ghanem A.H.M., 2002. Sustainable Management of Wadi Systems. Hydrology of Wadi Systems, UNESCO-IHP Series, Technical Documents in Hydrology, N° 55, UNESCO, Paris.

Shiklomanov, I.A., 1998. World Water Resources: a New Appraisal and Assessment for the 21st Century. UNESCO-IHP Publication, UNESCO, Paris.

The Hague, 2000. Second World Water Forum and Ministerial Conference, The Hague, The Netherlands, 17-22 March 2000.

WMO, 1997. Comprehensive Assessment of the Freshwater Resources of the World. UN-wide publication issued by WMO on behalf of participating agencies.

World Water Council, 2000. Final Report: The Hague - From Vision to Action. *Second World Water Forum and Ministerial Conference*, The Hague, 17-22 March 2000.

Salinity Management for the Landscapes and Rivers: The Murray-Darling Basin in Australia

ROBERT J. NEWMAN
Principal, Catchment Management Consulting, 57 Powell Street, Berri SA 5343, South Australia.
E-mail: bobnewman@ozemail.com.au

ABSTRACT: The Australian scene is typified by vast weathered landscapes in an arid climate – the wide brown land. Two centuries of agricultural developments have created massive changes in this landscape that have disturbed the hydrologic balance. Current land-uses are filling the landscape with excessive recharge bringing ancient salts to the surface. Salinity is evolving as a national crisis for the landscape and the rivers.

The flat terrain combines with low rainfall and high evaporation to create a natural trap for salts. There is little scope for natural discharge to the sea. The early Australian explorers searched for inland lakes and seas but found only vast salt lakes, which are most often dry. The Murray-Darling Basin in the South East of the continent drains $^1/_7$ of Australia. This Basin generates 40 % of the nation's agricultural production. The twenty major rivers of the Basin provide water for 3 million people and 75% of the nation's irrigated agriculture.

The paper describes the extent of the salinity problems considered over the next one hundred years. The early successes of collaborative institutional arrangements within a federated governance system are described. The early solutions have involved engineering works along rivers and in the irrigated regions, however it is increasingly recognised that broad-scale landscape change in the dryland, coupled with reducing demand for water will become necessary for sustainable outcomes over coming decades.

The solutions rely on taking a shared responsibility for offsetting the impacts of past planning decisions whilst ensuring full responsibility for future actions. Six governments have established inter-linked salinity strategies involving new funding commitments and onerous compliance conditions.

The Australian setting

Australia is the second driest continent after Antarctica. Australia is a single nation continent that should be able to enjoy effective natural resource decision-making. Within the last century we have exploited most of the readily available useful water resources of the temperate climate regions. The European settlers over the past two centuries presumed to understand the agricultural potential of these regions. The North of Australia is tropical and has not been developed to the same extent despite the abundant water resources

However the temperate regions of Australia are significantly different from those in Europe. The Australian landscape is ancient. It has not been rejuvenated by glaciation and uplift. The climate tends towards arid with low rainfall and high evaporation. The terrain is generally flat, low lying with deeply weathered sub-strata. Over geologic time salts from rainfall (cyclic salt) have accumulated within the landscape. Natural salt lakes (primary salinisation) are commonplace and the rivers naturally experienced variations in salinity levels. The climate varies significantly over seasons and also over periods of several years. The flow regimes consequently vary dramatically between floods and droughts. The natural vegetation had evolved to make the best use of the limited rainfall and in so doing concentrates the salts. The soils are mostly lacking in nutrients. Australia is a difficult place to farm and many mistakes have been made since the pioneering days. This paper concentrates on the future salinity implications of both the past and future land use changes.

European style farming developments tend to rely upon predictable climate and hydrologic cycles. In the early years of opportunistic agriculture farmers learned to cope with such variations but as closer settlement and increasing farm debt has occurred, farmers have become increasingly optimistic. This has placed increasing stress on the landscape capability.

It has now become apparent that the health of the inland landscapes and the associated river systems are in decline. Nowhere is this more evident than in the Murray-Darling Basin. We now abstract close to 75% of the natural stream runoff in the Basin. The Murray Mouth is now in artificially created drought conditions for half the time. That is the frequency of drought-like flows has increased from 5% under natural conditions to more than 50% under current development conditions. (Murray-Darling Basin Ministerial Council, 1995)

Water Resources Perspectives: Evaluation, Management and Policy. Edited by A.S. Alsharhan and W.W. Wood.
Published in 2003 by Elsevier Science, Amsterdam, The Netherlands, p. 141-154.

Natural Resource Management decision making

Following the first settlement in 1788, the earlier colonies and territories became federated in 1901 when Australia became a single nation. State and Federal Governments have legislative responsibilities under separate constitutions. Natural resource management decisions are essentially State obligations. The States have competed to exploit their natural resources whether it is for agriculture or minerals. Within two hundred years a very high proportion of the landscape was cleared and developed for agriculture. The salinity problems have emerged over the last fifty years. However it is noteworthy that Wood (1924) first recognised the potential for land and stream salinisation in 1924.

Water does not respect the political boundaries and consequently issues such as over exploitation and pollution have created tensions amongst the States. It is generally water issues that have encouraged a collaborative arrangement between these Governments. In fact the water sharing debate between New South Wales, Victoria and South Australia was a key driver towards federation in 1901.

Local Governments operate within the States' jurisdictional frameworks and are responsible for local planning decisions including land use. However the implications of landuse change occurs at a broader scale and is consequently not generally taken into account in the local government decision-making.

The National Water Reform Agenda

In 1994 the Council of Australian Governments (CoAG) national competition policy initiative examined issues relating to the supply of water and management of water resources. A national task force on CoAG water reform skilfully applied a form of competitive federalism (Aeuckens and Hoey, 1998) to maximise the synergies between the state jurisdictions, which have primary responsibility for water. The key elements of this water reform agenda include:

- Cost recovery and pricing
- Institutional reform
- Water allocation, property rights and trading
- The environment and water quality
- Public consultation and education

The CoAG water reform agenda throws down significant challenges to the water jurisdictions whilst promising the States significant benefits in both encouraging economic development and reducing the impacts of water use. In return the National Competition Policy offers significant financial incentives to the States. Significant progress has already been made. Nevertheless the environmental outcomes are falling behind. Property rights for water allocations have been assigned but the sustainable yield has not been well defined. The pressure to return water for environmental flows will mean that governments will need to consider buying back allocations to protect the riverine ecosystems.

Integrated Natural Resource Management for the 21st century

Recently there has been a trend towards regional scale planning based upon hydrologic catchments. Most States are developing regional catchment management organisations. The Federal Government is using its financial influence to foster these regional organisations. The latest federal funding package will be assigned through block funding to regional catchment organisations. Some States are resisting this pressure.

Regional catchment bodies are not intended to be another layer of governance, but rather a new coordination framework with the emphasis on natural resource management. In some States the catchment bodies have been granted the legal capacity to generate revenue and to establish regulatory frameworks whereas elsewhere they are purely coordinating bodies with limited agendas. State agencies are retaining control over the funding programs.

The Murray-Darling Basin Initiative *(the Initiative)* is a unique example of collaborative management at a Basin scale that crosses four state boundaries and includes six governments. The area covered is one seventh of the continent and includes four States and the Australian Capital Territory within which the national Capital, Canberra, is located. This is not a regional body but a collaborative initiative amongst the five Basin States and Territories and the Commonwealth Government. It includes twenty major regional catchments.

Similar arrangements are emerging for other major basins in Australia and elsewhere. An example is the Eyre Basin, which drains into the outback centre of Australia. However that basin is predominately extremely arid. Lake Eyre, an inland salt lake at the end of the system, has filled only twice in recent history.

These catchment management approaches are increasingly involving a higher level of community involvement in decision-making. The section on the institutional responses includes further description of the evolution in decision-making processes

Progressively, the natural resource management agenda is becoming interwoven with the regional development agenda. The days of the environment playing 'catch up' are nearly over. Nevertheless, the challenge for restructuring the landscape into fully sustainable uses is still before us. Needless to say there is a strong resistance by existing land and water holders to participate in changes to their

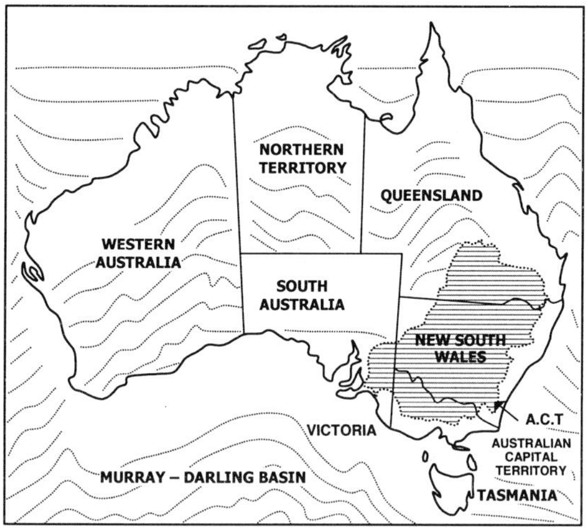

Figure 1. Salinity management for the landscapes and rivers of the Murray-Darling Basin in Australia

lifestyle even when productivity declines to the extent that the land use is no longer profitable. The tendency is for landholders to obtain supplementary income off-site. There is an increasing drive towards land stewardship arrangements which support the ongoing use of land in private ownership under mutually agreed arrangements that are more sustainable. However these arrangements are very difficult to secure over the longer term.

There is also increasing recognition of the need for a higher level of 'duty of care' to be applied to existing land and water users. Best management farming practices are being identified and promoted across the Basin. This is applying to both the irrigated and dryland regimes.

The Murray-Darling Basin

In this paper the author draws from experience in the Murray-Darling Basin. This Basin is one of the great river basins of the world. It drains one seventh of the area of Australia through twenty major rivers. The Murray-Darling Basin generates forty percent of the total agricultural production of the nation, including 75% of the irrigated production and provides water to twenty percent of the nation's population.

The River Murray is some two thousand five hundred kilometres long. However, when comparing with the great rivers of the world it becomes apparent just how arid is Australia (Table 1). The yield from the Basin is not only much less than all other basins but is also very variable from year to year.

Table 1. The World's Major Rivers.

River	Country	Length (km)	Catchment area (km^2)	Annual flow (GL/a)	Annual yield		
					ML/ha	ML/km^2	mm
Amazon	South America	6570	6,150,000	5,518,800	8.97	897.4	897.4
Yangtze	China	5980	1,827,000	1,014,700	5.55	555.4	555.4
Congo	Africa	4700	3,460,000	1,293,000	3.74	373.7	373.7
Yenisey/Selenga	Russia	5870	2,619,000	565,700	2.16	216.0	216.0
Zambezi	Africa	3500	1,330,000	220,752	1.66	166.0	166.0
Indus	India	2900	1,166,000	157,680	1.35	135.2	135.2
Mississippi/Missouri	USA	6020	5,980,000	405,100	0.68	67.7	67.7
Hwang Ho	China	4840	771,000	51,100	0.66	66.3	66.3
Nile	Africa	6690	2,802,000	88,500	0.32	31.6	31.6
Murray - Darling	Australia	2560	1,063,000	14,700	0.14	13.8	13.8

Source: Britannica 96 and Murray-Darling Basin Commission.

Clearly this makes industries that rely upon a uniform water supply year after year very vulnerable. Such industries are seeking the most reliable supply and require preference ahead of others during drought. They also seek water of consistent quality. The CoAG water reform agenda is driving water users into higher value production including permanent plantings for viticulture and horticulture.

Salinity in the Murray-Darling Basin

Salinity in the Murray-Darling Basin is essentially a natural phenomenon. When explorer Captain Sturt first encountered the Darling River in 1829 it was too salty to drink. It was a period of drought and the only stream flow was the base-flow resulting from groundwater seepage.

The Basin is a very effective trap for salt. The Basin contains in excess of 100,000 million tonnes of salt. Each year, three million tonnes is added by rainfall whilst only two million tonnes is exported to the sea. Over most of the Basin, the evaporation rate far exceeds the rainfall. Consequently the pure water is lost and the salt is concentrated. This provides the initial process whereby salt is concentrated in the landscape. The natural vegetation of the Basin had evolved to become very effective at drawing water from great depths, due to the need to survive long periods of severe drought. Consequently the water that percolates through the extended root-zone towards the water table contains all the cyclic (atmospheric) salt in only a tiny fraction of the original rainfall water.

A further concentration process occurs as the regional groundwater migrates slowly across the vast plains of the lower Murray and Darling rivers. The water, which is already moderately salty, is occasionally exposed to the atmosphere in discharge zones, which are sometimes salt marshes, and sometimes-salt lakes. The concentration of salt at the surface causes the water to become heavy and drop back through the groundwater regime, mixing with the lesser saline water in the deep, poorly consolidated, sediments (up to 600m deep) and continuing its slow flow, down gradient, towards lower reaches of the Basin. In this region, there are no tributaries, just braided anabranch systems. The river gradient is less than 100m in around 2000 km of river distance.

Consequently, over recent geologic time, the groundwater system in the lower Murray (Mallee zone) downstream of Swan Hill is now very saline, generally around half the salinity of the sea. The Mallee zone region stretches more than one thousand river kilometres from the Murray mouth. The region is typified by stunted Mallee vegetation that can thrive in the very arid regime with rainfall generally less than 350 mm. It is a dryland tree that can survive extended drought and high soil salinity. The floodplain is a relatively confined, geomophologically discrete, riverine corridor that survives on the natural flooding of the river. It tends to be between fifteen and two kilometres across and is a highly valued ecosystem that is now under significant threat due the changed river flow and salinisation.

The lower Murray is itself a natural groundwater drain, draining about one quarter of the whole Basin. Unlike most river basins in the world, the Murray-Darling Basin does not have a vast delta at its seaward end that allows some groundwater release to the sea. There is no other way that salt can reach the sea apart from through the mouth of the Murray at Goolwa in South Australia. The natural ecosystems of the floodplain had accommodated the variations in flow and consequent variations in salinity.

Whilst the upper Murray minimum discharge is 25% of the median, the value for the Darling River is just 0.04% (Crabb, 1997). However these systems cannot easily adapt to new changes over short time periods. As a natural drain, the River Murray always received regular annual flushes. Nowadays, these flushes are far less frequent due to the increased diversions across the Basin.

The upland tributary rivers of the Basin also suffer similar extremes of flow. Under natural conditions salinity in the landscape was not very evident because there is generally a higher rainfall regime and salt cannot accumulate so readily. Nevertheless under a closer examination there is salt in these uplands and some of the ephemeral streams experience significant salinity levels. Some of the major tributary systems, (eg the Paroo, Wimmera and Lachlan Rivers), do not reach the primary rivers but terminate in inland marshes. Whilst these are clearly susceptible to salinisation it has not been particularly evident. The relatively infrequent flushing events, which transport salt further downstream, must have been sufficient to create a low-salinity equilibrium. For example the Paroo has flowed into the Darling only once in a century. Other tributary systems have a groundwater outlet leading salt towards the lower Basin. This also maintains a flux of salt away from the terminal marshes. This is the case for the Wimmera River in the Victorian Mallee zone, which terminates at Lake Albercutya. These processes are entirely natural and are referred to as primary salinisation.

Salinity in relation to changed land and water uses - secondary salinisation

Secondary salinisation is the result of anthropogenic intervention in the landscapes. As the development of the river system occurred, through the construction of dams and weirs, increasingly regulated flows were introduced. The natural extremely low flows with associated high salinities were avoided. However as diversions increased the salinity levels rose again. The use of water for irrigation caused further mobilisation of the regional saline groundwaters. This exacerbated the induced salt loads into an already stressed river system.

The salinity regime for the Murray is further disturbed, due to the diversion of nearly 75% of the natural flow for consumptive use. Inadequate water is left to dilute and transport the natural salt accessions to the sea. At the same time, we are exacerbating the situation by inducing additional salt back to the river through inefficient irrigation and drainage systems. A fundamental indicator of the sustainability of irrigation ventures is that they should export only the salt that is drawn from its water supply, - no more and no less.

Irrigation regions export salt to the rivers through surface and sub-surface drainage systems and also through the groundwater systems. Current data suggests that some areas are discharging five times the salt imported whilst others are simply accumulating salt. From a salinity perspective, the limit to irrigation development is the ability of the rivers to remove the residual salt. That limit has been reached, or even exceeded, for the Murray-Darling Basin.

Salinity responses in the Murray-Darling Basin

When a severe drought occurred in 1967 the salinity levels in the lower Murray below Swan Hill in Victoria and throughout South Australia rose to levels that caused severe losses of permanent plantings of citrus and stone fruit.

This event caused the Basin States- South Australia, Victoria and New South Wales to initiate joint studies into the salinity processes for the Lower Murray. The (then) River Murray Commission engaged consultants Gutteridge, Haskins and Davey, (1970) to undertake the Murray Valley salinity investigations. Later another significant study (Maunsell and Partners, 1979) focused upon coordinated action. These studies placed a focus on the economic impacts of salinity to human users of the water for both irrigation and urban consumptions. The choice of remedies were assessed in terms of benefits and costs to water users.

The obvious salt inflows caused by piped irrigation drainage returns and floodplain drainage disposal basins were identified, and some of these diverted to 'out-of-valley' salt disposal basins. Salinity control schemes were constructed in Victoria and New South Wales. The Noora Drainage Diversion Scheme constructed by South Australia in 1984 involved a major pumping scheme to divert salt from irrigation drainage away from the river valley. These schemes were State initiatives.

The River Murray Commission constructed the Rufus River Salt Interception Scheme as the first joint works funded equally by the partner governments. This scheme addressed the salt intrusions caused by the operation of Lake Victoria as a regulating storage, which had mobilised the underlying regional saline groundwater.

The salinity responses of the lower Murray is described in many technical publications but the discussion paper by Meacham (1984) provides a lucid summary of the state of knowledge at that time. Whilst there was recognition of the implications of broad-scale dryland clearing on salinity, it is apparent that the main focus of attention was on the implications of irrigation and drainage close to the rivers.

Dryland salinity was emerging as a cause for concern to dryland agriculture across the Basin. The actions pursued were mostly associated with the local salinisation effects causing losses of dryland agricultural productivity. A report was prepared by the Murray-Darling Basin Ministerial Council in 1993, which described the fundamentals and several case studies, and examined management options and implementation arrangements. The implications for rivers, public and private infrastructure and regional economies were identified.

Having learnt to appreciate the salinity issues associated with irrigation and drainage in the lower Murray, the Initiative is beginning to address the broad-scale dryland salinity problems that are now emerging in the dryland farming regions around the Basin margins. The southern and eastern upland margins of the Basin are the source for the flows. However it is now apparent that this landscape is also underlain by salt. During earlier arid geologic periods, with lower sea levels, salt was blown inland by the westerly winds and deposited in the upper landscapes of the Basin uplands. Generally the salt is still retained in the groundwater systems, albeit at much lower salinity levels and in much more erratic spatial extent. There are some geologic situations where salt is stored in shallow clay profiles above the regional water tables. Such salt can still be mobilised through increased recharge, but it is the recharge water itself that picks the salt up and transports it laterally down-slope to discharge locally.

This salt is now being mobilised through dryland agriculture. Dryland salinity is a problem which has been initiated more than a century ago but which has only recently emerged as a recognised threat. The replacement of the native perennial vegetation with annual crops which use much less of the rainfall have caused increase recharge to raise groundwater levels. The landscape is now 'leaking ' more than under natural vegetation conditions. As watertables rise to within two metres of the surface, salinisation of the soils begins to occur even where the groundwater salinity is relatively low. The reduced productivity exacerbates the recharge causing salinisation to accelerate. As salt scalds develop, rainfall wash off takes the accumulated salt to the streams and rivers. Streams in the upland landscapes are becoming saline at an alarming rate.

Given such a gloomy picture how can we survive?

It is not clear that we will! We may have to learn to live with the consequences of salinisation and artificial drought. However at least we now understand the fundamental issues that we are facing and have now started to quantify the extent and scope of the problem. The recognision of a problem is the first step to its solution. Through the 1999 salinity audit, we now have a comprehensive first stage prediction of salinity in the Basin.

We are doing better than that; we are beginning to find the community pressure and political will to correct the situation. Over the last twenty years, the policy settings have begun to avert these threats.

The more recent responses to the increasing salinity threat are described in later sections.

Institutional responses; the Murray-Darling Basin 'Initiative'

Governments began to jointly address the river salinity problem, albeit after some bitter interstate debate attempting to assign blame. The debate even resulted in claims by the government of South Australia against the Government of New South Wales in the Supreme Court of New South Wales. That claim was eventually resolved politically, with the recognition that actions undertaken in one part of the Basin could impact further downstream. This concept was enshrined in the first Murray-Darling Agreement (1992).

Following the acrimonious debate of the early eighties, the Murray-Darling Basin Ministerial Council (The Council) was established in 1985, to take responsibility for the management of the whole Basin, dealing with one million square kilometres of land and water in four states. The Council replaced the earlier River Murray Commission, which had membership at the bureaucratic level not the political level. The River Murray Commission had developed the water resources of the Murray and Lower Darling through water sharing arrangements between the States since 1914 and the construction of dams and weirs.

The Ministerial Council's charter is:

"to promote and coordinate effective planning and management for the equitable, efficient and sustainable use of the water, land and other environmental resources of the Murray-Darling Basin".

The Council operates under an Agreement, which is empowered by legislation in each of the participating Governments. Queensland joined the Agreement in 1993. Decisions are made on a consensus basis.

A Community Advisory Committee drawn from a wide range of backgrounds across the Basin supports the Council. The Murray-Darling Basin Commission is the executive arm of the Council and is served through a modest office based in Canberra. Generally the States provide the major input to the Commission's operations through arrangements as contracting governments and constructing authorities.

In 1989 the Council prepared a Natural Resources Management Strategy that placed a strong emphasis on community awareness, education and involvement in natural resource programs. Initially communities of Common Concern were encouraged to form and to undertake environmental rehabilitation projects. These community groups grew into local action planning groups preparing more strategic resource management plans.

Over the past decade, a very significant effort of research and investigation has been mounted. In recent years the available data has been marshalled, so that the scale of issues can be quantified and large-scale options evaluated. The integrated catchment management approach is starting to take effect. A key feature of the integrated catchment management approach is the negotiation of targets for resource outcomes. In 1995 the MDBMC agreed to a Cap on water diversions across the Basin. This bold decision drove a need for enhanced integrated action, which has accelerated since then, (Blackmore 1994).

In June 2001, after extensive public consultation, the Council endorsed a policy statement on Integrated Catchment Management (ICM) for 2001 to 2010 (Murray-Darling Basin Ministerial Council, 2001a). This statement reinforces the need for strategic targeted action across a broad range of issues. It provides clarity on the characteristics of well-developed catchment strategies and action plans. It sets out a timetable for action that is constrained by the need to gather additional data and to undertake further research into solutions. Salinity is the first Basin strategy to be endorsed under this policy. The Council's Basin Salinity Management Strategy (Murray-Darling Basin Ministerial Council, 2001b) was released in September 2001, once again after extensive consultation.

The ICM policy statement provides a commitment that was difficult to negotiate:

"We, the communities and governments of the Murray-Darling Basin commit ourselves to do all that needs to be done to manage and use the resources if the Basin in a way that is ecologically sustainable."

An important value of the 'strategies' of the Council is that they are reported upon publicly each year. The Ministers of each participating Government are called to account. This tends to maintain the momentum behind each strategy and avoids them becoming yet another volume on the shelf! In just thirteen years, the Council have made some very challenging and courageous decisions.

National Environmental Programs

In 1989 the National Farmers Federation and the Australian Conservation Foundation initiated an unprecedented collaboration to encourage the Federal Government to participate in a 'Decade of Landcare'. For the first time the Commonwealth Government volunteered a major funding program. The National Landcare Program was initially set at $340 million over six years to natural resource management. A change of government in 1995 and the partial privatisation of the national telephone carrier (Telstra) provided a further $1.3 billion over the next six years (the Natural Heritage Trust). These programs were broad based community support initiatives that covered a wide range of natural resource outcomes including salinity.

A major outcome of these programs is the development of some 4000-community Landcare groups nationally. However there is some criticism that the physical programs have been ad-hoc and not strategically useful to provide broad-scale outcomes.

The prime initiators of these programs, Toyne and Farley (2000) criticised these programs for having become over politicised and involving cost shifting by all levels of government away from earlier core programs towards these grant based community programs. These authors have suggested that the cost of rehabilitating Australia's landscape will involve a large and strategic investment of public and private funds amounting to $65 billion dollars over ten years which might be partly funded by a national landcare tax of 1%.

To date Governments have not yet supported this scale of investment, however a new federal program, the National Action Plan for Salinity & Water Quality (NAPS&WQ) has been announced involving $700 million over seven years to be matched dollar for dollar by the States. The Federal Government is using its financial programs to promote regional catchment management organisations through block funding longer-term programs rather than funding individual projects. However at the time of publishing some States have yet to commit to the program.

The Murray-Darling Basin Commission Salinity and Drainage Strategy (1988)

One of the first actions of the Council in 1988 was the adoption of the Salinity & Drainage Strategy (Murray-Darling Basin Commission, 1989). This strategy has played a very significant role in the management of irrigation-induced salinisation of the rivers and the irrigated lands in the lower Murray region. The kernel of the Strategy (Salinity and Drainage Strategy, 1989) was that it made each jurisdiction accountable for the salinity impacts in the River Murray of any new actions or policy changes that were undertaken by those jurisdictions after the commencement of the Strategy.

This Strategy recognised that there would be further salinity impacts resulting from past practices, such as land clearing and intensive irrigation, and from the continued application of existing water allocation policies and it was therefore agreed that jurisdictions would not be held accountable for those impacts.

The Strategy allowed for a system of salinity credits and debits, the credits could be used to offset the salinity impacts from deliberate new actions for which the jurisdictions were accountable, (the debits). Under this system, jurisdictions could undertake actions, which impact on salinity, provided that they had previously attained the required credits. Undertaking actions or implementing new policies that improved salinity levels in the River Murray could attain credits.

Jurisdictions can attain credits through an initial program of engineering works that would reduce the average salinity at Morgan (South Australia) by up to 80 EC (EC) units. By contributing to the joint funding of those works, New South Wales and Victoria were each credited with up to 15 EC Units, as a salt disposal right. South Australia and the Commonwealth were also joint contributors to the funding of the initial program of works but did not take any credits. (An EC unit is a rapid measure of salinity. One EC unit equates to 0.6 mg/L total dissolved salts for the saline waters of the Basin).

The Murray-Darling Basin Commission maintains a register of the status of each jurisdiction's credits and debits. The register is regularly reviewed and endorsed annually. The impact of this strategy is illustrated on figure 2. Salinity Trends at Morgan. It is clear that the strategy has brought some respite but that the underlying trend is still daunting.

After a decade of working under the Salinity and Drainage Strategy, 1988 a comprehensive review was undertaken. The review demonstrated that the Salinity and Drainage Strategy had been very successful at controlling the impact of new actions after 1988. The objective of reducing the salinity at Morgan to below 800 EC units for 95% of the time has substantially been achieved. Peak salinities exceeding 800 EC have been reduced from 38% of the time to just 8%.

This success had been achieved despite the underlying trend being much greater than originally anticipated. The 1999 review of the Salinity and Drainage Strategy (1989) showed that the trend had been 4 EC per year rather than 1.5 EC per year as

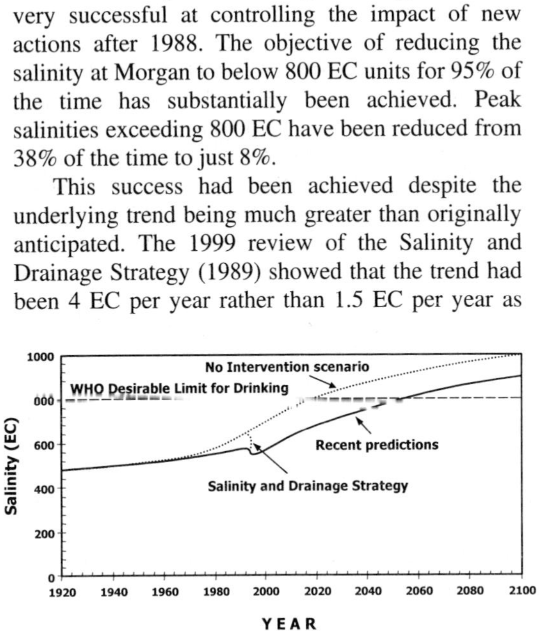

Figure 2. Forecast (Oct 1999) of Salinity at Morgan

previously thought. It was recognised that the emerging dryland salinity problem would be far more severe than considered twenty years earlier.

Murray-Darling Basin Ministerial Council Salinity Audit 1999

Since 1988 numerous local studies have been undertaken of salinity outbreaks in the landscape and streams across the Basin. In 1997 a study of the historic salt trends in the rivers had indicated that many streams were exhibiting increasing salinity levels. The Murray was one of only a few which exhibited decreasing trends.

The review of the Salinity and Drainage Strategy (1989) lead to the need to understand the salinity issues around the Basin more completely. A study of the future salt loads predicted to be mobilised into the tributary rivers of the Basin was broadened to become an audit of salinity Basin-wide. Around fifty per cent (50 million hectares) of the Basin was studied. This is the area that has suffered the most significant land-use changes and was likely to generate 80% of the future salinisation in the Basin.

This audit provided the first comprehensive assessment of salinity covering the rivers, landscapes, agricultural productivity, ecologic impacts in the rivers and dryland, and impacts on public and private infrastructure,

In summary, the Salinity Audit suggested that the extent of secondary land salinisation within the Basin would rise by around five-fold over the next fifty years. The overall extent would represent around 8% of the surveyed landscape with up to 40% of some catchments at risk. Much of the land at risk represents the prime agricultural land, which generally lies lower in the landscape. Six of the major tributary rivers were predicted to experience salinity levels that will be beyond the thresholds for human and ecologic tolerances.

Following the Murray-Darling Basin Salinity Audit, a National Dryland Salinity Audit (NL & WA, 2001) has since been undertaken which has provided further analysis of the infrastructure and environmental assets at risk.

The costs of salinity

Estimates of the cost of salinity to the wider society, especially for the predicted level of salinisation fifty years hence, are difficult to agree upon. Within the Murray-Darling Basin the latest estimates for the costs of the current impacts within eight priority catchments is $200 to $300 Million per year. Judgement would suggest that this might represent 60% to 80% of the current costs for the whole Basin. Guesstimates of the future costs at today's prices, for the level of salinisation anticipated fifty to one hundred years from now, suggest that this will rise to around $1,000 million per year.

These costs are borne differentially by all sectors of the community and governments. Irrigators and dryland primary producers are the most vocal affected group. However the data suggests that households and domestic water users suffer the largest share of the costs. As yet, there is no practical way to measure the financial costs to the environment.

But, have the right questions been asked?

Is it reasonable to expect that, when the landscape and the rivers have been so heavily exploited, no costs associated with salinity will occur? Salinity and its associated costs are just an inevitable outcome of the past land uses. The dilemma is that it occurs decades or even centuries after the causal action took place, and consequently affects future generations. It is more useful to consider the marginal costs and benefits of the opportunities to ameliorate the impacts of salinity.

This approach has been effectively used for the past thirty years in the Lower Murray, where engineering interventions have been implemented to divert irrigation drainage water or regional groundwater away from the River Murray. Conventional benefit/cost analyses have enabled an orderly program of works and measures to be pursued. The benefits have been justified by assessing the incremental costs to horticultural and urban water users. A range of cost functions have been developed which whilst questionable in detail have generally been accepted by Governments, to justify public expenditure. As a general rule the benefits or costs of a one EC (average) incremental change in salinity has been assessed at $100,000 per year. Engineering schemes have been constructed which can mitigate salinity at about half this cost. These schemes will respond relatively quickly generating benefits within a few years.

The problem of costing salinity management in the dryland is more intractable using conventional economic approaches. The landscape responds to changes over decades or centuries. Furthermore as the landscape changes so will the behaviour of the land and water users. Landholders don't just accept the penalty costs of salinity, they find alternative uses of their land, which are less sensitive to salt. In fact land use changes are occurring without consideration for the longer-term resource implications. Landholders respond to national and global terms of trade, which can often exacerbate salinity. Currently, Governments have little control over such decisions.

Water users will trade their water allocations to other users where the salinity is tolerable for the crops being grown. Urban water users respond by using products appliances that are resistant to the ambient salinity levels. Such responses can modify water and land use dramatically over a few decades; whilst the biological remedies might take much longer to take effect.

Conventional economics, preferred by today's governments, cannot value any benefits or costs after a few decades. Studies undertaken in the Basin by Australian Bureau of Resource Economics (Australian Bureau of Resource Economics, 2001) have indicated that the net present value of future salinity compared with the current baseline is about $600 million using a real discount rate of 5%. These estimates exclude certain infrastructure costs and all environmental costs and are therefore an under-estimate. Nevertheless they clearly demonstrate that the interventions that the scientists and politicians envisage cannot be justified by traditional economics alone.

The difficulty with evaluating costs and benefits associated with salinity management is the consideration of the intergenerational and ecologic parameters.

Murray-Darling Basin Ministerial Council Basin Salinity Management Strategy 2001

The Basin Salinity Management Strategy (BSMS) provides a Basin wide approach to salinity management, which emphasises the future working relationships between the governments of the Basin. It provides leadership to governments and communities to work together to control salinity. It is consistent with the recent State Strategies for salinity management and the Commonwealth Government's National Action Plan for Salinity & Water Quality.

The Principles of the Basin Salinity Management Strategy

The Basin Salinity Management Strategy adopts the key principles of the Council's ICM policy statement of April 2001. These include – commitment, targets, management arrangements, multiple mechanisms, monitoring, evaluating and reporting, roles and responsibilities, investment, and related initiatives.

Specifically for salinity this implies:
- Target driven decision making
- Accountability for 'future actions'
- A shared responsibility to address the 'legacy of history'
- Sharing benefits and costs
- Optimal investments & balanced decisions.

The BSMS has nine key elements which are:
- Developing capacity
- Identifying values and assets at risk
- Setting salinity targets
- Managing trade-offs
- Implementing salinity management plans
- Redesigning farming systems
- Targeting reforestation and vegetation management
- Joint salt interception works
- Basin-wide accountability and responsibility

Readers are advised to read the whole strategy (which is available at the Murray-Darling Basin Commission Web site). However some of the key features are described below.

End of valley targets

The rivers are the integrators of salinity across the landscape. The BSMS adopts in-stream 'end-of-valley' salinity targets as the principle-reporting framework. This acknowledges the prime common interest of the partners to the Initiative – the rivers of the Basin. However these targets are not ends in themselves but merely 'stakes in the sand' around which decisions will be made.

The notion of stream salinity targets recognises that there are thresholds of salinity tolerance for both human and environmental needs. However these targets are driven less by quantum standards or limits, than by realistic quantifiable decision-making. The first dilemma of such targets is defining the baseline conditions. It is only in the River Murray and some other rivers of the southern Basin that we have a reliable long-term record of salinity, flow and salt loads. Most of the selected targets are close to or slightly above the status quo. They are nevertheless quite difficult to achieve. This reflects the inevitability of 'living with salt' and the imperative to protect only key values and assets.

The target chosen for the end of the Basin at Morgan is that salinity shall remain less than 800 EC for 95% of the time; this is essentially the status quo. This level has been achieved after twenty years of aggressive intervention which has clawed salinity levels back from the much higher levels experienced in the 1970,s. Nevertheless a considerable effort will be still required to maintain this situation.

The target at Morgan also allows for a 10% increase in salt load without increasing peak salinity levels. In order to maintain this Morgan target over the next fifteen years, additional salinity control works and measures are needed of a similar order to those already implemented over the past fifteen years. About half of the new salt at Morgan will come from the upstream tributaries with the remainder coming from within the Mallee zone. The Mallee zone is particularly sensitive to the impacts of new irrigation development and wetland flushing. Accountability for future actions poses a particular challenge for this zone.

Salinity debits and credits

Under the BSMS, the system of inter-government salinity credits and debits developed for the Salinity & Drainage Strategy (1989) will continue to operate and in fact be expanded Basin-wide, albeit with the accountability remaining focussed on the lower Murray region. An important feature of the BSMS is the shared responsibility to offset the 'legacy of history'. In the salinity registers the legacy of history is represented by a cumulative debit clicking up each year. Over many years this becomes an onerous obligation.

The Initiative partners have agreed to differentiate the accountability for deliberate 'future actions' requiring a specific 'pollution offset' (recorded on the 'A register') from the less certain issues of the 'legacy of history'.

The Morgan target in the lower Murray does not in itself provide a driver for salinity intervention within the northern river catchments of the Basin. The most cost effective measures to deal with the lower Murray will arise within that zone. It is the end-of-valley salinity targets, representing the local catchment health that become the driver for action further afield. Nevertheless the cumulative effect of actions in the distant valleys will ultimately become measurable at Morgan and consequently will attract salinity credits.

In order to protect the value of credits generated through the engineering schemes, the Commission will operate two registers in harmony. A 'B register' will account for the more diffuse and less confident actions to offset the slower dryland salinity impacts.

Benefit and cost sharing for a joint works salinity control program

The new Basin Salinity Management Strategy, with its emphasis on shared responsibility for off-setting the Basin-wide 'legacy of history', together with accountability for future actions creates a more complex regime in which to negotiate future benefit and cost sharing arrangements. The Council has agreed to a further program of works providing a 46 EC reduction over seven years of which 31 EC will be assigned to the 'legacy of history' and 15 EC will provide offsets for future actions and be assigned equally to New South Wales, Victoria and South Australia. Queensland has not been included in this initial package because its contribution to salinity in the Southern Basin will be relatively modest. For the time-being the agreed cost sharing arrangements are the same as under the Salinity and Drainage Strategy (1989), The Commonwealth assists with capital works (quarter shares) with the States sharing the long-term operating costs (third shares).

The National Action Plan for Salinity & Water Quality has resulted in a 50%:50% partnership between the States and Commonwealth Governments to deal with legacy of history problems at a national scale. The benefit and cost sharing debate for the future joint works is yet to be fully resolved. The emphasis on the legacy of history aspects generates a blame debate between the Governments. This has been difficult to quantify given the differing development regimes within each State and consequently agreement has been reached to jointly share the responsibility for offsets. The Commonwealth Government rightly argues that it should bear no responsibility for offsets for future actions involving economic developments by the States.

Salinity intervention options

Salinity problems affecting water users and which originate within irrigated regions have been shown to be amenable to cost effective engineering solutions. Within the Murray-Darling Basin there is now more than thirty years of experience in dealing with salinity in the River Murray that has generated measurable benefits.

These interventions, have involved river management for improved dilution flows together with engineering salinity control schemes and drainage schemes in irrigation areas. Whilst engineering interventions can be readily justified, the opportunities are limited and most of the obvious schemes have been built. Also the schemes often have a limited life. Biological solutions involving landscape change are now being sought.

As secondary salinity takes hold in the dryland farming regions and its implications for the landscapes and rivers are acknowledged, some serious decisions are needed. What is the level of salinity with which we are prepared to live?

The community expectations are that future intervention actions will address the causes rather than the symptoms. The emerging recognition of dryland salinity as a major future cause of salinity within the Basin demands a biological solution. Improved dryland farming practices and revegetation of swathes of the landscape have been suggested to reduce recharge to the saline aquifers. However such solutions are difficult to implement at the scale that appears to be required. The responses are inevitably very slow.

More importantly, these solutions are difficult to assess. Will they actually solve the problem? For example the reduced recharge associated with enhanced evapotranspiration could also be associated with decreased run-off. The rivers might suffer less flow and consequently stream salinities might rise in the short term but improve over the longer term.

Applying science to the salinity problem

A major strength of the Murray-Darling Basin Initiative has been the ability to accelerate scientific study and to direct it towards effective policy development and decision-making. Site-specific investigations and engineering studies also play a large role in the ability to undertake effective interventions.

In general, the primary goals of the current strategies are to avoid further salinisation of the landscapes and the rivers. Clearly the measure of success of such strategies can only be compared with the scientific predictions of a 'no further intervention' scenario.

The ability to prepare predictions of the extent of landscape and river salinisation over the coming 20, 50 and 100 years has been particularly challenging. The need to relate the biophysical predictions to economic outcomes, agricultural risk assessments, stream salinity and infrastructure implications further compounds the scientific challenge.

The 1999 Salinity Audit was based upon a fairly crude extrapolation of recent historic groundwater trends using GIS special modelling based upon geologic regimes, groundwater salinity and the terrain. The available trend data was mostly around twenty years. Such an approach can attract criticism; in particular because it does not allow for an equilibrium situation to develop. Also in much of the Basin the density of groundwater trend data is very poor. The methodology involved considerable professional judgement to provide extrapolation into the data poor regions. However most of the other previous work in the Basin was either very locally based or use methodologies that could not be practically extended across the Basin.

Nevertheless by all peer review standards, there was a consensus that salinity is in fact a serious problem that justifies a significant level of public and private intervention. The Murray-Darling Basin Ministerial Council Salinity Audit, together with the National Land & Water Audit – dryland salinity theme, paints a very distressing picture. These predictions justify the need for a national scale response that has resulted in a $1.4 billion dollar seven year investment program. However such generic predictions should not be used as the basis of spending a single dollar on physical works. Each site-specific intervention requires a locally specific investigation using data and assessment methodologies that are appropriate to the scale of the local problem and the extent and scope of intervention that is envisaged.

Much of the recent scientific work within Australia has been focussed upon developing these locally specific modelling solutions. The National Dryland Salinity Program is an important Government initiative that has facilitated coordinated research and dissemination of dryland salinity research outcomes to the catchment communities. An example of the outcome of this program is the guideline for identifying and valuing the impacts of dryland salinity.

At the Basin-wide scale, recent work by the Cooperative Research Centre (CRC) for Catchment Hydrology has built upon some earlier work, by the Bureau of Rural Sciences (Coram et al., 2000), which characterises sub-catchments according to their ability to respond to land use changes. It uses the notion of groundwater flow systems. The new work, (Dawes et al., 2001), provides a water and salt balance leading towards a new hydrologic equilibrium for each catchment over time. This approach provides a more rational solution to salinity prediction than that used in the earlier salinity audit work. The audit investigations relied upon the extrapolation of groundwater trends across the landscape.

Agencies such as the Department for Land & Water Conservation of New South Wales are also developing new integrated modelling tools to predict salinity and its consequences. One of these tools is based upon fuzzy logic systems allowing the extension of a range of observational data at the land surface.

At the Basin scale, the current emphasis is towards establishing agreed historical benchmarks for flow and salinity regimes within the Basin's twenty primary tributary rivers. Once these benchmarks are established, it is reasonable to bring a wide range of salt mobilisation predictions of landscape salinisation processes to the individual nodes of these hydrologic models. This facilitates the assessment of the downstream implications of both the "no further intervention' scenario and any 'intervention option' under consideration.

Other work at the Basin scale is focussing on the ecological and social and economic implications of salinity management. This work is emerging as very challenging because the social consequences of salinity interventions such as large-scale revegetation can be just as unpalatable as with living with the consequences of salinisation resulting from no further intervention. The need for a multi-disciplinary scientific approach has been recognised and the response is accelerating.

CONCLUSIONS

Effective salinity management involves some deep philosophical debate. This debate is very dependant upon good science. Australia has been seriously entering into this debate over the past five to ten years. The relevant information and analysis is increasingly becoming available.

Scientists, bureaucrats and politicians are now strong advocates for revegetation of the landscape as a panacea for salinity. Whilst this option receives popular support by the broader community, those landholders who are asked to put their land back to trees resist it. In those parts of the landscape, which are most likely to suffer salinity in the future, the local communities and visitors are comfortable with the current sparsely treed landscape. It is often picturesque. Just 5% trees in the landscape are very evident when seen from the ground. Unfortunately, this picturesque landscape is now recognised as not sustainable. The remaining native vegetation at the mid-slope and valley floors will ultimately be lost. When travelling through the Basin the extent of dead tress and obvious land salinisation is increasingly evident. Nevertheless, the declining productivity alone is not sufficient justification for the aggressive landscape change necessary for stabilising salinity. It is the water users downstream that will bear the economic brunt of future salinisation.

Remedial measures involving sporadic tree planting have been pursued for more than a decade. Unfortunately such revegetation is not strategically planted and far too scattered to be of much effect. Communities and landholders have embraced the biologic solution but have not appreciated just how much is needed and where it should be placed.

The need for targeted revegetation is now recognised by scientists. A balance between reducing salt loads whilst maintaining stream flows for dilution is essential. This message has yet to be fully accepted by catchment communities and politicians.

The author's perception of social attitudes to salinity management, within the Murray-Darling Basin, is that the desirable community objective is to maintain the landscape and the rivers at about the same levels of salinity that we encounter today. There is a real concern that future generations should enjoy the benefits of the landscapes and rivers that we currently enjoy. The dilemma is that this is probably unachievable at any cost.

The economic productivity of the Basin's landscapes and rivers will continue to improve through improvements in farming systems including genetic engineering. Irrigation systems will continue to become more efficient.

Further engineering and flow management interventions can address the decline in river salinity that affects water users particularly in the lower Murray. Funds will be found for these actions, because the water is so valuable. However the opportunities for such interventions are limited further up the Basin's catchments.

On the other hand, there is a strong resistance to the retirement of both land and water allocations. Nevertheless these options will need to be considered. This will involve targeting salinity interventions at hotspots within the landscape and some difficult trade-offs will need to be negotiated. Blatant poor farming practices will be identified and possibly banned. However the bigger challenge will be the identification of those landscapes that will not be sustainable with any continuation of current land uses. Rezoning the landscape for new sustainable uses will involve much social and political debate. An increasing effort is being placed upon getting the commercial plantation investors involved in low rainfall farm forestry.

Where such interventions are shown to be absolutely essential a protracted debate will be inevitable. Some strong community and political leadership will be necessary. Scientists will need to be very convincing. The measure of success over the long term will be that salinity will only be a little worse than it is today rather than a lot worse as our current estimates suggest. How will we continue to provide funding whilst the situation is actually getting worse? The predictions that are being made will need far more validation. This will require much more field data that we have today and much more sophisticated models.

A more difficult issue will be the question of who pays. The issue of public and private benefit will involve some protracted debate. The role of a 'duty of care' will take a higher priority. Defining that will be quite a challenge.

ACKNOWLEDGEMENTS

The author has previously been employed by the Murray-Darling Basin Commission and is currently contracted to the Commission. The paper draws widely from the work of the Commission and the participating Governments together with research work from Cooperative Research Centres and studies by consultants.

Nevertheless, the opinions are those of the author. Increasingly community bodies at the local and regional scales are grappling with these processes and are now taking responsibility for sustainability decisions.

REFERENCES

Aeuckens, V. and Hoey, P., 1998. National water reform, an Australian experience; COAG. Presentation to the United Nations Commission on Sustainable Development, New York 1998.

Blackmore, D.J., 1994. Integrated Catchment Management-The Murray-Darling Basin Experience. Water Down Under , Adelaide, November 1994

Coram, J.E., Dyson, P.R., Houlder, P.A. and Evans, W.R., 2000. 'Australian groundwater flow systems contributing to dryland salinity'. Bureau of Rural Sciences project, undertaken for NL&WRA's Dryland Salinity Theme, Canberra. 2000.

Crabb, P., 1997. Murray-Darling Resources, Murray-Darling Basin Commission. Canberra, 1997.

Dawes W, Gilfedder M, Walker GR, Evans WR. 2001. Biophysical modelling of catchment scale surface and groundwater response to land-use change. In Proceedings of MODSIM 2001, Ghassemi F, Whetton P, Little R, and Littleboy M (eds.). Modelling and Simulation Society of Australia and New Zealand: Canberra; pp 535-540.

Gleeson, A.M., 2000. The Boyer Lectures: the rule of Law and the Constitution; ABC Books 2000.

Gutteridge-Haskins and Davey and Hunting Technical Services, 1970. Murray Valley Salinity

Investigations – a report to the River Murray Commission 1970.

Heaney, A., Beare, S., Bell, R., 2001. Targeting land & water use options for salinity management in the Murray-Darling Basin. Australian Bureau of Resource Economics, Oct 2001 (Available at the Murray-Darling Basin Commission, Website: www.mdbc.gov.au)

Maunsell and Partners, 1979. Murray Valley salinity & drainage, development of a coordinated plan for action a report to the River Murray Commission, 1979.

Meacham Ian, 1984. The River Murray Salinity Problem – A discussion paper by River Murray Commission.

Murray-Darling Basin Ministerial Council, 1993. Dryland Salinity Management in the Murray-Darling Basin; Murray-Darling Basin Ministerial Council; Canberra; June 1993.

Murray-Darling Basin Ministerial Council, 1995. An Audit of Water use in the Murray-Darling Basin; Canberra. June 1995.

Murray-Darling Basin Ministerial Council, 1999. The Salinity Audit of the Murray Darling Basin – a 100 year perspective Murray-Darling Basin Ministerial Council (*Available at the* Murray-Darling Basin Commission, Website: www.mdbc.gov.au).

Murray-Darling Basin Ministerial Council, 2001. Integrated Catchment Management in the Murray-Darling Basin 2001-2010 Murray-Darling Basin Ministerial Council June 2001.

Murray-Darling Basin Ministerial Council, 2001. Basin Salinity Management Strategy 2001-2015 Murray-Darling Basin Ministerial Council September 2001.

Newman, R.J. and Jessup, A., 1994. The Murray Darling Basin Salinity & Drainage Strategy – Is it meeting expectations? - A South Australian perspective, Australian Water & Wastewater 16th Federal Convention; Sydney; 1994.

Toyne, P. and Farley, R., 2000. The Decade of Landcare looking backward – looking forward. A discussion paper; the Australia Institute for a just, sustainable, peaceful future, (July 2000) www.tai.org.au.

Wilson, S., 1999. Guidelines and Instructions for identifying and valuing the impacts of dryland salinity National Dryland Salinity Program & Murray-Darling Basin Commission, Canberra 1999).

Wood, W.E., 1924. Increase in Salt in Soil and Streams following the Destruction of Native Vegetation, Journal of the Royal Society of Western Australia, Vol X, No 7, March 1924.

Zhang, L., Dawes, W., Walker, G.R., 1999. Predicting the effect of changes on catchment average water balance. CRC CH Technical Report No 99/12 November 1999.

Alternative Water Policies for the Gulf Cooperation Council Countries

WALEED K. AL-ZUBARI
Desert and Arid Zones Sciences Program, College of Graduate Studies, Arabian Gulf University, PO Box 26671, Bahrain.
E-mail: waleed@agu.edu.bh

ABSTRACT: The Gulf Cooperation Council countries are characterized by extremely arid climate and are in general devoid of reliable surface water supplies and depend primarily on groundwater (90%) and desalination (7.8%) of seawater to meet their water requirements. Since the early 1970s, these countries have experienced rapid population growth, and an accelerated social, agricultural, and industrial development, resulting from the sudden increase in oil revenues. Accompanying this growth has been substantial increase in water demands, which has placed great pressures on the region's limited water resources. Currently, all Gulf Cooperation Council countries are experiencing water deficits totaling more than 15 billion cubic meters (Bcm). Three scenarios are considered in this paper for the period 1995-2025: 1) business as usual, 2) supply augmentation, and 3) supply augmentation and policy remedies. In the third scenario, a major review and shift in water policies, emphasizing conservation and demand management, are suggested and outlined, with the overall objective of securing long term water supplies while meeting strict criteria for socio-economic, financial and environmental sustainability and public health requirements. The results of this study indicate that the Gulf Cooperation Council countries would continue to experience a deficit in their water resources for all three scenarios, although it is lesser in the third scenario, if their targets of maximum food security are adhered to, and their population growth rates continue as projected. However, the successful implementation of the proposed policies in the third scenario would be the milestone for paving the long intricate path towards coping with the water scarcity in this arid region.

INTRODUCTION

The Gulf Cooperation Council countries (Bahrain, Kuwait, Oman, Qatar, Saudi Arabia, and United Arab Emirates), comprising with Yemen the Arabian Peninsula, occupy an area of about 2.8 million Km^2 (UNSPD, 1997). Most of the region is characterized by an extremely arid climate that is devoid of reliable surface water supplies, and depend entirely on groundwater and desalination to meet their water requirements.

These countries have experienced an accelerated development growth since the early 1970s, due to the sudden increase in oil revenues, which have led to a rapid increase in their economic base and improvement in the standard of living. During the last five decades the total population in the countries of the region has increased about five folds, from about 5 million in 1950 to 25.5 million in 1995 (UNSPD, 1997). The average population growth rate in the Gulf Cooperation Council countries at present is over 3.5%, which is the highest in the world.

The rapid increase in population, along with social, agricultural, and industrial development, has resulted in substantial increase in water demands, placing great pressures on the region's limited water resources. Total annual water demand has increased from 6 billion cubic meters in 1980 (Al-Alawi and Abdulrazzak, 1994) to 26 Bcm in 1995. Food self-sufficiency has been a goal of the Gulf Cooperation Council countries and with the agricultural sector being the main water consumer (85%). Currently, the Gulf Cooperation Council countries are suffering from a huge deficit in their water resources reaching about 15 Bcm annually. This deficit is met mainly by over-drafting of groundwater resources, by desalinization of seawater, and by reuse of treated wastewater.

If current population growth rates and water use practices and patterns continue, annual water demand may reach 49 billion cubic meters by the year 2025. It is anticipated that this projected increase in water demand will be met mainly by further mining of groundwater. Unless major review and shifts in the current policies of population and adopted food self-sufficiency are made, and an appropriate and drastic measures in water conservation are implemented, the Gulf Cooperation Council countries are headed for a future major water crisis.

In the 21st century, it is anticipated that the Gulf Cooperation Council countries will experience a series of major environmental problems resulting from:

Water Resources Perspectives: Evaluation, Management and Policy. Edited by A.S. Alsharhan and W.W. Wood.
Published in 2003 by Elsevier Science, Amsterdam, The Netherlands, p. 155-167.

- Escalating water demands and lack of conservation measures;
- Slow rate of water resources augmentation due to limited opportunities;
- Continuous deterioration of water quality and reduction in the yield of heavily exploited aquifers;
- Modest programs for the treatment of wastewater from rapidly developing urban communities;
- Inefficient methods of wastewater treatment and solid waste disposal; and
- Lack of a solution to the issue of rapid population growth.

This study presents the available water resources in the Gulf Cooperation Council countries, discusses the problem of population growth and adopted agricultural policies and their impacts on water demands, and looks at the anticipated future water demands in these countries. Alternative policies towards sustainable development and use of water resources from a regional perspective are presented and discussed, and proposed policy remedies and other course of actions are outlined.

WATER RESOURCES

Prevailing Climate

The Arabian Peninsula is characterized by extremely arid climate. With the exception of the coastal strips and mountain ranges, the Peninsula is largely desert. It is characterized by irregular, scanty rainfall of less than 100 mm/y and high evaporation rates exceeding 3,000 mm/y (Table 1). More details on the Arabian Peninsula climate and its physiographic features are given elsewhere (e.g., Al-Alawi and Abdulrazzak, 1994).

Surface Water

The Gulf Cooperation Council countries have very little surface water resources. The only areas receiving enough rainfall to generate reasonable amounts of runoff are the southwestern parts of Saudi Arabia, the southern parts of United Arab Emirates, and the southern parts of Oman. Annual surface runoff volume in the Gulf Cooperation Council countries is estimated to average 4.8 Bcm annually, with very high temporal and spatial variation that it is typical of arid regions. The limited surface water that does occur is used in flood irrigation, and impounded behind dams, and recharge to alluvial aquifers beneath wadis.

Groundwater

The Gulf Cooperation Council countries depend primarily on groundwater to meet their water requirements. Groundwater is divided into two types. The first is the shallow aquifers, developed in the alluvial deposits along the main wadi channels and the flood plains of drainage basins. Shallow aquifers are the only renewable water source in the Gulf Cooperation Council countries, with an approximate annual recharge of 3.5 Bcm, with an estimated combined reserves of about 117.3 Bcm for the Gulf Cooperation Council countries (Al-Alawi and Abdulrazzak, 1994). The shallow alluvial groundwater is used extensively for domestic and irrigation purposes. Along coastal areas, seawater intrusion is a common problem.

Table 1: Available water rresources in the Gulf Cooperation Council countries (1995).

Country	Annual Rainfall (mm)	Annual Evaporation (mm)	Conventional (Mcm)			Non-Conventional (Mcm)				
			Surface Runoff[1]	Groundwater		Desalination		Treated Wastewater		Agricultural Drainage Water
				Recharge[2]	Used	Capacity	Produced	Capacity	Produced[3]	
Bahrain	80	1,650-2,050	-	-2	239	75	56	58.4	55.2	-
Kuwait	110	1,900-3,500	-	-2	405	475	402	193	95	-
Oman	20-300	1,900-3,000	1,450	955	1,223	51	34	>30	26	-
Qatar	75	2,000-2,700	-	50	286	185	126	>105	103	-
Saudi Arabia	70-500	3,500-4,500	3,210	3,850	17,000	>875	795	>667	526	30
UAE	89	3,900-4,050	150	125	1,615	704	385	>108	102	-
Total			**4,810[1]**	**4,980**	**20,768**	**2,365**	**1,798**	**>1,161**	**907**	**30**

(1) Most are seasonal flow of wadi systems; (2) Recharge to aquifers occurs by underflow from equivalent aquifers in Saudi Arabia, Bahrain = 90-110 Mcm/y, Kuwait = 58-117 Mcm/y; and (3) Reused treated wastewater volume is only 392 Mcm, used mainly in irrigation, the remains are used either for recharge purposes or disposed off to the sea. *Sources: compiled data from Al-Alawi and Abdulrazzak 1994; FAO 1997; Al-Zubari 1997; and official country reports for the period 1996-1999.*

Groundwater also occurs in the deep sedimentary aquifers. This deep groundwater is non renewable fossil groundwater that is thousands of years old dating from the rainy Pleistocene and Pliocene geological periods. Deep groundwater reserves are estimated at about 2,175 Bcm with the major part (1,919 Bcm) located in Saudi Arabia. Annual recharge to deep aquifers in the Gulf Cooperation Council is estimated at 1.5 Bcm per year (Al-Alawi and Abdulrazzak, 1994). The quality of the deep aquifers varies greatly, being suitable for domestic use in only few areas. Most of the water from these deep aquifers is used for agricultural purposes.

At present, annual groundwater abstraction from groundwater is about 21 Bcm (Table 1), with annual groundwater recharge of about 5 Bcm. Al-Turbak and Al-Dhowalia (1996) estimated that as of 1995 about 35% of the non-renewable groundwater resources in Saudi Arabia have been depleted. Mining of the groundwater has resulted in continuous and sharp declines in groundwater levels and severe quality deterioration due to seawater intrusion and deep brines up-flow (UNESCWA 1999a; FAO, 1997). In addition, over-irrigation and surface dumping of partially treated and untreated wastewater has generated large volumes of contaminated waters that have increased the pollution levels of shallow aquifers (Al-Zubari, 2000).

Desalination

Desalination of seawater began in the mid fifties and has expanded very rapidly. At present, municipal water supplies in the Gulf Cooperation Council countries rely mainly on desalinized seawater, which is used either directly or blended with groundwater. As of 1995 total capacity of desalination plants (including those under-construction) was more than 2.36 Bcm/y (Table 1), with total actual produced water of about 1.8 Bcm/y. Desalination costs range from 0.25-1.0 US$/m^3 for brackish groundwater and 1.4-8.0 US$/m^3 for seawater (UNESCWA, 1999b).

It is expected that the total desalination capacity of the region will be more than 3 Bcm by the year 2020 (Ismail, 1995). Desalinization has several major disadvantages: relatively high cost, short operational life (15–25 years) and dependence on fossil fuel. Additionally, desalination negatively impacts the environment through air pollution by emitted oxides and seawater and marine life pollution by rejected brines, which have elevated temperatures, increased salt concentration and may contain residual treatment chemicals and trace elements.

Treated Wastewater

Wastewater treatment in the Gulf Cooperation Council countries constitutes an increasing water source. In the early eighties, most of the Gulf Cooperation Council countries completed sewage water treatment facilities and urban sewage networks for most of the large cities. Almost all of the Gulf Cooperation Council countries now have modern treatment facilities with tertiary and advanced treatment capabilities. The total design capacity of the major wastewater treatment facilities is about 1.2 Bcm/y. This represents no more than 31% of the total municipal water usage of about 3.6 Bcm/y. Wastewater losses to the groundwater cause water table rise in urban areas (e.g., Riyadh and Kuwait Cities) and pollution of the shallow aquifer. Discharge of untreated wastewater represents a health hazards to the coastlines and the marine environment.

Total treated wastewater in 1995 was about 0.9 Bcm/y. Reused treated wastewater is approximately 0.4 Bcm/y, i.e., about 45% of total treated, and about 10% of all wastewater resources available. Treated wastewater are used mainly for urban uses (irrigating gardens, parks, and road ornamentals), fodder crops irrigation, and highway landscaping (Al-Zubari, 1997). The remainder is dumped at wadis to infiltrate the shallow aquifers or to the sea. All of the Gulf Cooperation Council countries have ambitious plans for the expansion in the utilization of reclaimed wastewater as a strategically alternative source to meet their future demands of irrigation water and to reduce groundwater abstraction (Al-Zubari, 1997).

Agricultural Drainage Water

In general, recycled agricultural drainage water is not included within the water resources management programs in the Gulf Cooperation Council countries. The exception is Saudi Arabia. At Al-Hassa Oasis in Saudi Arabia, about 30 Mcm/y of drainage water mixed with groundwater is reused for agriculture purposes. This has had the advantage that it helped in alleviating the stress on groundwater, increased water use efficiency, and helped in solving the problem of water logging in the Oasis (Al-Kuwaiti et al., 1999). Agricultural drainage water reuse has future potential. It is anticipated that more recycled drainage water will be used in the Gulf Cooperation Council countries.

STATUS OF POPULATION GROWTH AND WATER DEMAND

Population growth in the Gulf Cooperation Council countries is a major issue affecting all sustainable socio-economic development. The estimated total population of the Gulf Cooperation Council countries in 1995 was about 25.5 million (UNSPD, 1997), with an average growth rate of 3.7% (Arab Fund, 1995). This thought to be the highest in the world. The large increase in the population is due to marked improvement in the standard of living and better health of the population as well the influx of a large number of expatriates. Latest figures (1997) on expatriate population in the Gulf Cooperation Council countries range from 25% to 85% of the total population. Furthermore, population policies that have been adopted in some countries of the region during the last three decades encouraged population growth through various forms of subsidies and incentives. Social traditions, appropriate economic climate and religious beliefs played an important role in population growth, which became difficult to control at later stage. Population projections for the Gulf Cooperation Council countries over the period of 1995-2025 are given in Table 2. It is anticipated that the total population of the Gulf Cooperation Council countries will reach about 57 million by the year 2025 (UNSPD, 1997).

In the past two decades, food security, or food self-sufficiency desires have prompted decision-makers in the Gulf Cooperation Council countries to encourage increased agriculture production. For example, in Saudi Arabia the Agricultural sector grew at a rate of 12% per annum between 1979 and 1991 (Al-Sheikh, 2001). This growth was made possible through governmental subsidy and incentive programs. The water requirements to satisfy this increased level of agricultural production was met through groundwater mining of the deep aquifers. Unregulated pumping coupled with lack of enforcement of rules against unlawful drilling, poor irrigation practices (irrigation efficiency 30-50%), and lack of agricultural water tariffs has resulted in excessive agricultural water consumption.

Furthermore, population increase and urbanisation in the past three decades have caused municipal water needs to increase at rates that have exceeded the rate of increase of desalinized seawater production. These needs are exaggerated by the lack of effective conservation programs, inadequate tariffs and charges for water use, and excessive leakage from municipal supply networks, leading to high per capita water consumption rates in the domestic sector.

The total water used for all purposes in the Gulf Cooperation Council countries in 1995 amounted to about 26 Bcm (Table 3). Currently the agricultural sector takes about 85% of the available water resources, followed by the municipal sector at 14%, and industry accounting for less than 2%. In terms of resources share, these requirements are satisfied mainly by groundwater resources by about 90%, and are supplemented by desalinated water by about 7.8%, and treated wastewater by about 1.7%.

Table 2. Projected population in the Gulf Cooperation Council countries.

	Year							Percentage 2025/1995)
Country	**1995**	**2000**	**2005**	**2010**	**2015**	**2020**	**2025**	
Bahrain	557	618	671	717	766	816	863	155
Kuwait	1 691	1 966	2 192	2 390	2 576	2 752	2 904	172
Oman	2 207	2 717	3 302	3 986	4 752	5 610	6 538	296
Qatar	548	599	648	693	734	764	782	143
Saudi Arabia	18 255	21 661	25 255	29 222	33 483	37 919	42 363	232
UAE	2 210	2 444	2 660	2 869	3 049	3 185	3 297	149
Total	**25 468**	**30 005**	**34 728**	**39 877**	**45 360**	**51 046**	**56 747**	**223**

Source: UNSPD 1997.

Table 3. Water use in the Gulf Cooperation Council countries (1995).

Country	Municipal (Mcm)	Industry (Mcm)	Agriculture (Mcm)	Total (Mcm)
Bahrain	107	19	161	287
Kuwait	297	13	323	633
Oman	85	6	1,150	1,241
Qatar	85	17	337	439
Saudi Arabia	2,387	193	18,575	21,155
UAE	600	73	1,539	2,212
Total	**3,561**	**321**	**22,085**	**25,967**
Rate of water use (L/d/cap)	383	34.5	2,376	2,793
% of water use	**13.7**	**1.3**	**85.0**	

Sources: Data from verified Country reports, GCC 1996, UNSPD 1997, FAO 1997.

Except for Oman, all Gulf Cooperation Council countries have renewable water of less than 1000 m^3/y per capita, the benchmark established to identify chronic water scarcity (WRI/UNEP/UNDP/WB, 1996). The overall average for renewable water for the Gulf Cooperation Council countries is about 466 m^3/y per capita (Table 4) versus the per capita water use of 1,020 m^3/y (1995). The per capita water deficit is about 554 m^3/y provided mainly by mining groundwater reserves. The water stress index (defined as percentage of renewable water resources versus actually used), is over 100% in five of the six Gulf Cooperation Council countries. Estimated groundwater mining rates, in Mcm, are in Bahrain at 100, Kuwait 200, Oman 240, Qatar 140, Saudi Arabia 13,558, and United Arab Emirates 1,495 (FAO, 1997).

Comprehensive water policies and plans to assess, develop, protect, and manage water resources (with the exception of Oman) are still lacking in most of the Gulf Cooperation Council countries. The present policies are fragmented and stress economic development without consideration to water deficit issue. Agriculture was greatly expanded to promote food self-sufficiency, without regard to the limited availability of water. Furthermore, water supplies were developed and augmented without concurrent conservation programs to reduce water consumption, or cost recovery plans. This was compounded by the institutional weakness of water and land management authorities, multiplication and overlap of water agencies and lack of coordination among them, inadequate technical and financial resources, and lack of public participation. It should be noted that some countries have recently started preparing their national water resources management strategy plans (e.g., Saudi Arabia).

Table 4. Per capita water share and water use in the Gulf Cooperation Council countries (1995).

Country	Population (1000)	Available water resources (1000Mcm)	Exploited water resource (1000Mcm)	Per Capita water share (m^3/year)	Per Capita water use (m^3/year)	Deficit (m^3/year)	Water Stress Index (%)
Bahrain	557	0.243	0.287	437	515	-78	118
Kuwait	1,691	0.616	0.633	364	374	-10	103
Oman	2,207	2.486	1.241	1,126	562	564	50
Qatar	548	0.340	0.439	620	801	-181	130
Saudi Arabia	18,255	7.092	21.155	388	1,159	-770	300
UAE	2,210	1.087	2.212	492	1,001	-509	203
Total	**25,468**	**11.864**	**25.967**	**466**	**1,020**	**-554**	**219**

FUTURE WATER DEMAND

Projections of the total water demand in the Gulf Cooperation Council countries are that the 1995 total annual water demand of 26 Bcm will grow to about 49.4 Bcm by the year 2025 (Figure 1). Future available water resources in 2025 will under the most optimistic projections not exceed 17 Bcm (groundwater recharge = 5 Bcm/y; desalinated seawater = 3 Bcm/y; reuse of treated wastewater =4.2 Bcm/y assuming that 50% of treated wastewater is reused; surface water = 4.8 Bcm/y assuming that all surface waters will be harvested and utilized). It is clear that available water resources cannot satisfy future water demands, and that the water deficit will continue to increase and may reach about 31 Bcm by the year 2025, if the present trends of population growth and water consumption persisted.

Under these conditions, further over-exploitation and mining of groundwater resources will become inevitable to satisfy the increasing water requirements in the region, with the negative impacts of rapid depletion of aquifer storage and deteriorating water quality and salinization of agricultural lands, leading eventually to the loss of these resources. Under these circumstances it will be difficult, if not impossible, to maintain the present emphasis on maximum food production and self-sufficiency, and that widespread import of foodstuffs will become necessary.

Clearly, there is a need for a review of the present water resources policies in the region, to be followed by positive steps towards managing and rationalizing water demands, and imposing realistic controls on water use. The Gulf Cooperation Council countries have been making great efforts in developing additional water sources and augmenting their conventional and non-conventional water supply. These efforts are represented by the extensive installation of desalination plants, expansion in wastewater reuse, and constructing dams to capture and store surface runoff for irrigation and enhancing shallow aquifer recharge. Almost all the efforts to date have been on the supply side.

On the other hand, conservation and demand management have not received much attention. Conservation measures should be enforced on the agricultural sector, the largest water consumer in the Gulf Cooperation Council countries and where major and effective saving can be achieved. Irrigation efficiencies are very low at 30-45% levels, resulting from the widespread use of traditional irrigation practices, lack of monitoring and tariffs for irrigation water, and the planting of crops with high water requirements. In the domestic sector, present per capita water use is considered high, averaging 383 L/d, and reaches in some countries more than 740 L/d, mainly due to lack of public awareness and low water tariffs that do not discourage excessive water use or waste. These issues are aggravated by a general weakness among the institutions dealing with water affairs, inadequate technical capabilities and training, and from unsatisfactory coordination with other concerned water authorities. Table 5 summarizes the prevailing problems and critical issues in the water sector in the Gulf Cooperation Council countries.

Programs related to demand management and conservation as well as institutional reforms have been implemented recently by some Gulf Cooperation Council countries. Example conservation measures implemented include: in Saudi Arabia reduction of fuel and agricultural subsidies, and consolidating water resources agencies under one responsible authority; in Bahrain, metering of groundwater wells and approval of plans for implementing irrigation water tariff, changing crop pattern to higher water use efficiencies; in Oman, groundwater wells regulations and stringent permitting system. Additionally, most Gulf Cooperation Council countries subsidize modern irrigation techniques and protected agriculture in addition to public awareness campaigns. These demand management efforts remain as fragmented measures and are not consistent with the requirement of an integrated water resources management approach. Figure 2 illustrates the observed trends in per capita total water consumption for the period 1980-2000, as a percentage change to 1980 consumption rates. The figure indicates per capita total water consumption has stabilized in most Gulf Cooperation Council countries and has even dropped in Saudi Arabia and Oman. However, total water consumption continues to increase (Figure 1). There is an obvious need for water policy and strategy formulation as well as an appropriate drastic measures and major shifts in the current policies of population and food security.

Table 5. Summary of the prevailing problems and critical issues in the water sector in the Gulf Cooperation Council countries

Problem/Issue	Evidence	Source	Impact
Limited Water Resources	• Decrease of per capita available water with time	• Prevailing Aridity • Fast population growth • Adopted food self-sufficiency policies	• Slows down socioeconomic development plans
Inefficient Water Use	• Irrigation losses reaching up to 55% • Excessive leakage from supply networks • Wasteful water use in the domestic sector	• Extensive traditional irrigation practices/ Agricultural lands tenure system/ inappropriate cropping patterns • Lack of consumer awareness and participation • Lack of realistic pricing and stringent regulation • Old water supply networks/ No sufficient funds and programs for maintenance and leak detection	• Excessive losses of available water
Internal Water Allocation Conflicts	• Increasing competition among sectorial users, particularly between the agricultural and municipal sectors	• Growing sectorial water demands • Limited water resources	• Increasing water scarcity
Water Quality Deterioration with Time	• Groundwater levels decline • Groundwater quality deterioration by seawater and deep connate water intrusion • Contamination of shallow aquifers by untreated/ partially treated wastewaters and irrigation water agro-chemicals • Agricultural lands salinization	• Over-exploitation of groundwater • Discharge of domestic and industrial untreated/partially treated affluents • Discharge of industrial affluent in domestic wastewater networks • Lack of strict penalties on polluters • Insufficient studies for the management and protection of groundwater resources • Excessive irrigation techniques	• Loss of groundwater resources • Spread of water related diseases • Land degradation and loss • General negative impact on environment
Inferior Quality of Services	• Lack of sewage collection system in large urban areas and lack of treatment capacity • Shallow water table in urban areas • Water related diseases • Groundwater pollution	• Shortage of financial allocations • Growing population • Fast pace of Urbanization	• Negative general impact on development and welfare
Weak Water Institutions	• Duplication of efforts and conflict in decisions • Inefficient water development and management	• Fragmented water authorities and lack of coordination • Ill defined responsibilities • Incompetent water authorities, and inadequate technical capabilities and training	• Weak sectorial water policies • Poor water development plans and action programs • Inferior water services

ALTERNATIVE WATER POLICIES TOWARDS SUSTAINABLE DEVELOPMENT

Three alternative water management scenarios are analyzed in this paper and evaluated as to their effect on the water balance required to ensure sustainable water development in the Gulf Cooperation Council countries for the period 1995-2025. The year 1995 was used as reference year due to the availability of consistent data for the Gulf Cooperation Council countries. In the three scenarios the assumption is made that there will be no change in the population policies in the Gulf Cooperation Council countries. The UN population projections are used (UNSPD, 1997). The three scenarios are as follows:

Scenario 1- Baseline (Business as Usual)

- No change in the pattern of the per capita water use in the domestic and industrial sectors, i.e. the increase in the per capita water consumption in these two sectors will continue to grow.
- No further development of current water resources (1995).
- Continuation of agricultural policies aimed at maximum food production.
- Improving agricultural productivity per unit of water to achieve 20% saving in agricultural water demand by 2025.

Scenario 2 - Supply Augmentation

- Increase surface and groundwater resources for the Gulf Cooperation Council countries by 100 Mcm/y through artificial recharge and water harvesting yielding a total increase of 3 Bcm by the year 2025.
- Increase desalination capacity gradually to reach 3.26 Bcm/y by the year 2025.
- Increase recycled wastewater gradually to 4.2 Bcm/y by the year 2025 (50% of domestic water demand in 2025).
- Increase reuse of agricultural drainage water by 5% every 5 years, to reach about 40 Mcm by the year 2025.
- Other assumptions of water consumption rates increase in the domestic and industrial sectors, policies of maximum food self-sufficiency, improvement in agricultural productivity per unit of water, as in Scenario 1.

Scenario 3 - Supply Augmentation and Policy Remedies

- Development and increase of surface and groundwater resources as in scenario 2.
- Maintaining the 1995 rates of per capita water use in the domestic and industrial sectors, and achieving gradual decrease in per capita water use in the municipal sector to reach 300 L/d through rationalization measures, review of water tariffs, and awareness. This would result in a saving of 3.2 Bcm by the year 2025 as compared to Scenario 1.
- Further, gradual decrease in the agricultural sector water consumption is achieved by increasing the efficiency of irrigation, changing crop pattern to crops with less water requirements, metering and reviewing the price of irrigation water, and improving wastewater management. This would result in a decrease in the agricultural total water demand by about 5.2 Bcm by the year 2025 and a total water demand reduction in all sectors of about 8.4 Bcm by the year 2025.

Scenario 1 represents business as usual and is the most pessimistic scenario. It represents the worst-case scenario. Justification for its consideration is based on: 1) most of the Gulf Cooperation Council countries are located in an extremely arid climate and that most of the easy and promising water sources have been exploited. Additional water resource development will require heavy investments, laborious investigations and intensive research programs; 2) the Gulf Cooperation Council countries have been affected strongly by two regional wars in the last two decades (Iraqi-Iranian war, Iraqi-Kuwaiti war). No permanent settlement has been reached and these wars have drastically affected the economy of the region and have upset many socio-economic development plans. This has resulted in the postponement of many water development schemes (added to that the current US-Afghanistan war); and 3) The national economies of the Gulf Cooperation Council countries depend on oil and oil-related industries that are sensitive to the fluctuating international oil prices and which may impact negatively their socio-economic development planning. The major water saving in this scenario is a reduction of 20% in the agricultural water demand by 2025. This saving would be accomplished by agricultural research in the field of maximizing agricultural productivity per unit of water, and applications of appropriate technologies, including biotechnology.

Scenario 2 represents the supply augmentation strategy. It takes into consideration the intensive efforts in increasing and augmenting water supply in both conventional and non-conventional water resources in the Gulf Cooperation Council countries. These efforts could eventually yield an additional 7.8 Bcm (3 Bcm conventional water resources and 4.8 Bcm non-conventional water resources) by the year 2025.

Scenario 3 is the supply augmentation and policy remedies strategy. In scenario 3 water demands are reduced through formulated water policies that emphasis conservation and demand management. Scenario 3 requires extensive research, investigation, development, and reform programs for implementation of desired actions in the various areas of integrated water resources management (analysis and planning, legal and institutional, and economic considerations and tools), in addition to needed associated programs and projects (Table 6). The overall objective of the proposed water policies outlined in Table 6 is to secure long term water supplies while meeting strict criteria for socio-economic, financial and environmental sustainability and public health requirements.

It should be noted that all scenarios do not take into account the impact of groundwater overdraft on the quality of groundwater withdrawn. i.e., it assumes withdrawn water will be usable. Mining and over-exploitation of shallow and deep aquifers will be associated with an inevitable deterioration of water quality and loss of groundwater usefulness, in addition to salinization of agricultural lands.

The results of the three Scenarios are shown in Table (7). In Scenario 1 (Business as Usual), the Gulf Cooperation Council countries will suffer from acute water shortages. The water deficit will increase to about 36 Bcm by the year 2025. Percentage of deficit to water demand will increase from -47.8% in 1995 to -72.5% in 2025. In scenario 2 (Supply Augmentation), the water deficit will be slightly reduced to 28 Bcm in 2025, while the percentage deficit in the water balance to total demand will increase from -47.8% in 1995 to -56.7% in 2025. In scenario 3 (Supply augmentation and policy remedies), the water deficit will continue to experience an increasing water deficit, from 11.2 Bcm in 1995 to 16.5 Bcm by 2025. The percentage water balance to demand will, more or less, level off at 45%.

It is evident that scenario 3 is to be preferred and constitutes an appropriate base for an alternative strategy to secure a degree of sustainable development of water resources. Even so, it is clear that the Gulf Cooperation Council countries will continue to experience a deficit in their water resources if the goal of food security are adhered to, and population growth rates continue as projected. This will result in mining about 400 Bcm of shallow and deep groundwater resources during the study period. This will also result in the deterioration of pumped groundwater quality and may limit the uses of this water.

Table 6. Proposed policy matrix for the water sector in the Gulf Cooperation Council countries.

Policy	Planning and analysis Aspects	Legal and Institutional Aspects	Economic considerations and Instruments	Projects and programs
Priority of Water Supply for Domestic, Agriculture, and Industry, respectively	Water resources assessment & development; Population growth projection; Water demand projections; Water quantity and quality monitoring and prediction	Protection zones for surface water, domestic supply wellfields, springs, and catchment areas; Training and capacity building of staff; New competent authorities; Controlling illicit connections; groundwater wells regulation	Water viewed as an economic commodity as much as possible; Wide application of metering system; Increasing water tariffs for high segments of water demands; Provision funds for M&O and new water facilities; Strict regulation and penalties; Full stakeholders participation	Hydrological & hydrogeological studies; Continuous maintenance of water facilities and networks; Construction of water purification plants and desalination plants for future needs; Planned information and education campaign to change consumers behaviour
Rational Water Use	Mapping municipal and industrial water supply connections to control consumption and leakage; Mapping domestic wastewater collection and distribution networks; Determining and locating abstraction wells for agricultural, industrial, and domestic purposes	Regulation to support water saving and protection measures; Detection and control of illicit connections; Full application of metering system; Adequate institutions and staff training for efficient and proper water use control	Proper water pricing in all consuming sectors; Rotational check of metering system; Periodic review of water supply costing	Application of water saving technologies; Studies for rational water utilisation in different consuming sectors; Planned information and education campaigns to change consumer behaviour; Regular calibration of meters; Leakage detection and control; Regular maintenance and rehabilitation of network

Groundwater Management and Protection	Comprehensive monitoring of groundwater quality and water levels; Use of mathematical modeling studies in the management and protection of groundwater resources	Controlling pumping rate to prevent mining the reserve; Coordination between concerned institutions for efficient management; Application of wells metering system; Adoptation of detailed well drilling licensing; Stringent penalties for groundwater contamination; Training in the field of groundwater resources management and protection and mathematical modelling tools	Changing irrigation charge system to be based on volume of drawn water; Cancellation of subsidies for well drilling; Incentives for application of appropriate agricultural water-saving schemes (crops, techniques, and water use efficiency); Reviewing water tariff for commercial, domestic, and industrial abstraction	Application of agricultural water-saving schemes; Update studies on groundwater availability for sustainable water management; Artificial recharge studies and projects; Public awareness campaigns and educational programs for farmers (major groundwater users)
Increasing Water Use Efficiency	Continuous monitoring of Water use for different sectors; Conducting studies on consumptive use in agricultural sector; Extensive application of water saving agricultural schemes	Metering system for different water uses; Licensing system of well drilling and abstraction rates; Strict regulations and penalties to control illegal abstraction; User Group/community participation	Irrigation charges to be based on water use efficiencies; Incentives for modern irrigation techniques; Periodic review of irrigation charges; Increasing domestic water supply tariff for high segments of domestic and industrial water consumption	Choice of appropriate crops,; Application of water saving and reuse technologies; Increasing the efficiency of distribution networks & irrigation conveyance; Extension services and awareness programs; Re-use of wastewater and irrigation drainage water; Regular maintenance of water supply networks
Water Pollution Control and Environmental Protection	Standardisation of drinking water quality compatible with the region conditions, and industrial effluents discharged; Improving water quality monitoring systems; Provision of funds for studies	Stringent legislation for water pollution; Restrict activities in watershed areas and wellfield zones (defining wellhead protection areas); Regulations for the treatment of industrial effluents before discharge; Competent authority for water pollution control	Applying the "polluter pays" principle; Pollution charges to be assessed proportional to volume and quality of effluents discharged to the environment; Control of industrial intake & discharge points; Economic-environmental research on wastewater treatment and reuse	Expansion and improving of wastewater collection networks; Environmental impact assessment for water projects; Education and public awareness campaigns; Wastewater treatment plants in big communities & industrial complexes
Institutional Capacity Building	Human resources development and training of staff; Support research institutes & universities capabilities in the field of water resources; Application of new technologies for administration and management affairs; Modelling for water resources development & management	Consolidation of water authorities in one agency; Updating water legislation and enforcement; Coordination between water-related administrations; Defining clear responsibilities and duties of water administrations and minimizing overlaps; Integrated programs for water use management	Financial support for upgrading institutional capability; Modifying employment regulations, incentives should replace the current civil service rules; Improving Accounting and costing systems; Budget control systems	Training program for skilled technicians and professionals; Looking at options for privatising certain services to increase water use efficiency and public participation

Increasing Water Use Efficiency	Continuous monitoring of Water use for different sectors; Conducting studies on consumptive use in agricultural sector; Extensive application of water saving agricultural schemes	Metering system for different water uses; Licensing system of well drilling and abstraction rates; Strict regulations and penalties to control illegal abstraction; User Group/community participation	Irrigation charges to be based on water use efficiencies; Incentives for modern irrigation techniques; Periodic review of irrigation charges; Increasing domestic water supply tariff for high segments of domestic and industrial water consumption	Choice of appropriate crops,; Application of water saving and reuse technologies; Increasing the efficiency of distribution networks & irrigation conveyance; Extension services and awareness programs; Re-use of wastewater and irrigation drainage water; Regular maintenance of water supply networks
Water Pollution Control and Environmental Protection	Standardisation of drinking water quality compatible with the region conditions, and industrial effluents discharged; Improving water quality monitoring systems; Provision of funds for studies	Stringent legislation for water pollution; Restrict activities in watershed areas and wellfield zones (defining wellhead protection areas); Regulations for the treatment of industrial effluents before discharge; Competent authority for water pollution control	Applying the "polluter pays" principle; Pollution charges to be assessed proportional to volume and quality of effluents discharged to the environment; Control of industrial intake & discharge points; Economic-environmental research on wastewater treatment and reuse	Expansion and improving of wastewater collection networks; Environmental impact assessment for water projects; Education and public awareness campaigns; Wastewater treatment plants in big communities & industrial complexes
Institutional Capacity Building	Human resources development and training of staff; Support research institutes & universities capabilities in the field of water resources; Application of new technologies for administration and management affairs; Modelling for water resources development & management	Consolidation of water authorities in one agency; Updating water legislation and enforcement; Coordination between water-related administrations; Defining clear responsibilities and duties of water administrations and minimizing overlaps; Integrated programs for water use management	Financial support for upgrading institutional capability; Modifying employment regulations, incentives should replace the current civil service rules; Improving Accounting and costing systems; Budget control systems	Training program for skilled technicians and professionals; Looking at options for privatising certain services to increase water use efficiency and public participation

Table 7. Water balance in the Gulf Cooperation Council countries (Bcm), alternative water policies results.

Scenario / Year	1995	2000	2005	2010	2015	2020	2025
Scenario 1: Baseline scenario (business as usual)							
Available Water Resources	13.57	13.57	13.57	13.57	13.57	13.57	13.57
Total Water Demand	26.00	29.22	32.52	36.19	40.24	44.86	49.38
Water Balance	-12.43	-15.66	-18.96	-22.63	-26.67	-31.30	-35.82
% W.Balance/Demand	-47.81	-53.58	-58.29	-62.52	-66.29	-69.76	-72.53
Scenario 2: Supply Augmentation							
Available Water Resources	13.57	14.68	15.85	17.10	18.44	19.87	21.36
Total Water Demand	26.00	29.22	32.52	36.19	40.24	44.86	49.38
Water Balance	-12.43	-14.54	-16.67	-19.09	-21.79	-25.00	-28.02
% W.Balance/Demand	-47.81	-49.77	-51.26	-52.74	-54.16	-55.72	-56.74
Scenario 3: Supply Augmentation and Policy Remedies							
Available Water Resources	13.57	14.68	15.85	17.10	18.44	19.87	21.36
Total Water Demand	26.00	27.51	29.02	30.81	32.88	35.44	37.81
Water Balance	-12.43	-12.84	-13.17	-13.71	-14.43	-15.57	-16.45
% W.Balance/Demand	-47.81	-46.65	-45.38	-44.48	-43.90	43.94	-43.50

CONCLUSION AND RECOMMENDATIONS

Population pressures and current agricultural policies for food self-sufficiency constitute the core problem in the management of water resources in the Gulf Cooperation Council countries. Currently, all the countries of the Gulf Cooperation Council are experiencing large water deficits compensated for by mining of groundwater resources, installation of expensive desalination plant, and the expansion in the reuse of treated wastewater.

The present imbalance between the available water resources and water demands is chronic, and is expected to escalate in the future, unless positive and major positive steps are taken soon to rationalize and manage water demands, increase and augment water supply, and impose realistic controls on water use. A major review and shift in water policies in the Gulf Cooperation Council countries, emphasizing conservation and demand management, are suggested and outlined in this paper, with the overall objectives of securing long term water supplies while meeting strict criteria for socio-economic, financial and environmental sustainability and public health requirements.

The successful implementation of these policies would be a milestone for paving the long intricate path towards coping with the water scarcity in this arid region. However, failure to adequately address the problem of the growing water imbalance would result in severe negative impacts in both the quantity and quality of water supplies, widening food deficits, and deterioration of the standard of living. Furthermore, unless major shift in the population policies are implemented, the water problem will become one of the major constraints in future development of the countries of the region.

In this study, a general water resources policy with its main management components is outlined for the Gulf Cooperation Council countries. It is suggested that each country in the Gulf Cooperation Council appoint a national water council (if not existing yet) whose members are ministers responsible for water resources development, management, and utilization, to coordinate at the highest political level, and to formulate the overall national water policy of the country. A technical body (secretariat/committee), to be suggested by the council, and composed of senior scientists, economists, engineers, lawyers from concerned ministries and authorities, university and research institutes, and public agencies and stakeholders including (if the country political system permits) representatives of water users would help the council and ensure fair institutional and inter-ministerial coordination at the national level.

The water strategy should consider the water sector development, demand, protection, and sustainable use in the medium and long term (15-30 years). Data and information are needed to adequately form the base on which the strategy would be formulated with possible suitable options. The work plan would include several stages with a number of parameters relevant to each stage. These are: 1) Collection and review of existing data and information; 2) Review and evaluation of previous development studies; 3) Current conditions of water resources, population growth, socio-economic conditions, land use, and water utilization and related legislation in force; 4) Evaluation of water resources potential; 5) Water demands forecast; 6) Water resources development planning framework; 7) Water resources management, institutional reform, water facilities; 8) estimate of total cost; 9) Evaluation of strategy formulation: financial, economic, and social analyses. According to the evaluation results, the selected options and alternatives for strategy formulation could be decided and programs, water projects, tariffs, new technologies, and legal and institutional reforms could be recommended.

Water policies are usually projected over longer time periods and have normally unforeseen impacts which is difficult to measure in details and my not be implemented as proposed, but respond to current need and priorities. Water policies may require continuous monitoring for progress, re-evaluation or even revision of priorities that meet obstacles, and therefore, monitoring policy actions would benefit policy management.

ACKNOWLEDGEMENT

This study was prepared within the framework of UNEP Global Environmental Outlook (GEO) Project, and was funded by the United Nations Environment Program, The Regional Office for West Asia (UNEP/ROWA). The author would like to thank the water professionals in the Arab Center for Studies of Arid Zones and Deserts (ACSAD) and the Arabian Gulf University (AGU) for their contribution and help towards completing this study.

REFERENCES

Al-Alawi, J, and Abdulrazzak, M., 1994, Water in the Arabian Peninsula: problems and perspectives. In: Water in the Arab World, Perspectives and

Prognoses, Rogers, P, and Lydon, P (eds.), Division of Applied Sciences, Harvard University, 171-202.

Al-Kuwaiti, K, Al-Dhofar, M, and Al-Jughaiman, AR, 1999. Efforts on raising irrigation water use efficiency in Al-Hassa Region, Saudi Arabia. Proc. 4th Gulf Water Conference, Al-Zubari, W K, and Al-Sofi, AK (eds.), WSTA, p. 265-283 (in Arabic).

Al-Sheikh H M H., 2001. Water resources and development in Saudi Arabia. In: Water in the Arabian Peninsula, Problems and Policies, Mahdi K A (ed.), Ithaca Press, 295-328.

Al-Turbak, A S. and Al-Dhowalia, K H., 1996. Contribution of treated wastewater in solving water shortage problems in Saudi Arabia. Proc. Symposium on Wastewater Treatment Technology and Reuse, King Saud University, Saudi Arabia, 14-33 (in Arabic).

Al-Zubari W K., 1997. Towards the establishment of a total water cycle management and re-use program in the GCC Countries. In: Water in the Arabian Peninsula, Problems and Policies, Mahdi K A (ed.), Ithaca Press, 255-273.

Al-Zubari, W K, 2000. Guidelines for groundwater protection and pollution control in the Gulf Cooperation Council countries. UNESCWA EGM on Implication of Groundwater Rehabilitation for Water Resources Protection and Conservation, Beirut, 14-17 November, 2000, E/ESCWAS/ENR/2000//WG.3/12.

Arab Fund, 1995. the Economic Arabic Report, September 1995.

FAO, 1997. Irrigation in the Near East in Figures. Water Report No.9, Rome.

Gulf Cooperation Council, 1996, Economic Report. No. 11. GCC General Secretariat, Saudi Arabia.

International Desalination Association (IDA), 2000. IDA Worldwide Desalting Plants Inventory CD-PAM2000. Wangnick Consulting.

Ismail, N., 1995. Strategic projection for planning and management of water resources in the Gulf Cooperation Council countries. Attaawun, vol. 10, pp. 47-62 (in Arabic).

UNESCWA, 1999a. Groundwater Quality Control and Conservation in the ESCWA Region. Report No. E/ESCWA/ENR/1999/1, ESCWA, Beirut.

UNESCWA, 1999b. Development of Freshwater Resources in the Rural Areas of the ESCWA Region Using Non-Conventional Techniques. Report No. E/ESCWA/ENR/1999/16, UN Economic and Social Commission for Western Asia, Beirut.

UNSPD, 1997. World Urbanization Prospects: The 1996 Revision (Annex Tables). United Nations Secretariat Population Division. New York.

WRI/UNEP/UNDP/WB, 1996, World Resources 1996-1997. World Resources Institute, UNEP, UNDP, and the World Bank. Oxford University Press, New York and London.

Reallocation of Water and the Hydrological Effects of Climate Change: The Upper Rio Grande Basin, Southwestern USA

MICHAEL E. CAMPANA

Water Resources Program, Department of Earth and Planetary Sciences, University of New Mexico, Albuquerque, NM 87131 USA. E-mail: aquadoc@unm.edu

LOUIS SCUDERI, O. PAUL MATTHEWS, DAVID S. BROOKSHIRE, KATE KRAUSE, JANIE CHERMAK, BRADLEY CULLEN, SETH SNELL and KIRK GREGORY

Department of Earth and Planetary Sciences, Department of Geography, Department of Economics, University of New Mexico, Albuquerque, NM87131, USA.

ABSTRACT: The Southwestern United States is a semi-arid region that has experienced rapid population growth in the past few decades. This growth shows little sign of abating, so much so that there is concern among water planners and managers as to where additional water supplies will be found. The few large existing perennial streams – such as the Colorado River and the Rio Grande - are overallocated, meaning that additional surface water supplies cannot be developed. Ground-water supplies are also heavily used, leading to excessive water-level declines and incipient land subsidence; some areas also have ground-water quality problems (e.g., arsenic). To exacerbate matters, the federal government's Endangered Species Act (ESA) may require that water be left in some streams to preserve aquatic and riparian flora and fauna, thus reducing the amount available for agriculture and municipal and industrial uses. Preliminary research indicates that parts of this region are entering a drought, which may not reach its driest period until 2020 – 2030. The upshot is that reallocation of water is inevitable.

One approach to reallocation involves water marketing, which is already in effect in some areas. Water marketing will certainly not increase the total amount of water, but presumably will provide a market-based mechanism to sell or lease water rights. As an example, irrigated agriculture, the largest user of water, is perceived by some to be an inefficient and low-value use of water. Under the water market approach, a farmer might decide to retire some of her acreage, thereby freeing some water which could be sold or leased to a municipality or perhaps dedicated to instream ecological use.

Clearly, increased population, ESA considerations and impending drought bode difficult times ahead for water managers and planners who must reallocate precious water supplies among competing uses. Reallocation decisions will be politically sensitive and must be done with as much certainty regarding water supply and availability as possible (note that "availability" pertains to both hydrological and legal availability). Informed decisions with strong scientific bases can mitigate some of the political issues. Sophisticated models that can simulate the effects of global change, drought, and reallocation are clearly needed.

In response to that need, we propose a GIS-based (Geographic Information System) integrated physical hydrological - behavioral model of the upper Rio Grande Basin of New Mexico and Colorado. The model is unique in that it is process-based with the physical processes embedded within the GIS. Land-atmosphere interactions, surface water and ground water are represented, as is the legal availability of water. The physical model will determine the hydrological availability of water in the face of reallocation and climate change (drought); the behavioral model can assess stakeholders' responses to reallocation and reduced/increased hydrological flows. A preliminary model of a portion of the basin, the Conejos River Basin in the heavily agricultural San Luis Valley in southern Colorado, will be available by January 2003 for economic simulations.

Future plans call for the integration of water quality and instream ecological flow components into the basic physical hydrological-behavioral model.

Our approach will find applicability elsewhere, whether they be arid or humid regions; there is nothing inherent in the model that restricts its use to semi-arid climatic zones. The model will provide a stronger scientific foundation for the difficult water reallocation decisions facing humankind in the 21st century.

Water Resources Perspectives: Evaluation, Management and Policy. Edited by A.S. Alsharhan and W.W. Wood.
Published in 2003 by Elsevier Science, Amsterdam, The Netherlands, p. 169-181.

INTRODUCTION

Water availability in the semi-arid Rio Grande Basin (Figure 1) of the southwestern United States is a key component in the economic and environmental viability of the region. As in most semi-arid regions worldwide, precipitation is highly variable from year to year with the consequences of such variability readily seen in terms of significant environmental and economic impacts. Climatological research has shown that precipitation in this region is highly sensitive to variations in atmospheric circulation over the Pacific Ocean. Over short times scales a tight linkage to El Niño/La Niña events produces significant annual to decadal variability in precipitation totals and distribution (Cayan and Webb, 1989; Cayan and Peterson, 1989). Over longer time scales historical and paleoclimatic evidence (Stine, 1990) suggests that prolonged droughts of multidecadal to centennial length have occurred and that these droughts may have significantly altered the physical environment.

In the Rio Grande Basin, the institutions that evolved to cope with shortages have created a complex water allocation system (Baxter, 1997; DuMars et al., 1984; Meyer, 1996; Stevens, 1988). Because this system protects private property rights while recognizing public rights in the same water, reallocation in times of shortage becomes a difficult issue (Matthews, 1991; Matthews et al., 2001). Although reallocation is already occurring, it will accelerate and become a policy priority if long-term water shortages occur as a result of climate change, and shortages in this region may be imminent. If shortages occur as predicted, reallocation will be an absolute necessity because all the surface waters in New Mexico were completely allocated by the early part of the 20th century, and ground water is considered tributary to surface water (Clark, 1971; Clark, 1987). No "unused" water is available. The reallocation will pit agricultural uses against urban and industrial needs. Additionally, the traditional water paradigm that ignored public uses such as environmental protection, has changed (Wilkinson, 1985, 1991). Although these changes are being accommodated today, they must also be considered in future reallocations. The assessment tool proposed will allow those policy changes to be evaluated.

With institutional and economic changes already occurring, the ability to evaluate the impacts resulting from climate change becomes even more complex. The model must be able to distinguish between the changes that would have occurred anyway and the changes that can be attributed to climate change. The coupled model proposed will be a useful tool in evaluating the consequences of both natural climatic variability and human induced changes in water distribution in the region.

We will use coupled physical, biological, and human system models in an integrated Geographic Information System (GIS) framework to provide a quantitative assessment of the consequences of climate change and its impacts on water availability in the Rio Grande Basin north of Elephant Butte Reservoir. Elephant Butte was chosen as the cut-off because this is the control point for supplying water to Mexico under our treaty obligations and supplying water to Texas under an interstate compact. This integrated assessment will operate entirely within a GIS unlike most other assessment models that use GIS mostly for display. This approach permits efficient access and processing of the spatial data required for an integrated assessment. Because water law and economics will be integrated with physical and biological components, the assessment methodology can be used to evaluate the economic consequences of different environmental policies as well as physical and institutional changes. The model will allow us will evaluate the following questions.

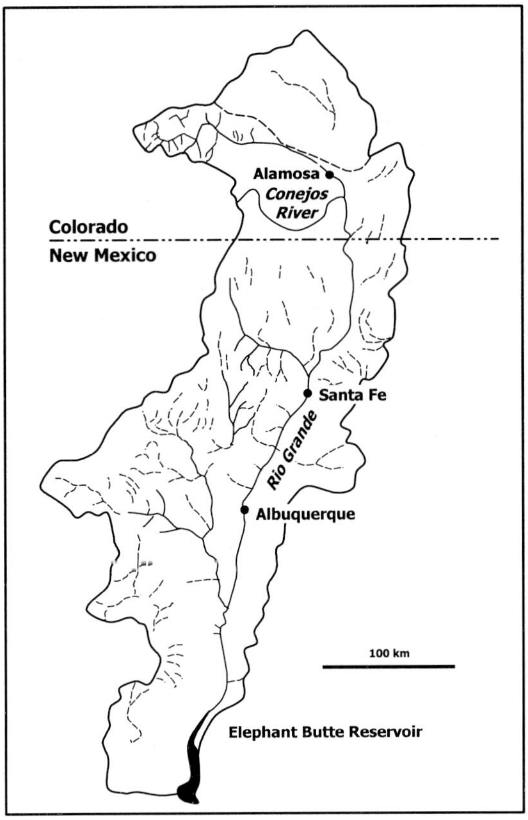

Figure 1. Upper Rio Grande Basin.

Physical Environment Questions and Economic / Institutional Impacts

1. What changes in water availability will be caused by El Niño/La Niña climatic variations? If shortages occur, which water users will be impacted because their water rights do not have a high enough priority under the appropriation doctrine? What are the economic impacts that will result from shortages or surpluses?
2. How do changes in vegetation and land use resulting from climate change, impact the amount of water available within a stream? What impact will that have on water users under the appropriation doctrine's priority system? What are the economic impacts of those changes?

Economic/Institutional Questions

1. If water shortages occur due to climate change, can the assessment method determine the impacts on other water users accurately enough to allow easier market reallocation between private parties? What is the economic impact of such changes?
2. If legislative reallocation occurs because of water shortages (i.e., minimum stream flows are set to protect endangered species), what is the economic impact?
3. If the evolution of water markets continues, what impact does climate change have on regional expansion (i.e., population and economic development)?

Modeling Research Questions

1. What is the effect of using different subsets of scale dependent processes as well as utilization of different resolution meteorological/climatic inputs on the modeling of total hydrologic output and model sensitivity?
2. Can the model distinguish ongoing institutional and economic changes from those that result from climate change?

Modeling Framework

The modeling framework uses a raster-based distributed water balance approach in which each raster element represents a bucket through which inputs and outputs may be routed. The Upper Rio Grande Basin (URGB) will be represented using a hierarchical resolution grid scheme based on a quadtree subdivision of the landscape. The raster data structure is designed to allow an infinite number of process resolutions on a as-needed basis (i.e., finer cells where detail is required, coarser cells where data preclude detail or processes operate on coarser scales) as well as the incorporation of remotely sensed imagery at several different spatial resolutions (Scuderi and Strome, 1988).

Specific grid-cell values (attributes) for a range of parameters will be georeferenced with respect to an underlying Digital Elevation Model with a minimum resolution within the database of 30 meters. Attribute values will be derived from existing mapped data sources (soils, geology), digital elevation models (slope, aspect), meteorological and climatic data (precipitation, temperature, snow cover), remotely-sensed information via image processing (vegetation, canopy cover, crop type), modeled attributes from within the model (crop water use, evapotranspiration, surface runoff, ground-water flow, total runoff) and economic and legal databases (water right priority, diversion points, land ownership, economic use). Finer resolutions are available for individual subcatchment and individual small-scale processes, but while possible, they will not be used in this study. At the coarse end, resolutions up to 2 kilometers will be used. This provides for a maximum of seven hierarchical layers with our model (Figure 2).

Water routing through the hydrologic network will allow determination of interaction effects between all database elements. One critical element from a policy perspective is water rights. Varying grid sizes represent the different scales at which the model will operate. The scale closest to the stream has a finer resolution because more detail is required while farther away the scale is coarser. This approach

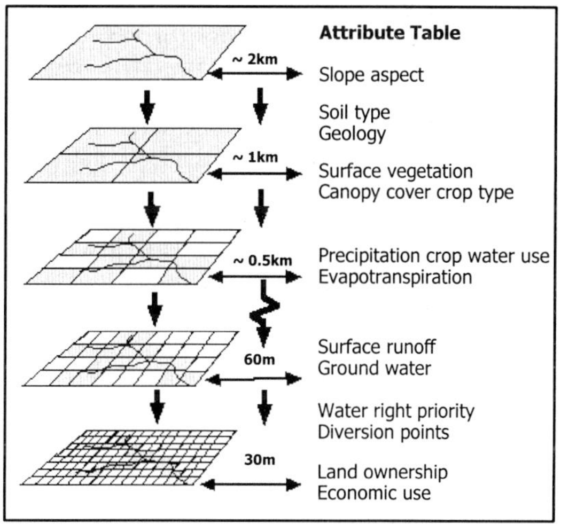

Figure 2. Hierarchical data structure and attributes. Note scale break between 0.5km and 60m - hierarchical levels at 250m and 125m are not shown.

allows the use of different resolutions in different areas and allows focusing in on smaller areas where specific subcatchment issues are to be addressed. Our approach allows for the evaluation of scaling effects on total hydrologic output using different subsets of scale dependent processes as well as the utilization and evaluation of different resolution meteorological/climatic inputs on model output and sensitivity.

The model is divided into four linked submodels that specifically deal with water rights, climatologic/vegetation inputs, hydrologic response, and economic utilization and valuation of the water resource.

Water Rights Submodel

The legal topology of water laws and treaties that act to constrain and modify water use and availability is included as topologically encoded layers and is part of the base system. Each submodel has, where appropriate, the ability to provide feedbacks to the other linked submodels.

Western water law is a combination of state and federal laws that define public and private rights in the use of water (Beck, 1991). These laws also control to a significant extent where, when and how water is used. State water law is based on the appropriation doctrine. This doctrine gives the first user of water a temporal priority over all who subsequently establish a right. As long as the water is put to a beneficial use, temporal priority is the controlling factor. If water shortages occur, the upstream water user with the lowest temporal priority must make up the shortage. This is true even if the person with the lowest priority can use the water more profitably or for more important purposes. Priority, not the nature of the use, controls water use. On top of the western states' appropriation doctrine is a layer of federal law. Although the federal government has deferred to the states on water allocation, federal laws that impact water are extensive. In New Mexico all surface waters are considered appropriated making reallocation a very real possibility.

Reallocation will require changes in water rights (Matthews, 1994). But changes are not always easy unless a model such as the one proposed here is used (McCormick, 1994). For example, senior priorities are protected against having their rights interfered with by junior users by virtue of their seniority. Junior users also have protection against senior users. Simply put, senior appropriators cannot change the way they are using water if junior water users will be harmed. Using ground water as an alternative may not be possible either because in New Mexico ground water and surface water are administered as a unit. Permits to develop new ground water sources may be denied unless a surface water use is retired. As a result, any change in the way water is to be used will have an impact on someone. If these upstream and downstream impacts are unknown a water market cannot operate, and the impacts of establishing minimum stream flows cannot be determined in advance. The example described below will illustrate how the system operates.

Consider a stream system with 10 users (Figure 3). The numbers are the priority each user has. The highest priority in the system is at the lowest point downstream. If number 1 does not have enough water, then 10 must cease using water until 1's right is satisfied. If 10 completely stop using water and 1's right is still not satisfied, then 9 must curtail water use. The process continues until 1 has enough water. Likewise, if 7 do not have enough water, then 10 must curtail use until 7's right is satisfied. If insufficient water is available from 10's curtailment, then 7's right remains unsatisfied. Seven cannot look to a lower priority (8 or 9) in another subcatchment, because there is no way for the water to get to where it can be diverted.

Suppose one of the water users wants to transfer (sell) water to a user outside the basin. For example, 10 has a right to divert 20 cubic feet per second of water during July, and she wants to sell it to a city outside the watershed. Senior appropriators downstream (7 and 1) could object to the sale if it will deprive them of water to which they are entitled. Suppose 2 wants to make a sale. If there is either surface or subsurface return flow, then 3, as a junior appropriator, could object because 3 is protected as to the conditions of the stream when 3 made the appropriation. All 2 can sell is the amount consumed.

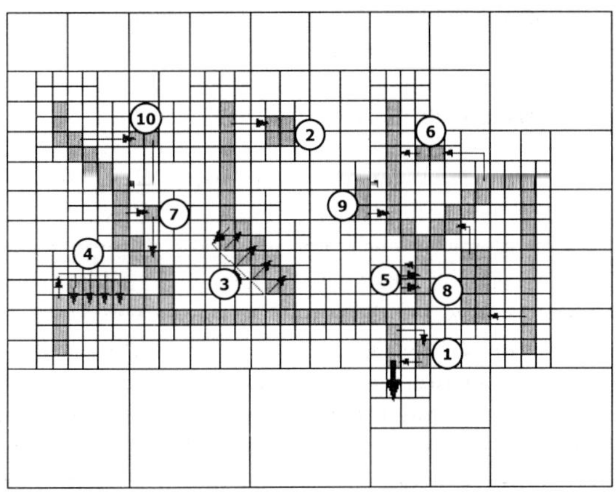

Figure 3. Schematic of a stream system with 10 users.

Suppose 6 wanted to change the point of return flow so it is downstream from 9's point of diversion. Nine could prevent this change, because junior rights are protected as to the conditions of the stream when their diversion was made.

The model being developed will account for this priority system by "turning off" the diversion of the most junior upstream appropriator when a downstream senior does not have enough water. This water will then be moved downstream. As it moves downstream through each grid cell, water that evaporates or infiltrates will be subtracted. If not enough water reaches the senior appropriator, the next most junior will be "turned off". This process will continue until the senior right is satisfied. A similar process can be used to determine the impact of changes in a water right that could occur because of a sale. Also, if the federal government determines the river has an endangered species and a minimum stream flow is required, junior appropriators could be "turned off" so that a specified volume of water is left in each of the stream's grid cells where protection is needed. If one part of the watershed becomes wetter and another drier, the model will show those differences and calculate their impact on water rights.

Climate/Vegetation Submodel

The primary physical driver of hydrologic systems is the availability of water in the drainage basin. While specifically a function of precipitation type, amount, intensity and duration, the spatial distribution of available water that can be channeled into a hydrologic flow (either surface or subsurface) is modified by the landscape as it responds to forcing energy balance factors and to surface characteristics, including; terrain, modified solar radiation receipt, air temperature, precipitation, cloud type, cloud cover (percent), water vapor pressure of air, wind velocity, water holding capacity of the soil, antecedent soil moisture levels, and vegetation (Bonan, 1989). Many of these variables are not commonly recorded and are rarely available over large areas. However, they can be estimated using standard energy balance approaches coupled with some knowledge of the meteorological, climatic, soil and vegetation characteristics of the region (Bonan, 1989; Blanchard et al., 1981; Choudhury and Blanchard, 1983).

The calculation of available water is a complex process and involves three linked submodels, namely: an atmospheric model derived from meteorological variables, an energy-balance model, derived from first principles of physics along with simplifying assumptions made to parameterize model initial conditions in the absence of measured variables, and a soil water model. Since we are starting without complete knowledge of the environmental conditions over the watershed and/or coarse scale meteorological data, the "missing" variables are estimated using alternate methods and by relying on some generalizing assumptions as described below.

We start with an energy balance model that has as its basis the net radiation equation. We calculate the shortwave fluxes of the net radiation equation utilizing methods outlined in commonly available energy balance models (Brock, 1981; Bristow, 1987; Choudhury, 1982; Oke, 1987). Daily meteorological conditions for sites within the study area are input directly from historic meteorological datasets. These sparse values are then interpolated to finer grid scales using an artificial neural network approach. These daily values are then used to estimate other elements of the model as described below.

Cloud density and cloud type effects on solar radiation receipt are first-order approximated with respect to precipitation by assuming that these variables are directly linked to precipitation totals (Oke, 1987). A cloud routine first evaluates the daily precipitation total and determines, based on climatologic norms for the region, the cloud type, cloud density and water vapor pressure of air normally associated with such precipitation totals. Air temperature is then used to determine whether precipitation is added directly to the soil moisture or snow store. In the case of non-precipitation days, cloud type and cover or cloudless conditions are randomly assigned to days based on climatic normals for these variables. These assignments can be modified with respect to monthly and or seasonal variations in these parameters, and will be augmented with meteorological radar and lightning strike data when available (Potter et al., 1998).

Surface albedo is derived with respect to surface cover. Growing season surface cover type is derived from 30 meter resolution multispectral LANDSAT imagery supplemented with existing maps and ground truthing. Winter season surface albedo, including the effects of snowfall events will be calculated from available meteorological data and prior day precipitation and/or snow levels (from the NOAA GOES satellite). Cover categories can range from dry or wet bare soil to fully vegetated cover during the height of the growing season. Snow cover/vegetative albedo values will change continuously and will be updated at each daily timestep.

We calculate the longwave fluxes of the net radiation equation utilizing the same energy balance model. The two primary longwave terms, sky counterradiation and surface reradiation, are functions of air and surface temperature respectively. Counterradiation is approximated using a modified version of Brunt's equation (Brunt, 1932). This approach has been shown by Hatfield et al., (1983) to provide hourly estimates to within an accuracy of five percent. Net downwelling longwave radiation is estimated utilizing a calculated surface temperature, which itself is a function of the amount of solar radiation received (Oke, 1987).

Total potential evapotranspiration is calculated directly from solar radiation receipt, air temperature, and empirically derived site constants using an approach initially developed by Priestly and Taylor (1972) and modified by Campbell (1977). In this approach the soil conductive flux is assumed to be zero when averaged over several days and the net surface reradiation is proportional to the incoming shortwave flux and air temperature. Total potential evapotranspiration can be calculated directly from solar radiation receipt, air temperature, and empirically derived site constants (Bonan, 1989).

Actual evapotranspiration depends on the evaporative demand of the atmosphere, transpiration from plants and available soil water. Moisture available for evapotranspiration can come from several layers in the soil, each of which must be modeled to arrive at the correct soil moisture content. These include an uppermost soil layer (<2 cm depth) in which evaporation occurs in a fairly unconstrained manner and a deep soil layer (>2 cm depth) in which evaporation is slowed by capillary action. In addition, snow layers, and losses from vegetation, surface runoff and deep runoff add additional complications.

To approximate the change in soil moisture content over time we adopt an Antecedent Precipitation Index (API) approach first suggested by Lindsey et al., (1949) and modified by Saxton and Lenz (1967). In their approach the API is a function of daily effective precipitation, the prior day's API value and a dimensionless recession or depletion coefficient (K) which is a function of soil characteristics. K is a negative exponential function of daily total evapotranspiration estimated with respect to the energy balance values calculated above, soil depth, field capacity and wilting point (Choudhury and Blanchard, 1983).

The actual calculation of the soil water levels utilizes the recession coefficient, the prior day's soil moisture content, and the maximum soil water capacity of the soil. Actual soil water loss (AWL) (Pastor and Post, 1984) is calculated with respect to daily meteorological conditions, with clear sky, high temperature, low humidity conditions placing the greatest potential demand on soil moisture. If available soil moisture exceeds the maximum capacity of the soil, the excess is routed to the hydrologic submodel (see below) and removed from the system before the evaporative losses are calculated. If precipitation occurs then the excess over evaporative demand is added to soil moisture store.

In addition to the above, the model needs to have the ability to be driven by different scale meteorological inputs. At one extreme we will utilize meteorological station data and spatial interpolation schemes to capture the spatial variability of the meteorological fields (primarily winter precipitation and winter and summer temperature). However, since summer precipitation is associated with convective cells and is difficult to evaluate using meteorological station data alone we will use lightning and radar data to capture its spatial distribution (Potter et al., 1998).

On the other extreme to capture output from General Circulation Models and to utilize this output at the proper spatial scale we will use an artificial neural network approach in which the network "learns" the distribution of temperature, precipitation and other meteorological fields over a landscape under different conditions and then "interpolates" this learned distribution to sites at sub-GCM cell resolutions.

Hydrologic Submodel

We will develop a physically based distributed hydrologic model capable of calculating daily water volume per grid cell. Traditionally, GIS and hydrologic models have been employed in loosely-coupled modeling approaches where the GIS is used to manipulate spatial data and extract model parameters for use in a separate hydrologic model. The model results are then displayed in the GIS. Our approach will develop a model that operates entirely within a GIS, thereby permitting efficient access and processing of the spatial data that are required for the modeling effort as well as to analyze and display model results. The model will assess the hydrologic inputs and outputs per cell on a daily time interval. Additionally, the exchange between surface water and ground water will be modeled. In this fashion, the model will allow for the calculation of water volume for any location within the grid, thereby permitting the identification of instream flow volumes at any point on a stream. A description of the spatial data layers and the physical model is given below.

Surface Water

In the model, each grid cell functions as a reservoir for which a daily water balance is computed by accounting for hydrologic gains and losses. Hydrologic gains are precipitation in the form of rain or snow, and later soil moisture movement. Hydrologic losses include evaporation, evapotranspiration, lateral soil moisture movement, and percolation losses. In the model, daily soil moisture accounting considers precipitation as rainfall or snow, evaporation and evapotranspiration, percolation, and runoff, in computing a water budget for a two-layer soil system. Percolation occurs at the rate of the hydraulic conductivity corresponding to the soil moisture content, and is calculated according to travel time through a layer (Arnold, et al., 1990). The soil profile is a two-layer system where the top layer represents a thin (<2 cm) zone of unconstrained evaporation and the lower layer corresponds to effective vegetative rooting depth where evaporation is governed by capillary action. Certain properties of the lower layer may vary, such as bulk density. Lateral soil water movement may occur when the storage in the lower layer exceeds field capacity after percolation and the travel time of later flow is computed using the ratio of saturated hydraulic conductivity and vertical hydraulic conductivity for the layer. Deep percolation will be counted as a loss to the surface water balance and will be used as input into the ground water component of our model, thereby permitting the computation of the exchange between surface water and ground water.

We propose to develop and evaluate two surface water models based upon (1) saturation excess and (2) infiltration excess. The saturation excess model will employ the excess precipitation (soil moisture excess above evaporative demand) computed by the soil moisture component of the climate submodel described above. For dry days, the water losses for each grid cell will be computed. For days with precipitation, interception losses will be subtracted from the precipitation depth based upon information in the GIS vegetation data layer. Additionally, soil moisture storage and percolation will be computed, and any excess precipitation determined. The excess will then be routed cell by cell through the basin. Snowmelt will be included in the water balance model. A thermal energy balance will be employed to track daily soil and any snowpack temperatures. Snowmelt is used to supplement rainfall and the depth to the uppermost frozen soil layer. Melted snow will be treated as rainfall. When a water diversion takes place, a diversion module will be used to determine the amount of available soil moisture capacity and the depth of runoff, or return flow, using the curve number method. The model will accommodate manual interval irrigation, in which a specified amount of water is applied on a regular schedule, as well as manual irrigation on irregularly spaced dates. The date and volume of irrigation requirements will be contained in the water allocation data layer, thereby permitting easy access.

The infiltration excess model will use the modified Soil Conservation Service (SCS), now the Natural Resources Conservation Service, curve number method (USSCS, 1972) to compute surface runoff from each grid cell. The method assumes that runoff occurs after the initial abstraction (Ia) of precipitation has been satisfied. Initial abstraction consists of interception, surface storage and infiltration. Initial abstraction (Ia) is expressed as 0.2S, where S is a surface retention parameter, which is a function of the curve number. A curve number is an empirical value that represents a measure of retention of water by a given combination of land cover, land use and soil characteristics, and ranges from 0 to 100. In order to produce runoff, precipitation must be greater than 0.2S. Runoff depth is influenced by the initial soil moisture retention parameter. Consequently, curve numbers will be adjusted to reflect the initial soil moisture conditions. For each dry day, the energy balance in the climate submodel described above will be used to compute the losses based upon initial soil moisture content. For days with precipitation, the model will first compute available soil moisture storage, then the retention parameter will be adjusted to reflect current soil moisture storage. In upland areas, excess precipitation will be routed sequentially from cell to cell as overland flow to receiving streams. The water equivalent from snowmelt will be used to adjust the curve number, and the melted snow will be treated as rainfall.

When a water diversion takes place, a diversion module will be used to compute the water balance within the cells where the water is applied. The module will compute the initial soil moisture capacity and the depth of runoff, or return flow, for either the saturation excess or precipitation excess model, depending upon which surface water model is selected. For the precipitation excess model, the curve number will be adjusted to reflect initial soil moisture conditions. The diversion module will accommodate manual interval irrigation, in which a specified amount of water is applied on a regular schedule, as well as manual irrigation on irregularly spaced dates. The date and volume of irrigation requirements will be contained in a water allocation data layer, thereby permitting easy access.

Routing of precipitation excess will be computed for overland flow and channel flow. The routing relationships will be based on distributed unit hydrograph theory, which computes the time required to reach steady uniform flow, and permits the use of multiple grid cell resolutions. Overland flow for non-channel cells will be computed as a function of cell area and excess precipitation. The time of travel within the cell will be computed based upon the time required to achieve equilibrium flow by the kinematic wave equation. This relationship uses cell parameters such as length of overland flow, Manning's roughness coefficient, excess precipitation, and the slope of the cell. Stream routing will be performed using the channel network to compute flow velocities and travel times between stream channel grid cells and the watershed outlet). Stream flow velocity will be computed using a combination of the Manning and continuity equations, which incorporate the slope of the channel cell. Discharge for a stream cell will be the mean of inflow from the upslope stream cell, plus direct upland overland flow, and outflow from the cell. Additionally, alluvial channels in semi-arid watersheds lose large quantities of stream flow. These transmission losses will be computed using a method described tailored for ephemeral streams.

Ground Water

Following our formulation of surface water flow, we will develop a ground-water submodel that operates entirely within a GIS. This will best be accomplished by developing a code virtually from scratch, as opposed to simply attempting to couple an existing code such as MODFLOW (McDonald and Harbaugh, 1988) to the surface water submodel. Their model did not operate entirely within a GIS, nor could it offer the resolution of ours (their grid cells were a constant 2.6 square km). Our approach will actually be more efficient in that it will permit a smoother integration among the climate, surface water and ground water submodels since all the models will have very similar foundations. We will use tested algorithms to describe ground-water flow and storage and surface water ground water exchange. What will be novel will be the nesting of these algorithms within a GIS framework and their linkage to the surface water submodel to produce a truly integrated physical model designed expressly for water allocation and climate change simulations.

Deep percolation from the surface water submodel will be used as input to the ground water submodel; boundary conditions can also be used to provide additional input, and pumping can be specified as a withdrawal. The ground-water reservoir will be treated as a system of grid cells; these can be of the same area as the surface water submodel cells, which can be as fine as 30 m x 30 m or as coarse as 2 km x 2 km, although in general, fine resolution (30 m x 30 m) is not required for ground water reservoirs. We will assume that ground-water flow is topographically driven and that the ground-water reservoir can be represented by a single layer of cells – essentially a large unconfined aquifer.

We will simulate flow by using a physically-based approach developed by Wigmosta et al. (1994): a grid-based, quasi three-dimensional saturated subsurface flow algorithm. They used it as the basis for their distributed hydrology-soil-vegetation model. Their method permits subsurface saturated flow to be calculated on a grid cell by grid cell basis, as opposed to the statistical-dynamical approach employed in TOPMODEL (Beven and Kirkby, 1979; Lamb et al., 1997). We chose the Wigmosta et al. (1994) approach over the statistical-dynamical one because we believe it to be more rigorous and, in a recent paper comparing the two methods, Wigmosta and Lettenmaier (1999) noted that the former yielded better results in transient simulations. In our research, transience is important as we specifically wish to treat the hydrologic changes induced by climate change. Wigmosta and Lettenmaier (1999) further noted that in models such as TOPMODEL, very little attention has been given to the representation of subsurface flow. Our efforts will devote much-needed attention to the representation of subsurface flow in integrated, GIS-based models by extending the Wigmosta et al. (1994) approach to larger systems involving more hydroclimatologic components.

Grid cells are centered at each elevation point; each cell can exchange water with its eight adjacent neighbors. We will assume that local hydraulic gradients can be approximated by water table slopes; in the upper reaches of the watershed, where such data may be unavailable, the topography can serve as a proxy for the water table slope. Darcy's Law is used to calculate flow from a particular cell to its neighbor cells; since the water table slope is assumed to approximate the hydraulic gradient, i.e., the Dupuit-Forchheimer (DF) assumptions, we can write:

$$Q = -Tw(\tan\beta) \qquad (1)$$

where Q is the volumetric flow rate from one cell to a neighbor cell; T = transmissivity in the direction of flow; w = width of flow; and $\tan\beta$ is the water table slope in the direction of flow (Bear, 1972). Use of the DF assumptions will introduce some error where water tables are steep, i.e., where $\tan\beta$ does not equal sin β, such as the in the vicinity of a pumping well. But since our model is not designed to provide detailed information on hydraulic heads in the vicinity of pumping wells, this more simplistic approach is entirely appropriate. We also note that even in cases where the DF assumptions are tenuous, the flow rates (i.e., values of Q) calculated from the DF assumptions are accurate (Bear, 1972).

Flow is partitioned from one cell to its neighbors using DEM data (see Wigmosta and Lettenmaier's (1999) development of equation 17 and their discussion in that paper's Appendix). The algorithm calculates depths to water tables, which can be converted to water table elevations. It also calculates exfiltration, or loss of water from the ground water zone to the surface. Should the grid cell above the ground water grid cell represent a stream, exfiltration can represent baseflow to the stream and "negative" exfiltration can represent infiltration from the stream - thus simulating surface water – ground water exchange.

Calibration will be effected by comparing our results to those of the various groundwater models constructed for portions of the watershed.

Economic Submodel

Many of the impacts that occur in the West from water reallocation are often analyzed from the perspective of a benefit-cost framework focusing primarily on the efficiency impacts. This often involves the determination of a set of market and non-market benefits and costs for a rather stylized set of tradeoffs. These analyses are relatively institution-free and do not easily allow for a clear understanding of all of the impacts of a water reallocation change that stems from a climate and/or institutional change.

In the West the central theme of the controversies of water reallocation, no matter the source, pertain to distributional effects. The controversies ask not what is the optimal amount of species habitat but in fact, what is the cost that must be incurred to minimally achieve a level of habitat to insure the survival of the species as a case in point. [Note: essentially, this is the setting of a safe minimum standard (SMS) approach which originates from uncertainty (see, for example Hanley et al., 1997). If society is unsure about the future cost of current environmental degradation, there are basically two courses of action that can be followed; (1) conserve the resource at a specific level, or (2) don't conserve the resource. Conserving is a risk-minimizing strategy and places the developer in the position of justifying development (Ciriacy-Wantrup 1952; Bishop 1978, 1993; Randall and Farmer, 1995). Thus, the question that is fundamentally at the root of legislation such as the Endangered Species Act or any other institutional adjustment and/or an analysis of a climate event is: what are the total and spatial economic impacts to all groups and economic sectors? In order to answer such a question, it is necessary to incorporate into the economic analysis the relevant physical and natural science setting and the appropriate institutional framework. Thus, in order to adequately model the impact of social or physical events on a watershed, events and impacts between the disciplines have to be assessed simultaneously. Rather than an event being considered as an exogenous impact to the system, a hybrid model allows for interactions between the physical and social worlds within the system.

Developing an integrated watershed simulation model that combines surface water and ground water considerations with an economic input-output (I-O) model and institutional water rights of stakeholders will allow us to estimate the economic impact of events (e.g., institutional changes or supply variations) in the watershed area, as well as estimate the impact on the watershed of economic or institutional decisions. An I-O model allows us a methodology to determine the value of production across sectors, given the available resources and applied technology. By combining an I-O model with the proposed distributed GIS model, the economic impact of events can be measured at a micro level that accounts for spatial variation in impact magnitudes. The importance of this is twofold: (a) the Rio Grande Basin is diverse in cultural heritage, population density, and industry and (b) events can impact both locally and regionally.

An I-O framework provides an estimation technique that defines the relationships between primary inputs (including water) and industry/household demand. When the level of a primary input changes, the impact of the change can be tracked through individual industries. By employing a micro-level analysis, we are able to estimate the impact across various industries and sectors of the economy. The proposed model will allow for accurate estimations of, among other things, transaction costs associated with institutional changes or the impact on the market economy of increased or decreased water supplies that is both spatial and accounts for divergent groups in the region.

The proposed spatial nature of an embedded I-O framework allows for "what if" questions. For example, consider investigating the spatial susceptibly of the agricultural sector to changes in physical (climate change) or institutional settings. In the Southwest, determination of these impacts is important in that a wide range of heritage groups are stakeholders in the overall water system. Thus, crop failure from climate change may not only hurt the rancher/farmer but may ripple through the economy. Given that different groups hold different shares of the overall economy, the possibility of disproportionate impact is real. Consequently, not only is there a multiplier effect that will ripple through the economy at the local, state and national level, but individual groups may share in the outcome differently. While it is difficult, from an economic perspective, to define social equity, the level of analysis afforded by this model also allows a comparison of impacts across communities and industries and it is possible that comparing the estimates can draw some inferences about social equity.

Because the model is integrated, the impact of economic decisions that result in changes in water-use patterns can also be assessed from both the economic perspective and the ecological perspective. Consider the farmer who is impacted by a reallocation, which reduces the available water to his farm. His current farming practices are no longer economic and so he is faced with the choice of: (a) switch to a crop that requires less water, but is also less profitable; (b) sell the land; or (c) if possible, sever and sell the water rights. While (a) has a potentially economic impact on the community and various industries, it does not change water-use patterns. However, (b) and (c) have the potential to impact the community economically, as well as the potential to impact water-use patterns in the watershed. This can result in an ecological impact on the watershed, as well as in an economic impact over a much larger area than that of (a).

Impacts also can be exacerbated by the fact that the Upper Rio Grande Basin has a cross-section of industries ranging from low-use to high-use, water sensitive industries. Whereas an impact on a low-use industry may have minimal effects, the impact on a high-use industry (e.g., Intel Corporation and the manufacturing of microchips) may have enormous ripple effects through the economy and through downstream water-use patterns.

The formal economic model will begin with a county level IMPLAN (Minnesota IMPLAN Group, Inc. 1997) based economic model. IMPLAN (short for Impact Analysis for PLANning) was developed in 1979 by the Forest Service of the U.S. Department of Agriculture in cooperation with the Federal Emergency Management Agency and the U.S. Department of the Interior's Bureau of Land Management. IMPLAN has two components: (1) software for an I-O model that provides an interface to describe an economy (at a national, state, county or zip code level), or predict impacts to the economy, given changes in the system; and (2) a database of economic and demographic variables. IMPLAN will provide the basis for the economic component of the hybrid model, as well as some of the necessary data.

SUMMARY

How can an integrated GIS help deal with reallocation? In the hypothetical example alluded to earlier each element of the water right can be incorporated spatially by assigning different attribute values to the required coverages. The time element can be simulated in a more realistic way by keeping the time step for calculating changes in flow short. For example, flow could be calculated on a daily or weekly time interval. In this way flow could be simulated and the movement of water within the system could be calculated on that basis. As more detailed information becomes available it could be incorporated into the different coverages to improve the quality of the database. The insecurity associated with water rights definition and identification may not be completely overcome, but tying the data to a spatial and temporal reference will make the existing data more usable.

In answering the question of how much water is needed or used at a site, models must be fully integrated and cannot look at just of few of the variables that influence determining consumptive use, duty of water, amount conserved or amount salvaged. The multiple attribute data provided within different GIS coverages allows for a more comprehensive approach. Site data can be used to determine evaporation, transpiration, infiltration,

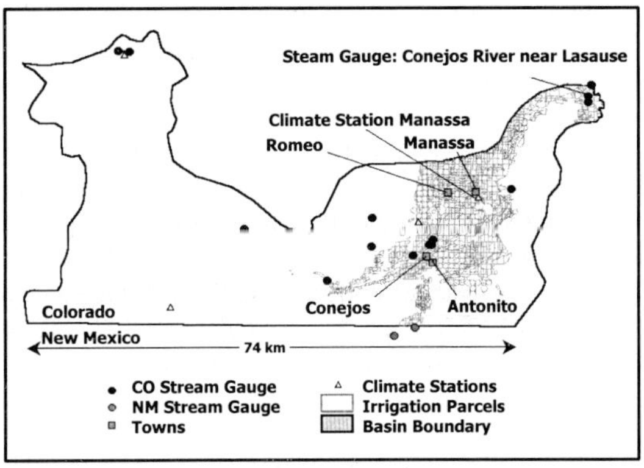

Figure 4. Pilot study area: Conejos river basin, Colorado (modified from Colorado Division of Water Resources' Decision Support System).

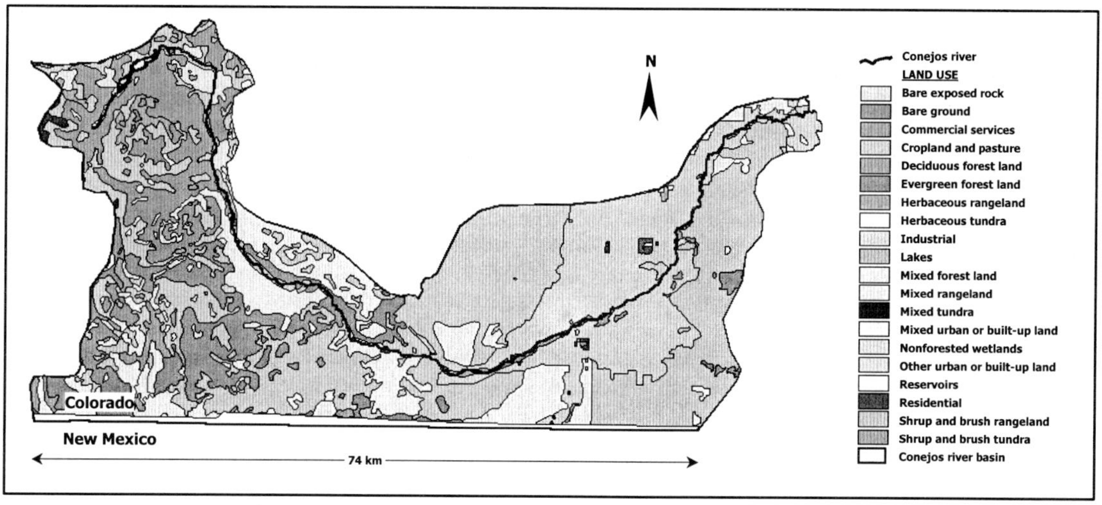

Figure 5. Conejos river basin, Colorado: land use (modified from Colorado Division of Water Resources' Decision Support System).

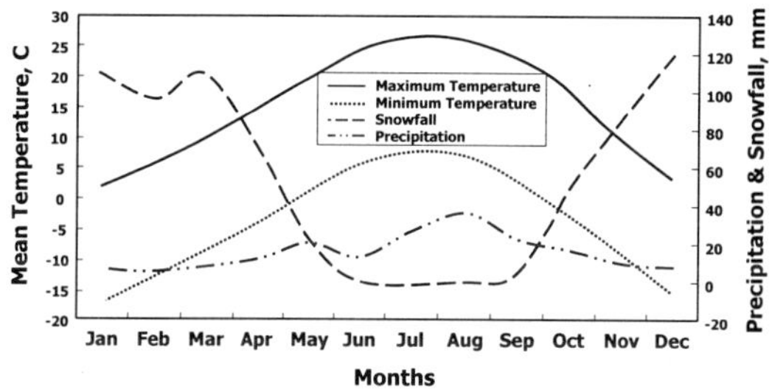

Figure 6. Climate data at Manassa, Colorado (1931-2000).

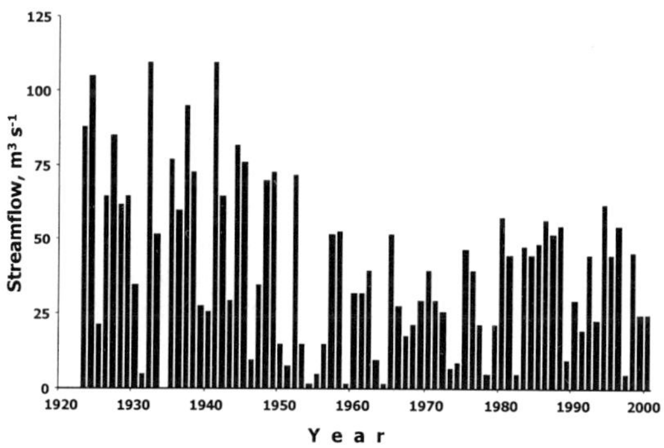

Figure 7. Conejos river near Lasauses, Colorado (1920-2000).

surface gains or losses from overland flow, and subsurface water movement for each grid cell. By changing the characteristics of different attribute data to simulate real world changes in site conditions, the impacts of those changes on water availability can be modeled (Matthews et al., 2001).

Just as the site variables can be modeled, the same process can be used to determine the impacts on third parties and the public. In this case, the water has to be routed from one grid cell to the next. For example, a specific site receives three inches of rain in two hours. Because of soil infiltration properties and slope only part of that water will infiltrate; the rest will become surface runoff. The GIS model will be able to "capture" the site specific infiltration and route the surface runoff across adjacent grid cells where it flows. If water is moved from one place of use to another the model will be capable of simulating the impacts this change would have (Matthews et al., 2001).

The proposed model, which will first be tested in the Conejos River Basin (Figures 4-7) can also distinguish differences in underlying consumer preference structures. Disaggregated consumer demand, coupled with a GIS modeling approach, allows for a model with spatially designated preferences. This micro-level analysis allows for a more robust analysis of consumer responses to changes in a water market. Additionally, the dynamic nature of such a model allows for a changing preference structure over time, if in fact preferences are influenced by expectations of climate change and long-term water availability (Matthews et al., 2001).

Theoretically, according to Matthews et al. (2001), all this is possible within a GIS, but some significant issues must be resolved before such a system is more than hypothetical. These problems, once resolved, will create a system which can more accurately answer the questions related to water reallocation and availability, climate change, legal availability and economics. Current methods will never provide adequate answers because they are not spatially or temporally detailed enough.

ACKNOWLEDGEMENTS

The assistance of Joshua Nims and Julie Anderson is greatly appreciated. This work is based in part upon a U.S. Environmental Protection Agency (USEPA) STAR grant (*An Integrated GIS Framework for Water Reallocation and Decision Making in the Upper Rio Grande,* Grant Agreement R-82807001-0); a National Science Foundation (NSF) grant (*A Quantitative Assessment of the Economic and Institutional Impacts of Climate Change on the Upper Rio Grande Valley Using an Integrated GIS Framework,* Agreement No. MMIA-9909140); and SAHRA (*Sustainability of semi-Arid Hydrology and Riparian Areas*, Agreement No. EAR-9876800) under the STC Program of the National Science Foundation. This article has not been subjected to the USEPA's required peer review policy and therefore does not necessarily reflect the views of the USEPA and no official endorsement should be inferred. Any opinions, finding, and conclusions expressed herein are those of the authors and do not necessarily reflect the views of SAHRA or the NSF.

REFERENCES

Baxter, John O., 1997. *Dividing New Mexico's Waters-1700-1912*, Albuquerque: University of New Mexico Press.

Bear, J., 1972. *Dynamics of Fluids in Porous Media*. New York: Elsevier, 764p.

Beck, Robert E. ed., 1991. *Water and Water Rights*, 6 volumes, Charlottesville, VA:The Miche Company.

Beven, K. J. and M.J. Kirkby, 1979. A physically based variable contributing area model of catchment hydrology. *Hydrology Science Bulletin*. 24: 43-69.

Bishop, R.C., 1978. Endangered species and uncertainty: the economics of a safe minimum standard. *American Journal of Agricultural Economics* 60: 10-18

Bishop, R.C., 1993. Economic-efficiency, sustainability, and biodiversity. *Ambio* 22(2-3): 69-73.

Bonan, G.B., 1989. A computer model of the solar radiation, soil moisture, and soil thermal regimes in boreal forests. *Ecological Modeling* 45: 275-306.

Bristow, K.L., 1987. On solving the energy balance equation for surface temperatures. *Agricultural and Forest Meteorology* 39: 49-54.

Brock, T.D., 1981. Calculating solar radiation for ecological studies. *Ecological Modeling* 14: 1-19.

Blanchard, B.J., McFarland, M.J., Schmugge, T.J., and Rhoades, E., 1981. Estimation of soil moisture with API algorithms and microwave emission.*Water Resources Bulletin* 17: 767-774.

Brunt, D., 1932. Notes on radiation in the atmosphere. *Quarterly Journal of the Royal Meteorological Society* 58: 389-420.

Cayan, D.R. and Peterson, D.H., 1989. The influence of North Pacific atmospheric circulation on streamflow in the west. *American Geophysical Union Monograph* 55: 375-398.

Cayan, D.R. and Webb, R.H., 1992. El Nino/Southern Oscillation and streamflow in the western United States. In: Diaz, H.F., and Markgraf, V., eds., *El Nino, Historical and Paleoclimatic Aspects*. Cambridge, Cambridge University Press.

Campbell, G.S., 1977. *An Introduction to Environmental Biophysics*. Springer: New York.

Choudhury, B.J., 1982. A parameterized model for global insolation under partially cloudy skies. *Solar Energy* 29: 479-486.

Choudhury, B.J. and Blanchard, B.J., 1983. Simulating soil water recession coefficients for agricultural watersheds.*Water Resources Bulletin* 19: 241-247.

Ciriacy-Wantrup, S., 1952. *Resource Conservation: Economics and Policy*. University of California Press: Berkeley.

Clark, I.G., 1987. *Water in New Mexico--A History of its Use and Management,* Albuquerque: University of New Mexico Press.

Clark, R.E., 1971. Water rights problems in the Upper Rio Grande watershed and adjoining areas. *Natural Resources Journal* 11: 46.

DuMars, C.T., M. O 'Leary, and A.E. Utton, 1984. *Pueblo Indian Water Rights--Struggle for a Precious Resource*. Tuscon: University of Arizona Press.

Hanley, N., Shogren, J.F., and B. White, 1997. *Environmental Economics in Theory and Practice*. Oxford University Press: New York.

Hatfield, J.L, Reginato, R.J. and Idso, S.B., 1983. Comparison of long-wave radiation calculation methods over the United States. *Water Resources Research* 19: 285-288.

Lamb, R., K. Beven and S. Myrabo, 1997. Discharge and water table predictions using a generalized TOPMODEL formulation. *Hydrological Processes* 11(9): 1145-1167.

Lindsey, R.K., Kohler, M.A. and Paulhus, J.L.H., 1949. *Applied Hydrology*. McGraw-Hill: New York

Matthews, O.P., 1991. Water is not 'real' property. *Water Resources Update* 85: 19-21.

Matthews, O.P., 1994. Changing the appropriation doctrine under the model state water code. *Water Resources Bulletin* 30: 189-196.

Matthews, O.P., L. Scuderi, D. Brookshire, K. Gregory, S. Snell, K. Krause, J. Chermak, B. Cullen and M. Campana, 2001. Marketing Western water: an a process-based geographic information system improve reallocation decisions? *Natural Resources Journal* 41(2): 329-371.

McCormick, Z., 1994. Institutional barriers to water marketing in the West. *Water Resources Bulletin* 30: 953 – 961.

McDonald, M.G. and A.W. Harbaugh, 1988. A modular three-dimensional finite-difference ground-water flow model. *U.S. Geological Survey, Techniques of Water-Resources Investigations*, Chapter 6, 576p.

Meyer, M.C., 1996. *Water in the Hispanic Southwest-- A Social and Legal History 1550-1850*, Tuscon: University of Arizona Press.

Minnesota IMPLAN Group, Inc., 1997. *IMPLANS Professional*. Minnesota IMPLAN Group, Inc.: Stillwater, MN.

Oke, T.R., 1987. *Boundary Layer Climates*. Methuen: London.

Pastor, J and Post, W.M., 1984. Calculating Thornthwaite and Mather's actual evapotranspiration using an approximating function. *Canadian Journal of Forest Research* 14: 466-467.

Potter, D.U., Gosz, J.R., Molles, M.C. Jr., and Scuderi, L.A., 1998. Lightning, precipitation and vegetation at landscape scale. *Landscape Ecology* 13: 203-214.

Priestly, C.H.B. and Taylor, R.J., 1972. On the assessment of the surface heat flux and evaporation using large-scale parameters. *Monthly Weather Review* 100: 81-92.

Randall, A. and M. Farmer, 1995. Benefits, Costs, and the Safe Minimum Standard of Conservation, in D. Bromley (ed.), *The Handbook of Environmental Economics*. Blackwell: Oxford.

Saxton, K.E. and Lenz, A.T., 1967. Antecedent retention index predicts soil moisture. *Proceedings of the American Society of Civil Engineers, Journal Hydraulics Division* 93: 223-241.

Scuderi, L.A. and Strome, W.M., 1988. Bringing remote sensing down to earth, *Advanced Imaging* 13: 44-50.

Stevens, A.S., 1988. Pueblo water rights in New Mexico. *Natural Resources Journal* 26: 535.

Stine, S., 1990. Late Holocene fluctuations of Mono Lake, eastern California. *Palaeogeography, Palaeo-climatology, Palaeoecology* 78: 333-381.

U.S. Soil Conservation Service, 1972. *National Engineering Handbook*. Section 4, Hydrology.

Wigmosta, M., Vail, L. and Lettenmaier, D., 1994. A distributed hydrology-vegetation model for complex terrain. *Water Resources Research* 30(6):1665-1679.

Wigmosta, M. and Lettenmaier, D., 1999. A comparison of simplified methods for routing topographically-driven subsurface flow. *Water Resources Research* 35(1):255-264.

Wilkinson, C. F., 1985. Western water law in transition. *University of Colorado Law Review* 56: 317-345.

Wilkinson, C. F., 1991. "In Memoriam: Prior Appropriation 1848-1991. *Environmental Law* 21: v-xviii

Water Management in the Arabian Gulf Region and a Partial Solution to Water Shortages

ALAN E.M. NAIRN
Environmental Science Research Institute, University of South Carolina, Columbia, SC 29208, South Carolina U.S.A.
E-mail: amenairn@hotmail.com

ABSTRACT: Even with the most stringent conservation measures on water use - agriculturally, domestically and commercially - there remains a deficit in all the Gulf countries. The most likely solution to overcome the deficit at the present time is the vastly increased production of desalinated water. Water management policies must be developed nationally to co-ordinate production, distribution and use with the necessary legislative background to deal with potential social, economic issues which can arise out of decisions on the distribution, allocation and charging for water.

One by-product of the hydrocarbons used as the energy source for desalination is the release of pollutants, largely carbon dioxide to the atmosphere. However the accession to significant volumes of pollutant gases runs contrary to the terms of the Kyoto accord. This can be avoided by use of solar energy as an alternate energy source. Of the techniques currently available, wind power has the greatest potential, and has been proven in the western US as well as on the plains of Northern Europe. It offers possibilities for both large and small scale operating units.

INTRODUCTION

Water shortages and water pollution problems are global in scale, but nowhere are they more clearly evident than in desert areas of the Arabian Gulf. The latter region is arid with no surface water flow save those following rare rain storms. The major part of surface water derives from beyond the borders of the Gulf Cooperation Council countries, in the Euphrates and Tigris basins. In consequence groundwater in the Arabian Peninsula accounts for 40 - 98% of the freshwater resource and only 1 - 10% of the meagre rainfall becomes available as soil moisture and ends up nourishing crops (Rogers and Lydon, 1994, Abdel Magid, 1995). Historically the region has supported low density, largely agriculture based communities in areas around oases, wells and springs.

The explosive growth in population and standard of living during the economic boom, which accompanied the discovery and exploitation of the rich hydrocarbon deposits, changed the situation fundamentally by placing increasing stress on the already inadequate water supplies. This has been exacerbated by the political and strategic demand for food self sufficiency. The water deficit was largely met by the ad hoc extraction of increasing amounts of groundwater, largely for agricultural use, and the construction of desalination plants to provide potable water. Extraction was initially done with little thought given to the consequences of excessive extraction and the attendant increase in salinity of groundwater supplies; however, the Gulf States have individually recognized that water is a national resource and have developed central water authorities to deal with all aspects of water management ranging from economic to health and social and legislative action. In recognition of the economics of water supply a scale of water charges was introduced, (see Table 1) though the charges fall far short of production and distribution costs. There is as yet no fully integrated plan in place.

Water policy has two distinct facets: the more technical issues of the planning and distribution of the actual water supply; and management dealing with the socio-economic and legal issues in the allocation and regulation of those water supplies. The principles are clearly set out by Abdel-Magid (1995). Water resource planning requires a strong institutional and legal framework to support effective management. Such management can be defined as the allocation of water resources through permits, regulations and standards set by competent authorities to ensure adequate water supplies for sustainable development to meet current and often competing demands, and anticipate future demands for water as determined within the frame of a water resource policy. Sustainable development must be taken to include a comprehensive and balanced economic and social system consistent with the strategic objectives of the state, rather than just the judicious and rational use of water resources. Conservation savings can be affected by economic means, or by charging an economic rate for water devoted to the growth of crops with high water

Water Resources Perspectives: Evaluation, Management and Policy. Edited by A.S. Alsharhan and W.W. Wood.
Published in 2003 by Elsevier Science, Amsterdam, The Netherlands, p. 183-189.

Table 1. Water Tariffs in the Gulf Cooperation Council countries (after Dabbagh et al., 1994).

Country	Monthly Tariff (cm)	Band width (cm)	Water Type	Price in $/m^3$
Saudi Arabia	0.15	1-100	Potable	0.04
	1.00	101-200	Potable	0.27
	2.00	201-300	Potable	0.53
	4.00	>300	Potable	1.07
Kuwait	0.80	1000 gals	Potable domestic	0.58
	0.25	1000 gals	Potable industrial	0.18
	0.10	1000 gals	Brackish domestic	0.07
	0.10	1000 gals	Brackish industrial	0.07
	0.20	1000 gals	Brackish agricult.	0.15
Bahrain	0.025	<60	Potable domestic	0.07
	0.08	61-100	Potable domestic	0.21
	0.20	>100	Potable domestic	0.53
	0.30	>450	Potable industrial	1.06
	0.002	<60	Brackish	0.01
	0.025	61-100	Brackish	0.07
	0.085	>100	Brackish	0.23
Qatar	4.40	Free for citizens	Potable	1.21
UAE	15	1000 gals	Potable	0.90
Oman	2.0	1000 gals	Potable domestic	1.14
	3.0	1000 gals	Potable industrial	1.71

demands and/or by changing the choice of crops, and by modifying the irrigation system as part of a general agricultural policy (Alsharhan et al., 2001).

The purpose of water resource planning and management is to provide a supply of water, now and for the future, since uncontrolled extraction of groundwater is not in the national interest. In the Gulf region and typically throughout the world, scientific and technological developments have outpaced policy development. What this means in terms of water is that the ability to extract and deliver supplies has proceeded without any clear policy decision on what limits or restrictions should be imposed. In the Gulf Cooperation Council countries, authority is typically divided between several ministries, the Ministry of Water and Electricity for potable water and commercial supplies, the Ministry of Agriculture and Fisheries for irrigation and farming, and the Ministry of Health responsible for drinking water standards. The allocation of water resources via permits, laws, regulations and standards is best set by a single competent authority within the frame or reference of a water resource policy. It requires a comprehensive treatment best provided by a single ministry which fully reflects upon the many uses of water, and the different economic value of water according to its use. For example, the economic return on water used for industry purposes is considerably greater than that used in irrigation. Thus water allocation is something which has to be ultimately established by political decision.

Water as a birthright of every individual is clearly established in the Koran, and incorporated in Islamic Law which is at the basis of the constitution of the individual states. It is therefore clear that the legislative backup to water policy decisions must be consistent with Islamic Law. As the governments of the states have generally undertaken the task of providing and distributing water, any policy decision which involves water involves the public so that it is essential that the plans be widely understood and their objectives agreed upon.

Water Policy

In the Gulf countries where the value of water is greater than that of the land, the governments have generally undertaken to provide and distribute water. Thus the formulation of a national management policy is essential not only to deal with questions of supply and distribution but also to deal with the complex social, economic and legal issues which can arise. In the attempt to develop a water management policy there are three main technical tasks. First and foremost is the need to assess the available supply after which ways of increasing that supply can be investigated. The search for new supplies includes the search for new sources and the optimization of the current sources, which includes the full range of conservation and economic restraints. The second arm of water policy concerns itself with the social, economic and fiscal issues (see section on Water Legislation and Alsharhan et al., 2001).

a) Assessment of water supplies

In the absence of surface water the assessment becomes an inventory of all wells and springs. This includes a documentation of all the relevant parameters as depth and movement of the groundwater table, water composition, covering both water quality and quantity, the rate of extraction, and aquifer characteristics. Also included should be full details of desalination units in operation, their cost of operation and the quantity and quality of water produced and how the residual ultra saline solutions and salts are disposed of. In a similar manner the treatment of wastewater, the amount and quality and the appropriate costs and current use of the treated water must be addressed. This is anticipated to become an increasingly important source. Even the contribution of dew in the coastal areas should not be ignored since it can maintain scrub vegetation some miles from the coast. Dew ponds, excavated shallow, clay lined hollows, in use since Neolithic times, can provide water locally for small communities and provide water for stock watering. The development of water retention dams for the flow from the rare storms, its percolation and recharge rate and the cost of construction and maintenance costs should be included.

Groundwater is still the principal source of supply, and the magnitude of it as a source can be seen in Table 2. Extraction and well head costs are low, but the costs of transfer and distribution are high through pipelines and pumping stations. Once the capital costs are met, running expenses are much less significant. The principal aquifers are well known, but facies changes can affect their porosity, permeability and transmissivity and production. In general, production from all the aquifers is well in excess of recharge with the result that more saline water is drawn from deeper horizons and sea water intrusion plays an increasingly significant role with the inevitable fall in water quality and increase in total dissolved solids. This effectively places a limit on the amount extracted without significant additional treatment prior to use beyond dilution with desalinated water to bring the total dissolved solids within acceptable limits.

As Table 2 shows, even with all the measures that are now applied, there is still a water deficit. The search for new groundwater sources continues, but the magnitude of the deficit is such that there is little chance of being able to develop adequate new sources. This leaves desalination as the most likely source of new supplies.

There are two common methods of desalination practised in the Gulf states:
1) multistage flash distillation, the traditional leader as the only method available in the mid sixties at the time of the growth in demand, and 2) the growing competition from reverse osmosis. In the 21st century it is anticipated that water produced by the growing competition from reverse osmosis will equal or exceed water produced by distillation, through the economies achieved by improved heat transfer, the use of chemicals to reduce scaling, more efficient membranes and reduced thermal power demands. An advantage of the membranes is that they can also provide a barrier to Cryptosporidium and Giardia cysts and are used in conjunction with microfiltration and ultrafiltration to remove not only suspended particles, but also bacteria.

As a review of Table 2 shows, even a doubling of the present output fails to meet the current requirements. Output can be increased by the construction of new plants, which in turn demands a major capital input. A key factor is the cost and amount of the energy requirements.

The only other substantial source of water is treated waste water although there are strong social sensitivities against its use except in agricultural areas where the social bias against it has declined. Wastewater is currently treated to secondary standards, viz. biological oxidation and disinfection, and it finds its main use in horticultural and agricultural projects, in major urban projects as fountains and green areas, and groundwater recharge of non-potable aquifers or wetland augmentation. Its value is enhanced in that it may contain a higher than usual organic content and may even provide the necessary salts and fertilizers depending upon the treatment to which it has been subjected. Its attraction is in the cost factor, since it costs only about one-fifth that of desalinated water. As urban areas expand, greater volumes of wastewater can be expected with a decline in the social bias against its use. When treated to tertiary standards, appropriate uses include landscape and golf course use, toilet flushing, car washing, crop irrigation and indirect potable use such as groundwater aquifer recharge and surface reservoir augmentation.

Wastewater is particularly useful in coastal areas where it can be used to create a barrier to saltwater intrusion. Treatment is becoming progressively more critical because of the high proportions of untreated domestic and industrial effluents discharged directly into water courses, irrigation canals and drainage ditches.

Alternative or additional suggestions to improve the supply of freshwater, such as the importing of freshwater by towing large plastic pods, or the use of tankers or, a much less feasible solution, towing icebergs, seem unlikely solutions in terms of the volume of water required, so there appears little alternative to desalination. However desalination raises the dual issue, water is needed, but the use of hydrocarbons as the energy source in the desalination process releases large quantities of carbon dioxide into the atmosphere which runs contrary to the Kyoto agreements. This question will be discussed in a subsequent section.

Table 2. Projected Water availability from all sources and the projected water demand for the years 2000 and 2010 based upon date drawn from Dabbagh et al., 1994 (Water measured in millions of cubic meters a year).

Country	Year 2000					Year 2010			
	Treated Water	Desali-nated	Ground-water	Projected demand	Surface Water	Treated Water	Desali-nated	Ground-water	Project demand
Saudi Arabia	710	1289	20212	23100	900	1000	1300	22100	25300
UAE	200	772	1185	2232	75	250	772	1359	2450
Oman	50	68	1072	1417	227	61	68	1229	1585
Kuwait	80	428	132	640	---	106	428	237	771
Bahrain	42	115	93	285	---	53	141	121	315
Qatar	43	216	75	334	0.4	43	216	129	388

The effective use of existing supplies can serve to extend the use of current supplies by means of a wide range of conservation techniques. The major techniques will require a major infusion of capital. In agriculture the replacement of open channel irrigation by such modern techniques as drip and sprinkler irrigation and by feeding water directly to the roots of crops combined with a change or reduction in the principal crops with high water demands (Dabbagh and Abderrahman, 1997) and concentrating more on crops with a lower water demand, but which yield better market prices (Ibnouf and Abdel-Magid, 1994). The cost may be a change in the policy of food self-sufficiency. Conservation by the better maintenance of piped supplies can reduce the water required domestically and commercially by up to 25%.

Historically water has been available at low cost such that there was never any impetus to save, but if now water is charged at an economic rate it reinforces recognition that water is a vital commodity, but available at a price which could or should encourage economies in its use. That price is a political decision and must take into account the many uses of water and the returns from the different uses as well as the costs of its supply.

b) Future supplies

Any reasonable projection for population growth incorporating increased living standards requires additional water supplies and underlines the inability of the system to meet those requirements under current conditions although the situation would be eased by the discovery of significant additional sources of groundwater. The most feasible source for the foreseeable future to meet the water deficiency lies in increased production of desalinated water, or the alternative to a major cut-back in agricultural production. The increased production of desalinated water entails the construction of new plants which in turn requires the infusion of major capital investment for their construction, distribution and maintenance. Despite many technological improvements which can improve efficiency and decrease power costs, the costs of running of new plants still remains appreciable even in terms of the economies of the Gulf hydrocarbon rich states (Alsharhan et al., 2001).

As mentioned earlier, the increased use of hydrocarbons in the desalination process poses a dilemma. Water is needed, but current techniques result in the release of large volumes of carbon dioxide contrary to the Kyoto agreements to reduce gas emissions. A way out of the dilemma may be by the use of solar energy as an alternate power source. A variety of ways exists to tap solar energy, but not all are equally practical. Photovoltaic cells, developed for use in satellites are the most attractive, but at the present time their high cost rules them out as a power supply for desalination units. Heliostats, focussed parabolic mirrors, have proved to have commercial applications and are obviously well suited to an area with 80% or more total sunlight hours, but experience with such plants is lacking.

The one technique which has shown major development in North and South America and Europe, from Scotland to Spain and the North German Plain, is wind power. Records from both California and Germany show that electricity generated from wind power under optimal conditions compares well with the costs of conventionally derived power (Flavin, 1981). If wind farms could be established and the electricity used to provide energy for desalination plants the dilemma might be resolved, and even the costs of the energy derived could represent a long term saving in water costs. In appendix some of the costs are outlined based almost entirely upon North American data.

Water Legislation

The ultimate object of water management is to provide a fiscally sound and socially acceptable policy which covers not only the operational and technical costs of providing water but can provide funds to cover future developments, perhaps under some form of 'public trust' doctrine to protect public rights and make some allowance for aquifer recharge. This means in effect that extraction should remain below some calculated safe yield. Legislation provides the unobtrusive underpinning to management and exists as a last resort to resolve issues which cannot be resolved in any other way. It exists as an arbiter according to set principles, consistent with the Shari'a and Islamic law since this is required by the constitutions of the Gulf countries. While any policy must be socially acceptable, it must also be sufficiently flexible to act as a policy guide concerning issues beyond the immediate scope of water management. For example, if it is decided that agricultural self-sufficiency is a major item of policy, then this directly affects water management. Several options are then open. Without a reduction in crop area, water could be charged at close to the economic rate with the inevitable consequence on food prices, and the danger of driving the small farmer out of business (see above), or more research could go into developing crop strains which can accept water of higher than usual salinity, or crops which have heavy water demand could be replaced by crops with lower water requirements, or major investments could be made into modern irrigation techniques. All of the suggestions are going to require significant infusions of capital for the equipment needed is in general not in place.

The laws must cover ownership rights and the transfer of ownership. They must control the function and jurisdiction of water and water related organizations in both the public and private sectors. There must be a code for the prevention, control and mitigation of harmful effects, including health hazards, arising from flooding, to pollution and salinization. There must also be laws relating to the treatment of waste water and the disposal of brines both from desalinization and oil field operations. They must provide for the establishment of water conservation zones and for research where this is needed. Above all the legal code must have a built-in enforcement code with a scheme of penalties of a sufficient magnitude to make compliance preferable to non-compliance. The laws must provide for appeals and adjudicators in questions of conflicting interpretations of the law.

The issues involve more than national policy since aquifers such as the Dammam aquifer in the East Saudi Basin serve not only Saudi Arabia, but also Kuwait, Bahrain, Qatar and, to a limited extent the United Arab Emirates. Yet no co-ordinated or regional policy exists (see Alsharhan et al., 2001).

In most countries the fundamentals of a system of laws have already been established under departments of the environment or water ministries.

In Oman prior to 1975, the Ministry of Agriculture and Fisheries supervised the use of irrigation water with the Ministry of Water and Electricity responsible for potable water supplies. Later that year the Water Resource Council under the chairmanship of the chairmanship of the Sultan of Oman was created to decide upon policies and programs and co-ordinate exploitation of water to ensure a balance between supply and demand. In 1979 the Public Authority for Water Resources was established under the Ministry of Water and Electricity to assist the Council for Water Resources. In 1985 The Ministry of the Environment took over water management, and the Department of Water Resources of the Ministry of Agriculture was transferred to the Ministry of the Environment. A separate committee was established in 1989 to oversee water use along the Batinah coast. That same year the Ministry of Water Resources shouldered the responsibilities for all the earlier councils and agencies.

In Bahrain decree 12/1980 called for the regulation of extraction and a reduction in the rate of quality deterioration, followed in 1982 by a further decree regulating extraction and the installation of water meters.

The first decrees regulating groundwater extraction in Saudi Arabia were also established in the 1980's.

In the United Arab Emirates in 1993 the Federal Environmental Agency was created by law 7/93 and assigned the role of protecting water resources and establishing water standards.

In 1994 in Qatar decree 13 water law 4 was amended, transferring the Environmental Protection Committee to the Ministry of Municipal Affairs and Agriculture.

In 1995 in Yemen a law was promulgated which called for a national water resource authority to establish policy, strategy and plans.

These few examples show an increasing awareness and concern with water problems although most of the decrees had limited coverage and relied upon regulations and the requirement of drilling permits as a means of controlling groundwater extraction. In effect there was/is no comprehensive water code. Metering extraction is still not widely practiced in the absence of water meters and penalties for non-compliance tend to be too low to act as effective deterrents.

Generation of new Sources of Water

Accepting as fact that the current sources of water, even under ideal conditions, involving maximum conservation, improved agricultural techniques and crop changes cannot meet even the short term water demand, the most feasible source for the necessary supply lies in the increased production of desalinated water. The costs for the production of major desalinated water supply, even with improved technology are still significant. One major aspect of such a major development with current techniques is the environmental one, namely the production of large volumes of exhaust gases, largely carbon dioxide. This creates a dilemma, more water or more polluting gases?

The Kyoto accord of the world governments was to set a timetable for the reduction of atmospheric carbon dioxide, yet the production of desalination exhaust gases runs contrary to that decision. The solution appears to be to seek alternate energy sources. In the Gulf region with its extremely thick sedimentary blanket the potential for significant geothermal sources is small, which leaves solar energy as the best alternative. There are many ways in which solar energy can be used, but most of the techniques are for passive conversion of solar energy and are inappropriate as power sources for desalination. The most obvious is the use of photovoltaic cells, but at the present time these are too expensive for wide scale use. Heliostats which focus solar radiation on a central tower to drive a turbine have proven to have a commercial application especially in a region with 80% or more daily sunlight hours, but experience in the large scale use of such plants is lacking. The one technique which has shown major developments in recent years is wind power. Plants have been extensively developed in California and on the North German Plain as well as in South America. They produce power which is closely comparable in cost with conventionally generated power and require relatively low maintenance.

In a desert area with dunes it is clear that wind velocities are appreciable, but it will require research programme to determine those areas which show relatively constant directions (see Breed and Reheis, 1999). There now exists a new generation of wind turbines which can generate 300-750 KW per turbine able to operate at a range of wind speeds and can thus be deployed in more moderate wind regimes. Such wind fields when established could provide the energy to drive a desalination unit. Smaller wind dynamos linked to the more economical reverse osmosis units could provide supplies for smaller and more isolated communities.

Two obvious localities in the UAE for wind powered desalination units are close to A1 Ain, important for its production of dairy products and in the Fujirah Emirate because of its production of vegetables. Both areas are close to the Oman Mountains so the chances of finding locations with wind velocities of sufficient magnitude and duration must berated high. In the Fujirah Emirate Bakhit and Nairn (1997) calculated that the emirate could balance its water budget with a 10% reduction in the cultivated area and the construction of an additional 4 million gallons a day desalination plant, the optimum capacity of a wind powered desalination unit. There is a potential for similar developments in active oilfield areas to deal with salt water produced in association with the oil, provided salinities are not excessively high and where potential gas supplies are available to maintain production during spells of low wind velocities.

Concluding Remarks

The future development of the Gulf countries depends upon their having access to adequate water supplies at an economic rate. This need is best met, at the present time, by an extensive increase in desalinated water, at least as great as the amount already produced. This implies the construction of major plants to supply the growing urban areas. Reverse osmosis is gradually replacing multi-flash distillation as the more efficient of the methods currently available. The power requirements for both up to this time have been derived from the use of hydrocarbons. Their waste products, the exhaust gases contain large volumes of carbon dioxide at a time when the Kyoto accord calls for a reduction in greenhouse gases. Switching to solar power would solve this problem. Of the available techniques, wind power has proven the most effective, but its application in the Gulf region will require a search for suitable locations to install the wind turbines (Tigges et al., 1999). If it is found to be practical there is a great future for wind power to generate energy to drive even small reverse osmosis plants for local usage, particularly in the more remote areas, as well as in major desalination plants.

The new generation of variable speed wind turbines can generate 300-750 KW per turbine rather than the 100 KW of the older turbines and are able to operate at a range of wind speeds and can thus be deployed in more moderate wind regimes (Millborrow, 1993, Davidson, 1991). Installation costs have been estimated at $1200 to $1000 per kilowatt which could drop to as low as $800 per kilowatt (these figures apply to costs in the US) particularly with the new generation of wind turbines. It is estimated that under optimal operating conditions the generating costs could then drop to 4c, and current planning, in the design stage, is to reduce the cost to 4c kwh. Maintenance and transmission costs may add another lc to the per kwh cost, however maintenance has not been a major problem in the Californian wind farms.

REFERENCES

Abdel Magid, Y., 1995. Planning water resources development in arid zones: An agenda for action in the Arab Region. Proceedings of the International Conference on Water Resource Management in Arid Countries, Muscat, Sultanate of Oman, 12-16 March 1995, 3: 47-54.

Alsharhan, A.S., Rizk, Z.A., Nairn, A.E.M., Bakhit, D.W. and Alhajari, S.A., 2001. *Hydrogeology of an Arid Region: The Arabian Gulf and Adjoining Areas*. Published by Elsevier B.V., The Netherlands, 331p.

Bakhit, D.W.M., Nairn, A.E.M., 1997. A proposed water conservation plan for the Fujairah Emirate. *Journal of Faculty of Science, UAE University* 9: 171-201.

Breed, C.S., Reheis, M.C., 1999. Desert Winds: Monitoring wind-related surface Processes in Arizona, New Mexico and California, United States Geological Survey Professional Paper 1598, 153p.

Dabbagh, T., Sadler, P., Al-Saqabi, A. and Sadeqi, M., 1994. Desalination: an emergent option. In: P. Rogers & P. Lydon (ed.), Water in the Arab World - Perspectives and Prognoses, Harvard University, Division of Applied Sciences, 203-242.

Dabbagh, A.E. and Abderrahman, W.A., 1997. Management of groundwater resources under various irrigation water use scenarios in Saudi Arabia. Arab Journal of Science and Engineering 22: 247-264.

Davidson, R., 1991. Performance up and costs down, Windpower Monthly, (Nov) 15p.

Flavin, C., 1981. Windpower: A turning point, Worldwatch Paper 45.

Flavin, C., Jenssen, N., 1994. Power Surge, 382p, Norton, New York.

Ibnouf, M.A.O. and AbdelMagid, I.M., 1994. Oman water resources: management, problems and policy alternatives. Proceedings of the 2nd Gulf Water Conference, 5-9 November, 1994, Bahrain, 2, 19-31.

Millborrow, D., 1993. Variable speed comes of age, Windpower Monthly 20-23. December.

Rogers, P. and Lydon, P. (ed.), 1994. Water in the Arab World: Perspectives and Prognoses, Published by the Division of Applied Sciences, Harward University, Harward University Press, 369p.

Tigges, R., Breed, C.S., Helm. P.J., 1999. Design and operation of gemet Stations of the Desert Winds Project : Monitoring Wind-related surface processes In Arizona, New Mexico and California, edit. Breed, C.S. Reheis, M. C., United States Geological Survey Professional Papers 1598: 135-193.

Appendix

The following estimates for the costs involved with a wind dynamo are largely derived from figures from North America (Flavin and Jenssen, 1994)

300-750 KW dynamo installation cost	$1000 to 12000 per KW, a figure which could drop to $800 per KW
Generating cost under optimal conditions	5c?kwh which could fall to 4c /kwh
Maintenance and transmission cost	1 c/kwh
Assumed minimal windspeed	5.8 km/s (=13 mph)
Cost for 500 MW plant	$500,000,000

Impact of Political, Scientific and Non-Technical Issues on Regional Groundwater Modeling: Case Study from Texas, USA

KEN RAINWATER, JEFF STOVALL, SCOTT FRAILEY and LLOYD URBAN
Texas Tech University Water Resources Center, Box 41022, Lubbock, Texas 79409-1022, U.S.A.
E-mail: ken.rainwater@ttu.edu

ABSTRACT: Recent legislation required regional "grass-roots" water resources planning across the entire state of Texas. The Texas Water Development Board (TWDB), the state's primary water resource planning agency, divided the state into sixteen planning regions. Each planning group developed plans to manage both groundwater and surface water sources and to meet future demands of various combinations of domestic, agricultural, municipal, and industrial water consumers. This presentation describes the challenges in developing a groundwater model for the Llano Estacado Regional Water Planning Group (LERWPG), whose region includes 21 counties in the Southern High Plains of Texas. While surface water is supplied to several cities in this region, the vast majority of the regional water use comes from the High Plains Aquifer System, often locally referred to as the Ogallala aquifer. Over 95 percent of the groundwater demand is for irrigated agriculture. The LERWPG had to predict the impact of future TWDB-projected water demands, as provided by the TWDB, on the aquifer for the period 2000 to 2050. If detrimental impacts were noted, alternative management strategies must be proposed. While much effort was spent on evaluating the current status of the groundwater reserves, an appropriate numerical model of the aquifer system was necessary to demonstrate future impacts of the predicted withdrawals, as well as the effects of the alternative strategies. The modeling effort was completed in the summer of 2000. This presentation concentrates on the political, scientific, and non-technical issues in this planning process that complicated the modeling effort.

INTRODUCTION

As the second largest of the United States in terms of both land area and population, Texas has challenges for water supply and demand that vary widely across the state. For example, annual precipitation varies from less than 25 cm/yr in the west to over 125 cm/yr in the east. The largest metropolitan areas, such as Houston, Dallas-Fort Worth, San Antonio, Austin, and El Paso, are growing quickly and require additional water sources. Agricultural areas, such as the High Plains, require irrigation for dependable yields and thus depend on groundwater pumpage that often exceeds recharge to the local aquifer system. An additional issue arises from the difference in the state's legal interpretation of water rights. Surface waters are classified as "waters of the state" and are controlled by an appropriative rights system enforced by the Texas Commission for Environmental Quality (TCEQ). Groundwater is owned by the overlying landowner, who has the "right of capture" to produce as much as desired. The only control on groundwater use can be administered by local groundwater conservation districts, which to date do not cover the entire state and are only enacted through voluntary cooperation of the local voting precincts. Water planning and financial support of those plans are administered by the Texas Water Development Board (TWDB). In recent decades, the TWDB generated statewide water plans on a five-year cycle, with the intent of providing useful data for local and regional planners and consumers. The state legislature can also participate through legislation that appropriates funds for projects or by directing the TNRCC or TWDB for selected actions.

In 1997, the 75th Legislature mandated statewide water resources planning under a bill that became known as Senate Bill 1 (SB1). The bill required a "grass-roots" approach for the planning effort, instead of the centralized TWDB approach of the past. The 254 counties in the state were divided into 16 planning regions as shown in Figure 1. The divisions were intended to approximate major river basins and regional aquifers, but the fits were fair at best. In addition, major metropolitan areas dominated their regions due to population and municipal water consumption. The bill also specified the makeup of the regional planning groups as eleven representatives of certain types of stakeholders. The stakeholder groups included the general public, county and municipal government, industry, small businesses, agricultural interests, environmental conservation, electrical utilities, river

Water Resources Perspectives: Evaluation, Management and Policy. Edited by A.S. Alsharhan and W.W. Wood.
Published in 2003 by Elsevier Science, Amsterdam, The Netherlands, p. 191-198.

authorities, water conservation districts, and water utilities. The regional planning groups were funded by the TWDB for a two-year period, during which time the group could hire consultants or perform their own studies to do the required tasks. The planning groups' responsibilities included identification of current water supply sources, estimation of present and future water demands by use type and county through the year 2050, and recommendation for future conservation or construction needs for support by the TWDB. The TWDB participated in the groups as a source of information, such as projections of future water demands, that could be used or superceded by the planning groups. At the end of the process, each planning group submitted a final report to the TWDB, which then compiled an overall planning document for the entire state.

This paper describes some of the special challenges answered by a research team from the Texas Tech University Water Resources Center (TTUWRC) in projecting the future of groundwater resources for the Region O planning group, which was also known as the Llano Estacado Regional Water Planning Group (LERWPG). As shown in Figure 1, Region O exists to the south of the Texas Panhandle, in an area often referred to as the Southern High Plains of Texas. Figure 2 shows the counties in Region O. Much of the region overlies the High Plains Aquifer system, known locally as the Ogallala aquifer due to the predominance of the Ogallala Formation in the productive layers of the subsurface. In Region O, only two cities (Lubbock [200,000] and Plainview [45,000]) have populations above 15,000, and 95 percent of the water consumption in the region is produced from the Ogallala for irrigation. The TTUWRC team was contracted by the LERWPG to simulate the future of the aquifer with a MODFLOW-based software system. The objective of this paper is to demonstrate how complexities of the planning process constrained the modeling effort.

Major Regional Water Issues

Two main water sources are utilized in Region O. Lubbock, Plainview, and several smaller cities utilize surface water provided by the Canadian River Municipal Water Authority (CRMWA). CRMWA manages Lake Meredith, a large reservoir north of Amarillo that is actually outside of Region O. Amarillo and a few other Region A cities also share in the CRMWA. Lake Meredith has been a very dependable supply since it opened in 1965, but each member city also uses nearby Ogallala well fields to supplement this surface water supply. Due to concerns about the relatively high total dissolved solids (>1000 ppm) in the Lake Meredith supply, the CRMWA has also purchased groundwater rights in Region A to allow blending of purer (TDS<350 ppm) Ogallala water with the lake supply. These issues benefit the member municipalities in Region O, but do not affect the groundwater pumpage in Region O. The second and larger water source in Region O is the Ogallala aquifer. All cities that are not CRMWA members either access much smaller reservoirs (White River Lake and Lake McKenzie) or have well fields inside or near their city limits. The petrochemical industry, electrical utilities, and other local industries depend on their own

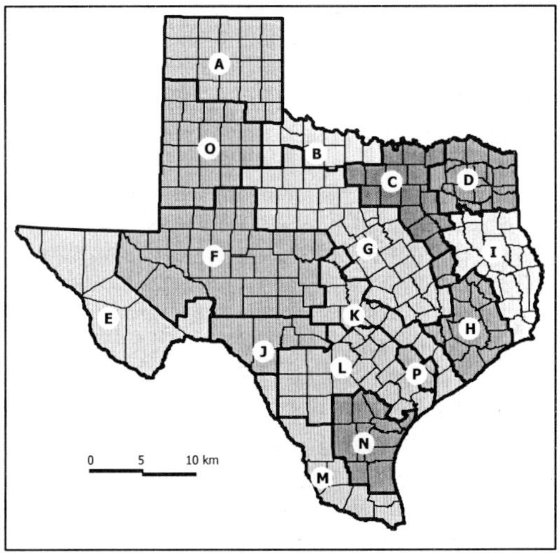

Figure 1. Geographic locations of planning regions in Texas.

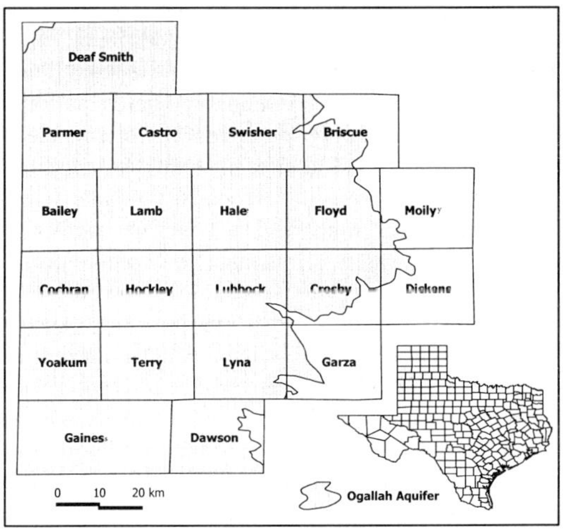

Figure 2. Counties in Region O (1 mile = 1.6 km).

groundwater supplies. Irrigated agriculture is almost completely supplied by wells owned and operated by the local farmers. Due to the heterogeneity of the alluvial Ogallala in Region O, irrigated acreage is nonuniformly distributed and is determined by the local productivity of the aquifer. Better production is found in areas with greater gravel content, with the best productivity in buried stream channels.

The state's differences in water rights for surface and groundwater also impact Region O. Perennial surface water sources, known as White River Lake, Lake Alan Henry, and Lake McKenzie, exist in Region O, but these sources are limited. Lake Meredith, built by the Bureau of Reclamation, is on the Canadian River, which begins in New Mexico and ends in Oklahoma after passing through the Texas Panhandle. The high TDS level in Lake Meredith is caused by saline springs in New Mexico, for which the CRMWA has enacted an interdiction project. In Region O, the minor surface water sources may play a larger role in future municipal water supply. The major conflict in groundwater resources management is the competition between the irrigating landowner's self-determination under the "right of capture" and the potential pumpage constraints that may be set by local groundwater conservation districts. The governing boards of the conservation districts are dominated by irrigated farmowners, who can use the districts' political connections to influence state legislators and agencies to protect the farmers' rights to use the aquifer, while discouraging other groundwater pumpage by municipalities, oil and gas exploration, and electrical utilities. One major responsibility of the groundwater conservation districts is to calculate annual water depletion allowances to reduce the income tax burden of farmers who are depleting their groundwater. "Conservation" has more often meant conversion from old furrow irrigation systems to modern center-pivot systems with low energy precision application devices. In practice, however, automated irrigation has made it easier to use more water, and most farmers appear to be driven more by annual decisions about which crop will pay better this year, rather than by protecting the long-term resources for their properties. Region O has one large conservation district, the High Plains Underground Water Conservation District No. 1 (HPUWCD), that includes all or parts of 15 counties in Region O and has existed for fifty years, and several single county districts that have been established for less than ten years. The districts all issue well permits and monitor depths to water and water quality in their jurisdictions, but they have been slow to computerize their databases.

As stated previously, SB1 required projections of wate supply and demands for 2000 to 2050. The current groundwater supply was estimated from maps of saturated thickness constructed from data provided by the groundwater districts for their areas and by the TWDB for areas outside the districts. The demands for groundwater in Region O were provided as county totals for use type (irrigation, municipal, rural domestic, range and livestock, industrial, steam electric power, mining, and feedlot) by the TWDB, and were based on historical trends and economic modeling by TWDB staff. SB1 also required consideration of drought impacts on supply and demand, without providing a specific definition of the hydrologic definition of drought for the region. The LERWPG was to then compare projected supplies and demands and thus identify potiential shortages. Obviously, any uncertainties in projected supply and demand values would impact uncertainty in the potential shortages. The LERWPG was finally responsible for proposing water management strategies to overcome the shortages. Only strategies proposed in the regional water plans will be eligible for future TWDB financial assistance.

Groundwater Modeling for Region

Model Assembly

The groundwater modeling software selected for this application was GroundWater Vistas™ (GWV), a proprietary product of Environmental Simulations, Inc. (1998). GWV is a pre- and post-processor for MODFLOW, the popular modular program by McDonald and Harbaugh (1988). GWV allowed interaction with two different calibration programs, and its telescopic mesh refinement (TMR) capability was useful for calibration of subareas within the region.

The TWDB and LERWPG provided county totals for groundwater pumpage for each groundwater use type, which then had to be distributed across the solution domain. The Ogallala aquifer was treated as a single unconfined layer, but the topography of its lower confining bed was considered. The contract with the LERWPG began in November 1999, and simulation results were required by July 2000.

One significant task in model assembly was updating the discretized geometry of the solution domain. The TWDB (Knowles et al., 1984; Peckham and Ashworth, 1993) had previously modeled the High Plains Aquifer system with a solution domain that included only the Southern High Plains of Texas. Harkins (1998) of the TTUWRC converted the TWDB input datasets for their own in-house computer model, known as GWSIM-III, to MODFLOW input datasets for the same discretized area. Those previous efforts considered the aquifer boundaries to be the Caprock escarpment on the east, north, and south, and the Texas-New Mexico state line on the west, and used grid cells measuring 4.67 km by 4.67 km. As part of the Regional Aquifer Simulation Analysis, Luckey et al. (1988) of the U.S. Geological Survey (USGS) modeled the High Plains Aquifer system in the Southern High Plains of Texas and New Mexico using an in-house computer model. Their solution domain set the aquifer boundaries as the Caprock escarpment on the east, north, and south, and the Pecos River erosional surface in New Mexico on the west. These boundaries much better represented the aquifer's areal extent. Weeks et al. (1988) had even larger cells in their discretized grid, with dimensions of 16.1 km by 16.1 km. The LERWPG required that the new model domain be discretized with the aquifer extent similar to that used by Weeks et al. (1988), but the size of the grid cells was to be 1.6 km by 1.6 km. The areal extent of the modeled domain is shown in Figure 3, and the area includes all or parts of 36 counties in two states.

There was no single data source for the hydrologic and hydraulic parameter required for the model. Values for the Texas side of the model domain were redistributed from the datasets used by Harkins (1998) using the TMR capability in GWV. The input values converted included hydraulic conductivity and specific yield. Elevations of the aquifer base were manually taken from contour

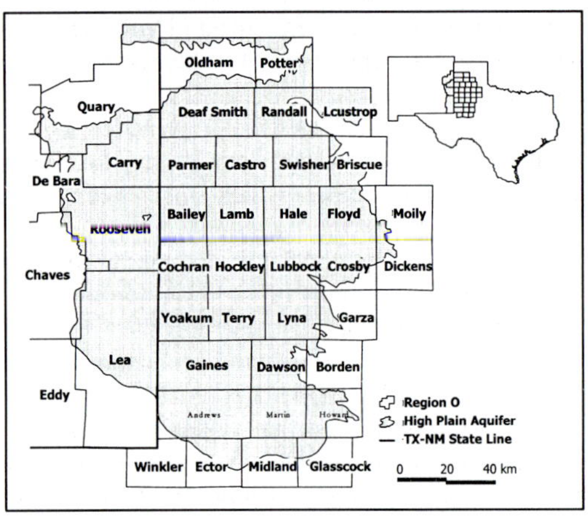

Figure 3. Areal extent of model domain (1 mile = 1.6 km).

maps provided by the groundwater conservation districts or from the TWDB database. For the New Mexico side, it was found that Weeks et al. (1988) and subsequent modelers (Luckey, personal communication) did not leave behind datasets that could be utilized for other modeling software. Their modeling group used another in-house computer program to generate discretized datasets for their aquifer simulation program from contour maps of the aquifer parameters. The effort to convert their data to the MODFLOW format was much more tedious. After the datasets were made on both sides of the state line, the two datasets were combined.

A second major challenge was generating contour maps of historical water levels across the solution domain. No single consistent data source was available. On the Texas side, the best source of data was from the local groundwater conservation districts. Each district measures water levels in select monitoring wells on an annual basis, and some districts publish contour maps of the distributions of water table elevation and saturated thickness. Unfortunately, the datasets for each district had different periods of record, some going back fifty years while some went back less than ten years. For the approximately one-third of the simulated area in Texas that was not in a groundwater conservation district, sparser data with more erratic timing from the TWDB was used to fill in the gaps. On the New Mexico side, the water levels were extracted from the even sparser USGS database. The short duration of the modeling project did not allow time for collection of better information from New Mexico water organizations. After creation of the two datasets on either side of the state line, they were combined and carefully reviewed for internal consistency.

The next challenge concerned the development of groundwater pumpage intensity and distribution in the historical period selected for calibration of the model. In Texas, due to the groundwater rights situation, irrigating farmers are not required to measure or report their groundwater consumption. However, to be eligible for federal crop insurance, farmers must report the size of their irrigated areas and crop type on those areas to the Farm Services Agency (FSA) in their county. This requirement does help, but it does not allow precise estimation of the amount of water used. An area is designated as irrigated whether it is only "pre-irrigated" to overcome initial soil moisture deficits, or if it is irrigated continuously throughout the growing season, or some condition between the two extremes. The FSA reports the area counts to the Natural Resource Conservation Service (NRCS).

NRCS field representatives in each county estimate the irrigation application rates for each crop in that county based on somewhat informal surveys of representative farms. These average application rates are then used with the FSA areas to calculate a single total irrigation volume for each county. That county number is then reported to the TWDB. It should be noted that the FSA and NRCS are both within the federal U.S. Department of Agriculture, while the TWDB is a state agency. For the New Mexico part of the model domain, the pumpage intensity and distribution was taken from the USGS modeling efforts, similar to the aquifer hydraulic parameters.

Obviously, the next step in building the model datasets was distribution of irrigation pumpage across the modeled domain in the Texas counties. The TWDB provided both historic and projected irrigation totals for each county. The TTUWRC team was required to determine the distribution of irrigated acres and local irrigation intensity. For the initial model setup, two methods were used to distribute the groundwater use in each county. For many of the counties, detailed irrigation survey maps were available from the groundwater conservation districts showing the locations of center-pivot irrigation systems in the county as of 1998. Groundwater use in counties for which maps were available accounted for over 96 percent of annual groundwater use for the modeled region. For these counties, groundwater demand was distributed in each county according to the distribution of center-pivot irrigation systems as shown on the map of that county. A transparent grid was overlaid on the map of irrigation systems, and the number of 0.4-km irrigation systems located within each grid cell was counted. Once all the irrigation systems had been counted, the county sum was calculated. The percent of county groundwater use in each grid cell was equal to the number of 0.4-km irrigation systems in the cell divided by the total number of irrigation systems in the county. For this purpose, a single 0.8-km irrigation system was assumed to be equivalent to four 0.4-km systems. In addition, it was assumed that the wells supplying water to the irrigation systems were located within the circle whose radius was defined by the length of the system. This preliminary method of assigning groundwater use did not initially account for the location of municipal well fields and domestic, industrial, and other agricultural users. However, the distribution of demand was refined during model calibration and adequate representation of these additional users was confirmed. In counties for which center-pivot irrigation system maps were not available, irrigation pumpage was distributed in the same spatial locations used by Harkins (1998), but scaled from their larger grid size to the smaller grids used in this study.

Calibration for 1985-1995

The LERWPG required the TTUWRC team to calibrate its regional model prior to predictive simulations. Selection of the best time period, with the best record of water level changes and irrigation pumpage, should be used was the subject of great debate within the LERWPG. The oldest groundwater conservation district, the HPUWCD, had historical water levels dating back over four decades, but the early number of monitored locations was much sparser than that used more recently. In addition, that district only covered 40 percent of the modeled area. Most of the single county districts were established after 1990. After much debate, the period selected by the LERWPG, not the TTUWRC team, was 1985 to 1995. This decade included a few abnormally wet years in the late 1980s and a relatively dry period in the early 1990s. Water level maps were provided by the HPUWCD for 1985, 1990, and 1995. Water level data outside the HPUWCD were taken from the TWDB's sparser monitoring well network and from the younger conservation districts' databases.

GWV had several capabilities that were employed in the calibration procedure. First, due to the large number (>30,000) of cells in the model and the presence of numerous internal areas with no initial saturated thickness, it was not possible to calibrate the entire domain at once. The TMR facility was used calibrate smaller subregions sequentially. The 1985 water levels were used as the initial conditions, while the 1995 water levels were set as the calibration targets. The cell parameters evaluated during the calibration procedure included hydraulic conductivity, specific yield, recharge, and pumpage distribution. The model was found to have relatively small sensitivity to hydraulic conductivity and specific yield, while recharge and pumpage distributions had to be balanced carefully to match the target water levels.

Predictive Scenarios

Once the TTUWRC model was calibrated, the LERWPG directed the TTUWRC to simulate several scenarios. Definition of the scenarios was quite challenging due to the diversity of the planning group. First, those stakeholders who were not employed in the water agencies had great difficulty comprehending the modeling work both in general and on specific issues. Second, some of the stakeholders who worked for local water agencies found it easy to question modeling assumptions, but rarely offered quantifiable suggestions to improve the work. At the request of the TTUWRC team, a smaller technical subcommittee was set up for direct oversight of the modeling, but that did not remove the second problem. After much discussion, the subcommittee and the TTUWRC team settled on a limited number of scenarios.

The first was the baseline simulation, used to predict the effects of groundwater pumpage using LERWPG water demand projections for Region O and TWDB water demand projections for counties outside Region O. The second scenario was a drought simulation to examine the effects of increased water usage on the aquifer. The drought was simulated as a four-year drought cycle starting first in 2015 and again in 2035 based on the observed historical droughts that have occurred in the 1930's, 1950's, 1970's, and 1990's. The drought of record was determined by Harkins (1998) to be a period in the 1950s during which the average annual precipitation on the South Plains was 16.5 cm below normal. The drought was represented in the model as 16.5 cm/yr of additional water demand for each irrigated acre in the region. The third scenario considered the possible positive impact of precipitation enhancement in the region. For purposes of this study, it was assumed that the precipitation enhancement program could result in an estimated annual increase in precipitation across the region of 5 cm in the future. It was assumed that this increase in precipitation resulted in a 2.5 cm reduction of groundwater pumpage and a 0.6-cm increase in recharge across the region. In the model, groundwater use by counties in Region O was reduced by 2.5 cm per irrigated acre, and recharge rates in these counties were increased by 0.6 cm/yr. The fourth scenario was a pumpage reduction simulation to examine the effects on the aquifer of reduced pumpage due to a combination of improvements in irrigation and application efficiency, plant genetics, cropping patterns, land management, and other agricultural advancements. The effects of these improvements were simulated as a reduction in the projected irrigation demands of 5% every five years to a maximum of 25% in the year 2020. It was hoped that the results of this simulation could be used to demonstrate the benefits and need for continued research in agricultural science. The final scenario was evaluated to determine the volume of groundwater available for use each year that results in a one percent annual decline in saturated thickness of the aquifer. The autocalibration feature of GWV was used to calculate the rate of withdrawal needed to produce the one percent annual decline of saturated thickness. The evapotranspiration package of MODFLOW was used to simulate pumpage for calibration purposes. It should be noted that this last scenario was evaluated based on a completely arbitrary demand of one of the groundwater conservation district managers. That stakeholder held that the "rule of thumb" for depletion of the Ogallala in the Region O area was an annual drop of one percent of the saturated thickness, and that "rule of thumb" would occur no matter what irrigation pumping was done in the future. The same individual also argued persistently for use of a single homogeneous value of specific yield for the entire region based on a similar tradition. Due to the arbitrariness of this scenario, its results are not presented in this manuscript.

Sample Modeling Results

The greatest time expense was for the calibration efforts. When compared to the 9,367 target 1995 head values in the primary HPUWCD district, the final calibrated dataset yielded a head distribution that had a mean error of 0.12 m and absolute mean error of 1.45 m. The percentages of the target heads that were matched within 1.5 m, 3.0 m, and 7.6 m were 62.7%, 89.7%, and 99.8%, respectively. For the entire regional model including 2,295 targets outside the HPUWCD area, the mean error was 0.19 m and the absolute mean error was 2.6 m. The percentages of the target heads that were matched within 1.5 m and 3.0 m were 43.7% and 68.5%, respectively. These calibration results were much better than those reported by the TWDB (Knowles et al., 1984) and USGS (Luckey et al., 1988).

Complete results of the predictive scenarios were provided to the LERWPG, and summarized highlights are presented here in two ways in Table 1. First, the volume of groundwater remaining in storage in the Region O counties at the end of the simulation in 2050, V_f, is compared to the initial groundwater storage for 1995, V_i, which was the time the water levels were reported completely. For reference, the initial storage in the Region O counties in 1995 was 1.634×10^{11} m^3. The number of counties that had more than 50 percent decline in storage by the end the simulation period, considered to be a critical depletion level, is also shown. The last column in Table 1 was derived to represent the impact of projected pumpage on areas that could not sustain those rates for the entire simulations. In MODFLOW, when the accumulated outflows for a grid cell exceed the accumulated inflows plus the amount in storage, the cell is shown as "dry" and no future pumpage is possible at future time steps. The satisfied demand percentage in Table 1 compares the total pumpage from grid cells that have not gone dry, D_s, to the total projected pumpage from all cells, D_p. The table compares the results for the different scenarios. The values in Table 1 show that while four counties have critical depletion after 2050 in the baseline and drought scenarios, both the precipitation enhancement and increased efficiency scenarios result in only two counties ending up with critical depletion. The impact of the drought scenario relative to the baseline scenario is relatively small because most of the major depletion in the region occurs in the first 15 years, prior to the occurrence of the first drought. Most of the grid cells at greatest risk of depletion due to the drought had already been depleted before 2015.

Table 1. Summarized Results of Predictive Scenarios.

Scenario	V_f/V_i	Counties $V_f/V_i<0.5$	D_s/D_p
Baseline	0.79	4	0.65
Drought impact	0.73	4	0.60
Enhanced precipitation	0.81	2	0.75
Increased efficiency	0.78	2	0.80

It should be noted that the simulation of the baseline scenario projects a much brighter future for Region O than the previous simulations by the TWDB (Knowles et al., 1984; Peckham and Ashworth, 1993) and the USGS (Luckey et al., 1988). The major finding of the calibration tasks that the calibrated recharge rate was much larger than that found in other studies. The typical average annual recharge rate used in the more recent modeling studies was 0.24 cm/yr, based on the work of Luckey et al. (1988), although Peckham and Ashworth (1993) did establish a few zones in the southern part of the study area with recharge rates as high as 5 cm/yr. The average total recharge rate found in this calibration study was 7.0 cm/yr, with the largest values in irrigated farm areas. This larger value reflects a combination of two processes. First, cultivation discourages runoff and encourages infiltration, increasing recharge. Second, irrigation return flow is very likely when part of the applied water moves down past the root zone. It is best to consider the total recharge rate as a sum of these two phenomena.

Lessons Learned from Regional Planning Group Control on Modeling Process

One major issue that complicated this work was the fact that aquifers do not recognize boundaries between states. The multiple political jurisdictions and bureaucratic data management procedures greatly complicated the assembly of information required for the modeling effort. Data sources with different mapping systems must be verified and correlated prior to combination into single datasets. The input data used in this work was resolved into values of the required parameters for over 30,000 grid cells, each 1.6 km by 1.6 km. This dataset is well suited to more modern database management systems, such as geographic information systems. The datasets were provided to the TWDB for future use. Cooperation across political boundaries is encouraged to make information more readily available. The TWDB should establish guidance for standard data collection and storage formats. In irrigated areas dependent on groundwater, the cooperation between federal and state agricultural and water agencies must be improved to more precisely estimate groundwater consumption. Interstate compacts for surface water bodies already exist due to recognition of shared water resources. It is time for formation of similar relationships for significant regional aquifer systems. Within Texas, it is time to establish conservation districts for all groundwater-producing areas. In 2001, the 77th legislature enacted what is known as Senate Bill 2 (SB2) for that purpose, and to strengthen the districts' regulatory authority. The districts' rules can now supercede the "right of capture" and regulate pumpage within their jurisdictions.

A second major issue that must be recognized is that each political subdivision or stakeholder group in the regional planning process has strong ties to its own agenda. The most directly involved agency is the local groundwater conservation district. Due to the connection between land ownership and the "right of capture," many of the district boards and managers see irrigation of farmland as the noblest use of groundwater, superceding the needs of municipal and industrial consumers. Many of the farmers served by these districts do all fiscal and operational planning on a short-term, annual basis. Depletion of their own water resources is a long-term issue that many of them prefer to treat like mining of oil and gas through their annual depletion allowances. Municipalities that depend on groundwater supplies must use long-term planning to spread out the costs of purchase, operation, and maintenance of their well fields. Due to deregulation of the electrical power generation industry, electrical utilities are expanding and building new facilities in this region. Most of these facilities will provide electricity for irrigation systems. These plants require significant water supplies that must also come from the local Ogallala aquifer. The competition for this valuable commodity signals that the "right of capture" approach with preference for agriculture will not serve Texas well today or in the future. The state must use groundwater conservation districts not only to collect data and encourage conservation, but also to regulate pumpage in their jurisdictions.

Finally, the formation of the regional planning group as a collection of stakeholders with common interests in water resources was well-intentioned but had limited success. The passage of SB2 in 2001 insures that the state's water planning is officially a work in progress. Quantification of water resources management strategies requires integrated efforts by engineers, hydrologists, hydrogeologists, economists, and planners. Public input is useful, but short-fused efforts funded poorly by state legislatures do not solve significant problems. The State of Texas should fund the TWDB to collect the proper mix of professionals to coordinate this type of effort with the cooperation of local groundwater conservation districts.

REFERENCES

Environmental Simulations, Inc., 1998. Ground Water Vistas, Version 2.30. Herndon, Virginia.

Harkins, D.K., 1998. The Future of the Texas High Plains Aquifer System: Modeling and Projections. Doctoral dissertation, Department of Civil Engineering, Texas Tech University, Lubbock, Texas.

Knowles, T., Nordstrom, P., and Klemt, W., 1984. Evaluating the Ground-Water Resources of the High Plains of Texas. Texas Department of Water Resources Final Report LP-173, Vols. 1-4, Austin, Texas.

Luckey, R.R., Gutentag, E.D., Heimes, F.J., and Weeks, J.B., 1988. Effects of Future Ground-Water Pumping on the High Plains Aquifer in Parts of Colorado, Kansas, Nebraska, New Mexico, Oklahoma, South Dakota, Texas, and Wyoming. US Geological Survey Professional Paper 1400-E.

McDonald, M.G. and Harbaugh, A.W., 1988. A Modular Three-Dimensional Finite-Difference Ground-Water Flow Model. USGS Water Resources Investigations Report 83-875, Reston, Virginia.

Peckham, D.S. and Ashworth, J.B., 1993. The High Plains Aquifer System of Texas, 1980 to 1990 Overview and Projections. Texas Water Development Board Report 341, Austin, Texas.

Weeks, J.B., Gutentag, E.D., Heimes, F.J. and Luckey, R.R., 1988. Summary of the High Plains Regional Aquifer-System Analysis in Parts of Colorado, Kansas, Nebraska, New Mexico, Oklahoma, South Dakota, Texas and Wyoming. U.S. Geological Survey Professional Paper 1400-A, 30p.

Sustainable Development and Management of Water Resources in the Arab Region

JEAN KHOURI
Director, Water Resources Division, The Arab Center for the Studies of Arid Zones and Dry lands, P.O.Box: 2440, Damascus, Syria. E-mail: john@scs-net.org

ABSTRACT: The Arab region, which occupies some 14 million square kilometer, encompasses Arab countries that lie essentially in semi-arid zones and many areas pass rapidly into arid and hyper-arid climatic environment. Renewable water resources are extremely low compared to other regions in the world. The population in the region has increased from about 78 million in 1950 to 292 million in 2000, thus water demand has increased nearly 4 times, due to improvement in the quality of life and adoption of the food security concept. Per capita availability has decreased from about 3100 m^3/year in 1950 to about 500 m^3/year in 2025, at the national level, water resources availability has dropped in 12 countries below the critical level 500 m^3/year.

Focus in the past century has been on sustainability. Sustainability requires integration of economic social, institutional as well as physical systems; As regards physical systems, the intricate relationship between surface and groundwater could be utilized to promote integrated water resources management. Governance need to be developed through greater participation and more transparency and cohesion. An important issue that should be addressed with regard to sustainability of hydrologic systems in arid and semi-arid zones concerns the spatial dimension and time scale of sustainability. Inadequate data, spatial and temporal variability, drought and climate changes add a dimension of uncertainty. For the foreseeable future water policy needs, therefore, to be made in a climate of uncertainty.

Under conditions of scarcity, not only intensive supply and demand management has been considered a fundamental need, but also a review of water sector policy is required when per capita water availability is less than 1000 m^3. As scarcity increases, marginal water resources and non-traditional sources are increasingly used. Water availability is considered a severe constraint on socio-economic development and environmental protection when per capita water resource availability drops below 500 m^3 . Water crises are likely to emerge and may attain serious proportion. Fortunately most countries with per capita water availability less than 500 m^3 in the Arab region are high income countries and are, in a position to respond to serious water scarcity through an increasing dependence on desalination.

Given the seriousness of the present water situation and risk of crises in the future, communities in the entire Arab region must strive for a situation where there is no further undermining of the natural resource base. People's attitude towards water must undergo a fundamental change. In view of the vital importance of water for sustaining life and promoting development, innovated approaches and policies are needed to deal with looming crises.

INTRODUCTION

The major challenge facing the Arab region in the first decade of the third millennium, and beyond, is that while physical availability of water is fixed and limited, demand will continue to increase steadily in the foreseeable future. Accordingly, the core problem is how to balance supply and demand under extremely difficult conditions: increasing scarcity and unpredictability.

With increasing awareness of water issues during the past decade, there has been a growing concern about the unsustainable use of water resources and the adverse impacts of water development. Continued growth in population, increased urbanization, industrial development and rising aspirations for improvement in living standards and attaining food self-sufficiency have all contributed to the substantial increase in water demand. At the present time, water demand is increasing at alarming rate, surpassing water availability in most countries. This imbalance is expected to continue in the 21st century.

Resources generated within the region are subject to high spatial and temporal variability. The resources of shared rivers, which constitute over 50% of the total resource base of the Arab region, experience increased pressures in their catchments in upstream countries.

These natural and human factors increase the water stress in the region and increasingly affect human activities and sustainability. Some 70 million

Water Resources Perspectives: Evaluation, Management and Policy. Edited by A.S. Alsharhan and W.W. Wood.
Published in 2003 by Elsevier Science, Amsterdam, The Netherlands, p. 199-220.

people still lack safe drinking water and the region imports about 50% of its food requirements. With the prospects of water crises emerging in the next decades, there is an urgent need to address existing and potential problems, with the aim to change attitude and approach to this vital resource. The lack of attention to the long-term sustainability of water resources development in the Arab region is gradually being replaced by growing concern as to the possible implications of increasing number of water issues. The countries of the region need to start working today for a more sustainable water future.

Water Resources

The trend of the water resources in the region has been towards water scarcity and in certain parts toward extreme scarcity. The per capita availability indicates a state of severe scarcity in the 21st century. At the present time the resource base comprises:

Conventional water resources
- renewable water resources
- non-renewable water resources

Non-conventional water resources
- desalinated water
- reused and recycled water

The total volume of precipitation in the Arab region is 2576 billion m^3 (Arab Center for the Studies of Arid Zones and Dry Lnads, 1997). It ranges from about 1800 mm in the Atlas and the Lebanon mountain ranges, to less than 10 mm or 100 mm in the Sahara and the Syrian Desert respectively (Fig.1). Similarly precipitation decreases from 1800 mm in south Sudan to less than 10 mm in the Nubian and Western Desert of Egypt. Average precipitation in the Arab region is estimated at 180 mm. The greater amount of rainfall, 1100 billion m^3 (bcm) (42%) falls on area not exceeding 11% of the total area of the Arab region. The greater part of the region (75%) receives an annual rainfall less than 200 mm. The remaining part (651 bcm) falls on 14% of the total area, which constitutes the rainfed agricultural belt. The total area of this productive zone is around 2 million m^3 (mcm). Due to water availability constraints, a food security plan could only be developed through an integrated management of rainfed and irrigated agriculture.

Perennial streams occur mainly in the Mashrek (East Mediterranean) and Maghreb (Fig. 2). In the former area the extensive occurrence of karstic features in carbonate rocks gives rise to intermittent streams in the coastal basins of Syria and Lebanon. The rift fault zone, which drains much of the groundwater in the Lebanon-Antilebanon massifs, sustains the flow of the Orontes and Jordan River which flow northwards and southwards in the rift valley. They are of medium size (1.5-2 bcm) compared to the coastal streams (less than one bcm). The average annual flow of Mediterranean river systems in the Mashrek and Maghreb do not normally exceeds one billion cubic meters. The Atlantic river system draining the Atlas Mountains, however, have higher discharge (Sbou 6.6 bcm, Oum El-Rabia 4.5 bcm). The average annual flow of larger international rivers, the Nile, Euphrates, Tigris and Senegal, is about 200 bcm. Of this total, the Arab countries can utilize 175 bcm, which forms some 62% of their total resources base. Annual flow from external sources is about 139 bcm. The remaining flow (36 bcm) is derived from internal sources. It is evident that high dependency of the region on international water resources is one of the most important issues that need to be addressed.

Major aquifer systems include alluvial, carbonate and sandstone aquifers. Porous alluvial aquifers occur in intermontane basins in the Lebanon-Antilebanon – Atlas mountain ranges. Sub-montane or deltaic aquifer systems comprise the Tihama coastal aquifer system in Yemen and Saudi Arabia, and Batina wadi aquifer system in Oman. Alluvial and deltaic aquifer systems in the Arabian Peninsula and North Africa are recharged mainly from wadi runoff or stream flow. They constitute the sole renewable water resources in the Arabian Peninsula, and play a major role in the sustainable development of these arid lands.

The Carbonate aquifer system, which extends along the Mashrek coast, is an extensive and highly productive groundwater system (Fig.3), surface–groundwater interaction in this mountainous terrain necessitates conjunctive use of surface and groundwater resources. Over-development of this "regional" aquifer system in the past decade has adversely affected the nature water regime.

Non-renewable water resources occur in extensive regional aquifer systems underlying the Sahara and hyper-arid deserts of northeast Africa and eastern parts of the Arabian Peninsula (Fig. 4). Their huge reserves have encouraged countries to develop their resources intensively in Libya, Egypt, Algeria, Tunisia and Saudi Arabia. Large water transfers "The Great Man-Made River" were adopted in Libya, whereas in Egypt, Saudi Arabia, in-site "oasis-type" development is the dominant

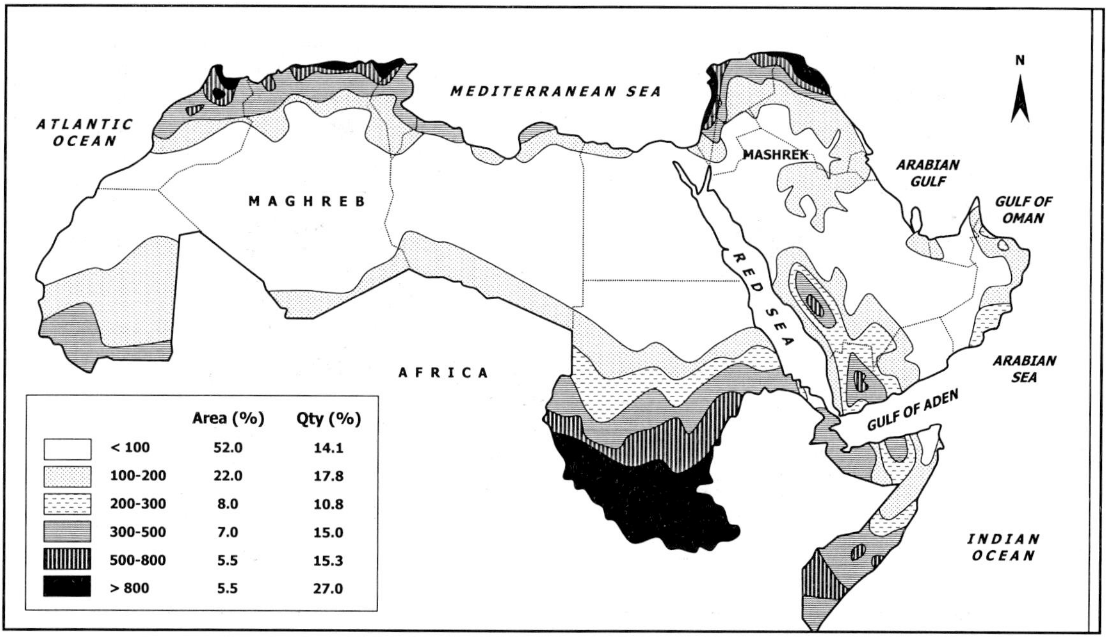

Figure 1. Annual rainfall distribution in the Arab region.

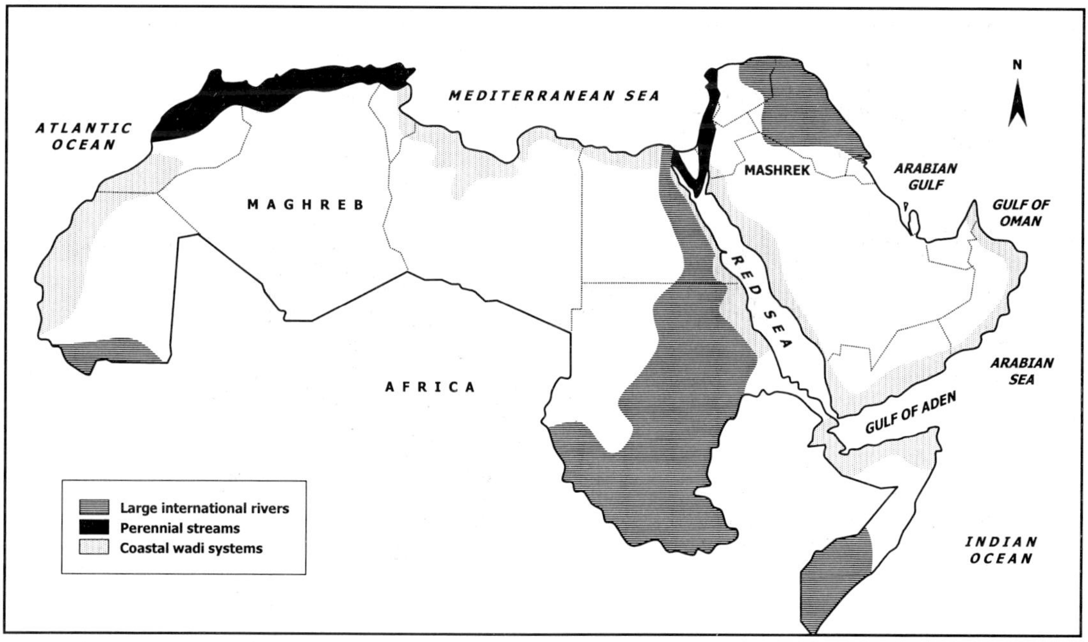

Figure 2. Perennial ephemeral river basins in the Arab region.

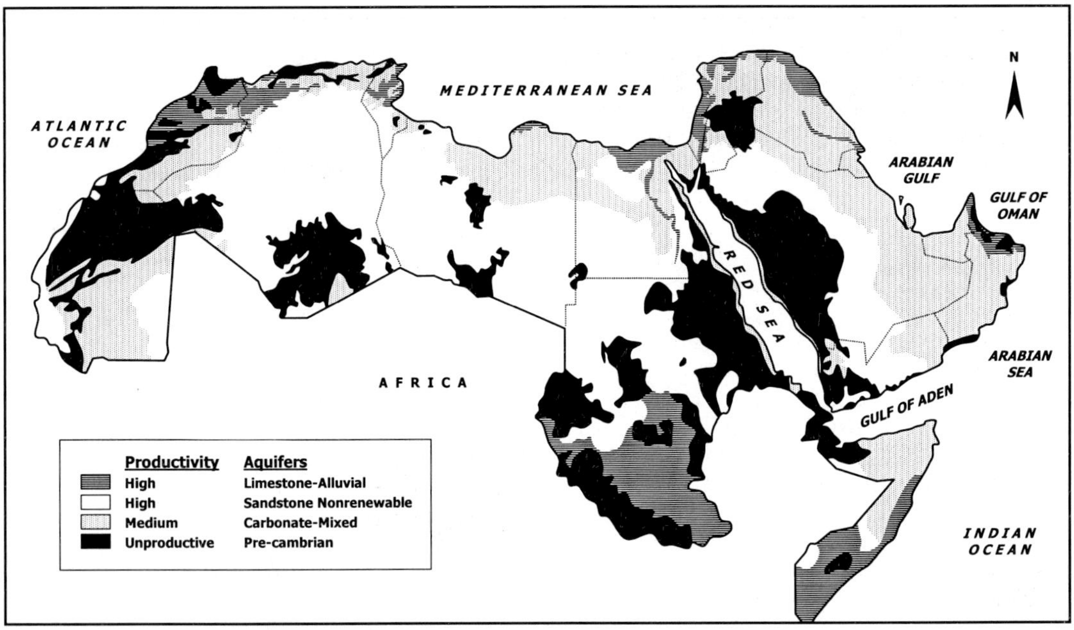

Figure 3. Productivity of the main aquifers in the Arab region.

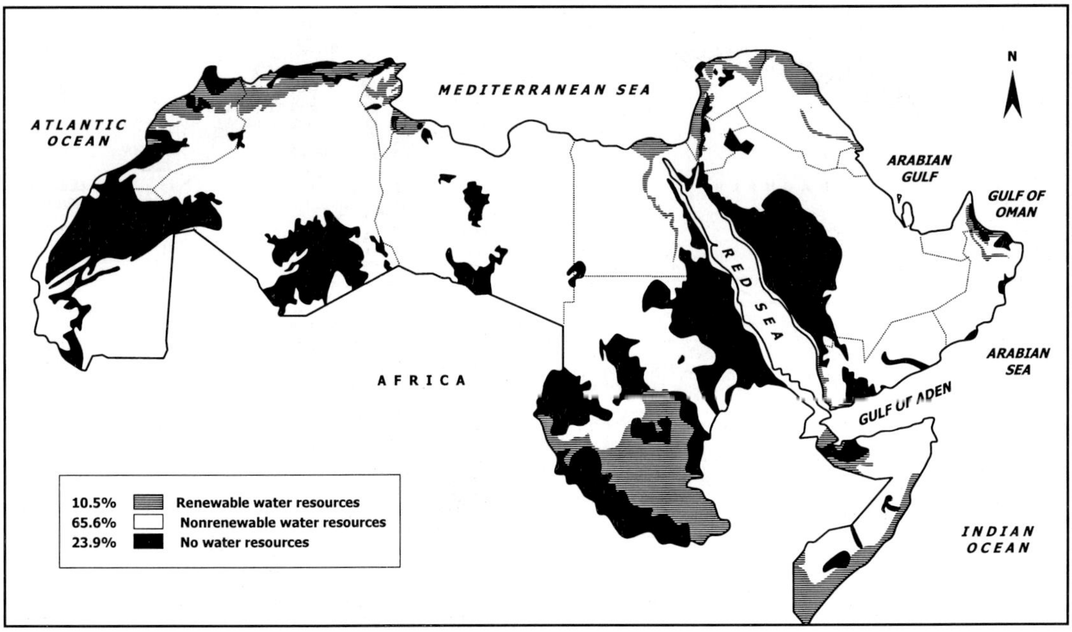

Figure 4. Renewable and nonrenewable water resources in the Arab region.

approach. On the basis of 50-year experience in groundwater irrigation in the Western Desert of Egypt, it was proposed that intensive cultivation in oases, should be replaced by small scale farming, utilizing an extensive approach for groundwater development. Alternatively, well fields could be designed on hydrogeologic criteria and their water may be transferred to collective cultivation (Hefny, 1991). This may help to address socio-economic, hydrogeologic, and contamination problems.

Renewable water resources in the region have been estimated at 246 bcm (Table 1). Of this total 209 bcm is surface water and 36.4 bcm groundwater. Inflow of surface water from adjacent regions (mainly via the Nile, Euphrates and Tigris) is about 139 bcm, whereas runoff general within the region is estimated at 86 bcm. If non-conventional sources and equitable shares from international rivers are added to these resources, the resource base could reach 275 bcm during the first decade of the 21st century.

Supply and Demand

There is a growing concern throughout the Arab region that scarcity of water resources, which is presently afflicting some parts of the region will become much more intense and widespread and water crises might ensue. The uneven distribution of water resources, the rapidly growing human population (Table 2), the deterioration of water quality and the accelerated demand for water are important factors, which impact the region's fragile environment in general and the vulnerable water resources in particular. Spatial and temporal variability, drought and global climate change in the long term adds a dimension of uncertainty to challenge of future action.

Table 1. Water Resources in the Arab Region.

Country	Conventional Water Resources (mcm)			Non -Conventional Water Resources(mcm)			Total (mcm)
	Surface Water	Ground Water	Subtotal	Desalination	Agricultural Drainage Reuse	Subtotal	
Jordan	475	275	750	2.5	61	63.5	813.5
Syria*	4276	5075	9351	-	1965	1965	11316
Iraq	60850	2000	62850	7.4	1500	1507.4	64357.4
Lebanon	2500	600	3100	47.3	21.5	68.8	3168.8
Palestine	30	185	215	0.5	2	2.5	217.5
Subtotal	**68131**	**8135**	**76266**	**57.7**	**3549.5**	**3607.2**	**79873.2**
U.A.E	185	130	315	405	108	513	828
Bahrain	0.2	100	100.2	75	17.5	92.5	192.7
Saudi Arabia	2230	3850	6080	795	131	926	7006
Oman	918	550	1468	47.3	21.5	68.8	1536.8
Qatar	1.4	50	51.4	131	33	164	215.4
Kuwait	0.1	160	160.1	338	30	418	578.1
Yemen	3500	1400	4900	9	52	61	4961
Subtotal	**6834.7**	**6240**	**13074.7**	**1850.3**	**392**	**2243.3**	**15318**
Egypt	55500	4100	59600	31.7	4790	4821.7	64421.7
Sudan	26000	1000	27000	0.6	-	0.6	27000.6
Somalia	8160	3300	11460	0.1	-	0.1	11460.1
Djibouti	200	50	250	0.15	-	0.15	250.15
Subtotal	**89860**	**8450**	**98310**	**32.55**	**4790**	**4822.55**	**103132.55**
Morocco	22500	7500	30000	1.2	350	351.2	30351.2
Algeria	13000	2000	15000	74.6	400	474.6	15474.6
Libya	397	650	1047	210	110	320	1367
Tunisia	2700	1940	4640	8.7	6	14.7	4654.7
Mauritania	5800	1500	7300	1.7	67.6	69.3	7369.3
Subtotal	**44397**	**13590**	**57987**	**296.2**	**933.6**	**1229.8**	**59216.8**
Total	**209222.7**	**36415**	**245637.7**	**2236.75**	**9666.1**	**11902.85**	**257540.55**

Source : ACSAD, 1997; ESCWA, 1999; ACSAD 1998.

* Water Resources of Syria: Except Equitable and Reasonable Share from the Euphrates River and Tigris River.

Table 2. Projection of population in the Arab region (thousand).

Country	1950	1990	2000	2010	2025
Jordan	1237	3423	5003	6747	9620
Syria	3495	12386	16125	20464	26292
Iraq	5158	18071	23280	31071	43482
Lebanon	1443	2554	3282	3723	4400
Palestine	1258	1839	2859	3972	5987
Subtotal	**12591**	**38273**	**50549**	**65977**	**89781**
U.A.E	97	1921	2442	2851	3234
Bahrain	116	490	613	689	789
Saudi Arabia	3201	16046	21930	31363	54030
Oman	456	1785	2518	3423	5020
Qatar	250	485	579	793	779
Kuwait	152	2143	2166	2339	2721
Yemen	4316	11590	18654	28661	44036
Subtotal	**8588**	**34460**	**48902**	**70119**	**110659**
Egypt	21834	56333	68523	80368	96462
Sudan	9190	24585	33064	42055	55553
Somalia	3072	8677	10916	14310	19302
Djibouti	256	632	991	1357	1962
Subtotal	**34352**	**90277**	**113494**	**138090**	**173779**
Morocco	8953	24334	28781	32959	38969
Algeria	8753	24935	32362	39197	48811
Libya	1029	4545	6562	8668	11871
Tunisia	3530	8080	9480	10904	12899
Mauritania	825	2003	2327	3114	4444
Subtotal	**23090**	**63897**	**79512**	**94842**	**116994**
Total	**78626**	**226907**	**292457**	**369028**	**491213**

Source : ACSAD 1997 , ESCWA 1999.

Table 3. Water Scarcity : Based on Per Capita Water Resources Availability.

Country	Population (thousand) 2000	Per Capita Water Availability (m^3/yr) 2000	Group	Water Scarcity
Kuwait	2166	74	Group I	Extreme
Palestine	2859	75		
Qatar	579	89		
U.A.E	2442	129	Group II	Severe
Jordan	5003	150		
Libya	6562	160		
Bahrain	613	164		
Djibouti	991	252	Group III	Very High
Yemen	18654	263		
Saudi Arabia	21930	277		
Algeria	32362	464		
Tunisia	9480	489		
Syria	16125	580	Group IV	High
Oman	2518	583		
Sudan	33064	817		
Egypt	68523	870		
Lebanon	3282	945		
Morocco	28781	1042	Group V	Moderate
Somalia	10916	1050		
Iraq	23280	2700		
Mauritania	2327	3137		

The water resources of Syria (9351 mcm) do not include the equitable and reasonable share from the Euphrates River and Tigris River.

There is already an increase throughout the region of the number of localities or countries where demand for water is outstripping available or potential water supplies (Table 3), and the resulting water shortage or "water stress" is likely to limit development.

Water use has increased substantially during the past two decades to meet rising demand resulting from high population growth, which has almost doubled (Table 2). Water withdrawal in 1985 was estimated at 162.4 bcm (Khouri et al, 1986). Total water use, from conventional and non-conventional sources, was about 172 bcm. Water withdrawal in the year 2000 is estimated at 200 bcm and total water use has reached 210 bcm, thus indicating that the region has reached a high degree of water stress (Table 4). Clearly withdrawal or present water resource development and use does not satisfy demand estimated at 238 bcm.

The volume of water currently utilized for agriculture is estimated at 188 bcm or 94% of total water use from conventional sources. Anticipated water demand for the 2025 is 380 bcm. Agricultural demand is about 300 bcm, far exceeding potential supplies. A significant part of groundwater abstractions comes from the reserves. In Saudi Arabia some 10 billion cubic meters is withdrawn from sandstone and limestone aquifers.

With increasing awareness about environmental issues during the past decade, many countries have initiated activities relating to environmental impacts assessments.

The impacts of intensive development on groundwater include excessive decline of water levels, depletion of aquifer storage and deterioration of groundwater quality. Depletion of reserves ranges from 25% in Saudi Arabia to proximately 50% in Bahrain and Qatar. Over-development in the Dakhla Oasis in Egypt has resulted in a continued lowering of groundwater levels of 1-4 meters per year (Khouri, 1999). The soils of Palmyra, Damascus (Syria) and the New Valley in Egypt are seriously affected by salinity. The dimension of the salinity problems changes substantially in the flood plains and deltas of large rivers. The high water table, low gradients, unfavorable drainage conditions and improper farming and irrigation practice have caused widespread water logging and salinization of soils.

Table 4. Water Stress in the Arab region, 2000, 2025.

Country	Renewable Water Resource (TR) 10^6m^3			Ground Water %	Total Demand (TD) 10^6m^3		TD/TR%	
	Surface	Ground	Total		2000	2025	2000	2025
Jordan	475	275	750	36.6	1242	1760	166	235
Syria	4276	5075	9351	54	16370	23555	175	252
Iraq	60850	2000	62850	32	66000	108000	105	172
Lebanon	2500	600	3100	19.3	1412	3850	46	124
Palestine	30	185	215	86	495	1290	230	600
Subtotal	**68131**	**8135**	**76266**	**11**	**85519**	**138455**	**112**	**181**
U.A.E	185	130	315	41.2	2180	3200	692	1016
Bahrain	0.2	100	100.2	99.8	282	609	282	608
Saudi Arabia	2230	3850	6080	63	17765	24200	292	398
Oman	918	550	1468	37.4	1847	2430	126	169
Qatar	1.4	50	51.4	2.7	347	485	670	943
Kuwait	0.1	160	160.1	99.9	590	1400	369	874
Yemen	3500	1400	4900	28.5	3571	4624	73	94
Subtotal	**6834.7**	**6240**	**13074.7**	**47.7**	**26582**	**36998**	**203**	**283**
Egypt	55500	4100	59600	6.9	68200	86300	115	145
Sudan	26000	1000	27000	3.7	22811	43672	85	162
Somalia	8160	3300	11460	28.8	1936	4580	17	40
Djibouti	200	50	250	20	517	1138	207	455
Subtotal	**89860**	**8450**	**98310**	**8.6**	**93464**	**135690**	**95**	**138**
Morocco	22500	7500	30000	25	13037	27703	44	92
Algeria	13000	2000	15000	13.3	7431	11000	50	73
Libya	397	650	1047	62	7501	16095	716	1537
Tunisia	2700	1940	4640	42	2909	12354	63	266
Mauritania	5800	1500	7300	20.5	1095	2162	15	29.6
Subtotal	**44397**	**13590**	**57987**	**23.4**	**31973**	**69314**	**55**	**121**
Total	**209222.7**	**36415**	**245637.7**	**15**	**237538**	**380457**	**97**	**155**

The water resources of Syria (9351 mcm) do not include the equitable and reasonable share from the Euphrates River and Tigris River.

SUSTAINABILITY ISSUES

Focus in the past decades has been on sustainability. The World Commission on Environment and Development (The Brundtland Commission) was concerned about how our actions today will influence the ability of future generations to meet their needs. The Brundtland report (Loucks, 1994) has defined sustainable development: "Humanity has the ability to make development sustainable to ensure that it meets the needs of the present without compromising the ability of future generations to meet their own needs". Sustainable water resource development and management, must address leading issues, to achieve a balance between water resources and water demand, under conditions of water scarcity. These issues comprise three categories: technical, managerial and financial issues.

Technical Issues

Technical issues in the Arab region are complex but manageable. A regional information and database is needed for better assessment and management of water resources. The data collection systems are often unsatisfactory. In only a few countries there are reasonable comprehensive time-series data available for rainfall, runoff, groundwater levels and water quality.

The water sector in the Arab region is characterized by inefficiencies in allocation among alternative uses, while significant wastage occurs in individual sub-sector use. In irrigation only a small part of water diverted by conventional gravity systems is available for plant use, normally about 25-30% compared with 60-75% in advanced systems such as drip and sprinkler systems. More significant is the overall efficiency level that encompasses losses from the irrigation network (conveyance and distribution) and field applications. Overall efficiency levels range between 20 percent for Yemen (spate irrigation) and 53 percent for Jordan, with Morocco achieving 42-49 percent (Sadik and Barghouti, 1994).

Although most countries in the region are confronted with high water stress (Table 4) leading to groundwater deterioration and depletion, extensive use of treated wastewater is not incorporated into national strategies. The need for reuse should be assessed, taking into consideration the availability of the resources, conventional or non-conventional and their potential and adequacy to meet present and future demands.

Uses of water have always economic, social and environmental consequences, some beneficial and others adverse. Groundwater irrigation in arid zones may create serious impacts unless it is managed in an environmentally sound manner. Salinization is undoubtedly one of the major environmental impacts of irrigation in large river basins such as the Nile, the Euphrates and the Tigris. The problem is more acute if groundwater is used extensively for irrigation in an arid environment. Groundwater development and use in oases, desert depressions and closed basins may cause serious environmental impacts.

Saline intrusion into freshwater aquifers is the common phenomenon that leads to the degradation of groundwater quality. In addition to the usual contamination sources such as seawater and saline water bodies in sabkhas (playas), gypsum and evaporate deposits may cause serious quality problems. The deeper parts of regional aquifer systems may contain brackish or saline water. Brackish groundwater is of extensive occurrence in the Western Sahara, Gulf countries and the Lower Jezira of Syria and Iraq (Fig. 5). Intensive exploitation of aquifer systems has the potential to lower the quality of the groundwater by drawing in lower quality waters from surrounding zones and from lower formations.

A complex hydrochemical regime exists in Bahrain. The regional carbonate aquifer system (the Dammam aquifer system) extends from Saudi Arabia to Bahrain. Upconing of saline water from deeper aquifers (Rus- Umm Radhuma aquifer) and intrusion of seawater resulting from excessive lowering of the potentiometric surface has caused substantial deterioration of water quality. The seawater / brackish water interface advances at a rate of about 100 m/yr (Zubari et al., 1993).

Groundwater has been intensively developed in several parts of the region and over-development of groundwater has caused rapid and excessive decline of water levels and depletion of aquifers. Qatar is an example where intensive development of the Dammam Carbonate regional aquifer system produced serious adverse effects. Total extraction from the carbonate aquifers in 1980 was estimated at about 80 mcm. Withdrawal increased to 244 mcm in 1997 (Al-Sulaiti, 1999) computations indicate an average depletion of aquifer storage at a rate of 20 mcm/yr. At such abstraction rate, it was estimated that the aquifer storage would be depleted in 20-30 years. Leading issues in the Arab region include water contamination problems facing most Arab countries, and affecting public health, ecosystems and quality of life. It may also render water resources in certain areas unusable without expensive treatment.

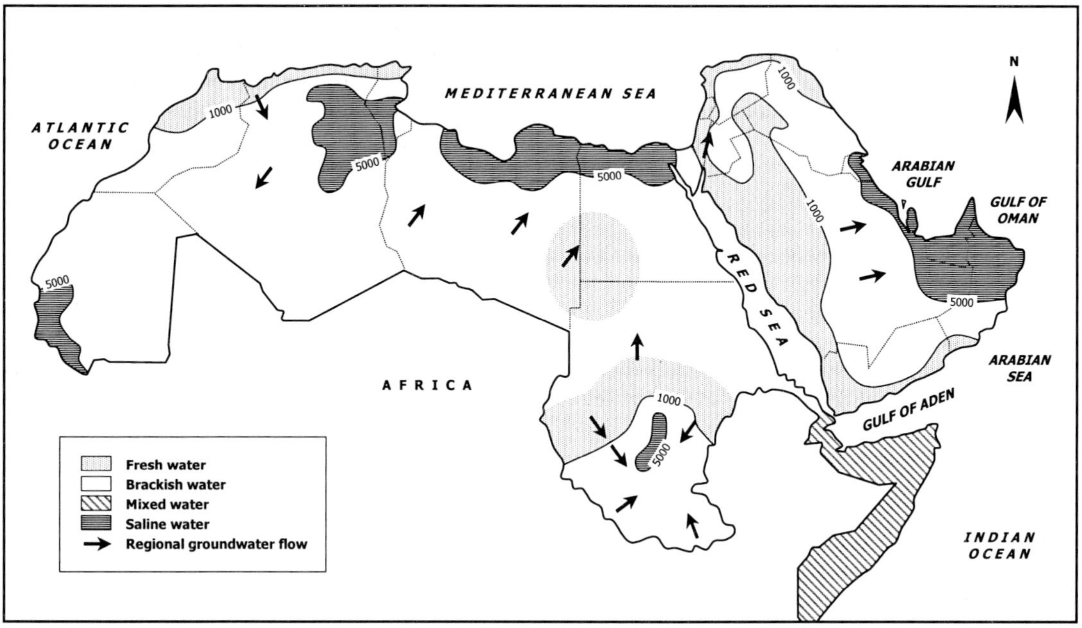

Figure 5. Ground water quality in the Arab region.

Managerial Issues

The need to have a regulating system for the management of water is widely recognized. A variety of statutory, legal, economic, administrative and other instruments and mechanisms have been developed, leading to a multiplicity of governmental, public and private bodies and non-governmental organizations (NGOs) of local, regional and national dimension.

Water management is in a satisfactory state in many Arab countries but there are several managerial issues that impede providing the desired level of service, and in some places services are deteriorating due to inadequate financial resources or appropriate methods.

The evolving conditions, both physical and socio-economic, may create problems that require a periodic review of water policies. There are two important factors that influenced water policies in the past two decades (1980s and 1990s). These are "Food security" and "Water security". The food security programme was launched in the late 1970s, whereas the water security programme was developed in the late 1980s (Khouri et al,1986). About one decade separated the adoption and implementation of these programmes, which were regarded as principal components of Arab League paramount strategies. Water policies, which were designed at the national level, have resulted in a continuous exhaustion of the naturally available water resources. The food security programme, however, aimed to enhance agricultural development in the Arab region, on the basis of studies of the "relative preference or distinction" of each area in the region in respect to climate water availability, soil and other factors required to yield maximum crop production per unit area under prevailing conditions.

In order to address the critical water problems in the region and deal with the complexity of the present situation, we need innovations in governance, new system for local and regional management of the limited, scarce, vulnerable and depleted water resources. In the Arab region water resources are "public" property susceptible to degradation or even destruction from over-use. Since the implementation of legislation is a major problem in most Arab countries, water is managed as a common property rather than public property and the consequences for the resource is "the tragedy of the "common" (World Humanity Action Trust, 2000).

Integrated water resource management necessitates the recognition of the inherently public nature of water resources. In a public property regime, a public institution is usually responsible for the most basic decisions about how water is to be used, but greater attention to economic incentives and tools will be essential.

Financial Issues

In most Arab countries financing is a principal constraint (Table 5). Investment for new water projects is becoming increasingly expensive because the easy access to water has reached the limit. The insufficient financial resources represent a major challenge to the conservation, development and management of water resources in the Arab region. The water sector requires at the present, as well as in the future, considerable investments, much larger than past and current investments. More funds are required for proper maintenance of water systems, and, in most Arab countries, major rehabilitation and modernization of water supply systems is needed.

Table 5. Per Capita Availability and Per Capita Income in the Arab region (United Arab Economic Report, 2000) .

Country	Per Capita Water Availability (m^3)	Per Capita Income $ U.S	Incomes Groups $ U.S.
Qatar	89	19026	High Income Group
U.A.E	129	17200	
Kuwait	74	13734	
Bahrain	164	9243	
Libya	160	6129	
Saudi Arabia	277	6059	
Oman	583	5537	
Lebanon	945	4754	
Iraq	2700	3649	
Tunisia	489	2051	Middle Income Group
Algeria	464	1565	
Jordan	150	1535	
Egypt	870	1441	
Morocco	1042	1277	
Syria	580	989	
Djibouti	252	930	
Sudan	817	424	Low Income Group
Mauritania	3137	416	
Yemen	263	330	

The water resources of Syria (9351 mcm) do not include the equitable and reasonable share from the Euphrates River and Tigris River

DEVELOPMENT OF ARAB WATER VISION

The establishment of an Arab Water Vision and development of actions that need to be taken to reach the vision is an approach that can help in exploring the future, enhancing the water situation and ensuring an acceptable level of sustainable development. The water vision should be developed in conformity with the aspiration, capabilities and potentials of the region. The vision development process involves a careful assessment of the water situation and water sector capabilities. Leading issues should be identified, analyzed and evaluated.

On the basis of the current status of water resources and their use, and by considering major constraints and challenges confronting water resource development and management, scenarios pertaining to the Arab Water Vision could be developed. They should aim primarily to achieve a sustainable water balance between future water demand, water potential and possible water supplies (World Water Council, 1999) (Table 6). The development of regional water scenarios requires a careful examination of two important elements: water availability and water demand.

Water resource assessment at the national and regional levels is influenced by several factors. These include upstream development in shared river or aquifer systems, spatial and temporal variability of rainfall, inadequate data on aquifer recharge and insufficient information on surface flow. Assessments, therefore, are made in a climate of uncertainty, due to the lack of reliable data on water resources and water use, and because of the complexity of the water systems.

Availability of data and water resource information in Arab region has improved in the last two decade. Estimates in the 1980s indicated water resource base of about 338 bcm (Khouri et al, 1986). In the late 1990s, estimates of renewable water resources in the Arab region were 265 bcm. In general, databases have improved during the 1990s in most Arab countries. On the basis of these national databases, the Arab Center for the Studies of Arid Zones and Dry Lands (ACSAD) established the "Arab Water Resources Data Bank" (AWRDB) (Arab Center for the Studies of Arid Zones and Dry Lands, 1998), which has contributed to the Arab region information base. Analyses of data from the AWRDB, within the framework of Arab Water Vision process (World Water Council, 1999), indicated that the maximum and minimum values of surface water is 303 and 155 bcm, respectively.

Table 6. Evolution of Per Capita Water Availability.

Country	Population (thousands)			Total Renewable Water Resources (millions)	Per Capita Water Availability		
	1950	2000	2025		1950	2000	2025
Jordan	1237	5003	9620	750	606	150	78
Syria	3495	16125	26292	9351*	2676	580	356
Iraq	5158	23280	43482	62850	12185	2700	1445
Lebanon	1443	3282	4400	3100	2148	945	705
Palestine	1258	2859	5987	215	171	75	36
Subtotal	**12591**	**50549**	**89781**	**76266**	**6057**	**1509**	**849**
U.A.E	97	2442	3284	315	4500	129	96
Bahrain	116	613	789	100.2	864	164	127
Saudi Arabia	3201	21930	54030	6080	1900	277	113
Oman	456	2518	5020	1468	3219	583	293
Qatar	250	579	779	51.4	206	89	66
Kuwait	152	2166	2721	160.1	1281	74	59
Yemen	4316	18654	44036	4900	1135	263	111
Subtotal	**8588**	**48902**	**110659**	**13074.7**	**1522**	**255**	**113**
Egypt	21834	68523	96462	59600	2730	870	618
Sudan	9190	33064	55553	27000	2938	817	486
Somalia	3072	10916	19802	11460	3730	1050	579
Djibouti	256	991	1962	250	977	252	128
Subtotal	**34352**	**113494**	**173779**	**98310**	**2862**	**866**	**566**
Morocco	8953	28781	38969	30000	3351	1042	770
Algeria	8753	32362	48811	15000	1714	464	308
Libya	1029	6562	11871	1047	1017	160	88
Tunisia	3530	9480	12899	4640	1314	489	360
Mauritania	825	2327	4444	7300	8848	3137	1643
Subtotal	**23090**	**79512**	**116994**	**57987**	**2510**	**729**	**495**
Total	**78626**	**292457**	**491213**	**245637.7**	**3126**	**840**	**500**

** Except equitable and reasonal shares from Euphrates and Tegris.*

Table 7. Water Resources of the Arab region According to Different Authors (in bcm).

	Khouri et al 1968	ACSAD 1997	Sadik & Barghouti 1997	Arar 1997	Khouri & Al-Weshah 2001
Surface Water	296.0	225.3	-	221.3	209.2
Groundwater	42.0	39.3	-	46.8	36.4
Total Conventional Water Resources	3380	264.6	244.6	268.1	245.6
Non- Conventional Water Resources	1.5	9.0	-	8	11.9
Total Resource Base	**353**	**273.7**	**-**	**276.9**	**257.5***

** Reuse equitable shares in international rivers will increase the resource base to 270 bcm.*

Estimates of groundwater resources ranged between 68 and 21 bcm. Accordingly, the maximum and minimum values of renewable water resources are 372 and 176 bcm, respectively. These values were considered to represent rainy and drought episodes (World Water Council, 1999). It can be concluded that water resource assessments in the Arab region need to be enhanced and data collection networks should be improved. Existing information on the status of water resources indicates, however, that earlier assessments are over-estimated. Most authors concluded that the resource base does not exceed 270 bcm. Few authors, however, evaluated accessible resources, i.e the resource which can be captured and converted to "real" water supplies (World Meteorological Organization, 1997).

Projection of water demand is even a more complicated issue. It has been considered the "core problem" in scenario analysis. The problem is centered around the projection criteria rather than the methodology. Significant driving forces and determining features for any consideration of future resource use in the Arab region is the population growth, income level and urbanization. Predicting water demand for the next 25 years requires, therefore, reliable forecasts of population size and economic growth at the country level.

Since such information is not readily available at the regional level and need to be compiled, most projections are based on simplified assumptions. Arab Center for the Studies of Arid Zones and Dry Lands projection of population size (Arab Center for the Studies of Arid Zones and Dry Lands, 1997), assuming current and decreasing growth rates, is 604 and 533 million inhabitants, respectively. The UN projection yielded 458 million population for 2025 (United Nations, 1986). In order to quantify the water scenarios, certain assumptions have to be made. Earlier attempts to develop water scenarios were based on two basic assumptions: food security (through self-sufficiency) and satisfaction of domestic industrial demand (Khouri et al, 1986). Agricultural water demands amounted to 378 bcm in 2030 while domestic and industrial demand was estimated at 55 bcm. It was soon realized that available water resources could not meet the projected demand of about 435 bcm. A revised vision for 2025 indicated that water demand could be reduced to about 386 bcm (Arab Center for the Studies of Arid Zones and Dry Lands, 1997). The balance (55 bcm) would be secured from raising the efficiency of water use and applying technical innovations. Food self-sufficiency will improve but food security will depend on the demographic transition and technological development in the field of biotechnology and water-related areas (Fig. 6).

Development of the Arab Water Vision within the framework of the Global Water Vision, has demonstrated that sustainable development is possible through a demand management and supply augmentation scenario (World Water Council, 1999). This conceived "sustainable scenario" indicates a sustainable water use is achievable at the regional level and cannot be achieved in certain sub-regions. The assumptions, which represent the collective view of a large number of experts from Arab institutions and regional and international organizations, are sound.

Implementation of action proposal, however, requires intensive efforts, large investments and closer cooperation at the regional, national and global level. The proposed supply augmentation to increase water supplies from the present level of about 210 bcm (World Water Council, 1999; Khouri, 2001) to the requirement of scenario IV (Table 8) could be achieved through improved efficiency of irrigated agriculture and reuse of treated wastewater.

In order to build on this concerted and collective efforts (World Water Council, 1999), the resource base was adjusted in order to become compatible with available resources. This has shown that the water deficit is larger than what has been estimated (Table 8) since the conceived "sustainable scarce" scenario III does not quantify food self-sufficiency, sufficiency level of 60% was proposed, and a fourth scenario was developed to attain the envisaged level of food self-sufficiency (Arab Center for the Studies of Arid Zones and Dry Lands, 1997).

Further adjustments are needed in the Arabian Peninsula and the Maghreb, in order to reach "Water Security" at the sub-regional level. The next step would be the development of national water visions, which should be consistent with the regional vision and help achieve a sustainable water use at the country level.

The vision exercise can only achieve tangible results if it is further developed on the basis of more reliable and accurate data and information base. The continuity of the planning process is needed because the impact of driving forces: demography globalization, technology, energy, biotechnology etc, would influence the basic assumptions and main criteria on which the present vision is developed.

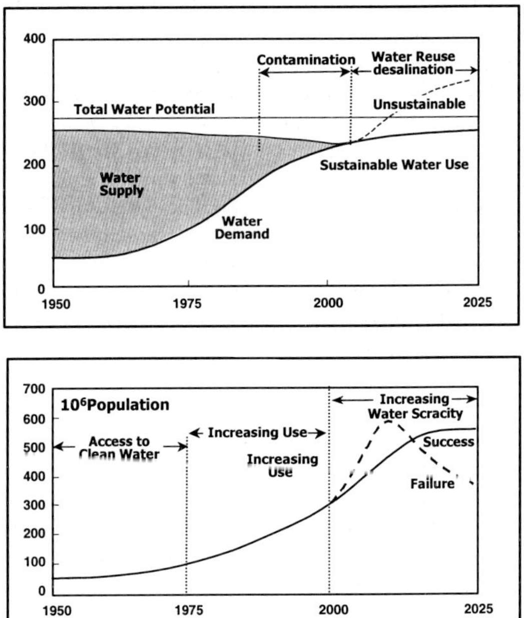

Figure 6. The demographic and freshwater transition in the Arab region

Table 8. Arab Water Vision: 2025 Scenarios (in bcm).

	Baseline	Scenario I 2025	Scenario II 2025	Scenario III 2025	Scenario IV 2025
Arabian Peninsula					
Water Availability	15.8	15.3	17.89	17.89	20.2
Water Demand	29.8	70.7	68.23	61.49	117.2
Water Balance	-14.5	-55.4	-50.34	-43.6	-27.6
Mashrek					
Water Resources	92.4	92.4	113.7	113.7	96.8
Water Demand	16.5	138.8	136.6	122.9	94.0
Water Balance	+25.9	-46.4	-22.9	-9.2	+2.8
Nile and East Africa					
Water Resources	10.3.1	103.1	111.67	111.67	107.6
Water Demand	83.4	165.0	159.30	143.41	109.8
Water Balance	+19.7	-61.9	-47.63	-31.74	-2.2
Maghreb					
Water Resources	59.2	59.2	60.48	60.48	63.14
Water Demand	25.7	57.3	53.80	48.44	37.0
Water Balance	+33.5	+1.9	+6.68	+12.04	34.4
Arab region					
Water Resources	270.0	270.0	306.48	306.48	288.0
Water Demand	210	431.8	418.07	376.26	288.0
Water Balance	+60	-161.8	-111.59	-69.76	000

Source: ACSAD, 1997; Khouri & AL-Weshah, 2001; World Water Council, 1999.

WATER RESOURCE MANAGEMENT: ACHIEVING SUSTAINABILITY

Sustainable water resource development should address leading sustainability issues, to achieve a balance between water availability and water demand, under conditions of water scarcity. Demographic, natural, social, economic and political factors interact to create often-unfavorable environment for sustainability. The challenge requires a new vision and innovated approach, working with and not against the natural behavior of water.

The general objectives of water resource strategies in the Arab region are:

- To reconcile the goals and programmes of water security with food security as defined by Arab organizations to achieve food self-sufficiency.
- Ensure that everyone has an access to safe drinking water.
- Integrate environmental considerations in water and land resource planning and development.
- Integrate the water sector with overall planning and management of the economy.

Unfortunately, it is rather difficult to design effective means of integrating environmental concerns at the national policy levels. Generally, priorities for water allocation are accorded to domestic use, agriculture and industry, in these descending order. Allocations among categories, however, are a political decision (Sadik and Barghouti, 1994) reflecting the broader, socio-economic and environmental objectives, which water as a national resources should properly serve. For achieving these broad objectives, it is imperative that the main responsibilities of water resource management lie with the state. The capabilities of government institutions need to be continuously developed and enhanced to enable them to undertake effectively and efficiently the functions of water resource development and management within a framework of policies, regulations and procedures.

Many planners tend to equate sustainability with resource preservation. This concept cannot be applied to non-renewable water resources, since preservation policies would substantially limit the economic and social development of arid and semi-arid areas, currently depending on these resources for their continued development and welfare. Countries in the region have chosen to continue mining their resources at various rates. The central question is how much is used now and how much should be used in the future, recognizing that the future is important but uncertain (Fig. 7). How should trade-off among various generations be made? Since future benefits, costs and losses, when

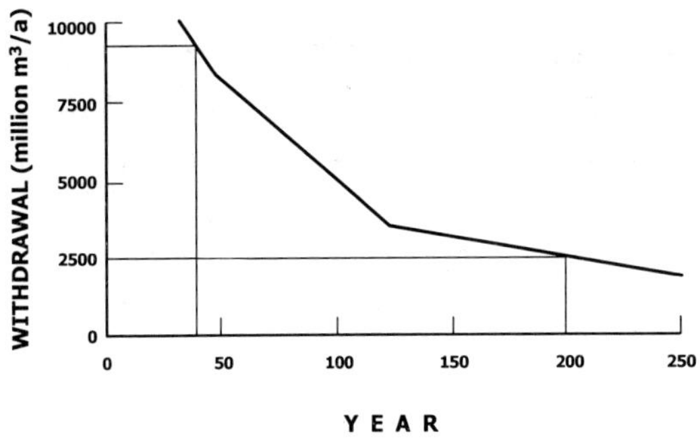

Figure 7. Relationship between total withdrawal and projected periods of water production (line-time of ground water) in Saudi Arabia (Khouri, 2001).

discounted, to the present time, decrease in value, any future loss far enough into the future will have minor effects on the present net benefits. Should our decisions today allow non-renewable water resources to be exhausted in the future (Loucks, 1994).

Another issue needs to be addressed with regard to sustainable water systems in arid zones. It concerns the spatial dimensions and time scale of sustainability. Conditions of sustainability vary for specific regions. Over what period should we expect to permit decreases in overall welfare due to depletion of water resources? Current understanding of water systems in arid zones is far from ideal. For the foreseeable future, policy at all levels of governance probably will need to be made in a climate of uncertainty. The "precautionary principle" is designed to avoid inertia. The principle states "where there are threats of serious or irreversible damage, lack of full scientific certainty should not be used as a reason for postponing measures to prevent environmental degradation" (World Humanity Action Trust, 2000).

Water resource development and management must be compatible with the limits and processes underlying natural water systems (Loucks, 1994). These processes could be environmental, ecological or hydrologic. Water resource development and management must interact with other sectors of economic and social activity. This entails managing the "whole range of development activities within a framework limited by the finite nature of the resource and optimizing the development strategy in terms of supply management, demand management, social equity, economic and environmental sustainability and natural capacities" (International Conference on Water and the Environment, 1992).

The conditions of sustainability require integration-the integration of economic, social and institutional as well as physical systems. The importance of such an approach was recognized in the Arab region and in many countries and was translated into national water legislations, since effective legislation is one of the basic instruments needed to ensure the rational use of water. Enforcement of water legislation, however, has not received the necessary consideration. Where governments have succeeded in developing water legislation, it often remains inadequately implemented.

Effective enforcement of water legislation needs participation, awareness-raising, and consultation among national institutions and interest groups. It is crucial to use participatory processes (through water user's associations) in order to manage the demand and to limit the time gap between the prevention mechanism and final decision-making.

Besides problems arising from the implementation and enforcement of modern water law, integrated water resource management in the Arab region has been confronted with major constraints and challenges, such as limited water resources and extreme rainfall variability, which are natural phenomena characterizing this arid and semi-arid region. These natural constraints have always existed in the region, and the Arab population has, for centuries, adapted to them. However, extreme events, such as recurrent and prolonged droughts, have set off the type of major outward migration that seems to have characterized certain periods in the history of the region. Under normal circumstances demographic pressures seem to have been relieved by movement within countries, sub-regions or the entire region. In modern times, population pressure has often led to demographic disequilibrium between rural and urban areas, creating acute urban water supply problems (Damascus, Amman, Riyadh, Sanaa). Pressure on water resources increased even in areas with abundant water resources and many areas or even countries moved to a status of high water stress (Table 4). This has compelled authorities to transfer water from sources that are further away from the sites of need. Accordingly, the cost of water supplies has substantially increased.

Water Policies

In most Arab countries the large allocation of water to agriculture has been driven by a strategic aspiration to attain food self-sufficiency at the national level (Sadik and Barghouti, 1994). Water policies, which were designed at the national level, have resulted in a continuous exhaustion of the naturally available water resources.

A structural approach to a policy review based on critical sustainability issues and quantified pressures on the water system has to be developed. Actions and measures to be taken to address over-development are country-specific, but guiding principles for policy review could be developed at the sub-regional or regional level. Water security is now high on the agenda. As regards food security, various scenarios have been developed at the national and regional level. Water resource institutions have realized that a pre-requisite to policy review and development is a continuous assessment of resources and demand. Matching supplies and demand is a basic issue for planning future withdrawal, but there are other issues that need to be addressed. Water contamination and depletion of aquifer storage need an urgent action, because of their threat to human health and sustainable development.

While each Arab country is attempting to tackle its situation according to its local needs, all face the imperative of formulating consistent national water policies as a matter of national priority, especially when addressing the allocation of water to agriculture. What is missing, usually, is an integrated water plan, which makes effective use of local resources taking into account different constraints in each country.

The national policies also need to address links between water and other sectors of the economy, such as agriculture, industry, energy and health, to account for the intersectorial competition for water demand. Often, there is also a failure to accommodate economic, social, cultural and ethical aspects in water development and management plans.

Kuwait and Saudi Arabia offer extreme cases on water policy formulation. In Kuwait (per capita water availability 74 m^3/yr), groundwater of the Dammam and Quaternary aquifers could only be used to blend desalinated water for urban water supply. Agriculture could reuse treated wastewater. Saudi Arabia (per capita water availability 250 m^3/yr) adopted in the past decade a water policy based on food self-sufficiency. The estimated agricultural water demand increased from about 2.0 bcm in 1980 to 17.8 bcm in 1995 (Al-Turbak, 1999; Al-Turbak and Dhowalia, 2001). Groundwater opstraction for agricultural purposes is expected to reach 22 bcm by 2010 (Al-Alawi and Abdulrazzak, 1994).

The key issue in policy formulation was identified when the gap between water availability and water uses increased substantially. It was realized that a large gap has developed between water availability and water uses, and that precious water is exported at minimal value. Accordingly, subsidies were reduced in 1993 reduction of cultivated areas. Groundwater levels recovered by 20 to 30%. Measures to manage water demands were also taken in urban areas.

Development of Governance

Governance is "the framework of social and economic systems and legal and political structures through which humanity manages itself" (World Humanity Action Trust, 2000). Recent trends in the area of governance focus on greater participation, more transparency and cohesion and on a new approach for re-organizing the water sector. In the eastern and southern Mediterranean countries, current approaches include de-centralization, corporatism and privatization in the form of mixed enterprise.

Experience has shown that in water -deficient countries, a holistic approach is a critical need. The proposed approach involve the management of freshwater as a finite and vulnerable resource and the integration of sectorial plans within the framework of national economic and social policy.

In most countries in the Arab region there is a fragmentation of authority in the water sector due to large number of institutions that deal with water resources as well as to lack of cooperation and coordination of activities. A coordinating body needs to be established to frame the overall policy of the country and direct the allocation of water, funds and manpower for different purposes. A priority need is to undertake action to implement effectively water legislation. Sound enforcement of water legislation is, however, country-specific. Coping with drought in the Arab region should comprise productive and reactive strategies (Albergel and Claude, 2000).

In an arid environment, water resource institutions should be empowered to act when emergency circumstances arise (Economic and Social Commission for Western Asia, 1999). They should be able to declare certain areas or basins protected, restricted or water rationed (Caponera, 1992). During a drought, strategies may include the imposition of limits on water withdrawal and prohibition of certain uses. Distinction must be made, however, between exceptional provisions extending power over a limited period and regular administrative control of water (Caponera, 1992).

Agreements for the use of water of international rivers are normally based on the Helsinki Rules and future agreements are expected to follow the provisions of the 1997 UN convention on the Law of the Non-navigational Uses of International Water Courses. There is a growing need for comprehensive legal instruments for international water bodies. Such agreements would promote the equitable use of shared water resources.

Institutions set up for the administration of international water resources comprise agencies or other forms of institutional arrangements, established for the purpose of dealing with its management. In the south Mediterranean region institutional development includes the establishment of commissions for the Nile and Senegal river basins. Committees have also been set up for the sustainable management of the Nubian Aquifer System and the major basins in the Northern Sahara, underlain by the continental sandstone aquifer system.

In order to achieve sustainable development, social and cultural changes would be required; people's understanding of water need to be improved. Water use is often changed. Nobody, even in countries endowed with abundant resources, has the right to waste water. Measures are needed to make socially and culturally acceptable the cooperation in the management of water resources and the introduction of practices leading to improvements of water productivity and water use. Measures are also necessary to promote stakeholder participation in water management.

People need knowledge about water in order to be involved in making decisions. There is a need to focus attention on education and training so that every person becomes aware of the importance of water. It must be highlighted that water saving is urgent need in the Arab countries, everywhere and in all seasons. The benefits are obvious, financial and environmental. It is the responsibility of governments, associations and local communities to promote this awareness.

Ethical issues need to be addressed to help in improving water management in the 21st century. The required ethics should be based on the concept "working with water". This concept should be the basis of laws and practices in order to achieve sustainable management of water resources and minimize the risk of "working against water" (World Humanity Action Trust, 2000); in managing water resources, the impact on ecosystems need to be taken into account.

The ethics of freshwater management must be based on a balance between traditional human values regarding conservation and the application of new technologies (Llamas, 2000). Women are rarely involved in the decision-making process regarding water resource management. Often, women are key water managers in small communities and villages. Studies often show that participation of women is both the ethical approach and also the pragmatic approach (Llamas, 2000).

In arid zones mining of non-renewable aquifer systems may be considered an ethical issue. Hydrogeologic evidence and modeling techniques could be used to generate scenarios on development options. Users and decision-makers should be provided with information such of resources on the impacts of development on the water levels and reserves.

Demand Management

Whereas planners should be aware of the possibilities of scientific breakthroughs, the scope for utilizing existing technological innovations for immediate effect is enormous; especially in the area of water use. The quest to conserve irrigation water should include the technological problems causing losses. Seepage, leakage, percolation, evaporation are the main causes of water losses in water distribution and application. Water can be conserved through better on-farm and system management. Modern on-farm irrigation methods such as sprinkler and drip irrigation can reduce water losses considerably. Efficiency rates of 70% for Jordan and 67% for Morocco are attributed mainly to the application of these methods (Sadik and Barghouti, 1994).

Strategies for water resource management are generally a combination of supply-and-demand oriented measures as dictated by physical, social and economic factors in each country. The sustainable use of water requires a global shift in strategy towards greater demand management. Demand management is defined as the management of the total quantity of the water abstracted from a source of supply, using measures to control waste and consumption.

Technical measures for water saving have financial and administrative implications. Hence, the character of demand management is multi-disciplinary and includes an array of instruments (World Humanity Action Trust, 2000):

- Technical: Water conservation, recycling, water saving technology (including retrofitting), leakage control, cropping patterns.

- Economic: Subsidies, incentives, taxes.
- Legal and administrative: Water law, water rights, licenses, regulations, penalties, enforcement, capacity building.
- Educational: Awareness, communication and education.

The combination of these instruments differs from country to country depending on physical, socio-economic and cultural environments.

Demand management requires decisions on where (that is, in which sector) and how water demand can be reduced. There will be conflicts between competing users, and trade-offs have to be made between the benefits obtained by different users within the context of the local and national economy. Conflicts commonly arise between agricultural use, on the one hand, and industrial and urban use on the other. Successful demand management also requires programmes of awareness-raising and promotion, and of education and training.

Even, in water-deficient countries such as those of the Arabian Peninsula, planning for the full utilization of treated effluent remains in the early stages. The total volume of reclaimed wastewater used in the Arabian Peninsula is estimated to be about 433 mcm. The reuse volumes, represent approximately 25 percent of the available treated wastewater.

Wastewater reuse ranges between 217 mcm in Saudi Arabia and 6 mcm in Yemen. In the region as a whole, reclaimed wastewater meets about 2 percent of total water demand or 14 percent of domestic and industrial demand. In Saudi Arabia, reclaimed wastewater is used for irrigation of non-edible crops, landscape irrigation, and for industrial cooling. In Kuwait, Bahrain, the United Arab Emirates, and Oman it is used for municipal irrigation of landscaped areas, while in Qatar it is used to irrigate animal food crops (Economic and Social Commission for Western Asia, 1999).

Wastewater irrigation based on up-to-date engineering technology and public health safeguards, is becoming a viable strategy and is often the least-cost solution for wastewater treatment and disposal. The new microbial guidelines provide a safe and rational basis on which countries can build a sound programme of wastewater reuse.

Supply Management

Using a diagnostic approach helps identify vulnerable areas that have the greatest risk to groundwater development (Zaporozec, 1994), and supports the efforts to address present exploitation problems and plan new development in less vulnerable areas. A diagnostic approach is an effective tool for planning future development when aquifer systems are untapped or subjected to a low degree of stress.

Methods of assessment of vulnerability of an aquifer system to contamination and susceptibility to exploitation side-effects have been proposed as an important tool for planning and management of groundwater resources (Adams and MacDonald, 1995; Vrba and Zaporozec, 1994). Assessment of groundwater vulnerability addresses mainly groundwater quality problems, whereas groundwater susceptibility deals mainly with aquifer sensitivity in term of quantity, such as decline of water levels and land subsidence. Groundwater susceptibility was presented as a potential adjunct to aquifer vulnerability in order to provide resource managers and planners with an additional method of evaluating potential aquifer degradation resulting from intensive development (Adams and MacDonald, 1995).

Adverse impacts on the environment in arid regions result mainly from water use under extreme climatic conditions, especially high temperature and high evaporation. Certain impacts could result from heavy pumpage. These include land subsidence and ecological impacts on aquatic ecosystems.

Susceptible areas are usually underlain by multi-layered aquifer systems containing fine-grained detrital sediments. Inelastic compaction of these argillaceous formations will not occur until water levels are lowered below a critical value. The continental and carbonate formations are susceptible to land subsidence in the North Sahara basins. Certain parts of Nubian aquifer system are also vulnerable in terms of land subsidence. Areas where the "pure" quartz sandstones pass northwards into mixed mairne-continental facies, are susceptible to land subsidence.

In susceptible areas, land subsidence causes engineering problems and failure of well casing. In regional carbonate aquifer systems, intensive development caused collapse of sinkhole, triggered by lowering of water levels. Several examples were reported in the western and northeastern parts of Syria during the 1950s and 1960s. To address these problems, a diagnostic approach could identify the degree of groundwater susceptibility, and a well-field design can mitigate possible impacts, nevertheless a strategic choice becomes necessary in extreme cases.

Impacts resulting from water use in arid zones can be minimized by improved management at the farm level. The most difficult problem in arid zones is disposal of drainage water in oases irrigated by groundwater. These depressed areas are usually cultivated because they offer more favorable water and land resources for agricultural development. Impacts resulting from intensive cultivation can only be mitigated if an appropriate plan for the disposal of drainage water and wastewater is implemented. In Ouorgla Oases in Algeria and Bahrya and as well as Siwa Oases in northern Egypt, complex groundwater development and use problems have risen due to natural factors and uncontrolled development and inadequate management practices. In order to address these environmental issues in the New Valley, in Egypt, the proposed groundwater development plan envisages limiting irrigation to relatively higher areas. The sabkhas and depressions could then be used as a discharge for drainage water, especially, when they act under natural conditions as discharge area to unconfined aquifer systems.

In irrigated areas, where return flow from the irrigation water recharges the aquifer, constituent salts leached from the soil or applied pesticides and fertilizers may enter the aquifer, thus degrading water quality. Management of groundwater quality in oases in arid zones need to reconcile two problems: leaching of salts from the soil profile and minimizing the movement of return flow into the aquifer system. In certain parts of aquifer systems and in even in major basins underlain by multi-layered aquifers (the Continental Intercalaire-Complex Terminal in the North Sahara Basin, the East Arabian Aquifer system, the Gezira Region aquifer systems in Syria and Iraq etc.), the major constraint on future development is degradation of water quality rather than limitation of water quantity.

Integrated Management

The natural unit for water resource management is the river basin or the aquifer system. Both units may be national or international. Integrated basin planning can lead to higher levels of sustained water use than more narrow focus on project or sectorial planning. The rapid growth of human population in the Arab region coupled with increasing per-capita water use and rising aspirations for attaining food self-sufficiency has resulted in state of high water stress in most countries (Table 4). Solutions to manage critical issues need to be found quickly. The likely impact of climate change on the hydrologic cycle adds to the urgency. There is a vital need to protect the environment. Ignoring the relationship between human population and the needs of the living world has resulted in the degradation of the environment.

In arid and semi-arid zones water resources are finite and often scarce. Water systems are vulnerable to natural and human impacts, so management requires sharing, rationing and moving quantities of water from place to place to overcome spatial variability of the resources. Using water effectively, protecting its quality and re-using it can effectively increase its availability. A variety of statutory, legal, economic and administrative instruments and mechanisms have to be developed. This has lead to the multiplicity of state-run, public and private bodies and Non-Governmental Organizations at the local, national and supranational levels.

In such an arid and semi-arid environment, an integrated management of resources must emphasize both supply and demand management including application of alternatives technologies for freshwater augmentation, which include desalination, reuse of wastewater, water harvesting and aquifer recharge. The concept of integrated approach as applied to water resource management, focuses on giving due consideration to economic, social, environmental and engineering requirements during planning. Integrated water resource management implies much more interactive and dynamic process requiring the implementation of a package of inter-related activities in a comprehensive manner (International Conference on Water and the Environment, 1992).

Recent development in this context is based on the application of a system approach and a diagnostic approach to address groundwater problems and minimize adverse impacts of water resources development (Khouri, 2000). The first approach would help improve understanding of the relationship between different components of the hydrologic cycle and the environment, atmosphere, surface runoff, soil and groundwater. The latter approach seeks to improve planning and protection of the resource by a-priori assessments. These assessments evaluate the sensitivity of the groundwater systems to possible pressures in terms of quality and quantity and can therefore play an important role in improving environmental management.

Achievement of the objectives of integrated water resource management requires the implementation of a wide range of interrelated activities. Focus in this respect is on demand management, but supply management continues to be an important component of countries water sector strategies. As new water sources become increasingly inaccessible, efficient supply management will require careful analysis of options that comprise both conventional and non-conventional source. The world total desalination capacity increased from 15 mcm/ d in 1991 to 20.3 mcm/d in 1995. Of this total, 11 million m^3/day is produced in the Arab region. They represent 51% of world capacity compared with 49% in 1991. Increase of water shortages has lead to greater attention to other non-conventional water resources such as wastewater reuse, re-cycling and water harvesting.

In spite of high spatial and temporal rainfall variability (Fig. 8), there is a large potential in many areas for increasing the supply of water from rainfall water management. Rainfall water management deals with all methods, which augment precipitation, manage rainfall and runoff and store them in the soil or underground for later beneficial use (Albergel, 2000). In dry climates sudden and violent floods characterize ephemeral stream-flow (Wadi flow). Devastating effects occur in low-lands and coastal plans. A flood warning network is a fundamental need to avoid or minimize flood damage and avoid loss of life, studying flash flood phenomenon and utilizing their water for beneficial purposes (Zebidi and Kallel, 1974). Rainfall water management techniques includes precipitation enhancement, water conservation, development of wadi systems through rainwater harvesting in upper reaches and floodwater harvesting and artificial recharge in the lower reaches.

Water Deficit		
600 mm	200mm	100 mm
Sub-humid	Semi-arid	Arid
(NW Syria) N		(NE Jordan) S

As water scarcity increases, rainfed areas receive high priorities in the present and future national development plans. In the Arab region, some 40 million hectares of rainfed croplands and about 500 million hectares of natural grazing lands could benefit from research findings of international research centers concerned with rainfed cropping and grazing systems. Supplementary irrigation could be introduced to stabilize the yields of rainfed winter crops through well timed applications of small amount of water. Response farming practices that aim at institutionalization of farmers role and incorporating information on rainfall expectations and crop response may produce immediate impacts.

One of the major water problems in the greater part of the Arab region is the management of non-renewable water resources. The extensive arid deserts and Saharas are underlain by regional aquifer systems shared between several countries. Non-renewable water resources occurring in the Mashrek are limited in quantities compared to the huge reserves of the North African aquifers. Non-renewable water resources in the Arabia Peninsula are limited but much greater in quantity than the Mashrek resources.

Since the 1992 UN Conference on the Environment and Development and the general acceptance of the concept of sustainability, plans have started to be made within a context that evaluates the environmental risks of proposals and takes account of environmental costs and benefits. Sustainable management of water resources includes not imposing the cost of cleanup onto future generations. Although the cost of prevention of water contamination is high, it is much less than the cost of cleanup: "Prevention is always better than cure".

Water resource development and use create impacts on the resource base and on the environment. These impacts have social and economic consequences. They are, however, often uncertain. Yet inadequate attention to the expected or predicted risks of adverse impacts can limit the effectiveness of even the best designed water resource management systems (Loucks, 1994).

Simulation models form the core of most groundwater assessments. The models have the power to integrate the complexities hydrogeologic settings, hydrogeologic processes and water utilization. The main steps required in the construction of groundwater models include the development of a conceptual model, construction of the model, calibration and verification. Some hydrogeologists have objected the use of the terms

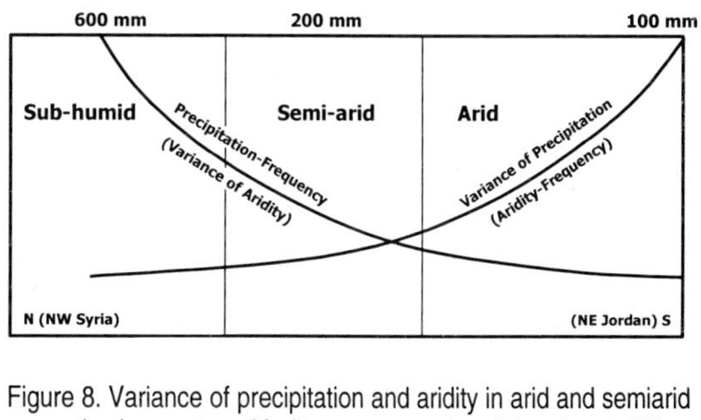

Figure 8. Variance of precipitation and aridity in arid and semiarid zones in the eastern Mediterranean region (Khouri and Miller, 1994)

verification and validation and suggest more meaningful terms such as model testing and sensitivity testing.

Simulation is considered a more realistic approach. It is widely used because it takes into consideration all possible development plans and their physical and chemical consequences on the aquifer system (lowering of water level depletion of reserves, deterioration of groundwater quality). It offers, therefore, elements of decision.

The wide occurrence of extensive regional aquifer systems in the Arab region underlying large basin, ranging from 500,000 to 2 million km^2, justifies the development regional models. Such regional models are intended primarily for regional assessments. The large extent of aquifers modeled and lack of data in large areas preclude the use of these models to address local and site-specific water resources issues (Sun and Johnston, 1994). To provide the greater resolution needed for a more detailed analysis of the effects of groundwater development, the regional flow model may, however, be further divided into sub-regional models. The regional flow models supply boundary conditions. For investigating site-specific groundwater problems, local models with small grid size (1 km^2 or less) are used.

The semi-arid areas of the Arab region are often confronted with recurrent droughts and floods. Since these areas are sensitive to natural impacts, water shortages caused by climatic extremes may become more critical by inadequate water use practices, especially in countries with high water stress.

The strategy for risk management has to provide for short-term solutions for the mitigation or prevention of damage from flood risks. As regards drought combat, long-term as well as short-term measures are required to reduce the negative impacts of drought. National drought policies should stress the creation of an infrastructure to supply the basic data and research needed for the assessment and response. A key part of the national policy should be the development of a general plan to prepare for and respond to drought episodes.

CONCLUSIONS

As result of prevailing dry climatic conditions, water resources are inherently limited in the Arab region. The arid part of the region is sensitive to even minor shift in energy balance. Variance of rainfall increases as aridity increases and, therefore, areas most subject to drought are those in which variations of annual rainfall are greatest. With the exception of the Lebanon and Atlas coastal mountains, these drought-prone areas encompass the greater part of the Mashrek and Maghreb countries. The zone of greatest unpredictability lies in these areas.

Population increased from about 80 million in 1950 to about 300 million in 2001. Populations increase and economic growth has spurred higher demand for the limited water resources. Scarcity of water resources that is presently afflicting some parts of region will become much more intense and widespread, and crises might ensue.

If water resource development is to be sustainable, principal sustainability issues should be addressed. Demographic, social, ethical, natural, economic and political factors interact to create often unfavorable environment for sustainability in the region. The establishment of an "Arab Water Vision" and development of actions that need to be taken to reach the vision is an approach that can help in exploring the future and ensuring an acceptable level of sustainable development.

Considering current status and trends of water resources in the Arab region and the influences of driving forces, policy development should be based on the following criteria and principles:

- Changing attitudes towards water, and acknowledging that water must be a fundamental issue, at the top of the agenda of governments, institutions and individuals.
- Reforming national water policies to address the major issues of the 21st century and reforming policy with regard to food security concept by adopting in water - deficient countries "a relative food security approach".
- Incorporating the environmental, social and cultural considerations into the various stages of water resource development and management.
- Focusing on sustainability and integration in developing the Arab strategy for water resource development and management.

Among the necessary actions that would enhance the water situation, the following are particularly important.

- Improve and/or optimize hydrologic monitoring networks in terms of quantity and quality and integrate hydrologic, meteorological and groundwater networks.
- Upgrade facilities and procedures used to store, process and analyse hydrologic data and make data and information derived from them available to users.

- Develop or strengthen groundwater protection programmes, on the basis of research on contamination, mass transport, modeling, vulnerability and risk assessment.
- Improve management of the resource from inception to reuse, making a better use of treated wastewater. Use aquifer storage recovery (ASR) technique to enhance strategic water resources and peak demand.
- Promote the incorporation of water demand management in national water strategies, by demonstrating the benefits associated with effective water demand management at the highest level of government.
- Improve and disseminate information to consumers and industries on water saving technologies, recycling, reuse and cleaner production processes.
- Establish incentives for the rehabilitation and modernization of installations and domestic devices. Remove disincentives to water utilities for promoting investments in water distribution networks.
- Establish and/or improve on-farm advisory services focussing on introduction of water-efficient irrigation techniques, and advice on water application and management and on crop varieties and markets.
- Develop strategies to improve sanitation and protection of surface and groundwater by establishing discharge standards, applying clean technologies and restricting use of pesticides.
- Encourage the use of groundwater conjunctively with surface water in wadis and other places where such joint use has a potential for increasing the available water supply. Control groundwater contamination.
- Improve coordination between different institutions and government agencies at the policy, planning and operational level.
- Promote decentralization of service delivery functions to specialized private or public companies through delegated management.
- Review, strengthen and re-structure water institutions in order to enhance their capacities in water related areas.
- Assess and revise, if required, existing legislation on the management and protection of water resources.
- Assess training needs and priority areas of training in the areas of water resource assessment, development and management.
- Develop a national training programme for the water sector based on the concept "continuing education".

REFERENCES

Arab Center for the Studies of Arid Zones and Dry Lnads, 1997. Water resources and their uses in the Arab region. Proceeding of the Second Symposium on Water Resources and Uses in the Arab region, Kuwait, 8-10 March, 1997. Published by the Arab Fund for Economic and Socio-Development, Kuwait.

Arab Center for the Studies of Arid Zones and Dry Lnads, 1998. Water resources data bank in the Arab region. Water Resources Division, ACSAD/WR/R 116, Damascus, Syria, 104p.

Adams B. and MacDonald A., 1995. A diagnostic approach to the determination of aquifer susceptibility to exploitation side-effects. IAH 26th International Congress, Edmonton, Alberta , Canada, 4-10 June, 1995.

Albergel J. and Claude J., 2000. Drought and water management in case of scarcity in North Africa and Middle East countries. International Seminar, Hydrology of the Mediterranean Regions, Montpellier, France, 11-13 October, 2000. IRD, France.

Albergel J., 2000. Small dams in the Mediterranean regions. International Seminar, Hydrology of the Mediterranean Regions, Montpellier, France, 11-13 October 2000. IRD, France.

Al-Sulaiti, M. Abu Yacob, 1999. A new vision for the water resources planning in Qatar. Proceedings of the 4th Gulf Water Conference, Bahrain, 13-17 February 1999, p. 15-34, Manama, Bahrain.

Al-Turbak A.S. and Al-Dhowalia K.H., 2001. Sustainable water management for the Kingdom of Saudi Arabia. Proceedings of the 5th Gulf Water conference, Doha, Qatar, 24-28 March 2001. The Secretariat General, The Gulf Cooperation Council (GCC), Riyadh, Saudi Arabia, 75-86.

Al-Turbak A.S. 1999. Future water supply and demand prediction in Saudi Arabia. Proceedings of the 4th Gulf Water Conference, Bahrain, 13-17 February 1999. Manama, Bahrain.

Arar A., 1997. Water use for agricultural purposes in the Arab region. Proceedings of the Second Arab Symposium on Water Resources and Their Uses in the Arab Region. Published by the AFSED, Kuwait, 277-348.

Caponera, D. 1992. Principles of water law and administration. A.A. Balkema, Rotterdam, The Netherlands, 260p.

Economic and Social Commission for Western Asia, 1999. Updating the assessment of water resources in the ESCWA member countries. Expert Group Meeting on Updating the Assessment of Water Resources in the Member States of the Economic and Social Commission of Western Asia (ESCWA), Beirut, 20-23 April 1999, 14 p.

Hefny, K., 1991. Planning for groundwater development of the Nubian sandstone aquifer for sustainable agriculture. Round Table Meeting on Planning for

Groundwater Development in Arid and Semi-Arid Region, Cairo, 5-9 October 1991. RIGW/IWACO, Cairo/Rotterdam, 113-124.

International Conference on Water and the Environment (ICWE), 1992. The Dublin Statement and report of the conference. Dublin, 26-31 January 1992. Published by World Meteorological Organization, Geneva, 55p.

Khouri, 2001. Arab water security: A strategy to achieve sustainability. Symposium on Water in the Arab World, Naples, Italy, 22-24 February 2001, 41p.

Khouri J. and R. Al-Weshah, 2001. Arab action plan on water resources for the 1st quarter of the 21 century. International Hydrologic Programme of UNESCO, 9th Regional Meeting of the Arab IHP National Committee, Muscat, Oman, 23-26 September 2001. UNESCO, Cairo.

Khouri J., 2001. Water resources and water demand in the Arab region during 2000-2025. Proceedings of the First International conference on Economical and Social Uses of Water in the Arab Countries, Beirut 18-21 June, World Water Council, RAMTAN, Riyadh, Saudi Arabia (in Arabic).

Khouri J., 2000. Integrated management of water resources in the arid and semi-arid regions of the Mediterranean Basin. International Seminar on the Hydrology of the Mediterranean Regions, Montpellier, France, 11-13 October, 36 p. Institute of Research for Development (IRD), France.

Khouri J., 1999. Impacts of intensive development on regional aquifer systems in arid zones. International Conference on Regional Aquifer Systems in Arid Zones, Managing Non-Renewable Resources, Tripoli, Libya, 20-24 November 1999. UNESCO, Paris, France.

Khouri J. and Miller J.C., 1994. Groundwater vulnerability in areas of climatic extremes. Guidebook on Mapping Groundwater Vulnerability, International Contributions to Hydrology, Verlag Heinz Heise, Hannover, 49-56.

Khouri J., Agha W.R. and Droubi A., 1986. Water resources in the Arab region, and future perspectives. Proceedings of the Symposium on Water Resources and Uses in the Arab region, Kuwait, 17-20 February. The Arab Fund for Economic and Social Development (AFESD), Kuwait.

Llamas R., 2000. Water and ethics. Special issue, USO Intensive de las Aguas de las Subterraneas, Aspecto Eticos, Technologies Y Economicos Series A, No. 5. Foundation Marcelino Batin, Madrid, Spain.

Loucks P., 1994. Water resources management: focusing on sustainability. UNESCO, Paris. IHP-IV Project M-4-3, 46p.

Sadik A.K. and Barghouti S., 1994. The water problems of the Arab World. Management of scarce resources. Harvard University Press. 1994.

Sadik A. K. and Barghouti S., 1997. The economics of water. Second Symposium on Water Resources and Uses in the Arab World, Kuwait, 8-10 March 1997. Published by the AFSED, Kuwait.

Sun R.J. and Johnston R.I., 1994. Regional aquifer analysis program of the U.S. Geological Survey, 1978-1992. U.S.G.S. 1099, Washington.

United Nations (UN), 1986. World population prospects. New York, UN 1986, 56-57.

Vrba J. and Zaporozec A., 1994. Guidebook on mapping groundwater vulnerability. International Contributions to Hydrogeology, Vol. 16, Verlag Heinz, Heise, Hannover, 131p.

World Humanity Action Trust, 2000. Governance sustainable –IV: Working with water. A report of the Humanity Action Trust, London, 52p.

World Meteorological Organization (WMO), 1997. Comprehensive assessment of the freshwater resources of the world. Published by WMO on behalf of UN Agencies and Stockholm Environmental Institute, New York, Geneva, 33p.

World Water Council (WWC), 1999. Arab countries vision consultations, Cairo-Marseilles-Beirut. WWC, UNESCO, ESCWA, 41p.

Zaporozec A. 1994. Concept of groundwater vulnerability. In: Guidebook on Mapping Groundwater Vulnerability. International Contributions to Hydrogeology, Vol. 16, Verlag Heinz Heise, Hanover, 3-8.

Zebidi H. and Kallel R., 1974. The operation of a flood warning system, its use for studying flash flood. Flash Floods Symposium, Proceedings of the Paris Symposium, IAHS, Publ. No. 112, 57-61.

Zubari W.K., Mubarak M.A. and Madani I.M. 1993. Development impacts on groundwater resources in Bahrain. *Water Resources Development*, 9(3): 263-279.

Water Management Strategies in the Arabian Gulf Countries

MOHAMED AL-SENAFY, KHALED AL-FAHAD and KHALED HADI
Water Resources Division, Kuwait Institute for Scientific Research, P.O. Box 24885, 13109 Safat, Kuwait.
E-mail: msenafy@kisr.edu.kw

ABSTRACT: Most of the Gulf countries are arid in nature. The scarcity of natural water resources and the growing gap between demand and supply of potable water in most of the Arabian Gulf countries force them to face their water challenge with wise policies and decisions. Until recently, increasing water demands were met by increasing the source capacity. The source of fresh water in many Gulf countries is desalination of seawater. The cost of desalination is high. The influx of oil wealth during mid 70's and 80's raised the standard of living considerably resulting in increased demands for fresh water. These water demands were met by the construction of more and more desalination plants. However in recent years, the drop in oil revenue due to world- wide recession and the rapid increase in per capita water consumption in the last 40 years, has forced many water supply authorities in Gulf countries to consider conservative alternatives to meet increasing water demands. One of the main alternatives is water conservation. Most appropriate water conservation strategies are: 1) Public awareness on water conservation, 2) Metering and pricing at appropriate rates, 3) Use of water saving devices and leak detection and prevention, and 4) Artificial recharge and storage of fresh water and treated wastewater. Of these, pricing of water is the most effective way to reduce water consumption. In general, water rates should be set at a level designed to provide necessary revenue for the water supply authority. However in most cases, water is highly subsidised by the government.

Water conservation under these highly subsidised pricing conditions, is extremely difficult to implement. Public education and even stringent legislation will not result in any significant reductions in water demand. Hence, for any significant reduction in water demand, the subsidies should be reduced or removed. Otherwise, water conservation can never be implemented and will remain a myth in most of the Gulf countries.

INTRODUCTION

This paper will concentrate on Kuwait as an example for the Arabian Gulf countries. Kuwait is an arid country with an average annual rainfall of about 120mm. The average daily evaporation rate is high and about 12mm (Al-Fahad and Al-Senafy, 2001). The average day temperatures vary from 18°C in winter months to about 45°C in summer months. Daily maximum temperatures in excess of 50°C are common in summer months.

Kuwait has few natural fresh water resources. The groundwater in Kuwait is mostly saline with zones of brackish water. The infiltration of rainwater is insignificant except in the northern region where unique geomorphologic conditions results in the rapid infiltration of rainwater during periods of short but heavy rain storm periods resulting in the formation of fresh ground water lenses. The safe yield from these lenses is about 6 mega litres per day (ml/d) (Al-Senafy, 2001) and hence is very small compared to the growing total demand of fresh water, which is about 1040 Ml/day (Ministry of Electricity and Water, 2000). The fresh water lenses are presently exploited for the production of bottled mineral water and other minor uses.

Most of the urban water demand for domestic, industrial and commercial users is met by water from the desalination of seawater using multistage flash (MSF) distillation units. Desalination technology was introduced in the mid-1950's and has developed very rapidly to compensate for the shortage in conventional water resources. The majority of the west Asia desalination plants are located in the Arabian Gulf (Food and Agriculture Organization, 2000). The cost of production of the desalinated water is about \$8.3/m^3 (Darwish et al., 1989). Brackish ground water with total dissolved solids of about 3000 ppm is used for urban landscaping and agricultural purposes. Occasionally fresh water is also used for landscaping and agricultural purposes.

Water Demand

Kuwait has dual distribution system, one for fresh water and the other for brackish water. The fresh water is a blend of desalinated water and brackish water (ratio of about 9:1). The brackish water is pumped from brackish ground water fields

Water Resources Perspectives: Evaluation, Management and Policy. Edited by A.S. Alsharhan and W.W. Wood.
Published in 2003 by Elsevier Science, Amsterdam, The Netherlands, p. 221-224.

and is supplied on alternate days. Table 1 lists price of water as charged to consumers in Kuwait.

It is clear from Table 1 that the price of desalinated water is highly subsidised. Consequently the per capita water consumption has increased substantially in the last 30 years. Table 2 presents the average per capita consumption of fresh and brackish water in Kuwait as a function of time.

Table 1. Price of potable and brackish water in Kuwait. (Source: Bushnak, 1990)

Consumer	Price in US $/m^3$	
	Fresh water	Brackish water
Domestic	0.67	0.08
Industrial	0.21	0.08
Agriculture	-	0.02
Tankers	0.25	Free

Table 2. Per capita water consumption in Kuwait. (Source, MEW-Kuwait, 2000)

Year	Population	Fresh water (l/c/day)	Brackish water (l/c/day)
1975	1,038,859	140	110
1980	1,400,917	210	100
1985	1,760,167	260	110
1990	2,125,053	280	90
1992	1,441,385	370	130
1995	1,721,968	440	140
1999	2,148,032	490	150

Groundwater used for agricultural purposes is not included in Table 2. It is estimated that about 120 ml/d was used for irrigation during the year 1989-90. The installed capacity of desalination plants increased from 27 ml/d in 1960 to about 1287 ml/d in 1999. The high production growth in the Arabian Gulf region exceeds the rate of water resource development. Consequently, the annual per capita share of water resources is decreasing, and at increasing rate. According to Global Water Assessment, the share per individual in Arab world has decreased during the last two decades from 2200 to 1100 m^3 per year (Food and Agriculture Organization, 2000). Table 3 shows the projections of total water demand over the period 1995-2015 in the Arabian Gulf Countries.

Water Conservation

Water conservation can be defined as making more efficient use of existing supplies through structural, operational, economic and socio-political means (Flack, 1981). Generally water conservation program is implemented on two occasions. The first is a permanent measure when the water demand exceeds the available water resources and the cost of procuring additional water resources is prohibitive. The second is a temporary measure implemented during periods of drought. In Arabian Gulf countries that are already suffering from a deficit in water resources the need for the implementation of water conservation measures from two reasons. These are (1) High cost of construction and operation of seawater desalination plants to meet the increasing water demands. In addition the present high cost of subsidy places increased burden on the government spending. (2) The increased use of brackish ground water for urban landscaping is depleting ground water reserves. A 1995 comparison between the Arabian Peninsula's annual groundwater recharge (7,200 million cubic meter) and groundwater abstraction rates (23,600 million cubic meters), indicates that the mining of groundwater reserves in the Arabian Peninsula is about 16,400 million cubic meters (Food and Agriculture Organization, 1997). Table 4 presents the groundwater depletion rates in the individual Arabian Gulf countries.

Table 3. Per capita total water demand projections in Arabian Gulf countries (1995-2015). (Source: Food and Agriculture Organization, 2000).

Country	Domestic water demand (L/c/d)					Industrial water demand (L/c/d)					Agricultural water demand (l/c/d)
	1995	2000	2005	2010	2015	1995	2000	2005	2010	2015	1995
Bahrain	540	555	570	585	600	95	100	105	110	115	792
Kuwait	540	555	570	585	600	15	20	25	30	40	168
Oman	130	135	140	145	150	15	20	25	30	40	1 428
Qatar	540	555	570	585	600	95	100	105	110	115	1 685
Saudi Arabia	356	370	375	380	390	30	35	40	45	50	2 788
United Arab Emirates	540	555	570	585	600	95	100	105	110	115	1 908

Table 4. Rates of groundwater depletion in Arabian Gulf countries. (Source: Food and Agriculture Organization, 1997)

Country	Groundwater depletion rate in million cubic meters
Bahrain	100.00
Kuwait	200.00
Oman	240.00
Qatar	140.00
Saudi Arabia	13, 558.00
United Arab Emirates	1, 495.00

In addition, the use of brackish water for landscaping in urban areas has resulted in the rise of subsurface groundwater levels, threatening the foundations of buildings and services as in Kuwait City.

Implementation of water conservation measures in Kuwait, and generally in most of the Arabian Gulf countries, faces certain problems due to the existing socio-economic conditions. These are (1) at present the community expects the availability of inexpensive water with no limitation on use, (2) Desalination of seawater creates an impression that the availability of treated water is unlimited because the source of water is sea, and (3) General desire to have trees, gardens and large landscaping areas. In addition to these, Gulf countries continually expand their industrial operations, resulting in the increased demand for water.

Implementation of Conservation Measures

Four conservation methods that are relevant to Arabian Gulf are (1) Public awareness (2) Use of water saving devices, leak detection and prevention, (3) Metering and billing at appropriate realistic rates, and (4) Artificial recharge and storage of fresh water and treated wastewater in groundwater aquifers.

Public awareness is the first and most important step in the implementation of water conservation program. Without public awareness, other measures of water conservation cannot be implemented. Public co-operation is vital for the successful implementation of conservation measures and public education is an important step in achieving public co-operation. Public education involves the development and distribution of brochures, posters, stickers etc. to the public. The most effective group targeted for public education is school children. This group is easy to access and very receptive to new ideas. In addition, a well-studied awareness program should be organised and publicised through the mass media.

Water saving devices attempt to decrease the amounts of water used in household functions without inconveniencing the user. Some of the devices are (1) low flush toilets, (2) dual flush toilets, (3) flow limiting showerheads and flow restrictors, etc. Leakage in water distribution systems is a significant "water use" category for many water supply authorities. The actual amount of system leakage is difficult to determine even in metered communities because of other unaccounted for water. Leakage detection and repair is a viable method of reducing the demand on the water system. Household leakage from plumbing fixtures presents a similar problem in that the precise amount of water use due to leakage is unknown but can account for a relatively large percentage of total household use. It is estimated that about 30% of water is lost or unaccounted for in major cities within the Gulf region (Bushnak, 1990).

The use of pricing policy is an effective tool for achieving water conservation. The economic incentive for using less amount of water is dependent upon consumer response as reflected in the elasticity of demand (Flack, 1981). The value of water to the consumer can be categorised as essential or less than essential to the consumer's well being. In house water use is considered essential and is not very responsive to price change. Outside use is less essential and is therefore more responsive to price changes.

Abdulrazzak and Khan (1990) undertook a conservation study on a sample population in Saudi Arabia. Some estimates for water savings and the related cost of implementation of conservation methods were obtained and presented in Table 5.

Table 5. Savings due to the implementation water conservation methods. (Source: Abdulrazzak and Khan, 1990)

Conservation method	Percentage savings	Cost of implementation
Public awareness	5 %	\$ 0.10/ m^3
Use of water saving devices and leakage reduction	10 - 20 %	\$ 1.05/ m^3
Metering and billing	20%	\$ 0.40/ m^3

One of the most important factors that control the effectiveness of water conservation methods is the price of water supplied to the consumer. If the water supplied to the consumer is highly subsidized or it is not related to the level of usage by the consumer, then whatever measure that is imposed to conserve water usage will be ineffective. In Gulf countries, cost of production of water range between \$0.5 -2.5 per cubic meter depending on the source of water. Much higher water costs exist for remote inland regions. Most of the Gulf countries subsidise water. For example in Kuwait, the cost of production of water is $8.3/kl and the water is supplied at a price varying between $ 0.25 and $ 0.67 per cubic meter. In Saudi Arabia, the price varies between $ 0.08 and $ 1.07 per cubic meter. Similar situation exists in other Gulf countries. Under these conditions, it is virtually impossible to implement any water conservation methods.

Artificial recharge and storage of fresh water in aquifers is likely to overcome some of the problems associated with the water resources management in the Arabian Gulf. The difference between summer demand and winter demand for fresh water in Kuwait is about 250 Mega liters per day (Viswanathan and Al-Senafy, 1998). This excess plant capacity in winter months could be used to produce additional water for artificial recharge and storage in aquifers. Storage of the fresh water is required to meet the seasonal and long-term uncertainties and crisis conditions. In the other hand recharging of treated wastewater through unsaturated zones of salty aquifers will improve the quality of wastewater to a level it can be used for unrestricted landscaping and gardening purposes. The usage of treated wastewater for irrigation will help to overcome the groundwater depletion problems in Arabian Gulf countries.

CONCLUSIONS

Arabian Gulf countries are facing a rapid increase in water consumption forcing the construction of additional seawater desalination plants in the near future. Highly subsidised water is one of the reasons for the increase in water usage. Water conservation methods are expected to reduce the present high demand levels. But under the existing socio-economic conditions and with highly subsidised water, the implementation of water conservation methods might prove to be difficult. A properly developed and implemented public education program coupled with correct pricing policy for water supplied, might overcome this obstacle. In addition to that, artificial recharge can play a major role in the management of water resources.

REFERENCES

Abdulrazzak M.J., and Khan M. Z. A., 1990. Domestic water conservation potential in Saudi Arabia. *Environmental Management* 14(2):167-178.

Al-Fahad, K. and Al-Senafy, M., 2000. Evaluation of groundwater quality in Kuwait. Proceedings of the groundwater quality conference, 2001. Sheffield, England, 164-166.

Al-Senafy, M., 2001. Geohydrology of fresh groundwater lenses in arid environment, Kuwait. Proceedings of the groundwater quality conference, 2001. Sheffield, England, 167-168.

Bushnak, A., 1990. Water supply challenge in the Gulf region. *Desalination* 78: 133-145.

Darwish, M.A., Jawad, M. A. and Aly, G. S., 1989. Technical and economical comparison between large capacity MSF and RO desalting plants. *Desalination* 76: 281-304.

Food and Agriculture Organization, 1997. Irrigation in the Near East in figures. Water report, No.9. Food and Agriculture Organization, Rome, Italy.

Food and Agriculture Organization, 2000. Alternative policy study: Water Resources Management in West Asia. ACSAD and Arabian Gulf University, Bahrain.

Flack, J.E., 1981. Urban water conservation. American Society of Civil Engineer, p.99.

Ministry of Electricity and Water, 2000. Statistical yearbook, 2000. Ministry of Electricity and Water, Kuwait.

Viswanathan, M.N. and Al-Senafy, M. N., 1998. Artificial recharge of groundwater. Proceedings of the third international symposium on artificial recharge of groundwater. Amesterdam. Netherlands, 29-33.

The Environmental Impact of Groundwater Exploitation in African Grasslands: Examples and a Case Study from the Kalahari Region

DAVID S. G. THOMAS
SCIDR, Department of Geography, University of Sheffield, Sheffield S10 2TN, UK.
E-mail: d.s.thomas@shef.ac.uk

ABSTRACT: In the Sahel and the Kalahari, the growth in groundwater extraction via simple wells and boreholes is contributing to an increased use of dryland natural resources, to competition for resources, and to environmental changes. Livestock production systems in particular have expanded through the utilisation of groundwater in areas devoid of perennial surface water sources. Environmental changes are most marked in vegetation systems that were utilised ephemerally and opportunistically in years of good rains prior to the all-year-round exploitation facilitated by groundwater. The nature of vegetation changes, notably biodiversity reduction and changed relationships between annual and perennial grasses and bush cover, is considered, including whether such changes represent environmental degradation and whether they are permanent. Process studies indicate that bush encroachment, a particularly marked form of vegetation change, is reversible at decadal to century time-scales, in the context of the natural dynamism of dryland ecosystems. At shorter time-scales such changes can even open up new resource opportunities that may relieve pressures on critical grazing resources at times of environmental stress. Ultimately however, available data on water availability suggest that the rate of groundwater extraction is probably unsustainable, and this may provide the ultimate limit on the wider environmental impacts that borehole-centred livestock systems have on the environment.

INTRODUCTION

In many semi-arid to arid grasslands of Africa, accessing groundwater via wells has been a traditional means of coping with water deficiencies in areas where rainfall is scarce or highly seasonal and surface water availability is extremely limited. In many areas, from the mid twentieth century onwards, more numerous and deeper boreholes and wells have been sunk or dug, sometimes replacing hand dug wells and increasing water extraction rates. Often reliant on simple technologies, borehole water extraction has contributed to marked transformations in pastoral and agricultural opportunities in many African countries, including Botswana, Ethiopia, Kenya, Mali, Namibia, Senegal, Somalia, South Africa and Sudan. A similar situation is reported for many Asian dryland areas, particularly steppe regions (Kharin et al., 1999), but is not specifically considered here.

In these dryland areas, access to water gives rise to *de facto* control over and use of the surrounding land. While extraction has a direct impact, often unsustainable, on the water resource itself, with groundwater reserves in many parts of the world under threat (Clarke *et al.,* 1996), it is also the root cause of other environmental impacts, which are frequently viewed as unsustainable or contributing to land degradation. When considering the use of water, particularly groundwater, resources in dryland environments, the effect of usage on water availability is only part of the environmental story. This paper focuses on the impacts that ground water use can have on the availability and nature of other natural resources, principally vegetation systems, focussing on the Kalahari region of southern Africa but also drawing upon examples from the Sahel region (Figure 1).

The expansion of boreholes and wells

Three processes have driven the expansion of wells and boreholes in drylands.

- Expanding cultivation, often commercially, in environments where intensive irrigation is needed (often commercially and using centre pivot irrigation systems). This has occurred for example in various dry regions in South Africa. An extreme example is the case of Saudi Arabia, where center-pivot irrigation has expanded dramatically in recent decades in very dry areas, and where the combined effects of urban and agricultural demand are expected to deplete groundwater reserves to 60% of their 1985 volumes by 2010 (Al-Ibrahim, 1991).
- Direct intensification of land use pressures, often involving increasing competition (and sometimes conflict) between pastoralists and crop farmers. This can lead to the sedentarisation of formerly mobile livestock systems, for example in eastern Sudan (Shazali and Ahmed, 1999).

Water Resources Perspectives: Evaluation, Management and Policy. Edited by A.S. Alsharhan and W.W. Wood.
Published in 2003 by Elsevier Science, Amsterdam, The Netherlands, p. 225-235.

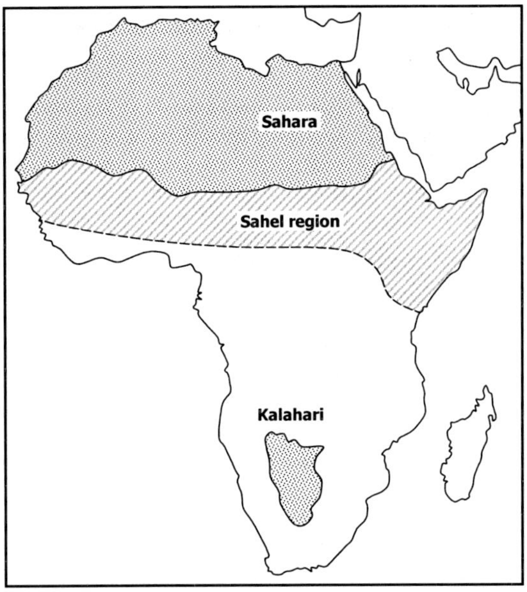

Figure 1. Areas considered in this paper

- Opening up previously little or under-used areas, or intensifying use of moderately used areas. In particular, this has been a notable process in the grasslands of central southern Africa, where dramatic increases in the number of borehole-centred cattleposts and ranches have occurred since the 1970s, allowing opportunistic grazing use to be replaced by sedentary, year round, livestock production (Cooke, 1985; Perkins, 1996).

In African drylands, the link between groundwater use and pastoralism is a particularly marked one, and is occurring significantly in marginal dryland areas both south of the Sahara, in East Africa, and central southern Africa.

Pastoralism and groundwater in African drylands

There are a number of environmental resource issues attached to the use of groundwater in dryland areas. First, it is necessary to consider the reasons that relate to enhanced groundwater use and the land use systems that this facilitates. Second, these land uses will themselves impact upon the environment. Intensified or changed resource impacts may ensue, and questions of sustainability may be raised. Third, the actual magnitude of groundwater extraction can impact upon no only future reserves, but on the very land uses that have been established through groundwater availability. These issues will be explored, with the first considered in the context of both the Sahel and central southern Africa and the latter two in more detail from research research in the Kalahari region of southern Africa.

Pastoralism and wells in the Sahel

The Sahel region lies immediately south of the Sahara, across the whole breadth of Africa from Senegal to Ethiopia, and is approximately bounded by the 150 mm and 600 mm mean annual isohyets. Drier northern areas are sometimes termed the Sahelian steppe and southern areas the Sudanian savanna (White, 1983). In many parts of the Sahel arable production and pastoralism have coexisted for centuries. Coexistence has been complementary, for example where each activity occurs in different ecological zones, or where coexistence uses the same territory but with specific 'articulation of activities across time and space' (Raynaut and Delville, 1997; p111), for example, pastoralists grazing their animals on fallow fields. Complex forms of land organisation may ensue where this occurs, for example in Mali (Gallais, 1984). While conflicts of interest have sometimes inevitably risen, they have tended to be resolved tolerably. In the region as a whole since 1940, however, the widespread northward expansion of arable production in the drier northern regions of the Sahel (Reynaut and Delville, 1997), as well as the growth in livestock numbers (Turner, 1999; Figure 2),

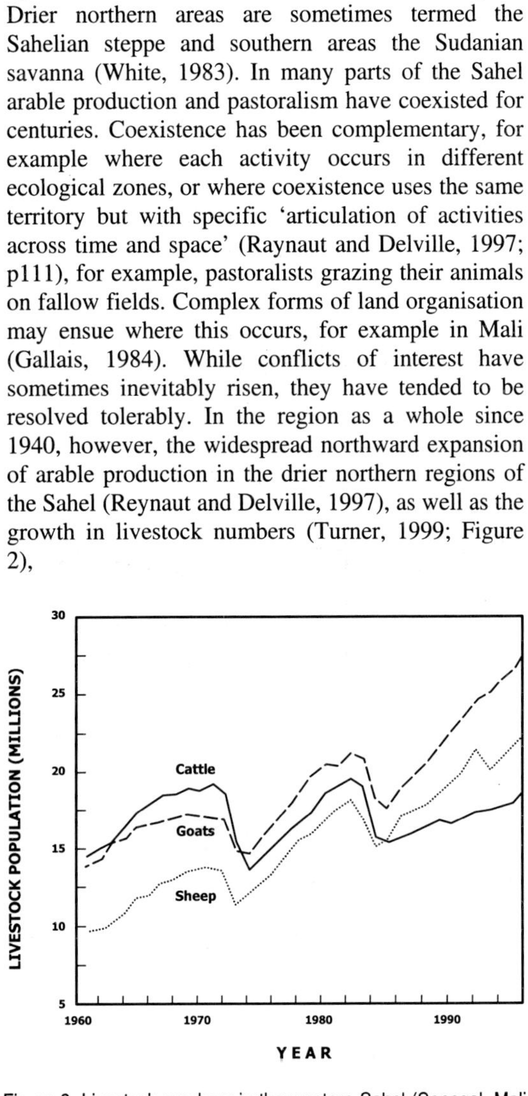

Figure 2. Livestock numbers in the western Sahel (Senegal, Mali, Burkina Faso, Niger, Chad). Over the 35 year period covered by the data, cattle numbers have risen by about 20%, but small stock numbers (Sheep and goats) have doubled. After Turner (1999). Turner identified complex reasons for the changing species' contributions, including small stock being better able to cope with degraded lands and the greater number of smallstock providing a better buffer against drought that the fewer cattle affordable by the total accumulated wealth.

including in the traditionally arable south (Ramisch, 1999), has occurred. It is also reported that the growth in total livestiock numbers incorporates an actual decline in the cattle population and an increase in small stock numbers (Reichelt, 1989; Turner, 1999). The overall growth in animal numbers has created pressures on land that have challenged this coexistence of arable and pastoral activities, and have in some instances led to violent conflict (Raynaut, 1997). Customary law has also traditionally managed wells. These have included the duration of residence permitted by a transhumance herder at an individual well, and has effectively also regulated the use of surrounding pastures. Increasingly, public wells have been sunk by governments, which have unrestricted access. Reichelt (1989), Reynaut (1997) and many others reports that this has contributed both to the number of people who now own livestock, and to the total number of livestock. Uncontrolled grazing in the pastures around these water points has been a major factor leading to land degradation particularly in the northern Sahel, which receives 200-250mm of rainfall per annum and has a dry season of up to ten months (Koechlin, 1997).

Pastoralism and groundwater in the Kalahari

The Kalahari region, centred on Botswana, is a surface water deficient area with mean annual rainfall varying from c200mm in the southwest to 600 mm in the northeast (Figure 3). The absence of perennial surface water, except in the major rivers at the northern and southern margins of the Kalahari, has exerted a considerable influence upon human activities and upon the development of the region and use of its natural resources. Proposals to irrigate the Kalahari by diverting south the waters of the Zambezi and Okavango Rivers, devised in South Africa during the early 20th century, (Schwatz, 1919, 1920), were never pursued but indicate the way in which the lack of permanent surface water has been perceived as a limit to development. Ephemeral accumulations of rainwater in pans and depressions may last days, weeks or even months into the dry season, and have been important for wildlife, hunters opportunist use of grazing by pastoralists more usually based in wetter areas to the east (Thomas, 2002).

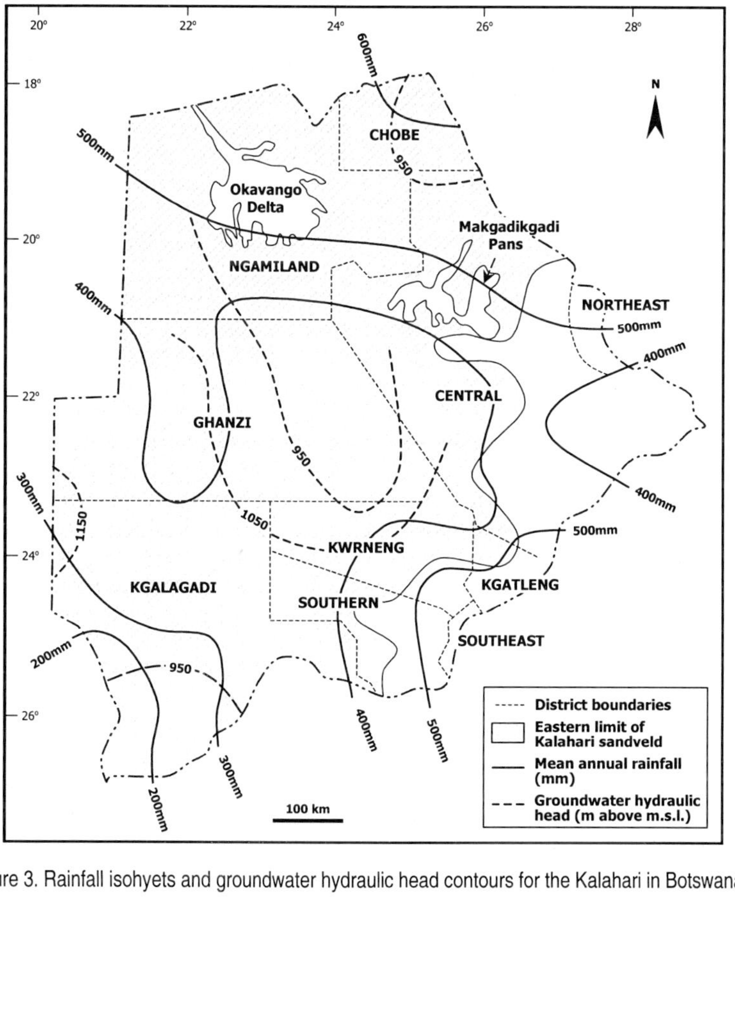

Figure 3. Rainfall isohyets and groundwater hydraulic head contours for the Kalahari in Botswana.

These surface water deficiencies have inevitably given rise to quests to utilise groundwater resources instead. Boreholes have been sunk in the Kalahari at an increasing rate from the mid-twentieth century onwards (Cooke, 1985). The total number of boreholes in the Kalahari is unknown but it is 'very high' (Hitchcock, 1978, Campbell 1981, Figure 4): they certainly total several thousand in the Botswanan Kalahari alone. Most tap groundwater aquifers in the Karoo sediments that underlie the younger unconsolidated Kalahari group sediments. Water tables are approximately 20 m below the surface around the basin margins and in excess of 100 m deep in central areas. A groundwater divide, coincident with the subsurface Bakalahari Schwelle (Thomas and Shaw, 1991) results in regional groundwater flow under most of the region trending northwest towards the Makadikgadi depression, with a hydraulic gradient of 0.03% (de Vries *et al.*, 2000; Figure 3).

Initially boreholes were sunk by the government for drought relief purposes, but, coinciding with the development of better drilling and pumping technologies from the 1950s, the Kalahari was viewed as an untapped grazing resource, with the potential to relieve pressure on resources in better watered eastern areas (Debenham, 1948, 1952). The formal development of the interior Kalahari for livestock production did not start to occur until the mid 1960s (Mazonde, 1994), notably from 1975 onwards in association with the Tribal Grazing Land Policy (TGLP) in Botswana and other opportunities that led to the sinking of boreholes. In the driest southwestern area of Botswana, in Kgaligadi District, the grazed area rose from c.13,000 km^2 in 1950 to c.32,000 km^2 in 1990 (White, 1993). This increase was associated with borehole numbers in the region rising from eight (plus seven wells) in 1955 to approximately 380 in 1990.

In the Botswanan Kalahari as a whole, the Tribal Grazing Land Policy created opportunities for the controlled establishment of leasehold ranches in a number of blocks. The opportunities to sink boreholes and fence land for the use of individuals have been further enhanced by the National Agricultural Policy of 1991, such that communal lands, administered successfully for decades by customary law, are increasingly subject to the effects of new legislation that affects the rights of individuals and access to land and water. Livestock numbers tripled, to 3.5 million, in the 30 years from 1966 (Wheelwright *et al.,* 1996). These changes have been seen as creating opportunities for resource conflicts, environmental degradation and the displacement of traditional livelihoods (White, 1993). The conflicts relate to wildlife versus livestock, subsistence versus commercial pastoralism, rural resident versus distant urban lessee, and sustainability versus short-term gains, and at least in part derive from the manner in which borehole lessees/owners gain *de facto* control over the surrounding environment, especially if boundary fences are erected (Thomas, 2002). The progressive fencing of ranch boundaries, constraining livestock to small areas in deference to the operation of traditional cattleposts (Perkins, 1996), and the sinking of extra boreholes at a density above the officially designated one per 8 km (Figure 4), are key issues indicating that access to groundwater impacts on the environmental resource base. There are a number of issues that arise from these wider environmental consequences, which are now considered.

Environmental consequences of groundwater use

Vegetation systems: degradation or change?

It has been widely cited that wells, boreholes and villages act as focal points for the development of bare degraded areas in drylands, resulting from high land use pressures, particularly linked to livestock grazing (e.g. Rapp, 1974). Scientific evidence does not support the notion that, other things being equal (i.e. it may be a different matter if the density of water points increases on the ground) these bare patches expand significantly over time. In other words, they do not act as foci for the expansion of desertification (e.g. Hellden, 1991). Indeed, remote sensing-based studies, by Ahlcrona (1988) in the Sudan and by Hanan *et al.* (1991) found no evidence of bare area growth around such point water sources over periods of up to 20 years. These studies do not however indicate that land use around water points does not impact on the environment. Instead of considering purely bare ground, it is necessary to consider more complex ways in which impacts may occur, particularly if land degradation is considered to be an effectively permanent change in the resource base that reduces productivity (*senso* Abel and Blaikie, 1989; UNEP, 1997). The changes could be represented by the near-permanent occurrence of bare ground that is then subject to erosion, but also includes factors such as vegetation biomass change and species composition change.

Research in all areas of dryland Africa has commonly identified these impacts as environmentally negative, with a substantial literature pointing towards the expansion of groundwater based livestock systems being unsustainable and leading to marked desertification (e.g. Nellis and Bussing, 1990; Ringrose *et al.*, 1990 and 1993). It is certain that borehole and well – centred grazing systems do lead to spatial changes in vegetation structure, composition, biomass and to localised soil erosion (Georgiadis and McNaughton, 1990; Perkins and Thomas, 1993). Because watering points act as the foci of grazing activity that radiates out from the water point, environmental impacts do however decrease in intensity with distance. This has been termed the 'piosphere effect' by Andrew (1988), based on research in arid Australia. The radial effect of environmental pressures, linked to decreasing 'herbivore use intensity' with distance away from water points, is an explanation of why bare 'deserts' do not generally expand spatially through time. What can result however is a zonation of impacts that can lead to annuli with different characteristics of vegetation structure and soil changes. Figure 5 shows a model of these impacts, based on empirical data from the eastern Kalahari (Perkins and Thomas, 1993). In practice the actual extent of the zones and the magnitude and nature of the vegetation and soil changes depends on a range of factors. These include total livestock numbers, duration of borehole use, and relationship between livestock induced changes and the natural vegetation system variability that characterises savanna ecosystems (e.g. Behnke *et al.*, 1993). In the piosphere model a bare sacrifice zone around the borehole is an inevitable consequence of high livestock pressures close to the borehole. At distance however (over about 2km from a relatively new borehole) herbivore pressures are commonly sufficiently diluted that impacts may be limited and a near-natural ecosystem results. Termed the grazing reserve in Figure 5, this area may only be threatened by environmental degradation if extremely large animal numbers use an individual borehole or if borehole spacing is reduced through new water points being tapped.

There are two widely noted vegetation changes that occur in borehole-centred grazing systems. First there is a progressive reduction in grass species diversity, with the most palatable perennial species being replaced by less palatable annual grasses, often with a single species (e.g. sour grass, *Schmidtia kalahariensis* in the Kalahari) dominating. Second is bush encroachment, whereby mixed savanna grassland is replaced by bush thicket. This is widely seen as one of the most negative aspects of borehole-dependant livestock impacts. In many Kalahari systems the encroaching bush layer is often dominated by *Acacia mellifera.* In the driest southwestern areas of the Kalahari a smaller shrub, *Rhygozum trichotomum,* is the dominant shrub species (Van Rooyen, 1998). In the bush encroached zone around boreholes the grazing resource is largely or totally replaced by bushy scrub, where predominantly only browse opportunities exist.

Bush encroachment has been attributed to a 'two-layer model' of soil hydrochemical change (Walker and Noy-Meir, 1982; Dougill et al., 1999). The reduction of grass cover under intense grazing pressure close to the borehole leads to a marked competative advantage being gained by deeper rooting plants, in terms of both soil nutrients and soil moisture availability and uptake (Figure 6).

Investigations of a range of borehole-dependant systems in southern Africa however show that while distinct vegetation changes do result from the intensive grazing of the environment around boreholes and wells, it is by no means clear that all such changes are permanent, or negative, in terms of their impacts on agricultural outputs and livelihood opportunities. Qualitative observations of the occurrence of recovery at abandoned boreholes (e.g. White, 1993) are supported by process-based studies of soil nutrients and soil water by Dougill *et al.* (1998). These do not suggest that soil hydrochemical changes associated with bush encroachment are significant, while natural disturbance events common in savanna ecosystems, notably droughts and fires, can reduce bush cover over time, at decadal to century scales (Dougill *et al.*, 1999).

Even at shorter time scales, the most negative aspects of vegetation change, the replacement of mixed perennial grassland communities with less palatable annual grasses and bush encroachment, can have positive dimensions depending on the adaptability of the land users. In a recent research project working amongst Kalahari pastoralists, many situations have been identified where pastoralists identify and exploit new opportunities within the vegetation changes that occur in association with borehole centred livestock systems (PANRUSA, 2001, 2002). These include exploiting the early growth of annual grass dominated areas to relieve pressure on perennial grasses, livestock browsing preferentially on the spring flowers of *Rhygosum*, again allowing grass growth to occur, and bush encroached *Acacia mellifera* acting as a critical browse resource for cattle and small stock during droughts. Together, these reinforce the observations of White (1993) and others that pastoralists and their animals can be highly resourceful at times of environmental stress. They also indicate that straightforward associations of vegetation change with land degradation should be treated with caution.

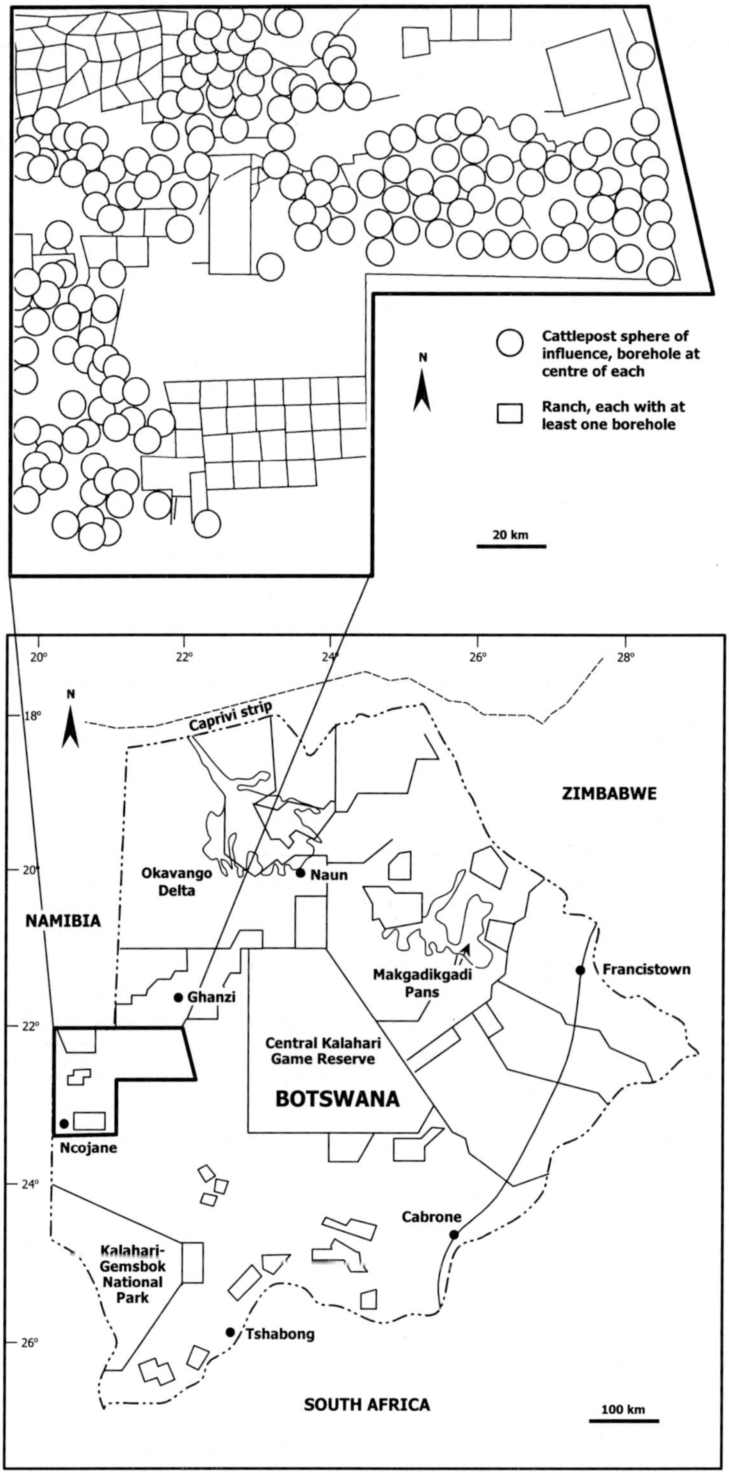

Figure 4. The density of boreholes in part of the wesetrn Kalahari, and their spheres of influence. Ciroloc chow the way in which the '8km spacing rule' has in some places been breached.

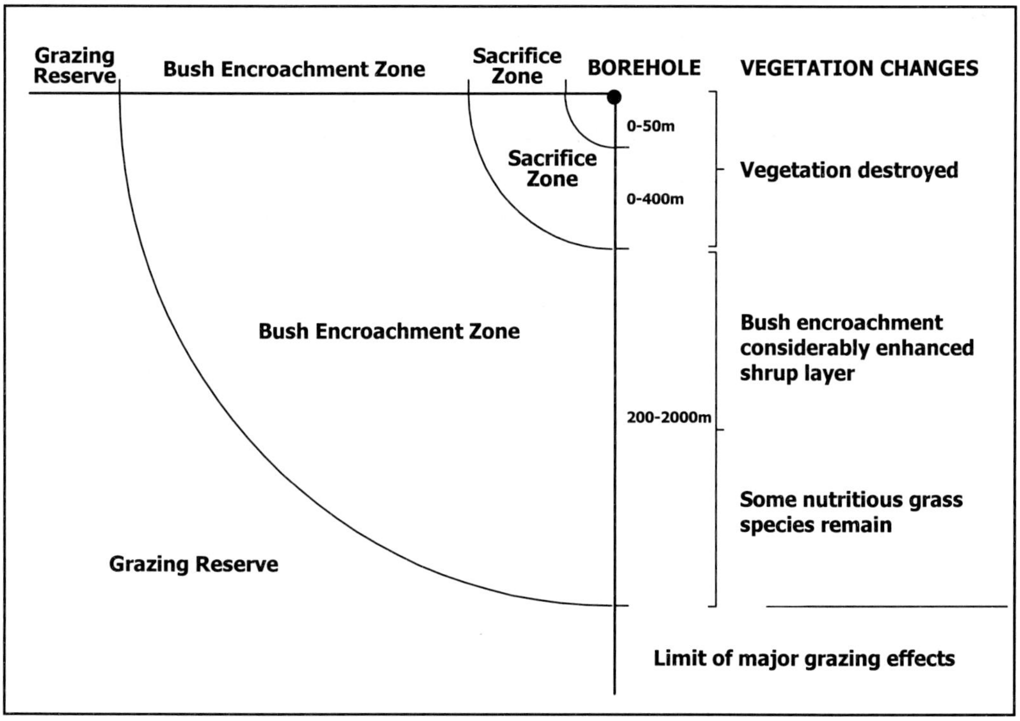

Figure 5. Schematic piosphere model developed from investigations in the eastern Kalahari. The precise compostion and size of each zone can vary spatially and with time, dependant of factors such as gazing intensity and age of borehole. After Perkins and Thomas (1993).

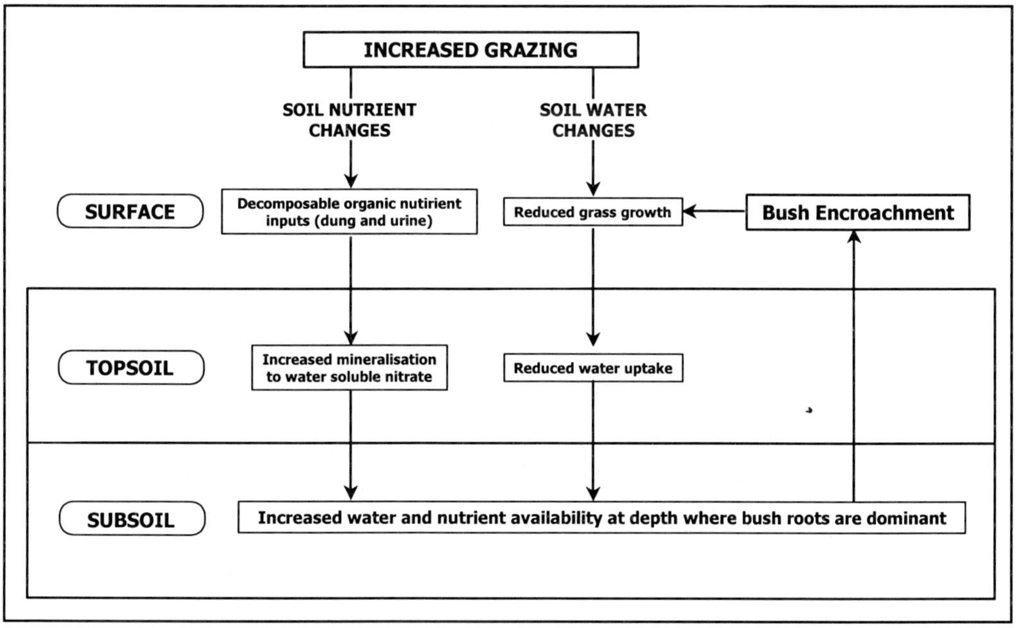

Figure 6. Two layer model of bush encroachment, showing hydrochemical changes that can result from high grazing pressure, leading to conditions favouring bush growth. After Walker and Noy-Mier (1982) and Dougill *et al.* (1999).

Livestock versus wildlife

A further significant impact of the expansion of groundwater based livestock production has been a concomitant reduction of large wildlife populations in the Kalahari (e.g. Williamson and Williamson, 1985; Wheelwright *et al.,* 1996). An additional issue is that many (e.g. Cooke, 1985) have advocated the substitution of livestock by wild ungulate farming in the Kalahari for several reasons. These include a lower potential for wildlife to degrade the environment, their ability to withstand drought, and lower water consumption demands (e.g. Brown, 1967). There are notable instances where wildlife ranching is expanding in the Kalahari, although start up costs is high. The likelihood of a large-scale replacement of livestock production, especially since the Botswanan livestock industry is closely linked to the export market is unlikely. It can also be noted that when wildlife are dependant on artificial watering points, such as in national parks, environmental impacts including the development of sacrifice zones and vegetation structural changes, do occur (Van Rooyen *et al.,* 1990; Thrash, 1998)

Kalahari groundwater: is the system sustainable?

The environmental impacts of borehole water use are highly variable and are as much a function of the land use system in place, which can depend on individual herders, as on the level of environmental change. Given uncertainties over the significance of ecological changes on livelihood systems, what may ultimately be more significant than the levels of environmental changes and whether these constitute degradation is that the rate of water depletion and the quality of groundwater may limit the persistence of these groundwater based systems.

There is much circumstantial evidence to suggest that Kalahari groundwater is being utilised at an unsustainable rate. The ever-growing rates of groundwater use raise issues of sustainability and recharge, though the lowering of Kalahari groundwater tables has been noted since the 19th century (Livingstone, 1858; Thomas and Shaw, 1991). Most water however comes from deep aquifers in the pre-Kalahari sediment basement, and many Kalahari borehole owners report a drop in water level, pumping rate and water quality after a few years operation of their bore. In the South African component of the extreme south west of the Kalahari, groundwater levels have dropped significantly in the last 20 years (Vegter, 1995). At Bokspits, (*pits* is a corruption of *puts*, meaning well) the most southwestern village in Botswana, a desalinisation plant was commissioned in 2001 because borehole water had become unpotable and health problems were ensuing.

There has been much speculation concerning whether recharge to deep aquifers occurs from rainfall percolation through the overlying unconsolidated Kalahari sediments (Foster, 1982; Mazor 1982; De Vries 1984; De Vries and Hoyer, 1988). Most studies have suggested that the thick sand over much of the basin and high summer evapotranspiration rates preclude recharge under current conditions. Isotopic data from the western Kalahari (Heaton *et al.,* 1983) suggest that the last major recharge events in the basin occurred during wetter climate intervals at 30-26 and 14-8 k a ago. Foster *et al.* (1982) indicate that presently recharge is only marked in areas of thin sand cover and during wet years, with greater recharge potentials where overlying sands are shallow or where the limited bedrock outcrops occur (de Vries and von Hoyer, 1988). More recently, by using both chloride tracer studies and hydraulic modelling, more precise recharge determinations have been produced (De Vries *et al.,* 2000).

In the central Kalahari, the recharge rate is currently determined at 1 mm per year where rainfall is up to 350 mm p.a., rising to 5 mm per year in wetter eastern locations. Further east in Botswana where rainfall is greater, the figure rises to 10-15 mm per year (Gieske, 1992). The overall interpretation of these studies is that under natural conditions groundwater depletion has occurred since the last major Pleistocene wet phase. Given the dramatic, though unquantified, enhancement above natural depletion levels, it is not difficult to conclude that currently usage is highly unsustainable. Alternative sources of water, such as from possible deeper and currently uncharted groundwater reserves, or piped from more distant sources, would have potentially dramatic and as yet uncharted financial implications for the functioning of Kalahari livestock systems.

CONCLUSIONS

From this assessment, which has drawn illustration from the Sahel and more detailed exemplification from the Kalahari, it is clear that the utilisation of groundwater has been a significant factor in the utilisation of natural resource use in major African grasslands over the last five decades. In conclusion, we can identify the following issues and concerns.

- Groundwater use has contributed significantly to the sedentarisation of activities, particularly livestock production, in areas where arable and pastoral production are increasingly in competition, as in many Sahel areas, or where year-round pastoralism was not previously possible, as in the Kalahari.

- The impact of groundwater use should not however be solely considered in terms of the use of the water itself. Access to groundwater is the key that unlocks access to other, particularly grazing, resources.
- Access to groundwater via wells and boreholes effectively gives those with rights to the water source *de facto* control over access to the surrounding environment. A concomitant demise in customary rights and laws can be generated, creating situations leading to disputes and conflict in parts of the Sahel and denial of access to natural resources previously utilised in the Kalahari.
- Environmental changes inevitably result from the biosphere effect of grazing pressures around wells and boreholes. These include the creation of bare sacrifice zones, changes in species composition, altered balances between annual and perennial grasses and bushy species) and biodiversity. Such changes are an inevitable function of intensive resource use.
- Change does not automatically mean degradation. There is no evidence to suggest that boreholes are the centres of expanding areas of desertification, although increasing bore densities could have this effect if sacrifice zones merge. It is easy to confuse natural vegetation variability with changes caused by grazing pressures, while process studies suggest that the latter may, significantly, be reversible at decadal to century scales.
- In the shorter term, vegetation changes perceived as negative can open up new resource opportunities that land users and their animals can exploit beneficially.
- Rates of groundwater recharge in the Kalahari are presently very low. While data on actual water extraction volumes are hard to come by, there is considerable circumstantial evidence to suggest that groundwater usage is currently unsustainable.

Ultimately, it may be this last point that limits the growth of groundwater use, and may lead to the current means of natural resource use in marginal dryland areas of Africa having to be reconsidered.

REFERENCES

Abel, N.O.J. and Blaikie, P.M., 1989. Land degradation, stocking rates and conservation policies in the communal rangelands of Botswana and Zimbabwe. *Land Degradation and Rehabilitation* 1: 101-123.

Ahlcrona, E., 1988. The impact of climate and man on land transformation in Central Sudan - Application of remote sensing. *Lund Universitets Geografiska Institurte Avhandlingar* 103, Lund University Press.

Al-Ibrahim, A.A., 1991. Excessive use of groundwater resources in Saudi Arabia: Impacts and policy options. *Ambio* 20: 34-37.

Andrew, M.H., 1988. Grazing impact in relation to livestock watering points. *Trends in Ecology and Evolution* 3: 336-339.

Behnke R. H., Scoones, I. and Kerven, C. (eds.), 1993. *Range Ecology at Disequilibrium.* London, ODI.

Brown, L. 1967. Wildlife *v.* sheep and cattle in Africa. *Oryx* I 10: 92-101.

Clarke, R., Lawrence, A.R., and Foster, S.S.D., 1996. *Groundwater: A threatened resource.* UNEP Environmental Library 15. Nairobi, UNEP.

Cooke H. J., 1985. The Kalahari today: a case of conflict over resource use, *Geographical Journal* 151: 75-85.

De Vries, J.J., Selaolo, E.T. and Beekman, H.E., 2000. Groundwater recharge in the Kalahari, with reference to paleo-hydrologic conditions. *Journal of Hydrology* 238: 110-123.

De Vries, J.J., 1984. Holocene depletion and active recharge of the Kalahari groundwaters- a review and an indicative model. *Journal of Hydrology* 70: 221-232.

De Vries, J.J. and von Hoyer, M., 1988. Groundwater recharge studies in senmi-arid Botswana: a review. In I. Simmers (ed) *Estimation of natural groundwater recharge.* Reidel, Dordrecht, 221-232.

Dougill, A. J., Heathwaite, A. L. and Thomas, D. S. G., 1998. Soil water movement and nutrient cycling in semi-arid rangeland: vegetation change and system resilience, *Hydrological Processes* 12:443-459.

Dougill, A. J., Thomas, D. S. G. and Heathwaite, A. L., 1999. Environmental change in the Kalahari: integrated land degradation studies for nonequilibrium dryland environments, *Annals, Association of American Geographers* 89: 420-442.

Foster, S.S.D., Bath, A,H., Farr, J.L. and Lewis W.J., 1982. The likelihood of active groundwater recharge in the Botswana Kalahari. *Journal of Hydrology* 55: 113-136.

Gallais, J., 1984. *Hommes du Sahel. Espaces-temps et pouvoirs. Le delta interieur du Niger 1960-1980.* Paris, Flammarion.

Georgiadis, N.J. and McNaughton, S.J., 1990. Elemental and fibre contents of savanna grasses: variations with grazing, soil type, season and species. *Journal of Applied Ecology* 27: 623-634.

Gieske, A., 1992. Dynamics of groundwater recharge: a case study in semi-arid eastern Botswana. PhD thesis, Vrije University, Amsterdam.

Hanan, N.P., Prevost, Y., Diouf,A. and Diallo, O., 1991. Assessment of desertification around deep wells in the Sahel using satellite imagery. *Journal of Applied Ecology* 28: 173-186.

Heaton, H.E., Talma, A.S. and Vigel, J., 1983. Origin and history of nitrate in confined groundwater in the western Kalahari. *Journal of Hydrology* 64: 232-262.

Hellden, U., 1991. Desertification- time for an assessment. *Ambio* 20: 372-383.

Hitchcock, R. K., 1978. *Kalahari Cattle Posts: A Regional Study of Hunter-Gatherers, Pastoralists and Agriculturalists in the Western Sandveld Region, Central District, Botswana.* Gaborone, Government Printer.

Kharin, N., Tateishi, R. and Harahsheh, H., 1999. *Degradation of the drylands of Asia.* Chiba University Center for Environmental Remote Sensing, Chiba, Japan.

Koechlin, J., 1997 Ecological conditions and degradation factors in the Sahel. In: C. Reynaut (ed) *Societies and nature in the Sahel.* London, Routledge, 12-36.

Livingstone, D., 1858. *Missionary Travels and Researches in South Africa.* London, Deighton Bell.

Mazonde I. N., 1994. *Ranching and Enterprise in Eastern Botswana,* Edinburgh, Edinburgh University Press.

Mazor, E., Verhagen, . Th., Sellschop, J.P.F., Jones, M.T., Robins, N.E., Hutton, L.G., and Jennings, C.M.H., 1982. Northern Kalahari gorundwaters: hydrologic, isotopic and chemical studies at Orapa, Botswana. *Journal of Hydrology* 34: 203-234.

Nellis, M.D. and Bussing, C.E. 1989. Desertification and sustainable development issues in Botswana. *Papers and Proceedings of Applied Geography Conferences, State University of New York at Binghampton* 12: 99-106.

PANRUSA, 2001. Briefing notes. http://www.sheffield.ac.uk/panrusa

PANRUSA, 2002. Poverty, policy and natural resource use in southern African drylands. DFID project R7074, Final technical report. Department of Geography, University of Sheffield. http:www.sheffield.ac.uk/panrusa

Perkins J. S. and Thomas D. S. G., 1993. Spreading deserts or spatially confined environmental impacts? Land degradation and cattle ranching in the Kalahari Desert of Botswana. *Land Degradation and Rehabilitation* 4: 179-194.

Perkins, J. S., 1996. Botswana fencing out the equity issue. Cattleposts and cattle ranching in the Kalahari Desert, *Journal of Arid Environments* 33: 503-518.

Rapp, A., 1974. A review of desertization in Africa- water, vegetation and man. *Secretariat for International Ecology, Report* 1, Stockholm.

Reichelt, R., 1989. Desertification in the Sahel: The exposing of the 'old erg' of an earlier Sahara. In Ernst, W., Hohnholz, J.H. and Bittner, A. (eds) *Natural Resources and Development: a biannual collection of recent German contributions concerning the exploration and exploitation of natural resources.* Tubinggen, Institut fur Wissenschaftliche Zusammenarbeiut, 104-113.

Reynaut, C., 1997. The transformation of social relations and the management of natural resources. 1. The birth of the land question. In: C. Reynaut (ed) *Societies and nature in the Sahel.* London, Routledge, 235-261.

Reynaut, C. and Delville, P.L., 1997. A shared land: complementary and competing uses. In: C. Reynaut (ed) *Societies and nature in the Sahel.* London, Routledge, 109-137.

Ringrose, S. Matheson, W., Tempest, F. and Boyle, T., 1990. The development and causes of range degradation features in southeast Botswana using multitemporal Landsat MSS imagery. *Photogrammetic Engineering and Remote Sensing* 56: 1252-1262.

Ringrose, S., Vanderpost, C. and Matheson, W., 1995. The use of integrated remotely sensed and GIS data to determine causes of vegetation change in southern Botswana. *Applied Geography* 15: 225-242.

Schwatz, E. H. L., 1920. *The Kalahari, or Thirstland Redemption.* Cape Town: Maskew Miller.

Schwatz, E. H. L., 1919. The progressive desiccation of Africa: the cause and the remedy, *South African Journal of Science* 15: 139-190.

Shazali, S. and Ahmed, A.G.M., 1999. Pastoral land tenure and agricultural expansion: Sudan and the Horn of Africa. *IIED Issue Paper* 85, London, IIED.

Thomas, D.S.G. and Shaw, P.A., 1991 *The Kalahari environment.* Cambridge, Cambridge Univesrity Press.

Thomas, D.S.G., 2002. Sand, grass, thorns and…cattle. The modern Kalahari environment. In Sporton, D. and Thomas D.S.G. (editors) *Sustainable livelihoods in Kalahari Environments: contributions to global debates.* Oxford, Oxford University Press 21-38.

Thrash, I., 1998. Impact of water provision on herbaceous vegetation in Kruger National Park, South Africa. *Journal of Arid Environments* 38: 437-450.

Turner, M.D., 1999. Merging local and regional analyses of land-use change: the case of livestock in the Sahel. *Annals of the Association of American Geographers* 89: 191-219.

Van Rooyen, A.F., 1998. Combatting desertification in the southern Kalahari: connecting science with community actions in South Africa. *Journal of Arid Environments* 39: 285-298.

Van Rooyen, N., Bezuidenhout, D., Theron, G.K. and Bothma, J. du P., 1990. Monitoring of vegetation around artificial watering points (windmills) in the Kalahari Gemsbok National park. *Koedoe* 33: 63-88.

Walker, B.H. and Noy-Meir, I., 1982. Aspects of the stability and resilience of savanna ecosystems. In Huntley, B.J.and Walker, B.H. (eds) *Ecology of tropical savannas.* Berlin, Springer-Verlag, 556-590.

Wheelwright, N.T., Dangerfield, J.M., Flyman, M. and Tjibae, M., 1996. Strengthening biological theory in wildlife research and management: a case study for Botswana. *South African Journal of Science* 92: 173-177.

White, F. 1983. *The vegetation of Africa.* Paris, UNESCO.

White. R., 1993. *Livestock Development and Pastoral Production on Communal Rangeland in Botswana.* Gaborone, Botswana: The Botswana Society.

Williamson, D. and Williamson, J., 1985. Botswana's fences and the depletion of Kalahari wildlife, *Parks* 10: 5-7.

From Vision to Action: Towards a National Policy for Integrated Water Management in Sudan

OSMAN MIRGHANI MOHAMED ALI
Institute of Environmental Studies, University of Khartoum, Sudan.
E-mail: osmirghani@yahoo.co.uk

ABSTRACT: An overview of the water resources in Sudan and the patterns of their utilization is given. Through an evaluating perspective, the problems and constraints are outlined. Sudan's efforts in the realm of water resources conservation and rational exploitation are reviewed. From an oblique reference to an environmental policy in the 1973 Constitution to the proposed National Water Policy of 1999, the legislation, policies and plans were followed. Prior to 1992 policies and legislation were fragmented. In the National Comprehensive Strategy for Development (1992-2002) the water resources sector was considered as a sub-sector of the agricultural sector. Currently, the water sector in Sudan is characterizes by overlapping legal regimes and the situation is further complicated by the federal system of governance.

The proposed Water Policy covers a wide variety of issues. It is built upon twelve overarching policy principles and objectives set out to embrace areas of water resources, utilization, water and the environment, regional water issues, socio-economic issues, disaster management and public safety and institutions, capacity building and technical assistance.

It is envisaged that the adoption and implementation of the National Water Policy would mark a milestone in the history of Sudan's development. The policy is reviewed and appraised. Specific aspects where the proposed policy needs strengthening were highlighted.

INTRODUCTION

Sudan is the biggest country in Africa, covering one-eighth of the continent with an area of 2.5 million km^2, and sharing its border with nine countries. It comprises several climo-ecological zones: desert, semi-desert, savanna, tropical forests, flood region (swamps), equatorial and montane. The population of the county is estimated at 30 million with an annual rate of growth of 2.8%. More than 50% of the population lives on just 15% of the land along the river Nile and its tributaries. The remaining population lives in towns away from the Nile and depends primarily on deep groundwater.

Water Resources in the Sudan

There are four main resources of water in the Sudan: rainfall, surface, ground water and the Red Sea. Rain represents the largest water source available in Sudan. Despite the fact that almost 50% of the country is desert and semi-desert, the annual rainfall is estimated at 1.07 trillion m^3, almost 45% of the total annual precipitation of the whole Arab region (AOAD, 1989). Rainfall increases from almost zero in the north to 1600 mm in the equatorial region in the south (Table 1).

The country has high potential evapotranspiration rates exceeding, 1450 mm in most areas. Considering the year as a whole, the annual water deficit is above 600 mm in most of the country. However, water deficit generally decreases during the rainy season. During the months July and August there is practically no water deficit to the south of latitude 12° N (Satakopan, 1965). The availability of rainwater is further complicated by the fact that great areas of the country are sandy soils.

Table 1: Distribution of annual rainfall and water input in various regions of Sudan (modified from IHP, 1988)

Climatic region	Area (10^6 km^2)	Average annual rainfall (mm)	Length of season (month)
Equatorial	0.14	1200-1600	7-9
Tropical forests and swamps	0.46	800-1200	6-7
Savanna	0.68	300-800	3-6
Semi-desert	0.50	75-300	1-3
Desert	0.73	3-75	0-1

The Nile system represents the main surface water source in Sudan. The main axis of the Nile system including the White Nile runs for more than 300 km within Sudan's borders. The combined length of the Blue Nile, the tributaries of the White Nile and the seasonal rivers is over 6000 km (Beshir

Water Resources Perspectives: Evaluation, Management and Policy. Edited by A.S. Alsharhan and W.W. Wood.
Published in 2003 by Elsevier Science, Amsterdam, The Netherlands, p. 237-244.

and El Moghraby, 1980). The average annual discharge of the Nile is estimated at 84 km^3 measured at Aswan in Egypt. Over 80% of the Nile's annual flow take place between August and October. Some of the seasonal wadis and khors in the eastern and western regions do not reach the main Nile. At their maximum discharge their annual flow is about 5-7 km^3 a year (SRFAC, 2001).

Sudan is unique among the ten countries of the Nile basin as it is a sink, a path and a resource. All the five sub-basins making the Nile basin (The White Nile, Sobat, The Blue Nile, Atbra and the River Nile) are represented within Sudan's border, which embraces 60% of the Nile system (MOIWR, 1999). The swamps of the upper reaches of the Nile could be one of the largest in the world with 8,000 km^2 permanent swamps and 80,000 km^2 seasonal swamps (Thompson, 1985).

The construction of dams along the Nile has resulted in the creation of large water impoundments such as Lake J. Aulia, Sennar, El Roseiris, Khashm el Girba and Nubia. Away from the Nile system, there are a number of non- nilotic lakes. Some of these are freshwater lakes such Kundi, Abyad, Keilak, and Turdat el Rahad. Saline lakes in the country are either crater lakes represented by the two lakes of Dariba in the massif volcanic caldera of Jebel Marra, and the Malha Lake in the Medoub series- both in western Sudan- or non-crater lakes such as Nikheila, Natroon and Salima oases in the northern region. The freshwater lakes are small, in areas sparsely populated and are mainly used as watering points for livestock.

Available records indicate that Sudan has a large storage of groundwater. The area of land suitable to bear groundwater is 1.3 million km^2, with an annual recharge of two milliard m^3. The main aquifers of the Sudan are formed of the Mesozoic Nubian Sandstone Formation; the detrital Quaternary to Tertiary deposits (of Um Ruwaba Formation, El Gezira Formation and Al Atshan Formation) and the Alluvial wadi-fill deposits (SNC/UNESCO, 2000) (Table 2). The Nubian formation is the most extensive and largest aquifer in the Sudan. Besides these inland and ground water resources, Sudan has 750 km of Red Sea coastline. However, currently this marine water is not utilized for drinking or irrigation, as desalination is not practiced on a large scale yet.

Utilization Patterns

Water consumption for domestic purposes is estimated at 100 l/day, albeit no authentic figures are available. Industry utilizes only 1% of the total water consumption. Irrigation is the principal consumer of the water in the Sudan with over 80% of the water withdrawal from the Nile being used for irrigated agriculture. Hydropower is generated from the Blue and River Atbra rivers. The present installed capacities are around 200MW. The Nile in the Sudan is used for transportation and the navigable water courses are estimated at 5000 km (Beshir and El Moghrabi, 1980). The Nile is renounced for its rich and diverse fisheries especially in the man-made lakes behind the dams.

Table 2: Groundwater basins in Sudan (SNC/UNESCO, 2000).

Formation	Main water basins	Annual recharge (10^6 m^3)	Annual abstraction (10^6 m^3)
The Nubian Sandstone	The Sahara Basin The Nile Basin The Central Darfur Basin The Atbra Basin The Nahud Basin The Sag El Naam Basin	1074	636
The Detrital Quaternary Tertiary Deposits	The Bara basin The Atshan Basin The Baggara Basin The Sudd Basin	585	148
Alluvial Wadi-filled depressions	El Gash Wadi Azum Wadi Nyala	376	160
Total		**2035**	**944**

The Vision: Evaluation and Perspectives

The strive of Sudan to institute a water policy has been a continuous process hinging on and spurred by a vision stemming from evaluation of the current status and future perspective of water needs

- Remarkable variations in the spatial and temporal distribution in quantities and duration of the rains. With high potential evapotranspiration, the rainwater surplus is practically nil for the whole country except in the extreme south.
- The Nile normally witnesses high seasonal and annual periodicity in its discharge.

- Storage capacities of surface water are limited and with decreasing efficiencies. More water flows out of the country (120 km^3) than is stored in.
- Most of the surface and groundwater resources are shared with neighboring countries.
- Sudan can use only 22% of the Nile water.
- Sudan is the only country within the Nile basin that has a potential to increase the natural flow via conservation of the river flow through the wetland areas of the upper stretches of Nile. The Jonglei Project's phase I and phase II are planned to provide 4.7 and 9.0 km^3 annually, respectively. If the amounts envisaged to be available from River Baro (Sobat swamps) and the Machar marshes (4.0 km^3 each) are added, then 21.7 km^3 could be added to the current discharge. Although these aliquots will be shared with Egypt, Sudan can increase its within-border waters by heightening El Roseiris dam -adding 7.0 km^3, by construction of Hamadab Dam at the fourth cataract and by more utilizing groundwater (Table 3).
- Although 52% of the country's area overlies water bearing formations, the exact hydrogeological situation – groundwater in storage and groundwater recharge- is yet to be elucidated and documented. Huge incongruities are reported in the literature.
- Although huge quantities of groundwater estimated as 1678 milliard m^3 are believed to be stored in the Nubian Sandstone basin, much of it is paleo-water (Kheir, 1986; Abdel Salam, 1986, cited by SNC/UNESCO, 2000) and thus non-renewable.
- Non-nilotic seasonal streams and wadis are not receiving equal attention as the Nile system. They are not sufficiently studied nor are they properly utilized.
- There are tremendous losses of water in the swampy regions of the upper Nile, estimated at 42 km^3 annually (El Moghrabi and Ali, 1989).
- The actual cultivated area is only about 29% of total arable land.
- Present hydropower utilization is only 2% of the potential power from the Nile and its tributaries.
- The per capita available water is declining progressively. As cited by SRFAC (2001), available water has decreased progressively from 5.3 thousand m^3 per year in 1970 to 2.8 thousand m^3 in 1995. The forecast for the year 2025 is 1.3 thousand m^3 per year. The present trends in availability, allocation and use of water resources point to unsustainable water resources, exploitation and demand (SRFAC, 2001).
- Although Sudanese water resources are comparatively clean, signs of contamination are manifested in both the Nile and the groundwaters and pollution is imminent (Dirar, 1986; Abdellatif *et al.,* 1993; Ali, 2000; SNC/UNESCO, 2000). Urbanization, agricultural practices, industrial waste and the newly discovered oil resources represent the main threat to water resources.
- Water quality does not receive due attention within the current institutional structure.
- The present water administrative structure ignores non-conventional resources such as sewage, irrigation runoff waters (AOAD, 2000) and desalinization.
- Inadequacy of water supply is affecting the sanitation standards. Only 30% of the rural population has access to adequate water supply while nearly 40% of the urban population and 90% of people living in rural areas lack basic sanitation services (MOIWR, 1999). On the other hand, almost 90% of the epidemic diseases are water induced or water related.
- The development of water resources cannot be viewed in isolation from the overall social and economic development in the country.
- It is a top priority to review and revise existing legal regimes to remove redundancies and inconsistencies.
- A major gap in water resource management includes data and information management, water quality monitoring and cross-sectoral coordination.
- Exodus of trained and qualified engineers and difficulties in training and qualifying younger engineers abroad due to financial constraints
- The country is suffering from the long civil strife, which impinges upon its political stability and development strategies.
- The Nile could be both a source of conflict and a basis for collaboration among the riparian countries.
- Management of water resources is best carried out on a regional or catchment-wide basis
- Under globalization, no vision of sound policies and planning could be developed without an international dimension.

Table 3: Availability and Constraints of Water in Sudan (modified from MOIWR, 1999).

Water Resources	Quantity (km^3/year)	Constraints
Sudan present share from the Nile (1959 Agreement[a])	20.5	Seasonal pattern coupled with limited storage facilities
Non-Nile streams	5.4	Highly variable, short duration flows difficult to monitor or harvest. Some are shared with neighbours.
Renewable Groundwater	4.1	Deep water entailing high cost of exploitation and pumping. Remote areas of weak infrastructure.
Present Total	**30**	
Expected share from reclamation of swamps	10.25	Capital intensive with considerable social and environmental cost. Security dimension due civil strife.
Heightening of El Roseiris Dam[b]	7.0	Capital intensive
Total	**47.25**	

a) Agreement signed by Egypt and the Sudan in 1959 whereby the former receives 55.5 km^3 and the later receives 18.5 km^3 of the 84 km^3 Nile annual discharge. 10 km^3 are regarded as lost in Lake Nassir-Nubia by evaporation.

b) On the Blue Nile. Major source of hydropower and irrigation water in the Sudan.

The Action

Water Management History

The draft document of the Sudan National Water Policy (MOIWR, 1999) divided the development of water policy and legislation in the Sudan into two eras: pre-1992 and post 1992. During the first era the main water regulations in the Sudan were based on the 1951 Regulations which are licensing regulations for pumping water from the Nile according to the Nile Pump Control Act (NPCA) of 1939. This was recently modified in the Water Resources Act of 1995. In addition there are about ten other pieces of fragmented legislations. There is no policy for selling water in the country.

The second era began with the National Comprehensive Strategy for Development of 1992 (1992-2002) when the irrigation and water resources sector was embodied as a sub-sector of the agricultural sector. The strategy sets a number of policy objectives and strategies oriented towards solving the water resources problems and developing the resource. At the close of the NCSD decade, most of those objectives have not been fulfilled. In 1995 the National Council for Water Resources (NCWR) was established headed by the Minister of the Federal Ministry of Irrigation and Water Resources (FMOIWR) with representatives from major suppliers and users of water at the national and state levels. Its mandate included formulation of the general policy for water resources as a long-term federal plan for optimal and balanced use of water resources, and determination of priorities. As for drinking water, the National Water Corporation Act of 1995 has established a corporation with policy powers in the drinking water (MOIWR, 1999).

A significant development resulted from the 1998 Constitution whereby the federal government is responsible for planning, regulating and executing inter-state waters and national electricity projects. The major federal development projects were considered by the Constitutional Decree No. 14 as national wealth, to be managed at the federal level.

Currently the water sector in the Sudan is governed by three overlapping legal regimes: The Nile Pump Control Act of 1939 and the 1952 Regulations made there under, the Irrigation and Drainage Act of 1990 and the Water Resources Act of 1995. The last Act has repealed the NPCA of 1939 while keeping the 1951 Regulations but it has not repealed the Irrigation and Drainage Act of 1990 (MOIWR, 1999). As for the ground water legislation gaps include the oil resources development and its impact on groundwater resources, and the groundwater basins shared with neighboring countries.

Proposed National Water Policy

For a water policy to be effective in Sudan, it has to cope with a set of extremes: drought and deluge episodes; Nilotic and non-Nilotic systems; marine and fresh waters, surface and sub-surface resources, Federal and State authorities; health and ailment; national and regional interests. The Proposed Water Policy (PWP) lays a framework composed of a set of over-arching policy issues that apply to the water sector as a whole and to most of the individual components. All the main specific areas of concern in water development, management and protection are to be tested against these principles and objectives. The overarching policy principles and objectives laid by the proposed policy are highlighted below:

1. Water is a scarce and valuable resource.
2. Access to water for human needs is the highest priority in the development of water resources.
3. Development of water resources must be demand-driven and management should be undertaken at the lowest possible level.
4. Development and management of water resources, and the operation and maintenance of water services must be economically sustainable through the recovery of cost from those who benefit.
5. All water should be managed in an integrated manner as part of the hydrological cycle.
6. Water resources management affects everybody and should be undertaken with the participation of relevant stakeholders
7. People are stakeholders and the national government is the custodian of all water in Sudan for the equitable benefit of all.
8. The gathering and management of accurate information for the investigation and continuos monitoring of water resources is essential for proper development, management and protection of the resource.
9. The environment needs to be protected in order to ensure sustainable utilization for present and future generations.
10. The development of water resources will be undertaken in order to maximize its benefits in the public interest whilst ensuring minimum adverse impact on the environment.
11. Public institutional arrangements at federal and state levels shall be integrated, efficient and transparent whilst avoiding duplication of functions and responsibilities.
12. Water and water related issues are an integral part of the wider economy and have direct effects on many other sectors that require inter-departmental and inter-sectoral communications and co-operation.

All the subordinate policies related to the development, management and protection of water resources are tested against the above principles and objectives. In formulating the surface-water resources policy, six issues and problems associated with their development and protection are addressed. These are: flood and drought spells; storage facilities for the highly variable flow; maintenance of irrigation infrastructure; supply and demand management; surface-water quality management and human resources. The policy principles and objectives emphasize that surface- water planning and development must be integrated at all levels. The protection of surface water is the obligation of all sectors and, therefore, the development policy had to be clear and accessible to the involved communities. A reliable data and information system is a pre-requisite for sound assessment, planning, management and development of surface-water. To ensure continuity of services, the operation and maintenance of surface-water should be based on the "user pays principle". To meet the increasing demand, the storage capacity has to be increased through heightening of El Roseiris Dam and via building new dams and barrages. Optimum and equitable use of surface-water should be promoted through the cooperation between the national water users. The final principle emphasizes the regulatory function of the Federal Government in ensuring that appropriate standards of service quality, sustainability and environmental friendliness are met by both water suppliers and users.

The Proposed Water Policy summarizes the key issues pertaining to groundwater in the aspects of monitoring and information base; quality, pollution and environmental degradation; development and utilization; and institutional arrangements. The policy principles and objectives stress that groundwater resources are an indivisible part of the hydrological cycle, the national water balance and the natural resources base. The Proposed Water Policy draws from the Islamic principle which regards water as a common property and spells it quite clear that groundwater resources are a national property, the equitable use of which is common to all subject to national authority control. Proper planning, assessment, development and management of water resources cannot be achieved without strengthening the information base at the national and the state levels. The Proposed Water Policy recognizes the present and future users right of access to clean and unpolluted groundwater resources and an un-degraded environment. Groundwater in the proposed policy is recognized as having a social and economic value based on the cost of its development, operation and maintenance. Concerning sustainability of groundwater resources, the Proposed Water Policy urges that abstraction shall be based on recharge and safe yield concepts. Sustainability of groundwater development and supply services shall be planned for and considered as part of development and management policy of the water resources systems.

The Proposed Water Policy is acclaimed for stressing the importance of the unconventional water resources. It draws the attention that research, planning and development of unconventional water resources, such as domestic sewage and desalinated waters, will be undertaken in order to use such alternative resources for Sudan.

In the realm of sanitation, the Proposed Water Policy identifies certain issues and problems such as supply-demand imbalance, sustainability of water supply systems, sanitation and water disposal, water supply and sanitation sector institutional reform. The reform is required to ensure that, *inter alia*, water supply and sanitation sectors and the associated projects should be integrated, adopting a community-based approach. Further, clear identification is needed for roles, responsibilities and obligations of federal and state governments, NGO's, local communities, the private sector and the National Council for Water Resource. Towards such extent, the policy and objectives stress that access to adequate water supply and sanitation is a basic necessity. The achievement of sustainable and financially viable water supply and sanitation services must be the objective of service providers.

In the agriculture and landuse sectors, the Proposed Water Policy points to specific key issues and problems. Wind and water erosion has resulted from overgrazing and over-cultivation. Crop yields are declining due to the ensuing environmental degradation, successive drought episodes, animal death and migration of rural people to towns. Other issues included are excessive removal of vegetation cover, concentration of agricultural activities in central Sudan, wildlife depletion, soil degradation and poor traditional landuse practices. To ameliorate the situation, the Proposed Water Policy principles and objectives urge that the improvement of water use efficiency in agriculture shall be a priority. The licensing of water use must include a rigorous assessment of water resources to be utilized. The water pricing must promote equitable and efficient water use. The use of water in irrigation projects should be monitored and evaluated properly.

The Proposed Water Policy reviews all the problems related to hydropower and sets suitable principles and objectives. It rates hydropower as a clean energy form, relatively cheap to produce and remains a priority for investment. The thermal backup shall be secured to fill the gap. In order to optimize the use of the water stored for different purposes, dam operation should be co-coordinated at all levels through appropriate institutional arrangements. Beneficiaries participation in management of hydropower must be institutionalized. Public awareness, clarity and transparency of energy policy are needed to gain acceptance of customers. While advocating private sector to be encouraged in the development of hydropower sector, the Proposed Water Policy maintains that the government must oversee the equality of services provided to the consumers.

As for industry sector, the Proposed Water Policy recognizes the associated problems and the expected increase in water requirements (5% of total water consumption by 2025) and sets definite principles and objectives. The use of water to transport and dilute waste and to act as a coolant has to be linked with treatment. The Precautionary Principle of pollution control should be adopted with financial penalties for polluters. Adequate legislation should prevail and be enforced to prevent pollution of the environment at large and fresh sources in particular.

The Proposed Water Policy regards river and sea navigation as constituting an important form of transportation in Sudan that needs to be promoted and adequately regulated in coordination and cooperation with the concerned bodies. Fisheries are catered for in the Proposed Water Policy and the principles and objectives pertaining to this important economic activity call for the identification and implementation of appropriate water resources management measures.

The environment is featured prominently in the Proposed Water Policy. The issues dealt with are: the environment as the resource base; pollution control; catchment degradation; biodiversity and wildlife. Four principles and objectives are highlighted: The necessity to sustain the aquatic environments; the realization that people and the consequences of their development are part of the environment; the impact of development and use of water resources should not compromise the long-term sustainability of aquatic environments; and a balance must be maintained between water resources development and utilization on the one hand and their protection on the other.

The Proposed Water Policy, aware that water does not recognize political boundaries, calls for Sudan to seek to cooperate with other countries for the development, optimum use and protection of international waters wherever possible in its national interests and without compromising its sovereignty. It regards watershed management inside the country and in neighbouring countries as a pre-requisite for efficient water use and for strengthening political relations.

For a water policy to be effective, it has to incorporate both recurring flood and drought phenomena. Within this context, the Proposed Water Policy sets as a priority the development of a National Disaster Management Plan to enable both the avoidance of and effective response to disasters. International co-operation is critical for proper and adequate response to natural and other disasters. Sudan will seek to participate in and contribute to international efforts such as the Program to Combat Desertification.

The Proposed Water Policy emphasizes that the development of water resources can not be seen in isolation from the overall social and economic development. Towards that end, the policy stresses that accommodating social and economic issues in water resources development projects is very important and often determines the difference between success and failure.

The Proposed Water Policy regards the major challenge in relation to institutional development of the water sector in the Sudan is to develop appropriate institutions for effective management, allocation and protection of the country's water resources. There is an urgent need to develop modern data collection and data information management processes and techniques through the application of computerization, system analysis, GIS, remote sensing telemetry and mathematical modeling. There is a need also for structured research on the impact of water policy. The proposed policy envisages that building the capacity of the water sector at large in the country is one of the fundamental requirements for achieving proper resources management. Finally, the PWP emphasizes that external donor assistance will be needed to support capacity building programmes, especially those that focus on users' group.

SUMMARY AND RECOMMENDATIONS

Summary

The paper gave an overview of the water resources in the Sudan and their patterns of utilization. Through evaluating the current status of the water resources in the Sudan and the inherent pitfalls in managing the water resources, a vision is perceived heralding the issues, policies and principles of the proposed water policy. The salient features of the proposed national water policy for Sudan, regarded as a milestone in the water sector in the country's endeavour to develop, prosper and promote the well being of its people, are highlighted. The proposed draft, the first of its kind in the country, is an excellent document reflecting the strenuous effort exerted in its production.

Recommendations

The proposed water policy needs to be subjected to more discussion and to be reviewed by as a wide a sector of people as possible. Certain areas have to be fine-tuned in particular the gender issues. Women in most parts of the Sudan are involved in all activities pertaining to water use such as agriculture, husbandry and in fetching water, sometimes from distant locations. As such, they are impacted and have an impact on water as a resource. Although any water policy has to reflect and be tailored to a specific country, it would be beneficial to draw from other countries experiences – especially the Nile basin countries - in developing and implementing water policies.

The relation between federal and state authorities in utilizing, developing and investing in inter-state rivers and groundwater basins has to be clarified and stated clearly in all decrees and regulations. The Proposed Water Policy should embrace issues of shared groundwater basins with emphasis on neighbouring activities that would adversely impact the water quality of the basin. Coordination is also needed in the institutions responsible for the legislation and implementation of water quality. Finally, the water policy needs to be strengthened by legislative power and to be backed by research capabilities. The Environment Protection Law of 2001, the Regulations for Environmental Impact Assessment and for the Environment Protection in the Petroleum Industry -both yet to be passed- will augment the former need while the latter could be achieved by a specialized national center for water resources research and technology.

REFERENCES

Abdellatif, E., Ali, O.M., Khalil, I.F. and Nyonje, B. M., 1993. Effects of sewage disposal into the White Nile on the plankton community. *Hydrobiologia* 259: 185-201.

Ali, O. M., 2000. Towards an Arab Region Water Security – The water resources in the Sudan: patterns of use and constraints to utilization. *The Arab Journal for Irrigation Water Management* 3(2): 45-50.

AOAD (Arab Organization for Agricultural Development), 1989. Utilization of Water for Agricultural Purposes and the Future Trends. Science and Technology, Institute of Arabic Development, Lebanon.

AOAD (Arab Organization for Agricultural Development), 2000. Study of Promotion of Institutional and Organizational Structures for the Management of Water Resources in the Arab Countries. Arab Organization For Agricultural Development, Khartoum. 132p.

Beshir, M. and El Moghraby, A.I., 1980. Ecological studies on the Sudanese Nile System. *Water Supply and Management* 4: 25-28.

Dirar, H., 1986. Coliform bacterial counts in the Nile water. *Environmental International* 12: 571-576.

El Moghraby and Ali, O. M., 1989. The Aquatic Environment in the Sudan. In: " Problems of Environment and Legislation in the Sudan". By (eds.)

Mahmoud, Z. N.; Abdellatif, E. and Ali, O. M. Sudan Environment Conservation Society, Khartoum. 83p.
IHP, 1988. Water Resources Assessment in the Arab Region. ACSAD, UNESCO/IHE. 398p.
MOIWR, 1999. Sudan National Water Policy. Draft Document.
Satakopan, V., 1965. Water Balance in the Sudan. Sudan Meteorological Service. Memoir No. 5, 33p.
SNC/UNESCO, 2000. Vulnerability of Groundwater Resources of Sudan to Pollution Risks. Khartoum. The Sudanese National Commission for UNESCO. 104p.
SRFAC, 2001. Integrated Water Resources Management. Scientific Research Foreign Affairs Centre. Sudan University for Science and Technology. Technical Paper Series No. (2). 64p.
Thompson, K., 1985. Emergent plants of permanent and seasonally flooded wetlands. In: P. Denny (Edit.), The Ecology and Management of African Wetland Vegetation. Dr Junk, Dordrecht, The Netherlands: 43-107.

Water Resources in the United Arab Emirates

ZEIN S. RIZK
Institute of Environment and Water, Ajman University of Science & Technology, P.O.Box: 346, Ajman, United Arab Emirates. E-mail: zeinrizk@hotmail.com

ABDULRAHMAN S. ALSHARHAN
Faculty of Science, United Arab Emirates University, P.O.Box: 17551, Al-Ain, United Arab Emirates. E-mail: sharhana@emirates.net.ae

ABSTRACT: The United Arab Emirates rely on non-conventional water resources, in addition to conventional resources, to meet the ever-increasing demands for water. Conventional water resources include seasonal floods, springs, falajes and groundwater. Non-conventional resources are the desalinated water and treated-sewage water. The existing conventional water resources in the United Arab Emirates include 125 Mm^3/yr (million cubic meter per year) from seasonal floods, 3 Mm^3/yr from permanent springs, 22 Mm^3/yr from seasonal springs, 20 Mm^3/yr of falaj discharges, 109 Mm^3/yr of aquifer recharge. The existing non-conventional water resources include 475 Mm^3/yr of desalinated water and 150 Mm^3 of reclaimed water. Future developments may increase the discharge of seasonal springs to 40 Mm^3 and falaj discharges to 40 Mm^3. Parallel increases in the volumes of desalinated water and treated wastewater may also be achieved. The most important water-related problems in the United Arab Emirates are the depletion of aquifers in several areas, such as at Al Ain and Al Dhaid; saline-water intrusion, and water quality degradation, such as that associated with the oil industry and agricultural activities.

Improvement of water-resource management in the United Arab Emirates can lead to water conservation, maintenance of better quality water, and restoration of deteriorating aquifer systems in many areas. The use of advanced irrigation technologies, construction of groundwater-recharge dams, and growing salt-tolerant crops are suitable agricultural approaches. Development of human resources is a priority in the United Arab Emirates, to provide trained national experts in water-related fields. Establishment of data banks and the application of advanced groundwater modeling and isotope hydrology techniques are powerful water-resources management tools.

INTRODUCTION

Fresh water resources are unevenly distributed on the Earth's surface, some countries suffer from excess water while others experience severe scarcity of water including the absence of permanent streams or lakes. Most of the Arab countries, including United Arab Emirates, belong to the second group of countries that have low rainfall, high temperatures and extremely high evaporation rates. In the United Arab Emirates, the management of scarce and vital water resources is an important national challenge. At present, the country does not make the best use of available water resources. Most of the water is used in irrigation, yet despite this, more than 50% of national food requirements are imported.

The United Arab Emirates lies in the southeastern part of the Arabian Peninsula between latitudes 22° 40' and 26° 00' N, and longitudes 51° 00' and 56° 00' E. The United Arab Emirates has mainland surface area of about 77,700 km^2. It is bounded to the north by the Arabian Gulf and the Sultanate of Oman, to the east and northeast by the Sultanate of Oman and the Gulf of Oman, and to the south and west by the Kingdom of Saudi Arabia (Fig. 1).

Geomorphic features have a major role in controlling the movement of both surface and groundwater. Geology controls the patterns of outcrop of aquifers and confining units, stratigraphic sequences, and structural zones. These greatly influence runoff volumes, infiltration rates, and the quality of surface and groundwater. In general, rainfall increases towards the north and east and decreases in the south and west. Estimates of the total annual volume of rainfall range from 700 to 1,480 Mm^3. Khalifa (1995) suggested that about 10% of this amount recharges groundwater every year. This gives an average annual recharge of 109 Mm^3. The authors, however, believe that only 2 to 3% of rainfall recharges groundwater in such an arid area, giving an average annual recharge of 21.8 to 32.7 Mm^3.

Water Resources Perspectives: Evaluation, Management and Policy. Edited by A.S. Alsharhan and W.W. Wood.
Published in 2003 by Elsevier Science, Amsterdam, The Netherlands, p. 245-264.

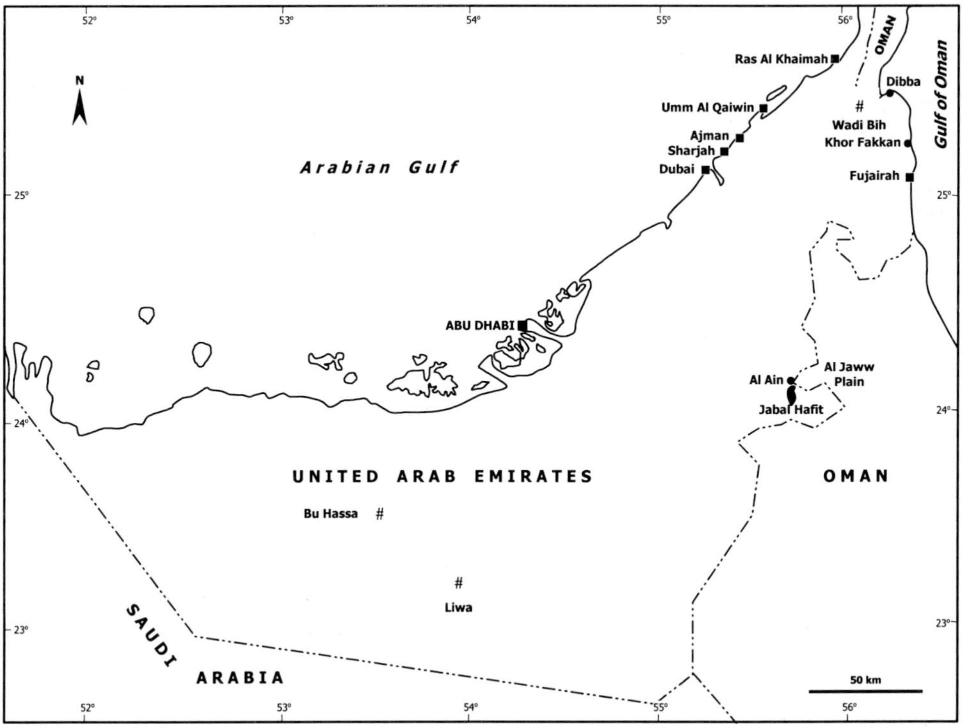

Figure 1. Location map of the study areas in the United Arab Emirates.

CONVENTIONAL WATER RESOURCES

The conventional water resources in the United Arab Emirates include seasonal floods, springs, falajes and groundwater. Because these resources do not meet the national demands for water, non-conventional resources are now intensively used, especially desalinated water. There are no perennial surface water resources in the United Arab Emirates, such as rivers or lakes. However, a small numbers of springs and several falajes provide a limited renewable supply of water.

Seasonal Floods

Desert areas are noted for flash flooding. In the United Arab Emirates, floods mostly occur in the eastern region, in association with strong, short lasting, rain storms. Because the porosity and permeability of the prevailing igneous and metamorphic rocks are low large volumes of rainwater move over the land surface as surface runoff. This flow begins in the mountainous areas in the east near the proximal end of watersheds and moves towards the Gulf of Oman in the east or towards the desert in the west. The western region is dry and lacks surface runoff, because of low rainfall (40 mm/year) (Rizk et al., 1997), high natural evaporation (3,360 mm/year) (Ministry of Agriculture and Fisheries, 1993), scarce vegetation cover, and high porosity and permeability of dune-forming sands.

Topographic maps, aerial photographs, and satellite images show that the mountain ranges of the eastern region have 70 drainage basins, 58 of them within the United Arab Emirates (Fig. 2). The catchment areas of these basins vary between 5 km^2 (Wadi Dhednah, Al Fujairah) and 500 km^2 (Wadi Al Bih, Ras Al Khaimah) (Al Shamesi, 1993). Some large wadis have more than one runoff event per year, others may have surface runoff once in several years, and the rest of the wadis may remain dry for even longer periods.

The annual contribution of seasonal floods to water resources of the United Arab Emirates, as estimated as the Ministry of Agriculture and Fisheries in 1993, is 125 Mm^3. The Ministry has constructed 45 dams across the outlets of several main wadis, providing a total surface-water storage capacity of 75 Mm^3 (Table 1). By the end of 2002, the total number of dams in the United Arab Emirates reached 88 dams with total storage capacity of 100 Mm^3. These dams protect against flood and assist the recharge of groundwater.

Table 1. Main dams constructed in the United Arab Emirates.

No.	Name	Capacity (Mm^3)
1	Bih	7.5
2	Ham	7.0
3	Hadf	0.3
4	Zikt	3.5
5	Tawyyaien	19.5
6	Hatta	4.5
7	Shuaib	20.0
8	Shi	3.0
9	Warraiyaa	5.5
10	Gulfa	0.125
11	Eden	0.05
12	Gheli	0.12
13	Siji	0.750
14	Sufini	0.460
15	Burak	0.280
16	Shawkah	0.272
17	Dalm	0.272
18	Safad, Thyib	0.260
19	Ghalilah	0.250
20	Merbih, Kadfaa	0.242
21	Kidaa	0.220
22	Shaam	0.152
23	Ramth	0.134
24	Mai	0.113

Al Asam (1994) summarized the advantages of dams in the United Arab Emirates as follows:

1. Groundwater recharge.
2. Conservation of floodwater lost to the sea or desert.
3. Storage surface water for irrigation.
4. Preservation of agricultural soil and prevention of soil and plant scourage by floodwater.
5. Protection against flood damage.
6. Accumulation of silt in artificial reservoirs of these dams and improvement of soil properties.

Rizk et al. (1998a) used topographic maps and satellite images, in addition to the records of 5 stream gauges and 11 meteorological stations, to estimate surface runoff and to assess the flood potential of the major drainage basins in the Al Ain area (Fig. 3). The calculated minimum annual rainfall needed to produce surface runoff in drainage basins of the Al Ain area is about 75 mm in the Northern Oman Mountains, and 90 mm in Jabal Hafit (Rizk et al., 1998b). The average annual runoff

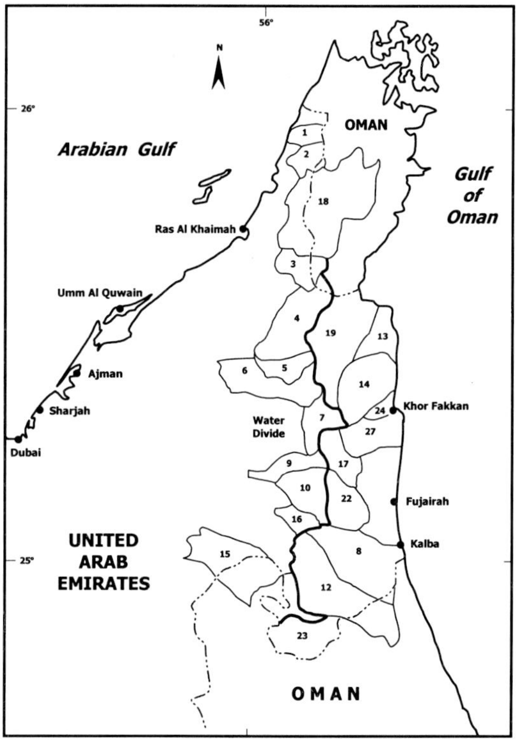

Figure 2. Major drainage basins in United Arab Emirates.

volume ranges from 0.25 Mm^3 in southwest Al Ain, to 3 Mm^3 in the northeast. The average annual runoff depth for the 1981-1991 period ranges from 5 mm in southwest Al Ain, to 20 mm in the northeast. The percentage of rainfall as runoff varies between 3% in Jabal Hafit basins and 18% in the basins of the Northern Oman Mountains. Based on the values of infiltration rate and length of overland flow, Wadi Sidr of the Northern Oman Mountains and Wadi Ain Al Faydah of Jabal Hafit appear to have the highest flood potential, while Wadi Muraykhat and Wadi Al Ain West have the lowest flood potential.

Springs

A spring is a groundwater "outcrop" (Meyboom, 1966) or concentrated groundwater discharge at the ground surface (Todd, 1980). Accordingly, they are used as indicators of discharge areas in arid and semiarid regions. The United Arab Emirates springs tend to discharge from local and intermediate groundwater-flow systems (Rizk and El-Etr, 1997). These include mineralized and thermal springs with therapeutic value. In the United Arab Emirates, several springs, such as Khatt (Ras Al Khaimah),

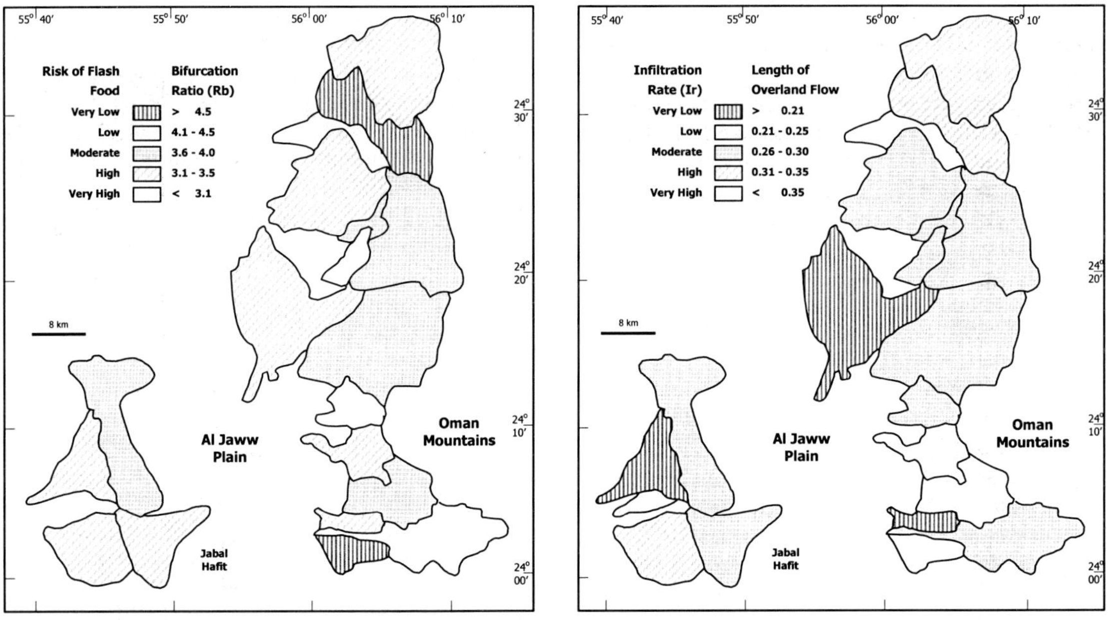

Figure 3. Flash flood and infiltration potential of major drainage basins in the Al Ain area, United Arab Emirates.

Maddab (Al Fujairah) and Bu Sukhanah or Ain Al Faydah (Al Ain), belong to this category, and have been utilized as recreational and touristic sites. The Khatt springs are located about 13 km east of Diba and 15 km south of Ras Al Khaimah, within limestone rocks of the Musandum Formation. Siji spring lies about 50 km west of Al Fujairah and 75 km east of Al Sharjah, at the contact between the ophiolite sequence and the western gravel plains (Alsharhan et al., 2001). Bu Sukhanah spring is about 4 km west of Jabal Hafit, and south of Al Ain town, the spring issues from Miocene gypsum and clay layers through thin loose Quaternary sediments (Fig. 4) (El-Shami, 1990).

The annual discharge of permanent springs in the United Arab Emirates was estimated by the Ministry of Agriculture and Fisheries (1993) as 3 Mm^3. Records of spring discharge for the period 1984-1991 indicate that the Bu Sukhanah spring has the highest discharge (2.50 Mm^3/yr), whereas Siji spring has the lowest discharge (0.06 Mm^3/yr). Discharge of all springs shows wide variations during the 1984-1991 period, with a net increase of Khatt south and Bu Sukhanah springs, and a net decrease of Khatt north spring. The discharge of Siji spring remained constant during the 1987-1991 period because of the construction of the Siji dam, which maintained a steady and continuous groundwater recharge for the spring. According to

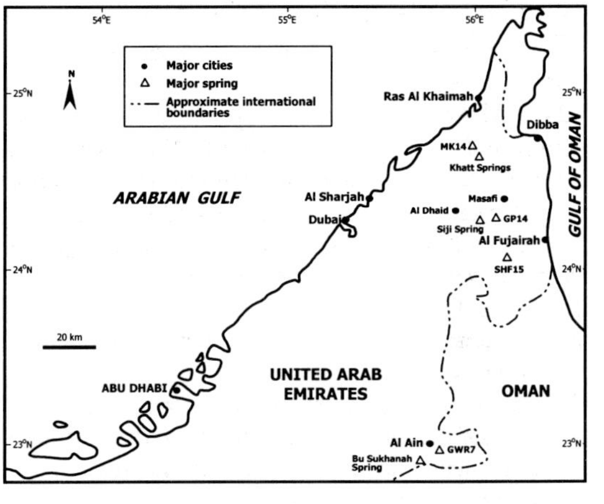

Figure 4. Location of major springs in United Arab Emirates.

their discharge and Meinzer's classification (1923), the studied United Arab Emirates springs belong to the second (Bu Sukhanah), fourth (Khatt south and Khatt north) and fifth (Siji) orders. The Khatt north, Khatt south, Siji and Bu Sukhanah springs can be divided into two categories: springs with discharges that are directly related to rainfall (Khatt north and Khatt south) and springs with discharges that are indirectly related to rainfall (Siji and Bu Sukhanah).

High-temperature water represents the most important physical property characterizing the United Arab Emirates springs (≈40°C). It is believed that the high-temperature water of springs is related to deep circulation of groundwater. Waters of springs discharging from local groundwater-flow systems are normally of low salinity with a temperature close to the mean annual air temperature. In contrast, water of the springs discharging from regional groundwater-flow systems may be highly mineralized and have elevated temperatures (Fetter, 1988). The Bu Sukhanah spring derives most of its water through a deeply circulating intermediate groundwater-flow system that recharges to the east in the Northern Oman Mountains. This could explain the spring's high-temperature water and high salinity.

Terratest (1975) has identified black organic deposits with 21 ppm uranium along the joints and bedding plains of some calcareous rocks of Jabal Hafit. The lack of these deposits in other layers is attributed to leaching by water which discharges into the Bu Sukhanah spring. The heat associated with radioactive decay of uranium may contribute to the rise in the water temperature of the spring. Khalifa (1997) reported high radon activity in this spring.

Except for Siji spring, the Total Dissolved Solids content of water from the studied springs has increased from 1968 to the present by 10% (Khatt south) and 50% (Bu Sukhanah). This increase is mainly related to intensive groundwater extraction in recharge areas of these springs and low rainfall during the last few years. However, the increase in Total Dissolved Solids content, from 5,500 mg/l in 1977 to 10,228 mg/l in 1994 for the Bu Sukhanah water, occurred at a time of increasing discharge, from 0.96 Mm^3/yr in 1984 to 2.50 Mm^3/yr in 1991, and needs further study. The dissolution of evaporite deposits of the Miocene, through which the spring water moves, may contribute to the high-salinity water of the spring.

Concentrations of the major ions vary from one spring to another, according to their local geologic and hydrogeologic conditions. Chemical analysis of water samples have indicated a small increase in concentrations of all ions within the same spring from 1991 to 1994. But, there was a great variation in concentrations of the same ions among different springs. Despite the fact that both the Khatt and Bu Sukhanah springs drain limestone, chemical analyses of spring water samples show that the concentration of calcium ion varies between 60 mg/l at Khatt south and 1,100 mg/l at Bu Sukhanah in 1991. This reflects the difference in the groundwater-flow systems feeding the springs. The water of the Khatt spring has a rapid circulation and response to rainfall and water-table fluctuations. In contrast, water chemistry of the Bu Sukhanah spring results from slow circulation and is insensitive to rainfall and water-table fluctuations. Sodium-ion concentration varied between 2 mg/l at the Siji spring and 1,600 mg/l at the Bu Sukhanah spring in 1991, reflecting the differences in the hydrogeologic conditions between the two springs (Alsharhan et al., 2001).

According to Chebotarev's (1955) sequence, the high sulphate-ion concentration of Bu Sukhanah spring suggests an old water source. In contrast, the high bicarbonate-ion concentration of Khatt springs may suggest a recent water source. The relatively low bicarbonate-ion concentration in the Khatt springs (200 mg/l), in 1994, can be explained by a decrease in recharge rates, especially for the Khatt north spring. The decrease in recharge reduces the more rapid, low solute local flow leaving a larger component of deeper, higher solute regional flow. Although there is a wide variation in sulphate-ion concentration, the chloride-ion content shows only an insignificant increase in the Bu Sukhanah water between 1991 (4,000 mg/l) and 1994 (4,040 mg/l).

The results of chemical analyses of water samples from the United Arab Emirates springs show that none the spring waters are potable. A plot of the specific conductance in μS/cm at 25°C against the Sodium Adsorption Ratio indicate that, except for the water of the Siji spring, water of studied springs are also not suitable for irrigating traditional crops. However, the waters can be used for irrigating specific crops in well- drained soils.

Falajes

Until recently, falajes represented the main arteries of life in the eastern United Arab Emirates. At their outlets palm oases flourished, permanent communities were established, with the agricultural way of life dependent upon their water. At the present time, many United Arab Emirates falajes are dry because of excessive groundwater pumping. However, several falajes still flow and supply water to large palm oases (Fig. 5). The design, construction, and maintenance of falajes are interesting topics and relics of historical water-distribution systems still exist near Al Hili archeological garden in the Al Ain area (Rizk, 1998; Alsharhan et al., 2001).

During the period 1978-1995, the total falaj discharges in the United Arab Emirates varied between a minimum of 9.0 Mm^3/yr, in 1994, to a maximum of 31.2 Mm^3/yr in 1982 (which represents 2.8 to 9.7 % of the total water use in the United Arab Emirates). Discharge varies from one falaj to another depending on the location of the source well(s), nature of the source aquifer, the amount of seepage from tunnel sides, and the mean annual rainfall. Some falajes cease to flow during periods of low rainfall while others are relatively unaffected. Average discharges of United Arab Emirates falajes for the period 1984-1996 varied from 2.38 l/sec in Falaj Habeesa to 89.77 l/sec in Falaj Bithnah. Falajes are classified according to discharge into three types, locally designated as Al Gheli, Al Daudi and Al Hadouri (Al Aidrous, 1990). Al Gheli-type falajes carry water only in winter and their discharge is directly dependent on rainfall. Despite their limited supply, water of Al Gheli falajes is renewable and has good quality. In fact, because most of the United Arab Emirates falajes belong to the Al Gheli type, discharges show a clear response to rainfall events. Because aquifers are the main water source, Al Daudi falajes have a permanent discharge. Al Daudi falajes, exhibit very little change in their discharge rates throughout the year, because groundwater storage is the main source of water for these falajes, and maintains their flow during the whole year. Al Hadouri falajes are typically connected to deep artesian aquifers. These falajes intercept the groundwater as it moves upward through fissures and fractures. A plot of the total annual discharge of United Arab Emirates falajes versus the mean annual rainfall on the eastern mountain ranges and gravel plains, shows a direct correlation (Fig. 6) (Rizk, 1998). Because of heavy groundwater pumping in many areas of the United Arab Emirates, several falajes have ceased to flow. Historical discharge records of Al Ain falajes (Fig. 7) were obtained from Halcrow and Partners (1969), Gibb and Partners (1970) and the Ministry of Agriculture and Fisheries (1993). These records indicated that Al Jimi, Al Mu'waji, and Al Qattarah falajes were dry in 1979, whereas Al Mu'Tarrad, Maziad and Al Hili falajes were dry in 1981-1982. Because of continuous maintenance, extension and pumping groundwater into their channels, the Al Aini and Al Daudi falajes in the Al Ain area are still active at the present time (Alsharhan et al., 2001).

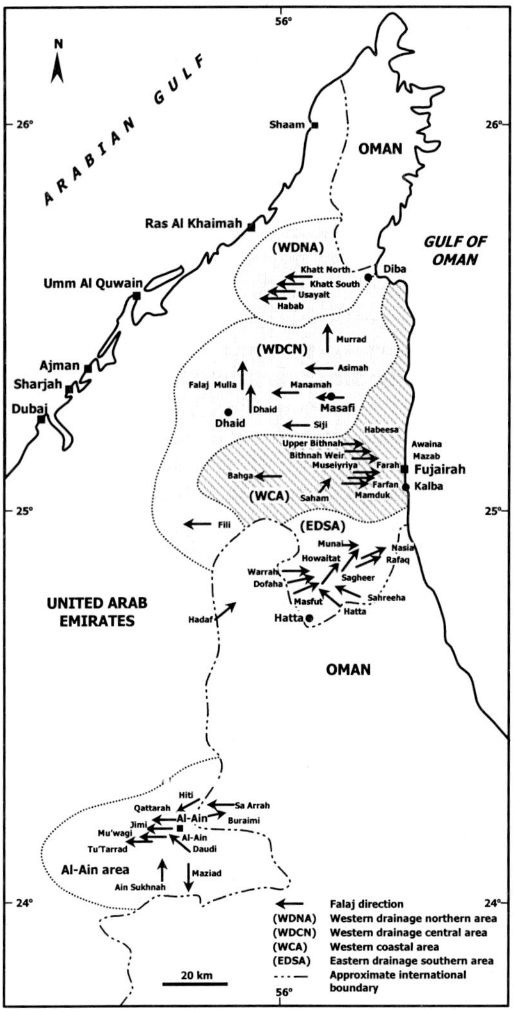

Figure 5. Location and drainage areas of the United Arab Emirates falajes.

The Electrical Conductivity of falaj waters varies between 450 µS/cm in Falaj Asimah (Al Fujairah) and 10,940 µS/cm in Falaj Ain Sukhnah (Al Ain). Generally, the Electrical Conductivity values are low for water samples collected from the falajes which drain ophiolite rocks (east of Al Ain and Al Fujairah areas), indicating low water salinity. In contrast, the Electrical Conductivity values are higher in water samples from the falajes draining limestone rocks in Ras Al Khaimah and west of Al Ain areas. The iso-Electrical Conductivity contour map shows that the Electrical Conductivity values of falaj waters are low near the watershed of the eastern mountains, and increase further east and west, with distance from the recharge area (Fig. 8). The iso-Electrical Conductivity contours also show that the groundwater salinity increases from the watershed towards the east and west. A water sample from Falaj Ain Sukhnah in Al Ain area has Electrical Conductivity of 10,940 µS/cm, constituting an exceptionally high salinity for a falaj water. A possible reason is that this particular falaj penetrates the evaporite deposits of the Miocene Fars Formation, which flank Jabal Hafit on both

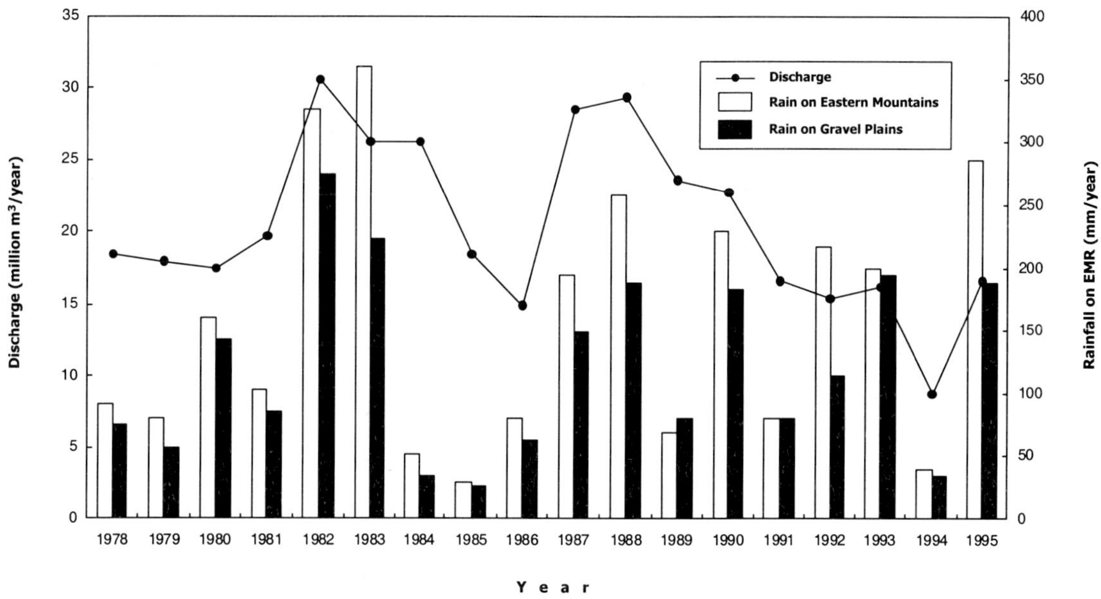

Figure 6. Total annual discharge (Mm3/y) of United Arab Emirates falajes versus the mean annual rainfall (mm) on the eastern mountain ranges and gravel plains.

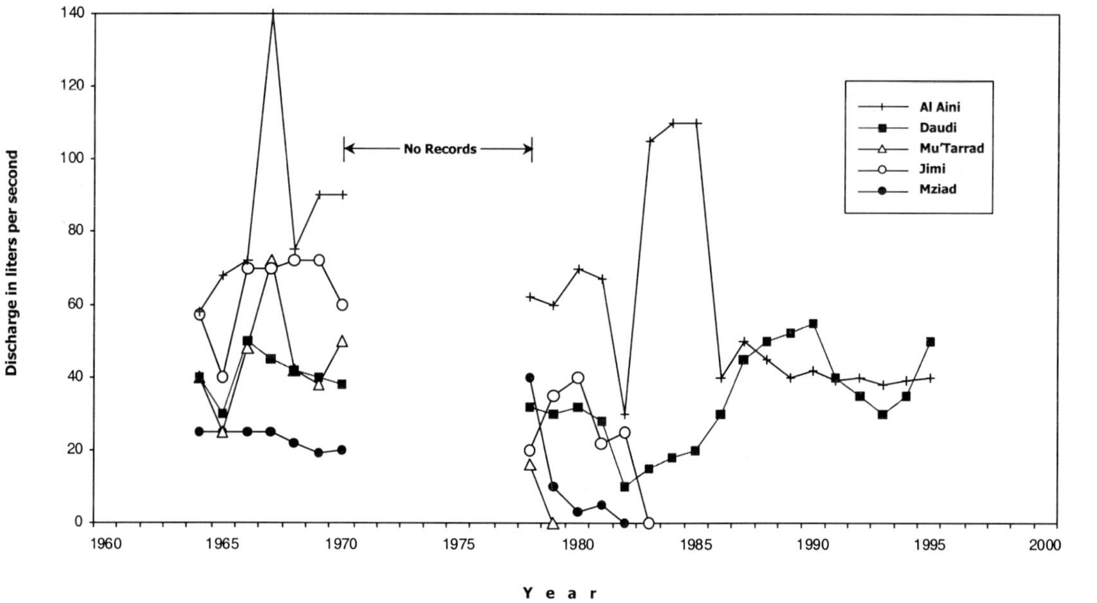

Figure 7. Hydrographs of Al Ain falajes for the period 1964-1996 (compiled data based on studies of Gibb and Partners, 1970; Ministry of Agriculture and Fisheries, 1993; and the authors).

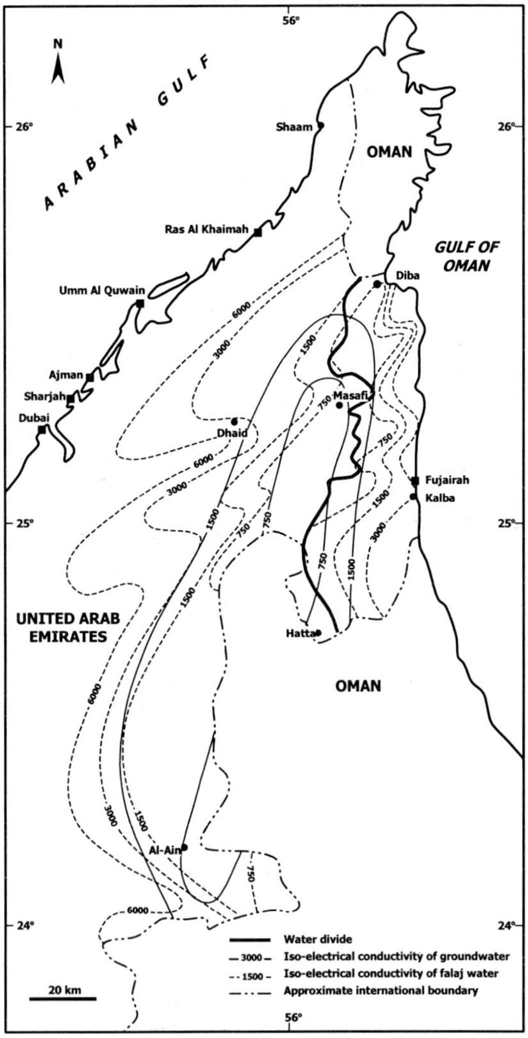

Figure 8. Iso-electrical conductivity (mS/cm) contour map of groundwater and falaj water during early 1996 in northeastern region of the United Arab Emirates.

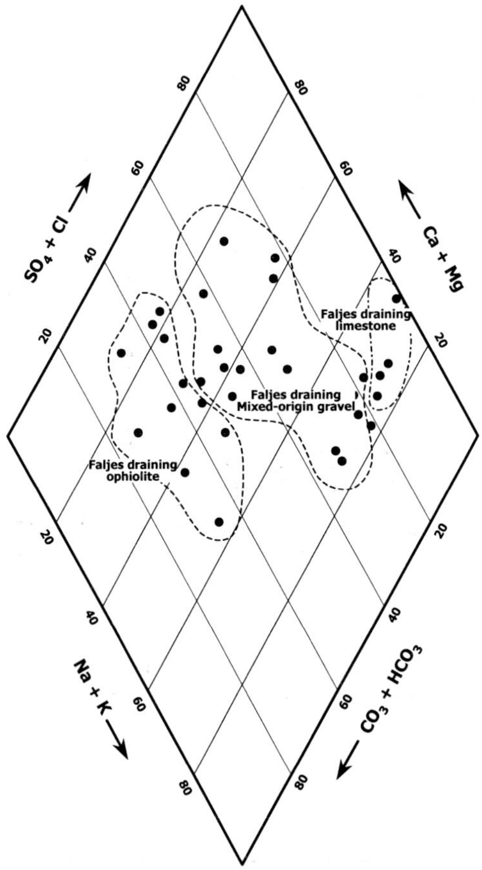

Figure 9. The trilinear plot of chemical analyses of water samples collected from United Arab Emirates falajes during early 1996.

sides. Dissolution of these evaporites may significantly increase the salinity of the falaj water. Jabal Hafit, the main recharge area of the falaj, is mainly composed of limestones containing some gypsum. Upon dissolution, gypsum can contribute to the high Electrical Conductivity of the water in the falaj. The plot of the Electrical Conductivity (μS/cm) versus falaj length shows that the Electrical Conductivity of water in open-channel falajes (Al Gheli type), increases with increasing falaj length (Rizk, 1998). The Electrical Conductivity increase results from the high natural evaporation rates from falaj channels and interaction between water and the bedrock. The longer the channel, the larger the contact surface with the bedrock, leading to larger Total Dissolved Solids contribution from the bedrock. Old falaj channels were constructed narrow and rather deep to minimize natural evaporation from falaj waters. In tunnel-type falajes (Al Daudi falajes), the Electrical Conductivity does not correlate with the falaj length because of the variation in rock type and source of water in these falajes. Despite its short length (0.5 km), Falaj Hubhub has a high Electrical Conductivity (2, 760 μS/cm). In contrast, Falaj Al Aini (6 km) is longer than Falaj Hubhub and has a much lower Electrical Conductivity (620 μS/cm). A trilinear plot of the chemical analyses of water samples collected from the United Arab Emirates falajes shows that most of the samples lie in the upper triangle of the diamond-shaped field (Fig. 9), pointing to the dominance of Na-Mg and chloride and bicarbonate water types. Water of the United Arab Emirates falajes is enriched in Mg^{2+} dissolved from Mg-rich ophiolitic and dolomitic rocks. Because the water of the falajes is mainly used for irrigation purposes, a United States Salinity Laboratory Staff (1954) diagram was used to plot values of Electrical Conductivity and Sodium Adsorption Ratio. The water of all the United Arab Emirates falajes, except Khatt South and Hubhub falajes, are good to fair for irrigation purposes.

Groundwater

The main aquifers in the United Arab Emirates include the limestone aquifers in the north and east, fractured ophiolite rocks in the east, gravel aquifers flanking the eastern mountain ranges on the east and west and sand dune aquifers the south and west (Fig. 10).

The Limestone Aquifers

Two important limestone aquifers exist in the United Arab Emirates; the northern limestone aquifer (or Wadi Al Bih Aquifer) in Ras Al Khaimah area and Jabal Hafit limestone aquifer, south of Al Ain city.

The northern limestone aquifer is predominantly composed of fractured limestone and dolomite. At Wadi Al Bih, the aquifer belongs to the Ru'us Al Jibal Group of Permian to Triassic age. The rocks forming the aquifer are stratified, hard, dense, and non-porous at the surface, with extensive internal karstification. The aquifer is recharged from a catchment of 475 km^2, at an average elevation of 1,050 m above sea level (Abu Al Enien, 1996). The average annual discharge from Wadi Al Bih and Al Burayrat well fields during the period 1990-1995 is 57.7 Mm^3. The aquifer's average hydraulic conductivity is 50 m/d, and the hydraulic head ranges from -5 m at Al Burayrat well field in the

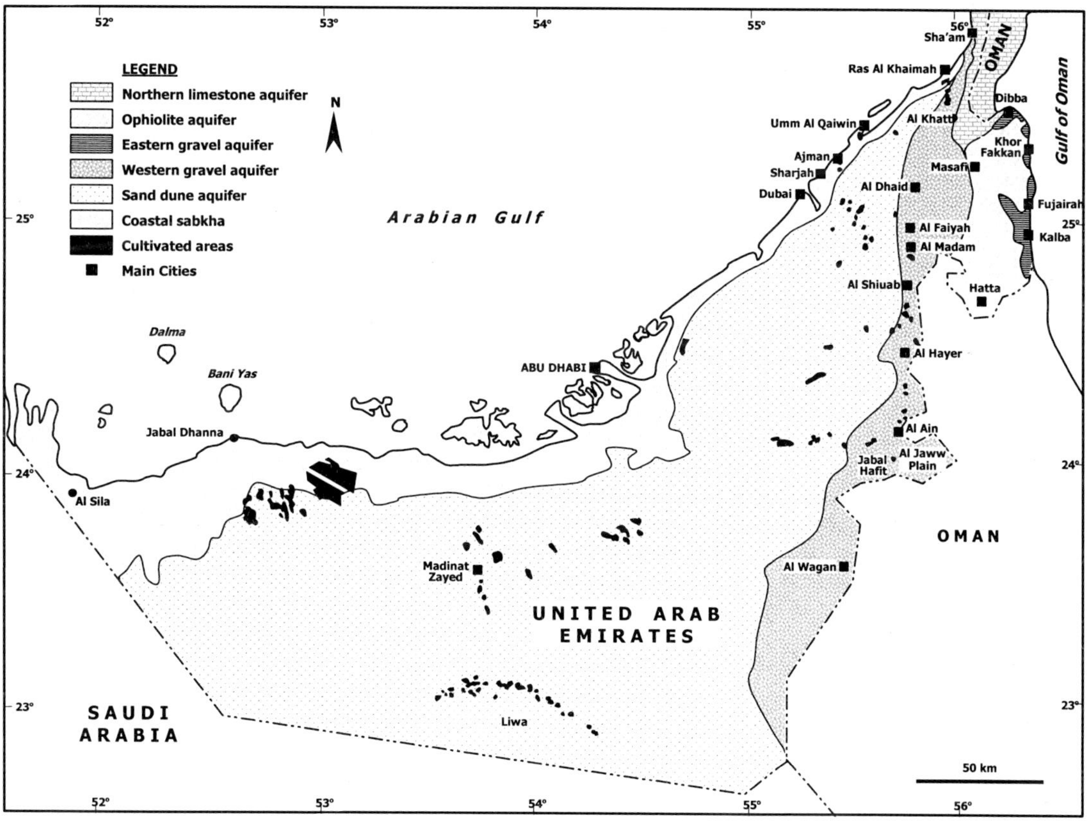

Figure 10. Main aquifers in the United Arab Emirates.

west and 50 m at Wadi Al Bih main dam, indicating that the groundwater flows from east to west, toward of the Arabian Gulf. Most of isotopes for the groundwater in this aquifer plot close to the global meteoric water line with a deuterium excess (d) of +20 (Akiti et al., 1992). The groundwater data which plots to the right of this meteoric water line may result from evaporitic effects, mixing of different water types or salt-water intrusion from the Arabian Gulf. The aquifer contains groundwater with an average isotopic composition (^{2}H = -10.8 ‰ and ^{18}O = -3. 7 ‰) being lighter than the average values of the stable isotopic composition for rainwater in the United Arab Emirates (^{2}H = 12.4 ‰ and ^{18}O = 0.8 ‰). This difference indicates that the possible source of water is from the Mediterranean region, which provides the meteoric water recharge of groundwater at high elevations.

Jabal Hafit is composed of a 1,500 m thick sequence of interbedded limestone and marl with gypsum and dolomite and evaporite formations of Lower Eocene to Miocene age. Limestone of the Middle Eocene Dammam Formation constitutes an aquifer in the Jabal Hafit area. The aquifer is characterized by extensive dolomitization and is affected by numerous faults and fractures. Porosity is virtually nil except for infrequent unfilled fractures, vugs, and heterogeneous secondary porosity (Whittle and Alsharhan, 1994). In a well field at the northern end of Jabal Hafit, fifteen wells (90 to 200 m deep) produce 7.7 Mm^{3} each year of thermal brackish water, with high radium-226 and radon-222. Calculations of the geothermal gradient based on silica thermodynamic data indicate that the depth of water circulation ranges from 1,500 to 2,100 m (Khalifa, 1997). Well yields of over 4,600 m^{3}/d are attributed to fracture permeability and dissolution of the aquifer. Water from the water well field now supplies a recreational spa and irrigation. The Middle Eocene limestone of Jabal Hafit has a black organic precipitate along joints and bedding planes that contains about 21 ppm of uranium (Terratest, 1975). The radioactive decay of uranium creates daughter elements including radium and radon. Radon is a gas and will escape into the air as water is exposed to the atmosphere. The recommended maximum level for radium-226 in drinking water is 5 picocuries per liter (World Health Organization, 1971). No maximum contaminant level has been set for radon-222. Because the salinity of water produced from Jabal Hafit well field exceeds the maximum level for drinking water, it is not suitable for human consumption. The caliper, temperature, and conductivity logs indicate the possibility of a confined-flow carbonate system. A color television inspection of a well indicated that most of the rock consists of massive limestone and marl with very low porosity, except for the interval between 101 and 112 m, which is fractured limestone. Fracture widths in this interval are as great as 9 cm, indicating that the interval could be responsible for most of the water yielded by the well. A flow meter survey for the same well indicated that high-temperature water was flowing into it from a fractured interval between 93 and 102 m (Khalifa, 1997). The temperatures in the wells range from 36.5 to 51.4°C. The water is slightly alkaline, sodium chloride rich, and has a Total Dissolved Solids content between 3,900 and 6,900 mg/l. Results of chemical analyses indicate two geochemically different water types. The first type of water is relatively low in temperature and Total Dissolved Solids, and the second type of water is relatively high in temperature and Total Dissolved Solids. In both types, sodium is the dominant cation and chloride is the dominant anion. The high chloride concentration at Jabal Hafit well field may indicate upwelling of salt-water from a deep source. Available sources of magnesium and sulfate include the dolomite, gypsum and anhydrite of the carbonate and evaporite rocks at Jabal Hafit. The environmental isotopes (^{2}H, ^{3}H and ^{18}O) in Jabal Hafit thermal water indicate that at least part of the water pumped from Jabal Hafit well field comes from recent recharge. Depletion of ^{2}H and ^{18}O relative to standard reflects recharge at high elevation. A conceptual model devised by White (1977) can be used to explain the presence of three water-bearing zones in the Jabal Hafit area: a fresh water zone replenished by meteoric water; a mixing zone where fresh water mixes with brackish water; and a deep saline water zone. The model involves the mixing of two different sources: fresh water from rains falling on the jabal and saline water moving upward from 2,000 m depth by hydraulic head. Brackish water forms as the two types of water mix together. The ^{2}H and ^{18}O isotopes in groundwater from Jabal Hafit well field support this model. These isotopes are lighter than other groundwater in the surrounding area, suggesting local recharge at high elevation. The tritium levels, while relatively low, still indicate mixing with recent locally recharged water (Alsharhan et al., 2001).

Ophiolite Aquifer

The ophiolite sequence is jointed and fractured and has been subjected to faulting. The main fault system runs in a northwest-southeast direction. Electrowatt (1981) described the Semail ophiolite suite as being typically medium-grained gabbro and fine to medium-grained diorites. The jointing, faulting and weathering of the Semail and the Hawasina beds of the Northern Oman Mountains give rise to good aquifers. Groundwater in this area occurs in joints and fractures.

Gravel Aquifers

The largest reserve of fresh groundwater in the United Arab Emirates occurs in the alluvial deposits of the piedmont plains bounding the eastern mountains from the east and west. Two aquifers can be distinguished. These are the eastern gravel aquifer and the western gravel aquifer.

The eastern gravel aquifer is composed of a series of alluvial flats filling the embayments between promontories of rock spurs extending into the Gulf of Oman. South of Khor Fakkan (Fig. 1), the flats and wadi fans coalesce to form an almost continuous littoral strip between the mountains and the sea. The fans near the mountains comprise rock and coarse gravel that become progressively finer in grain size as the distance from the mountains increases. The aquifer contains fresh groundwater that drains from the wadi fans towards the sea. Stable isotope from groundwater from the eastern gravel aquifer plots on the meteoric water line. The low chloride concentrations suggest younger water, in hydrogeological terms. Consequently the wells probably draw from a local groundwater-flow system. The stable isotope contents are relatively depleted compared with the waters from the western sand and gravel aquifer. The deuterium excess 13.6 suggests that this region partially receives recharge from two air masses, the winter precipitation from

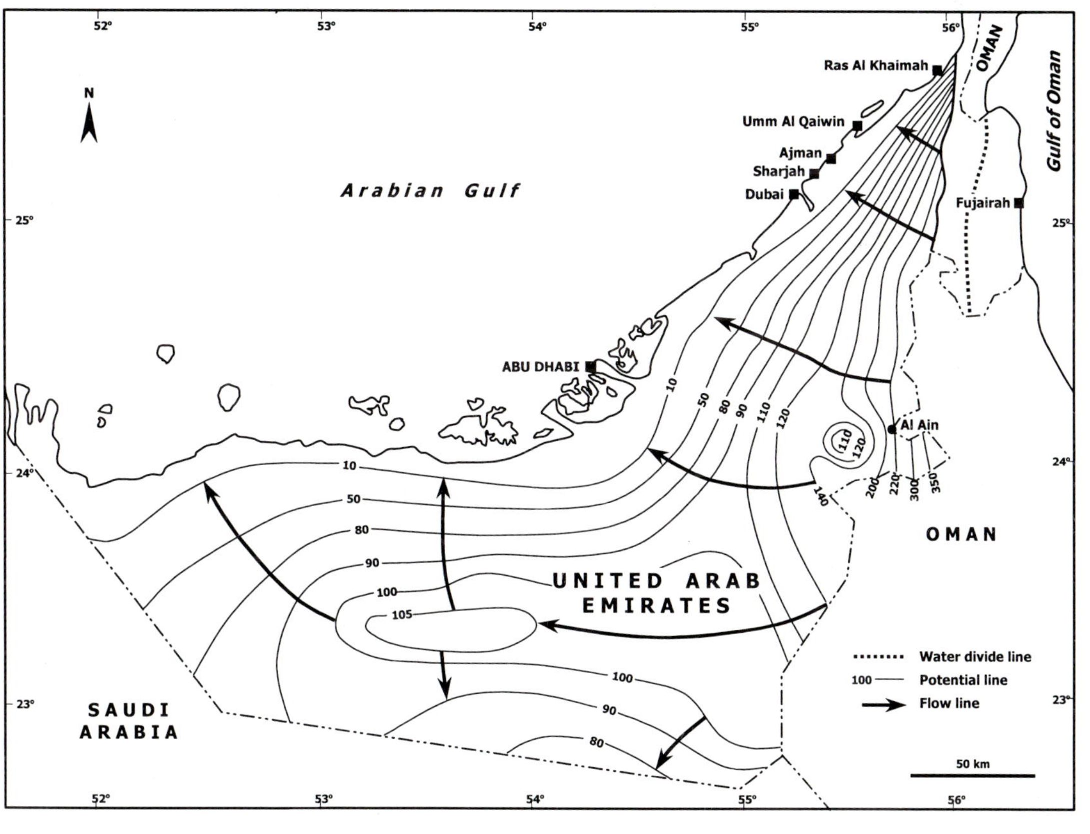

Figure 11. Hydraulic head map, in meters above sea level, for the sand and gravel aquifer, 1995, United Arab Emirates (modified from US Geological Survey-Groundwater Research Project, 1996; Hutchinson, 1996).

the Mediterranean Sea and the Monsoon rains of the Indian Ocean (Yurtsever, 1992). The tritium concentration in groundwater of the eastern gravel aquifer is higher than that of present-day rainfall. The water was probably recharged after 1972 (an exceptionally wet year) with tritium decaying over time during groundwater circulation. Groundwater in this aquifer contains the highest activities of ^{14}C of Total Dissolved Inorganic Carbon in the United Arab Emirates. The ^{14}C ages of groundwater range from modem to 7,000 years B.P. This agrees with the high ^{3}H content in the aquifer and confirms that this aquifer is receiving modem recharge.

Quaternary alluvium of the western gravel aquifer is composed of an approximately 60 m sequence of sand and gravel with thin interbeds of silt and clay. Most of the alluvium was derived from the ophiolitic Oman Mountains. In the north and west of Al Ain, present-day wadis are located between NE-SW trending sand dunes. Recharge of the gravel aquifer in the Al Ain area comes from rain, which falls on the western flank of the Northern Oman Mountains and runs through wadis where it infiltrates and recharges the aquifer. Gibb and Partners (1970), Hydroconsult (1978), and German Engineering (1982), prepared water-table maps for the western gravel aquifer in the Al Ain area. Woodward and Menges (1991) used oil exploration uphole seismic data to construct a water-table map of the gravel aquifer in this area. The latter map shows a buried paleodrainage network, containing saturated alluvial fill, which may constitute major fresh water aquifers in the Al Ain area. Analyses of groundwater samples collected from the western gravel aquifer generally plot to the right of the meteoric water line, indicating enrichment during infiltration. This enrichment could be a result of the residence of water in surface depressions before recharge. The clay in the alluvium does not permit rapid infiltration, and therefore promotes isotopic enrichment as the water moves. The high chloride

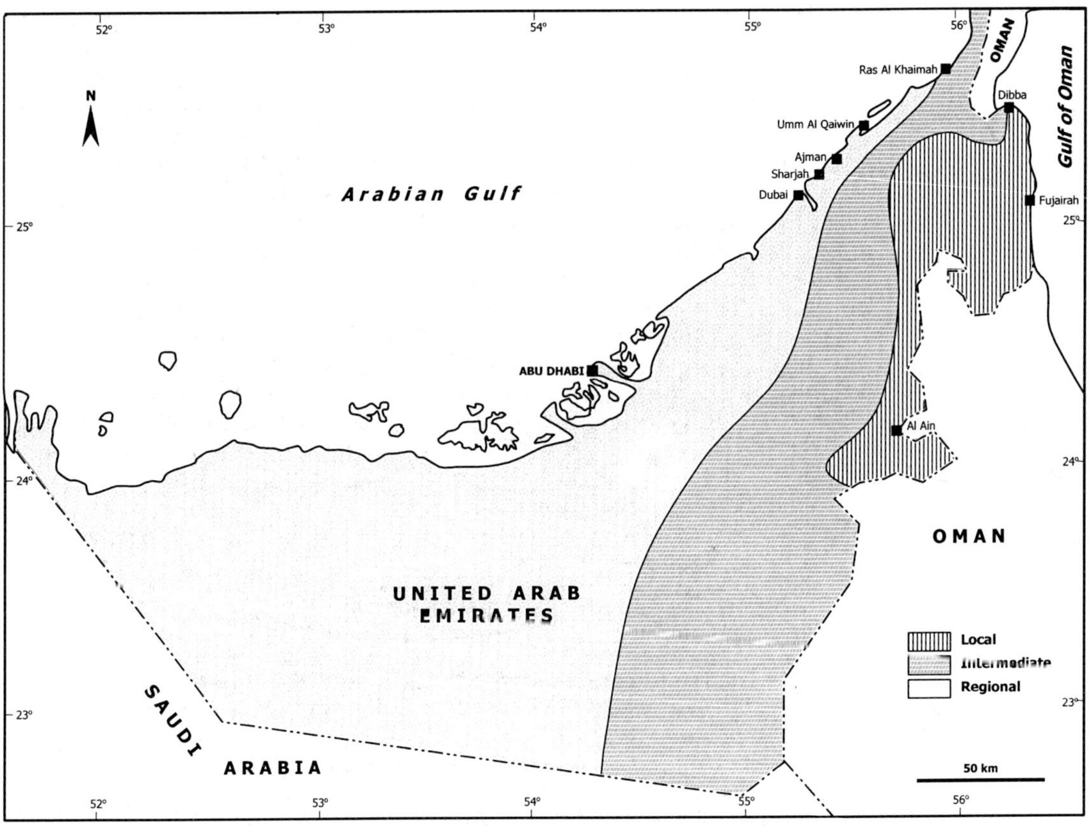

Figure 12. Approximate distribution of groundwater flow systems in the United Arab Emirates.

content and enriched stable isotopes confirm the effect of groundwater flow and its dissolution of salts during flow. Evaporation from groundwater lowers the value of the deuterium excess. It is also possible that the western gravel aquifer receives old water that mixes with rainwater percolating downwards through fractures (Rizk and Alsharhan, 1999). Present-day recharge is restricted to the mountainous areas and the areas adjacent to the mountains. A general increase in groundwater salinity of the western gravel aquifer occurs as water moves from the recharge area in the east towards the discharge area in the west. The hydraulic head map for the sand and gravel aquifers shows that the eastern mountains are the main recharge areas for groundwater in the United Arab Emirates, whereas the Arabian Gulf and the Gulf of Oman are the main discharge areas (Fig. 11). Close to these mountains, low-salinity groundwater (230-1,000 mg/l) exists in the Al Jaww plain, Masafi and Al Shuayb areas. Groundwater salinity of 1,000-3,500 mg/l exists in the Al Ain, Diba, Hatta, Khatt and Al Fujairah areas, whereas salinity of 3,500-6,500 mg/l was recorded in the Ras Al Khaimah, Madinat Zayed, Liwa and Dubai. Groundwater with > 10,000 mg/l of total dissolved solids is found at Al Dhaid, west and south of Al Ain and Kalba areas. Local natural discharge areas are encountered in association with inland sabkhas at Al Ain and Liwa. The western coastal sabkhas may also represent discharge areas of regional flow systems. The study of groundwater flow in the United Arab Emirates points to the presence of local, intermediate and regional groundwater-flow systems (Fig.12). Local groundwater flow systems are limited to the eastern mountains where the hydrologic cycle is relatively fast and groundwater has a short residence time. Water of these systems has a low salinity and belongs to the HCO_3^- water type. Groundwater of local flow systems has a good quality, such as those of Masafi and Al Jaww plain areas. It appears that the Khatt (Ras Al Khaimah) and Maddab (Al Fujariah) springs discharge from local groundwater flow system. The inland sabkhas represent the main discharge areas for groundwater of the intermediate system. Groundwater is mainly brackish, has a moderate residence time and belong to the SO_4^{2-} water type. Because it has relatively high salinity, groundwater of this system has limited uses. The Ain Al Faydah (Al Ain) spring appears to discharge from this system. Groundwater of regional flow systems is mainly saline, has a long residence time and belongs to the Cl^- water type. Coastal sabkhas represent the main discharge areas for groundwater of this system (Hutchinson, 1996). Centers of high groundwater temperatures are observed in springs and wells of Khatt (Ras Al Khaimah), Al Dhaid (Al Sharjah), Hatta (Dubai) and south Al Wagan (Abu Dhabi) areas (Rizk et al., 1997). These centers lie along a NNE-SSW striking thrust fault that represents the western boundary of the Northern Oman Mountains. It is believed that high-temperature groundwater is related to the deep circulation of groundwater. The electrical conductance of groundwater samples collected from the sand and gravel aquifer during early 1996 varied between 252 μS/cm on the Al Jaww plain, east of the Al Ain city, and 173,000 μS/cm in a sabkha area along the Al Ain-Abu Dhabi road. Salt-water intrusion as a result of heavy groundwater pumping is noticed south of Dubai, west of Suweyhan and southwest of Al Ain. It is also observed north of Khor Fakkan and west of Kalba along the eastern coast. The aquifer recharge is lower than the discharge rates. This point must be considered for water-resources planning purposes. Wells in the immediate vicinity to the mountains contain high tritium, whereas the wells at the western edge of the western gravel aquifer contain little or no tritium. The activities in ^{14}C for Total Dissolved Inorganic Carbon are very low, suggesting a great age for the groundwater. The ^{14}C age of groundwater in Abu Dhabi area ranges from modem to 15,000 years B.P. The iso-electrical conductance contour map shows that the groundwater in the gravel aquifer flanking the eastern mountains is mainly fresh and can be used for all purposes. However, because of the excessive pumping, groundwater in several areas is now suffering from salt-water intrusion not only from the sea but from deeper horizons of the same aquifer, and possibly from nearby sabkha deposits. The groundwater is very hard in the northeast at Al Dhaid, Kalba, Al Khaznah and along the western coast, while groundwater in the eastern part of the sand and gravel aquifer does not have a hardness problem and can be used for domestic purposes. The calculated Sodium Adsorption Ratios show that the groundwater in the northern and eastern parts of the country has little harmful effect on plants and soils. Groundwater along the western coast, west Al Ain and east Liwa has high Sodium Adsorption Ratios values and can be very harmful to plants and soils when used for irrigation.

Sand Dune Aquifer

Sand dunes cover about 74% of the total area of the United Arab Emirates and gradually increase in elevation from sea level at the western coast until they reach 250 m above ground level in the Liwa-Al

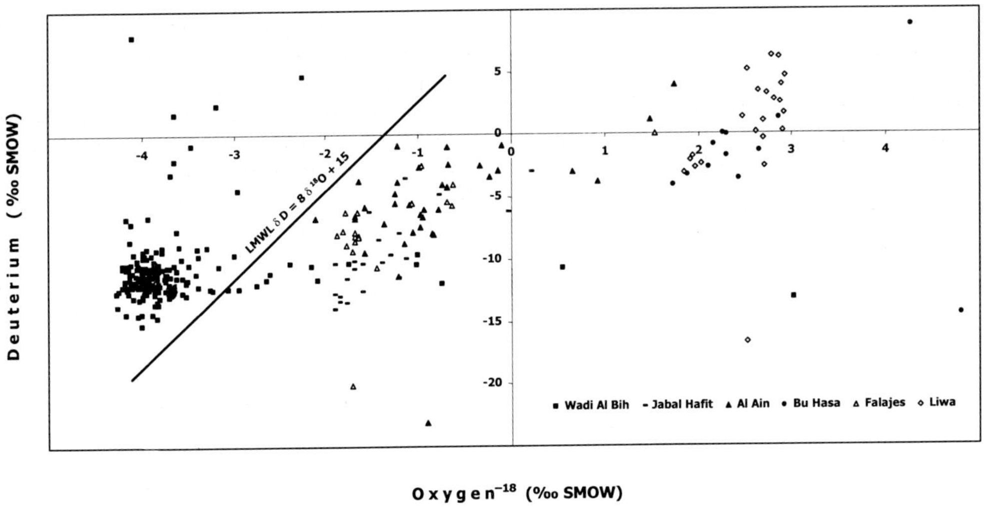

Figure 13. Stable isotqpes of Wadi Al Bih aquifer, Jabal Hafit aquifer, Western Quaternary gravel aquifer (Al Ain), Al Ain falajes, and Liwa Quaternary sand aquifer

Batin basin, south central part of the United Arab Emirates. Sand dunes are the least studied aquifer system in the United Arab Emirates. Fresh water occurs in the Quaternary sand dunes between Liwa and Madinat Zayed. Preliminary hydrogeological investigations of sand dunes in the Bu Hasa oil field indicate the presence of a similar fresh water mound. Exploration of the sand dunes between Al Wagan and Liwa may lead to the discovery of similar fresh water lenses. The ^{2}H and ^{18}O isotopes in the groundwater of sand dune aquifers plot very far from the global meteoric water values, indicating enrichment before or during groundwater recharge (Fig. 13). The wells tapping groundwater in sand dunes contain practically no detectable tritium. However, the samples analyzed in August 1996 contain about 4 TU, indicating the possibility of irrigation return flow after exposure to the atmosphere (W.W. Wood, personal comm., 2000).

NON-CONVENTIONAL WATER RESOURCES

Desalination Water

Two main types of desalination processes are now available commercially: distillation processes and membrane processes. The most important distillation methods are multi-stage flash distillation and multi-effect distillation. Both involve the evaporation of saline feed water and its condensation back into fresh water, leaving dissolved substances in the waste brine. Water desalination in the United Arab Emirates was initiated in 1973, in Abu Dhabi at an annual production rate of 7 Mm^3, reaching 395 Mm^3 in 1997. Since 1974, over sixty desalination plants have been built in the United Arab Emirates. Most of the plants are on the coast or on islands, although a small number were built inland, such as Al Burayrat (Ras Al Khaimah) and As Surrah (Umm Al Quwein), where brackish groundwater is desalinated. The number of desalination plants in the United Arab Emirates rose from one station at Abu Dhabi, during 1976, to 65 stations in the 1995, with each Emirate having at least one desalination plant. The daily production of desalinated water in the Abu Dhabi Emirate, for example, increased from 12,500 m^3 in 1969 to 590,000 m^3 in 1989 (Alsharhan et al., 2001).

Desalinated water represents one of the most important water sources in the United Arab Emirates at the present time, and needs to expand in order to meet the increasing demands for water. It is also important to develop local industries, which can provide chemicals, and equipment parts needed for water desalination plants operating in the United Arab Emirates and possibly other Gulf countries. Support for scientific research towards the use of solar energy for water desalination is also needed.

Records of the Ministry of Agriculture and Fisheries is indicated that there are 76,000 groundwater-production wells used for domestic purposes. These wells produced 137.5 Mm^3 in the year 2000 (Table 2).

Table 2. Volumes of groundwater and desalination water production for domestic purposes in the United Arab Emirates during 2000, and predicted water demands in 2005 and 2010 (data from the UAE Ministry of Electricity and Water).

Emirate	Water Production (Mm^3/yr)				
	2000				
	Groundwater	Desalination	Total	2005	2010
Abu Dhabi	33	358	391	954.7	1208
Dubai	12	200.5	212.5	364.6	527
Sharjah	45	51	96	152.8	168.9
Northern Emirates	47.5	25.5	73	139.9	163.4
Total	137.5	635	772.5	1612	1295.5

Treated Wastewater

In principle, no higher-quality water should be used for a purpose that can tolerate a lower quality of water. As a substitute for fresh water in agriculture and industry, treated wastewater has an important role to play in water resources management in the United Arab Emirates.

The estimated volume of sewage water treated annually in the United Arab Emirates is 150 million m3. It is used in the irrigation of public parks and beautifying streets and roundabouts of the major cities. The discharge of treated sewage water has exceeded 175 Mm^3 in the year 2000. It is worth mentioning that the sewage water after primary, secondary and tertiary treatment can be used in agricultural irrigation especially since after treatment it is purged of toxic materials and bacteria. Sewage water could be used for industrial needs, and to recharge aquifers. It could be used as additional source of drinking water after careful confirmation of its quality for such purpose.

The four sewage treatment plants in the United Arab Emirates are located in Abu Dhabi, Dubai, Al Ain, and Al Sharjah. The first sewage treatment plant, constructed in Abu Dhabi during 1973, had a daily capacity of 4,000 m^3. The capacity of the plant reached 120,000 m^3/d in 1994. The Al Aweir plant in Dubai has a daily capacity of 110,000 m^3. The plants provide primary, secondary (biological treatment) and tertiary (advanced) treatments, which improve the quality of the water and render it suitable for reuse in irrigation, because it does not contain heavy metals. Disinfection of treated sewage water by chlorination during the tertiary treatment kills bacteria and viruses that are health hazards.

The economic feasibility of the treatment of sewage water and its reusage depend on many factors, such as the cost of treatment and the degree of required treatment in comparison to the cost of producing alternative water resources for the same purpose. The cost of producing one m^3 of desalinated water is 5 United Arab Emirates Dirhams (US$ 0.27 per Dirham), whereas the cost of producing one m^3 of treated sewage water is 2 Dirhams (Hamouda, 1995). A question still remains about the possible use of treated sewage water in agricultural irrigation, and chemical and biological effects on the plants, soils and groundwater, in addition to the possible health hazards associated with the use of treated-sewage water in irrigation. It is possible to avoid such consequences by safe treatment, which ensures the elimination of pollutants, and purification of the water to be used for crop irrigation. It is essential to make periodical analyses and field studies to promote the safe use of treated sewage water for irrigation, and to educate the community appropriately to avoid any adverse effects on the environment.

WATER-RELATED PROBLEMS

Depletion

Comparison between the annual groundwater recharge and groundwater abstraction rate (most of which are related to the demands of the agricultural sector) indicates that most of the groundwater resources in the United Arab Emirates are over-drafted and are becoming depleted. In other words, water is being drawn from groundwater storage, with no significant recharge to replace the amounts withdrawn. The authors estimated annual recharge for groundwater in the United Arab Emirates is range of 21.8 to 32.7 Mm^3, while groundwater abstraction, mainly by the agricultural sector, is 880 Mm^3 per year. This imbalance has resulted in a continuous decline in groundwater levels and quality deterioration in most of the well fields due to seawater intrusion and uprising of highly-saline formation. Cones-of-depression ranging from 50 to 100 km in diameter now exist in the Al Dhaid, Hatta, Al Ain and Liwa areas.

Salt-Water Intrusion

The possible sources of salt-water intrusion are: salt-water intrusion from the sea in coastal regions, lateral movement of saline water from a nearby sabkha-dominated areas, or upwelling of saline water from lower stratigraphic units into shallow fresh water aquifers. The $Cl/(CO_3+HCO_3)$ ratio is used to evaluate salt-water intrusion. The chloride ion (Cl^-) is a dominant anion in salt water and normally occurs in small amounts in groundwater. The bicarbonate ion (HCO_3^-) is the most abundant anion in groundwater in the United Arab Emirates because of dissolution of limestone and dolomite (Hutchinson, 1996; Alsharhan et al., 2001). The groundwater in most of the country, except for the central part of the ophiolite aquifer, is seriously affected by salt-water problems. Salt-water intrusion into relatively fresh groundwater of the Quaternary aquifer in the Al Ain area is now taking place as a result of excessive groundwater pumping for different purposes (Garamoon, 1996). Salt-water intrusion in this area is also indicated by the dominance of $MgCl_2$ and NaCl salts. The Na/CI ratio in seawater is less than unity (0.85), while groundwater has Na/Cl ratios greater than unity (Hounslow, 1995). Therefore, it is used to outline areas suffering from salt-water intrusion. Salt-water intrusion problems exist in Ras Al Khaimah, Al Dhaid, Dibba, Kalba, Dubai Jabal Al Dhanah, Madinat Zayed, Liwa and Al Ain (Fig. 14).

Water Quality Problems

Between 1964 and 1985, the groundwater salinity increased in the Bu Hasa camp area from 3,500 mg/l to 5,500 mg/l. The iso-salinity contour map constructed with the field data collected by the authors in April 1996 indicates that the salinity of groundwater in the southeastern part of the Bu Hasa area has tripled during the period 1985-1996, probably because of the disposal of oil-field brines in unlined pits in this area. The analyzed groundwater from the Quaternary Liwa aquifer is free of hydrocarbon pollution. In contrast, brine in the Miocene clastic aquifer contains high concentrations of aliphatic and aromatic compounds. If this water reaches the Quaternary aquifer, it will result in a serious contamination problem, especially for the domestic water-well field. The plot of ^2H (‰) versus ^{18}O (‰) for water samples from the Liwa and Bu Hasa areas suggests that the oil-field water and water of the Quaternary Liwa aquifer are distinct, indicating the possibility of sealing properties of the Miocene clastics in the Bu Hasa area. This seal prevents upward movement of wastewater from the Miocene clastic sediments into the Quaternary Liwa aquifer. Groundwater in the Bu Hasa area has a stable isotopic composition identical to that of groundwater in the Liwa area, indicating a common source.

According to Freeze and Cherry (1979), nitrate (NO_3^-) is the most commonly identified contaminant in water. The WHO (1971) recommended limits for nitrate in drinking water are 10 mg/l as nitrate-nitrogen and 45 mg/l as nitrate (NO_3^-). Centers of high NO_3^- are encountered in Wadi Al Bih, south of Dubai, Al Ain, Al Khaznah, Madinat Zayed and Liwa. Nitrate-ion (NO_3^-) concentrations as high as 1000 mg/l in shallow groundwater of the United Arab Emirates were measured west of Al-Kaznak and Liwa areas. Because of the close correlation between high NO_3^- contents and areas of intensive farming, it appears that agriculture is the main source of nitrates in shallow groundwater of the United Arab Emirates (Fig. 15).

SUMMARY

The United Arab Emirates relies on non-conventional water resources, in addition to conventional ones, to meet the ever-increasing demands for water. The conventional water resources include seasonal floods, springs, falajes and groundwater. The non-conventional resources are desalinated water and treated-sewage water. The existing water resources in the United Arab Emirates are summarized in Table 3.

A significant fraction of floodwater is now tapped by 88 groundwater-recharge dams with a total storage capacity of 100 Mm^3. Several dams are under construction at present and several others are planned for the future.

The United Arab Emirates permanent springs provide about 3 Mm^3 of water per year. While the discharges of Khatt springs is directly related to rainfall, the discharges of Bu Sukhanah spring (Ain Al Faydah) is independent of rainfall. During the 1984-1991 period, the salinity of springs increased between 10% and 50% as a result of low rainfall and heavy groundwater pumping in spring recharge areas. High spring-water temperatures (≈40°C) are related to deep groundwater circulation. The high HCO_3^- concentration in Khatt springs suggests a recent age, whereas the high SO_4^{2-} content in Bu Sukhanah spring water indicates old age. The Sodium Adsorption Ratio of springs water varies between 0.82 and 15.01, indicating that it is not always suitable for irrigation of traditional crops.

Despite their limited discharge, falaj water is a renewable resource directly related to rainfall. During 1978-1995, the total falaj discharge in the United Arab Emirates varied between 9 Mm^3/yr in 1994 and 31.2 Mm^3/yr in 1982, which represents 2.8

to 9.7% of the total water use in the country. The electrical conductance of falaj waters varies between 450 micro-Seimens per centimeters (μS/cm) in Falaj Asimah (Al Fujairah) and 10,940 μS/cm in Falaj Ain Sukhnah (Al Ain). An iso-electrical conductance map indicates that the salinity of falaj waters is minimal near the watershed of the Northern Oman Mountains and increases to the east and west. In the Al Gheli falajes, the salinity increases with increasing falaj length, but in the Al Daudi falajes, the salinity is generally low regardless of falaj length. According to Electrical Conductance and Sodium Adsorption Ratio, water of the Al Gheli falajes is good for irrigation, whereas the waters of the Al Daudi and Al Hadouri falajes are fair to poor for irrigation purposes.

The annual recharge for groundwater in United Arab Emirates is 109 Mm^3, whereas the present groundwater extraction rate averages 880 Mm^3/yr. This highly unbalanced situation resulted in aquifer depletion in many areas such as Al Ain and Al Dhaid; a steady increase in groundwater salinity in well fields over the entire country; dryness of many shallow wells; and salt-water intrusion problems. Due to excessive groundwater pumping, cones-of-depression ranging from 50 to 100 km in diameter now exist in the Al Dhaid, Hatta, Al Ain and Liwa areas. The Northern Oman Mountains in the United Arab Emirates represent the main recharge area for groundwater. Close to these mountains, low-salinity groundwater (230-1,000 mg/l) exists in the Al Jaww plain, Masafi and Al Shuayb areas. Groundwater salinity of 1,000-3,500 mg/l exist in the Al Ain, Diba, Hatta, Khatt and Al Fujairah areas, whereas salinity of 3,500-6,500 mg/l was recorded in the Ras Al Khaimah, Madinat Zayed, Liwa and Dubai. Groundwater with > 10,000 mg/l of dissolved salts is found at Al Dhaid, west and south of Al Ain and Kalba areas.

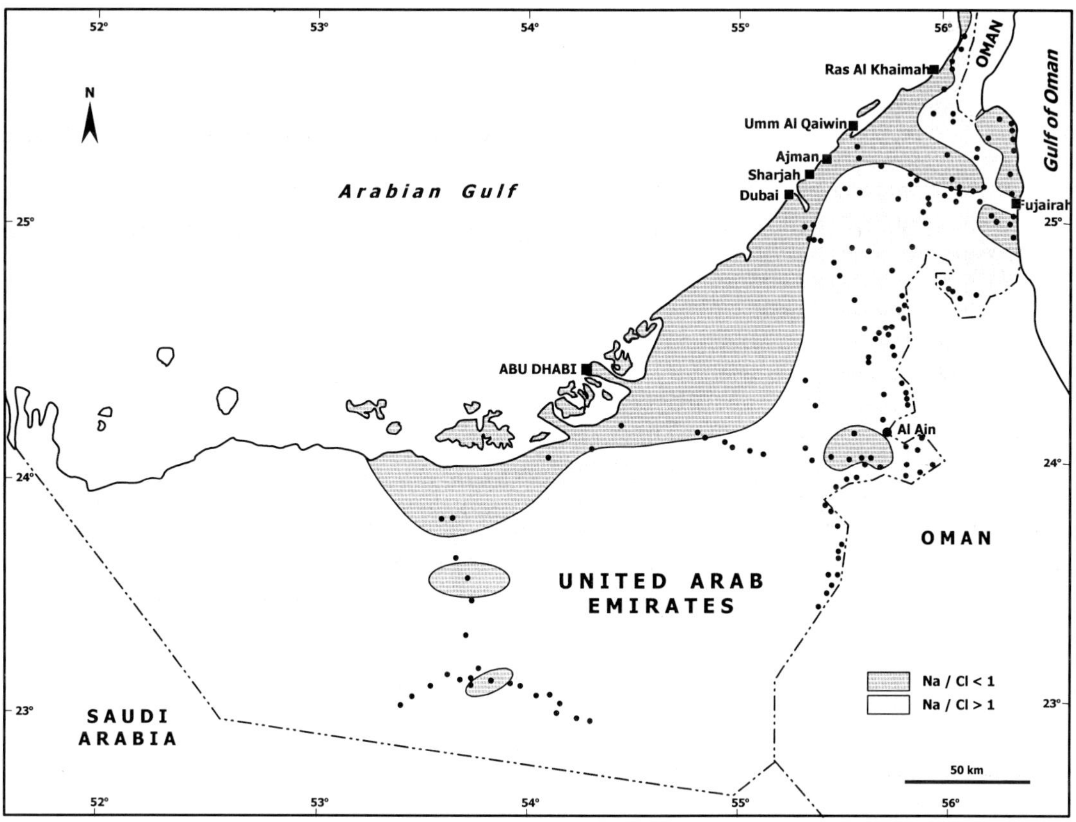

Figure 14. Hydrochemical coefficient Na/Cl in groundwater from the sand and gravel aquifer during 1996, United Arab Emirates.

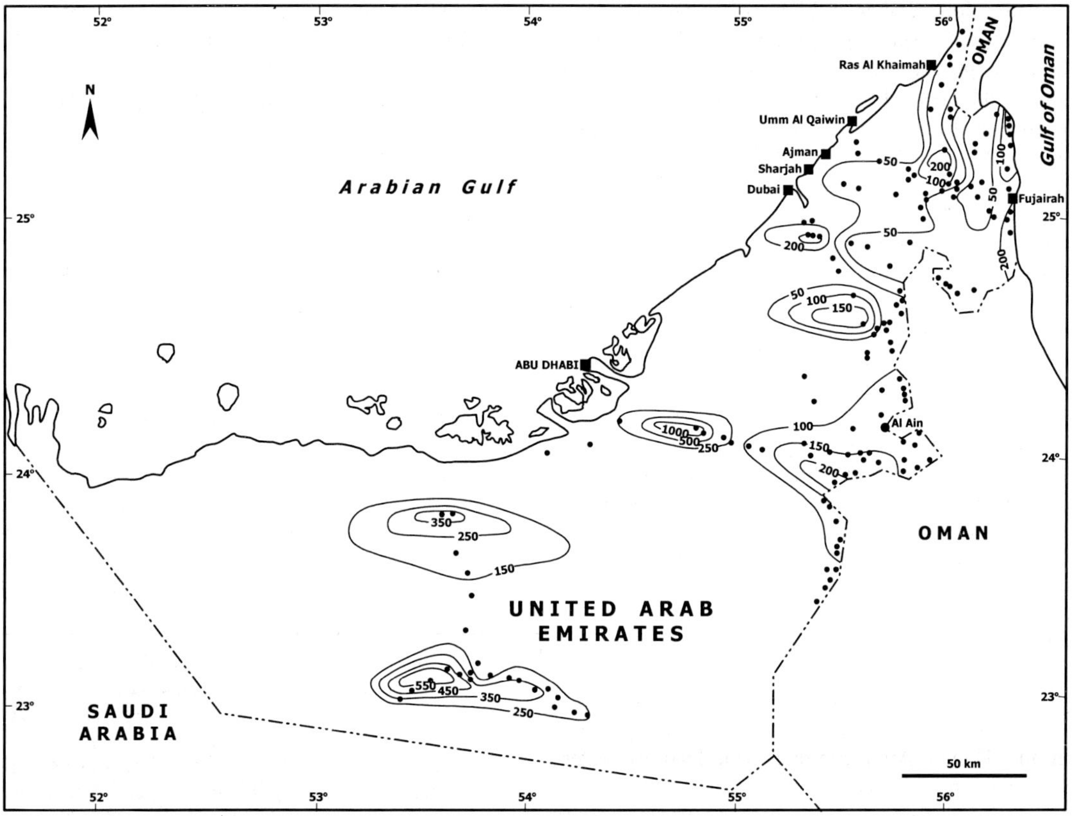

Figure 15. Iso-concentration contour map of the nitrate ion (mg/l) in groundwater from the sand and gravel aquifer during 1996, United Arab Emirates.

Table 3. Summary of conventional and non-conventional water resources in the United Arab Emirates, derived from different sources.

Resource	Existing	Potential	In use	Source
Conventional Resources (Mm^3/y)				
Seasonal floods	125	125	125	Al Asam, 1996
Perennial springs	3	6	3	Rizk and El-Etr, 1997
Seasonal springs	22	40	---	Ministry of Agriculture and Fisheries, 1993
Falajes	20	40	20	Rizk, 1998
Aquifer recharge	120	120	120	Khalifa, 1995
Groundwater	---	---	880	Ministry of Agriculture and Fisheries, 1993
Non-Conventional Resources (Mm^3/y)				
Desalinated water	----	----	694	Ministry of Electricity and Water, 1998
Reclaimed water	150	----	150	Hamouda, 1995

The annual production of desalinated water has increased from 7 Mm^3 in 1973 to 694 Mm^3 in 1999. In 1985, the desalination plants in the United Arab Emirates produced 204 million m^3 of water, which represents 60% of the domestic water needs. In 1998, the production of desalinated water reached 527 Mm^3, which is 76% of the water used for domestic purposes. The percentages of desalinated water production in 1997 was 57% in Abu Dhabi, 35% in Dubai, 5% in Sharja and 3% in the northern Emirates.

The sewage water discharge in the United Arab Emirates increased from 1.5 Mm^3 in 1973 to 142 Mm^3 in 1994. There is about 10% annual increase in sewage water production as a result of increasing population, increasing per capita water use and extension of sewage network to serve about 70% of the population. The sewage water discharge in the United Arab Emirates has exceeded 175 Mm^3 in the year 2000.

The most important water-related problems in the United Arab Emirates are the depletion of aquifers in several areas, such as Al Ain and Al Dhaid, saline-water intrusion and water quality deterioration such as those associated with oil industry or agricultural activities.

Improvement of the present water management in the United Arab Emirates can lead to water conservation, maintain better-water quality and restoration of deteriorating aquifer systems in many parts of the country. The use of advanced irrigation technologies, construction of groundwater-recharge dams and growing salt-tolerant crops are proper agricultural approaches. Development of human resources provides trained national experts in water related fields. Establishment of data banks and application of advanced groundwater modeling and isotope hydrology techniques represent powerful management tools.

REFERENCES

Abu Al Enien, H., 1996. Wadi Al Bih alluvial fan water resources: Al Kuwait University, Geography Department, Publication No. 198: 78p.

Akiti, T. T, Gonfiantini, R., and Mutawa, A., 1992, Aspects of isotope hydrology of the United Arab Emirates: Ministry of Electricity and Water internal report, 50p.

Al Aidrous, M.H., 1990, Falajes of the Al Ain-First Edition: Al Motanabi Publication, Abu Dhabi, 109p. (in Arabic).

Al Asam, M.S., 1994. Dams in the United Arab Emirates and their role in groundwater recharge: Proceedings of the Second Gulf Water Conference, Manama, Bahrain, 203-218.

Al Asam, M.S., 1996. United Arab Emirates water resources use in agriculture and conservation: IDA conference of Desalination, Abu Dhabi, 625-637.

Al Shamesi, M.H., 1993. Drainage basins and flash flood hazards in Al Ain area, United Arab Emirates: M. Sc. Thesis, Faculty of Science, United Arab Emirates University, 151p.

Alsharhan, A.S., Rizk, Z.A., Nairn, A.E.M., Bakhit, D.W. and Alhajari, S.A., 2001. *Hydrogeology of an Arid Region: The Arabian Gulf and Adjoining Areas.* Published by Elsevier B.V., The Netherlands, 331p.

Chebotarev, I.I. 1955. Mechanisms of natural waters in the crust of weathering: Geochim. *Cosmochim Acta* 8: 22-48, 137-170, 198-212.

El-Shami, F., 1990. The hydrochemistry of the spring at Ain bu Sukhanah, United Arab Emirates. *Arab Journal of Scientific Research* 8 (1): 33-49.

Electrowatt, 1981. Wadi Bih dam and groundwater recharge facilities: Ministry of Agriculture and Fisheries, United Arab Emirates, 67p.

Fetter, C. W., 1988. Applied hydrogeology: Macmillan Publishing Company, New York, second edition, 592p.

Freeze, R.A., and Cherry, J.A., 1979. Groundwater: Prentice-Hall, Englewood Cliffs, New Jersey, 604p.

Garamoon, H.K., 1996. Hydrogeological and geomorphological studies on the Abu Dhabi - Al Ain - Dubai rectangle, United Arab Emirates: Ph. D. Thesis, Ain Shams University, Cairo, Egypt, 277p.

German Engineering, 1982. Um Ghafa water project, Al Ain, United Arab Emirates: Preliminary Consultant Report for Water and Electricity Department, 11p.

Gibb, Sir Alexander and Partners, 1970. Water resources survey, supplement to interim report, Subsurface investigations in Al Ain area: Department of Development and Public Works, Abu Dhabi.

Halcrow, Sir William and Partners, 1969. Report on the water resources of the Trucial States. *Engineering report for the Trucial States Council* 1: 161 p.

Hamouda, M.F., 1995. Reuse of sewage water and its role in water resources development in United Arab Emirates: Culture and Science Society, Meeting on Water Balance in United Arab Emirates, Dubai, 10p.

Hounslow, A. W., 1995. Water quality data analysis and interpretation: CRC Press Inc., Lewis Publishers, New York, 397p.

Hutchinson, C., 1996. Groundwater resources of Abu Dhabi Emirate: U. S. Geological Survey Administrative Report, 136p.

Hydroconsult, 1978. Reconnaissance report and development proposals, Abu Dhabi, United Arab Emirates, eastern region water resources: Government of Abu Dhabi, Ministry of Petroleum and Mineral Resources report, 126p.

Khalifa, A.A., 1995. Surface water and Groundwater resources in United Arab Emirates: Culture and

Science Society, Meeting on Water Balance in United Arab Emirates, Dubai, 12p.

Khalifa, M.A., 1997. Hydrogeology of the geothermal fractured-rock well field at Jabal Hafit, Abu Dhabi Emirate: Proceedings of the Third Gulf Water Conference, Muscat, Sultanate of Oman, 125-140.

Meinzer, D.E., 1923. The occurrence of groundwater in the United States, with discussion of principles: U. S. Geological Survey Water Supply Paper, 489p.

Meyboom, P., 1966. Groundwater studies in the Assiniboine River drainage basin-I. The evolution of a flow system in south central Saskatchewan. *Geological Survey Canada Bulletin* 139: 65.

Ministry of Agriculture and Fisheries, 1993. Hydrology, v. 3, 1980-1991: Department of Soil and Water, Ministry of Agriculture and Fisheries, United Arab Emirates, 294p.

Rizk, Z.S., and El-Etr, H.A., 1997. Hydrogeology and hydrogeochemistry of some springs in the United Arab Emirates: *The Arabian Journal for Science and Engineering* 22 (1C): 95-111.

Rizk, Z.S., Alsharhan, A.S., and Shindo, S.S., 1997. Evaluation of groundwater resources of United Arab Emirates: Proceedings of the Third Gulf Water Conference, Muscat, Sultanate of Oman, 95-122.

Rizk, Z.S., 1998. Falajes of United Arab Emirates: Geological Settings and hydrogeological characteristics. *The Arabian Journal for Science and Engineering* 23(1C): 3-25.

Rizk, Z.S., Garamoon, H.K., El-Etr, H.A., 1998a. Morphometry, surface runoff and flood potential of major drainage basins of Al Ain area, United Arab Emirates. *The Egyptian Journal of Remote Sensing and Space Sciences* 1(1): 391- 412.

Rizk, Z.S., Garamoon, H.K., El-Etr, H.A., 1998b. Application of remote sensing techniques to a hydrogeological investigation of sand dunes southwest of Al Ain city, United Arab Emirates. *The Egyptian Journal of Remote Sensing and Space Sciences*, 1 (1): 369-390.

Rizk, Z.S., and Alsharhan, A.S., 1999. Application of natural isotopes for hydrogeologic investigations in United Arab Emirates: Proceedings of the Fourth Gulf Water conference, Manama, Bahrain, 197-228.

Terratest, 1975, Abu Dhabi mineral survey - Stage II. Detailed investigation of promising areas, Final report. Phase II. Ain bu Sukhanah (unbuplished report).

Todd, D. K., 1980. Groundwater hydrology: John Wiley and Sons, New York (second edition), 535 p.

United States Salinity Laboratory Staff, 1954. Diagenesis and improvement of saline and alkali soils: U.S. Dept. Agri., Agricultural Handbook no. 60: 60-160.

White, W.B., 1977. Conceptual models for carbonate aquifers: Revisited In: Dilamarter, R.R. and Csallany, S.C., editors, Hydrologic problems in karst regions, U.S., Bowling Green, Kentucky, Western Kentucky University, 176-187.

Whittle, G.L., and Alsharhan, A.S., 1994. Dolomitization and chertification of the Early Eocene Rus Formation in Abu Dhabi, United Arab Emirates. *Sedimentary Geology* 92, 272-285.

World Health Organization (WHO), 1971. International standards for drinking water: Third edition, Geneva, Switzerland.

Woodward, D.G., and Menges, C.M., 1991. Application of uphole data from petroleum seismic surveys to ground water investigations, Abu Dhabi, United Arab Emirates. *Geoexploration* 27: 193-212.

Yurtsever, 1992. Preliminary evaluation of isotope results from reconnaissance samples collected in Kuwait: IAEA Technical Cooperation project, Isotope Hydrology of the Middle East, 12p.

SCIENCE AND ENGINEERING ADVANCES IN SUPPORT OF WATER RESOURCES MANAGEMENT

Renewable and Non-renewable Groundwater in Semi-arid and Arid Regions

W. MIKE EDMUNDS
School of Geography and Environment, Oxford University, U.K. E-mail: wme@btopenworld.com

ABSTRACT: For thousands of years humans have adapted to live within the limits of the water cycle. Groundwater has generally provided reliability in supplies in semi-arid and arid regions through periods of drought, but now, within a generation, a global water crisis has developed, due to poor resource management since the scientific evidence has been clear, that water withdrawals exceed natural rates of renewal. For groundwater it is vital that finite amounts of any renewable resource can be recognised and quantified. In areas with low rainfall geochemical techniques are well-suited to recharge estimation. Unsaturated zone data using the chloride mass balance, linked to regional studies are described from sandy areas of the African Sahel, which can provide reliable estimates of renewable amounts of water, taking into account also the spatial variability. Results using this approach from around the world indicate that a cut-off in significant recharge at the regional scale occurs at around 200 mm/yr mean annual rainfall, although some preferential recharge may still occur via river channels. Under more arid conditions mining of non-renewable palaeowaters is being widely practiced. Chemical and isotopic indicators are able to confirm the timing and mode of emplacement of these waters as a guide to their non-renewability.

Water quality issues also emerge which can threaten and further diminish the amounts of the renewable reserves. Salinity encroachment from excessive pumping (in addition to salinisation due to agriculture) poses considerable threats. Other issues considered include naturally occurring high nitrate concentrations related to past vegetation cover and in some cases the high baseline concentrations of certain elements arising from natural geological environments. A sound scientific basis is yet required for improved water management – for improved definition of the renewable resource and indications of non-renewable waters, water quality distribution and stratification and in providing early warning of change and the design of monitoring strategies. Above all an integration of available scientific evidence into policy making is badly needed.

INTRODUCTION

For many thousands of years human populations in arid and semi-arid regions have relied on and adapted to natural oscillations in the hydrological cycle. Groundwater remained a largely hidden resource, but natural discharges from springs provided reliable perennial resources, which controlled both migration patterns and settlement of early populations (Hassan, 1997). Human ingenuity led to intervention in the hydrological cycle with construction of dams, irrigation schemes and, in the case of groundwater, tunnelling to intercept and lower the water tables through the construction of qanats, aflaj and galleries. Springs were artificially contained and shallow wells were created to exploit shallow groundwaters, permitting early societies to offset cycles of drought. This small-scale intervention in the natural cycle persisted with little detriment until the 20th century. It is only in the last one or two generations at the most that the groundwater potential of arid and semi-arid regions has been exploited with such dramatic consequences.

Many aquifers in arid and semi-arid lands (as well as in temperate regions) contain evidence of palaeowaters which were recharged during prolonged, much cooler periods of the late-Pleistocene, when the global climates and recharge patterns were significantly different to those of today (Edmunds and Milne, 2001). In coastal regions, groundwater circulation was enhanced by the lowering of sea level by up to 130 m (Hathaway et al., 1979; Edmunds and Milne, 2001). The period of maximum glaciation in northern latitudes was characterised in many arid low-latitude regions by much drier climates. During the Holocene, coincident with rapid warming and global climatic adjustments, some extreme wet periods occurred especially in parts of northern Africa, which resulted in important late-stage groundwater recharge. Since 4500 BP greater aridity has characterised the modern era creating the present-day distribution of arid and semi-arid regions, beneath which large reserves of mainly fresh groundwater may be preserved. Small climatic perturbations at the century to decadal scale such as the "Little Ice Age" or the prolonged drought of the

Water Resources Perspectives: Evaluation, Management and Policy. Edited by A.S. Alsharhan and W.W. Wood.
Published in 2003 by Elsevier Science, Amsterdam, The Netherlands, p. 265-280.

previous millennium (800-1000BP) may also have had local impacts on water availability but probably did not have any long term effects on aquifer depletion.

It is against this background that current and projected groundwater resources and their exploitation must be considered. Current estimates of renewable global freshwater resources are weighted towards wetter regions and contain inadequate information on renewable groundwaters, as well as the quantity and quality of the palaeo-reserves now being drawn (Shiklomanov, 1998; 2000; Gleick, 2001). Sharply falling water tables in arid and semi-arid regions testify to significant over-development of groundwater, specifically that the rates of groundwater abstraction are exceeding the rates of natural replenishment from current rainfall by large amounts and that the aquifer response to massive withdrawals cannot cope; in some places transient conditions may exist, where water level decline is proportional to the hydraulic diffusivity (transmissivity/storage) of the aquifer (Custodio, 1992). On the basis of available scientific evidence there are clear limits to the exploitation of these fragile resources. Questions arise relating to both the quantity and quality of the groundwater. Several important issues emerge where scientific evidence is needed as a basis for underpinning resource management :

- is modern recharge occurring and if so, how much;
- if non-renewable resources are proven, what is their extent and the timing and mechanism of the recharge event(s); are we mining fossil reserves?
- What are the controls on water quality both spatially and with depth; are these entirely natural?
- how may the water quality change with time as development continues;

In this paper these and related issues are examined in detail, some conclusions are drawn and priorities proposed for water management in arid and semi-arid regions aiming where possible at the sustainable development of the finite but renewable water resources and their surrounding ecosystems, where it may be necessary to rethink water usage within stringent limits.

PRESENT DAY RENEWAL OF AQUIFERS

Considerable uncertainties exist in estimates of the renewable groundwater resources in different countries and groundwater recharge assessment remains an immensely complex problem. Techniques developed for recharge estimation in high rainfall areas are inadequate when transposed to semi-arid and arid regions of the world. In addition there are several factors, especially the heterogeneity of the soils and the near surface geology, vegetation and land use change, that confound these estimations. Simple recharge estimates based on empirical formulae are inadequate for low rainfall areas with high evapotranspiration (Gee and Hillel, 1988; Allison et al., 1994). It is however possible using geochemical techniques to estimate rates of diffuse recharge in many semi-arid landscapes underlain by sandy sediments. This approach may underestimate total recharge in certain areas, especially where preferential flow additionally takes place through fractures or drainage channels. Minimum estimates are nevertheless critical in water resources planning.

The use of the chloride mass balance in the unsaturated zone is now being widely adopted for diffuse recharge estimation in semi-arid regions (Allison et al., 1994; Gaye and Edmunds, 1996; Wood and Sandford, 1995). Conservative rainfall solutes, especially chloride, are concentrated by evapotranspiration and the consequent build-up in salinity is inversely proportion to the recharge rate. This method is therefore ideally suited to the measurement of low recharge rates; it is also attractive since, in sharp contrast to instrumental approaches, decadal or longer average rates of recharge, preserved in thick unsaturated zone profiles may be obtained. Under favourable circumstances unsaturated zone profiles also preserve a record of the recharge history through the downward movement by piston displacement of the moisture. Such profiles have been used to examine the recharge history over the decadal to century scale (or longer) in several parts of the world, as well as to reconstruct the related climate change (Edmunds and Tyler, 2002).

The results of recent applications of the chloride mass balance approach to derive recharge assessments in different parts of the northern hemisphere (Africa and Asia) are shown in Figure 1, with the data summaries in Table 1. Some of these are single data points but for some areas multiple sites are represented for the same area; only those profiles are shown which are ≥8m in length, where steady state moisture and chloride are recorded, indicating net downward flux of moisture. The sources of data are given in the caption. Several interesting features emerge;

Table 1. Recharge estimation from profiles in various countries in Africa and Asia. For sources of data see Figure 1.

Country	Site Ref	Profile length (m)	P (mm)	Chloride Concentration	Direct Recharge (mm/yr)	Annual Recharge ($m^3/km^2/yr$)
CHINA	CN-97	15.2	360	-	47	47000
	CN-98	15.5	550	-	68	68000
CYPRUS	AK2	-	406	119	56	56000
	AK3	-	406	122	55	55000
	AK4	-	406	83	80	80000
	AK5	-	406	200	33	33000
	AK6	-	406	126	53	53000
	AK7	-	406	650	10	10000
	AK8	-	406	70	94	94000
	AK9	-	406	70	74	74000
EGYPT	RA1	20	300	-	24	24000
	RA2	20	300	-	18	18000
INDIA	S1	15	240	-	14.5	14500
	S2	15	240	-	18.2	18200
	SD1	12	240	-	20	20000
JORDAN	S1	10.5	65	-	0.03	
	S2	21.5	480	-	28.4	28400
NIGERIA	GM1	15.5	320	14.1	40	40000
	MD1	22.5	320	21.3	27	27000
	MN1	16.5	320	41.6	14	14000
	N-TM	18.8	320	11.5	49	49000
	W-WGR	19.3	320	17.6	32	32000
	MG	16.3	400	29.5	25	25000
	KA1	15.5	320	27.9	20	20000
SAUDI ARABIA	CRP-1	18	133	-	1.8	1800
SENEGAL	L2	29	290	27.9	29.1	29100
	L3	22.5	290	81	10.1	10100
	L5	11	290	73	11.1	11100
	L6	12.5	290	80	10.1	10100
	L11	13	290	89	9.1	9100
	L12	13	290	175	4.6	4600
	L18	34	290	23.6	34.4	34400
	L9	11	290	95	8.6	8600
	L10	10.5	290	1660	0.49	490
	L14	8.5	290	50	16.2	16200
	NL1	45	290	3.2	18	18000
	NL2	22	290	-	9.8	9800
	L23	37	290	-	78.3	78300
	K1	19	545	747	2.5	2500
	K2	12.5	730	327	108	108000
	K3	7.5	545	158	18.1	18100
	K4	10	545	99	72.1	72100
SYRIA	I	20.6	220	-	1.9	1900
TUNISIA	TOZ3	19	100	322	1.55	1550
	TOZ4	14	100	786	0.64	640
	TOZ5	13.5	100		0.55	550

i) There is a cut off point around 200 mm mean annual rainfall which signifies the lower limit for effective recharge.

ii) Above this threshold there is the possibility for recharge as in Cyprus where 100 mm recharge is measured from 400 mm rainfall. Nevertheless, the technique records small but finite recharge rates in some areas with rainfall as high as 550 mm.

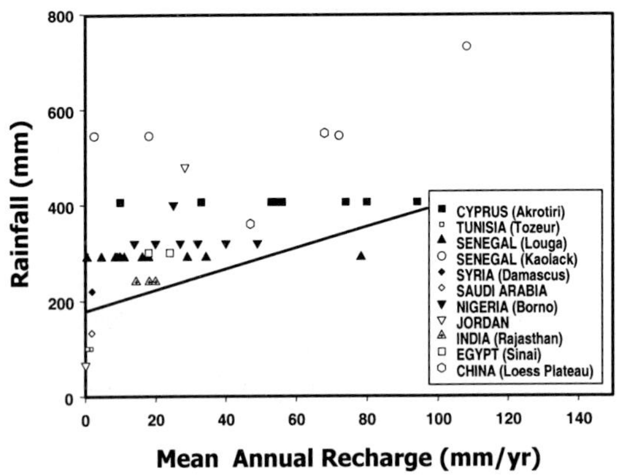

Figure 1. Estimations of point source recharge using chloride mass balance from several countries in Africa and Asia. Data are from Cyprus (Edmunds et al., 1988); Tunisia (Zouari et al., 2001); Senegal (Edmunds et al., 1987; Gaye and Edmunds, 1996; IAEA, 2001); Syria (Abou Zakhem and Hafez, 2001); Saudi Arabia (Al-Sagaby and Moallim, 2001); Nigeria (Edmunds et al., 2002); Jordan (Amro et al., 2001); India (Navada et al., 2001); Egypt (Hussain, 2001); China (Ruifen and Keqin, 2001).

iii) At any one site a large range of recharge may occur in a small area implying that spatial variability is likely to be considerable, due to soil, geology or vegetation controls, since seasonal rainfall variability is likely to be evened out over the decadal timescales of most profiles.

iii) Finite, if low, recharge rates still occur even in those areas with mean annual rainfall below 200mm (eg. Tunisia, Saudi Arabia). However these low rates also signify moderate salinity in the profiles; with the absence of recharge, salinity accumulation will be continuous as found in southern Jordan.

Most of the profiles are from sandy terrain (sandstones or unconsolidated dune sands). Few examples are shown of clay-rich sediments. An exception is the Damascus Basin profile which is predominantly clay-loam (Abou Zakhem and Hafez, 2001) and which shows very low recharge. Two profiles from China are from loess sediments and these show modest rates of recharge. The recharge phenomenon is examined further with reference to the data from Senegal where a wide range of recharge rates has been measured across the Sahel region (Figure 1) with rainfall ranging from 290-720 mm/yr.

Point source recharge estimates

Four representative profiles of chloride from the unsaturated zone near Louga north-west Senegal (Figure 2) illustrate the small-scale variability of

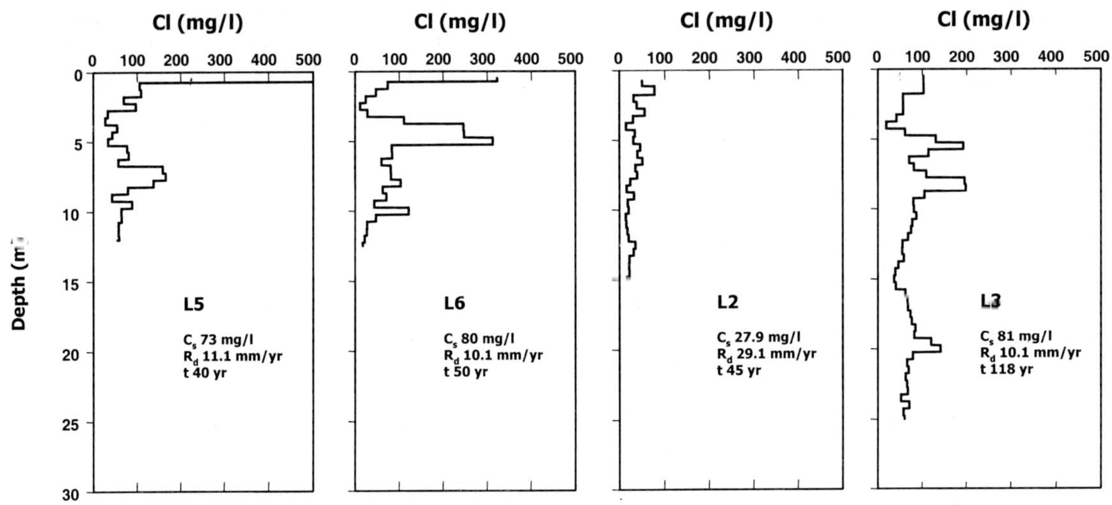

Figure 2. Four profiles (L2, L3, L5, L6) of chloride in the unsaturated zone at Louga, northern Senegal with mean chloride concentrations (C_s) and derived estimates of direct recharge (R_d).

recharge. The four profiles selected are within one site (of area $0.1m^2$), where the water table was at 35m (area A, Figure 3). A total a total of 13 profiles over 8.5m have been obtained from this site together with several of shorter length (Gaye and Edmunds; 1996). All were obtained by hand auguring from fixed Quaternary dune sands where the long-term (100 year) rainfall is 356 mm/yr, falling by 36% to 223 mm/yr from 1969 to the early 1990s during the Sahel drought. The mean concentrations of chloride (C_s) in these four profiles range from 28 to 81mg/l, which correspond to mean direct recharge from 10 to 25 mm/yr. These represent mean recharge rates over periods of several decades (a range of 40-118 years determined from the chloride accumulation rates). Microtopographical and textural features of the dune sands can explain the variability of recharge, where soil thickness and vegetation density vary at the plot scale.

The oscillations in Cl in the point-source unsaturated zone profiles (Figure 2) record the recharge history at the decadal scale and provide a record of climate change over the past century. The record of L3 profile, for example, corresponds well to the sequence of drought and wet periods contained in the instrumental records of rainfall and river flow, which have been discussed elsewhere (Edmunds et al., 1992; Edmunds and Tyler, 2002).

Thus, point source estimates in semi-arid areas can provide important long term recharge estimates which cannot be provided by classical methods. However it is clear that several unsaturated profiles are needed to overcome heterogeneity at the plot scale, even in homogeneous terrain in regions such as Senegal and the Sahel, where fixed dune sands are common. The recharge estimates of all the unsaturated zone profiles measured in the Louga region range from 0.5 to 78 mm/yr and it is clear that the variations in profile estimates must be understood in terms of the wider regional landscape.

Spatial variability- regional scale

One way to determine the spatial variability of recharge using geochemical techniques is to use data from shallow dug wells. Water at the immediate water table in aquifers composed of relatively homogeneous porous sediments represents that which is derived directly from the overlying unsaturated zone. Recharge estimates may then be compared with those from unsaturated-zone profiles to provide robust estimates of renewable groundwater, taking into account the three dimensional moisture distribution and based on timescales of rainfall from preceding decades. A series of 120 shallow wells in a 1600 km^2 area of NW Senegal were used to calculate the areal recharge from the same area from which the point source recharge estimates were obtained (Edmunds and Gaye, 1994). The topography of this area has a maximum elevation of 45m and decreases gently towards the coast and to the north; the natural groundwater flow is to the north-west. The dug wells are generally outside the villages and human impacts on the water quality are regarded as minimal; the area is either uncultivated savannah or has been cleared and subjected to rainfed agriculture. The Cl concentrations in the shallow wells have been contoured, and range from <50 to >1000 mg/l Cl (Figure 3a). These are proportional to recharge amounts which vary from >20 to <1mm a^{-1} (Figure 3b) The spatial variability may be explained mainly by changes in sediment grain size between the central part of the dune field and its margins where the sediments thin over older, less permeable strata. Over the whole area the renewable resource may be calculated at between 13 000 and 1100 $m^3/ km^2/yr$.

Regional comparisons

It is evident from the Senegal data therefore that the range in recharge estimates for a given rainfall regime indicated in Figure 1 can be explained by small-scale local differences in soil properties, geology and vegetation. Recharge rates may also vary in proportion to short-term climatic events, even the recent 20 year Sahel drought, but the estimates of renewable resources allow for this and probably slightly underestimate the resource since they are based on the lower rainfall of the late 20th century. In addition, the progressive clearance of natural vegetation in this and other areas may also enhance rates of groundwater recharge. It is possible to extrapolate to some extent the results of this study to other areas of the African Sahel where comparable geology – sandy deposits overlying bedrock, or other porous sediments – form the main landform. Indeed, recent comparable studies in similar terrain in Nigeria (Edmunds et al., 2002) support the Senegal conclusions.

Natural vegetation differences between regions may however lead to a non-linearity between rainfall and recharge. This is illustrated by the recharge data for Kaolack (Figure 1 and inset "K" Figure 3) lying in the Sudanian vegetation zone at rainfall values of 550-750 mm/yr, indicate a wide range in recharge. This region lies some 400 km to the south of Louga, which is in the Sahelian vegetation zone, where the rainfall is approximately half that of Kaolack. Thick

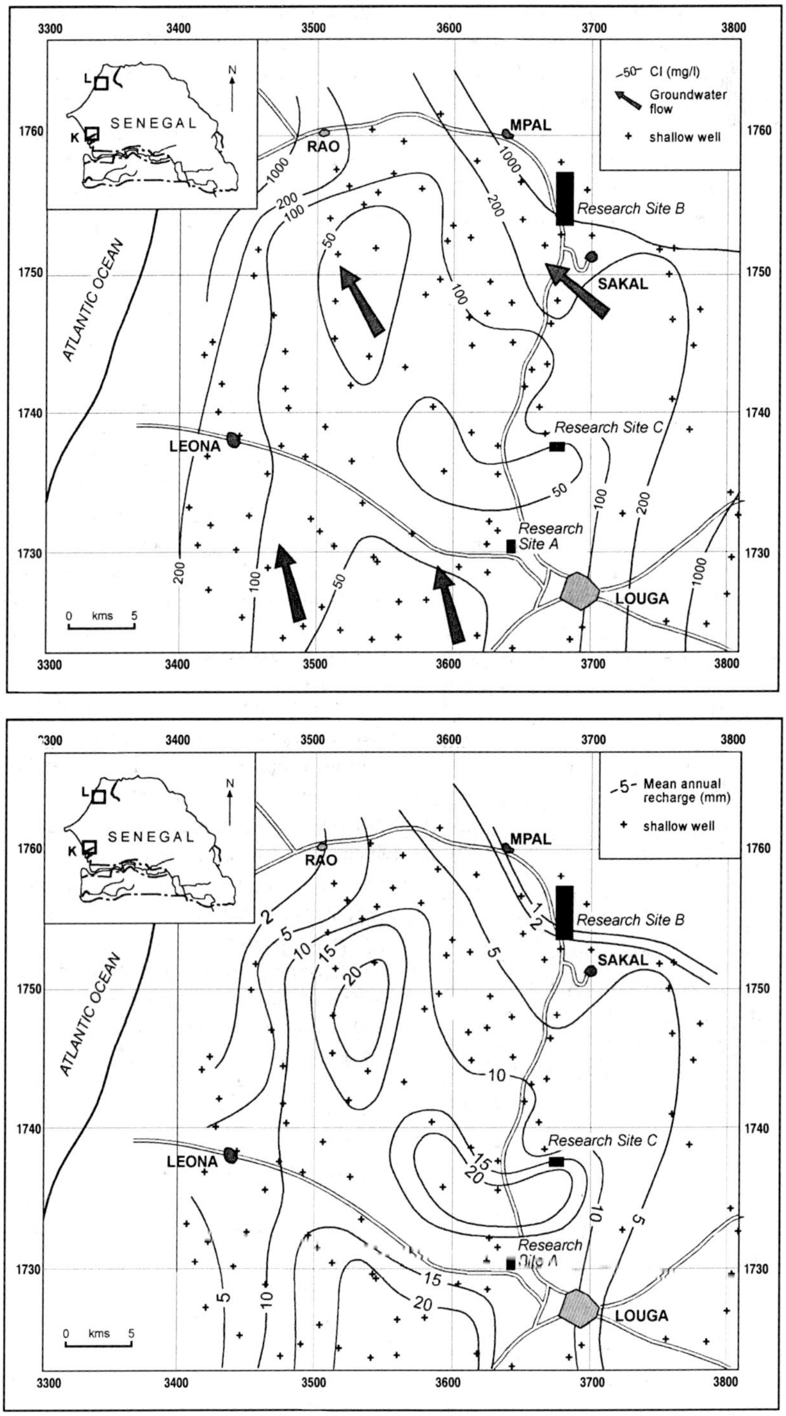

Figure 3. Spatial variability of Chloride (above) and derivation of recharge estimates (below) for the Louga region of NW Senegal. The location of three areas where point source estimates of recharge have also been made are indicated.

unsaturated zones in fixed dune sand deposits also characterise the Kaolack region, but the greater water use by the denser vegetation means that recharge to groundwater is similar to or less than that in Louga region. It is only where clearance for agroforestry has occurred that significant recharge (in excess of 50 mm/yr) is recorded. Thus it is important not to assume that higher rainfall necessarily leads to greater renewal of groundwater.

Where rainfall becomes too low for recharge in any but the most porous media, the unsaturated zone becomes saline, geochemical reactions lead to the formation of minerals in the soil zone and the development of indurated crusts. This situation has been described from the Butana region of Sudan where the mean annual rainfall prior to 1980 was 225 mm but where clay-rich soils inhibited recharge (Darling et al., 1991). The accumulation of salinity in this region is proportional to at least 2000 yr, probably dating to the last period of wetter climate. Similar high concentrations of chloride in the unsaturated zone corresponding to low recharge rates have been recognised in Australia (Allison and Hughes, 1983) and in southern and western USA (Phillips, 1994; Tyler et al., 1996).

PALAEOWATER RESERVES

Deep drilling for groundwater took place widely in north Africa and the Middle East from the 1960s onwards mainly as a consequence of the exploration of large sedimentary basins for hydrocarbons. Many of these drilling programmes intercepted freshwater at depths of 2km or more in certain basins. Popular explanations of these freshwaters linked them with recharge from the Nile or with recharge over impossibly long distances from tropical areas to the south. However these discoveries coincided with the earliest application of radiocarbon and stable isotope methods which linked the groundwater emplacement to pluvial episodes related to past wetter climates (Degens, 1962). The term palaeowaters was first used when it was realized that the climatic conditions of recharge of these fossil waters, as initially identified using stable isotopes of water, $\delta^{18}O$ and $\delta^{2}H$, was colder than at present. It is now widely agreed that these huge freshwater accumulations are unconnected to the modern hydrological cycle and that exploitation of these deep reserves amounts to mining. It is important to promote a better understanding of their occurrence and emplacement as the basis for an informed management of non-renewable supplies, and to recognize the difference between modern, renewable resources and the palaeowaters.

Timing of palaeowater recharge

Isotopic methods have been used widely in arid and semi-arid regions to demonstrate the presence of palaeowaters as well as the timescales of recharge using in particular the activity of radiocarbon in groundwaters (see Fontes and Edmunds, 1989; Clark and Fritz, 1997). From a purely statistical viewpoint it is clear from the evidence now available from some hundreds of radiocarbon analyses that distinct episodes of groundwater recharge related to fundamental reorganisations of climate must have characterised the Holocene as well as the Pleistocene (Sonntag et al., 1978; Edmunds et al., in press). Stable isotope ($\delta^{18}O$ and $\delta^{2}H$) signatures have been used in many studies to distinguish between water from the Holocene and present day from that of the Late Pleistocene. Light isotopic signatures, usually at least 2 per mil more negative in $\delta^{18}O$ characterise the main phase of recharge of most Sahara/Sahel aquifers during cooler climates of the last glacial period.

Evidence is found widely in unconfined aquifers in the Sahara and Sahel that groundwater at the immediate water table is of Holocene age (Fontes et al., 1991). A good example is provided in the Sirt Basin in east-central Libya (Edmunds and Wright, 1979). During exploration studies in the mainly unconfined post-Middle Miocene (PMM) aquifer a distinct body of very fresh groundwater (<50 mg l^{-1}) around 100 m deep was found cross-cutting the general NW-SE trend of salinity increase (Figure 4). Because of the good coverage of hydrocarbon exploration wells (with supporting water supply wells) in this region a three-dimensional impression can be gained of the water quality. This feature, around 10 km in width, may be traced in a roughly NE-SW direction for around 130km where the depth to the water table is currently around 30-50 m. It is clear that this feature is a channel that must have been formed by recharge from an ancient wadi. No obvious traces of this river channel were found in this area which had undergone significant erosion, although neolithic artefacts and other remains testify to settlement during the Holocene. Whereas the regional, more mineralised, groundwaters gave radiocarbon values of 0.7-5.4% modern carbon (pmc), the fresh waters gave values from 37.6-51.2 pmc and also were distinctive in their hydrogeochemistry. These younger waters gave ages ranging from 5000-7800 yr.

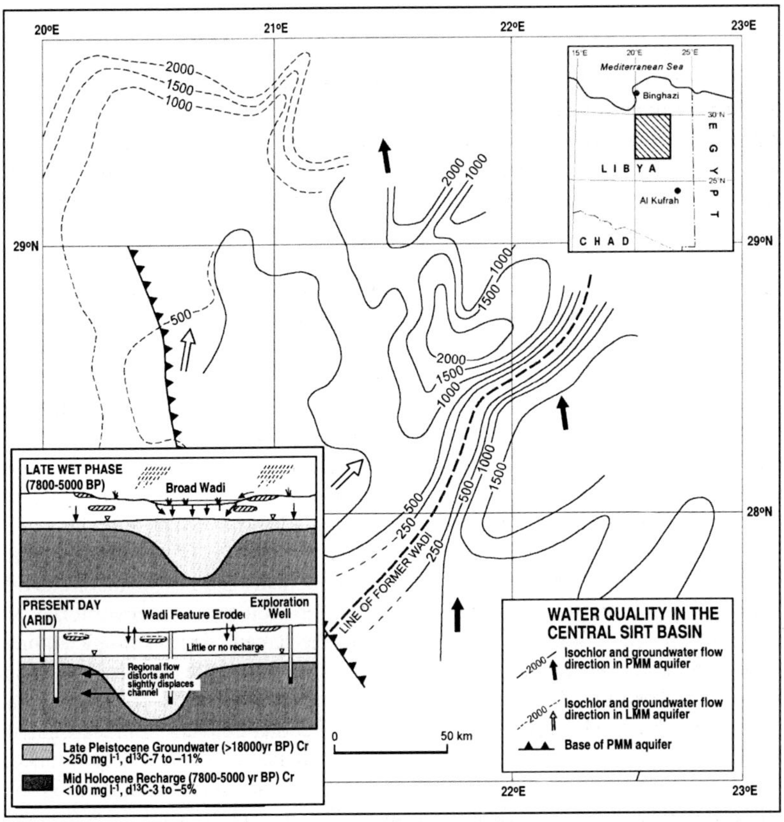

Figure 4. Distribution of Chloride in the phreatic aquifer system in north-central Libya identifying the presence of a freshwater body. The configuration of the freshwater system is also shown in cross-section.

Radiocarbon evidence from shallow wells in the vicinity (Figure 4) proved that fresh, younger water was also present at the water table indicating that, simultaneously with the river flows, direct recharge was occurring at a regional scale (Edmunds and Wright, 1979). Coinciding with the hydrogeological work, palaeoenvironmental studies traced the line of a large, now-inactive river system from the Tibesti mountains (Pachur, 1974; 1980). Thus, converging evidence from both above and below ground demonstrate that in the middle Holocene a significant wet phase occurred. Recharge to the aquifer and perennial river flow probably only continued until around 5000 yr BP. Palaeontological evidence including that of large mammals shows that the wadi system was active until at least 3500 BP (Pachur, 1975) after which time the present arid conditions (<20mm yr) commenced with no evidence so far recorded of modern recharge having reached the water table. Thus the present-day configuration of shallow groundwater quality beneath desert areas can still reflect the relict drainage network.

Groundwater recharge during the Holocene, as indicated by ^{14}C data, may have occurred quite widely from flooding as the result of higher river levels or from rivers changing their courses more frequently. In addition to the example from Libya, this is probably the case for the Nile valley in northern Sudan as suggested by Pachur et al. (1987) and can be well demonstrated in the wetter Gezira region where dated palaeochannels (Williams and Adamson, 1980) correspond to areas beneath which groundwaters of Holocene and Late Pleistocene age may be recognised (Malmberg and El Shafi, 1975; Haggaz and Kheirallah, 1988). Another example of selective (Holocene) recharge via no-longer-active river channels is provided from the arid region of north-eastern Kenya (Pearson and Swarzenki, 1974). Similarly there is evidence of a northern extension of the Niger river in Mali during the Holocene and radiocarbon evidence confirms that this led to groundwater recharge (Fontes et al., 1991) to a distance of some 150 km northwest of the present Niger River bend.

The palaeo-modern interface

Sustainable management of groundwater requires an identification of that part of the hydrological cycle that is replenished from modern rainfall. This may prove difficult in terms of classical water balance or in terms of water quality (although it is suggested that a step change in salinity accompanies climate change). The way forward is to use locally suitable methods to estimate the modern recharge (the chloride mass balance approach being suited to regional or diffuse recharge estimation in many semi-arid regions) and to understand this in relation to the hydrogeology and the landscape as a whole. In areas where regional recharge is likely to be negligible (below the 200 mm/yr threshold), selective recharge may take place via high permeability zones especially river channels (Wheater, 1996). Thus in the region of Sudan described above where soil and unsaturated zone salinity indicate negligible recharge, renewable freshwater sustaining the local population is found in linear water bodies beneath the river systems.

An additional example is given of northern Nigeria (Figure 5a, b) where the modern regime of renewable groundwater has been interpreted in relation to the deeper groundwater system (Edmunds et al., 1999). The Quaternary Chad Formation (Figure 5a) contains three aquifers (Upper, Middle and Lower) of which the latter two are confined beneath thick clays. The upper aquifer is more heterogeneous and is overlain by colluvial, lacustrine and dune deposits of Holocene to Recent age. The regional landscape is shown conceptually in terms of the sediments and contained waters, which have different ages. The mean annual rainfall in this area is 320 mm (1970 – 1990). The chloride mass balance has been used to estimate the modern recharge for unsaturated zone profiles both in the dunes and in the colluvial inter-dune and flood plain sediments. A range of 16-30 mm is found (Edmunds et al., 2002). Extrapolating using Cl data for 360 shallow wells a spatially-averaged recharge rate for the region (18,000 km^2) is found to be 43 mm/yr-slightly higher than unsaturated zone estimates. The overall

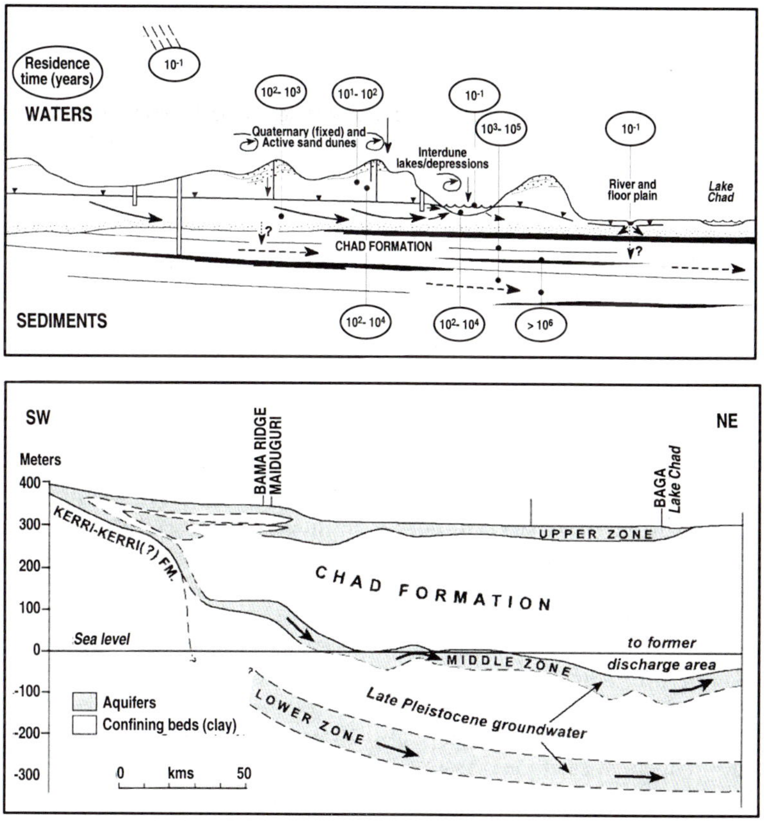

Figure 5. Landscape elements of the Manga Grasslands and adjacent areas of northern Nigeria near to Lake Chad (a) where Holocene dune sands overly the Upper Zone of the Chad Formation and (b) the cross section of the Chad Formation indicating groundwater ages. This illustrates the interface between the shallow aquifer with modern recharge and the beeper palaeowater system.

estimates compare favourably however with water balance estimates for the areas of the grasslands (Carter and Alkali, 1996).

The confined groundwaters in the Middle and Lower Zones of the Chad Formation are of good quality (Cl below 100 mg/l) and have been investigated using radiocarbon, noble gases and stable isotopes. Nearly all groundwaters sampled gave consistent ages in the range 19,000-24,000 yr BP. Noble gas recharge temperatures indicate that the groundwater recharge took place under a climate approx 7°C cooler than at the present day and was confined to a period of high rainfall prior to the glacial maximum in Europe, following which an arid period with dune accretion took place with no further recharge. At the time of recharge surface drainage was to the north east of the present Lake Chad and groundwater was able to discharge into this depression (Pays Bas). At around 12,000 years BP a wetter climate began with the formation of lake Megachad, covering the former discharge area and preventing new recharge to the aquifers of the Chad formation (except at outcrop, discharging to the lake).

With the onset of the semi-arid climate of the modern era, commencing around 4,000 yr BP, recharge once more took place although there is as yet no evidence that the deeper aquifers have been renewed since that time. An unanswered question, mainly due to the absence of intermediate depth boreholes, is to what extent the Upper aquifer is being renewed. Lake Megachad also deposited new clay sediments over some of the area, creating semi-confined or confined conditions in parts of the Upper aquifer, restricting the modern renewable groundwater regime to the areas of superficial sediments.

WATER QUALITY ISSUES

Although palaeowaters found in large sedimentary basins generally constitute a valuable resource of high quality water, a number of issues related to the specific geological or hydrogeological situation, residence times or climate change may need to be considered. The main quality issues are i) salinity, ii) build up over long time periods of harmful accumulations of some elements, and iii) the occurrence of high nitrate concentrations. High pCO_2 in some freshwater aquifers in sandstones may also give rise to severe corrosion problems (Barnes and Clarke, 1969) and scaling problems are also significant in many large sedimentary basins where thermal waters, saturated with regard to certain minerals (notably calcite and gypsum), are produced as a result of deep circulation. These issues need consideration alongside human-induced quality impacts.

Groundwater salinity in semi-arid and arid regions

The build-up of groundwater salinity occurs primarily through the acquisition of solutes from input sources, principally rainfall. The accumulation and the release of this salinity from the near surface takes place naturally as a result of climatic cycles. As described earlier, high salinity in the unsaturated zone may indicate that the minimum threshold for continuous recharge to occur is not exceeded at the present day. Progressive water-rock interaction with evaporite-containing sediments or mixing with formation waters provides significant secondary salinity sources. In coastal regions marine intrusion creates significant salinity problems in regions with high population densities; the problem may also be excerbated by palaeo-marine waters trapped near the coasts from former sea water advances. Edmunds and Droubi (1998) have recently published an overview of salinity in arid zone groundwaters.

Human intervention through groundwater development especially during the past 100 years and in semi-arid and arid regions has however greatly disturbed the distribution of the natural salinity. Thus the prolonged drawdown of groundwater may introduce water from lower permeability strata and, clearance of natural vegetation my enhance recharge but displace salinity from the unsaturated zone. The latter problem of dryland salinity is best illustrated and has been most widely documented from Australia. In Western Australia, Victoria and the Murray Basin saline interstitial waters, representing solute accumulations from rainfall over hundreds of thousands of years and contained in the unsaturated zone of the deeply weathered soil profiles, have been released over a period of decades due to the clearance of native vegetation (Williamson et al., 1987; George et al., 1997). Groundwater levels have risen by more than 30 m and aquifers exist where none existed before clearing. Salinity in streams is increasing at a rate of 10-90 mg/l/yr.

The combined use of chloride and the stable isotopes of water ($\delta^{18}O$, $\delta^{2}H$) provide a powerful technique for studying the evolution of groundwater salinity as well as recharge/discharge relationships (Fontes, 1980). It is also possible to distinguish rainfall derived from geological sources of salinity using Br/Cl ratios as illustrated in Figure 6 using data from various sources world-wide. In the Continental

Intercalaire in Algeria (Edmunds et al., 2003a) along a section from the Saharan Atlas to the discharge area in the Chotts in Tunisia chloride increases progressively from around 200 to 800 mg l^{-1}. The initial Br/Cl ratio is much below sea water ratio and the these values are maintained for some 400m along the flow lines indicating that dissolution of halite predominates (any rainfall salinity signature is overprinted). Towards the discharge area of the Chotts the Br/Cl ratios increase, indicating increasing influence of marine formation waters, possibly the merging of different flow lines. In the East Midlands aquifer at low salinity a slight Br/Cl enrichment relative to sea water is related to changes in Br/Cl in the rainfall in groundwaters dated to the Holocene and late-Pleistocene. In the Milk River aquifer (Fabryka-Martin et al., 1991), uniform Br enrichment occurs throughout the aquifer; this is related to the organic-rich nature of the sandstone, where most of the Br enrichment occurs early in the flow line. In the Great Artesian Basin of Australia the dilute waters derived from the NE of the basin and located along the flow path of some 1200 m all show similar slight Br enrichment and are considered to represent long term inputs of marine aerosol-derived salts; by contrast, the saline waters are Br depleted and have an evaporite origin (Herczeg et al., 1991). In Mali the origins of two sets of groundwaters from the unconfined Azaouad Basin (Continental Intercalaire) may also be separated using the Br/Cl ratios (Fontes et al., 1991). Slightly enriched values in the Saharan (northern) groundwaters are attributed to rainfall origins whereas waters with greater enrichment are likely to be derived from the Niger river (containing organic material) when it extended to the north during the Holocene.

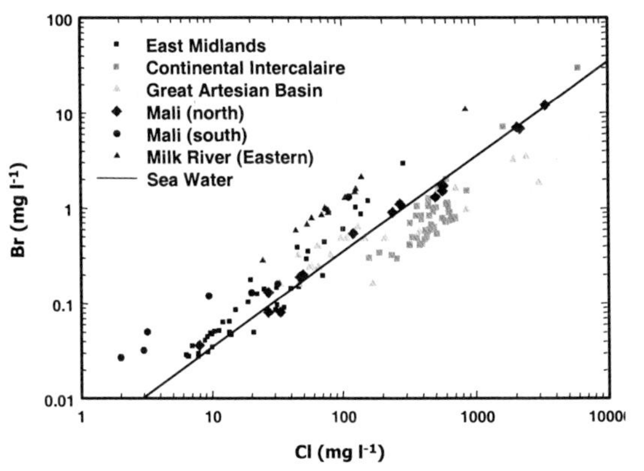

Figure 6. Bromide/chloride ratios in different aquifers indicating different sources of salinity/chloride.

Baseline quality

Groundwater quality in large sedimentary basins and in semi-arid regions to some extent retain features of the initial rainfall composition, modified by evapotranspiration, but also evolves as a result of time-dependent geochemical reactions along flow paths. The baseline or background quality must first be determined before pollution from human sources can be defined (Langmuir 1997; Edmunds et al., 2003a). In addition, elevated concentrations of naturally-occurring solutes (F, Se, As, Cr, NO_3) that may be regarded as harmful may be present. Distinct major ion chemistries will develop according to the geology of the aquifer, with major differences in baseline chemistry being found in particular between carbonate and non-carbonate systems. Sequential changes in chemistry including trace element behaviour, may be recognised in many aquifers and these are strongly influenced by the redox conditions (Smedley and Edmunds, 2002). Aerobic conditions prevail in most semi-arid regions allowing the stabilisation of some metals such as Cr, U, Se, As as oxyanions and these may reach concentrations above international guidelines for drinking water (Edmunds et al., in press).

High nitrate concentrations are widely preserved under aerobic, water-table, phreatic conditions in arid and semi-arid climates of the Sahara and Sahel (Edmunds and Gaye, 1997; Edmunds, 1999) and are also found in other continents (Heaton, 1984; Barnes et al., 1992; Hartsough et al., 2001). The available data from five countries is summarised in Figure 7. In Libya, Niger, Sudan and Algeria many of the dated groundwaters contain nitrate close to or above the 11.3 mg l^{-1} limit. The frequency of high nitrate groundwaters appears to be equivalent through the Late Pleistocene and Holocene. Moreover these concentrations are close to the baseline values observed for the region in modern groundwaters. The

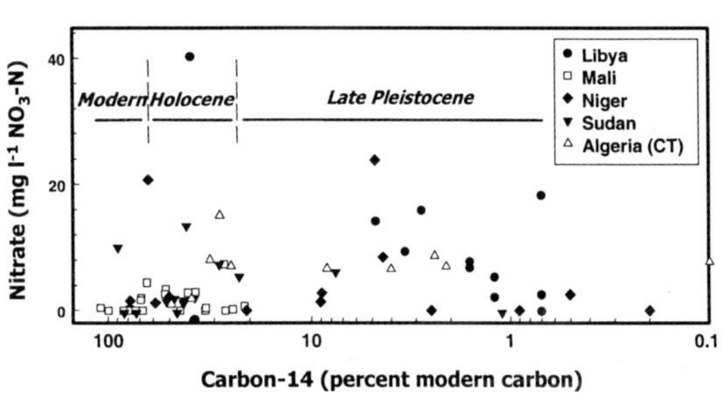

Figure 7. Nitrate concentrations in dated North African groundwaters.

origin of the high nitrate is attributed to natural leguminous vegetation under former more humid conditions; at the present day the cultivation of leguminous crops may also lead to high NO_3 concentrations. Therefore the presence of high nitrate in groundwaters in arid regions is more likely to be due to natural causes rather than pollution from agrochemicals.

Human impacts

Human-induced contamination for large basins is a limited problem at the regional scale and even for unconfined areas, since population densities are generally low and in any case a degree of protection is afforded against diffuse or point source pollution due to low rates of recharge. Drawdown of saline waters from discharge areas (oases and sebkhat areas) as well as contamination from evaporites or saline formation waters in multi-layered aquifers, as discussed above, present hazards to the abstraction of fresher palaeowaters which are found in the most permeable horizons. However, with the widespread aerobic conditions at shallow depths, little attenuation capacity is provided and agricultural chemicals and other wastes will tend to accumulate rather than be degraded. Significant threats come however from irrigation and from wastewater discharges where transit times are likely to be rapid and where return waters will increase salinity. In populated areas unconfined aquifers may be degraded and there is a priority to avoid wholescale losses of upper, renewable freshwater layers, since long delays may occur in the arrival of pollution fronts. Waste-water reuse and artificial recharge are not considered in this paper.

CHALLENGES FOR WATER RESOURCE MANAGEMENT AND POLICY

This paper has drawn attention to examples which illustrate how modern renewable water may be identified and distinguished from non-renewable palaeowater and how in favourable circumstances estimates may be made of the renewable resource, although elsewhere quantification of the renewable resource remains an intractable problem. It is also shown that quantitative issues may be intermixed with quality problems. These may be exacerbated by improper well design or management where fresh and saline waters are mixed together, or where contaminated waters insidiously invade scarce reserves of renewable groundwater or fresh palaeowaters.

Reliable water balance estimation

Understanding the storage and flows of freshwater in the hydrological cycle is absolutely fundamental to discussions of water problems and yet at the start of the 21st century, despite all the scientific work to date, some major knowledge gaps remain as shown by recent reviews (Gleick, 2001; Shiklomanov, 1993; 2000). This deficiency is very real for groundwaters especially in semi-arid regions and, since estimates either do not take account of the non-renewable palaeowater component or, are likely have inaccurate estimates of the renewable modern component. For many semi-arid regions, small renewable resources (a minute fraction of the global renewable estimates) are critical since populations generally in these areas have adapted to low daily consumptions. The question of recharge estimation is therefore an area of concern and current activity by hydrogeologists in general (Simmers, 1997; Scanlon and Cook, 2002).

Improved estimations of recharge

Nevertheless, regional estimation of renewable resources is possible, using chemical mass balance, as illustrated above for Senegal and Nigeria. In Senegal for example, the minimum available resource (obtained using rainfall of 223 mm/yr from the drier years) lies between 13,000 and 1,100 $m^3/km^2/yr$. The target minimum human daily water requirement (drinking, sanitation, bathing, food preparation) has been estimated at 50 litres (Gleick, 1996) or approx 18 m^3/person per year. Thus in sandy areas with 20 mm/yr recharge (13,000 $m^3/km^2/yr$) a population of around 750 living off rain-fed agriculture could in theory be sustained (assuming the aquifer specific yield is close to the porosity of the aquifer). This well-exceeds the present population density in most rural areas, where water is drawn from dug wells, although in some marginal areas described above, demand exceeds renewal. The use of uncontrolled pumping with electric pumps in such areas would severely alter the water balance. Whilst these calculations allow for low rainfall periods, they do not account for losses to deeper aquifers for which the water table aquifer acts also as the only present-day recharge source. Development of the deeper confined or semi-confined groundwaters with communal pumped wells is often favoured in areas formerly served by several village dug wells, but the same amount of renewable water as calculated for shallow phreatic aquifer systems must be viewed as the renewable component.

The modern interface and water quality stratification

To achieve sustainable water resource conditions in semi-arid regions revised groundwater abstraction procedures are needed. These include the construction and design of wells to take account of the natural stratification in recharge history and associated groundwater quality variations. Results of scientific investigations indicate that present day recharge amounts may be very low (negligible below 200 mm mean annual rainfall) and that this renewable groundwater may be restricted to thin freshwater layers overlying fresh or saline palaeowaters. No account is taken of this in well design such that, in addition to over-withdrawal, modern and palaeowaters are often developed from the same well or borehole. It is concluded that this is the general explanation for widespread decline in water tables in many part of the world. It is necessary therefore to explore the vertical stratification, both in groundwater age and quality in the water column, in existing and new wellfields in order better to define and understand the interface between renewable groundwater and fossil water inherited from earlier periods in the history of the hydrological cycle. Hydrogeophysical logging combined with hydrochemical and isotopic investigations can help provide the necessary basis for understanding the age/quality relationships with depth (Buckley et al., 2001).

Management of non-renewable groundwaters

Groundwater exploitation strategies therefore must be based upon clear evidence of the proportion of the renewable and non-renewable components (Custodio et al., 2001). In many cases water resources exploitation is proceeding in the clear knowledge that these are non-renewable and are being mined as in Libya (Pim and Binsariti, 1993; UNESCO, 2002) and in many other countries. An upper limit is foreseen to the life of wellfields and anticipation is made that technological solutions (eg desalination) will come on line to meet long-term needs. Water conservation, especially in the agricultural sector, and public awareness are essential as part of good management practice. Elsewhere, states in semi-arid areas with limited access to fossil fuels and unable to implement energy consuming, large-scale desalination schemes are forced to re-evaluate the water balance and the inventory of renewable vs. non-renewable water resources. Limited modern recharge becomes a vital strategic resource. This is illustrated above in the case of northern Nigeria. The large artesian reserves in the Chad basin, discovered in the 1960s are now after 40 years sub-artesian, deep wells are badly corroded and populations are being forced to reconsider reliance on renewable water resources and traditional water abstraction methods. Sustainable settlement will increasingly involve careful management of the shallow aquifer and available surface waters (subject to drought cycles and abstraction by upstream users) with the remaining artesian water acting as a strategic reserve.

Facing up to realities

A significant gap persists between scientists and policy makers where water is concerned, although there are signs that this gap may be closing (Howsam and Carter, 1996). Nevertheless it usually requires the shock of dramatic falls in water levels and associated social disruption for action to be taken. Sufficient scientific data has accrued over the past few decades to alert global attention to the water crisis that initially is affecting the semi-arid and arid areas of the world. For many countries that are affected, renewable resources, sufficient to sustain livelihoods and natural ecosystems are available, but mis-management leads to a secondary scarcity. A scientific basis for management of water resources remains essential and application of new technologies including diagnostic tools such as stable and radioisotopes, full use of remote sensing and GIS approaches and the development of appropriate integrated modelling approaches (linking hydraulics, water quality and management) need to form part of the resource assessment. For the success of the scientific delivery a commitment is also required to less glamorous activities such as regular field monitoring, establishment of ground truth and data management. A great failure of the scientific onslaught of the twentieth century has been the general lack of follow-through and application of scientific conclusions to water management. Scientific research typically has a three-year life span-the average PhD or donor project-and the results (with some notable exceptions) often lie crumbling unknown in archives. A new improved dialogue between scientists and those involved in water management is now needed.

ACKNOWLEDGMENTS

Although this paper has drawn in part from work conducted at the British Geological Survey, the opinions expressed are those of the author.

REFERENCES

Abou Zakhem, B. and Hafez, R., 2001. General hydroisotopic study of direct infiltration and evaporation process through the unsatrurated zone in Damascus Oasis, Syrian Arab Republic. 131-170 In: Isotope based assessment of groundwater renewal in water scarce regions. TECDOC 1246, IAEA, Vienna.

Allison, G.B. and Hughes, M.W., 1978. The use of local recharge to an unconfined aquifer. *Australia Journal of Soil Research* 16: 181-195.

Allison, G.B. and Hughes, M.W., 1983. The use of natural tracers as indicators of soil water movement in a temperate semi-arid region. *Journal of Hydrology* 60: 157-173.

Allison, G.B., Gee, G.W. and Tyler, S.W., 1994. Vadose-zone techniques for estimating groundwater recharge in arid and semi-arid regions. *Soil Science Society of America Journal* 58: 6-14.

Al-Sagaby, A. and Moallim, M.A., 2001. Isotope based assessment of groundwater renewal and related anthropogenic effects in water scarce areas: sand dunes study in Qasim area, Saudi Arabia. P. 221-229 In: Isotope based assessment of groundwater renewal in water scarce regions. TECDOC 1246, IAEA, Vienna.

Amro, H., Kilani, S., Jawawdeh, J., Abd El-Din, I. and Rayan, M., 2001. Isotope based assessment of groundwater renewal and related anthropogenic effects in water scarce areas: a case study in Jordan. P.171-220 In: Isotope based assessment of groundwater renewal in water scarce regions. TECDOC 1246, IAEA, Vienna.

Barnes C.J., Jacobson, G. and Smith, G.D., 1992. The origin of high nitrate groundwaters in the Australian arid-zone. *Journal of Hydrology* 137: 181-197.

Buckley, D.K., Hinsby, K, and Manzano, M., 2001. Application of geophysical borehole logging techniques to examine coastal aquifer palaeohydrogeology. In: Edmunds and Milne (eds.) Palaeowaters of Coastal Europe; evolution of groundwater since the late Pleistocene. *Geological Society, London, Special Publication* 189: 251-270.

Carter, R.C. and Alkali, A.G., 1996. Shallow groundwater in the northeast arid zone of Nigeria. *Quarterly Journal of Engineering Geology* 29, 341-355.

Barnes, I. and Clarke, F.E., 1969. Chemical properties of groundwater and their corrosion and encrustation effects on wells. *US Geological Survey Professional Paper, 498-D*, 58p.

Clark, I.D. and Fritz, P., 1997. Environmental Isotopes in Hydrogeology. Lewis, Boca Raton, 328p.

Custodio, E., 1992. Hydrogeological and hydrochemical aspects of aquifer overexploitation. In I. Simmers (ed) *Selected Papers on Aquifer Overexploitation: Proceeding sof the international congress IAH, Tenerife, 15-19 April 1991.* Hannover: Heise.

Custodio, E., Edmunds, W.M. and Travi, Y., 2001. Management of coastal palaeowaters. In: Edmunds and Milne (eds.) Palaeowaters of Coastal Europe; evolution of groundwater since the late Pleistocene. *Geological Society, London, Special Publication* 189: 313-327.

Darling, W.G., Edmunds, W.M., Kinniburgh, D.G. and Kotoub, S., 1987. Sources of recharge to the basal Nubian sandstone aquifer, Butana region, Sudan. In *Isotope Techniques in Water Resources Development*: 205-224.Vienna. IAEA.

Degens, E.T., 1962. Geochemische Untersuchungen von Wässern aus den ägyptischen Sahara. *Geologische Rundschau* 52: 625-639.

Edmunds, W.M., 1999. Groundwater nitrate as a palaeoenvironmental indicator. In: Geochemistry of the Earth's Surface. Proceedings of 5th International Symposium on the Geochemistry of the Earth's Surface. Reykjavik, Iceland. Balkema, Rotterdam. 35-38.

Edmunds, W.M. and Wright, E.P., 1979. Groundwater recharge and palaeoclimate in the Sirte and Kufra basins, Libya. *Journal of Hydrology* 40: 215-241.

Edmunds, W.M., Darling, W.G. and Kinniburgh, D.G., 1988. Solute profile techniques for recharge estimation in semi-arid and arid terrain. In: Estimation of Natural Groundwater Recharge. NATO ASI Series. Reidel, Dordrecht, 139-158.

Edmunds, W.M., Darling, W.G. and Kinniburgh, D.G., 1987. Estimation of aquifer recharge using geochemical techniques. Final Report of the Lower Atbara River Basin Project. Report WD/OS/87/1. British Geological Survey.

Edmunds, W.M. and Gaye, C.B., 1994. Estimating the spatial variability of groundwater recharge in the Sahel using chloride. *Journal of Hydrology* 156: 47-59.

Edmunds, W.M. and Gaye, C.B., 1997. High nitrate baseline concentrations in groundwaters from the Sahel. *Journal of Environmental Quality* 26: 1231-1239.

Edmunds, W.M. and Droubi, A., 1998. Groundwater salinity and environmental change. Pp 503-518 In: Isotope Techniques in the Study of past and current Environmental Changes in the Hydrosphere and Atmosphere. IAEA. Vienna.

Edmunds, W. M., Gaye, C.B. and Fontes, J-Ch., 1992. A record of climatic and environmental change contained in interstitial waters from the unsaturated zone of northern Senegal In: *Isotope Techniques in Water Resources Development, 1991,* 533 549 Vienna. IAEA.

Edmunds, W.M. and Milne, C.J., 2001. Palaeowaters of Coastal Europe; evolution of groundwater since the late Pleistocene. *Geological Society, London, Special Publication* 189: 327p.

Edmunds, W.M., Fellman, E. and Goni, I.B., 1999. Lakes, groundwater and palaeohydrology in the Sahel of NE Nigeria: evidence from hydrogeochemistry. *Journal of the Geological Society of London* 156:345-355.

Edmunds, W.M., Fellman, E., Goni, I.B. and Prudhomme, C., 2002. Spatial and temporal distribution of recharge in northern Nigeria. *Hydrogeology Journal* 10: 205-215.

Edmunds, W.M. and Tyler, S.W., 2002. Unsaturated zones as archives of past climates: towards a new proxy for continental regions. *Hydrogeology Journal* 10: 216-228.

Edmunds, W.M., Shand, P., Hart, P and Ward, R.S., 2003a. The natural (baseline) quality of groundwater in England and Wales: a UK pilot study. Science of the Total Environment. (in press).

Edmunds, W.M. Guendouz, A.H., Mamou, A., Moulla, A.S., Shand, P. and Zouari, K., 2003b. Groundwater evolution in the Continental Intercalaire aquifer of Southern Algeria and Tunisia: trace element and isotopic indicators. Applied Geochemistry, 18: 805-822.

Edmunds W.M. Dodo, A., Djoret, D., Gasse, F., Gaye, C.B., Goni, I.B, Travi, Y. Zouari, K. and Zuppi, G.M. (in press).Groundwater as an archive of climatic and environmental change – the PEP III traverse. (PEP III) Conference Aix-en Provence, August 2001.

Fabryka-Martin, J., Whittemore, D.O., Davis, S.N., Kubik. P.W. and Sharma, P., 1991. Geochemistry of halogens in the Milk River aquifer, Alberta, Canada. *Applied Geochemistry* 6: 447-461.

Fontes, J-Ch., 1980. Environmental isotopes in groundwater hydrology. In: Handbook of Environmental Isotope Geochemistry, Fritz, P and Fontes, J Ch (eds). Elsevier, Amsterdam, Vol. 1, 75-140.

Fontes, J-Ch and Edmunds, W.M., 1989. *The use of environmental isotope techniques in arid zone hydrology. Technical Documents in Hydrology.* Paris:UNESCO.

Fontes, J-Ch, Andrews, J N, Edmunds, W M, Guerre, A and Travi, Y., 1991. Palaeorecharge by the Niger River (Mali) deduced from groundwater geochemistry. *Water Resources Research* 27: 199-214.

Gaye, C.B. and Edmunds, W.M. 1996. Groundwater recharge estimation using chloride, stable isotopes and tritium profiles in the sands of northwestern Senegal. *Environmental Geology* 27: 246-251.

Gee, G.W.and Hillel, D., 1988. Groundwater recharge in arid regions:review and critique of estimation methods. *Journal of Hydrological Processes* 2: 255-266.

George, R. et al., 1997. Salinity threatens the viability of agriculture and ecosystems in Western Australia. *Hydrogeology Journal* 5: 6-21.

Gleick, P.H., 1996. Basic water requirements for human activities: meeting basic needs. *Water International* 21: 83-92.

Gleick P.H., 2001. The World's Water 2000-2001. Island Press. Washington DC.

Haggaz, Y.A.S. and Kheirallah, K.M., 1988. Palaeohydrology of the Nubian aquifer northeast of the Blue Nile, near Khartoum, Sudan. *Journal of Hydrology* 99: 17-125.

Hartsough, P, Tyler, S.W., Sterling, J., and Walvoord, M., 2001. A 14.6 kyr record of nitrogen flux from desert soil profiles as inferred from vadose zone pore waters. *Geophysical Research Letters* 28: 2955-2958.

Hassan, F. A., 1997. Nile floods and political disorder in Early Egypt. In: Third Millennium BC Climate Change and Old World Collapse. *NATO ASI Series. Series 1*, 49: 1-24.

Hathway, J.C., Poag, C.W., Valentine, P.C. et al. 1979. US Geological Survey core drilling on the Atlantic shelf. *Science* 206: 515-527.

Heaton, T.H.E., 1984. Sources of the nitrate in phreatic groundwater in the western Kalahari. *Journal of Hydrology* 67: 249-259.

Herczeg, A. L., T. Torgersen, A. R. Chivas, and M. A. Habermehl, 1991, Geochemistry of ground waters from the Great Artesian Basin, Australia. *Journal of Hydrology* 126: 225-245.

Howsam, P. and Carter, R. (eds)., 1996. Water Policy. Allocation and management in practice. Spon. London. 384pp.

Hussain, M.F., 2001. Water flow and solute transport using environmental isotopes and modelling. PP 231-271. In. Isotope based assessment of groundwater renewal in water scarce regions. TECDOC 1246, IAEA, Vienna.

IAEA, 2001. Isotope based assessment of groundwater renewal in water scarce regions. TECDOC 1246, IAEA, Vienna.

Langmuir, D. Aqueous Environmental Chemistry, Prentice Hall. New Jersey, 1997, 600 pp.

Malmberg, G.T. and El Shafi, S., 1975. Application of environmental isotopes to selected hydrologic studies in Sudan. *Unpublished Report.* Vienna. IAEA.

Navada, S.V., Nair, A.R., Sinha, U.K., Kulkarni, U.P. and Joseph, T.B., 2001. Application of isotopes and chemistry in unsaturated zone in arid areas of Rajasthan, India. P. 119-130 In: Isotope based assessment of groundwater renewal in water scarce regions. TECDOC 1246, IAEA, Vienna.

Pachur, H-J., 1974. Geomorphologische Untersuchungen im Raum der serir Tibesti (Zentralsahara). *Berlin Geographische Abandlungen.* 17, 58p.

Pachur, H-J., 1975. Zur spätpleostozäne und holozäne Formung auf der Nordabdachung des Tibest-Gebirge. *Die Erde,* 106, 21-46.

Pachur, H-J., 1980. Climatic history in the Late Quaternary in southern Libya and the western Libyan desert. *In M.J. Salem & M.T. Busrewil (eds) The Geology of Libya*, 781-788. London. Academic Press.

Pachur, H-J., Röper, H-P., Kröpelin, S. and Goschin, M., 1987. Late Quaternary hydrography of the eastern Sahara. *Berlin geowiss. Abh.* A (75/2) 331-384.

Pearson, F.J. and Swarzenki, W.V., 1974. ^{14}C evidence for province, Kenya. *In Isotope Techniques in Groundwater Hydrology* Vol.II 95-109. Vienna. IAEA.

Pim, R.H. and Binsariti, A., 1994. The Libyan great man-made river project: the water resource. *Proceedings of the Institute of Civil Engineers, Water Maritime and Energy.* 106: 123-145.

Phillips, F.M. 1994. Environmental tracers for water movement in desert soils of the American southwest. *Soil Science Society of America Journal* 58: 15-24.

Ruifen, L and Keqin. W., 2001. Environmental isotope profiles of the soil water in loess unsaturated zone in semi-arid areas of China, p. 101-108 In: Isotope based assessment of groundwater renewal in water scarce regions. TECDOC 1246, IAEA, Vienna.

Scanlon, B.R. and Cook, P.G. (eds)., 2002. Hydrogeology Journal Special Issue.

Shiklomanov, I.A. 1998. World Water Resources. A new appraisal and assessment for the 21st century. UNESCO.1998. 37p.

Shiklomanov, I.A., 2000. Appraisal and assessment of world water resources. *Water International* 25: 11-32.

Simmers, I. (editors)., 1997. Recharge of phreatic aquifers in semi-arid areas. IAH International Contributions to Hydrogeology 19. Balkema, Rotterdam. 277pp.

Sonntag, C., Klitsch, E., Lohnert, E.P., Munnich, K.O., Junghans, C., Thorweihe, U., Weistroffer, K. and Swailem, F.M., 1978. Palaeoclimatic information from D and ^{18}O in ^{14}C-dated North Saharian groundwaters; groundwater formation from the past. In: Isotope Hydrology, IAEA, Vienna, p 569-580.

Smedley, P.L. and Edmunds, W.M. 2002. Redox patterns and trace element behaviour in the East Midlands Triassic sandstone aquifer, UK. *Groundwater* 40: 44-58.

Tyler, S.W., Chapman, J.B., Conrad, S.H., Hammermeister, D.P., Blout, D.O., Miller, J.J., Sully, M.J., Ginani, J.N. (1996). Soil-water flux in the southern Great Basin, United States: Temporal and spatial variations over the last 120,000 years. *Water Resources Research* 32: 1481-1499.

UNESCO, 2002. IHP-VI Technical Documents in Hydrology No. 42 (UNESCO, Paris, 2001).

Wheater, H., 1996. Wadi hydrology: Process response and management implications. In: Mourits, L.J.M. et al. (eds.) Wadi Hydrology and Groundwater Protection, UNESCO IHP-V, *Technical Documents in Hydrology* 1: 1-13.

Williams, M.A.J. and Adamson D.A., 1980. Late Quaternary depositional history of the Blue and White Nile rivers in central Sudan. *In M.A.J. Williams and H.Faure (eds) The Sahara and the Nile* 281-304. Rotterdam: Balkema.

Williamson, D.R.et al., 1987. Response of input and output of water and chloride to clearing for agriculture. *Journal of Hydrology* 94: 1-28.

Wood, W.W. and Sanford, W.E., 1995. Chemical and isotopic methods for quantifying groundwater recharge in a regional semi-arid environment. *Ground Water* 33: 458-468.

Zouari, K, Maliki, M.A., Moumni, M. and Aranyossy, J.F., 2001. Chemical (Cl) and isotopic (^{18}O, ^{2}H and ^{3}H) study of the unsaturated zone in the arid region of Nefta, south Tunisia, p. 71-84 In: Isotope based assessment of groundwater renewal in water scarce regions. TECDOC 1246, IAEA, Vienna.

Groundwater Recharge Estimates in Arid Areas Using Channel Morphology and a Simulation Model

WAITE R. OSTERKAMP and LEONARD J. LANE
Geological Survey, United States Department of the Interior Research Project Office, 1675 West Anklam Road, Tucson AZ 85745, U.S.A. E-mail: wroster@usgs.gov

ABSTRACT: In water-deficient areas, ground-water recharge by transmission loss along ephemeral-stream channels typically is an unmeasured water-budget residual. This value is poorly constrained and alternative approaches are sought. By combining channel-morphology techniques inferred from down-channel changes in streamflow with a simple distributed-parameter simulation model, estimates of recharge from transmission loss are made from direct measurements of watershed variables. Inputs for the simulation model, including rainfall, may be determined indirectly using outflow estimates from selected basins or by estimating downstream changes in discharge as related to channel geometry. Recharge also can be estimated by routing discharge of an index storm down-channel, accounting for channel losses, and adjusting for inter-channel recharge during low-frequency storms.

BACKGROUND

Transmission loss is the abstraction, or loss, of flow in ephemeral- or intermittent-stream channels as discharge migrates downstream. Water loss occurs by infiltration into normally dry (unsaturated) sediment forming the channel bed and banks. Thus, part of the abstraction becomes ground-water recharge. In arid and semiarid regions, where most streamflow is ephemeral and is an important process for recharge, most additions to the ground-water reservoir occur by transmission loss (Lane, 1985; Osterkamp et al., 1995). This part of the arid-lands water balance, which becomes available for recharge, typically is very small compared to evapotranspiration but represents a component of precipitation that is potentially usable for water supply. Hence, the measurement of transmission loss and rates of ground-water recharge are of vital concern to water managers in arid and semiarid regions.

Recharge generally is computed as a remainder after other water-budget components are estimated. This approach is not advised in arid areas where rapid evaporation of rainfall accounts for most water loss and a small error in its estimation may lead to a large error in the estimate of recharge. Hence, alternative techniques to evaluate transmission loss directly, and recharge indirectly, have been proposed in recent decades. A method to determine transmission loss from channel measurements as input to a distributed-parameter simulation model is described here. Examples of its use are given for channels of Abu Dhabi (UAE), and the Amargosa River Basin (Western USA).

Previous strategies to evaluate transmission loss directly rely on the downstream routing of inflows or the computation of flow from precipitation records. Examples include loss-rate equations for known inflows to reaches of ephemeral-stream channels (Burkham, 1970a; 1970b), regression equations relating volumes of inflow, outflow, and flow loss (Lane et al., 1971; Sinha and Sharma, 1988), techniques of storage routing as cascading leaky reservoirs (Lane, 1972; Wu, 1972; Peebles, 1975), application of a kinematic-wave model that yields estimates of infiltration (Smith, 1972), and simple differential equations describing rates of transmission loss (Jordan, 1977; Lane, 1980). The approach of Lane (1980) led to the simulation model described here.

APPROACH

This paper describes an empirical water-balance technique to estimate recharge by transmission loss along ephemeral-stream channels. It relies on a distributed transmission-loss model for channelized flow (Lane, 1982; 1985) and evaluates losses using calibrations that relate measures of channel shape and drainage-basin area to discharges, especially flood magnitudes. The input variables for the transmission-loss model, including rainfall, if data are lacking, are geomorphically estimated outflows from selected drainage basins. These input variables are applied to model calculations for other drainage basins. The model inputs for outflow are determined, if feasible, from channel morphology to infer an index event of specified recurrence for the normally flashy streams.

Water Resources Perspectives: Evaluation, Management and Policy. Edited by A.S. Alsharhan and W.W. Wood.
Published in 2003 by Elsevier Science, The Netherlands, p. 281-286.

The index-event volume is related to long-term flow volume as a means of estimating average-annual recharge. Thus, recharge is derived by routing index discharges, if necessary from an index storm, downstream, accounting for channel losses during the flows, and adjusting for interchannel recharge during low-frequency precipitation events.

Theory

The differential equation describing transmission loss in a channel reach is based on the assumptions that (1) the volume of loss in a reach is proportional to the volume of upstream inflow, (2) the loss rate is constant, and (3) the rate of lateral tributary inflow is uniform along the channel length during the flow event. The first-order equation is:

$$dV(x,w)/dx=-wc^*-wk^*V(x,w)+V_l/x \qquad (1)$$

in which V(x,w) is flow volume in a channel segment of length x and mean width w, V_l is volume of lateral inflow, and c^* and k^* are empirical parameters. The solution to equation 1 is:

$$V(x,w)=a^*(x,w)+b^*(x,w)V_u+F^*(x,w)V_l/x \qquad (2)$$

in which V(x,w) becomes the outflow volume, V_u is upstream inflow volume, and $a^*(x,w)$, $b^*(x,w)$, and $F^*(x,w)$ are coefficients. In the absence of lateral inflow, the upstream volume, V_u, is greater than $a^*(x,w)/b^*(x,w)$ or all the inflow is lost in the channel segment and V(x,w) equals 0. If lateral inflow occurs, there must be finite outflow and V(x,w) is greater than 0.

Relations between the coefficients and the parameters are derived as:

$$a(x,w)=[a^*/(1-b^*)][1-b^*(x,w)] \qquad (3)$$

in which a^* and b^* are intercept and slope, respectively, for a unit channel,

$$b^*(x,w)=\exp(-k^*xw) \qquad (4)$$

$$F(x,w)=[1-b^*(x,w)]/k^*w \qquad (5)$$

$$c^*=-k^*a/(1-b^*) \qquad (6)$$

The values of a^*, b^*, and k^* have been determined for specific measures of effective steady-state hydraulic conductivity (K), mean duration of flow to the reach, and mean volume of inflow to the reach (Lane, 1982). Analyses of data from experimental watersheds in Arizona (Murphey et al., 1977) yielded regression equations to estimate mean duration of flow and mean runoff volume from a watershed area.

Hydrographs for 127 flow events in the western USA were used to calibrate the transmission-loss equation (Lane, 1980). For channels of Abu Dhabi and the Amargosa River, separate calibrations were made for a^*, b^*, and k^*, and inflow durations and volumes were based on flow and rainfall records (Osterkamp et al., 1994, 1995). Because the effective saturated conductivity, K, is the constant hydraulic conductivity of bed sediment under natural conditions of entrapped air and suspended sediment, it is as little as a tenth the conductivity estimated with infiltrometers and clear water. Effective conductivities were derived from total losses for an event divided by length and width of the channel reach and the upstream duration of flow. The estimates were averaged over all flow events for a channel segment to derive mean effective hydraulic conductivity. Values of K based on bed-material classes (Lane, 1982), coupled with values obtained during field activities, provided input values of K for the example studies.

Application

The structure of the model considers a channel network from upland areas to an outlet (Fig. 1). Each primary channel receives input from an upland-flow and two lateral-flow areas (Fig. 1A). Secondary channels of complex watersheds receive input from two, or perhaps one, upstream channels and one or two lateral-flow areas (Fig. 1B). Each upland- and lateral-flow area can have different rainfall and infiltration inputs. Thus, the model is distributed in that a watershed may have discrete upland and lateral-flow areas.

Each primary watershed is described by area above the point of interest (A, Fig. 1A), length of channel segment (AB), average channel width, effective hydraulic conductivity of the bed, upstream area, and lateral areas. Outflow from a channel segment becomes input to a downstream segment until the outlet is reached. Transmission losses for the primary channel segments accumulate to yield loss for an entire basin. The surface-water balance is computed by accounting for all rainfall, infiltration on the upland and lateral-flow areas, transmission loss, and runoff.

Streamflow Estimates

Records defining long-term precipitation in arid areas are scarce and discharge records of sufficient length to give reliable flow indices for ephemeral-stream channels are rare. Two indirect methods to evaluate runoff are the use of channel widths to estimate discharge, and development of area-dependent rainfall/runoff relations for arid regions. For both approaches, either mean discharge is estimated or event-based discharge volumes are estimated that allow computations of mean discharge. Estimates for mean discharge are presumed to result in transmission loss and recharge equal to the sum of flow losses that typically occur in a year. Elaboration is in Osterkamp et al. (1994, 1995).

Measurement of channel width to estimate discharge assumes that alluvial channels adjust to the flows that they convey (Leopold and Maddock, 1953). By measuring channel properties at many sites of known discharge, flows of specified frequency are related to geometry by the continuity equation for stream discharge:

$$Q_i = WDV \qquad (7)$$

in which Q_i is instantaneous discharge, W is flow width, D is mean water depth, and V is mean velocity for a flow at the section. Expanding equation 7 to power form yields:

$$Q_i = kW^bD^fV^m \qquad (8)$$

in which k' is a coefficient and exponents b, f, and m depend on drainage-basin properties including the sizes and flux of fluvial sediment. Equation 7 can be expanded to:

$$Q_i = rW^b \qquad (9)$$

$$Q_i = sD^f \qquad (10)$$

$$Q_i = tV^m \qquad (11)$$

in which r, s, and t are coefficients. Because equations 9, 10, and 11 require measurement of width, depth, and velocity for an instantaneous discharge, their use is impractical for ungauged sites in arid areas. To avoid flow-specific measurements, therefore, discharge estimates are made by considering the channel geometry and the sizes of bed and bank material.

The most reliable relations for arid areas yield mean discharge or a flood discharge from channel widths grouped by properties of sediment, climate, or vegetation (Hedman and Osterkamp, 1982). Using channel-width data from numerous gauged sites, power functions relating width and measures of discharge lead to estimates of streamflow at ungauged sites (Hedman and Osterkamp, 1982). Thus, a flow of specified frequency can be estimated for an ungauged site. If basin area is known and other data required to calculate runoff from a 2-year rainfall are estimated, input values can be evaluated iteratively to replicate the field-determined 2-year flow. Because the model computes water balance as volumes of rainfall, runoff, channel loss, and overland loss (largely evapotranspiration), output too includes volume of streamflow, as estimated by geomorphic techniques and expressed as single flow events, rather than as peak discharge.

Where channels are disturbed and cannot be measured well, application of the geomorphic/distributed-parameter simulation approach can be modified to rely on area-dependent rainfall/runoff relations. Data from ephemeral-stream channels of southwest USA (Osterkamp et al., 1995) gave envelope relations between unit flood discharges and basin areas. Included is a curve for $Q_{2.33}$, a flow of return period 2.33 years that, in desert areas, is assumed to result from the mean-annual precipitation event also of return frequency 2.33 years. For ground-water recharge studies in Abu Dhabi, therefore, available precipitation records were used to define an index storm having a return period of 2.33 years.

Flow data from gauged channels of southwest USA suggest that the volume of $Q_{2.33}$ passing an ephemeral-stream site approximates the volume of mean annual discharge. The mean-annual rainfall (index) event, therefore, is among input data to compute a flow volume about that of mean-annual discharge. Input values, including the index-storm magnitude, are adjusted to yield an outflow volume equaling the mean-annual discharge indicated by the curve for $Q_{2.33}$. Simulation-model output includes flow duration, in hours, as determined by an approach suggested by Ardis (1973) and related to basin area by Murphey et al. (1977); recharge, then, is estimated by routing the 2.33-year index-storm discharge down channel, accounting for transmission loss.

EXAMPLES

Streamflow, transmission loss, and recharge were estimated for channels of Abu Dhabi, (UAE), and the Amargosa River Basin, western USA, using the geomorphic, distributed-parameter simulation approach. Summary estimates are in Table 1.

Table 1. Estimated drainage area (DA, km^2), average annual precipitation (P, mm), transmission loss (TL, Mm^3), ground-water recharge (G, Mm^3), and recharge as percent precipitation (G%), Al Ain area (Abu Dhabi) and Amargosa River Basin.

	Estimated drainage area (km^2)	Average annual precipitation (mm)	Transmission loss (mm^3)	Groundwater recharge (mm^3)	Recharge precipitation (%)
Al Ain area	6290	130	46.1	55.3	6.8
Amargosa Basin	3980	180	18.4	20.5	1.6

Determinations of recharge were computed for 17 basins of Abu Dhabi; the computations were summed to estimate total recharge for the area. In the Amargosa Basin, recharge was estimated by determining channel change along a 130-km reach and for 90 km of main tributaries. The distributed-parameter simulation model was also used by Lane (1982) and Nichols (1999).

Abu Dhabi (United Arab Emirates)

Investigations to evaluate and model ground-water resources near Al Ain, Abu Dhabi, UAE, required modeling of a water budget, including estimates of ground-water recharge. Most stream channels (wadis) of the area, west of the Oman Mountains, are altered and thus measures of shape do not indicate discharges. In bottomlands of the mountains, highly braided channels are typically floored with coarse debris and the hillslopes often expose bedrock but little vegetation (Fig. 2). Many flow-input volumes to the simulation model, therefore, were computed from relations of rainfall and runoff for simple or complex watersheds (Fig. 1). Transmission loss was assumed complete downstream, where no channel definition was apparent; hence, all ground-water recharge was inferred to occur above these sites.

The study area had 17 basins discharging water from gaps along the western edge of the Oman Mountains at 26 sites; seven basins had two or more primary channels and were modeled as complex watersheds (Fig. 1B). To generate input volumes of streamflow and rates of transmission loss above and below gaps, a rainfall/runoff relation was developed from isohyetal data for the Oman Mountains:

$$R_a = 0.755P_p - 10.6 \qquad (12)$$

in which R_a and P_p, in mm, are runoff volume and annual peak precipitation (Osterkamp et al., 1995).

Model results of average annual ground-water recharge (millions of cubic meters, Mm^3) were adjusted to account for inter-channel additions by high-intensity storms (Lane and Osterkamp, 1991).

The study indicated that about 55 Mm^3 are recharged annually to the ground-water reservoir in the Al Ain area, about 93% of which occurs above gaps in the Oman Mountains. Computed average-annual recharge, as water depth, was 18 mm in basins of the mountains (2870 km^2) and 1.1 mm on fans (3420 km^2) west of the mountains. Assuming average-annual precipitation of the area to average 130 mm, recharge is nearly 7% of rainfall (Table 1).

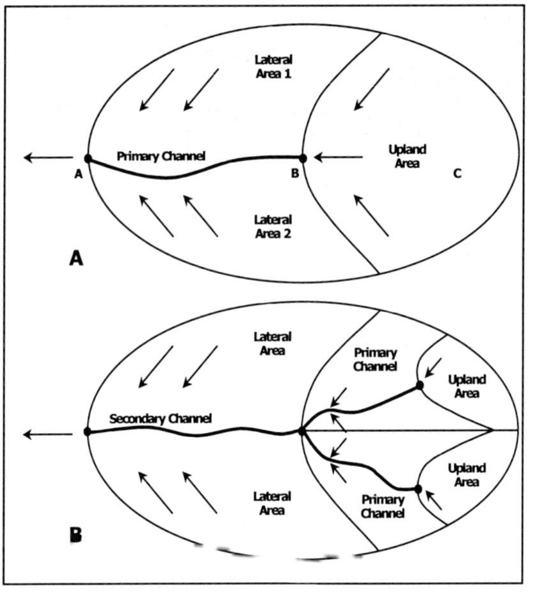

Figure 1. Schematic diagram illustrating physical elements for model input to a simple watershed (A) and a complex watershed (B). Arrows indicate streamflow and runoff directions from contributing parts of the watershed to a primary or secondary channel.

Figure 2. Photograph showing coarse alluvial deposits of a stream channel and bedrock sideslope, western Oman Mountains.

Figure 3. Photograph showing headwaters area of Amargosa River Basin, Nevada, USA.

Amargosa River Basin in Nevada, USA

As part of safety programs for the Nevada Test Site, USA, transmission losses were computed in support of a water-balance model for the Amargosa River (Osterkamp et al., 1994). As for the Al Ain area, the goal was to estimate transmission loss and recharge along ephemeral-stream channels of the basin.

Channels of the Amargosa Basin are generally undisturbed; thus flows based on channel widths, and calibrated by records from 12 stream gauges, were input directly to the simulation model. Accounting for inflows from lateral areas (Fig. 1A) based on rainfall and inferred infiltration potential, discharges were routed along channels of the Amargosa River and its main tributaries. Reductions in mean discharge, as suggested by net decreases in channel size, led to computed volumes of recharge through transmission loss. Total recharge included additions by infiltration in inter-channel areas during low-frequency storms (Lane and Osterkamp, 1991).

Average-annual recharge in the modeled area (3980 km^2) was estimated to be 20.5 Mm^3 (Table 1), suggesting mean-annual recharge of 5.2 mm. Study results indicate that net abstraction of water occurs downstream along most channel reaches. Where a channel is near mountains with elevated rates of rainfall and runoff, the mean rate of streamflow may increase and thus recharge by transmission loss also increases. As for the Al Ain area, recharge rates appear to be much greater in mountainous parts of the basin (Fig. 3) than in the lower interior. About 90% of recharge in the Amargosa Basin occurs by abstraction along ephemeral-stream channels, the rest results from infrequent rainfalls and infiltration through soil of upland and inter-channel areas.

CONCLUSION

Model results for ground-water recharge in the Al Ain area and the Amargosa River Basin differ much but may not be contradictory. Computed depths of average annual recharge for the two areas are, respectively, 8.8 and 5.2 mm despite evidence that mean annual rainfall in the Al Ain area is 50 mm less than in the upper Amargosa Basin. Al Ain watersheds have few plants to transpire soil moisture, and headwater areas of the Oman Mountains have much bare-rock surface from which runoff readily flows to alluvium of high hydraulic conductivity. The upper Amargosa Basin has greater vegetation cover, less bare-rock area to enhance runoff, and more soil areas of lower hydraulic conductivity than does the Al Ain area. Because the transmission-loss model accounts for rainfall, runoff, and infiltration, results obtained for the two example areas may offer reasonable inputs for the larger water-budget studies.

The simulation model upon which ground-water recharge was computed for the Al Ain and Amargosa areas are event-based and originally unintended for long-term estimates of recharge. The work described here, with field-based (geomorphic) techniques for its use, permits integration to larger scales of space and time and, therefore, inputs to models of large ground-water systems. Extension of these techniques could include addition of sediment-transport algorithms to estimate sediment discharge by otherwise unsampled flows in ephemeral-stream channels.

REFERENCES

Ardis, C.V., 1973. A double triangular model for generating storm hydrographs. *Proceedings of the First World Congress on Water Resources*, 4: 350-360.

Burkham, D.E., 1970a. A method for relating infiltration rates to streamflow rates in perched streams. *United States Geological Survey Professional Pap*er 700-D: 266-271.

Burkham, D.E., 1970b. Depletion of streamflow by infiltration in the main channels of the Tucson Basin, southeastern Arizona. *United States Geological Survey Water-Supply Pap*er 1939-B.

Hedman, E.R. and W.R. Osterkamp, 1982. Streamflow characteristics related to channel geometry of streams in western United States. *United States Geological Survey Water-Supply Pap*er 2193.

Jordan, P.R., 1977. Streamflow transmission losses in western Kansas. *Journal of Hydraulics Division* ASCE 103: 905-919.

Lane, L.J., 1972. A proposed model for flood routing in abstracting ephemeral channels: Hydrology and Water Resources in Arizona and the Southwest. *Arizona Academy of Science, Hydrology Section* 2: 439-453.

Lane, L.J., 1980. Transmission losses, National Engineering Handbook, NEH 4: USDA SCS.

Lane, L.J., 1982. Distributed model for small semiarid watersheds. *Journal of Hydraulics Division* 108: ASCE, 1114-1131.

Lane, L.J, 1985. Estimating transmission Losses. Proceedings of Development and Management Aspects of Irrigation and Drainage Systems: San Antonio, Texas. *Journal of Irrigation and Drainage Division*, ASCE, 106-113.

Lane, L.J., M.H. Diskin, and K.G. Renard, 1971. Input-output relationships for an ephemeral-stream channel system. *Journal of Hydrology* 13: 22-40.

Lane, L.J. and W.R. Osterkamp, 1991. Estimating upland recharge in the Yucca Mountain area. *Drainage and Irrigation Division Proceedings*, ASCE, 170-176.

Leopold, L.B., and T. Maddock, Jr., 1953. The hydraulic geometry of stream channels and some physiographic implications. *United States Geological Survey Professional Pap*er 252.

Murphey, J.B., D.E. Wallace and L.J. Lane, 1977. Geomorphic parameters predict hydrograph characteristics in the Southwest. *Water Research Bulletin* 13: 343-348.

Nichols, M.H., 1999. The spatial distribution of sediment sources and sinks at the watershed scale in semiarid areas. Unpublished PhD dissertation Las Cruces: New Mexico State University.

Osterkamp, W.R., L.J. Lane and C.S. Savard, 1994. Recharge estimates using a geomorphic/distributed-parameter simulation approach, Amargosa River Basin. *Water Resources Bulletin* 30: 493-507.

Osterkamp, W.R., L.J. Lane and C.M. Menges, 1995. Techniques of ground-water recharge estimates in arid/semi-arid areas, with examples from Abu Dhabi. *Journal of Arid Environments* 31: 349-369.

Peebles, R.W., 1975. Flow recession in the ephemeral stream. Unpublished PhD dissertation Tucson: University of Arizona.

Sinha, B.P. and S.K. Sharma, 1988. Natural groundwater recharge estimation methodologies in India: Estimation of Natural Groundwater. *Dordrecht: Reidel*. 301-311.

Smith, R.E., 1972. Border irrigation advance and ephemeral flood waves. *Journal of Irrigation and Drainage Division* 98, ASCE: 289-307.

Wu, I-pai, 1972. Recession flows in surface irrigation. *Journal of Irrigation and Drainage Division* 98, ASCE: 77-90.

Enhanced Salinisation

ANDREW S. GOUDIE
School of Geography and the Environment, University of Oxford, Mansfield Road, Oxford, OX1 3TB, UK
E mail: andrew.goudie@geog.ox.ac.uk

ABSTRACT: This paper reviews the five main classes of enhanced salinity: irrigation salinity, dryland salinity, urban salinity, salinity brought about by inter-basin water transfers, and coastal zone salinity. It also draws attention to potential future changes in salinity that could be brought about by climate change and rising sea-levels.

INTRODUCTION

Human activities cause enhanced or secondary salinisation in drylands in a variety of ways (Goudie and Viles, 1997). In Table 1 these mechanisms are grouped into five main classes: irrigation salinity; dryland salinity; urban salinity; salinity brought about by inter-basin water transfers; and coastal zone salinity. Increasing salinity has a whole series of consequences that include a reduction in the availability of potable water (for humans and/or their stock), deterioration in soil structure, reduction in crop yields and decay of engineering structures and cultural treasures. It is, therefore, a major environmental issue.

Human-induced salinisation affects about 77 million ha on a global basis, of which 48 million ha are in susceptible drylands (Middleton and Thomas, 1997) (Table 2).

Table 1: Enhanced Salinisation.

1	Irrigation salinity	• Rise in groundwater • Evaporation of water from fields • Evaporation of water from canals & reservoirs • Waterlogging produced by seepage losses
2	Dryland salinity	• Vegetation clearance
3	Urban salinity	• Water importation and irrigation • Faulty drains & sewers
4	Inter-basin water transfers	• Mineralisation of lake waters • Deflation of salts from desiccating lakes
5	Coastal zone salinity	• Over-pumping • Reduced freshwater recharge • Sea-level rise • Ground subsidence

Table 2: Global extent of human-induced salinisation in the susceptible drylands (million hectares).

Continent	Light	Moderate	Strong	Extreme	Total
Africa	3.3	1.9	0.6	-	5.8
Asia	10.7	8.1	16.2	0.4	35.4
South America	0.9	0.1	-	-	1.0
North America	0.3	1.2	0.3	-	1.8
Europe	0.8	1.7	0.5	-	3.0
Australasia	-	0.5	-	0.4	0.9
Global total	16.0	13.5	17.6	0.8	47.9

Source: GLASOD; Middleton and Thomas 1997, Table 4.17

Irrigation Salinity

In recent decades there has been a rapid and substantial spread of irrigation across the world. The irrigated area in 1900 amounted to less than 50 million hectares. By 2000 the total area amounted to five times that figure. During the 1950's the irrigated area was increasing at over 4% annually, though this figure has now dropped to only about 1%. This spread of irrigation has brought about a great deal of salinisation and waterlogging.

The amount of salinised irrigated land varies from area to area (Table 3), but in general ranges between 10 and 50% of the total. It needs to be noted, however, that there is a considerable range in these values according to the source of the data (compare Table 4) and this may in part reflect differences in the definition of the terms 'salinisation' and 'waterlogging' (see Thomas and Middleton, 1993).

Water Resources Perspectives: Evaluation, Management and Policy. Edited by A.S. Alsharhan and W.W. Wood.
Published in 2003 by Elsevier Science, Amsterdam, The Netherlands, p. 287-293.

Table 3: Salinisation of irrigated cropland.

Country	Percentage of irrigated lands affected by salinisation
Algeria	10-15
Australia	15-20
China	15
Colombia	20
Cyprus	25
Egypt	30-40
Greece	7
India	27
Iran	<30
Iraq	50
Israel	13
Jordan	16
Pakistan	<40
Peru	12
Portugal	10-15
Senegal	10-15
Sri Lanka	13
Spain	10-15
Sudan	<20
Syria	30-35
USA	20-25

Source: FAO data as summarised in *World Resources* (1987, 1988).

Table 4: Global estimate of secondary salinisation in the world's irrigated lands.

Country	Cropped Area (Mha)	Irrigated area (Mha)	Share of irrigated to cropped area (%)	Salt-affected land in irrigated area(%)	Share of salt-affected to irrigated land(%)
China	96.97	44.83	46.2	6.70	15.0
India	168.99	42.10	24.9	7.00	16.6
CIS	232.57	20.48	8.8	3.70	18.1
USA	189.91	18.10	9.5	4.16	23.0
Pakistan	20.76	16.08	77.5	4.22	26.2
Iran	14.83	5.74	38.7	1.72	30.0
Thailand	20.05	4.00	19.9	0.40	10.0
Egypt	2.69	2.69	100.0	0.88	33.0
Australia	47.11	1.83	3.9	0.16	8.7
Argentina	35.75	1.72	4.8	0.58	33.7
South Africa	13.17	1.13	8.6	0.10	8.9
Subtotal	842.80	158.70	18.8	29.62	20.0
World	1473.70	227.11	15.4	45.4	20.0

Source: Ghassemi *et al.* 1995. Table 18

Irrigation causes secondary salinisation in a variety of ways (Rhoades, 1990). Firstly, the application of irrigation water to the soil leads to a rise in the water-table so that it may become near enough to the ground surface for capillary rise and subsequent evaporative concentration to take place. When groundwater comes within 3m of the surface in clay soils, and even less for silty and sandy soils, capillary forces bring moisture to the surface where evaporation occurs. There is plenty of evidence that irrigation does indeed lead to rapid and substantial rises in the position of the water-table, as shown by the data in Table 5. Rates typically range between 0.2 and 3m per year.

Table 5: Increase in level of water tables due to irrigation.

Irrigation project	Country	Water-table height (m)	
		Original depth	Rise per year
Nubariya	Egypt	15-20	2.0-3.0
Beni Amir	Morocco	15-30	1.5-3.0
Murray-Darling	Australia	30-40	0.5-1.5
Amibara	Ethiopia	10-15	1.0
Xinjang Farm 29	China	5-10	0.3-0.5
Bhatinda	India	15	0.6
SCARP 1	Pakistan	40-50	0.4
SCARP 6	Pakistan	10-15	0.2-0.4

Source: Tolba and El-Kholy 1992. Table 5:94

Secondly, many irrigation schemes, being in areas of high temperatures and rates of evaporation, suffer from the fact that the water applied over the soil surface is readily concentrated in terms of any dissolved salts it may contain. This is especially true for crops with a high water demand (e.g. rice) or in areas where, for one reason or another, farmers are profligate in their application of water.

Thirdly, the construction of large dams and barrages creates extensive water bodies from which further evaporation can take place, once again leading to the concentration of dissolved solutes.

Fourthly, notably in sandy soils with high permeability, water seeps both laterally and downwards from irrigation canals so that waterlogging may occur. Many irrigation canals are not lined, with the consequence that substantial water losses can result.

Dryland Salinity

A prime cause of dryland salinity extension is vegetation clearance. The removal of native forest vegetation, by reducing interception and evapotranspirational losses, allows a greater penetration of rainfall into deeper soil layers which causes groundwater levels to rise, thereby creating conditions for the seepage of sometimes saline water into low-lying areas. This is a particularly serious problem in the wheatbelt of Western Australia and in some of the Prairie areas of North America. In the case of the former area it is the clearance of Eucalyptus forest that has led to the increased rate of groundwater recharge and to the spreading salinity of streams and bottom-lands. Salt "scalds" have developed. The speed and extent of groundwater rise following such forest clearance is shown in Figure 1. Until late 1976 both the Wights and Salmon catchments were forested. Then the Wights was cleared. Before 1976 both catchments showed a similar pattern of groundwater fluctuation, but after that date there was a marked divergence of 5.7m (Peck, 1983). The process can be reversed by afforestation (Bari and Schofield 1992). Revegetation policies could also provide increased carbon sinks and so could provide synergistic value (Pittock and Wratt, 2001).

Groundwater levels have increased some tens of metres since clearance of the natural vegetation began. For instance, historical observation indicates that they have increased by up to 30m since the 1880's in south eastern Australia and by about 20m in parts of south west Australia. In some of the upland areas of New South Wales, groundwater levels have increased by up to 60 m over the last 70 to 80 years (http:www.agso.gov.au/information/structure/egg/mb/salinity2.html) while in New South Wales, the area of land affected by dryland salinity is currently reported to be about 120,000 hectares. However, if current land use trends and groundwater rise continues, this figure has the potential to increase to as much as 7.5million hectares by 2050 (http://nccnsw.org.au/veg/context/salinity_fs.html). In Western Australia there is already an estimated 1.8 million ha of farmland that is salt-affected. This area could double in the next 15 to 20 years, and then double again (http://www.agric.wa.gov.au/progserv/natural/trees/salinity/salwa.html) before reaching equilibrium. In all, some 6.1 million hectares has the potential to be affected by dryland salinity.

Dryland salinity is also a major problem on the Canadian prairies. In Alberta, approximately 0.65 million ha are affected by secondary salinity, with an average crop yield reduction of 25%. In Saskatchewan 1.3 million hectares are affected, and in Manitoba 0.24 million hectares (http://www.agric.gov.ab.ca/sustain/soil/salinity/).

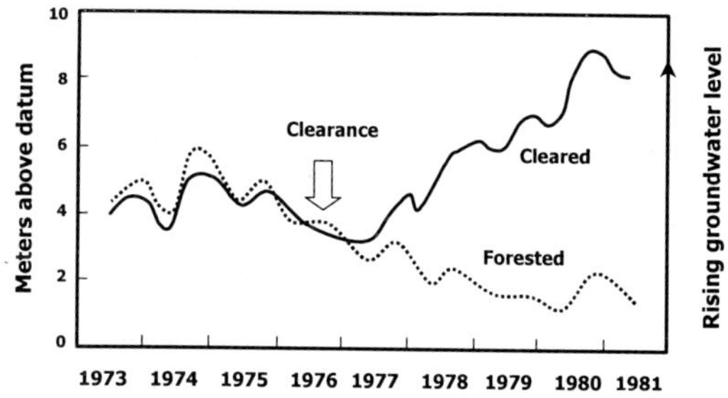

Figure 1. Comparison of hydrographs recorded from the boreholes in Wights (solid line) and Salmon (broken line) catchments in Western Australia. Both catchments were forested until late 1976 when Wights was cleared. (Modified after Peck, 1983, figure 1).

Rising water tables resulting from land use changes are now being identified in other areas. For example, there has been a marked rise in the water table of the Continental Terminal in southwest Niger (Leduc et al., 2001). The rates have been between 0.01 and 0.45 million year^{-1}. The reason for this is the replacement of natural woodland savanna with millet fields and associated fallows. This has promoted increased surface runoff which concentrates in temporary endoreic ponds and then infiltrates to the water-table.

Urban Salinity

Recent decades have seen a great growth in urban areas in drylands. This is, for example, the case with some of the Gulf States of the Middle East. The city of Abu Dhabi had a population of around 8,000 in 1960. By 1984 it had reached 243,000, a 30-fold increase, and by 1995, 928,000, a 116-fold increase. The city of Dubai had a population of around 45000 in 1960. By 1984 this had reached 266,000, representing an almost 6-fold increase and by 1995, 674,000, a 15-fold increase.

Urbanisation can cause a rise in groundwater levels by affecting the amount of moisture lost by evapotranspiration. Many elements of urbanization, and in particular the spread of impermeable surfaces (roads, buildings, car parks, etc) interrupt the soil evaporation process so that groundwater levels in *sabkha* (salt plain) areas along the coast of the Arabian Gulf rise at a rate of 40 cm yr^{-1} until a new equilibrium condition is attained; the total rise from this cause may be 1-2 m (Shehata and Lotfi, 1993). This can require the construction of horizontal drains.

Urbanisation can lead to other changes in groundwater conditions that can aggravate salinisation. In some large desert cities the importation of water, its usage, wastage and leakage, can produce the ingredients to feed this phenomenon. This has, for example, been identified as a problem in Cairo and its immediate environs (Hawass, 1993). The very rapid expansion of Cairo's population has outstripped the development of an adequate municipal infrastructure. In particular, leakage losses from water pipes and sewers have led to a substantial rise in the groundwater level and have subjected many buildings to attack by sulphate- and chloride-rich water. The medieval Qalawun Complex is one example.

There are other sites in Egypt where urbanization has been identified as a major cause of accelerated salt weathering of important monuments. Smith (1986), for example, described the damage to tombs and monuments in Thebes and Luxor in the south of the country.

Inter-Basin Water Transfers

A further reason for increases in levels of salinity is the changing state of water bodies caused by inter-basin water transfers. The most famous example of this is the shrinkage of the Aral Sea, the increase in its mineralisation, and the deflation of saline materials from its surface and their subsequent deposition downwind. Although estimates of the amount of salty material being deflated from the desiccating floor of the sea vary, they agree that some tens of millions of tonnes of salt are being translocated by dust storms (Figure 2) each year (Saiko and Zonn, 2000). The sea itself has had its mineral content increased more than threefold since 1960. Data on recent salinity enhancement in the USA, Asia and Australia is provided by Williams (1999) (Table 6).

Table 6: Enhanced salinity of lake basins.

Lake	Period	Change (gL^{-1})
Mono (California)	1941-1992	48 to 90
Pyramid Lake (Nevada)	1933-1980	3.75 to 5.5
Dead Sea	1910-1990s	200 to 300
Aral Sea	1960-1991	10 to 30
Qinghai (China)	1950s-1990s	~6 to 12
Corangamite (Australia)	1960-1990s	35 to 50

Source: Data in Williams, 1999

Another important illustration of the effects of inter-basin water transfers is the desiccation of the Owens Lake in California. Diversion of water to feed the insatiable demands of Los Angeles has caused the lake to dry out, so that saline dust storms have become an increasingly serious issue (Gill, 1996). Sampling of airborne dust in the area has shown that 20-70% of the dust is soluble salts (Tyler et al., 1997).

Coastal Zone Salinity

Another prime cause of the spread of saline conditions is the incursion of sea water brought about by the over-pumping of groundwater. Salt water displaces less saline groundwater through a mechanism called the Ghyben-Herzberg principle. The problem presents itself on the coastal plain of Israel, in parts of California, on the island of Bahrain, and in some of the coastal aquifers of the United Arab Emirates. A comparable situation has also arisen in the Nile Delta, though here the cause is not necessarily solely groundwater over-pumping, but may also be due to changes in water levels and freshwater recharge caused by the construction of the Aswan High Dam. Figure 3 shows the way in which chloride concentrations have increased and spread in the Llobregat Delta area of eastern Spain because of the incursion of seawater.

A major future threat is posed by sea-level rise associated with global warming and ground subsidence. In 1995 the best estimate for global

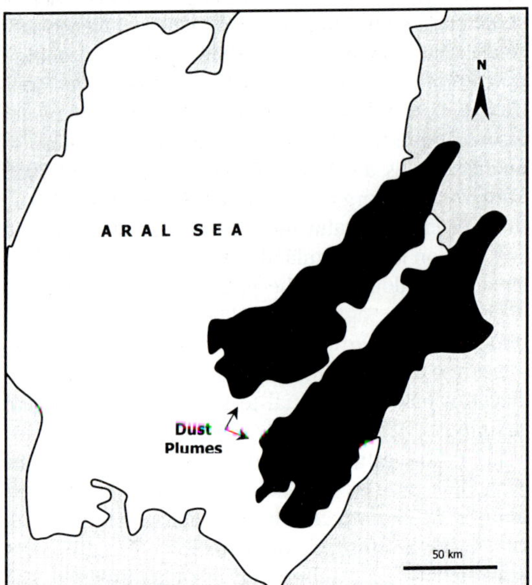

Figure 2. Dust plumes caused by the deflation of salty sediments from the drying floor of the Aral sea as revealed by a satellite image.

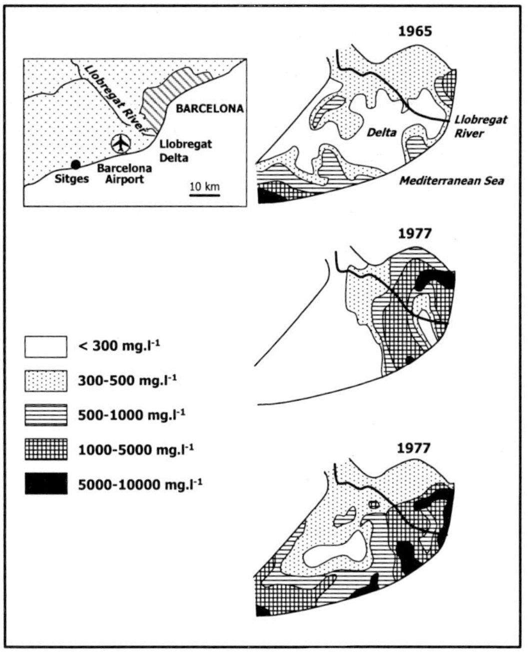

Figure 3. Changes in the chloride concentration of the Llobregat Delta aquifer, Barcelona, Spain as a result of sea water incursion caused by the over-pumping of groundwater. (Modified from Custodio *et al.* 1986).

sea-level change by 2100 compared to the present was a rise of 38-55cm (Houghton et al., 1996), but in subsiding areas this would be exacerbated. Although such rates of rise might appear modest they could have serious implications in sensitive low-lying environments such as deltas, estuaries, atolls and sabkhas. This is especially true where there are coastal freshwater aquifers, as along the Batina coast of Oman or on the seaward margins of the Nile Delta. The balance between fresh aquifers and sea water incursion is a very delicate one. Recent modeling, using the Ghyben-Herzberg relationship, suggests that in the Nile Delta, where subsidence is occurring at a rate of 4.7mm per year, over pumping is taking place, and less freshwater flushing is being achieved by the dammed Nile, a 50 cm rise in Mediterranean level will cause an additional intrusion of salt water by 9.0 km into the Nile Delta aquifer (Sherif and Singh, 1999).

This is made evident when one considers the Ghyben-Herzberg relationship between fresh and saline groundwater (Figure 4). Fresh water has a lower density than salt water, such that a column of sea water can support a column of fresh water approximately 2.5 per cent higher than itself (or a ratio of about 40:41). So where a body of fresh water has accumulated in a reservoir rock or sediment which is also open to penetration from the sea, it does not simply lie flat on top of the salt water but forms a lens, whose thickness is approximately 41 times the elevation of the piezometric surface above sea-level. This is called the Ghyben-Herzberg principle. The corollary of this rule is that if the hydrostatic pressure of the fresh water falls as a result of overpumping in a well, then the underlying salt water will rise by 40 units for every unit by which the fresh-water table is lowered. A rise in sea-level can cause a comparably dramatic alteration in the balance between fresh and salt water bodies. This is especially serious on low islands (Figure 5) (Broadus, 1990).

An example of concern about salinisation following on from sea-level rise is the Shanghai area of China (Chen and Zong, 1999). The area is suffering subsidence as a result of delta sedimentation (at 3 mm year) and groundwater extraction (now less than 10 mm year). It is very low-lying and in some counties the groundwater table is at 1.0 – 1.2m. Thus a rise in sea level will raise the groundwater level, prolong waterlogging and cause a greater dominance of saline water in the deltaic area.

In any one location the causes of seawater incursion may be complex. A particularly good exemplification of this is provided by the coastal plains of northern Australia. Here Mulrennan and Woodroffe (1998) have assessed the potential role of such factors as sea-level change, rainfall variability, boat erosion of creeks, the activities of feral water buffalo and sediment compaction.

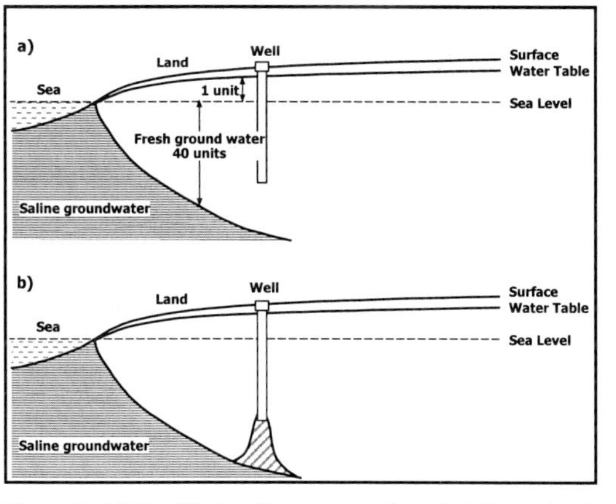

Figure 4. (a) The Ghyben-Herzberg relationship between fresh and saline groundwater (b) The effect of excessive pumping from the well. The diagonal hatching represents the increasing incursion of saline water (Modified after Goudie and Wilkinson, 1977, figure 63).

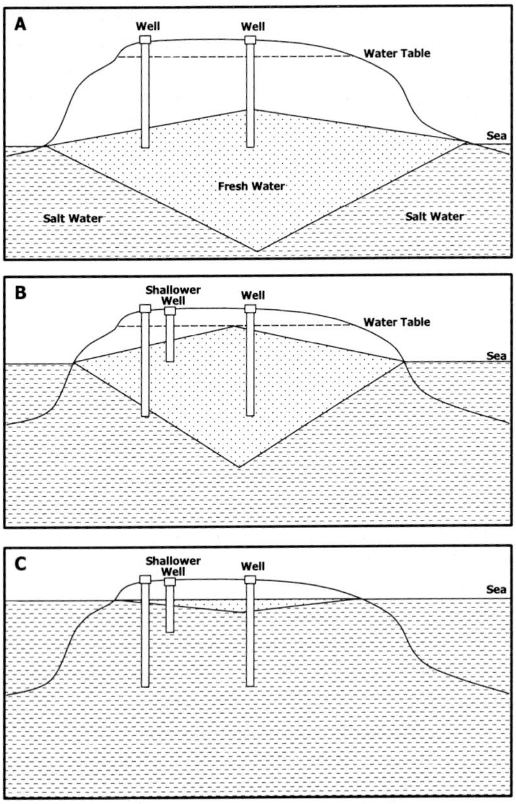

Figure 5. Impact of sea level rise on an island water table. Note: The freshwater table extends below sea level 40cm for every 1 cm by which it extends above sea level. (A) For islands with substantial elevation a 1 m rise in sea level simply shifts the entire water table up 1 m, and the only problem is that a few wells will have to be replaced with shallower wells (B). However, for very low islands the water table cannot rise due to runoff, evaporation and transpiration. A rise in sea level would thus narrow the water table by 40 cm for every 1 cm that the sea level rises (C) effectively eliminating groundwater supplies for the lowest islands. (Modified after Broadus, 1990).

The Future: Climate Change

Global climate change, brought about by the enhanced greenhouse effect could exacerbate current trends in salinisation (Table 7).

First, many arid areas may become still more arid. Moisture deficits will increase and streamflows will decrease, partly because of increased evaporative loss promoted by higher temperatures, and partly because of decreased precipitation inputs. Lakes and reservoirs will suffer more evaporative losses and so could become more saline. In many areas stream flow will decline by 60-70% (Goudie, 2000). Greater moisture stress may lead to an expansion in the quest for irrigation water, there may be less freshwater inputs to flush out estuarine zones, and rates of evaporative concentration of salt may increase. Such tendencies could lead to an increase in salinisation (Imeson and Emmer, 1992; Szabolcs, 1994). Conversely, in some areas falling groundwater levels may follow on from decreased recharge, and this could lead to improvements in the incidence of groundwater-induced salinisation.

Table 7: Global warming and salinisation.

Negative effects	Positive effects
• Increased moisture stress will lead to a greater need for irrigation. • Higher evaporation losses will lead to increased salt concentrations. • Lesser freshwater inputs will reduce estuarine flushing and decrease replenishment in coastal aquifers. • Higher sea-levels will cause change of salinity in estuaries and in coastal aquifers.	• Groundwater levels may fall

In addition, as this paper has shown, sea-level rise poses a potential threat to coastal aquifers and to estuarine ecosystems, and so is a serious concern for areas like the Nile Delta.

REFERENCES

Bari M. A. and Scholfield N. J., 1982. Lowering of a shallow, saline water table by extensive eucalyptus reforestation. *Journal of Hydrology* 133: 273-291.

Broadus J. G., 1990. Greenhouse effect, sea level rise and land use. *Land Use Policy 7*: 138-153.

Chem X. and Zong Y. 1999. Major impacts of sea-level rise on agriculture in the Yangtze Delta area around Shanghai. *Applied Geography* 19: 69-84.

Custodio E, Iribar V, manzane B. A. and Galofre A., 1986. Evolution of sea water chemistry in the Llobregat Delta, Barcelona, Spain. In Proceedings of the 9th Salt Water Chemistry Meeting, Delft.

Ghassemi F, Jakeman A. J. and Nix H. A., 1995 Salinisation of land and water resources. Wallingford: CAB International, 536p.

Gill T. E., 1996 Eolian sediments generated by anthropogenic disturbance of playas: human impacts on the geomorphic system and geomorphic impacts on the human system. *Geomorphology* 17: 207-228.

Goudie A. S and Wilkinson J. C., 1977. The Warm Desert Environment. Cambridge: Cambridge University Press.

Goudie A. S., 2003. The impacts of global warming on the geomorphology of arid lands. In: A.S. Alsharhan, W.W. Wood, A.S. Goudie, A. Fowler & E.M. Abdellatif (editors). Desertification in the Third Millennium. A.A. Balkema, Rotterdam, The Netherlands.

Goudie A. S. and Viles H. A., 1997. Salt weathering hazards. Chichester: Wiley

Hawass Z., 1993 The Egyptian monuments: problems and solutions. In: M. J. Thiel (ed.) Conservation of stone and other materials. London: Spon, 19-25.

Houghton J. T, Meira Filha L. G, Callender B. A, Harris N, Kaltenberg A. and Maskall K. (eds.). Climate Change 1995: The Science of the climate change. Cambridge: Cambridge University press.

Imeson A. and Emmer I. M., 1992. Implications of climate change on land degradation in the Mediterranean. In: L. Jeftic, J. D. Milliman and G. Sestini (eds.). Climate Change and the Mediterranean. London: Arnold, 95-128.

Leduc C. Favreau G. and Schroeter P., 2001 Long term rise in a Sahelian water-table: the Continental Terminal in South-West Niger. *Journal of Hydrology* 243: 43-54.

Mainguet M., 1995. L'Homme et la Secheresse. Paris: Masson.

Middleton N. J. and Thomas D. S. G., 1997. World Atlas of Desertification (2nd Edition). London: Arnold.

Mulrennan M. E. and Woodroffe C. D., 1998. Saltwater intrusion into the coastal plains of the Lower Mary river, Northern Territory, Australia. *Journal of Environmental Management* 54: 169-188.

Peck A. J. 1983. Response of groundwater to clearing in western Australia. Proceedings of International Conference on Groundwater and Man, 327-335.

Pittock A.B. and Wratt D., 2001. Australia and New Zealand. In: J.J.McCarthy, O.F.Canziani, N.A.Leary, D.J. Dokken, and K.S. White (eds.). Climate change 2001: Impacts, Adaption and Vulnerability. Cambridge: Cambridge University Press, 591-639.

Rhodes J. D., 1990. Soil salinity – causes and controls. In: A. S. Goudie (eds.). Techniques for desert reclamation. Chichester: Wiley, 109-134.

Saiko T.A.and Zonn I.S., 2000. Irrigation expansion and dynamics of desertification in the Circum-Aral region of Central Asia. *Applied Geography* 20: 349-367.

Shehata W. and Lotfi H., 1993. Preconstruction solution for groundwater rise in Sabkha. *Bulletin of the International Association of Engineering Geology* 47: 145-150.

Sherif M. M. and Singh V. P., 1999. Effect of climate change on sea water intrusion in coastal aquifers. *Hydrological Processes* 13: 1277-1287.

Smith S. E., 1986. An assessment of structural deterioration on ancient monuments and tombs in Thebes. *Journal of Field Archaeology* 13: 503-510.

Szabolcs I. 1994 State and perspectives on soil salinity in Europe. *European Society for Soil Conservation Newsletter* 3: 17-24.

Thomas D. S. G. and Middleton N. J., 1993. Salinisation : new perspectives on a major desertification issue. *Journal of Arid Environments* 24: 95-105.

Tolba M. K. and El-Kholy O. A., 1992. The world environment 1972-1992. London: Chapman and Hall.

Tyler S. W, Kranz S, Parlange M. B, Albertson J, Katul G. G, Cochram G. F, Lyles B. A. and Holder G., 1997. Estimation of groundwater evaporation and salt flux from Owens Lake, California, USA. *Journal of Hydrology* 200: 110-135.

Williams W.D., 1999. Salinisation: a major threat to water resources in the arid and seni-arid regions of the world. *Lakes and Reservoirs: Research and Management* 4: 85-91.

Timing of Recharge, and the Origin, Evolution and Distribution of Solutes in a Hyperarid Aquifer System

WARREN W. WOOD
U. S. Geological Survey, MS 430, 12201 Sunrise Valley Drive, Reston, VA 20192, USA. E-mail: wwwood@usgs.gov and School of Geography and the Environment University of Oxford, Mansfield Road, Oxford OX1 3TB UK.

ZEIN S. RIZK
Institute of Environment and Water, Ajman University of Science & Technology, P. O. Box 346, Ajman, United Arab Emirates. E-mail: zeinrizk@hotmail.com

ABDULRAHMAN S. ALSHARHAN
Faculty of Science, United Arab Emirates University, P. O. Box 17551, Al Ain, United Arab Emirates. E-mail: sharhana@emirates.net.ae

ABSTRACT: Examination of an aquifer system in the Liwa Crescent/ Bu Hasa area of the Emirate of Abu Dhabi on the southeastern edge of the Rub al Khali, Arabian sub-continent, provides insight into the timing of ground-water recharge and the origin and evolution of solutes in a representative hyperarid area. Ground-water flow in the aquifer system is radially outward from the center of two ground-water mounds, corresponding to two 110 m-thick sand deposits. The isotopic data from ancient ground waters from the Liwa Oasis with its unusual "d" (deuterium excess) of approximately -15 indicate that Holocene moisture derived from previously evaporated water on the surface of the Indian Ocean rather than from the Mediterranean/Arabian Gulf. Such a source is consistent with a summer monsoonal circulation. Hydraulic heads in underlying aquifers are lower than those of the Liwa/ Bu Hasa aquifer; thus, there is no advective water or solute input from these lower units. Solute diffusion from underlying aquifers provides a small solute flux, but it is inadequate to account for the observed solute mass, nor is the ionic ratio consistent with this source. There are no laterally adjacent aquifers; thus, there is no influx of solutes from these sources. Dissolution of the aquifer framework provides for only a minor fraction of most of the observed solutes. Most solutes in this aquifer system are derived from atmospheric precipitation. Salts contained in rain were stored on the surface and in the unsaturated zone during the hyperarid time interval between the ends of the Pleistocene recharge event (26,000 years BP) and the beginning of the Holocene recharge (9000 years BP). During the Holocene recharge era (9000 to 6000 years BP), these stored salts were mobilized and transported to the ground water. The initial solute distribution has been slightly modified as the solutes and water were transported along the flow path. As ground water moves away from the apex of the mound, it encounters a series of interdunal sabkhat. Water is evaporated from the sabkhat leaving soluble chloride and nitrate minerals on the surface and retrograde carbonate and sulfate minerals in the unsaturated zone. When recharge occurs through the interdunal sabkhat, which are the only areas of recharge in the aquifer, soluble salts on the surface, modern ^{14}C, and tritium are added to the aquifer. There is vertical mixing within the aquifer because the density of the recharged water is greater than the density of the ground water in this nearly homogeneous and isotropic aquifer. Relatively insoluble carbonate and sulfate minerals are retained in the unsaturated zone of the interdunal sabkhat, causing a change in the solute ratios downgradient. Only the portion of the aquifer not associated with interdunal sabkhat retains the original solute composition determined by mobilization of stored salts. These findings, although specific to this aquifer, provide a useful model by which to evaluate other aquifer systems in hyperarid environments.

INTRODUCTION

Arid areas occupy 25 percent of the continental surface, excluding polar regions (Thompson and Turk, 1999). As world population is predicted to increase from 6 to 9 billion in the next 50 years (Lutz et al., 2001), these previously marginal areas will undergo water-resource development. Most of this development in arid areas will be ground water; thus, knowledge of the origin, distribution, and evolution of solutes and source and timing of recharge is essential for the implementation of appropriate water-management practices (Gleick, 2002).

Our experimental design consisted of selecting an aquifer system in a hyperarid environment with

Water Resources Perspectives: Evaluation, Management and Policy. Edited by A.S. Alsharhan and W.W. Wood.
Published in 2003 by Elsevier Science, Amsterdam, The Netherlands, p. 295-312.

well-defined chemical and flow properties that minimized the number of potential solute and water sources. Our site-selection criteria were met in an area north and northwest of the apex of the Liwa Crescent on the northeastern edge of the Ar Rub al Khali, or "Empty Quarter," approximately 150 km southwest of the city of Abu Dhabi (Figure 1). The aquifer in this area is essentially homogeneous and isotropic with respect to flow; has a well-defined flow system; is mineralogically homogeneous; experiences low rainfall flux; receives no water or solutes from underlying or adjacent aquifers; lacks vegetation; and is not contaminated on a regional scale by human activity. Furthermore, the aquifer contained wells for sampling solute chemistry, isotopes, and hydraulic head; and cooperative landowners assured access by motor vehicle.

Hydrogeology

High sand dunes (130 - 150 m above mean sea level) with relief up to 50 m dominate the topography of the area. Nearly flat (typically less than 1 m of relief) valley floors (interdunal sabkhat) separate the dunes over much of the area (Figure 2). The surface elevations of these sabkhat are controlled by a dynamic equilibrium between the water table and eolian processes (Wood and Imes, 1995). That is, as the water table drops over geologic time, the sabkhat surfaces are lowered by wind erosion. As water level rises, sediment is trapped, and the sabkhat surfaces rise.

Dunes forming the Liwa aquifer are Quaternary as dated by Optically Stimulated Luminescence (OSL) (Table 1). Dunes are deposited on thin, less than 30 m, essentially flat-lying clastics of Miocene age (Alsharhan et al., 2001), the top of which is near (± 10 m) sea level. These clastics have lower hydraulic conductivity than the Liwa aquifer and contain brackish water. The top of these clastics forms the base of the active flow system of the Liwa aquifer. Below the clastics is a Tertiary and pre-Tertiary sequence of carbonate and evaporite formations with low hydraulic conductivity, containing brines with greater than 50,000 mg/L TDS (Table 2).

The Liwa aquifer is composed largely of quartz, equal amounts of calcite and dolomite, and minor amounts of anhydrite, feldspar, and opaque and heavy minerals (Ahmed et al., 1998; Hadley et al., 1998). Porosity of fine eolian sand typically varies between 35 to 40% (Driscoll, 1986), and laboratory evaluation of three samples averaged 38% with little variance. Grain-size analyses (Figure 3, Table 3) show a high degree of uniformity, similar to the dunes in other areas of the Emirates (Ahmed et al., 1998). Specific yield of this aquifer is assumed to be 0.2, based on analysis of a similar aquifer on a coastal sabkha of the Arabian Gulf (Sanford and Wood 2001). Aquifer hydraulic conductivity is approximately 1.1 m/d (meters/day), based on the arithmetic average of seven regionally distributed aquifer tests in the Liwa area (Wood and Imes, 1995), and on similar tests in eolian sediments near the coast (Sanford and Wood, 2001). Hydraulic conductivities calculated from grain-size analyses (Taylor, 1948) are consistent with these values and range between 0.52 and 1.67 m/d (Table 4). Because of the uniform grain size, the eolian origin, and the general lack of cementation, the hydraulic properties are assumed to be homogeneous and isotropic.

The Quaternary aquifer has a maximum saturated thickness of approximately 110 m that thins away from the center of the dune complex (Figure 4). The ground-water surface forms two mounds reflecting the topography of the area: one approximately 20 km north of the apex of the Liwa Crescent, and one near Bu Hassa, approximately 90 km northwest of the apex (Figure 4). The water table at the top of both of the mounds is approximately 110 m above MSL (mean sea level), and ground-water flow is radially away from the mounds (Figure 4). A north/south-generalized cross section through the Liwa ground-water mound illustrates probable flow patterns (Figure 5). Heads in underling formations are approximately 70 m below those in the Liwa aquifer; thus, there is no advective transport of water or solutes to the Liwa aquifer from this source. The altitude of the paleo-ground-water table at the center was 128 m above MSL at approximately 28,000 years BP and again at 6000 years BP (Wood and Imes, 1995).

Historical records of annual rainfall in the Liwa Crescent area are limited but document extreme rainfall variability averaging 40 mm/y (millimeters/year) (National Atlas, U.A.E., 1993). Potential pan evaporation exceeds 3.5 m/y (National Atlas, U. A. E. 1993). Wood and Imes (1995) and Wood and Imes (this volume) showed that paleo-rainfall between 32,000 and 26,000 years BP and again between 9000 and 6000 years BP was approximately 200 ± 50 mm/y. Rainfall in the intervening hyperarid interval is believed to be approximately 40 mm/y, similar to the present. Estimates of wet and dry intervals are consistent with other findings in the area (McClure 1976; Clarke and Fontes, 1990; Burns et al., 1998; Burns et al., 2001; Neff et al., 2001; and Stokes et al., this volume).

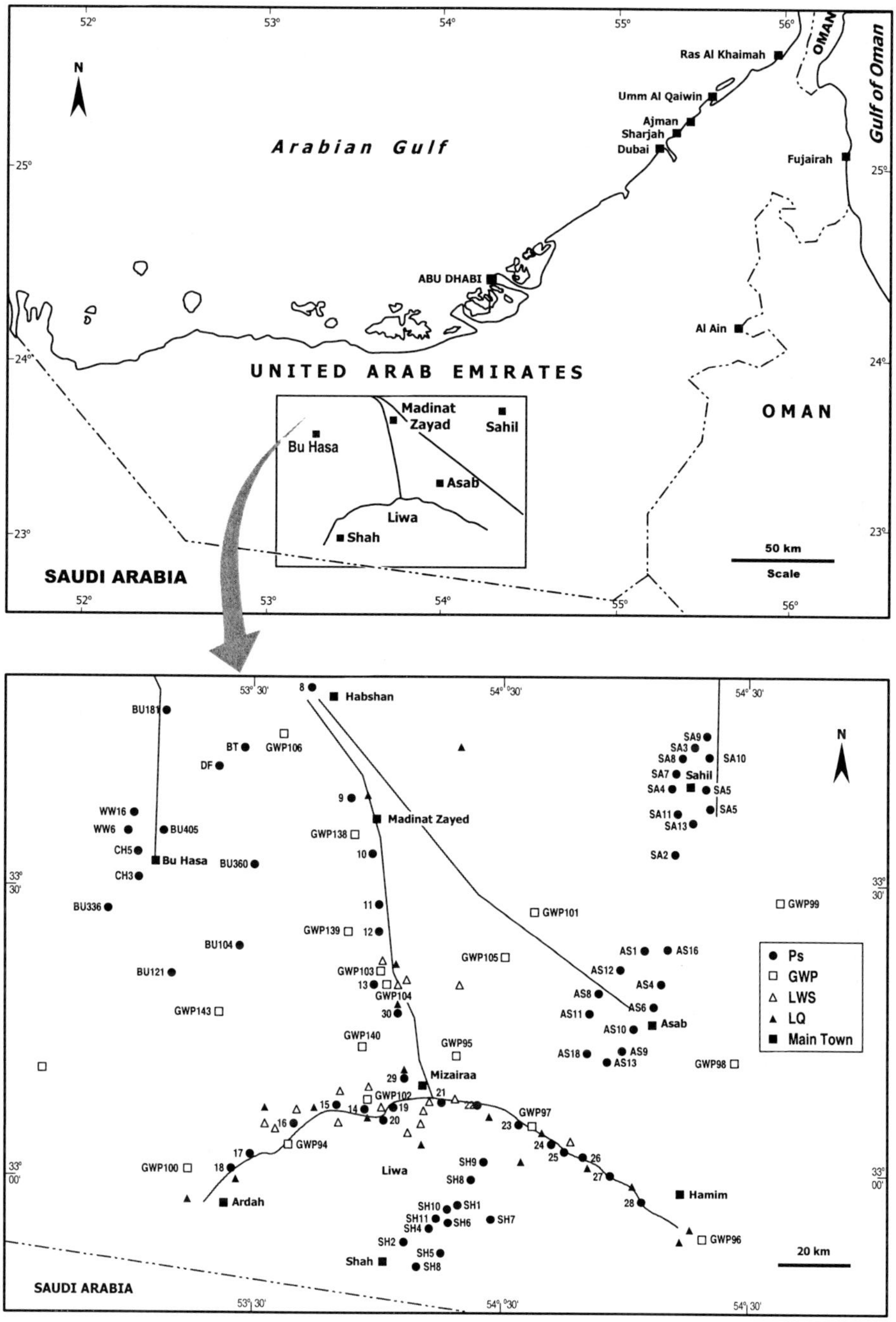

Figure 1. Map of the study areas showing roads, cities, and locations of wells sampled. Present study (● PS), Ground Water Project (□ GWP), Water and Electricity Department (△LWS), and water quality monitoring (▲LQ).

Table 1. Dates in thousands of years (ka) from OSL (Optically Stimulated Luminescence). "Wet" assumes that sample was at or below water table since deposition, "dry" assumes the sample was above water table since deposition. Analyses by A.G. Wintle and G.A.T. Duller, University of Wales, UK.

Sample ID	LAB No.	Location UTM Zone 39			Measured % Water Content	K% Bulk	Th (ppm)	U (ppm)	Equivalent Dose (Gy)	Dry Burial		Wet Burial		Description
										Dose Rate (Gy/ka)	Age (ka)	Dose Rate (Gy/ka)	Age (ka)	
LM-2 (.5-.6m)	124	746340E	2534641N	Zone 39	4.01	1.37	2.06 ± 0.38	1.04 ± 0.12	323.5 ± 28.1	2.40 ± 0.152	134 ± 14	2.023 ± 0.135	160 ± 8	Paleo interdunal sabkha
LM-4 Road Cut (240-257 cm)	125	751528E	2545266N	Zone 39	0.41	1.16	1.61 ± 0.33	0.44 ± 0.10	320.2 ± 39.2	1.950 ± 0.106	164 ± 22	1.640 ± 0.098	195 ± 27	Paleo interdunal sabkha
LM-6 near surface	126	219156E	2547039N	Zone 40	1.95	1.10	1.28 ± 0.32	0.87 ± 0.10	112.4 ± 8.6	2.063 ± 0.119	54 ± 5	1.749 ± 0.109	64 ± 6	Paleo interdunal sabkha
LM-7 near surface	127	223630E	2537931N	Zone 40	2.15	1.22	1.46 ± 0.32	0.70 ± 0.10	87.7 ± 2.1	1.947 ± 0.116	45 ± 3	1.623 ± 0.104	54 ± 4	Paleo interdunal sabkha
190193-4	130	783757E	2579561N	Zone 39	2.51	1.19	0.64 ± 0.18	0.68 ± 0.06	102.9 ± 9.0	2.031 ± 0.118	51 ± 5	1.720 ± 0.108	60 ± 6	Cemented forest dune
210193-2	131	798176E	2545448N	Zone 39	0.21	1.24	0.94 ± 0.24	0.69 ± 0.08	126.6 ± 3.1	2.113 ± 0.132	60 ± 4	1.785 ± 0.118	71 ± 5	Cemented forest dune
210193-4	192	76[illegible]152E	2618558N	Zone 39	3.73	1.22	1.26 ± 0.33	1.42 ± 0.11	243.8 ± 30.2	2.164 ± 0.122	113 ± 15	1.802 ± 0.109	135 ± 19	Cemented forest dune
Near Liwa Hotel	202	781787E	2558287N	Zone 39	1.16	1.36	1.37 ± 0.25	0.81 ± 0.08	7.7 ± 0.4	2.395 ± 0.140	3.2 ± 0.3	2.012 ± 0.127	3.8 ± 0.3	Holocene dune

Table 2. Analyses in mg/L for brine underying the Liwa and Bu Hasa aquifer.

Source	Name	Depth in meters	UTM east	UTM north	UTM zone	Date sampled	Ca	Mg	Na	K	Sr	SiO_2	Alk as HCO_3	Cl	SO_4	Br	NO_3	Dissolved Solids	Density at 20°C	pH	$\delta^{34}S$	$\delta^{2}H$	$\delta^{18}O$
Tertiary	GWP-287A	121	600,328	2,647,191	39	Jan-98	4,800	2,600	35,000	1,300	85	30	109	69,500	2,850	140	1,625	117,857	1.0798	7.9	17.73	-12.6	3.36
Tertiary	GWP-287B	127	600,329	2,647,220	39	Jan-98	4,500	2,500	34,000	1,300	85	25	73	67,600	2,950	140	1,500	114,505	1.0749	7.9	18.24	-14.2	2.95
Tertiary	GWP-288A	89	687,801	2,666,649	39	Jan-98	1,450	950	10,500	300	25	36	55	19,700	4,050	11	33	37,078	1.0258	8.0	19.80	-6.2	2.37
Tertiary	GWP-289	119	777,587	2,655,712	39	Jan-98	1,150	675	5,600	120	23	15	49	10,300	4,000	5	8	21,922	1.0146	7.8	19.84	0.7	2.92
Tertiary	GWP-290A	88	210,989	2,655,365	40	Feb-98	1,500	950	15,500	450	27	20	50	25,000	6,750	15	22	50,251	1.0369	7.4	17.72	5.9	4.09
Tertiary	GWP-291A	142	205,481	2,670,574	40	Feb-98	2,200	1,350	31,500	950	40	18	39	55,100	4,650	20	35	95,867	1.0686	7.8	18.42	5.0	4.64
Tertiary	GWP-292	119	236,909	2,671,579	40	Feb-98	3,600	2,400	48,000	1,500	62	23	34	87,300	3,300	6	33	146,239	1.0971	7.4	17.82	7.5	5.19
Tertiary	GWP-306A	137	208,465	2,668,318	40	Mar-99	2,750	1,250	34,500	1,100	50		64	62,500	4,800	12	50	107,040	1.0820	7.7	-	-	-
Pre Tertiary	Bu Hassa oil field	>1000	-	-	-	Nov. 95	16,400	4,900	60,500	2,100	725	50	180	131,500	500	310	650	217,437	1.1415	6.4	-	-2.8	3.19
Pre Tertiary	Ramaitha oil field	>1000	-	-	-	April-96	14,200	1,600	57,900	1,350	1.000	22	210	110,000	350	620	1	186,553	1.0289	6.2	30.82	-5.3	6.47
Pre Tertiary	Ham Yaphour oil field	>1000	-	-	-	April-96	9,700	1,200	40,000	600	200	20	142	62,000	200	345	0	114,008	1.0291	6.6	30.15	7.1	7.12

Table 3. The results of grain size analysis of six sediment samples from boreholes (depth ≈ 30 m) and sand dunes in the Liwa area. Sample locations are shown on Figure 1.

Sample	GWP-95				GWP-96				GWP-100			
Sieve Size (mm)	Weight (gm)	Weight (%)	Cumulative weight (%)	Finer by weight (%)	Weight (gm)	Weight (%)	Cumulative weight (%)	Finer by weight (%)	Weight (gm)	Weight (%)	Cumulative weight (%)	Finer by weight (%)
1.17	0.00	0.00	0.00	100.00	0.00	0.00	0.00	100.00	0.00	0.00	0.00	100.00
0.85	0.26	0.89	0.89	99.11	0.27	0.95	0.95	99.05	0.50	1.20	1.20	98.80
0.60	0.09	0.31	1.20	98.80	0.56	1.96	2.91	97.09	0.38	0.92	2.12	97.88
0.43	0.23	0.79	1.99	98.01	1.44	5.05	7.96	92.04	0.81	1.95	4.07	95.93
0.30	1.38	4.74	6.74	93.26	1.92	6.73	14.69	85.31	3.09	7.44	11.52	88.48
0.21	6.37	21.90	28.64	71.36	7.98	27.98	42.67	57.33	8.61	20.74	32.26	67.74
0.15	9.24	31.76	60.40	39.60	10.30	36.12	78.79	21.21	12.87	31.00	63.26	36.74
0.06	11.52	39.60	100.00	0.00	6.05	21.21	100.00	0.00	15.25	36.74	100.00	0.00

Sample	GWP-102				GWP-105				Sand Dune			
Sieve Size (mm)	Weight (gm)	Weight (%)	Cumulative weight (%)	Finer by weight (%)	Weight (gm)	Weight (%)	Cumulative weight (%)	Finer by weight (%)	Weight (gm)	Weight (%)	Cumulative weight (%)	Finer by weight (%)
1.17	0.00	0.00	0.00	100.00	0.00	0.00	0.00	100.00	0.03	0.02	0.02	99.98
0.85	0.12	0.80	0.80	99.20	0.84	3.36	3.36	96.64	0.42	0.21	0.23	99.77
0.60	0.06	0.40	1.19	98.81	2.09	8.36	11.72	88.28	19.59	10.01	10.24	89.76
0.43	0.40	2.65	3.85	96.15	3.47	13.87	25.59	74.41	19.27	9.84	20.08	79.92
0.30	0.72	4.78	8.63	91.37	3.14	12.55	38.14	61.86	18.89	9.65	29.73	70.27
0.21	3.02	20.04	28.67	71.33	4.42	17.67	55.82	44.18	29.05	14.84	44.57	55.43
0.15	5.57	36.96	65.63	34.37	4.76	19.03	74.85	25.15	78.50	40.10	84.67	15.33
0.06	5.18	34.37	100.00	0.00	6.29	25.15	100.00	0.00	30.00	15.33	100.00	0.00

Table 4. Calculated uniformity coefficient (C_u) and hydraulic conductivity (K) of six sediment samples from the Quaternary Liwa aquifer in the Liwa area.

Map No.	d_{10}	d_{50}	d_{60}	Uniformity coefficient	K_{Hazen} (m/d)
GWP-95	0.078	0.17	0.19	2.44	0.52
GWP-96	0.100	0.19	0.23	2.30	0.85
GWP-100	0.080	0.18	0.20	2.50	0.54
GWP-102	0.085	0.18	0.19	2.24	0.61
GWP-105	0.090	0.24	0.30	3.33	0.70
Sand Dune	0.140	0.20	0.23	1.64	1.67

d_{10} = grain size that is 10% fine by weight (mm).
d_{50} = grain size that is 50% fine by weight (mm).
d_{60} = grain size that is 60% fine by weight (mm).
K_{Hazen} = hydraulic conductivity calculated by Hazen method (m/day).

Figure 2. Photograph of dunes and interdunal sabkha near Liwa Crescent, Emirate of Abu Dhabi, United Arab Emirates.

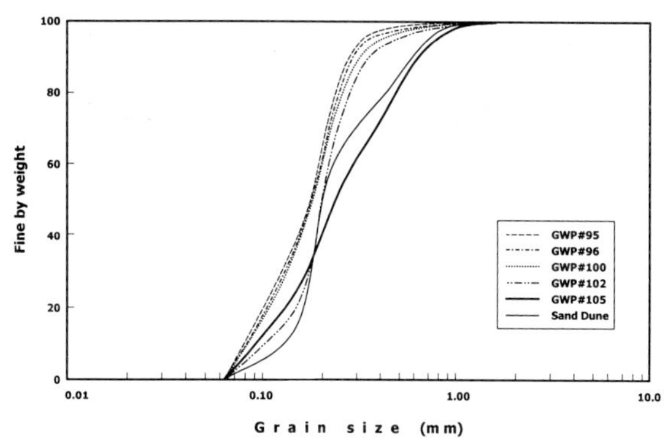

Figure 3. Grain-size analyses of six samples collected from the Quaternary Liwa aquifer.

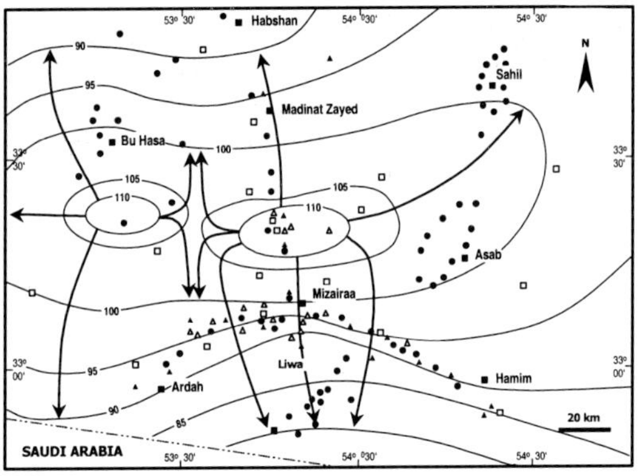

Figure 4. Contour map of hydraulic head and saturated thickness of the Quaternary Liwa aquifer, in meters above mean sea level. Ground water flow path is indicated with arrow headed lines. February 1999.

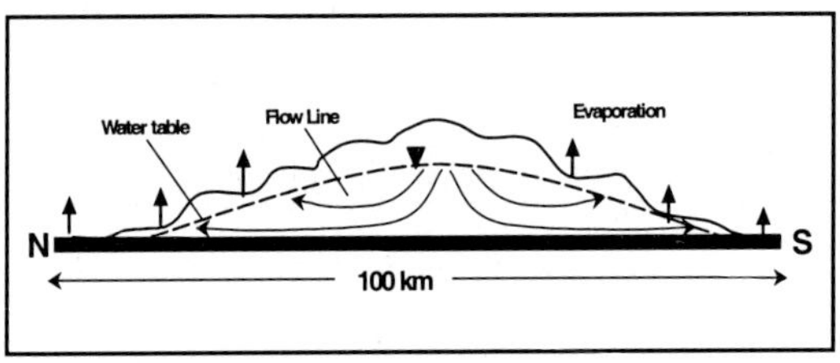

Figure 5. Diagram of a north/south-generalized cross section illustrating probable flow patterns through the Liwa ground-water mound.

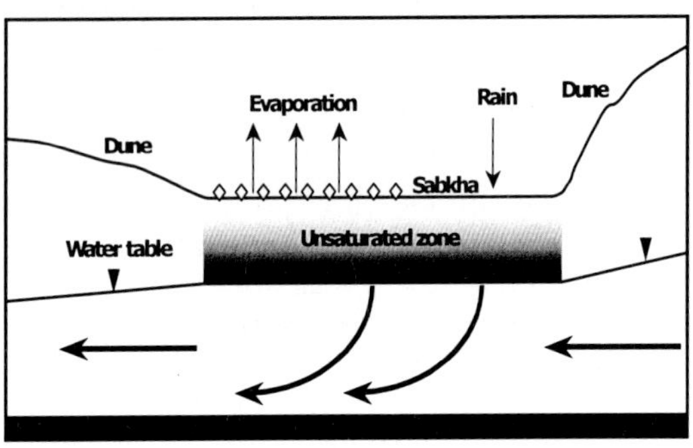

Figure 6. Diagram illustrating recharge and discharge from inland sabkhat in the Liwa area.

Wood and Imes (this volume) show that much of the water with low dissolved solids was recharged approximately 6000 years BP. Under pre-development conditions, most ground water from the Liwa aquifer was discharged by evaporation from the top of the capillary fringe in the interdunal sabkhat (Figure 6). Little evaporative discharge occurs beneath the dunes because the water table is many meters below the ground surface.

Current ground-water recharge to the Liwa aquifer is unknown, but believed to be small. The lack of observed water-level rise on continuous water-level recorders after rainfall events and the lack of tritium in the center of the ground-water mound beneath the dunes (Wood and Imes, this volume) are consistent with the lack of significant ground-water recharge through the dunes. Recharge, however, almost certainly occurs in the 5 to 10% of the area covered by interdunal sabkhat. Sanford and Wood (2001) and Wood et al. (2002) demonstrated that over 90% of rainfall is recharged in a similar hyperarid area on a coastal sabkha of the U. A. E. This unusually high percentage of precipitation recharge occurs because the flat topography reduces runoff, a capillary fringe extending to the surface minimizes the need for "wetting" of the formation after rainfall, and the low thermodynamic activity of water in saline environments reduces evaporation of water on the surface. There is net water loss by evaporation in the sabkhat relative to that gained by recharge. Recharge flux to the system, however, is small because the annual rainfall flux is low, and the interdunal sabkhat cover less than 10% of the surface in this area.

Wood and Imes (1995) demonstrated that recharge must have been approximately 1.4 mm/y during the two late Quaternary wet periods. If it is assumed that the Holocene wet period lasted 3000 years (9000 to 6000 years BP) and that the steady-state rate of recharge was 1.4 mm/y, then the total water recharged was approximately 4.2 m. Accepting a value of 0.2 for specific yield dictates that the water level at the center of the mound must have risen approximately 21 m during this recharge episode. Because the saturated thickness is 110 m, it is concluded that this Holocene water must be superimposed on Pleistocene water.

The dunes have little or no vegetation (Figure 2); however, interdunal sabkhat near the apex of the Liwa Crescent have extensive halophilic vegetation, largely *Zygophyllum hamiense* (common name, Schweint). Interdunal sabkhat are devoid of vegetation as one moves in any direction away from the Liwa and Bu Hassa mounds (Figure 2). This vegetation pattern is assumed to be a function of the concentration of dissolved solids in the ground water (Figure 7).

Source and timing of paleo-precipitation

Isotopes of water provide some insight into the hydrology of this system by identifying the probable

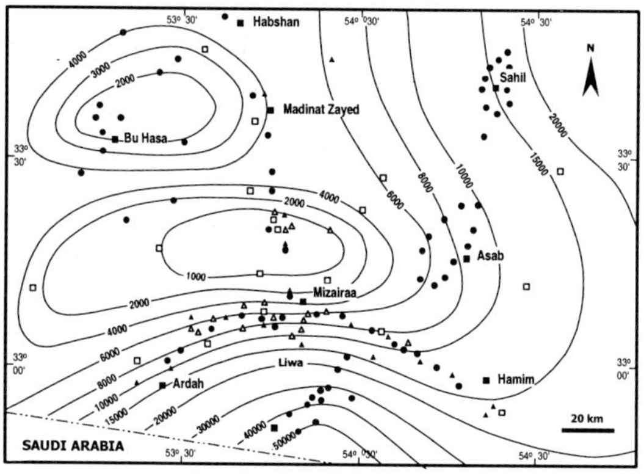

Figure 7. Contour map of total dissolved solids, in milligrams per liter, in the Quaternary Liwa aquifer, February 1999.

source and timing of precipitation. Analysis of carbonates in speleothems, travertines, and lacustrine and capillary-zone deposits indicates that the current hyperaridity prevalent across much of the Arabian Peninsula has been punctuated by past episodes of increased rainfall. Absolute dating of these and other climate-sensitive deposits documents episodes of increased precipitation during the late Pleistocene (32-26 ka) and Holocene (9-6 ka) (McClure, 1976; Clarke and Fontes, 1990; Wood and Imes, 1995; Burns et al., 1998; Burns et al., 2001; Neff et al., 2001; and Wood and Imes, this volume). Although the age of these deposits and water is well established, the source of the moisture and seasonal timing of the precipitation remains elusive.

Summer monsoon-associated precipitation during the mid-Holocene in southern Arabia is thought to be the result of northward extension of the ITCZ (intertropical convergence zone) (McClure, 1976; Clarke and Fontes, 1990; and Burns et al., 1998) with moisture derived from the Indian Ocean during the summer. This contrasts with intensification of the present "shamals," or northern rain, which has an Arabian Gulf moisture source and winter precipitation. The monsoon hypothesis is consistent with other sectors of the Old World paleomonsoon belt (deMenocal et al., 2002; Zhisheng et al., 2000).

Ground water from Liwa exhibits an unusual "d" (deuterium excess) of approximately -15 relative to the +10 of the GMWL (Global Meteoric Water Line) and the +22 of typical Eastern Mediterranean precipitation (Figure 8, Table 5). Deuterium excess is expressed as

$$d = \delta^2H - 8\,\delta^{18}O, \qquad (1)$$

and δ is: $[(R_X/R_S)-1]\,1000$. R_X is the isotopic ratio of the sample, and R_S is the isotopic ratio of Vienna Standard Mean Ocean Water (VSMOW) in parts per thousand (permil, ‰). Deuterium excess is controlled by the temperature-dependent equilibrium of water vapor with the water source and relative humidity (Merlivat and Jouzel 1979). The Global Meteoric Water Line is expressed as

$$\delta^2H = 8.17\,\delta^{18}O + 10.35. \qquad (2)$$

Table 5. The stable isotopes ^2H and ^{18}O per mil, in groundwater samples from the Quaternary Liwa and Bu Hasa aquifer. Analyses were made at International Atomic Energy Agency (IAEA) Central Laboratories in Vienna, Austria and referenced to VSMOW (Vienna Standard Mean Ocean Water).

Liwa			Bu Hasa		
Map No.	^{18}O (‰)	^2H (‰)	Map No.	^{18}O (‰)	^2H (‰)
8	2.53	-16.7	WW-02	-	-
9	1.96	-2.7	WW-06	2.86	1.3
10	1.85	-3.1	WW-07	-	-
11	1.92	-2.1	WW-11	2.65	-1.4
12	1.94	-1.8	WW-15	-	-
13	2.03	-2.4	WW-16	-	-
14	2.62	0.1	WW-17	-	-
*15	10.51	43.1	WW-18	-	-
16	2.70	-0.4	WW-19	-	-
17	2.70	1.0	WW-20	-	-
18	2.88	2.6	WW-21	-	-
19	2.92	1.7	CH-03	2.44	-3.6
20	2.71	-2.6	CH-05	2.11	-2.7
21	2.78	6.3	CH-06	-	-
22	2.64	3.5	CH-07	-	-
23	2.73	3.3	BU-104	4.27	8.7
24	2.89	4.0	BU-116	-	-
25	2.93	4.7	BU-121	2.31	-0.1
26	2.90	0.2	BU-173	1.73	-4.1
27	2.47	1.4	BU-177	1.89	-3.3
28	2.81	2.8	BU-181	2.31	-1.8
29	2.52	5.2	BU-360	6	8.5
30	2.86	6.2	BU-465	2.26	0

**15 is a water sample collected from an interdunal sabkha.*

It is extremely improbable that stable isotopes from the Liwa ground waters are representative of evaporation of the winter shamals because the delta deuterium values are isotopically much more negative than those that would occur by evaporation of the current "shamals" sampled at Bahrain (Figure 8) (International Atomic Energy Agency, 1992). It is not likely that the water was derived from eastern Mediterranean precipitation because of the large positive value of "d" in this source. Nor is it likely that evaporation of the current Oman monsoons (Clarke et al., 1987) could produce the observed distribution because the slope of Liwa isotope data is greater than (8.2), not less than, these data (5.6) (Figure 8). Physics dictates that the most likely explanation for the large negative "d" value is that the precipitation was derived from a previously evaporated water source (Merlivat and Jouzel, 1979).

With increased precipitation, significant relatively fresh water would have been discharging from the Red Sea to the Indian Ocean. This input could account for the observed isotopic anomaly because it would have been both less dense (due to the freshwater component) and isotopically enriched in ^{2}H and ^{18}O due to evaporation. This water would have floated over cold, upwelling Indian Ocean water and been driven northeastward along the coast of Oman under monsoonal influence, becoming the moisture source for precipitation in southern Arabia during the late mid-Holocene and presumably earlier episodes of monsoonal intensification. The amount of water discharging from the Red Sea would have been greater than at present and dominated the isotopic signature of the surface of the Indian Ocean in this area. The surface-water circulation and stable water isotopes of the Red Sea therefore have been linked by feedback processes to precipitation in the catchment area. An example of negative "d" precipitation can be seen in the shaded area of Figure 8 that represents the first day of precipitation of the new monsoon. That is, this precipitation is derived from surface waters that are presently discharged from the Red Sea or Oman coast. Because there is little current discharge, the volume is small, and the effects are eliminated after the first day or two of monsoonal activity. The remainder of the annual precipitation is derived from upwelling water once the surface water has been swept eastward.

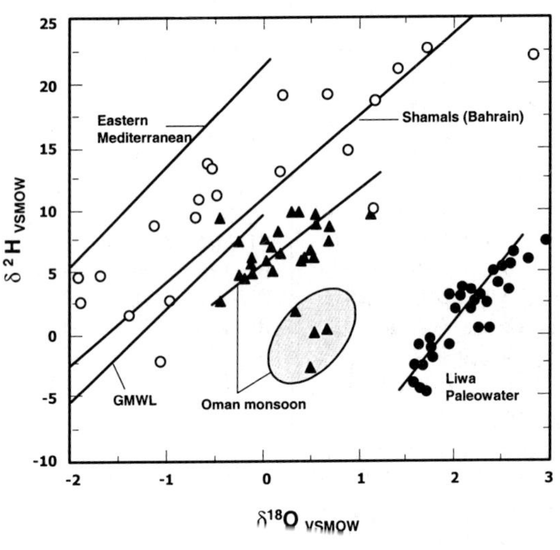

Figure 8. Graph of ^{18}O and ^{2}H ‰ values relative to VSMOW from the Liwa ground water, GMWL (Global Meteoric Water Line), current eastern Mediterranean line, weekly averages of shamal precipitation from Bahrain (International Atomic Energy Agency, 1992), and July, 1987 Oman monsoon (Clarke et al., 1987). The shaded data, collected the first and second days of a new Oman monsoon season, are consistent with the proposed isotope model of previously evaporated water as the source of ancient ground water in the Liwa area.

The isotopic data from ancient ground waters from the Liwa Oasis indicate that Holocene moisture sources derived from the Indian Ocean rather than the Mediterranean/Arabian Gulf. Such a source is consistent with a proposed summer monsoonal circulation (McClure, 1976). This information provides important deterministic constraints on the paleoclimatic and paleoenvironmental dynamism of the Arabian Peninsula.

Solute Geochemistry

Solutes in aquifer systems generally originate from reaction of water with the solid framework of the aquifer, from advective or diffusive transport from underlying or adjacent aquifers, or from atmospheric precipitation. In vegetation-free areas, solutes are removed from aquifer systems by water discharge to the surface, by discharge to adjacent or underlying formations, and by precipitation of solid phases. The Liwa aquifer system was selected because there is no upward or lateral advective flow into the aquifer. Natural solutes in the Liwa aquifer, therefore, must originate by dissolution of the solid aquifer framework, from diffusion from underlying formations, or from atmospheric precipitation.

The distribution of dissolved solutes clearly shows increasing concentration away from the core of the ground-water mound along the direction of flow (Figure 7). Individual ions (Figures 9, 10, and 11; Tables 6, 7, and 8), with the exception of bicarbonate (Figure 10c), follow the same general pattern (see below). Any model for the origin of these solutes and water must address this distribution. Because of the nearly homogeneous and isotropic nature of the aquifer, the solute transport in this system can be estimated by modification of Darcy's Law:

$$q = -k\,(dh/dl), \qquad (3)$$

where q is the Darcy flux (L/T), k is hydraulic conductivity (L/T), and dh/dl is the gradient (dimensionless) where L is length and T is time. To calculate velocity of a particle of water or conservative solute, porosity (θ) (dimensionless) is incorporated thus:

$$v = -k\,(dh/dl)\theta, \qquad (4)$$

where v (L/T) is the particle velocity. Using values of 1.1 m/d for hydraulic conductivity, a gradient of 30 m in 80 km (0.000375) (Figure 4), and a porosity of 0.38, a particle of water or conservative solute will have moved approximately 350 m since cessation of major recharge 6000 years BP. With this small distance, it is clear that the observed distribution of the solutes is not a function of horizontal advective transport.

Solutes from weathering of the aquifer framework

Solutes can be gained or lost from solution by interaction of water with the aquifer framework. Consider the common carbonate/sulfate system. Rainfall acquires bicarbonate and hydrogen ions by the interaction of atmospheric and soil carbon dioxide with water:

$$CO_2 + H_2O = HCO_3^{-1} + H^{+1}. \qquad (5)$$

As the water moves along the flow path, H^{+1} generated by Equation 5 dissolves calcite ($CaCO_3$), adding additional bicarbonate to solution:

$$H^{+1} + CaCO_3 = HCO_3^{-1} + Ca^{+2}. \qquad (6)$$

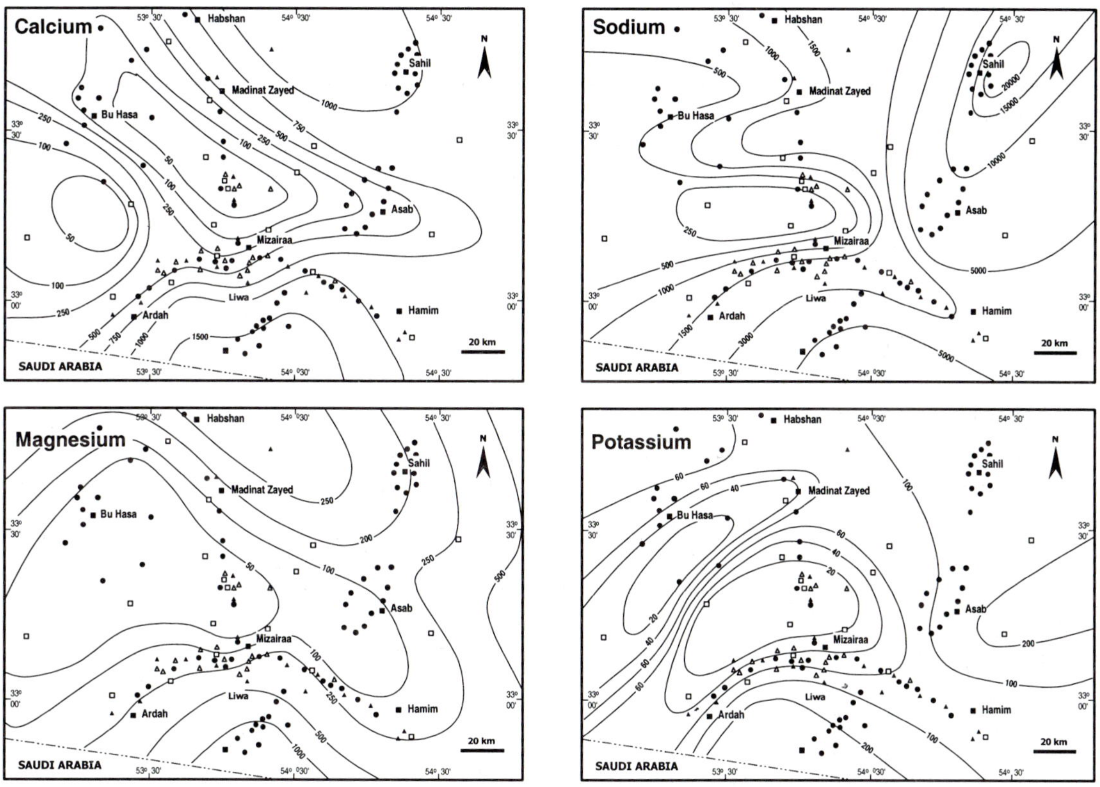

Figure 9. Contour maps of concentrations of major cations in the Quaternary Liwa aquifer, (A) calcium, (B) magnesium, (C) sodium, and (D) potassium, February 1999.

The dissolution of gypsum ($CaSO_4{\cdot}2H_2O$) adds calcium and sulfate ions to the solution:

$$CaSO_4{\cdot}2H_2O + H_2O = Ca^{+2} + SO_4^{-2} + 3H_2O. \quad (7)$$

The addition of calcium (Equation 7) forces thermodynamic supersaturation of the solution with respect to calcite, a mineral of lower solubility, and calcite precipitates. This combined reaction is:

$$CaSO_4{\cdot}2H_2O + HCO_3^{-1} = CaCO_3\downarrow + H^{+1} + SO_4^{-2} + 2H_2O. \quad (8)$$

The net result is that bicarbonate is removed, and the solution becomes more acidic (addition of H^+ ions) in a downgradient direction. Because the concentration of bicarbonate is small, the mass of calcium removed by calcite precipitation is small, and the calcium concentration continues to increase with the increasing gypsum dissolution.

The thermodynamic saturation index (SI), (ratio of the logarithm of equilibrium ion activity product K_{iap} to the equilibrium constant K_{eq}) is used to assess the potential for solubility or precipitation of a mineral (Plummer et al., 1988). The SI of gypsum changes from a negative value (undersaturated) toward 0.0 (equilibrium) as the water moves away from the center of the ground-water mound (Wood and Imes, this volume), consistent with the addition of Ca^{+2} and/or SO_4^{-2} ions. Calcite is essentially at equilibrium in all the samples evaluated, suggesting that equilibrium is obtained before the water has moved far. Thus, relatively little calcite is dissolved.

Trace elements are likely derived from weathering of the aquifer framework. The abduction of the oceanic mantle in the formation of the Oman Mountains provides a likely source of Cr-rich pyroxene (Lippard et al., 1986). Chromium in solution (Cr^{+6} or hexavalent) (Figure 11c, Table 8) almost certainly represents weathering of pyroxene

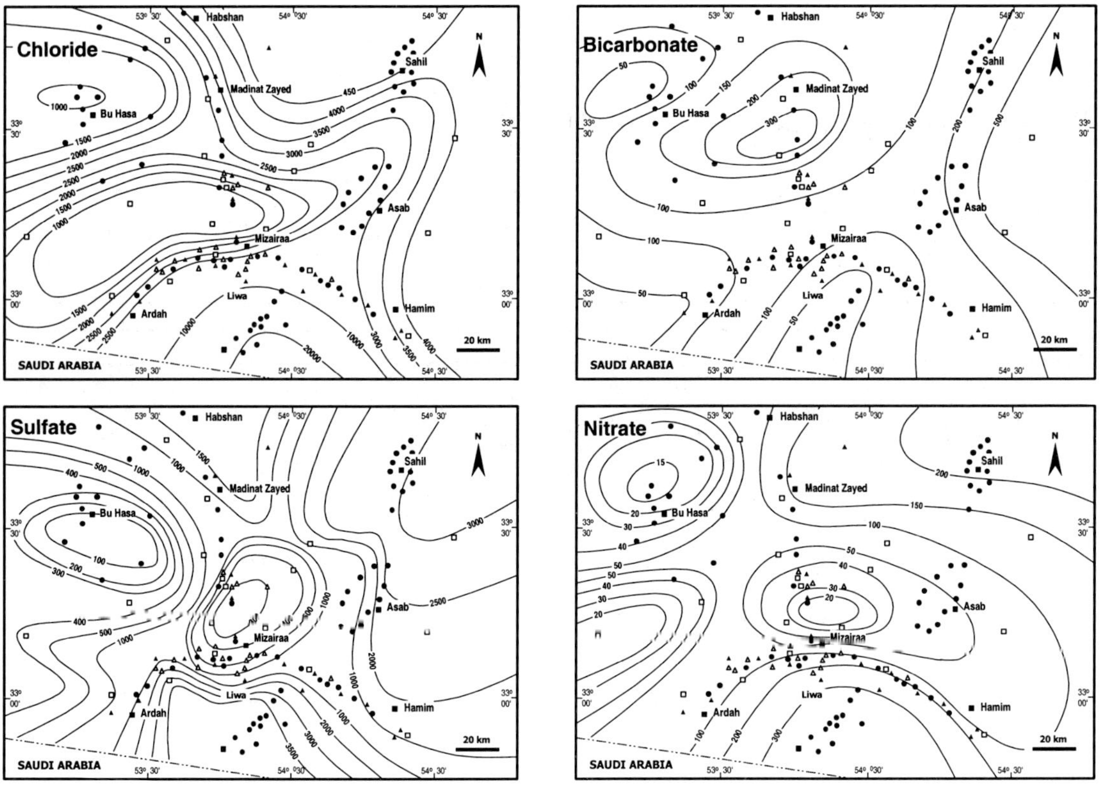

Figure 10. Contour maps of concentrations of major anions in the Quaternary Liwa aquifer, (A) chloride, (B) sulfate, (C) bicarbonate, and (D) nitrate, February 1999.

incorporated in the opaque fraction of the sand. Pyroxene from the ophiolites contains a significant amount of chromium in the +3 valence state, substituting for iron in the crystal lattice. As the pyroxene weathers rapidly in this environment, Cr^{+3} becomes available and is likely oxidized from +3 to +6 by the reduction of manganese. In the hexavalent state, chromium is mobile and moves with water by advective transport. Chromite, although present in significant quantities in the opaque fraction of the Liwa aquifer, is chemically stable in this environment and not likely to dissolve and provide chromium in solution.

Dissolved zinc is probably derived from weathering of the mineral sphalerite [ZnS] in the heavy mineral fraction of the sands. Fluoride likely is derived from the weathering of the mineral fluorapatite [$Ca_5(PO_4)_3F$] in the heavy mineral fraction. Other trace elements are undoubtedly added to the solute mass by weathering of the framework. While weathering provides a rationale for increasing concentration of some elements downgradient, the mass added is small. Nor does weathering address the origin of the large mass of sodium, magnesium, chloride, and nitrate, because there are no minerals in the aquifer framework to provide these elements.

Diffusion

Large concentrations of calcium, sodium, magnesium, and chloride exist in the brines in the underlying formations (Table 2); thus, diffusion of the solutes will occur from these brines to the lower TDS waters of the Liwa aquifer. The basic equation controlling the diffusion into geologic material under one-dimensional, steady-state conditions is given in Fick's First Law:

$$q = -D\ (\Delta C/\Delta X), \qquad (9)$$

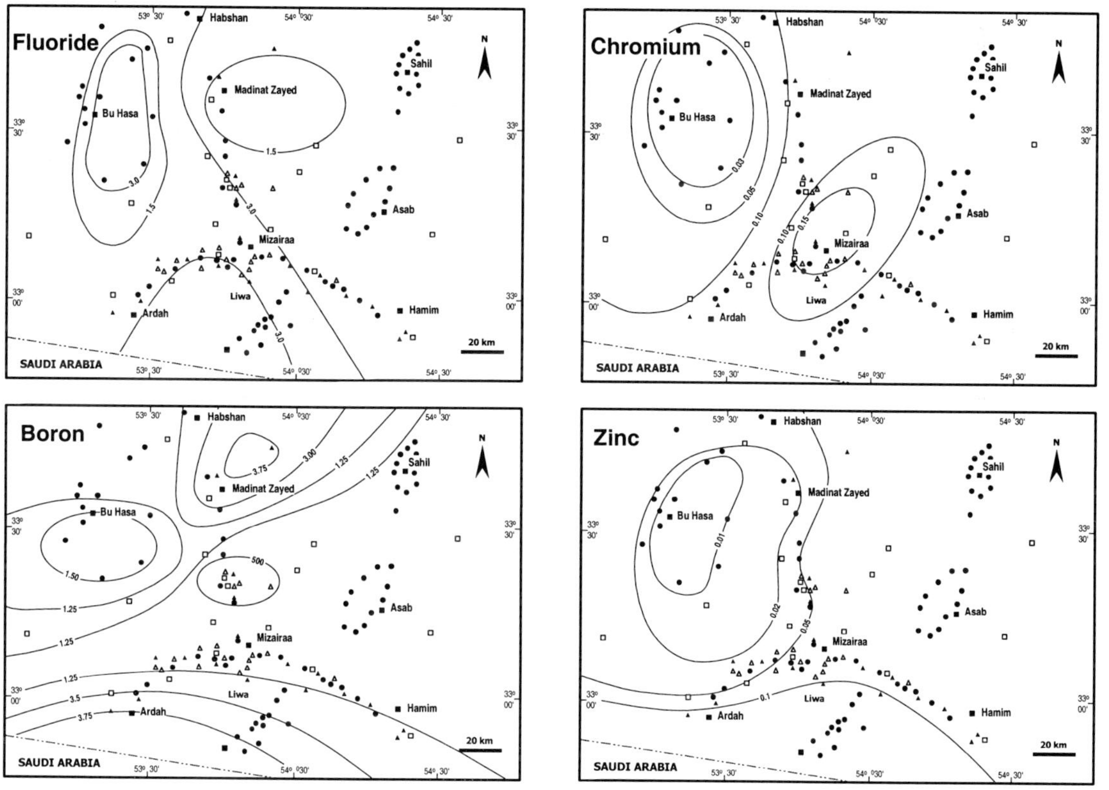

Figure 11. Contour maps of concentrations of selected trace elements in the Quaternary Liwa aquifer, (A) fluoride, (B) boron, (C) chromium, and (D) zinc, February 1999.

Table 6a. Chemical analyses of ground water collected from the Quaternary Liwa aquifer during 1999. Electrical conductance (EC in μS/cm), total dissolved solids (TDS in mg/l), and major ions (mg/l).

No.	pH	EC	TDS	Ca^{++}	Mg^{++}	Na^{+}	K^{+}	HCO_3^{-}	Cl^{-}	SO_4^{-}	NO_3^{-}
8	7.6	18,100	13,600	760	290	2900	93	98	5300	1700	380
9	7.4	10,200	7700	590	160	1600	58	49	2600	1300	160
10	7.4	9300	7000	420	100	1500	55	49	2400	1000	160
11	7.6	8200	5800	170	83	1500	40	85	2300	2500	-
12	7.4	5600	3600	62	35	1100	33	307	1400	1300	-
13	7.5	2300	1500	72	10	440	16	92	480	570	55
14	7.6	9100	6600	290	100	1700	35	92	2200	2200	220
15	7.6	56,700	50,700	480	720	15900	400	171	19800	12000	210
16	7.4	11,500	8600	580	150	2100	100	165	2600	1700	180
17	7.7	9400	6900	590	180	1400	85	159	1800	1900	590
18	7.5	8900	6300	590	170	1400	83	63	1700	2000	-
19	7.7	15,300	11,500	730	260	2800	95	140	4000	2000	320
20	7.4	12,300	9200	620	190	2200	68	123	3200	1500	480
21	7.3	7400	5100	550	200	1000	45	79	1300	1500	22
22	7.5	11,800	8800	710	240	2000	68	73	2800	2000	358
23	7.4	9000	6400	710	190	1500	35	55	2100	1700	-
24	7.6	11,800	8800	780	220	2500	53	123	2800	2100	400
25	7.6	13,100	9900	740	230	2900	65	57	3400	2100	270
26	7.3	13,900	10,400	630	280	2400	55	85	3500	2500	260
27	7.5	6800	4500	270	100	1600	38	110	1700	1200	260
28	7.4	12,400	9300	600	160	2000	80	49	2900	2200	230
29	7.3	3000	1900	140	41	580	21	58	500	870	220
30	7.8	1300	810	21	7	330	13	104	210	190	65

Table 6b. Chemical analyses of ground water collected from the Quaternary Liwa aquifer during February 1999. Electrical conductance (EC in μS/cm), total dissolved solids (TDS in mg/l), and major ions (mg/l).

Map No.	pH	TDS	Ca^{++}	Mg^{++}	Na^{+}	K^{+}	HCO_3^{-}	Cl^{-}	SO_4^{-}	NO_3^{-}
WW-02	7.3	3700	140	65	990	45	150	1600	640	7
WW-06	7.4	2000	71	31	540	23	180	820	320	9
WW-07	7.5	3300	73	54	920	38	240	1500	440	10
WW-11	7.6	5000	240	96	1110	49	91	2500	900	12
WW-15	7.1	2200	63	30	490	15	110	1000	420	9
WW-16	7.4	2900	120	47	640	25	110	1400	550	13
WW-17	6.9	2000	59	25	450	17	110	900	390	17
WW-18	7.6	3800	160	67	990	42	120	1600	730	13
WW-19	7.4	2200	75	27	600	20	130	890	390	14
WW-20	7.5	2500	90	40	540	21	140	1200	470	15
WW-21	7.7	2800	87	37	810	26	110	1100	590	13
WW-22	7.5	2100	43	16	480	12	110	670	400	12
WW-23	7.2	1900	49	20	440	16	130	870	340	10
CH-03	7.6	5400	190	77	850	35	84	2700	1500	13
CH-05	7.6	2400	67	27	490	15	110	1100	550	19
CH-06	7.8	6200	250	102	1200	50	89	2900	1500	12
CH-07	7.5	2600	74	29	650	20	110	1200	520	18
BU-104	-	5800	110	73	1700	50	83	3800	18	-
BU-116	-	1800	21	10	540	15	32	1200	8	3
BU-121	-	1800	4000	5	440	8	160	1000	100	-
BU-173	-	1900	6	3	590	10	160	1100	7	-
BU-177	-	2319	7	0.2	660	10	84	1500	30	-
BU-181	-	4258	137	60	960	35	100	2000	940	15
BU-185	-	1348	10	2	340	9	81	760	130	-
BU-329	-	3135	214	60	590	45	-	1500	580	7
BU-360	-	2829	1	0.1	800	10	310	1500	160	-
BU-377	-	1939	11	7	510	8	77	1100	270	-
BU-465	-	2692	79	33	700	21	120	1200	490	9
AL DHAFRA	-	2770	114	47	650	34	97	1300	550	17
BIN THANI	-	3398	1298	490	770	34	100	1600	650	16

Table 7. Chemical analyses, (in mg/l), of ground water from the Quaternary Liwa aquifer.

A. Chemical analysis of groundwater samples from Groundwater Project wells in the Liwa area.

Map No.	Location	pH	TDS	Ca^{2+}	Mg^{2+}	Na^+	K^+	HCO_3^-	Cl^-	SO_4^-	NO_3^-
GWP-94	Lagain	7.5	16,800	700	350	4200	150	60	7900	3500	83
GWP-95	Mizairaa	7.5	990	65	16	250	12	110	230	350	20
GWP-96	Hameem	7.3	16,200	850	350	4600	83	220	7600	2700	75
GWP-98	Al Hamra	7.7	12,800	720	270	3600	220	510	5600	2500	53
GWP-99	Al Zafrah	7.2	15,100	1100	350	4100	98	1500	5600	3800	130
GWP-100	Huwaila	7.2	1400	62	12	430	33	82	460	330	44
GWP-101	Asab Northwest	7.0	9700	600	200	2700	98	18	3800	2300	53
GWP-102	Bousadeen	7.3	3400	88	54	1100	25	390	1250	670	57
GWP-103	Qafah	7.7	910	9	3.0	330	7	130	300	180	35
GWP-104	Qafah	7.7	1100	120	20	220	17	51	170	540	6
GWP-105	Heea	7.6	3000	87	32	990	44	82	1400	450	41
GWP-138	Jarub	-	3300	85	3.0	850	30	110	1100	500	-
GWP-139	Mussama	7.6	1800	20	5.0	600	19	140	500	300	-
GWP-140	Khurais	7.6	1100	30	6.0	230	13	93	200	200	35
GWP-142	Huwaila West	7.8	2000	75	30	450	26	85	650	300	18
GWP-143	Bu Hasa South	7.4	1300	15	5.0	350	12	120	350	200	-

B. Chemical analysis of ground water from the Asab oil field.

Map No.	pH	TDS	Ca^{++}	Mg^{++}	Na^+	HCO_3^-	Cl^-	SO_4^-
AS-1	7.5	6500	600	160	1400	61	2100	2100
AS-4	8.2	9500	500	200	2500	68	3200	2800
AS-6	8.1	9000	330	90	2800	81	3700	1900
AS-8	7.9	4300	240	75	1200	85	1500	920
AS-9	8.0	6900	330	110	1800	84	2600	1500
AS-10	8.0	9000	370	120	2400	61	3100	2200
AS-11	7.9	6900	250	69	1900	110	2400	1500
AS-12	7.8	6400	290	71	1700	76	2200	1600
AS-13	7.6	9900	360	130	2700	220	3200	2600
AS-15	7.7	18,900	1000	420	5200	41	9000	2800
AS-16	8.2	22,800	1100	450	6600	450	11,300	2600

C. Chemical analysis of ground water from the Sahil oil field.

Map No.	pH	TDS	Ca^{++}	Mg^{++}	Na^+	HCO_3^-	Cl^-	SO_4^-
SA-1	7.6	6000	450	94	1500	85	2100	1600
SA-2	7.4	13,200	770	250	3600	73	5400	3000
SA-3	7.6	14,200	920	280	3700	73	5700	3300
SA-4	7.7	15,500	1200	100	4300	61	6700	3000
SA-5	7.8	16,700	1400	92	4500	61	7400	3100
SA-6	6.8	18,000	930	290	5000	81	7600	3500
SA-7	6.9	27,500	1000	220	8900	61	13,500	3700
SA-8	6.6	13,200	910	300	3300	73	5200	3300
SA-9	6.7	13,200	900	310	3300	73	5200	3200
SA-10	7.2	15,100	1100	350	4100	51	5600	3800
SA-11	7.6	6000	450	94	1500	85	2100	1600

D. Chemical analysis of ground water from the Shah oil field.

Map No.	TDS	Ca^{++}	Mg^{++}	Na^+	HCO_3^-	Cl^-	SO_4^-
SH-1	53,500	1600	1000	9800	44	18,800	2800
SH-2	36,300	1500	920	9600	61	18,100	2700
SH-3	33,100	1300	840	9900	49	17,400	3600
SH-4	47,700	1600	1200	14,800	53	25,900	4200
SH-5	51,000	1800	1200	15,700	42	28,000	4200
SH-6	40,100	1600	880	11,900	40	21,600	2800
SH-7	28,800	1100	150	9500	63	14,600	3300
SH-8	27,200	1300	570	8100	57	18,900	3250
SH-9	40,800	1800	880	1100	49	20,700	2900
SH-10	47,600	2300	1500	13,500	57	26,000	4300
SH-11	26,300	1100	620	7900	55	13,800	2900

Table 8. Chemical analyses of trace chemical constituents (in mg/l) in the groundwater from the Quaternary Liwa aquifer in the Liwa and Bu Hasa areas, February 1999.

Liwa							Bu Hasa						
No.	Sr	F^-	Fe	B	Cr	Zn	No.	Sr	F^-	Fe	B	Cr	Zn
8	-	-	0.06	-	-	0.04	WW2	24	3.0	0.03	2.3	0.02	0.01
9	19	-	0.05	3.1	-	0.03	WW7	-	3.8	0.13	2.1	-	0.01
10	0.3	0.1	0.04	-	-	0.02	WW19	13	2.9	0.07	1.8	0.03	0.01
11	4.5	-	0.03	1.2	-	0.02	WW21	11	3.0	0.10	2.1	0.02	-
12	2.6	-	0.06	0.8	-	0.06	WW6	-	2.7	0.03	1.8	0.03	0.02
13	2.0	3.0	0.03	1.1	0.14	0.02	WW18	21	2.4	0.11	2.3	-	0.02
14	32	3.0	0.03	1.6	0.17	0.02	WW11	31	3.1	0.08	2.0	0.02	0.02
15	19	6.2	0.24	1.1	0.14	0.05	WW17	10	3.0	0.03	1.5	-	0.01
16	18	4.2	0.06	1.0	0.12	0.04	WW16	-	2. 9	0.08	1.1	0.02	0.01
17	14	-	0.04	2.1	-	0.02	WW20	18	2.5	0.07	1.6	0.02	0.01
18	16	-	0.05	3.9	-	0.04	WW15	19	3.0	0.07	1.3	0.02	-
19	21	4.2	0.07	1.3	0.19	0.03	WW23	13	3.0	0.11	0.9	-	-
20	20	6.0	0.05	1.4	0.18	0.02	WW22	-	3.0	0.07	1.0	-	0.02
21	20	4.8	0.04	1.5	0.19	0.18	CH7	9	2.6	0.07	1.0	-	0.01
22	23	3.0	0.06	1.1	0.12	0.09	CH6	-	2.7	0.02	1.3	-	0.01
23	25	2.1	0.05	1.1	0.13	0.05	CH5	10	2.5	0.05	0.6	-	0.01
24	16	-	0.06	3.0	-	0.08	CH3	23	3.4	0.15	0.9	-	-
25	9	-	0.05	1.3	-	0.04	BU-377	-	3.2	0.07	0.8	-	0.01
26	11	1.9	0.07	1.2	0.15	0.03	BU-173	-	3.3	0.23	0.8	-	-
27	21	-	0.03	1.7	-	0.02	BU-465	-	3.6	0.03	1.1	0.02	-
28	28	-	0.05	3.1	-	0.03	BU-181	-	3.1	0.18	0.9	0.05	-
29	20	4.8	0.03	1.5	0.19	0.42	BU-177	-	3.5	0.10	0. 6	-	0.01
30	21	-	0.03	0.9	-	0.05	BU-329	-	0.5	-	1.1	-	0.16
31	-	-	0.01	-	-	0.35	Al Dhafra	21	0.4	0.13	0.8	-	0.01
32	-	-	-	-	-	-	Bin Thani	45	1.9	0.07	0.6	0.02	0.02
33	-	-	0.02	-	-	0.02	BU-185	-	2.6	0.18	0.6	-	0.01
34	-	-	0.03	-	-	0.01	BU-360	-	2.2	0.13	1.4	0.02	0.01
35	-	-	0.01	-	-	0.01	BU-104	-	1.3	0.07	0.7	-	0.01
46	-	-	0.00	-	-	0.02	BU-121	-	2.6	0.05	1.0	-	0.01
47	-	-	0.07	-	-	0.03	BU-336	-	1.4	0.13	0.7	0.06	-
48	-	-	0.03	-	-	0.01	BU-116	-	0.7	3.07	0.7	0.01	0.01

Table 9. Soluble salts derived from placing a known weight of surface material in a known quantity of distilled water.

Field No.	UTM	Ca µg/g	Mg µg/g	Na µg/g	K µg/g	Sr µg/g	Si µg/g	Fe µg/g	Mn µg/g	Cl µg/g	SO_4 µg/g	Br µg/g	NO_3 µg/g	Zn µg/g	Cr µg/g	Ba µg/g	B µg/g	V µg/g
LQ-1	767256E	60	13	66	11	4.2	11.4	1.44	0.12	123	129	0.34	14.61	0.21	1.40	0.28	0.24	0.06
LQ-2	773785E	19	3.9	17	6.6	1.2	7.5	0.33	0.03	10	35	0.93	1.78	0.15	1.35	0.25	0.02	0.01
LQ-5	783606E	45	11	24	9.3	2.2	7.5	0.36	-	102	17	0.00	1.22	0.18	1.20	0.23	0.04	0.03
LQ-6	784547E	45	7.8	45	8.4	1.5	9.0	1.35	-	67	71	0.20	5.24	0.13	1.01	0.20	0.08	0.02
LQ-8	785931E	93	23	69	16	2.1	10.2	1.83	0.13	258	96	0.41	6.69	0.14	1.01	0.19	0.26	0.01
LQ-9	219995E	90	20	33	16	1.7	11.7	1.74	0.14	216	11	0.00	1.67	0.19	1.13	0.22	0.08	0.05
LQ-10	221270E	90	20	14	14	2.1	10.2	2.13	0.23	180	9.9	0.00	0.00	0.22	1.11	0.21	0.05	0.05
LQ-11	203470E	78	17	9.6	12	1.7	7.8	1.44	0.11	167	8.8	0.24	1.00	0.13	1.18	0.22	0.00	-
LQ-12	806671E	69	14	7.8	8.7	1.4	6.6	1.05	0.11	131	14	0.27	0.00	0.18	1.13	0.18	0.04	-
LQ-13	789180E	81	17	6.6	12	1.7	10.2	1.8	0.21	152	11	2.04	7.56	-	-	-	0.05	0.02
LQ-14	799720E	375	20	9.6	13	3.9	12.0	1.5	0.21	129	660	-	0.34	-	-	-	0.11	-
LQ-18	770754E	45	6.6	4.2	8.4	1.7	7.5	0.9	0.21	21	16	-	-	-	-	-	0.05	-
LQ-19	764298E	22	3.9	2.8	6.3	0.7	9.9	1.8	0.21	5.3	20	-	-	-	-	-	-	-
LQ-20	774827E	60	11	6.0	9.6	1.3	7.8	1.2	0.21	97	12	-	-	-	-	-	-	-

where q is solute flux ($(M/L^2)/T$), D is the diffusion coefficient (L^2/T), C is concentration (M/L^3), and M is mass (L and T were defined previously). This equation can be used to estimate the maximum effect of mass transport by diffusion. Consider the following initial conditions using two horizontal, porous slabs one meter on a side: the lower one is infinite in thickness and contains a solute with a constant concentration of 26,000 mg/L; the upper slab is 100 m thick and contains a constant concentration of 1000 mg/L. Both slabs are homogeneous and isotropic with a porosity of 38%. Using representative values of $D = 1x10^{-11}$ m^2/s, a concentration difference of 25 g/m^3 over a distance of 100 m would result in a steady-state mass flux to the upper slab of $0.25 \times 10^{-11} g/m^2/s$. Because there are 2.83×10^{11} seconds in the 9000 years since Holocene recharge started, the total mass flux to this water over that time period would be approximately 0.9 g/m^2 or 900 mg/m^2. Because the upper slab is 100 m thick and has a porosity of 38%, the total pore volume of the upper slab would be 38 m^3 or 38,000 L. The average mass contribution by diffusion would thus be approximately 0.024 mg/L, an insignificant amount relative to the total concentration. In the above analysis, the horizontal velocity in the upper slab is assumed to be zero. Although that is not strictly true, the velocity is very low, and the assumption would have little impact on the transport of solutes.

Although there is a low concentration of NO_3 in the underlying aquifer, the concentration of NO_3 in the upper aquifer is high. The conservative chloride/bromide ratio is different in the upper and lower aquifers (Tables 2 and 9). Thus, the chemical data are also inconsistent with the underlying aquifers provide a significant source of solutes.

Atmospheric precipitation as a source of solutes

Based on the above analysis, most of the solutes in this aquifer system must be derived from atmospheric precipitation. Eolian transport of sea aerosols in precipitation on the continents is well established (Junge and Werby, 1958; Mandel and Shiftean, 1981; and Hingston and Gailitis, 1976), as are the effects of these aerosols on surface waters and soils (Person and Fisher, 1971; Dethier, 1979; Eisenreich, 1980; Wiman and Agren, 1985; and Wiman et al., 1990). Accumulation of solutes over geologic time has been shown to be the major source of solutes in the arid areas of Australia (Wiman et al., 1990). Because of the infrequent nature of rainfall, we were unable to collect rainfall samples during the duration of this study; however, based on data from other arid areas, the input TDS is generally believed to be approximately 10 mg/L (Junge and Werby, 1958). Soluble salts on the surface of the dunes, however, were collected (Table 9) and are believed to represent the ionic ratio, but not concentration, of those salts stored between the Pleistocene and Holocene recharge events. In an arid area, salt input from adjacent sabkhat would significantly increase the load (Wood and Sanford 1995).

The following conceptual model accounts for the observed solute concentrations and distribution in the ground water of the Liwa aquifer. Between approximately 26,000 years BP (the end of the late Pleistocene recharge event) and approximately 9000 years BP (beginning of the Holocene recharge event), salts from precipitation accumulated on the surface and in the unsaturated zone. That is, rainfall was insufficient to transport them to the water table. In this 17,000 year interval, each square meter received 10 mg/L TDS in rainfall with an average annual flux of 40 mm/y, resulting in an accumulation of $6.8x10^6$ mg/m^2. When precipitation increased between 9000 and 6000 years BP, recharge occurred, and these stored salts were mobilized and transported to the ground water.

The solute distribution in the Holocene water is believed to be largely a function of the thickness of the unsaturated zone before recharge. That is, with a thick unsaturated zone associated with the topographic apex of the aquifer, storage is large, and sufficient recharge water flushes through the system to remove most of the stored soluble salts. When the unsaturated zone is thin, as on the flanks of the aquifer, storage is small, and only a small portion of the recharge enters the aquifer. The rest is rejected and is evaporated. Furthermore, because the water table is shallow, significant evaporation can occur, which increases the solute concentration. Thus, the solute concentration is greater on the flanks of the sand mound than near the apex because it has not been diluted. The observed "halo" solute distribution surrounding the mound is a function of the pre-recharge thickness of the unsaturated zone. That is, it is assumed that rainfall and salts stored on the surface would have been relatively uniform on this topographically low-lying aquifer system. Controls on solute distribution during the Pleistocene recharge event where probably similar to those during the Holocene recharge event. The present concentration distribution has not reached a steady state.

If it were assumed that recharge during the 3000 year wet interval of the Holocene was approximately 1.4 mm/y (Wood and Imes, 1995), 4.2 m of total recharge would have occurred. If all the stored salts from the previous 17,000 years ($6.8x10^6$ mg/m^2) were mobilized, one would expect to observe concentrations of approximately 1620 mg/L in the core of the aquifer. This expectation is confirmed within the limit of error and thus consistent with the proposed model (Figure 7, Tables 6 and 7).

High concentrations of Mg^{+2}, Na^{+1}, Cl^{-1}, and NO_3^{-1} are also consistent with an atmospheric source (Junge and Werby, 1958) because there are no significant mineral sources for these elements in the aquifer framework (Ahmed et al., 1998; Hadley et al., 1998). Nitrate (NO_3^{-1}) is formed by the oxidation of ammonia (NH_4^{+1}) formed in the atmosphere by the action of lightening on atmospheric N_2. Because of a lack of vegetation in this system, there is little nitrogen take-up, and it accumulates by evaporative concentration, like other solutes. There is significant nitrate that has accumulated on the surface since the end of the Holocene recharge interval (Table 9). The presence of the ubiquitous sabkhat increases the potential mass of salts on the surface (Wood and Sanford, 1995).

The initial solute distribution is modified by the presence of interdunal sabkhat. Capillary forces bring water and solutes to the surfaces of these sabkhat; water is evaporated, and soluble chlorides and nitrate salts are deposited on the surface. The relatively insoluble calcite, gypsum, and anhydrite that exhibit retrograde solubility are precipitated in the unsaturated zone as a function of increasing temperature toward the surface. When recharge occurs, the soluble salts are returned to the ground water while the relatively insoluble carbonates and sulfates remain in the unsaturated zone. There is a net loss of water by evaporation relative to recharge; thus, the concentrations of the solutes increase beneath the sabkhat. The ground water and solutes move downgradient toward the next interdunal sabkha, where the solutes are further concentrated by this process. Vertical mixing is promoted by the high density of recharging water relative to the density of water in the aquifer, and by the nearly isotropic hydraulic conductivity of the aquifer. This downgradient transport is coupled with the removal of some calcium, sulfate, and carbon, reflecting precipitation of relatively insoluble sulfates and carbonates in the unsaturated zone (Figure 6).

Summary and Conclusions

A Quaternary sand aquifer near the Liwa Crescent on the northeastern edge of the Ar Rub al Khali, United Arab Emirates, was selected to test several hypotheses on the seasonal timing of ground-water recharge, and on the origin, evolution, and distribution of solutes in a hyperarid area. Examination of this aquifer system provides insight into the timing of ground-water recharge and the origin and evolution of solutes in a representative hyperarid area. Ground-water flow in the aquifer system is outward in all directions from the center of two ground-water mounds, corresponding to two thick sand deposits. Hydraulic heads in underlying aquifers are lower than those of the Liwa/ Bu Hasa aquifer; thus, there is no advective water or solute input from this source. Solute diffusion from underlying aquifers provides a small solute flux, but it is inadequate to account for the observed solute mass; nor is the ionic ratio consistent with this source. There are no laterally adjacent aquifers; thus, there is no influx of solutes from these sources. Dissolution of the aquifer framework provides only for a minor fraction of most of the observed solutes. Therefore, most solutes in this aquifer system are almost certainly derived from atmospheric precipitation. Salts contained in rain were stored on the surface and in the unsaturated zone during the hyperarid time interval between the end of the Pleistocene recharge event (26,000 years BP) and the beginning of the Holocene recharge event (9000 years BP). During the Holocene recharge era (9000 to 6000 years BP), these stored salts were mobilized and transported to the ground water.

Initial solute distribution was controlled by the thickness of the pre-recharge unsaturated zone. With a thick unsaturated zone, aquifer storage is large, and recharge water flushes through the system, removing most of the stored soluble salts. Where the unsaturated zone is thin, the aquifer storage space is small, and only a small portion of the recharge enters the aquifer. The rest is rejected and lost to evaporation. Thus, the solute concentration is greater on the flanks than at the center because it has not been diluted by recharge.

The initial solute distribution has been modified as the solutes and water are transported along the flow path. Calcium and sulfate have been added by the dissolution of gypsum, as have certain trace elements, and bicarbonate has been reduced by precipitation of calcite. As ground water moves away from the apex of the mound, it encounters a series of interdunal sabkhat. Water is evaporated from the sabkhat, leaving soluble chloride and nitrate minerals on the surfaces and retrograde carbonate and sulfate minerals in the unsaturated zone. When recharge occurs through the interdunal sabkhat, soluble salts on the surfaces, modern ^{14}C, and tritium are added to the aquifer. There is vertical mixing within the aquifer because the density of the recharged water is greater than the density of water. This mixed water is then carried downgradient toward the next interdunal sabkha. Relatively insoluble carbonate and sulfate minerals are retained in the unsaturated zone of the interdunal sabkhat, causing a change in the solute ratios in a downgradient. The portion of the aquifer near the center of the mound which is not affected by interdunal sabkhat retains the original solute composition determined by mobilization of stored salts. It should be noted, however, that the water and solute movement is very slow, and much of the solute distribution is controlled by the initial thickness of the unsaturated zone rather than by ground-water transport.

Controls on solute distribution during the Pleistocene recharge event where probably similar to those during the Holocene recharge event. The system is not at steady state with respect to solute distribution. The isotopic data from ancient ground waters from the Liwa Oasis with its unusual "d" (deuterium excess) of approximately -15 indicate that Holocene moisture derived from previously evaporated water on the surface of the Indian Ocean rather than from the Mediterranean/Arabian Gulf. Such a source is consistent with a summer monsoonal circulation first proposed by McClure (1976). These findings, although specific to this aquifer, provide a useful model by which to evaluate other aquifer systems in hyperarid environments.

ACKNOWLEDGMENTS

We thank the International Atomic Energy Agency for isotope analyses and the United Arab Emirates University for support for this study. We thank Jeffery Imes and Ward Sanford for reviews that enhanced both the clarity of thought and the presentation. The authors, however, assume all responsibility for any errors.

REFERENCES

Ahmed, E. A., Soliman, M. A., Alsharhan, A. S., and Tamer, S., 1998. Mineralogical characteristics of the Quaternary sand dunes in the eastern province of Abu Dhabi, United Arab Emirates. *In:* Quaternary Deserts and Climatic Change: A. S. Alsharhan, K. W. Glennie, G. L. Whittle, and C. G. St. C. Kendall, (editors), A. A. Balkema, Rotterdam, 85-90.

Alsharhan, A. S., Rizk, Z. A., Nairn, A. E., Bakhit, D. W., and Alhajari, S. A., 2001. Hydrogeology of an Arid Region - The Arabian Gulf and Adjoining Areas: Elsevier, 331p.

Burns, S. J., Matter, A, Frank, N., and Mangini, A., 1998. Speleothem-based paleoclimate record from northern Oman. *Geology* 26(6): 499-502.

Burns, S. J., Fleitmann, D., Matter, A., Neff, U., and Mangini, A., 2001. Speleothem evidence from Oman for continental pluvial events during interglacial periods. *Geology* 29(7): 623-626.

Clarke, I. D., Fritz, P., Quinn, O. P., Rippon, P. W., Nash, H., and bin Ghalib al Said, B., 1987. Modern and fossil groundwater in an arid environment, a look at the hydrogeology of Southern Oman, in Use of Isotopes in Water Resources: International Atomic Energy Agency, Symposium 299, Vienna, p. 167-187.

Clarke, I. D., and Fontes, J. C., 1990. Paleoclimactic reconstruction in Northern Oman based on carbonates from hyperalkaline ground waters. *Quaternary Research* 33: 320-336.

deMenocal, P., Ortiz, J., Guilderson, T., Adkins, J, Sarnthein, M., Baker, L., and Yarusinsky, M., 2002. Abrupt onset and termination of the African humid period - Rapid climate responses to gradual insolation forcing. *Quaternary Science Reviews* 19: 347-361.

Dethier, D. P., 1979. Atmospheric contributions to stream-water chemistry in the North Cascade Range, Washington. *Water Resources Research* 15:787-794.

Driscoll F. G., 1986. Groundwater and Wells: Johnson Filtration Systems, Inc., St. Paul, 1089p.

Eisenreich, S. J., 1980. Atmospheric input of trace metals to Lake Michigan. *Water, Air, Soil Pollution* 13: 287-301.

Gleick, P., 2002. The world's water - The biennial report on freshwater resources: Island Press, Washington, DC, 334p.

Hadley, D. G., Brouwers, E. M., and Bown, T. M., 1998. Quaternary paleodunes, Arabian Gulf coast, Abu Dhabi Emirate - Age and paleoenvironmental evolution. *In:* Quaternary Deserts and Climatic Change: A. S. Alsharhan, K. W. Glennie, G. L. Whittle, and C. G. St. C. Kendall, eds., A.A. Balkema, Rotterdam, 123-140.

Hingston, F. J., and Gailitis, V., 1976. The geographic variation of salt precipitation over Western Australia. *Australian Journal of Soil Research* 14: 319-335.

International Atomic Energy Agency, 1992. Statistical treatment of data on environmental isotopes in precipitation, Technical Report Series No. 331: International Atomic Energy Agency, Vienna, 781p.

Junge, C. E., and Werby, R. T., 1958. The concentration of chloride, sodium, potassium, calcium and sulfate in rainwater over the United States. *Journal of Meteorology* 15(5): 417-425.

Lippard, S. J., Shelton, A. W., and Gass, I. G., 1986. The ophiolite of Northern Oman: The Geol. Society Memoir # 11, Blackwell Scientific. Pub., Oxford, UK, 178p.

Lutz, W., Sanderson, W., and Scherbov, S., 2001. The end of world population growth. *Nature* 412: 543-545.

Mandel, S., and Shiftean, Z. L., 1981. Groundwater Resources: Academic Press, New York, 269p.

McClure, H. A., 1976. Radiocarbon chronology of late Quaternary lakes in the Arabian desert. *Nature* 263: 755-756.

Merlivat, L., and Jouzel, J. J., 1979. Global climatic interpretations of the deuterium-oxygen 18 relationship for precipitation. *Journal of Geophysical Research.* 84: 5029-5033.

National Atlas of the United Arab Emirates, 1993. United Arab Emirates University, 164p.

Neff, U., Burns, S. J., Mangini, A., Fleitmann, D., and Matter, A., 2001. Strong coherence between solar variability and monsoon in Oman between 9 and 6 kyr ago. *Nature* 411: 190-293.

Person, F. J. Jr., and Fisher, D. W., 1971. Chemical composition of atmospheric precipitation in the northeastern United States: U.S. Geological Survey Water Supply Paper 1535-P, p.1-23.

Plummer, L. N., Parkhurst, D. L., Fleming, G. W., and Dunkle, S. A., 1988. A computer program incorporating Pitzer's equations for calculation of geochemical reactions in brines: U. S. Geological Survey Water-Resources Investigation Report 88-4153, 310p.

Sanford, W. E., and Wood, W. W., 2001. Hydrology of the coastal sabkhat of Abu Dhabi, United Arab Emirates. *Hydrogeology Journal* 9(4): 338-366.

Stokes, S., Bray, H., Wood, W. W., and Goudie, A. S., (this volume), Later Quaternary paleorecharge events in the Arabian Peninsula. *In*: A.S. Alsharhan & W.W. Wood (editors). Water Resources Perspectives: Evaluation, Management and Policy. Elsevier Science, The Netherlands.

Taylor, D. W., 1948. Fundamentals of soil mechanics: John Wiley and Sons, New York.

Thompson, G. R., and Turk, J., 1999. Earth Science and the Environment - 2nd Ed.: Harcourt College Publisher, Fort Worth, TX, 589p.

Wiman, B. L. B., and Agren, G. L., 1985. Aerosol depletion and deposition in forests - A model analysis. *Atmospheric Environment* 19: 335-347.

Wiman, B. L. B., Unsworth, M. H., Lindberg, S. E., Bergkvist, B., Jaenicke, R., and Hanson, H. C., 1990. Perspectives on aerosol deposition to natural surfaces-Interactions between aerosol residence times, removal processes, the biosphere and global environmental change. *Journal of Aerosol Science* 21: 313-338.

Wood, W. W., and Imes, J. L., 1995. How wet is wet? Constraints on late Quaternary climate in southern Arabian Peninsula. *Journal of Hydrology* 164: 263-268.

Wood, W. W., and Imes J. L. (this volume), Dating of Holocene ground-water recharge in the Rub al Khali of Abu Dhabi - Constraints on global climate-change models. *In*: A.S. Alsharhan & W.W. Wood (editors). Water Resources Perspectives: Evaluation, Management and Policy. Elsevier Science, The Netherlands.

Wood, W. W., and Sanford, W. E., 1995. Eolian transport, saline lake basins, and groundwater solutes. *Water Resources Research* 31: 3121-3129.

Wood, W. W., Sanford, W. E., and Al Habschi, A. R. S., 2002. The source of solutes in the coastal sabkha of Abu Dhabi. *Bulletin of the Geological Society of America* 114(3): 259-268.

Zhisheng, A., Porter, S. C., Kutzbach, J. E., Xihao, W., Suming, W., Xiaodand, L., Xiaoqiang, L., and Weijian, Z., 2000. Asynchronous Holocene optimum of the East Asian monsoon. *Quaternary Science Review* 19: 743-762.

Fully-integrated Modelling of Surface and Subsurface Water Flow and Solute Transport: Model Overview and Application

EDWARD A. SUDICKY, JOEL E. VANDERKWAAK, JON P. JONES, JON P. KEIZER, ROBERT G. MCLAREN
Department of Earth Sciences, University of Waterloo, Waterloo, Ontario, Canada N2L 3G1.
E-mail: sudicky@sciborg.uwaterloo.ca

GEORGE B. MATANGA
U.S. Bureau of Reclamation, MP-700, Sacramento, California 95825, USA.

ABSTRACT: A physically-based, fully-integrated model has been recently developed that considers water and solute transport on the two-dimensional land surface and three-dimensional dual continua (i.e. porous medium-macropore) subsurface under variably-saturated conditions. The two-dimensional form of the nonlinear diffusion-wave equation, together with Manning's equation to compute overland flow velocities, are employed on the surface, while Richards' equation and Darcy's law are assumed to hold in the subsurface. Two- and three-dimensional forms of the advection-dispersion equation are used to describe solute transport on the surface and within the subsurface, respectively. Linkage between the various continua is through first-order, physically based flux relationships or through pressure head and concentration continuity assumptions. Full coupling of the flow regimes is achieved by solving and assembling one system of discrete algebraic equations such that surface water depths and subsurface pressure heads, as well as water fluxes between continua, are determined simultaneously. Likewise, one system of advective-dispersive transport equations is solved to obtain the solute concentrations in the surface and subsurface flow domains. Robust and efficient discretization and solution techniques are utilized to achieve a high degree of computational efficiency. The model is applied to the Laurel Creek watershed in Southern Ontario, which is about 75 km^2 in area and includes both urban and rural land uses. Through past and ongoing field studies, the watershed is reasonably well characterized and is being monitored on an ongoing basis. In addition to diverse land use and surface cover and more than 100m of topographic relief, the watershed is underlain by a complex interconnected sequence of sand and gravel aquifers used for water supply that are separated by discontinuous clayey aquitards.

INTRODUCTION

This study was undertaken to demonstrate the feasibility of applying the Integrated Hydrology Model (*InHM*) described by VanderKwaak (1999) and to a larger scale problem than had been previously attempted. The model uses an integrated approach to solve the surface/subsurface flow equations by assembling and simultaneously solving one set of discrete algebraic equations arising from the application of the Control Volume Finite Element (CVFE) approach. On the surface, the two-dimensional form of the nonlinear diffusion-wave equation is employed together with the Manning equation to represent flow over the land surface and in streams. The three-dimensional mixed form of Richards' equation is used to describe variably-saturated subsurface flow in the porous medium. The mixed form of Richards' equation applies in both the saturated and unsaturated zones. The dynamic fluid exchange fluxes between the surface and subsurface continua are computed implicitly in a fully-coupled manner. The *InHM* is different in this respect from most other models that either treat surface/ subsurface flow as two independent systems or as being weakly coupled through a common flux term. In the latter case, a simple iterative solution strategy that is prone to convergence difficulties is often employed to handle the nonlinear nature of the flux coupling term. These difficulties are avoided with the fully-coupled, fully-implicit approach used in *InHM*.

The Laurel Creek watershed, which covers 75 km^2 in Ontario, Canada (see Figure 1), was selected as a candidate site for a large-scale application of the model. It contains areas of both urban and rural land uses, with highly complex subsurface stratigraphy of glacio-fluvial origin and more than 100 m of topographic relief. As well, over 10 years of hydrological and ecological studies have been carried out on the watershed.

The scope of this paper is to present preliminary modeling results for a conceptual model whose surface and subsurface characteristics are strongly heterogeneous and are based on extensive field data.

Water Resources Perspectives: Evaluation, Management and Policy. Edited by A.S. Alsharhan and W.W. Wood.
Published in 2003 by Elsevier Science, Amsterdam, The Netherlands, p. 313-318.

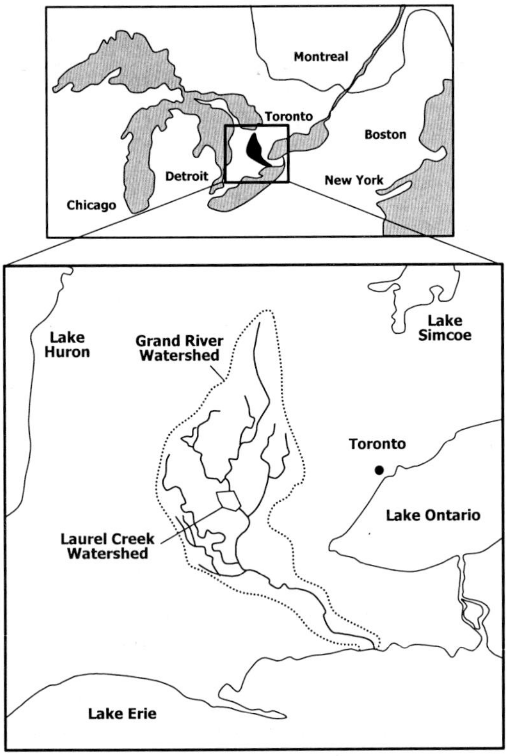

Figure 1. The Laurel Creek watershed.

While not yet fully calibrated to the transient hydraulic response of the Laurel Creek watershed due to individual precipitation events, the modeling results demonstrate that *InHM* is a powerful new, physically-based simulator that holds promise to represent watershed responses in a holistic three-dimensional framework. Although *InHM* has the capability to simulate advective-dispersive solute transport in both the surface and subsurface flow regimes, this aspect will not be considered here.

GOVERNING EQUATIONS

Three-dimensional variably-saturated subsurface flow in the porous medium is described in *InHM* by

$$\nabla \cdot q \pm q^b \pm q_{ps}^e = \frac{\partial \theta S_w}{\partial t} \tag{1}$$

where q is the Darcy flux, q^b is a specified or internally computed source/sink representing water injection, withdrawal or evapotranspiration, q_{ps}^e is the rate of water exchange between the porous medium and surface continua, θ is porosity, S_w is water saturation, and t is time. While *InHM* is capable of describing subsurface flow in a dual-porosity, dual-permeability context through the use of a form of (1) applied to macropore flow, this feature is not needed for the Laurel Creek application and therefore will not be described. The Darcy flux in (1) is given by:

$$q = -k_{rw} \frac{\rho_w g}{\mu_w} k \nabla(\Psi_p + z) \tag{2}$$

where k_{rw} is the relative permeability, ρ_w is the density of water, g is the acceleration due to gravity, μ_w is the viscosity of water, k is the intrinsic permeability vector, Ψ_p is the pressure head, and z is the elevation head. Capillary pressure relationships are described in this study using a reference curve (van Genuchten, 1980) scaled to accommodate spatially-variable porous medium properties as a function of local permeability and porosity (Leverett, 1942). Capillary pressure hysteresis, while supported in the model, is not considered in the simulations presented in this study.

Transient flow of water on the land surface (both overland and open channel) is described by the nonlinear diffusion-wave approximation of the depth-integrated shallow water equations. Two-dimensional surface flow is conceptualized as a second continuum that interacts with the underlying variably-saturated porous medium through a thin interfacial boundary layer of thickness a_s. Assuming a negligible influence of inertial forces and a shallow depth of surface water, Ψ_s, the conservation of water on the land surface is described by

$$\nabla \cdot \Psi_s^{mobile} q_s \pm q^b \pm q_{sp}^e = \frac{\partial (S_w h_s + \Psi_s^{store})}{\partial t} \tag{3}$$

where q_s is the surface water velocity, q_b is a source/sink term (e.g. rainfall), and $q_{sp}^e (= -q_{ps}^e)$ is the surface-porous medium water exchange rate. Surface saturation S_w is defined as (VanderKwaak, 1999),

$$S_w = \Psi_r^{2(1-\Psi_r)} \tag{4}$$

$$\psi_r = \min[1, \max[\Psi_s / h_s]] \tag{5}$$

where h_s is the average height of the non-discretized surface microtopography. The stored, Ψ_s^{store}, and mobile, Ψ_s^{mobile}, water depths are defined as (VanderKwaak, 1999),

$$\Psi_s^{store} = \max[0, \Psi_s - h_s] \tag{6}$$

$$\psi_s^{mobile} = \max[0, \Psi_s - \Psi_s^{pond}] \tag{7}$$

where Ψ_s^{pond} is the depth of ponded water that does not participate in overland flow (e.g. depression storage). Surface water velocities are calculated utilizing a two-dimensional form of the empirical, though well-established, Manning water depth/friction-discharge equation given by

$$q_s = -\frac{(\Psi_s^{mobile})^{2/3}}{n\Phi^{1/2}} \nabla(\Psi_s + z) \tag{8}$$

where n is the Manning's surface roughness tensor and Φ is the friction (or energy) slope approximated as

$$\Phi = \left[\left(\frac{\partial(\Psi_s + z)}{\partial x}\right)^2 + \left(\frac{\partial(\Psi_s + z)}{\partial y}\right)^2\right]^{0.5} \tag{9}$$

The linkages between the different components of *InHM* are through first-order, physically-based flux relationships driven by pressure-head gradients (VanderKwaak, 1999).

Surface-subsurface water exchange rates, q_{sp}^e, are approximated as

$$q_{sp}^e = \alpha^e(\Psi_p - \Psi_s) = -q_{ps}^e \tag{10}$$

where the nonlinear water exchange coefficient, α^e, is given by

$$\alpha^e = k_{rw}^e \frac{\rho_w g}{\mu_w} \frac{k_p^{zz}}{a_s} \tag{11}$$

where ρ_w and μ_w are the density and viscosity of water, k_p^{zz} is the intrinsic permeability of the soil surface in the vertical direction and a_s is a characteristic boundary-layer thickness at separating the surface and subsurface flow domains.

The interface relative permeability, k_{rw}^e, is defined in the upstream continuum

$$k_{rw}^e = \begin{cases} k_{rw_p} : \Psi_p \geq \Psi_s \\ k_{rw_s} : \Psi_s > \Psi_p \end{cases} \tag{12}$$

which ensures that the discretized solution is monotone. The surface pseudo-relative permeability, k_{rw_s}, is defined as being equivalent to the surface saturation (4). This function effectively reduces the surface area associated with each node on the surface in coupling the surface and subsurface flow continua, thus restricting rapid infiltration during concentrated ponding in rills (Dunne et al., 1991; Woolhiser et al., 1996). Equation (10) describes both rates of surface water infiltration.

$$(\Psi_p - \Psi_s) < 0 \Rightarrow q_{sp}^e \leq 0; q_{ps}^e \geq 0 \tag{13}$$

and groundwater seepage (or exfiltration),

$$(\Psi_p - \Psi_s) > 0 \Rightarrow q_{sp}^e \geq 0; q_{ps}^e \leq 0 \tag{14}$$

Infiltration and exfiltration rates are determined by spatially and temporally variable subsurface pressure head gradients and surface water depths, and by spatially-variable porous medium properties. The flow of water in both the surface and subsurface continua is therefore intimately coupled and the rates of water exchange between continua are not specified, but evolve in time and space as a function of local hydrodynamics.

The governing equations, (1) and (3), are discretized in space using the control volume finite-element method, which combines the geometric flexibility of finite elements with the local conservation characteristics of control volumes. Each coupled system of nonlinear equations in an *InHM* simulation is solved implicitly using Newton iteration (VanderKwaak, 1999). Efficient and robust iterative sparse matrix methods are used to solve the large sparse Jacobian systems (VanderKwaak et al., 1997) such that three-dimensional simulations involving grids with several hundred thousand nodes can be performed on well-equipped Pentium-class computers.

PHYSICAL SYSTEM

A 25 m Digital Elevation Model (DEM) obtained from the Grand River Conservation Authority in Ontario, Canada was used to define a two-dimensional triangular-element mesh (see Figure 2) representing the top of the model domain (ground surface) as well as the lateral extents of the watershed. The elevations were mapped from the 25 m DEM onto a finite element mesh generated for *InHM*. The *InHM* is coarser than the DEM and thus some detail is lost when mapping elevations onto it.

The surface mesh was divided into five distinct land-use classes with an appropriate Manning's coefficient assigned to each class. The stream channels were incised into the surface mesh using stream cross-sectional data where available, and a modified version of the Rosgen stream classification was used to estimate channel geometries elsewhere (Annable, 1995).

In *InHM*, surface water flow is simulated using the two-dimensional mesh, which is superimposed on a three-dimensional finite-element mesh used to simulate subsurface flow. The top of the three-dimensional mesh is coincident with the two-dimensional mesh such that dual nodes exist over the ground surface. Twenty layers separate the surface and the base of the three-dimensional mesh, which is defined by the bedrock surface.

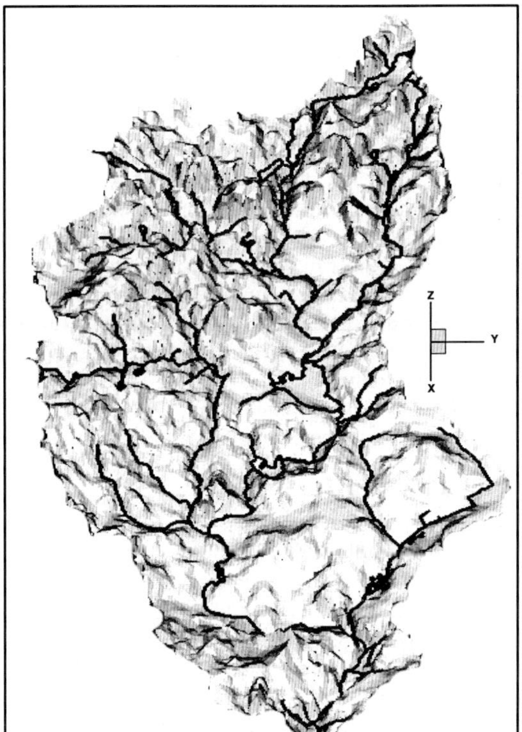

Figure 2. Laurel Creek surface drainage network and 25m resolution DEM.

In this study, the conceptual model consists of four aquifers between four aquitards (Figure 3). Hydrostratigraphic surfaces delineating boundaries between the aquifer and aquitard units, including the bedrock surface, were extracted from over 1100 borehole logs provided by the Ontario Ministry of the Environment. This permitted unique hydraulic conductivity values to be assigned to each hydrostratigraphic unit. A veneer of surficial soils has been incorporated into the conceptual model (along with their pertinent wetting and drying characteristics) based on soil survey studies.

The bottom and side boundaries of the domain were assumed to be impermeable with respect to both surface and subsurface flow. In order to arrive at a quasi-steady initial condition prior to the application of individual precipitation events, a net annual average rainfall of 20 cm/yr was applied uniformly to the two-dimensional surface mesh over the entire watershed. This value was calculated from the long-term average baseflow measured at the watershed outlet. Outflow is allowed in six surface nodes in the two-dimensional surface mesh, which were chosen along the segment of the domain where Laurel Creek discharges into the Grand River. A nonlinear critical-depth boundary condition is applied at these outflow nodes that neither constrains the flow rate nor the surface water depth.

MODEL RESULTS

The model was run in transient mode using the net annual average precipitation rate as input until a steady-state equilibrium was achieved both in the surface and subsurface flow regimes. At the start of

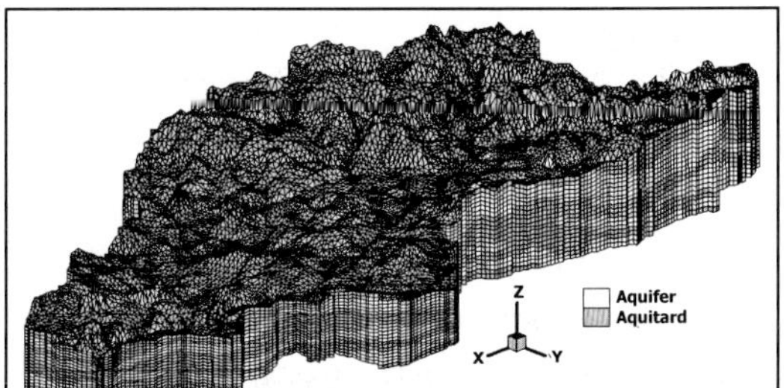

Figure 3. The *InHM* finite element mesh and subsurface stratigraphy.

the simulation, no surface was assumed to exist on the surface, and the water table was initially positioned at the land surface. During the approach to steady state, the model formed the surface drainage network, which includes Laurel Creek and its tributaries, because of groundwater discharge into the topographic low points defining the locations of the initially dry channels. Steady state was achieved after about 1800 days. The simulation required several days of computer time on a Pentium III 800 processor with 1024 megabytes of RAM memory. It should be noted that the model does not *a priori* assume any rainfall-runoff generation mechanisms. Instead *InHM* determines where water infiltrates, exfiltrates or forms surface water in drainage channels or wetlands. Moreover, if the model is correctly parameterized, the computed steady-state drainage network should be similar to the observed drainage network. Figure 4 provides this comparison. The simulated surface-water drainage network is seen to be similar to the observed network with the exception of some ponding along the edges. This ponding is due to a combination of using a relatively coarse surface mesh containing localized topographic irregularities and the fact that the lateral boundaries of the domain are assumed to be zero-flux boundaries, both in the surface and subsurface systems. Although not shown here, the calculated three-dimensional pattern of the hydraulic head distribution in the subsurface exhibits a reasonable correspondence with the observed hydraulic head data obtained from monitoring wells, although the model is not yet calibrated.

Once steady-state conditions were achieved, a 125-day record of actual precipitation events was input in the model in order to assess the ability of *InHM* to compute streamflows due to individual, discrete rainfall events at a gauged station where continuous monitoring data exist. For this preliminary simulation, evapotranspiration was neglected in the model for the sake of simplicity. The rainfall record is shown in Figure 5. The computed and observed hydrographs are compared in Figure 6. It can be seen that the results from the *InHM* model consistently over-predict the observed streamflow, although the timing of the peak flows and the shapes of the recession portions of the curves are similar. The over-prediction of the streamflow is mainly because of the neglect of evapotranspiration in this preliminary simulation, and likely because the vertical permeability assigned to the surface soils is too low such that too little rainfall infiltrates into the subsurface.

CONCLUSIONS

The preliminary results indicate that it is feasible to apply *InHM* in three dimensions to a hydrologic system of the scale of the Laurel Creek watershed and larger. As well, the robust nature of the fully-coupled numerical solution approach and the use of an efficient sparse-matrix solver allows simulations to be carried out on Pentium-class computers. While the model is not yet calibrated, early results indicate that *InHM* is capable of further tasks with representing the transient hydraulic

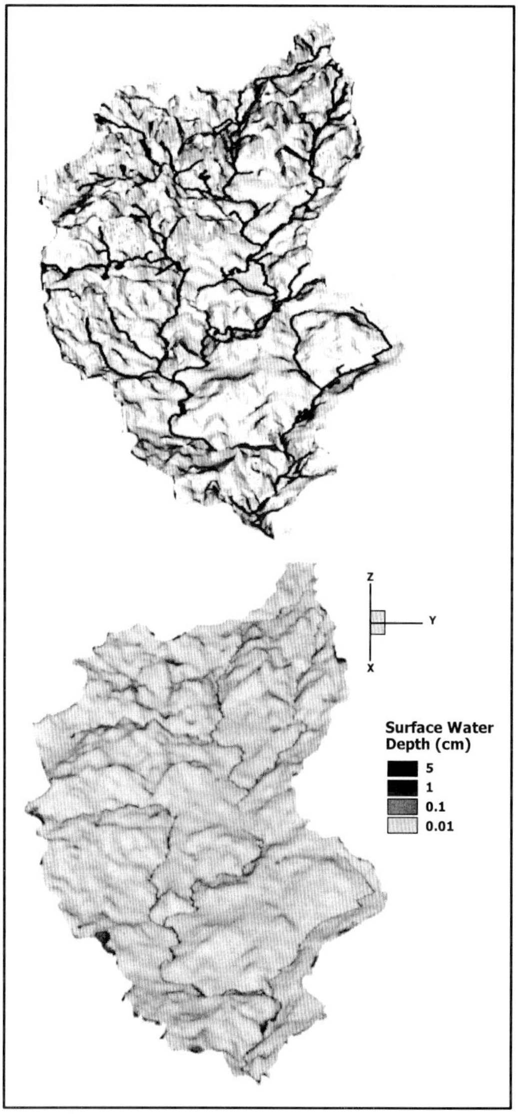

Figure 4. Observed versus computed surface drainage network under steady-state baseflow conditions.

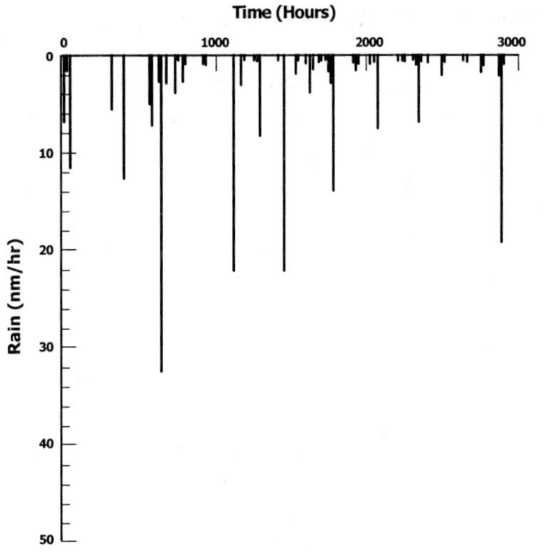

Figure 5. Record of input precipitation events for a three-month period.

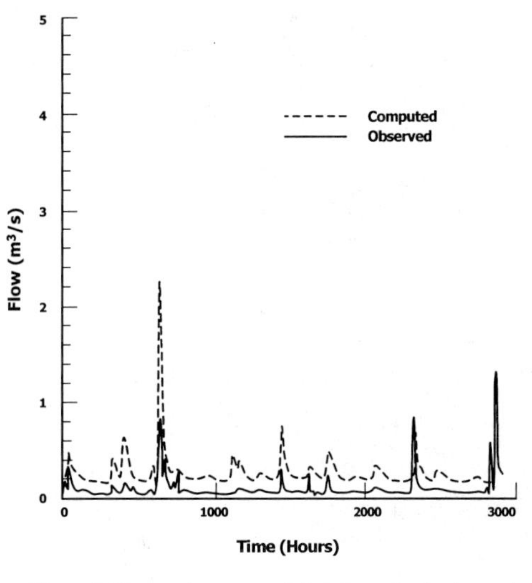

Figure 6. Observed versus predicted streamflow hydrograph at Erbsville Road recording station.

response of both the surface and variably-saturated subsurface flow regimes *simultaneously* in a fully-integrated manner. Although the application of *InHM* to the Laurel Creek watershed involves a temperate climate setting in which rainfall is plentiful, its design is well-suited to simulate coupled surface/subsurface flow in arid climates because both the saturated and unsaturated zones are accounted for in the subsurface through the use of Richards' equation.

respect to its application to the Laurel Creek watershed will consist mainly in refining the surface and subsurface properties of the physical system in accordance with the available data, and to include evapotranspiration processes. Once calibrated, the next step that is planned is to assess the ability of *InHM* to simulate the migration of contaminants through the watershed from both point and non-point sources.

ACKNOWLEDGEMENTS

A NSERC Strategic Research Grant in partnership with the Grand River Conservation Authority, the Regional Municipality of Waterloo and the City of Waterloo funded this study.

REFERENCES

Annable, W., 1995. Morphological relationships of rural water courses in southwestern Ontario for use in natural channel designs. *M.Sc. Thesis, University of Guelph, Guelph, Ontario, Canada.*

Dunne, T., 1978. Field studies of hillslope flow processes and their significance. In: M.J. Kirkby (ed), *Hillslope Hydrology,* Wiley Interscience, New York, 227-293.

Leverett, M.C., 1941. Capillary behavior in porous soils. *Transactions of the American Institute Mineralogy and Metallurgy, Petroleum Engineering* 142: 152-169.

VanderKwaak, J.E. 1999. Numerical simulation of flow and chemical transport in integrated surface-subsurface hydrologic systems. *Ph.D. Thesis, Department of Earth Sciences, University of Waterloo, Waterloo, Ontario, Canada.*

VanderKwaak, J.E., P.A. Forsyth, K.T.B. MacQuarrie and E.A. Sudicky, 1997. User's Guide.

Van Genuchten, M.Th., 1980. A closed-form equation for predicting the hydraulic conductivity of unsaturated soils. *Soil Science Society of America Proceedings* 44: 892-898.

WatSolv sparse matrix iterative solver package, Version 2.16. *Waterloo Centre for Groundwater Research, University of Waterloo, Waterloo, Ontario, Canada,* 23p.

Woolhiser, D.A., R.E Smith and J.-V. Giraldez, 1996. Effects of spatial variability of saturated hydraulic conductivity on Hortonian overland flow. *Water Resources Research* 32: 671-678.

From Drainage to Recharge to Discharge: Some Timelags in Subsurface Hydrology

PETER G. COOK, IAN D. JOLLY, GLEN R. WALKER
CSIRO Land and Water, Private Bag 2, Glen Osmond SA 5064, Australia. E-mail: peter.g.cook@csiro.au

NEVILLE I. ROBINSON
CSIRO Mathematical and Information Sciences, Australia.

ABSTRACT: In arid areas, low rates of groundwater recharge and thick unsaturated zones mean that significant time lags can occur between changes in drainage below plant root zones and changes in aquifer recharge. The magnitude of the timelag can be approximated using simple analytical expressions, or more accurately determined using numerical models of soil water movement. It is related to the thickness of the unsaturated zone, the magnitude of the drainage rates, and the soil hydraulic properties. In southern Australia, in areas where water tables are 30 m below the land surface, timelags of 30 - 200 years occur between clearing of native vegetation for agriculture and an increase in groundwater recharge. It is predicted that similar timelags would occur between revegetation and a reduction in recharge.

The timelag between a change in groundwater recharge and the change in groundwater discharge is related to the distance between the recharge and discharge areas, and the aquifer transmissivity and specific yield. For large groundwater basins with low recharge fluxes, this timelag can be many hundreds or thousands of years. In some cases, this allows groundwater extraction at rates well in excess of recharge rates to continue for a number of years before the impact of this policy will be seen. On the other hand, in southern Australia revegetation to reduce recharge and hence lower watertables and reduce saline groundwater flows to one of Australia's most important rivers (the Murray River) are being considered. The timelag between the reduction in recharge and the consequent reduction in discharge limits the effectiveness of this management option in areas further than a few kilometres from the river.

INTRODUCTION

Over long time periods, the impacts of development on water resources can be predicted from simple water balance calculations. For example, groundwater extraction will cause natural discharge from the aquifer to be reduced by an amount equal to the rate of extraction (Sophocleous, 1997). In the case of agricultural development, changes in aquifer discharge will be of the same magnitude as the total change in drainage beneath these areas.

Large groundwater systems, however, can take many years to fully respond to development, and so water balance calculations are of limited value in predicting aquifer responses. In arid areas, low rates of groundwater recharge and thick unsaturated zones mean that significant time lags can be associated both with unsaturated and saturated responses (e.g., Abdulrazzak and Morel-Seytoux, 1983; Morel-Seytoux, 1984). Where a large net increase in soil drainage occurs, such as from a change in land use (e.g., clearing of native vegetation, or the change from dryland to irrigated agriculture), there will be a time lag before this increase in drainage becomes an equivalent increase in groundwater recharge. Conversely, where there is a large decrease in drainage, there will be a timelag before recharge decreases. Similar timelags occur between changes in groundwater recharge and consequent changes in groundwater discharge. In this paper we present some simple equations that can be used to estimate the magnitude of these time delays. Their application is illustrated with respect to land salinisation processes in southern Australia, and groundwater resource management.

FROM DRAINAGE TO RECHARGE

In this section, we present a simple model for vertical water movement through a homogeneous unsaturated zone. This model is then used to quantify timelags between changes in water flux below the plant root zone (drainage) and consequent changes in water flux at the watertable (recharge).

In the case of steady drainage through a homogeneous unsaturated soil, the soil water velocity will be the same at all depths, and so the

Water Resources Perspectives: Evaluation, Management and Policy. Edited by A.S. Alsharhan and W.W. Wood.
Published in 2003 by Elsevier Science, Amsterdam, The Netherlands, p. 319-326.

drainage rate below the soil surface will be equal to the aquifer recharge rate (at the water table). The matric potential profile is given by:

$$z = z_{WT} + \int_0^{\psi} \frac{d\psi}{1 + D/K(\psi)} \quad (1)$$

where D is the drainage rate, $K(\psi)$ is the hydraulic conductivity, z_{WT} is the unsaturated zone thickness, ψ is matric potential and z is depth below the land surface. For large z_{WT} and small z, $\partial\psi/\partial z \to 0$, and so $K(\psi) = D$. The volumetric water content profile, $\theta(z)$, can be determined from the matric potential profile, $\psi(z)$, using the soil moisture characteristic, $\psi(\theta)$.

Step Increase in Drainage

Where a step increase in drainage occurs, there will be a time delay before the aquifer recharge rate increases. This time delay is related to the time for the soil water content profile to increase from that in equilibrium with the initial drainage rate to that in equilibrium with the new (higher) drainage rate. The time delay is determined by the initial and final drainage rates, the thickness of the unsaturated zone, and the soil moisture characteristic and hydraulic conductivity functions.

Where a large increase in drainage occurs, the timelag (t_L) between the increase in drainage and the increase in recharge can be approximated by

$$t_L = z_{WT}(\theta_2 - \theta_1)D^{-1} \quad (2)$$

where D is the change in drainage rate, θ_l is the mean water content of the soil under the initial drainage rate, and θ_2 is the mean soil water content in equilibrium with the new drainage rate (Jolly et al., 1989; Cook et al., 1989). Figure 1 shows predicted increases in aquifer recharge as a function of time, for an increase in drainage rate of 25 mm/yr, (θ_2-θ_1) = 0.06 and watertable depths between 10 and 50 m. For the case of a 50 m watertable depth, the timelag between the increase in drainage and the increase in recharge is estimated to be 120 years.

However, Cook et al. (1989) noted that the rise in the watertable at any particular point is not due to aquifer recharge occurring at that point, but rather reflects recharge occurring over a larger area. Where drainage is spatially variable, then the recharge flux to the unconfined aquifer on a regional scale as a function of time is given by

$$R(t) = \int_{z_{WT}(\theta_2-\theta_1)/t}^{\infty} y f(y)\,dy \quad (3)$$

where $R(t)$ is the mean regional groundwater recharge rate at time t and $f(y)$ is the probability density function that defines the spatial variability of drainage (Cook et al., 1989; Allison et al., 1990; Figure 2).

Step Decrease in Drainage

A similar time delay occurs between a step decrease in drainage, and the consequent decrease in aquifer recharge. In this case, the time delay is determined by the time required for the soil to drain from the initial (higher) water content, to the final (lower) water content. Jolly and Cook (2001) calculated time delays between decreasing drainage and the subsequent decrease in aquifer recharge using a numerical solution to the one dimensional Richards' Equation:

$$\frac{\partial\theta}{\partial t} = \frac{\partial}{\partial z}\left(K(\psi)\left(1 - \frac{\partial\psi}{\partial z}\right)\right) \quad (4)$$

The lower boundary condition represented the water table:

$$\psi = 0 \quad at \quad z = z_{WT} \quad (5)$$

and the upper boundary condition was for zero recharge (no flux):

$$\frac{\partial\psi}{\partial z} = -1 \quad at \quad z = 0 \quad (6)$$

The initial condition was for the steady state profile, given by Equation (1).

Figure 3 depicts the aquifer recharge rate as a function of time since the decrease in drainage for a loamy sand soil and water table depths between 10 and 50 m. For these simulations, the initial condition was for steady state drainage (and recharge) of 25 mm/yr. Drainage decreases to zero at time t = 0. In all cases, aquifer recharge continues unchanged for a period of time, before beginning to slowly decrease. For these simulations, the time taken for the recharge rate to reduce to half of its initial value is approximately proportional to the unsaturated zone thickness, increasing from 8 years for a water table depth of 10 m, to 36 years for a water table depth of 50 m.

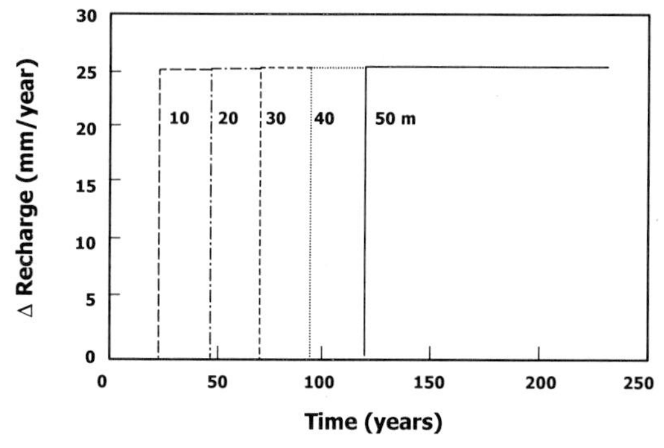

Figure 1. The increase in aquifer recharge as a function of time since an increase in drainage, for an increase in drainage rate of 25 mm/yr, (θ_2-θ_1) = 0.06, and water table depths of 10, 20, 30, 40 and 50 m.

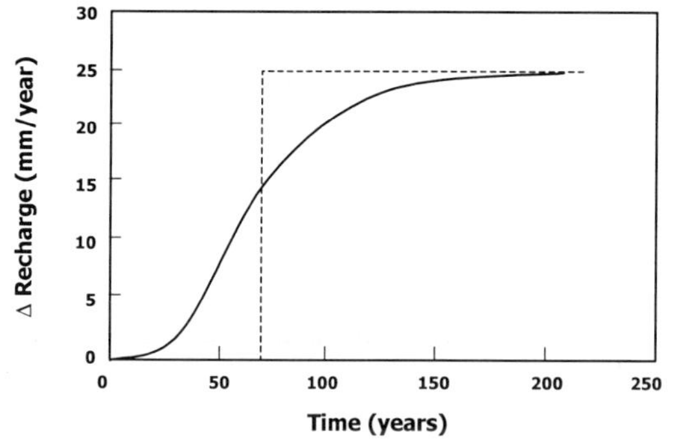

Figure 2. The increase in aquifer recharge as a function of time since an increase in drainage. The broken line depicts aquifer recharge for a uniform increase in drainage rate of 25 mm/yr, (θ_2-θ_1) = 0.06, and water table depth of 30 m. The solid line is for spatially variable drainage, defined by a log-normal distribution with mean of 25.1 and median 22.2 (other parameters identical).

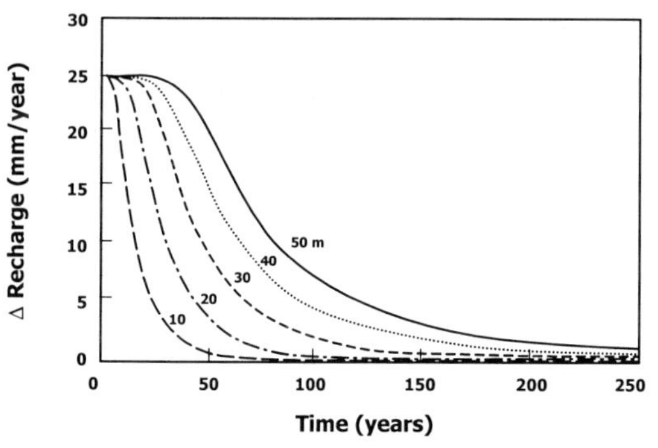

Figure 3. Predicted changes in aquifer recharge following cessation of drainage for five watertable depths, based on numerical modelling and using the soil moisture characteristic for a loamy sand. From Jolly and Cook (2001).

Sisson et al. (1980) present an approximate analytical solution to describe redistribution of water in a uniform soil under gravity drainage. By making a unit gradient assumption, the authors show that the drainage rate below depth z following cessation of infiltration can be approximated by

$$R(t) = \begin{cases} D_1 & t \le \dfrac{z}{A} \\ D_1\left(\dfrac{z}{At}\right)^{\frac{1}{1-n}} & t > \dfrac{z}{A} \end{cases} \tag{7}$$

where D_1 is the initial steady drainage rate,

$$A = \frac{D_1}{n(\theta_1 - \theta_2)} \tag{8}$$

and

$$\frac{K(\theta)}{D_1} = \left(\frac{\theta - \theta_2}{\theta_1 - \theta_2}\right)^{1/n} \tag{9}$$

is the Brooks and Corey (1964) hydraulic conductivity – water content function over the region of interest, θ_1 is the soil water content at drainage rate D_1, and θ_2 is the soil water content at zero drainage. Equation (7) was developed to approximate water redistribution in a soil draining under the influence of gravity in the absence of a watertable. Nevertheless, in areas of deep watertables it also allows approximation of the timelag between changes in drainage and changes in aquifer recharge. This approximate solution shows that the time required for the recharge rate to reduce to one half of its initial value is linearly related to the watertable depth, and will be less where the initial drainage rate is higher. Figure 4 compares the numerical results of Jolly and Cook (2001) for a 30 m watertable depth with the analytical solution produced using the same soil parameters. The difference between the two curves is due to the influence of the capillary fringe on the drainage process, which is not included in the analytical approximation.

FROM RECHARGE TO DISCHARGE

In this section, we present a simple model for groundwater flow in a homogeneous aquifer. The model is then used to quantify the timelags between changes in aquifer recharge and changes in aquifer discharge.

Consider an unconfined aquifer of length, L, and uniform width, with uniform porosity, ε, specific yield, S, and hydraulic conductivity, K, overlying a horizontal, impermeable aquiclude (Figure 5). The aquifer is bounded upgradient:

$$\frac{\partial h}{\partial x} = 0 \quad at\ x = L \tag{10}$$

where $h(x,t)$ is hydraulic head, and x is distance; and discharges to a stream downgradient, which is represented by a constant head boundary:

$$h = k \quad at\ x = 0 \tag{11}$$

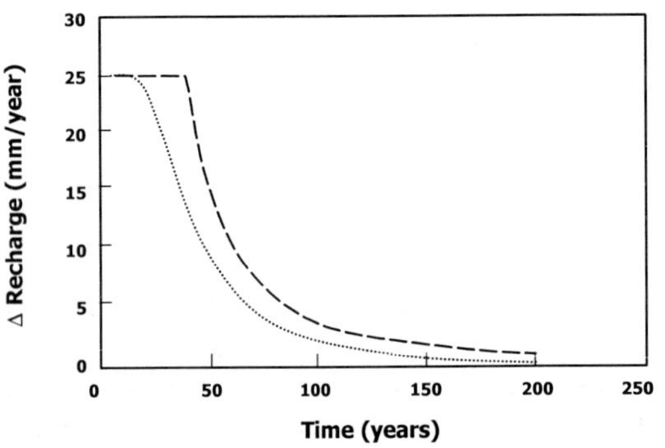

Figure 4. Comparison of analytical (dashed line) and numerical (dotted line) models for prediction of changes in aquifer recharge following cessation in drainage. Simulations are for an initial drainage (and recharge) rate of 25 mm yr^{-1} and a water table depth of 30m.

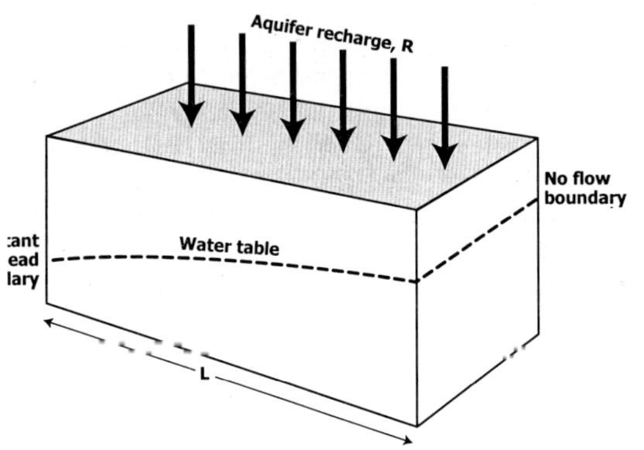

Figure 5. Schematic representation of a simple aquifer system, of constant width and length L, discharging to a river (represented by a constant head boundary).

If fluctuations in the water table are small compared to the aquifer thickness, then the flow of groundwater can be described by the linearised Boussinesq equation:

$$S\frac{\partial h}{\partial t} = K\bar{h}\frac{\partial^2 h}{\partial x^2} + R(x,t) \tag{12}$$

where $R(x,t)$ is the recharge rate, $\bar{h}$ is the mean water table depth, and t is time. This approach assumes that the aquifer transmissivity, $T=Kh$, is uniform. We define

$$Q(t) = T\frac{\partial h}{\partial x} \quad at\ x = 0 \tag{13}$$

to be the discharge to the river. The position of the water table under steady state conditions can be determined by solving Equation 12 with $\partial h/\partial t = 0$. If the aquifer recharge rate is uniform, then under steady flow conditions, the position of the water table is given by

$$h(x) = k + \frac{RL^2}{T}\left(\frac{x}{L} - \frac{1}{2}\left(\frac{x}{L}\right)^2\right) \tag{14}$$

and the discharge rate to the river is equal to the aquifer recharge rate:

$$Q = RL \tag{15}$$

To simulate changes in aquifer discharge arising from changes in groundwater recharge, we have used an analytical solution to Equation 12. Using the linearised form of the Boussinesq Equation, timelags for step increases and decreases in recharge can be represented by the same mathematical equations. For simplicity, we have defined the transmissivity to be $T = Kk$, so that the assumption of constant transmissivity is most accurate (i.e. closer to the aquifer transmissivity, Kh) close to the river, which is the area of most interest. The initial condition is given by the steady state water table position for recharge rate R_1 (Equation 14). For an instantaneous change in recharge rate from R_1 to R_2 at time t = 0, the solutions for the hydraulic head $h(t,x)$ and aquifer discharge $Q(t)$ become

$$h(t,x) = k + \frac{R_2}{T}\left(Lx - \frac{x^2}{2}\right) + \frac{2L^2}{T}(R_1 - R_2)\sum_{n=0}^{\infty}\frac{\sin\left[\frac{\beta_n x}{L}\right]}{\beta_n^{\ 3}}\exp\left[-\frac{\beta_n^2 T}{L^2 S}t\right] \quad (16)$$

and

$$Q(t) = LR_2 + 2L(R_1 - R_2)\sum_{n=0}^{\infty}\frac{\exp\left[-\frac{\beta_n^2 T}{L^2 S}t\right]}{\beta_n^2} \quad (17)$$

where $\beta_n = (2n+1)\frac{\pi}{2}$. Where the instantaneous change in aquifer recharge at time $t = 0$ is restricted to the region between x_1 and x_2, the solution for the aquifer discharge becomes:

$$Q(t) = LR_1 - (R_1 - R_2)(x_2 - x_1) + 2L(R_1 - R_2)\times \sum_{n=0}^{\infty}\frac{1}{\beta_n^2}\left(\cos\frac{\beta_n x_1}{L} - \cos\frac{\beta_n x_2}{L}\right)\exp\left[-\beta_n^2\frac{T}{L^2 S}t\right] \quad (18)$$

Because Equation 18 is linear, solutions for more general step changes in $R(t,x)$ (over various distances and times) can be readily calculated by summing solutions of this form.

Figure 6 shows the increase in groundwater discharge as a function of time for increases in aquifer recharge that are limited to bands of 1 km width parallel to the river which drains the aquifer. An increase in recharge rate of 10 mm yr^{-1} at time $t = 0$ is simulated (K = 5 m/day, S = 0.1, k = 50 m, L = 100 km). In all cases, under steady-state conditions, the increase in groundwater discharge corresponding to this increase in aquifer recharge is 10 m^2 yr^{-1}. This diagram illustrates that (for the above aquifer parameters), recharge that occurs further than 20 km from the river has little impact on groundwater flow rates to the river for more than 200 years.

APPLICATIONS

Land Salinisation

In southeastern Australia, clearing of native Eucalyptus (mallee) vegetation for dryland agriculture has led to land and river salinisation. The native vegetation is extremely water-efficient, and its roots may extend to 20 m depth. Several studies have shown that rates of groundwater recharge under native mallee vegetation are less than 0.1 mm/yr (Allison et al. 1990). Clearing of the native vegetation, and its replacement with shallow-rooting crops and pastures results in increases in the amount of water draining below the plant root zone. The increase in drainage, once transmitted to the water table, causes the water table to rise. Where the watertable rises to close to the land surface, concentration of salts during evaporation causes land salinisation. The rising water table also causes increased hydraulic gradients towards the Murray River. Because the groundwater adjacent to the river is saline, this causes increased flows of saline groundwater into the river, and hence increases in river salinity.

Because the unsaturated zone over much of the area is thick (up to 50 m), and also because groundwater flowpaths towards the river are long (up to 200 km), the salinisation process takes a long time to develop. Based on a drainage rate under dryland agriculture of 10 mm/yr, soil volumetric water contents under mallee and agriculture of θ_1 = 0.06 and θ_2 = 0.12, and depth to water table of z_{WT} = 30 m, it is estimated that there will be a time lag of 180 years before the groundwater recharge rate increases, and the water table begins to rise. Of course, spatial variability in drainage rate, depth to water table and soil properties mean that the magnitude of this timelag also displays large spatial variability.

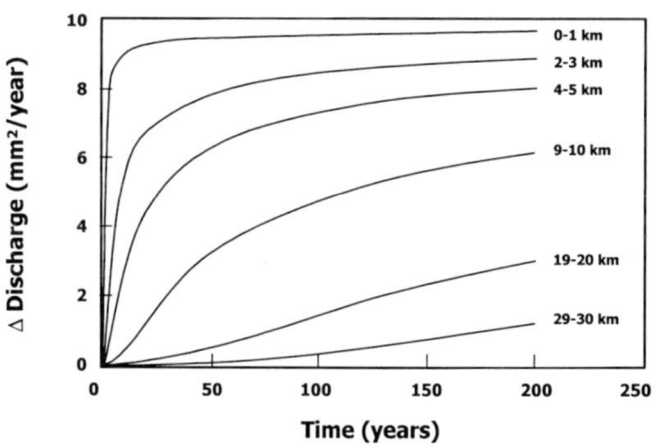

Figure 6. Change in groundwater discharge to the river as a function of time since a change in recharge over 1 km wide strips parallel to the river. Simulations are for a change in recharge rate of 10 mm/yr at time t=0, T = 250 m²/day, S = 0.1, L = 100 km.

Jolly et al. (1989) and Walker et al. (1991) showed that matric potential profiles could be used to identify the position of the pressure front created by the increase in drainage. In the example shown in Figure 7, the drainage rate is estimated to be 45 mm/yr, and the water table is at approximately 60 m. In this case the velocity of the pressure front is estimated to be 0.83 m/yr (7.5 m in 9 years), and so it is estimated that there will be a timelag of 72 years before the aquifer recharge rate increases.

Initially, the rate of water table rise will be $\frac{\partial h}{\partial t} = (R_2 - R_1)/S$. For an increase in recharge of 10 mm/yr, and specific yield of $S = 0.1$, this equates to an initial rate of water table rise of 0.1 m/yr. However, the rate of water table rise will slow as discharge to the river increases. Assuming a catchment length of L = 100 km and transmissivity of T = 200 000 m^2/yr, the rate of water table rise 320 years after the recharge rate increases (500 years after land clearance), is estimated to be 0.08 m/yr at x = 50 km, and 0.01 m/yr at x = 5 km (Equation 16).

Once the water table begins to rise, groundwater flow to the river will begin to increase. Using the same aquifer parameters as above, the groundwater discharge rate to the river is predicted to increase by 130 m^3/yr per metre width of aquifer 70 years after the aquifer recharge rate increases (250 years after land clearance) and by 280 m^3/yr per metre width of aquifer 320 years after the aquifer recharge rate increases (500 years after land clearance). If the groundwater salinity is 20 000 mg/L, then the increase in salt load to the river 500 years after clearing is predicted to be 16 tonnes/day per km width of aquifer (i.e. per km length of river). For the approximately 500 km of river length for which these conditions apply, this represents approximately 8000 tonnes/day of salt in 500 years. The groundwater flow rate (and hence salt load) to the river is predicted to continue to increase for more than 5000 years.

More accurate assessment of the future increase in groundwater flow to the river, including consideration of spatial variations in drainage, watertable depth and aquifer transmissivity, and incorporating river geometry has been determined using a groundwater flow model (Allison et al., 1990; Barnett, 1990). The results of this extensive study were for a increase in salt load to the river of approximately 2000 tonnes/day after 500 years, which, while less, is of the same order as that determined with the simple equations presented here.

One of the management options being considered to reduce the flow of saline groundwater to the river is revegetation to reduce recharge. However, the timescales associated with this strategy may be quite long. Figure 8 shows the reduction in groundwater discharge to the Murray River as a function of time, resulting from revegetation along a 5 km strip adjacent to the river, for a sandy loam soil and water table depths before revegetation of 10 m and 40 m. The model assumes that revegetation prevents additional drainage below the surface, but does not allow plants to extend roots deeply into the soil to 'mine' existing soil water.

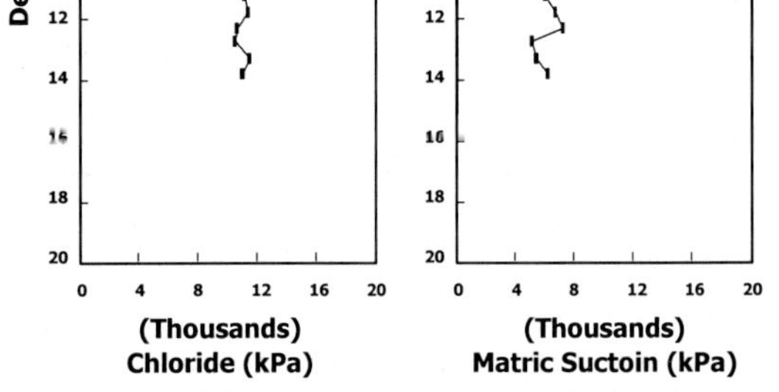

Figure 7. Chloride and matric suction profiles beneath land that was cleared 9 years before sampling. The chloride front has been displaced to between 4 and 5 m depth, and the pressure front has moved to 7.5 m. From Jolly et al. (1989).

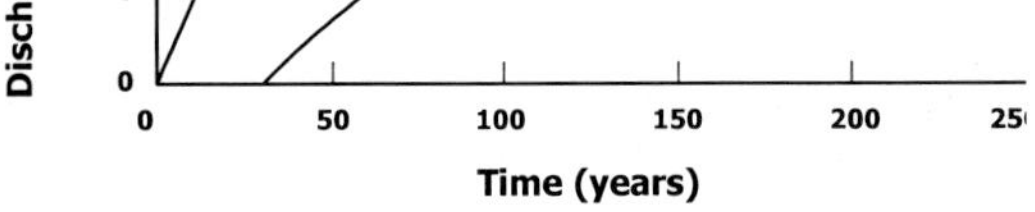

Figure 8. Reduction in groundwater discharge to the river as a function of time since revegetation, for revegetation of a 5 km strip adjacent to the river. Simulations incorporating the timelag between revegetation and the reduction in aquifer recharge for a sandy loam soil with water table depth of 10 m and 40 m are compared. From Cook et al. (2001).

The results include the timelag between cessation of drainage and the reduction in aquifer recharge, and the timelag between the reduction in aquifer recharge, and the consequent reduction in aquifer discharge. The initial condition for this simulation was the steady-state water table configuration for a recharge rate of R_1 = 0.1 mm yr^{-1}. At time t = 0, it was assumed that recharge to the aquifer everywhere increased to R_2 = 10 mm yr^{-1}. At time t = 100 years, drainage decreased to zero. Solution of the Richards' Equation was used to determine the aquifer recharge rate as a function of time since the reduction in recharge, and this information was then used as input to Equation 18. For a sandy loam soil with 40 m water table depth, aquifer recharge is 10 mm yr^{-1} until approximately t = 120 years (20 years after cessation of drainage), after which it begins to decrease. Fifty years after revegetation (t = 150 years), aquifer recharge is approximately 6.5 mm yr^{-1}. The simulations clearly show that the time required to achieve a reduction in discharge increases as the unsaturated zone thickness increases (Figure 8). Eventually, the discharge reduction is equal to 50 $m^2 yr^{-1}$, which is equal to the width of the revegetation strip (5 km) multiplied by the recharge reduction (10 mm yr^{-1}). This represents a decrease in salt load to the river of 2.7 tonnes day^{-1} km^{-1}. For early time periods, there are diminishing benefits in extending the revegetation strip beyond 5 km from the river (Figure 6).

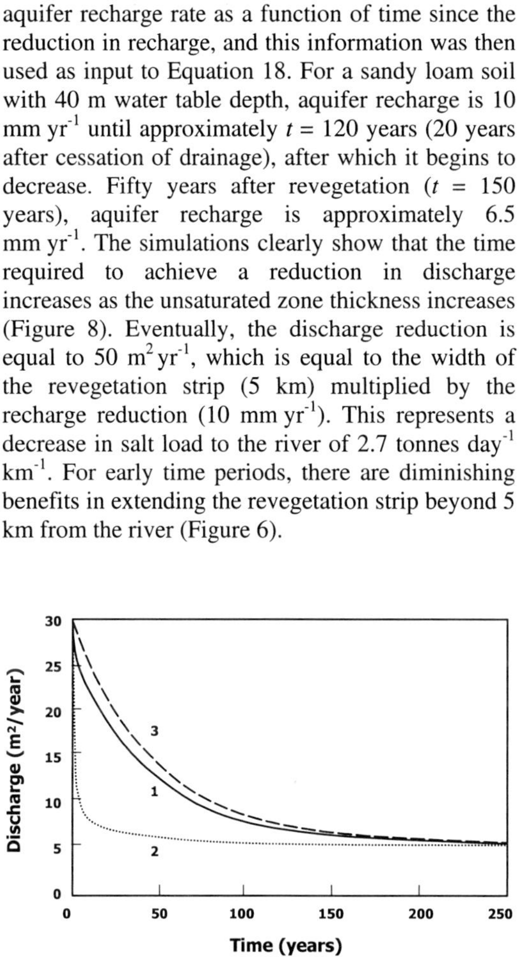

Figure 9. Impacts of groundwater pumping on groundwater discharge to a river. 1: Groundwater extraction distributed evenly throughout catchment; 2: Groundwater extraction concentrated with 1 km of river; 3: Groundwater extraction excluded from 2 km buffer zone adjacent to river. Simulations are for aquifer parameters: L = 10 km, S = 0.1, T = 10^5 m^2/yr, R_1 = 50 mm/yr, and a groundwater extraction rate of 500 m^2/yr.

Impacts of Groundwater Extraction

The 'safe yield' of an aquifer has been defined as: "the attainment and maintenance of a long-term balance between the amount of groundwater withdrawn annually and the annual amount of recharge" (Sophocleous, 1997). This definition, however, ignores natural groundwater discharges. More recently, there has been a concern that groundwater use should not impact the environment, and should allow for groundwater flows to rivers to be maintained. Groundwater extraction represents an increase in aquifer discharge. Provided that the aquifer recharge rate remains unchanged, then groundwater extraction must ultimately reduce the natural aquifer discharge by an equivalent amount. The ultimate reduction in natural discharge will then depend on the total groundwater extraction. However, the timelag between the commencement of groundwater extraction and the reduction in natural discharge will depend on the location of the wells.

The timelag between groundwater extraction and the reduction in aquifer discharge can be modeled using Equations 17 and 18, where R_2 is the net recharge rate (recharge minus groundwater pumping). Consider an aquifer of length L = 10 km, transmissivity T = 10^5 m^2/yr, specific yield S = 0.1, with uniform recharge rate R = 50 mm/yr. Under natural conditions, the aquifer discharges to a stream, the discharge rate of 500 m^2/yr being equal to the product of the recharge rate and the aquifer length. Figure 9 shows the reduction in groundwater discharge to the stream that occurs following the commencement of groundwater extraction of 250 m^2/yr at time t = 0. In particular, the figure compares situations where groundwater extraction is evenly distributed across the aquifer, where all groundwater extraction occurs within 1 km of the river, and where groundwater extraction is excluded from a 2 km buffer zone adjacent to the river. While all of these groundwater extraction scenarios ultimately have the same impact on the river (a halving of the groundwater inflow component), the early-time impact can be controlled by regulation of bore location. In this case, the use of a 2 km buffer zone results in reduced impact on the river for more than 100 years.

CONCLUSIONS

In large groundwater basins, the impacts of land use changes can take many years to manifest. In areas of thick unsaturated zones, changes in drainage due to changes in vegetation cover may not alter aquifer recharge rates for many decades. The processes that determine this timelag are different in the case of increasing or decreasing drainage, although both can be estimated from simple analytical or numerical models. There is also a timelag between changes in groundwater recharge, and the consequent change in groundwater discharge. The magnitude of this timelag is most strongly affected by the aquifer transmissivity and the distance of the recharge perturbation from the discharge area, and can also be estimated using a simple analytical model. For dryland salinity management in southern Australia, these timelags mean that the full impacts of clearance of native vegetation are unlikely to be manifested for many thousands of years. However, while these timelags have resulted in a relatively slow rate of land degradation, they also mean that ameliorative actions may take a long time to be effective. In the case of groundwater resource management, increasing distances between groundwater extraction and rivers can significantly delay the impacts of development.

REFERENCES

Abdulrazzak M.J. and Morel-Seytoux H.J., 1983. Recharge from an ephemeral stream following wetting front arrival to the water table. *Water Resources Research* 19: 194-200.

Allison G.B., Cook P.G., Barnett S.R., Walker G.R., Jolly, I.D. and Hughes M.W., 1990. Land clearance and river salinisation in the western Murray Basin, Australia. *Journal of Hydrology* 119: 1-20.

Barnett S.R., 1990. Murray Basin hydrogeological investigation - Mallee region groundwater modelling exercise. Department of Mines and Energy, Adelaide. Report 90/36.

Brooks R.H. and Corey A.T., 1964. Hydraulic properties of porous media. Hydrology Paper 3. Colorado State University, Fort Collins.

Cook P.G., Walker G.R. and Jolly I.D., 1989. Spatial variability of groundwater recharge in a semiarid region. *Journal of Hydrology* 111: 195-212.

Cook P.G., Leaney F.W. and Jolly I.D., 2001. Groundwater recharge in the Mallee Region, and salinity implications for the Murray River. CSIRO Land and Water, Technical Report 45/01.

Jolly I.D. and Cook P.G., 2002. The time lags in salinity control: controlling recharge and halting watertable rise. *Natural Resource Management* 5(2): 16-21.

Jolly I.D, Cook P.G, Allison G.B. and Hughes M.W., 1989. Simultaneous water and solute movement through an unsaturated soil following an increase in recharge. *Journal of Hydrology* 111: 391-396.

Morel-Seytoux H.J., 1984. From excess infiltration to aquifer recharge: a derivation based on the theory of flow of water in unsaturated soils. *Water Resources Research* 20: 1230-1240.

Sisson J.B., Ferguson A.H. and van Genuchten M.Th., 1980. Simple method for predicting drainage from field plots. *Soil Science Society of America Journal* 44: 1147-1152.

Sophocleous M., 1997. Managing water resources systems: why "safe yield" is not sustainable. *Ground Water*35(7): 561.

Walker G.R., Jolly I.D and Cook P.G., 1991. A new chloride leaching approach to the estimation of diffuse recharge following a change in land use. *Journal of Hydrology* 128: 49-67.

Environmental Tracers Applied to Quantifying Causes of Salinity in Arid-Region Rivers: Results from the Rio Grande Basin, Southwestern USA

FRED M. PHILLIPS, SUZANNE MILLS, M.H. HENDRICKX
Earth & Environmental Science Department, New Mexico Tech, Socorro, NM 87801 USA.
E-mail: Phillips@nmt.edu

JAMES HOGAN
Department of Hydrology & Water Resources, University of Arizona, Tucson, AZ 85721 USA.

ABSTRACT: The upper Rio Grande extends ~1,200 km between its headwaters in southern Colorado (USA) and the USA/Mexico border region. Over this distance the total dissolved solids content of the water increases from ~40 mg L^{-1} to over 2,000 mg L^{-1}. This increase has previously been attributed to evapotranspirative concentration and to flushing by irrigation water of salts accumulated by pre-irrigation evapotranspiration. We employed environmental tracers, including $\delta^{18}O$, $\delta^{2}H$, Cl^-, and the Cl/Br ratio to help identify the causes of salinization. Both $\delta^{18}O$ and $\delta^{2}H$ are enriched with flow distance due to evaporation, but not sufficiently to explain most of the salinization as caused by evapotranspiration. The Cl/Br ratio increases from about 50 in the headwaters to over 1,000 at the distal end of the river basin, indicating influx of subsurface saline waters, which are commonly characterized by high Cl/Br ratios. The Cl^- concentration and the Cl/Br ratio increase in a step-wise fashion and are typically localized at the southern ends of the sedimentary basins comprising the Rio Grande Rift, suggesting that the salts are from discharge of deep ground water where it is forced to the surface by bedrock highs, rather than due to flushing by irrigation water.

INTRODUCTION

Water scarcity is commonly considered the major water-resources problem in arid regions, but in fact, poor water quality is in many cases as important a problem in these regions, or an even greater one. Although toxic contaminants have received a very large share of both the public and the scientific interest in water quality issues, the buildup of ordinary salinity in arid-region surface and ground water is generally of much greater practical significance. Understanding of the causes of salinization is a prerequisite to its management and possible mitigation. Although many studies have been conducted on specific salinization problems, most of these are focused on limited geographical areas that are usually only small fractions of entire river basins. Although such approaches may be locally effective, they ignore the large-scale linkages that cause progressive salinization at the basin scale. This paper will deal with salinization problems at the basin scale, using the Rio Grande basin in the southwestern United States as an example.

The solute burden of rivers can be divided into three major categories. The first is cyclical and surficial salts. These originate from atmospheric deposition (both in precipitation and as dry fallout) and from weathering reactions of precipitation with surficial rocks and soil in the watershed. The spatial origin of such salts is very diffuse and they are virtually impossible to limit or control. The second category is the inflow of subsurface-origin saline waters. Such waters may be derived from solution of evaporite deposits, or may be of connate, diagenetic, or geothermal origin. If the points of introduction of such saline waters are localized, it may be possible to intercept them or otherwise mitigate their effects. The third category is anthropogenic salt inputs. These may arise from industrial and domestic waste streams, road salt, dietary salt, fertilizers and soil amendments, petroleum extraction, and many other human sources. Such anthropogenic sources are in principle controllable, but their limitation may in fact be economically or socially difficult. In arid and semiarid river basins the effects of evaporation and transpiration subsequently concentrate salts from all three categories above. Evaporative increases in concentration may tend to increase the importance of the more conservative solutes (e.g., Cl^-, Na^+) in the solute burden, while decreasing the proportion of the more reactive ones (e.g., Ca^{2+}, HCO_3^-). The extent to which increases in dissolved solids can be attributed to evapotranspirative concentration as compared to solute addition can be difficult to quantify.

Water Resources Perspectives: Evaluation, Management and Policy. Edited by A.S. Alsharhan and W.W. Wood.
Published in 2003 by Elsevier Science, Amsterdam, The Netherlands, p. 327-334.

Traditional approaches to investigating river salinization have relied on water quality measurements made at river gaging stations to calculate salt burdens. Relatively constant burdens with distance down river are taken to imply that concentration increases can be attributed to evapotranspiration, whereas increasing burdens imply some additional source of salt. However, such studies are limited to gaged sites, which are frequently sparse, and they do little to identify sources of salt influx. In this study, we employ environmental tracers in addition to traditional salt-burden calculations. The tracers can be measured at any desired spatial frequency in order to pinpoint locations of salt addition, and can be used to fingerprint sources of salinity.

Rio Grande Basin

The headwaters of the Rio Grande are in the southern Rocky Mountains, in the state of Colorado, USA (Figure 1). Most of the runoff that supplies the river originates as snowfall at elevations of 3,000 to 4,000 m in the San Juan, Sangre de Cristo, and Jemez mountain ranges. The river flows south through the states of Colorado and New Mexico before turning southeast to form the boundary between the United States and Mexico and ultimately flowing into the Gulf of Mexico. This study was limited to the northern portion of the drainage basin, down to a point 130 km south of the cities of El Paso and Ciudad Juarez. This point is 1,150 km downstream from the river's headwaters. Further downstream the river is frequently dry, although eventually additional tributaries restore perennial flow. The elevation at this point is 1,100 m and the area of the drainage basin above it is 120,000 km^2.

Hydrogeologic Framework

The relatively linear course of the Rio Grande as it flows south is largely due to structural control by the Rio Grande rift. This is an intracontinental extensional feature similar in origin to the better-known East African rifts. Rifting was initiated at ~30 Ma and was initially rapid, but has slowed with time (Keller and Cather, 1994). The rift consists of a series of elongate basins that are usually bounded on either side by Precambrian crystalline or Paleozoic sedimentary basement rocks (see Figure 1). The basins are formed by full or half grabens that are filled with up to 4,000 m of alluvial and lacustrine sediment (Keller and Cather, 1994; Wilkins, 1998). The grabens are terminated at either

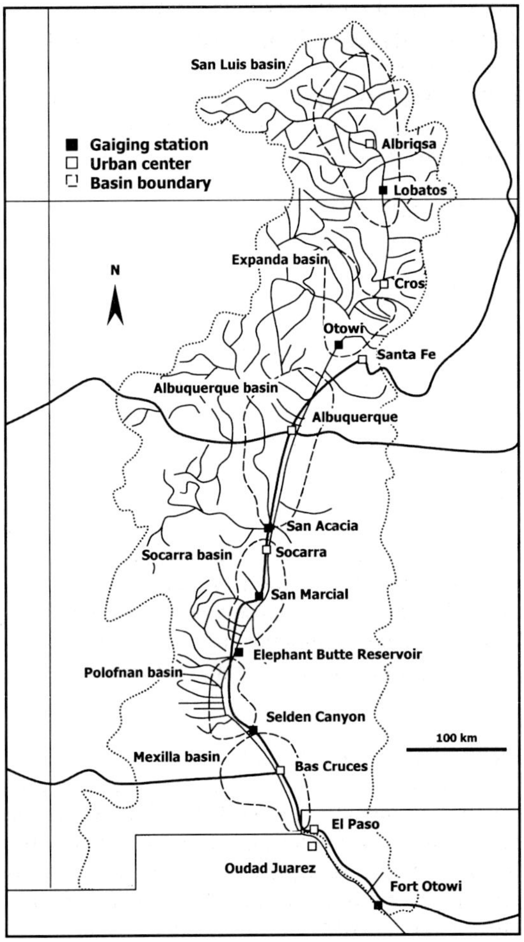

Figure 1. Map of the upper Rio Grande, showing population centers, gaging stations mentioned in text, and sedimentary basins that form the Rio Grande rift.

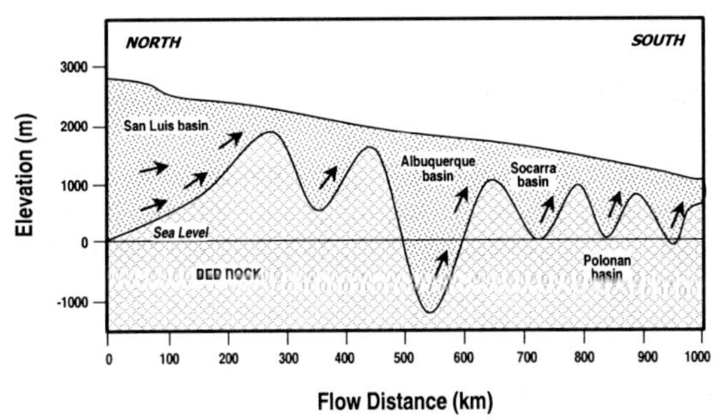

Figure 2. Schematic hydrogeologic cross-section of the Rio Grande rift, parallel to the path of the river. Arrows indicate direction of groundwater flow. Basin fill depth dotted where not well-constrained. Basin structure is based on data from Anderholm (1987), Hawley (1984), Keller and Cather (1994), and Wilkins (1998).

end by transform structures that bring basement rocks close to the surface. Figure 2 shows a cross-section of the basin structure parallel to the course of the river.

Precipitation and Runoff

Although precipitation is as high as 1,300 mm yr^{-1} over small areas in the mountainous headwaters, the majority of the drainage basin ranges from semiarid to arid. Most of the basin receives less than 300 mm yr^{-1} and the driest areas only 160 mm yr^{-1}. Summer temperatures are high and over most of the basin potential evapotranspiration varies from 1,000 to 2,000 mm yr^{-1}.

Within the study area the river discharge first increases as tributaries flow in, then decreases, partly because no further significant tributaries join the river, and partly because of evapotranspirative losses from riparian vegetation and reservoirs, as well as irrigation and municipal diversions. At the Colorado-New Mexico border, about 270 km from the headwaters, the mean annual discharge is 17 m^3 s^{-1}, although the natural flow would be about 40 m^3 s^{-1} without agricultural diversions (discharge and water use data are from Ellis et al., 1993). The river reaches its maximum discharge of 49 m^3 s^{-1} at Otowi, New Mexico, 430 km down the river. This would probably be closer to 70 m^3 s^{-1} under natural conditions. At Albuquerque, New Mexico, 550 km down river, the discharge is 44 m^3 s^{-1}. Above Elephant Butte Reservoir (800 km downstream and by far the largest reservoir on the river) the discharge amounts to 40 m^3 s^{-1}; below the reservoir it is about 33 m^3 s^{-1}. The difference is largely attributable to evaporation from the reservoir. At El Paso (1,025 km) this discharge has been further reduced to 20 m^3 s^{-1}, mostly due to agricultural diversions.

Water Use

There are five major population centers within the drainage basin: Santa Fe, New Mexico (450 km from the headwaters, population 70,000), Albuquerque, New Mexico (550 km, population 420,000), Las Cruces, New Mexico (950 km, population 75,000), and the twin cities of El Paso, Texas (1,025 km, population 600,000) and Ciudad Juarez, Chihuahua, Mexico (1,025 km, population 1,300,000). Most municipal water supply is from wells, but Santa Fe and El Paso do use significant amounts of surface water as well. Use of surface water by municipalities totals about 1 m^3 s^{-1}.

Total irrigated area in the basin is approximately 370,000 ha, of which 260,000 ha are in Colorado, near the headwaters of the river. Withdrawals for irrigated agriculture amount to about 70 m^3 s^{-1}, of which about half returns to the river. Losses due to open-water evaporation (especially from Elephant Butte Reservoir) and riparian evapotranspiration are difficult to quantify, but, based on balancing the various river discharges and consumptive uses given above, are probably about 18 m^3 s^{-1}.

Water Quality

The dissolved solids content of the Rio Grande increases markedly as it flows from its headwaters in Colorado to the southern limit of the study area (Figure 3). The total dissolved solids content in the headwaters region averages about 40 mg L^{-1}. By the time the river reaches El Paso the average value is about 750 mg L^{-1}, and at Fort Quitman, south of El Paso, it commonly attains values in excess of 2,000 mg L^{-1}. Not only the solute concentration, but also the solute burden increase downstream, from about 100,000 tons yr^{-1} at Lobatos to about 600,000 tons yr^{-1} south of San Marcial. The concentrations of dissolved solids vary seasonally, and during the winter, when base flow is not diluted by releases of irrigation water from dams, can greatly exceed the average values given above (Figure 4). Water quality in the vicinity of El Paso approaches the

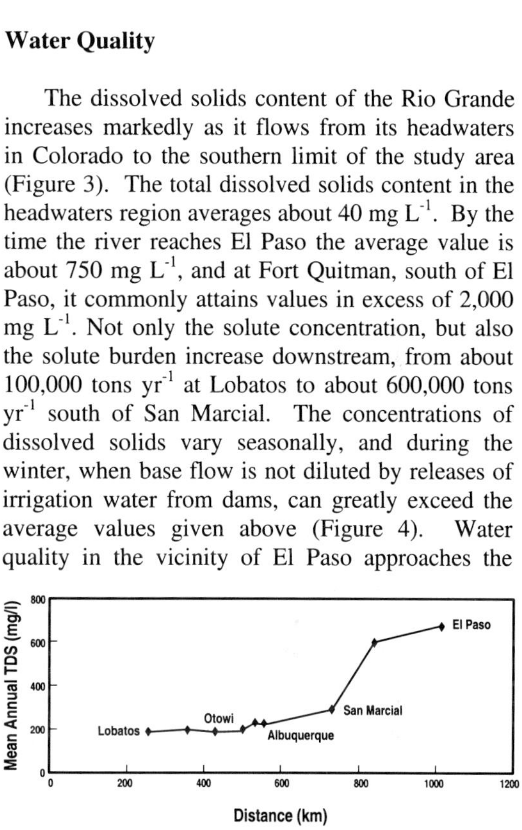

Figure 3. Variation of mean annual total dissolved solids (TDS) content of the Rio Grande with flow distance (data from U.S. Geological Survey gaging station records).

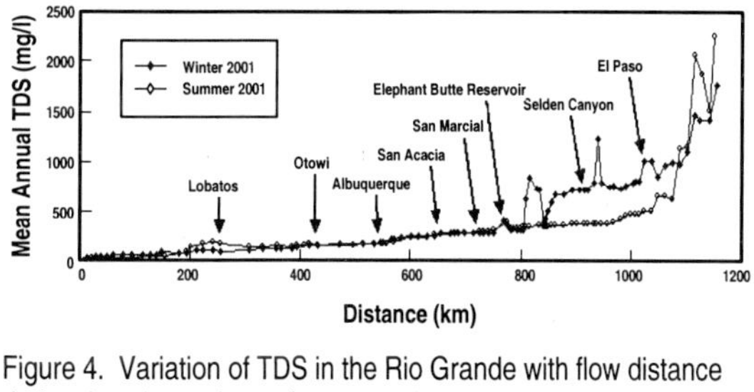

Figure 4. Variation of TDS in the Rio Grande with flow distance during the winter of 2000/2001 and the summer of 2001.

limits advised for both irrigation and drinking-water use. South of El Paso it is not usable for most purposes. Analysis of historical records shows that during periods of extended drought, such as the mid-to-late 1950's, salinity consistently exceeded the present-day averages. Population, and thus demand on the water resources, has greatly increased since the 1950's, and a repetition of such a long-term drought could have disastrous consequences with regard to water quality.

The causes of this prominent increase in salinity with flow distance have never been fully explained. Evapotranspiration during irrigation and by riparian vegetation clearly plays a role, but cannot explain the observed increase in solute burden. The work of the National Resource Committee (1938) attributed it principally to displacement of natural saline ground water by infiltrating irrigation water. Hayward (1956) and Wilcox (1957) laid more emphasis on increases in the proportion of the more soluble salts during evaporation in irrigated soils.

Application of Environmental Tracers

Studies employing environmental tracers to investigate the sources of salinity have been pursued as part of the Rio Grande Basin project carried out through the Center for Sustainability of Semi-Arid Hydrology and Riparian Areas, funded by the U.S. National Science Foundation. High spatial-resolution sampling campaigns (~10 km sampling intervals) have been conducted by the authors twice per year since the winter of 1999/2000: during late summer when irrigation water is being released from the reservoirs, and during mid-winter when there are no irrigation releases and flows are minimal. The main tracers employed to date have been the stable isotopes of oxygen and hydrogen in the water molecule, chloride, and bromide.

Stable isotopes of oxygen and hydrogen

The stable isotopes of oxygen ($^{18}O/^{16}O$) and hydrogen ($^{2}H/^{1}H$) in the water molecule are a commonly-employed tool for understanding the origins and circulation of natural waters (Coplen et al., 1999; Gat, 1996). The stable isotope content of precipitation at middle and northern latitudes is most strongly controlled by the temperature at the time of the precipitation event, with isotopically lighter precipitation being associated with colder temperatures (Dansgaard, 1964). Inasmuch as most of the runoff for the Rio Grande originates as snow at high elevation, it can be expected to be isotopically light.

One of the most important controls on the stable isotope composition of surface water is evaporation. The heavier isotopic species ($H_2{}^{18}O$ and ^{2}HHO) are less volatile than their lighter (and more common) counterparts and thus tend to be retained during evaporation. This causes progressive isotopic enrichment during the process of evaporation. In contrast, transpiration does not affect the isotopic composition of the residual water because plants transmit water from the soil to the atmosphere without isotopic separation. Measurement of the stable isotope composition of salinized water is a useful method for discriminating the cause of the salinity because water that is saline due to evaporation will be isotopically more enriched than the source water, whereas water that is saline due to salt addition or transpiration will not change isotopic composition.

The variation of stable isotope composition of the river water is illustrated in Figure 5. When plotted against each other (Figure 5a), the two isotopes form a linear array with a slope of 5, which is considered diagnostic of evaporation. As expected, the initial $\delta^{18}O$ and $\delta^{2}H$ are relatively light (-14 and -100‰, respectively) and are consistent with winter precipitation for the region (Adams et al., 1995). The plot of $\delta^{2}H$ against flow distance in Figure 5b shows obvious trends that correspond to features of the river system. The peak in $\delta^{2}H$ at about 250 km is explained by inflow of evaporated drainage water from the Closed Basin Project in the San Luis Valley. This isotopically heavy inflow is diluted downstream by isotopically light tributaries. The major step in $\delta^{2}H$ at 800 km is at Elephant Butte Reservoir, which is known to be the single largest evaporative sink in the entire river system. Downstream of Elephant Butte Reservoir the isotopic composition is surprisingly stable, given the large amount of diversion for irrigation and inflow of return waters in the Mesilla Valley. This may indicate that most water loss in this stretch is by transpiration.

A simple Rayleigh distillation model (Campbell and Larson, 1998) can be used to estimate the degree of evaporation necessary to produce the observed enrichment. Assuming an effective isotopic enrichment factor for ^{2}H of –89 ‰ during evaporation, the Rayleigh model indicates that about 35% of the river flow has been evaporated by the time the river reaches El Paso. This differs from the water balance based on gaging that was described above, which indicates that about 75 % of the water is lost to evapotranspiration above El Paso. However, transpiration does not fractionate ^{2}H or ^{18}O. The stable isotopes thus indicate that about half

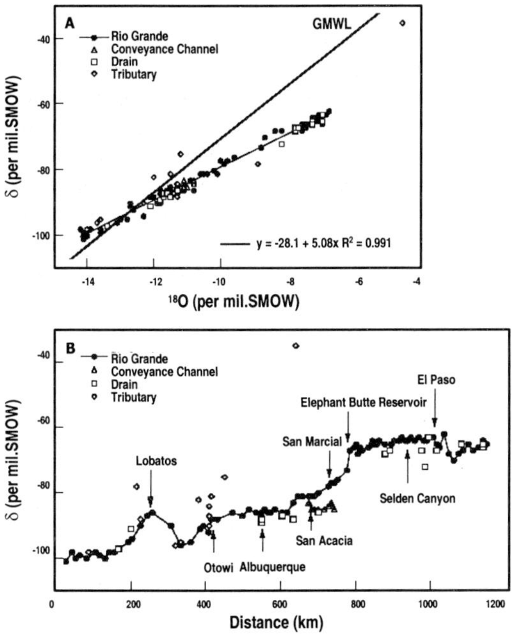

Figure 5. Stable oxygen and hydrogen isotopes in waters of the Rio Grande during August 2001. "GMWL" is the global meteoric water line. (a) d18O versus d2H. Values are in per mil relative to the SMOW standard. (b) d2H in per mil as a function of flow distance.

of the river losses are due to evaporation and the other half due to transpiration from irrigated fields and riparian areas.

Chloride/Bromide Ratio

Chloride has many uses as a tracer in natural water (Feth, 1981). It is normally present only in the form of Cl^- and does not participate in precipitation or adsorption reactions. Chloride is thus highly conservative under natural conditions. Bromide generally exhibits similar geochemical characteristics, and thus the Cl/Br ratio is a conservative indicator for the origin of natural waters (Davis et al., 1998). Meteoric water is usually characterized by a Cl/Br mass ratio of 150 to 50. In contrast, most subsurface sources have substantially larger Cl/Br ratios, ranging from ~200 for many oilfield brines up to over 10,000 for solutions from the dissolution of marine evaporites (Davis et al., 1998; Whittemore, 1995). Many anthropogenic sources of chloride, such as sewage, also have high Cl/Br ratios (Vengosh and Pankratov, 1998). If the progressive increase in salt burden of the Rio Grande with flow distance can be attributed to influx of subsurface or anthropogenic salts, it should be accompanied by a similar progressive increase in the Cl/Br ratio.

Figure 6 shows the change in the Cl^- concentration and the Cl/Br ratio with flow distance, sampled in August 2001. Chloride concentration in the headwaters is about 0.5 mg L^{-1}. It increases to about 9 mg L^{-1} at Albuquerque, 50 mg L^{-1} at Elephant Butte, and 100 mg L^{-1} at El Paso. The Cl/Br ratio at the headwaters is about 50, at the lower end of the meteoric range. Were all of the increase in Cl^- concentration attributable to evapotranspirative consumption of the water, this ratio would remain relatively constant downstream. (Minor geographical variations in the Cl/Br ratio of atmospheric deposition and possible effects of preferential uptake of Br by plants could cause small downstream variations in the river Cl/Br ratio). In fact, it increases to about 200 at Albuquerque and 500 at El Paso. Thus it is clear that the salinization is caused, at least in part, by addition of saline water having high Cl/Br ratios, most likely of subsurface origin.

In order to help quantify the mechanisms of salinization, the Cl^- and Br^- concentrations were modeled using a simple two-component mixing and evaporation model:

$$m_i = e\left[m_{s,i}f + m_{a,i}(1-f)\right]$$

where m is salt concentration, i indicates Cl^- or Br^-, a is the atmospherically-derived end member, s is the subsurface end member, f is the fraction of water originating from subsurface inflow, and e is inverse evaporative loss (i.e., e is equivalent to V_o/V_x where V_o is the original volume of runoff water and V_x is

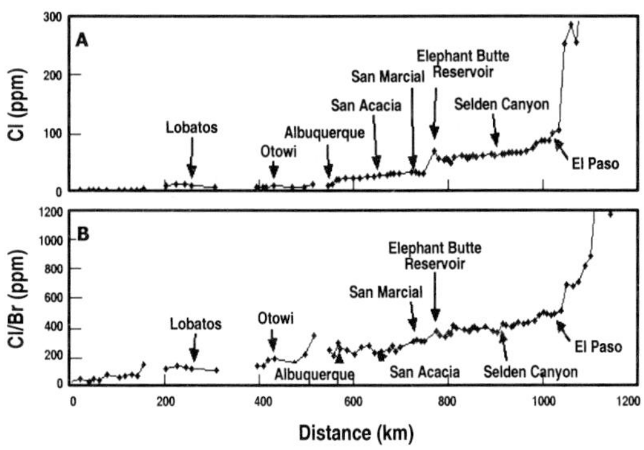

Figure 6. Chloride and bromide in waters of the Rio Grande. (a) Chloride as a function of flow distance during August 2001. (b) The Cl/Br ratio as a function of flow distance.

the volume remaining after evapotranspirative loss at some distance x down the river). These two equations (for Cl^- and Br^-) were solved simultaneously for e and f using the Cl^- and Br^- measured at each point down the river. The meteoric end-members were taken as the Cl and Br data for the next station upstream from each measurement point. The end-member concentrations for the subsurface (i.e., saline groundwater) were chloride, 280 mg L^{-1} and bromide, 0.5 mg L^{-1}. These values were taken from the most concentrated ground waters identified by Bexfield (2001) in the "discharge zone" of the Albuquerque Basin, which is located at its southern end.

The results of the mixing/evaporation model calculations are shown in Figure 7. The model employed is very simplistic and the results should be considered as only semi-quantitative (for example, anthropogenic sources of Cl^-, effects of varying composition of the subsurface end member, or mixing of tributaries of differing compositions are not considered). However, it does illustrate the major trends of salinization. The evaporative fraction (i.e., the fraction of the original runoff remaining after evapotranspiration) shown in Figure 7a decreases from 1 in the headwaters to about 0.4 at El Paso. The basin water balance based on stream gaging, given above, indicates that about 75% of the flow at Otowi is evaporated by the time the river reaches El Paso, thus the two estimates are in general agreement. The fluctuations around the regional trend are strongly influenced by variations in the Cl^- and Br^- concentrations of the various inflows and probably do not accurately reflect variations in evaporation at the local scale.

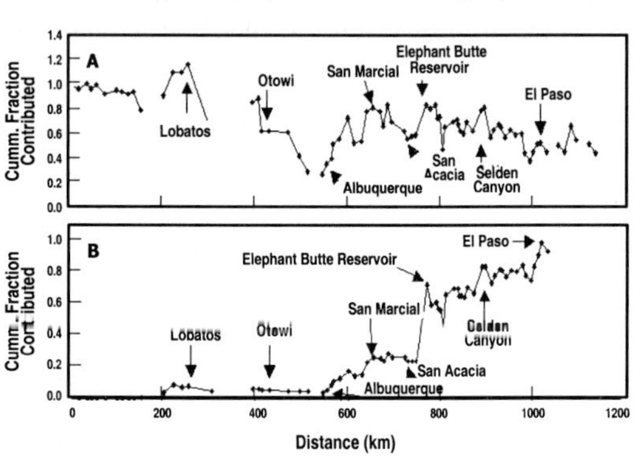

Figure 7. Preliminary interpretation of effects of evaporation and salt addition on chloride concentration of the Rio Grande, based on August 2001 data. (a) Evaporative fraction (volume remaining after evapotranspiration/original volume) as a function of flow distance. (b) Relative chloride addition (as fraction of final chloride added) as a function of flow distance.

The calculated relative salt additions are shown in Figure 7b. This graph shows that Cl^- is added to the river in a series of distinct steps, rather than the relatively uniform distribution of Cl^- concentration increase caused by evapotranspiration. Six localities of apparent salt addition can be identified. The first step is south of Alamosa, the second at the south end of Albuquerque, the third just north of San Acacia, the fourth at Elephant Butte Reservoir, the fifth at the head of Seldon Canyon, and the sixth north of El Paso. Most of these have a similar geological setting. The first step is at the south end of the San Luis Basin. The third is at the south end of the Albuquerque Basin. The fourth is at the southern termination of the Socorro Basin and the fifth at the southern end of the Palomas Basin. Finally, the last is at the southern end of the Mesilla Basin. The only step that does not correspond to this pattern is the second one, and this increase in Cl^- may be related to discharge from the Albuquerque sewage treatment facility. In general, these salinity inputs do not appear to be closely related to agricultural drainage. Most of the major drain outflows were sampled and those north of Elephant Butte Reservoir showed only slight enhancement of Cl^- concentrations over the diverted river water. One notable exception is a drain near Socorro having both high Cl^- and a high Cl/Br ratio. The source of this saline water is not known. South of Elephant Butte most of the drains have moderately elevated Cl^- and slightly elevated Cl/Br. One drain at the far south end of the Mesilla Basin is strongly elevated in both parameters. This drain is not in a heavily irrigated area and appears to be intercepting mostly natural ground-water discharge.

CONCLUSIONS

The Rio Grande presents a classic case study of arid-region river salinization. At the headwaters of the river both the TDS and the Cl^- concentrations are low (40 and 0.5 mg L^{-1}, respectively). These both increase progressively during flow through approximately 1,000 km of desert environment, to values of about 750 and 100 mg L^{-1}, respectively, at El Paso/Ciudad Juarez, the last major population center on this section of the river. Both conventional river gaging and stable isotopes of oxygen and hydrogen indicate that only approximately 30% of the inflow across the Colorado/New Mexico border escapes evapotranspiration to reach El Paso. However, evapotranspirative concentration of the cyclic salts from the headwaters area can account for less than 20% of the increase in total dissolved solids.

The most common explanations of the additional salt input have been that infiltration of excess irrigation water raised the water table and flushed soils and shallow ground water made saline by natural, pre-irrigation evapotranspiration in the desert environment, or that water logging of irrigated fields prior to the installation of drains accumulated salts that were flushed after drainage was installed. Were either of these the case, the salinity inputs should be localized at agricultural drains and the Cl/Br ratio should remain approximately constant (since the salt was derived from pre-irrigation evapotranspiration of river water or direct precipitation). However, the increase of the Cl/Br ratio with flow distance indicates that most of this additional salinity accrues from influx of saline subsurface waters. The salinity increase is not gradual, but rather stepwise, with most of the steps at the southern ends of the major sedimentary basins that form the Rio Grande Rift. These observations indicate that most of the salinization of the Rio Grande is probably due to regional discharge of ancient ground water that has followed deep flow paths through the basins and acquired high salinity due to diagenetic or geothermal reactions. Additional studies are planned to more definitively identify the sources and locations of these discharges. This may ultimately enable some amelioration of the salinization of the Rio Grande.

ACKNOWLEDGEMENTS

This work was supported in part by SAHRA (Sustainability of Semi-Arid Hydrology and Riparian Areas) under the STC Program of the United States National Science Foundation, Agreement Number EAR-9876800, and in part by National Science Foundation grant EAR-9614509. We thank Donald Whittemore and two anonymous reviewers for suggesting improvements to the manuscript.

REFERENCES

Adams A. I., Goff F., and Counce D., 1995. Chemical and isotopic variations of precipitation in the Los Alamos Region, New Mexico: Los Alamos National Laboratory, *Report LA-12895-MS,* Los Alamos, NM, 35p.

Anderholm, S. K., 1987. *Hydrogeology of the Socorro and La Jencia basins, Socorro County, New Mexico: U.S. Geological Survey Water-resources investigations report 84-4342*, 62p.

Bexfield L. M., 2001. Occurrence and sources of arsenic in ground water of the Middle Rio Grande Basin, central New Mexico: M.S. Thesis, New Mexico Institute of Mining and Technology, Socorro, 143p.

Campbell A. R. and Larson P. B., 1998. Introduction to stable isotope applications in hydrothermal systems: In: *Techniques in Hydrothermal Ore Deposits* (ed. J.P. Richards and P. B. Larson), *Reviews in Economic Geology,* Vol. 10, Society of Economic Paleontologists and Mineralogists, 173-193.

Coplen T. B., Herczeg A. L., and Barnes C., 1999. Isotope engineering - Using stable isotopes of the water molecule to solve practical problems: In *Environmental Tracers in Subsurface Hydrology* (ed. P. Cook and A. L. Herczeg) Kluwer Academic, Dordrecht, 79-110.

Dansgaard W. (1964) Stable isotopes in precipitation: *Tellus* 16: 436-468.

Davis S. N., Whittemore D. O., and Fabryka-Martin J., 1998. Uses of chloride/bromide ratios in studies of potable water: *Ground Water* 36: 338-350.

Ellis S. R., Levings G. W., Carter L. F., Richey S. F., and Radell M. J., 1993. Rio Grande Valley, Colorado, New Mexico, and Texas: *Water Resources Bulletin* 29: 617-646.

Feth J. H., 1981. Chloride in natural continental water: *U.S. Geological Survey Water Supply Paper 2176*, 36p.

Gat J. R., 1996. Oxygen and hydrogen isotopes in the hydrologic cycle: *Annual Review of Earth Planet Science* 24: 225-262.

Hawley, J. W., 1984. Hydrogeologic cross sections of the Mesilla bolson area, Dona Ana county, New Mexico and El Paso County, Texas: *New Mexico Bureau of Mines and Mineral Resources Open-File Report 190*: 10p.

Hayward H. E., 1956. The salinity factor in the reuse of waste waters: In *The Future of Arid Lands*, Publication No. 43 (ed. G. F. White) American Association for the Advancement of Science, Washington, D.C., 279-290.

Keller G. R. and Cather S. M., 1994. Basins of the Rio Grande Rift: Structure, Stratigraphy, and Tectonic Setting: *Geological Society of America Special Paper 291,* Boulder, Colorado, 304p.

National Resource Committee, 1938. *Regional Planning: Part VI, The Rio Grande Joint Investigation in the Upper Rio Grande Basin, Colorado, New Mexico, and Texas, 1936-1937:* U.S. Government Publishing Office, 566p.

Vengosh A. and Pankratov I., 1998. Chloride/bromide and chloride/fluoride ratios of domestic sewage effluents and associated contaminated ground water: *Ground Water* 36: 815-824.

Whittemore D.O., 1995. Geochemical differentiation of oil and gas brine from other saltwater sources contaminating water resources: Case studies from Kansas and Oklahoma: *Environmental Geosciences* 2: 15-31.

Wilcox L. V., 1957. Analysis of salt balance and salt-burden data on the Rio Grande: In *Problems of the Upper Rio Grande: An Arid Zone River*, Publication No. 1 (ed. P. C. Duisberg) U.S. Commission for Arid Resource Improvement and Development, Socorro, New Mexico, 39-44.

Wilkins D. W. (1998) Summary of the Southwest Alluvial Basins Regional Aquifer-System Analysis in Parts of Colorado, New Mexico, and Texas: *U.S. Geological Survey Professional Paper 1407-A*, 49p.

Climate Outlooks for Water Management Adaptation to Climate Change in the Middle East

RICHARD WASHINGTON and MATTHEW SWANN
School of Geography and the Environment, University of Oxford, Mansfield Rd., Oxford, OX1 3TB, United Kingdom, Email: richard.washington@geography.ox.ac.uk

ABSTRACT: Planning and adaptation management strategies are clearly needed for water management in the arid subtropics. Such strategies need to embrace a very wide range of time scales since human induced climate change is unique in that the full consequences of human actions will be felt for many centuries (Houghton, et al., 2001). This paper argues for a new way of approaching adaptation to climate change by considering the use of climate predictions at a continuum of time scales (from seasonal to multi-decadal) to improve resource management. We focus on the Middle East and ask three key questions:

a) In what way can we conceptualise adaptation to climate change on a continuum of time scales in a way that embraces knowledge of natural patterns of climate variability?
b) What are the key patterns of natural climate variability in the Middle East and what impact do they have on rainfall?
c) What are the limits of climate predictability?

1. INTRODUCTION

The detection and attribution of human induced climate change as a result of rising levels of Greenhouse gases in the atmosphere is now widely accepted (Houghton, et al., 2001). Climate change will undoubtedly bring costs and benefits to many regions of the globe, but regions already marginal for humans are at great risk from possible negative impacts. The Intergovernmental Panel on Climate Change (IPCC) Third Assessment Report indicates a likely drop in observed Northern Hemisphere sub-tropical precipitation of the order of 0.3% per decade during the 20th Century.

Planning and adaptation management strategies are clearly needed for water management in the arid subtropics. Such strategies need to embrace a very wide range of time scales since human induced climate change is unique in that the full consequences of human actions will be felt for many centuries (Houghton, et al., 2001). This paper argues for a new way of approaching adaptation to climate change by considering the use of climate predictions at a continuum of time scales (from seasonal to multi-decadal) to improve resource management. We focus on the Middle East and ask three key questions:

a) In what way can we conceptualise adaptation to climate change on a continuum of time scales in a way that embrace knowledge of natural patterns of climate variability?
b) What are the key patterns of natural climate variability in the Middle East and what impact do they have on rainfall?
c) What are the limits of climate predictability?

2. CLIMATE ADAPTATION ON A CONTINUUM OF TIME SCALES

Typical research into climate change adaptation has begun with climate change scenarios generated by global climate models for time slices in the 21st century, such as the 2020, 2050 and 2080. This is a poor basis for understanding adaptation: the scenarios are not probabilistic predictions and such long time scales are generally beyond the time frame of resource management. In most cases, the impacts and adaptation strategies are more sensitive to assumptions about socio-economic conditions than the climate change scenarios.

This papers argues for a new way of approaching adaptation to climate change by using climate predictions at a continuum of time scales (from seasonal to multi-decadal) to improve resource management.

In this paper we propose that climate change adaptation may be considered as a process of learning - using information about future climates at a variety of time scales. Resource managers are concerned with seasonal variations in operational decision making, with interannual and interdecadal variations and variability in strategic planning, and possibly long-term climate change as a risk for large investments. This range of time scales of climate

Water Resources Perspectives: Evaluation, Management and Policy. Edited by A.S. Alsharhan and W.W. Wood.
Published in 2003 by Elsevier Science, Amsterdam, The Netherlands, p. 335-348.

change and variability—seasonal, interannual, interdecadal and climate change—is a natural component of resource management in the learning process of climate change adaptation since resource managers are faced with climate impacts on an ongoing basis.

The justifications for this novel approach are numerous and include:

1. Climate change and climate variability are not separate processes. Natural climate variability and climate change are closely linked in the operation of the climate system. In Europe, for example, climate variability on interannual to interdecadal time scales is dominated by the North Atlantic Oscillation (NAO) which imposes a strong signal on Northern Hemisphere temperature trends as a whole (Hurrell, 1995). Elsewhere it has been argued that climate change may occur through the operation of patterns of climate such as the North Atlantic Oscillation (Corti et al., 1999; Stone et al., 2001).
2. Climate anomalies during extreme episodes of the North Atlantic Oscillation or related to the ENSO (El Niño-Southern Oscillation) are larger than many long-term climate change anomalies projected by model scenarios, especially for the next 20 to 50 years. Adaptation strategies which are geared to cope with large climate anomalies associated with climate variability therefore embrace the envelope of change expected from long term climate change.
3. Improvements in climate modelling and predictability of global climate, particularly on seasonal and decadal time scales (Sutton and Allen, 1997; Rodwell et al., 1999) are being developed in ongoing international programs such as CLIVAR and formerly TOGA. Current work suggests that there is sufficient skill in climate outlooks to be of use to resource managers.
4. Adaptation to current climate is a more immediate concern to resource managers than responding to hypothetical threats in 2050. Connecting the near term and long term provides a means to engage stakeholders in adaptive resource management strategies.

Water management, particularly irrigation and allocation of water among competing users, should be a high priority in the process of climate change adaptation. Present resource management is sensitive to climatic variations and large investments are at risk if adverse climate is realised.

Many resource managers and researchers, including the World Meteorological Organisation, US Office of Global Programs and the World Bank, have endorsed the idea of using seasonal, interannual and interdecadal climate outlooks as a strategy for adapting to climate change. However, the presumed benefits of climate outlooks for adapting to climate change have not been systematically evaluated and the predictability of climate is not known for several key regions of the world. Many scientific and technical issues need to be addressed and our new integrated methodology requires demonstration and evaluation before the presumed utility of using climate outlooks can be fully accepted.

3. KEY PATTERNS OF NATURAL VARIABILITY

In this section we consider the key patterns of natural variability in the Northern Hemisphere and evaluate their role in modulating Middle Eastern climate, particularly water related variables.

3.1. Northern Hemisphere Modes of Variability

The definition of modes of atmospheric variability in the Northern Hemisphere draws on a rich history of research beginning with Walker and Bliss (1932). Interest in Northern Hemisphere teleconnections was revived in the late 1970s (e.g. van Loon and Rogers 1978, Rogers and van Loon 1979) but it was the work of Wallace and Gutzler (1981), Horel (1981), Esbenson (1984) and later that of Barnston and Livezey (1987) that saw much of the definitive progress.

The North Atlantic Oscillation (NAO) is one of the dominant modes of Northern Hemisphere climate variability (Walker and Bliss, 1932; van Loon and Rogers, 1978; Wallace and Gutzler, 1981) and is present in all seasons (Barnston and Livezey, 1987). The signature of North Atlantic Oscillation is a north-south dipole of anomalies, with centres over Greenland/Iceland and the Azores. The positive phase is defined as below normal pressure in the region of the Icelandic low but above normal pressure in the Azores (corresponding with a steepening of the geostrophic gradient). This leads to an enhanced Atlantic storm track and anomalously high transport of moisture and heat by transient eddies (Hurrell, 1995) corresponding to changed precipitation and temperature patterns extending from eastern North America to western and central Europe, and into Scandinavia, Siberia and central Russia (Walker and Bliss, 1932; van Loon and Rogers, 1978; Rogers and van Loon 1979; Hurrell, 1995).

Atlantic variability in the Autumn, Winter and Spring is also controlled by the East Atlantic (EA) pattern (Barnston and Livezey, 1987). Similar to the North Atlantic Oscillation, this pattern consists of a north-south dipole but extends across the entire North Atlantic basin with a line of separation through England and France (Barnston and Livesey, 1987). The EA pattern is distinct in that the anomaly centres are concentrated near the nodal lines of the North Atlantic Oscillation dipole, resembling a southward shifted North Atlantic Oscillation so that the subtropical component is part of the subtropical ridge of the North Atlantic (Barnston and Livezey, 1987).

Eurasia is influenced by the East Atlantic/ West Russian (EA/WR) pattern (Barnston and Livezey, 1987) which in Winter, has two main anomaly centres located over the Caspian Sea and western Europe. This simple pattern is replaced by a more complicated structure in Spring and Autumn, so that two main anomaly centres, of opposite sign, are found over western or north-western Russia and over northwestern Europe. The third centre, (in phase with the Russia centre) is located off the Portuguese coast in Spring, but moves westward in Autumn.

The main mode of variability over Scandinavia, Siberia and the Arctic Ocean is known as the Scandinavia pattern (SCA) which is active in all months except the high Summer (Barnston and Livezey, 1987). Blocking events over Scandinavia and Russia are objectively measured by this mode. The East Pacific (EP) pattern is evident in all months except August and September, and reflects a north-south dipole of height anomalies over the eastern North Pacific and eastern Alaska (Barnston and Livesey, 1987). Characteristics of strong positive phases of the East Pacific pattern include an anomalous northeastward extension of the Pacific jet stream toward western North America, with enhanced westerlies over the Pacific Northwest States, and northern California.

The North Pacific (NP) pattern appears mostly in Spring and Summer consisting of a main anomaly centre spanning the western and central North Pacific, and a smaller centre of opposite sign across eastern Siberia, Alaska and central North America. Overall. The Pacific North American (PNA) pattern is perhaps the best known mode of Pacific based variability. It appears in all months except June and July. The Pacific North American pattern reflects a quadrupole pattern of geopotential anomalies, with anomalies of similar sign located south of the Aleutian Islands and over the southeastern United States.

The Tropical / Northern Hemisphere (TNH) pattern (Mo and Livezey, 1986; Barnston and Livesey, 1987) consists of one primary anomaly centre over the Gulf of Alaska and a separate anomaly centre of opposite sign over the Hudson Bay. A weaker area of anomalies, having similar sign to the Gulf of Alaska anomaly, extends across Mexico and the extreme southeastern United States.

The Polar/Eurasian (POL) pattern is a tripolar mode of geopotential height variability, with the main centre of activity over the north polar region, and two centres of opposite sign, the strongest over northern Japan and the Sea of Okhotsk and a weaker one over central/southern Europe (Barnston and Livezey 1987). These anomaly centres link the Polar/Eurasian to the strength of the circumpolar circulation, and appear most prominently between December and February. The positive phase of the pattern is represented by below-average heights in the polar centre, and above-average heights in the mid-latitude centres. The Polar/Eurasian pattern varies on both interannual and interdecadal time scales, with a tendency towards the same state from one winter to the next.

3.2.Northern Hemisphere Teleconnection Indices

The procedure used to identify the teleconnection patterns discussed in Section 3.1 is outlined in detail in Barnston and Livesey (1987). They used Rotated Principal Component Analysis (RPCA) to isolate the primary teleconnection patterns in the Northern Hemisphere circulation for all months. Here the results from an Rotated Principal Component Analysis technique applied to monthly mean 700-mb height anomalies between January 1964- July 1994 are used (Bell, 1998). The indices are the time coefficients (or amplitudes) of the Rotated Principal Component Analysis modes which have been calculated over the standard seasons December to February (DJF), March to May (MAM) etc. The data used to define these indices derive from the NCEP-NCAR Reanalysis Project which entails the initialisation of a State of the Art climate model with observed atmospheric conditions every 6 hours to yield a data set of globally complete yet physically reasonable fields. Indices reflecting the amplitudes of all the patterns discussed in Section 3.1 have been used in this study.

3.3. Northern Hemisphere Teleconnection Indices and Middle Eastern Winter Rainfall

In this section we relate the temporal indices of the climate variability patterns discussed in Section 3.2 to Middle Eastern water related atmospheric parameters. Only the December to February (northern winter) season is considered as these months include the dominant rainfall season for the Middle East. We start with an overview of the patterns related to the annular mode of northern hemisphere zonal flow.

Recent work on Northern Hemisphere atmospheric variability on interannual and interdecadal time scales has focussed on mid-latitude zonal flow. On a hemispheric scale this is best represented by the Arctic Oscillation (AO), or Northern Hemisphere annular mode (NAM) (Wallace, 2000; Thompson, 2001). In addition, Barnston and Livezey's key (1987) work on low frequency atmospheric circulation patterns highlights regional patterns of atmospheric mass distribution, discussed in Sections 3.1 and 3.2 above. We start with a consideration of the link between the Northern Hemisphere annular mode and winter rainfall in the Middle East, and use Barnston and Livezey's work on regional patterns to support the investigation of links between the teleconnection indices discussed in Section 3 and Middle Eastern winter rainfall.

The NCEP/NCAR reanalysis project provides the most reliable global gridded dataset for the purposes of correlating regional rainfall with Northern Hemisphere Teleconnection Indices (NHTI). Both the rainfall data and the Northern Hemisphere Teleconnection Indices are derived from the NCEP/NCAR reanalysis data, and the correlations in 3.4 were carried out at the Climate Diagnostics Center website (http//:www.cdc.noaa.gov/ correlations) Since rainfall in the Middle East is limited to the winter months, only the months of December, January and February (DJF) were used. This also corresponds with the season when the majority of the Northern Hemisphere Teleconnection Indices are most prominent.

3.4. RESULTS

Figure 1 illustrates the relationship between the Arctic Oscillation index (Northern Hemisphere annular mode) and 700mb zonal winds. The link between this index and the hemispheric circulation is clear, with the strongest correlations over the North Atlantic/European region. The link between the Arctic Oscillation and Middle East is of lesser significance, with a negative centre over the Mediterranean basin. Further south over the Arabian Peninsula however, the link becomes weakly positive, as part of a weak hemispheric tropical/sub-tropical connection with the Arctic Oscillation. An examination of the link between the AO and Middle Eastern winter rainfall (Figure 2) shows that there is a highly significant negative relationship centred over southern Turkey, Syria and Northern Iraq.

The North Atlantic Oscillation shows a negative relationship with 700hPa zonal winds as far east as western Iran but of lesser importance and centred on the northern sub-tropical Atlantic. The North Atlantic Oscillation precipitation relationship is also centred further west (Figure 4), although the area of weaker negative correlations extends as far east as Iraq, and there is a further area of negative loadings extending from the tropics to the southern Arabian Peninsula. This is spatially similar to the North Atlantic Oscillation -700mb zonal wind pattern.

The Polar/Eurasian pattern displays a clear link with winter rainfall in the Middle East. Figure (5) illustrates the zonal wind relationship, with an area of high negative correlations over the Mediterranean basin. There is a small area of positive correlations over the Gulf of Oman, which appears to be part of a hemispheric connection from the tropics to the main centres of Polar/Eurasian activity.

Middle Eastern precipitation appears to be linked most strongly with the Polar/Eurasian Pattern (POL). An area of strong negative correlations is centred over Turkey and Syria, extending into Iran, Iraq, northern Saudi Arabia and Jordan. This is the main centre of negative precipitation loadings associated with the Polar/Eurasian index globally.

The East Atlantic pattern show similar spatial patterns to the North Atlantic Oscillation and Polar/Eurasian Pattern when correlated with 700mb zonal winds (Figure 7). The southern negative correlations are centred over north Africa, and extend as far as Syria and Jordan. The relationship between the East Atlantic index and precipitation (Figure 8) follows a similar spatial domain to the East Atlantic-zonal wind relationship. However the southern centre of negative correlations extends further east to India centred on 30°N.

The East Atlantic/West Russian pattern index displays two main centres of high correlations with 700mb zonal winds, with the main positive centre off the coast of Norway, and the negative centre over southern Europe (Figure 9). A second positive centre exists over Egypt/Sudan and Saudi Arabia. In general, it appears that the East Atlantic/West Russian index is less linked to zonal flow in the southern mid-latitudes than the other indices.

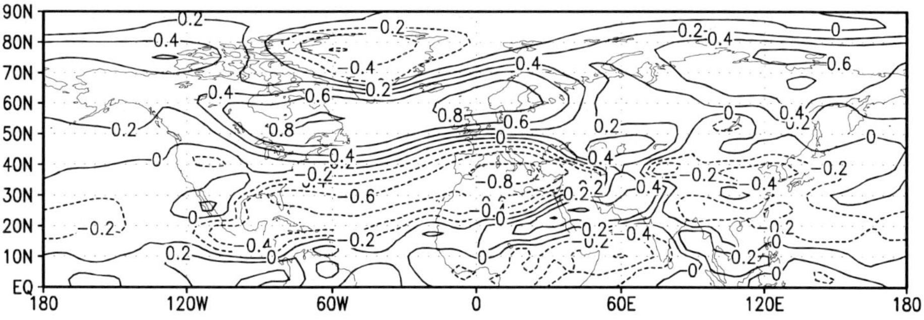

Figure 1. Correlations between DJF AO and 700hPa Zonal Flow, 1958-2000.

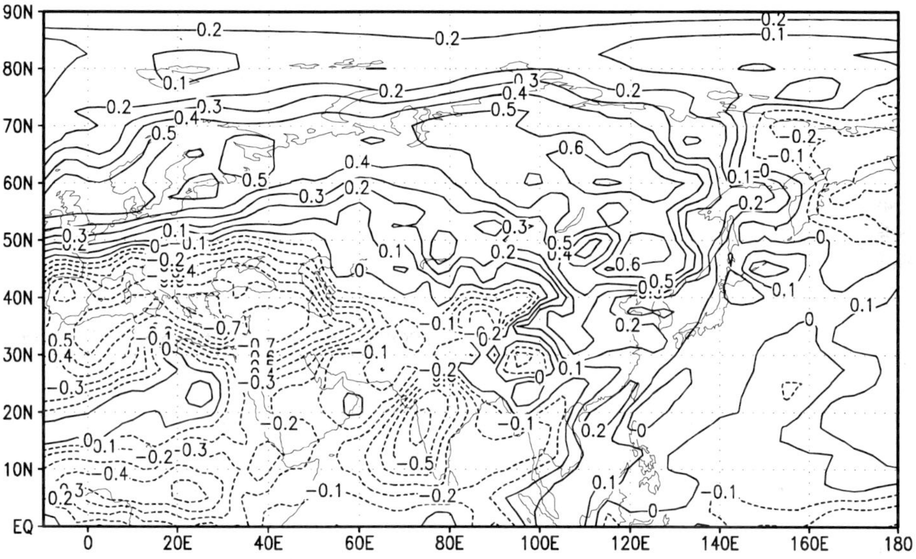

Figure 2. Correlations between DJF AO and Surface Precipitable Water 1958-2000.

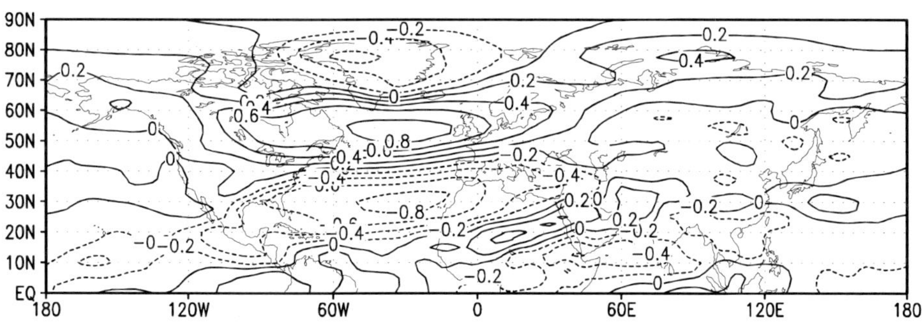

Figure 3. Correlations between DJF NAO and 700hPa Zonal Flow 1950-2000.

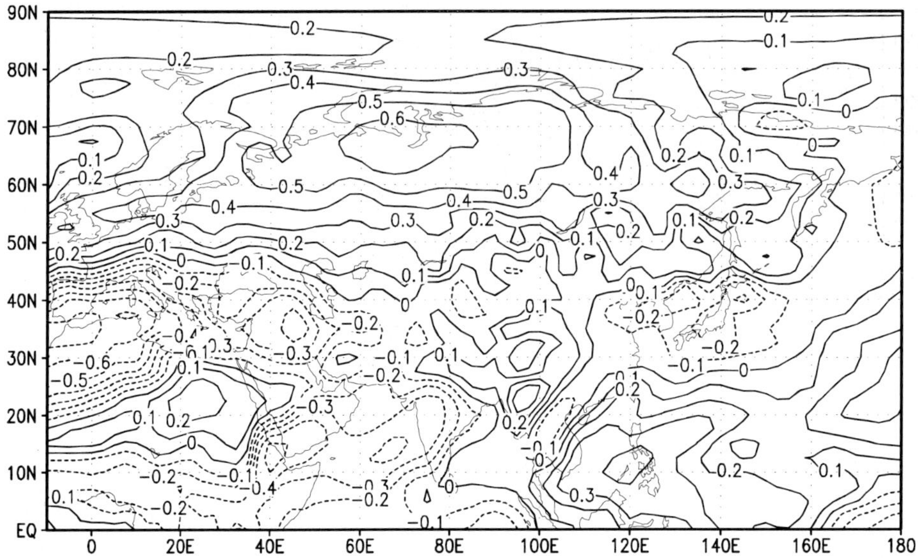

Figure 4. Correlations between DJF NAO and Surface Precipitable Water 1950-2000.

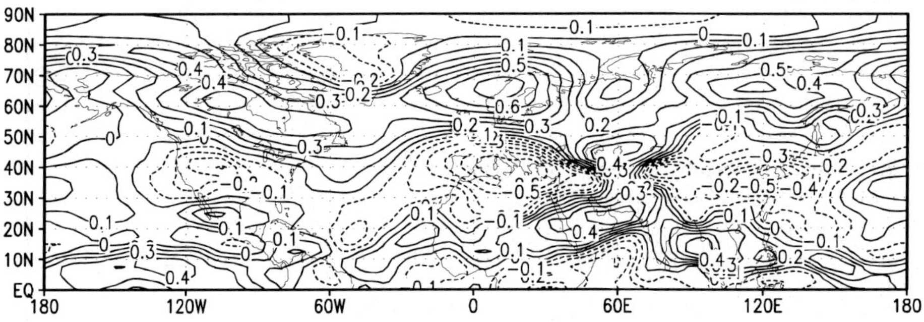

Figure 5. Correlations between DJF POL and 700hPa Zonal Flow 1950-2000.

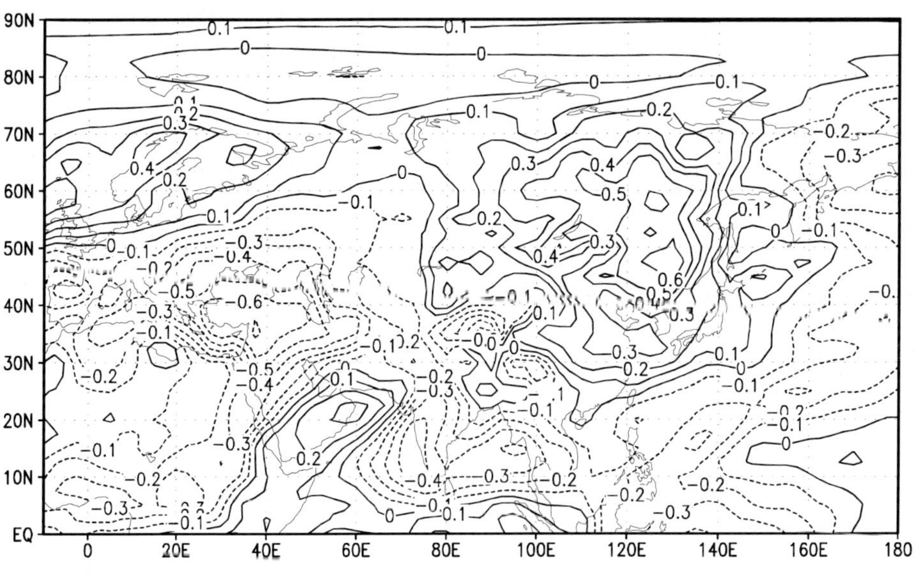

Figure 6. Correlations between DJF POL and Surface Precipitable Water 1950-2000.

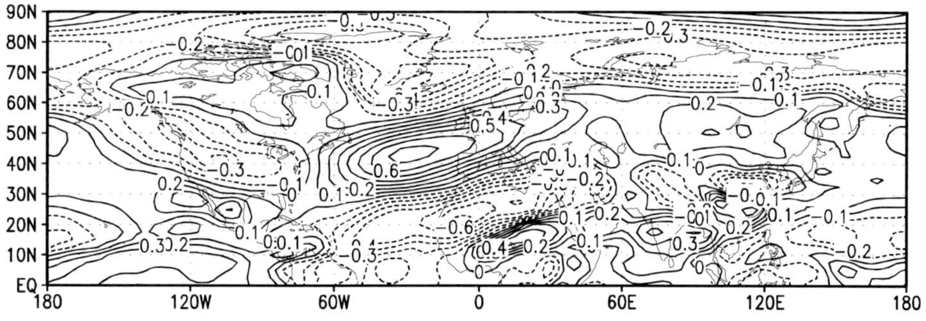

Figure 7. Correlations between DJF EA and 700hPa Zonal Flow 1950-2000.

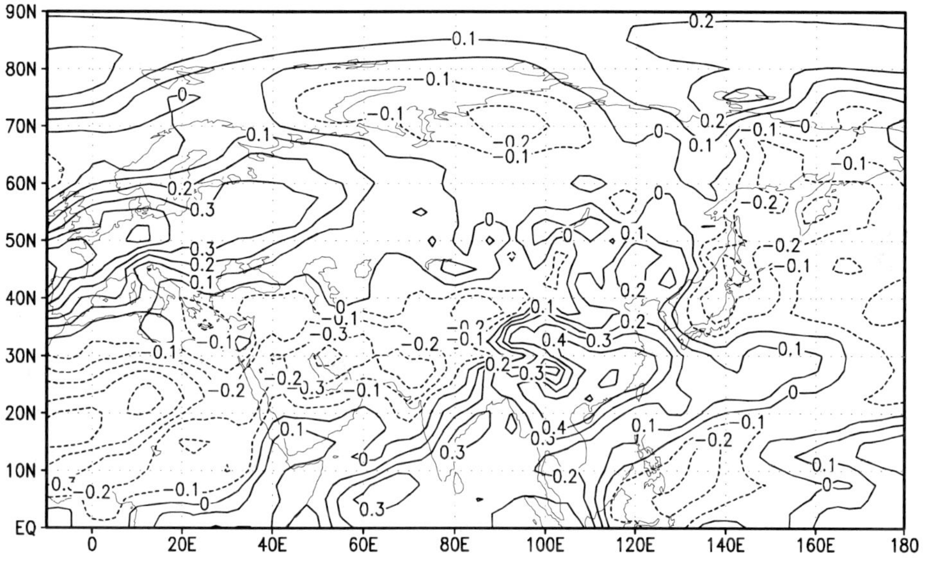

Figure 8. Correlation between DJF EA and Surface Precipitable Water 1950-2000

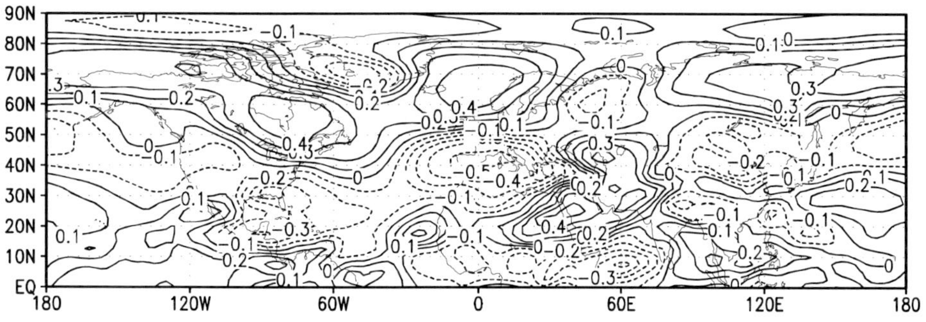

Figure 9. Correlations between DJF EA/WR and 700hPa Zonal Flow 1950-2000.

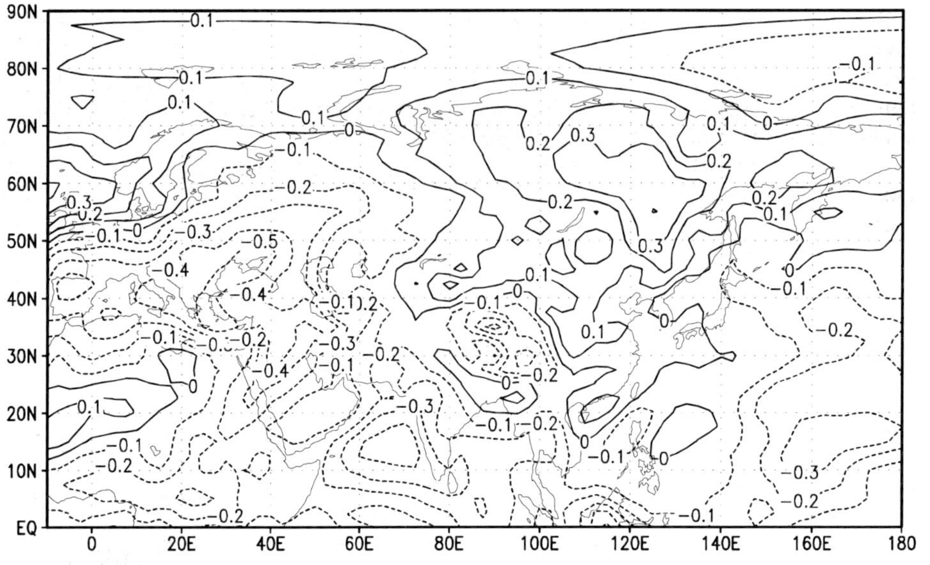

Figure 10. Correlations between DJF EA/WR and Surface Precipitable Water 1950-2000.

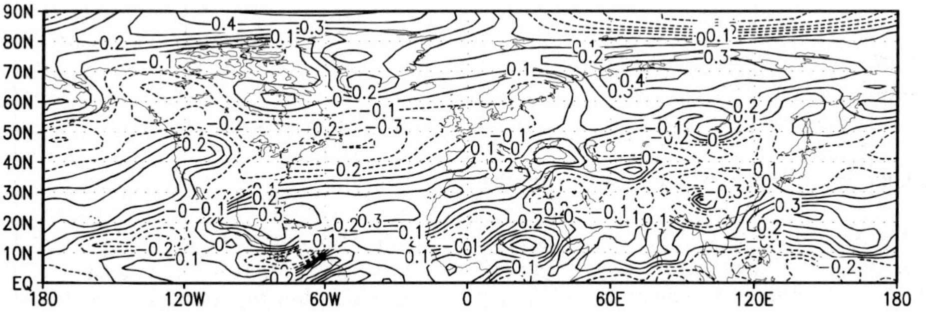

Figure 11. Correlations between DJF AT and 700hPa Zonal Flow 1958-1999.

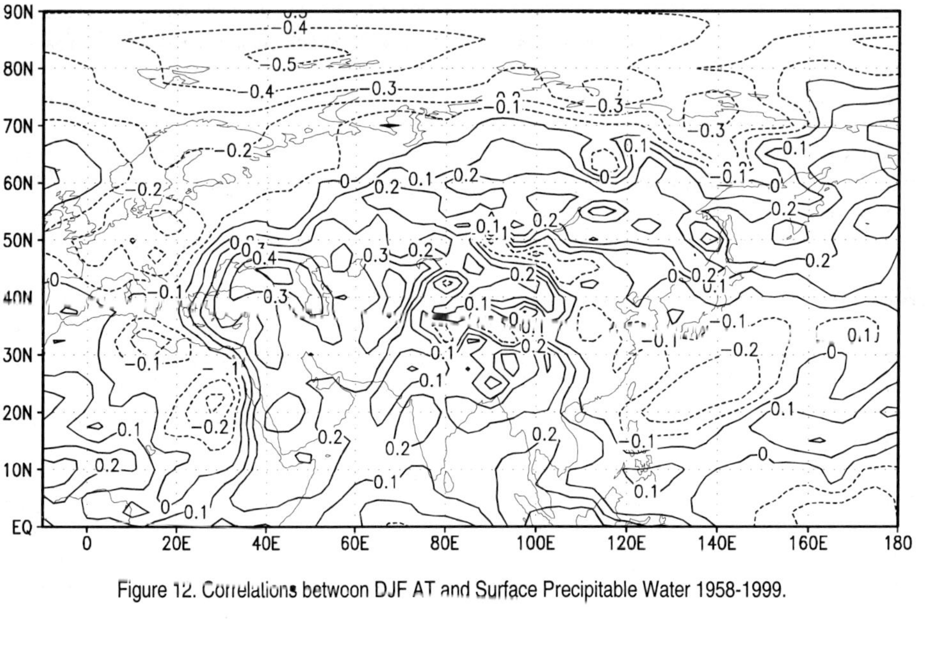

Figure 12. Correlations between DJF AT and Surface Precipitable Water 1958-1999.

The East Atlantic/West Russian index-precipitation correlation shows a much more widespread region of negative loadings centred on southeastern Europe (Figure 10). It appears to remain of some significance in much of the Middle East, and similarly to the Polar/Eurasian Pattern - precipitation correlation (Figure 6), extends south-west into Africa.

The Atlantic Tripole sea surface temperature pattern indicates that there is a weak positive relationship with this index and winter rainfall in the Middle East. (Figure 12). It has been argued that this signature of sea surface temperature, which is thought to be most coherent beneath the surface, is a major forcing factor in the sign and amplitude of the annular mode in the North Atlantic region (Sutton and Allen, 1997). Given the slow evolution of sea surface temperature anomalies, it is plausible that predictability of this mode on multi annual timescales is within reach. Modelling studies have further demonstrated this potential (Rodwell et al., 1999).

4. INHERENT PREDICTABILITY OF ATMOSPHERIC MODES

The usefulness of applying climate predictability on a range of time scales in the context of climate change adaptation rests in the degree of predictability present in the atmosphere. Clearly, if a region is subject to a large degree of interannual or interdecadal variability of a key climate variable, but the modes of atmospheric variability have no inherent predictability, then adaptation process to climate change via adaptation to climate variability, will be severely constrained. In the previous section it was shown that patterns of northern hemisphere climate variability do impact on the region of interest in this paper. But it remains difficult to determine how predictable those patterns are. Fortunately, there are theoretical means to estimate this. One such scheme is introduced in this section.

4.1. Sensitivity to Initial Conditions

The atmosphere is known to be sensitive to initial conditions (Lorenz, 1963). Precisely the same forcing can produce hugely divergent evolutions of the climate system given infinitesimally small differences in initial conditions (Lorenz, 1963). The concept of initial conditions is easiest to understand when described in the context of a climate model which is assumed to be a realistic representation of the real atmosphere. To start the model, the values of all the primary variables (for example temperature, pressure, moisture) need to be specified for each point in the atmosphere. Since computing power prevents all points in the atmosphere from being represented, these models usually treat the atmosphere as a collection of boxes on a grid. The equations which the model uses to simulate the atmosphere are then integrated forward in time using the specified starting or initial conditions of the primary variables once, but thereafter calculating the values of these variables on the basis of the equations. Experiments indicate that a very small change in the starting values over a small area in the atmosphere is sufficient for the patterns of weather to evolve completely differently in time across the model atmosphere. This has been termed the sensitivity of the atmosphere to initial conditions (Lorenz, 1963; 1993).

In the real world the atmosphere is always initialised (if we insert some time point as a nominal starting condition) by unique or very nearly unique conditions since the weather from one day to the next, one hour to the next or one second to the next is unique. As outlined in the previous paragraph, one of the few opportunities we currently have to test atmospheric sensitivity to initial conditions is in a climate model whereby only one condition, here the initial values, may be arbitrarily changed while any forcing function remains constant from one model run to the next. Multiple integrations of such an experiment, with each integration subjected to identical forcing through time from say solar radiation and ocean temperatures, yet each being uniquely defined by arbitrarily small differences in starting conditions of the model atmosphere, would yield a set of data for which a quantitative scheme could be used to partition the data variance into components essentially common to all integrations and variance unique to each integration. When taken across the globe as a whole, we might label these the Forced and Free Manifolds respectively.

4.2. Quantifying Predictability

The concept of quantifying sensitivity to initial conditions is a necessary extension of Lorenz's (1963) work, but has taken time to appear in the literature (Rowell, 1998; Ward and Navarra, 1997). The process of identifying the forced and free model response consists of separating the variance within and between the model integration for each grid box (or spatial location) in the model for any particular atmospheric field. The free response, corresponding to unforced variability (internal variability) may be measured by the variance of each observation from the ensemble mean for each year of the model integration. Rowell *et al.* (1995) specify this as follows for a single variable at one grid box in the model (i.e. the simplest case):

$$\sigma_{\text{int}} = \frac{1}{N(n-1)}\sum_{i=1}^{N}\sum_{j=1}^{n}(x_{ij} - \bar{x})^2$$

where N is the number of years in the integration, n is the number of ensemble members (i.e. the number of times the experiment is run with unique initial conditions) and $\bar{x}$ is the ensemble mean for the ith year. This component of variability in the atmosphere therefore consists of how different the atmosphere is from one experiment to the next. Since the one and only factor distinguishing one experiment from the next is the arbitrary initial conditions, the one and only factor that can account for this component of the variance is precisely those initial conditions. If this unforced component of variance is very large, as will be shown for certain parts of the atmosphere in later sections, then the atmosphere's evolution or the trajectory of weather and climate, will have been critically influenced by the arbitrary state of the atmosphere at the beginning of the experiment. If, on the other hand, this component is very small, the influence of different starting conditions can be deduced to be small since the atmosphere is influenced more by the forcing which is common to all the experiments (Washington, 2000).

In order to define the forced variance, the ensemble mean variance must be calculated. This is given by:

$$\sigma_{en} = \frac{1}{N-1}\sum_{i=1}^{N}(\bar{x} - \bar{\bar{x}})^2$$

where $\bar{\bar{x}}$ is the ensemble mean of all data, other terms being defined earlier.

The ensemble mean, $\bar{\bar{x}}$, is given by:

$$\bar{\bar{x}} = \frac{1}{Nn}\sum_{i=1}^{N}\sum_{j=1}^{n}x_{ij}$$

The value of the forced variance is then calculated as the difference between the ensemble mean variance and the internal variance. The total variance is the sum of the internal variance and the forced variance. Ratios between the internal and total variance and between the forced and total variance, when calculated for each of the model grid boxes for a particular field, yield the Free and the Forced Manifold respectively. In full the calculation is as follows:

$$\sigma_{SST} = \frac{1}{N-1}\sum_{i=1}^{N}(\bar{x} - \bar{\bar{x}})^2 -$$
$$\frac{1}{n}\left(\frac{1}{N(n-1)}\sum_{i=1}^{N}\sum_{j=1}^{n}(x_{ij} - \bar{x})^2\right)$$

The higher the value of the forced response (i.e. the component common to the sea surface temperature forcing, hence the sea surface temperature subscript), the larger the predictability of the climate. Since this quantity can be calculated for each point in a climate model, the predictability of regional climate may easily be calculated.

Forced Response and Predictability in the United Kingdom Meteorological Office Model

The example given here uses the United Kingdom Meteorological Office (UKMO) Hadley Centre atmosphere general circulation model (HADAM2A). HADAM2A has been successfully used in studies of anthropogenic climate impacts (Folland *et al.*, 1998). Features include a resolution of 2.5° latitude by 3.75° longitude, with 19 hybrid layers in the vertical (5 representing boundary layer processes). Physical parameterisations include a gravity wave drag scheme and a radiation scheme that computes fluxes in several parts of the shortwave and longwave bands and which responds to cloud growth occurring dynamically in the model. Large scale cloud amount, convective cloud amount, liquid water concentrations and ice concentration are the prognostics used in the cloud scheme. A surface hydrology scheme includes parameterised surface moisture storage, surface and sub-surface runoff and a prescribed vegetation canopy that interacts with precipitation and runoff. A detailed description of the model physics is available in Phillips (1994).

Several critical assumptions relating to the model performance are necessary for the method adopted here to be successful in quantifying forced and free modes. These include:

1) That a 'mean' or quasi-equilibrium state exists both in nature and in the model.

2) That the model converges to the correct solution. Whether this solution is regarded as correct cannot be determined analytically.
3) That the model is a realistic representation of the real atmosphere.

HADAM2A is an improved version of the model submitted by the United Kingdom Meteorological Office as part of the Atmospheric Model Intercomparison Project (Gates, 1992), which identified the United Kingdom Meteorological Office model as amongst the best four models currently available. More recent reviews and model intercomparisons have again shown the performance of HADAM model to be comparable with the best available models (Srinivasan *et al.*, 1995; Airey *et al.*, 1996; Gates *et al.*, 1996; Joubert, 1997). The climatologies of HADAM1 (the version submitted to AMIP) and HADAM2 are compared in Gallani (1994).

The suite of experiments analysed in this section consist of an ensemble of four integrations forced throughout with observed sea surface temperatures (sea surface temperature), sea ice, changing equivalent carbon dioxide and the changing tropospheric sulphate aerosol data set parameterised as in Mitchell *et al.* (1995). The sea surface temperature data used to force HADAM2A derive from the sea surface temperature (Global Sea Ice and Sea Surface Temperature) data set. sea surface temperature is the result of developments covering some two decades of research at the United Kingdom Meteorological Office (UKMO). These data set generation schemes are thoroughly documented elsewhere (Bottomley *et al.*, 1990; Colman, 1992; Parker *et al.*, 1995; Folland and Parker, 1995; Rayner *et al.*, 1996). Each integration runs from October 1903 to December 1994. Ensemble members differ in terms of arbitrarily chosen initial conditions from recent United Kingdom Meteorological Office operational analyses. The first two months of the experiment were discarded to allow the model atmosphere to reach near equilibrium with the boundary layer forcing after imposition of the arbitrary initial atmospheric conditions. Sea Surface Temperature data were interpolated to the model grid and linearly interpolated to 5-day means to ensure a smooth evolution of the boundary layer.

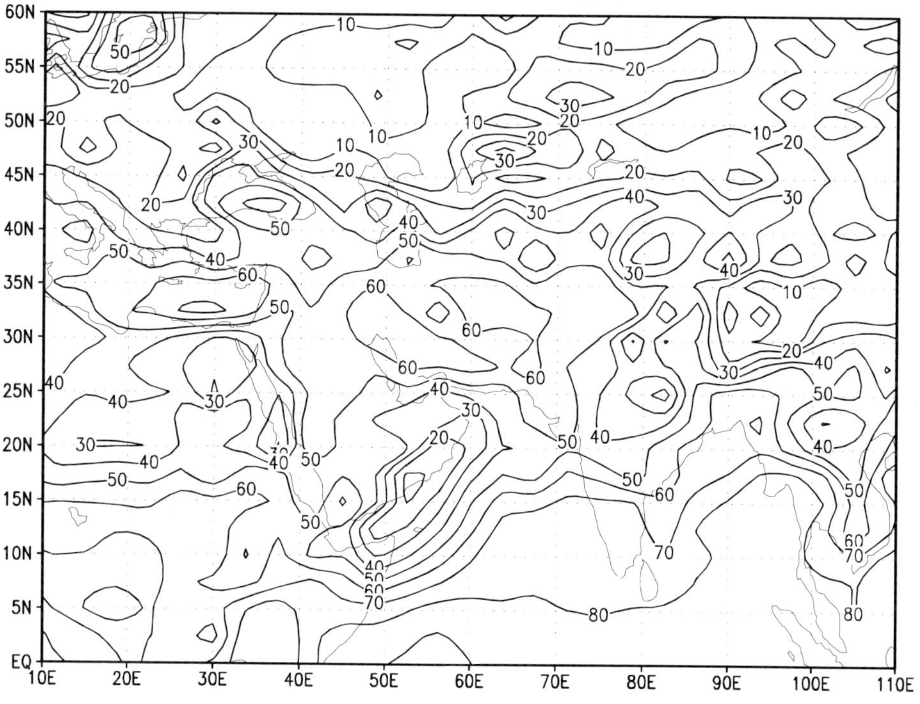

Figure 13. Forced vs. Free manifold DJF Rainfall.

Predictability of Winter Rainfall

The Forced and Free Manifolds determined from the statistical formulation outlined earlier, are presented and described in this section for the domain 0°N-60°N, 10°E – 110°E. for December to February (DJF) model rainfall (Figure 13). The statistics have been calculated over the complete period of model integration (1903-1994). The most striking result to emerge is the extremely high component of the forced rainfall response in the tropics but the extremely low values found in the midlatitudes where transient dynamics are known to dominate (Grotjahn, 1993). In essence this indicates that the tropical atmosphere, as represented by rainfall which usefully integrates many atmospheric processes over several vertical layers in the model atmosphere, responds to sea surface temperature forcing in a similar manner regardless of the initial conditions. The component of internal variability or chaos is therefore very small. Put differently, sea surface temperature forcing is far more important than atmospheric internal variability throughout most of the tropical atmosphere. The opposite is true in the midlatitudes. Here the atmospheric state with which the model was initialised is still important many decades later, despite the forcing of the boundary layer by sea surface temperature. While the tropical atmosphere is predictable on the basis of sea surface temperature forcing, the midlatitude atmosphere, on the whole, is not. This result is not new (Charney and Shukla, 1981; Palmer and Anderson, 1994). Predictability is generally lower over the tropical and subtropical continents than over the oceans at the same latitude (Washington, 2000).

One particularly interesting feature is the abrupt shift from forced to free response between the Eastern Mediterranean – Iraq, Iran region and the region parallel with the Gulf of Aden along the coasts of Yemen and Oman. Interesting links may exist with the Monsoon outflow dynamics. What Figure 13 is unable to show is whether the predictability is due to particular modes of variability which the model integrations have captured.

DISCUSSION AND CONCLUSION

Climate could be thought of as a non-linear system whose variability is determined principally by a number of quasi stationary regimes between which a state vector frequently shifts. In the case of the tropical atmosphere, the attractor which describes the trajectory of climate evolution is influenced by systems with inherently slower time scales, in this case that of the oceans. When the atmospheric system locates in one particular part of the attractor phase space, the atmospheric system is likely to be stable and predictable. Put differently, there are regions of the attractor where neighbouring trajectories do not rapidly diverge, so that in these parts of the phase space the atmosphere is insensitive to initial conditions. Taking El Niño-Southern Oscillation as an example, either El Nino or La Nina pose as stable modes, yet when the system is in transition, that predictability declines drastically. Since the stable modes persist for some time, particularly in the ocean, predictability on seasonal time scales is possible. Therein lies the greatest application of knowledge of chaos, namely seasonal forecasting. Figure 13 suggests that such predictability may exist in the case of the Middle East.

Based on the methods presented in the previous section, the midlatitude atmosphere is highly chaotic compared with the tropics. One of the reasons for this is that individual weather systems in the tropics do not feed back significantly on the basic flow. Yet historically it was the patterns of climate variability in the middle latitudes that captured a large portion of theorists attention, in particular, the interrelationships between fluctuations in the zonal mean wind and the quasi-stationary waves in the atmosphere through the Index Cycle (Rossby et al 1939; Rossby, 1941). The sinusoidal perturbations of the planetary scale circulation as represented by the Index Cycle may be thought of as a chaotic process (Lorenz, 1986). A key to medium and long range (seasonal) forecasting lies in understanding and predicting the state of the Index Cycle. Nowhere in the middle latitudes is this clearer than in the North Atlantic where the North Atlantic Oscillation (NAO) is the dominant mode of variability. In essence the North Atlantic Oscillation is a local expression of the Index Cycle so the key to seasonal forecasting for the North Atlantic sector lies in the predictability of the North Atlantic Oscillation together with a sound knowledge of the teleconnections with which it is associated. Encouragingly, the results discussed in section 3.3. suggest that the phase of the annular mode whose local expression in the domain studied is the Polar-Eurasian Pattern, may in part be forced by the oceans, a result which has been argued for the local expression of the annular mode in the North Atlantic, namely the North Atlantic Oscillation (Sutton and Allen, 1997; Rodwell *et al.*, 1999).

The problem of climate change has also been posed within the perspective of chaos (Palmer,

1998; Mitchell and Hulme, 1999). For example, given a small imposed forcing (such as changes in greenhouse gas concentration), climate change may be viewed in terms of changes to the probability distribution functions (PDF) of climate regimes such as phases of the North Atlantic Oscillation (Palmer, 1998). This notion is beginning to influence climate change detection and attribution experiments such that long model integrations forced by greenhouse gas are now being run in ensemble mode, with experiment members differing by arbitrary starting conditions (Mitchell and Hulme, 1999). More importantly, as set out in this paper, there is a plausible link between the operation of modes of climate variability and climate change. Predictability of climate on a continuum of time scales from interannual to interdecadal, may well provide an inroad for a response and adaptation mechanism to climate change. The evidence discussed in this paper indicates that the Middle East may be one such region. A careful evaluation of the modes of variability and their impact on climate in greenhouse-gas climate model experiments is the next step in the process suggested here.

REFERENCES

Airey, M. J., M. Hulme and T. C. Johns, 1996. Evaluation of simulations of terrestrial precipitation in UK Met. Office/Hadley Centre climate change experiments. *Geophysical Research Letters* 23: 1657-1660.

Barnston, A. G., and R. E. Livezey, 1987. Classification, seasonality and persistence of low-frequency atmospheric circulation patterns. *Monthly Weather Review* 115: 1083-1126.

Bell, G. D., 1998. http://nic.fb4.noaa.gov/data/teledoc/telecontents.html

Bottomley, M., C. K. Folland, J. Hsiung, R. E. Newell and D. E. Parker, 1990. *Global Ocean Surface Temperature Atlas*. HMSO, London.

Climate Diagnostics Center website: http//:www.cdc.noaa.gov/Correlations

Charney, J. G. and J. Shukla, 1981. Predictability of monsoons, in *Monsoon Dynamics*, eds J. Lighthill and R. P. Pearce, CUP, Cambridge.

Colman, A. W., 1992. Development of worldwide marine data eigenvectors since 1985. *Climate Research Technical Note CRTN* 28: Hadley Centre, UK Meteorological Office.

Corti S., Molteni F. and Palmer T. N., 1999. Signature of recent climate change in frequencies of natural atmospheric circulation regimes. *Nature* 398: 799-802.

Esbensen, S. K., 1984. A comparison of intermonthly and interannual teleconnections in the 700mb geopotential field during the Northern Hemisphere winter. *Monthly Weather Review* 112: 2016-2032.

Folland, C. K. and D. E. Parker, 1995. Correction of instrumental biases in historical sea surface temperature data. *Quarterly Journal of the Royal Meteorological Society* 121: 319-367.

Folland, C. K., D. M. H. Sexton, D. J. Karoly, C. E. Johnson, D. P. Rowell and D. E. Parker, 1998: Influences of anthropogenic and oceanic forcing on recent climate change. *Geophysical Research Letters* 25: 353-356.

Gallani, M. L., 1994. *Comparison between 1st and 2nd versions of the Climate Model, Hadley Centre.* Internal Note no 58.

Gates, W. L., 1992. AMIP: The Atmospheric Model Intercomparison Project. *Bulletin of the American Meteorological Society* 73: 1962-1970.

Gates, W. L., A. Henderson-Sellers, G. J. Boer, C. K. Folland, A. Kitoh, B. J. McAvaney, F. Semazzi, N. Smith, A. J. Weaver and Q-C. Zeng, 1996. Climate models – evaluation. In: Houghton, J. T., L. G. Meira Filho, B. A. Callander, N. Harris, A. Kattenberg and K. Maskell. (Eds). *Climate change 1995: The science of climate change*, Cambridge University Press, Cambridge.

Grotjahn, R., 1993. *Global atmospheric circulations*, Oxford University Press, Oxford.

Horel, J. D., 1981. A rotated principal component analysis of the interannual variability of the Northern Hemisphere 500mb height field. *Monthly Weather Review* 109: 2080-2092.

Houghton, J. T., Y. Ding, D. J. Griggs, M. Noguer, P. J. van der Linden, X. Dai, K. Maskell and C. A. Johnson, 2001. *Climate Change 2001: The Scientific Basis.* Cambridge University Press, Cambridge.

Hurrell, J. W., 1995: Decadal trends in the North Atlantic Oscillation: Regional temperatures and precipitation. *Science* 269: 676-679.

Joubert, A., 1997. Simulation by the Atmospheric Model Intercomparison Project of the atmospheric circulation over southern Africa. *International Journal of Climatology* 17: 1129-1154.

Lorenz, E. N., 1963. Deterministic non-periodic flow. *Journal of Atmospheric Science* 20: 130-141.

Lorenz, E. N., 1986. The Index Cycle is Alive and Well. Namias Symposium, J. O. Roads (ed), *Scripps Institution of Oceanography Reference Series* 86-17: 188-196, University of California, San Diego.

Lorenz, E. N., 1993: *The essence of chaos*, Univ. Washington Press, 227p.

Mitchell, J. F. B., T. C. Johns, J. M. Gregory and S. F. B. Tett, 1995: Climate response to increasing levels of greenhouse gases and sulphate aerosols. *Nature* 376: 501-504.

Mitchell, T. D and M. Hulme, 1999: Predicting regional climate change: living with uncertainty. *Progress in Physical Geography* 23: 57-78.

Mo, K. C., and R. E. Livezey,1986. Tropical-extratropical geopotential height teleconnections during the Northern Hemisphere winter. *Monthly Weather Review* 114: 2488-2515.

Palmer, T. N. and D. L. T. Anderson, 1994. The prospects for seasonal forecasting - a review paper. *Quarterly Journal of the Royal Meteorological Society* 120: 755-793.

Palmer, T. N., 1998. Nonlinear dynamics and climate change: Rossby's legacy. *Bulletin of the American Meteorological Society* 79: 1411-1423.

Parker, D. E., C. K. Folland and M. Jackson, 1995. Marine surface temperature: observed variations and data requirements. *Climatic Change* 31: 559-600.

Phillips, T. J., 1994. A summary documentation of the AMIP models. *PCMDI report,* 18, University of California, Lawrence Livermore Laboratory, California.

Rayner, N. A., E. B. Horton, D. E. Parker, C. K. Folland and R. B. Hackett, 1996. Version 2.2 of the global sea-ice and sea surface temperature data set, 1903-1994. *Climate Research Technical Note CRTN,* 74, Hadley Centre, UK Meteorological Office.

Rodwell, M. R., D. P. Rowell, and C. K. Folland, 1999. Oceanic forcing of the wintertime North Atlantic Oscillation and European climate, *Nature* 398: 320-323.

Rogers, J. C. and van Loon, H., 1979. The seesaw in winter temperatures between Greenland and northern Europe. Part II: Some ocean and atmospheric effects in middle and high latitudes. *Monthly Weather Review* 107: 509-519.

Rossby, C-G., 1941. The scientific basis of modern meteorology. *U.S. Yearbook of Agriculture, Climate and Man,* 656-661.

Rossby, C-G., and co-authors, 1939. Relation between variations in the intensity of the zonal circulation of the atmosphere and the displacement of the semi-permanent centres of action. *Tellus* 2: 275-301.

Rowell, D. P. C. K. Folland, K. Maskell, and M. N. Ward, 1995. Variability of summer rainfall over tropical north Africa (1906-92): Observations and modelling. *Quarterly Journal of the Royal Meteorological Society* 121: 669-704.

Rowell, D. P., 1998. Assessing potential seasonal predictability with an ensemble of multidecadal GCM simulations. *Journal of Climate* 11: 109-120.

Srinivasan, G., M. Hulme and C. G. Jones, 1995: An evaluation of the spatial and interannual variability of tropical precipitation as simulated by GCMs. *Geophysical Research Letters* 22: 2139-2142.

Stone, D. A., A. J. Weaver, and R. J. Stouffer, 2001. Projection of climate change onto modes of atmospheric variability. *Journal of Climate* 14: 3551-3565.

Sutton, R. T. and M. R. Allen, 1997. Decadal predictability of North Atlantic sea surface temperature and climate. *Nature* 388: 563-567.

Thompson, D. W. J., and J. M. Wallace, 2000. Annular modes in the extra-tropical circulation. Part I: Month-to-month variability. *Journal of Climate* 13: 1000-1016.

Van Loon, H. and J. C. Rogers, 1978. The seesaw in winter temperatures between Greenland and northern Europe 1. General Description. *Monthly Weather Review* 106: 296-310.

Walker, G. T. and E. W. Bliss, 1932. World Weather V, *Memoirs of the Royal Meteorological Society* 4: 53-84.

Wallace, J. M., 2000. North Atlantic Oscillation/annular mode: Two paradigms-one phenomenon. *Quarterly Journ. Royal Met. Soc.* 126: 791-805.

Wallace, J. M., and D. S. Gutzler, 1981. Teleconnections in the geopotential height field during the Northern Hemisphere winter. *Monthly Weather Review* 109: 784-812.

Ward, M. N. and Navarra, A. 1997. Pattern analysis of sea surface temperature-forced variability in ensemble GCM simulations: Examples over Europe and the tropical Pacific. *Journal of Climate* 10: 2210-2220.

Washington, R., 2000: Quantifying Chaos in the Atmosphere. *Progress in Physical Geography* 24: 4499-514.

Lenses, Plumes and Wedges in the Sultanate of Oman: A Challenge for Groundwater Management

PHILIP G. MACUMBER
Department of Earth Sciences, University of Melbourne, Victoria, Australia. E-mail: pmacumber@optusnet.com.au

ABSTRACT: Groundwater in eastern Arabia is commonly associated with relatively flat-bedded aquifer systems in Tertiary marine limestone. However in the Sultanate of Oman, there are a number of unusual flow systems in which large freshwater lenses and plumes occur as discrete bodies within, but separate from the broader aquifer system. Each of the occurrences requires special management considerations as the specifics of the partial flow system, rather than those of the aquifer in general, determines the nature, extent, and at times, the survival of the resource. Examples of such occurrences are the large and exceptionally fresh groundwater lenses (TDS 400-160 mg/l) overlying saline regional systems in hyper-arid central Oman. The larger lenses represent modern day recharge, and as such run counter to more conventional wisdom about recharge potential under such adverse conditions. The lenses are a renewable resource but require careful management to avoid upcoming, saline pollution and eventual destruction.

A similar requirement occurs on the Al Khawd Fan at the eastern edge of the Batinah coastal plain of northern Oman where water tables are at or close to sea level, 6 km inland near important wellfields. While the potential for upconing of saline water would appear to be acute, this is less precarious than would normally be the case. High groundwater pressures occurring at depth within the aquifer, generated from upbasin within the regional and intermediate flow systems, counter the inland migration of the intruded seawater wedge and thus partially protect the resource. Westward, on the Batinah coastal plain, a further type of flow system sees large plumes of groundwater originating from recharge areas high (>2000 m) on the Jabal Akhdar, and then passing across the Batinah plain to the coast. In the case of the Wadi Maawil, the high altitude isotopic signature of the Maawil plume is still present at the coast after crossing the Batinah plain, showing that there is very little low altitude recharge on the plain along the path of the plume. In a more general sense this demonstrates that recharge from precipitation is insignificant on the Batinah coastal plain. The presence of the plume across the deepest part of the Eastern Batinah, an area with the highest groundwater potential in the coastal plain alluvial aquifer, requires special consideration in management of the aquifer.

INTRODUCTION

This paper examines non-conventional groundwater flow systems in Oman, which have broad implications for other regions. Two different settings are provided, one in the coastal alluvial aquifers of northern Oman (Al Khawd Fan, Eastern Batinah coastal plain, and the adjacent high relief Jabal Akhdar-Fig.1, area 1), and a second in Tertiary limestone aquifers of the hyper-arid Al Wusta Region of central Oman (Ma'abar Depression-Fig.1, area 2, and the Wadi Rawnab catchment-Fig.1, area 3).

In the Eastern Batinah and in Al Wusta (Wadi Rawnab and the Ma'abar Depression) much of the understanding of the lenses and plumes has come from the use of what is generally considered an outdated technique, cable tool drilling. While considerably slower than rotary drilling, and limited in depth capacity, the collection of good data was paramount and this approach made it simple for field staff to sample. It was adopted in order to provide a simple, reliable method for detailed close-interval sampling (2m to 5m) of profiles where fresh groundwater overlies saline water with an intervening transitional zone. The technique permitted sampling for stable isotopes and chemical analysis thereby enabling chemical and isotopic profiling through the freshwater zone, the transition zone, and into the underlying saline regional groundwater or intruded seawater in the case of the coastal plain.

As the boreholes deepened, casing which closely followed the drill down the hole sealed off the overlying parts of the profile. One advantage of this program was it enabled the recording of potentiometric head changes with depth, which in the case of the upper Rawnab gorge showed the presence of a strongly developed recharge profile. In the Eastern Batinah seawater intrusion investigation, it permitted quasi permeability tests at each sampling interval in the alluvial aquifer by adopting

Water Resources Perspectives: Evaluation, Management and Policy. Edited by A.S. Alsharhan and W.W. Wood.
Published in 2003 by Elsevier Science, Amsterdam, The Netherlands, p. 349-370.

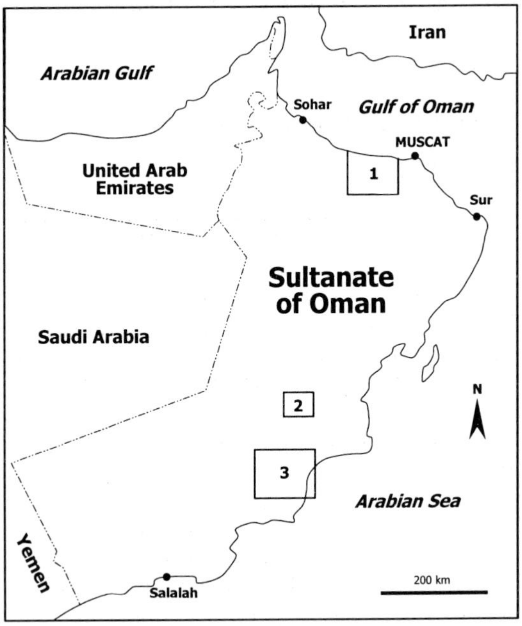

Figure 1. Locality plan, Sultanate of Oman.

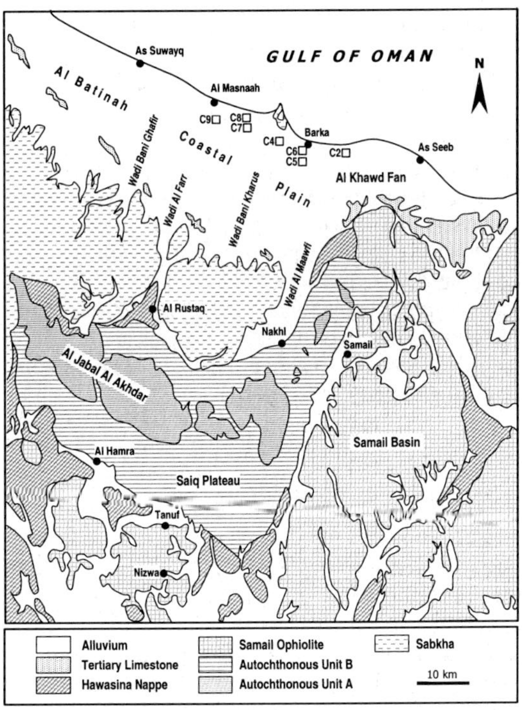

Figure 2. Geology of the Jabal Akhdar and Eastern Batinah-Al Khawd Region.

a method of short term extraction-recovery tests (Macumber, 1998). This in turn, when linked with the isotopes and salinities recorded at each depth, provided data on preferential pathways used for the migration of intruded water into the aquifer.

The program initially used on the Ma'abar Depression, was later extended to cover the large wadis draining to the Al Jazir coastal plain, focusing on the 80 km long course of the Wadi Rawnab. Termed the Cable Tool Program (Macumber, 1998) for the Eastern Batinah seawater intrusion investigation, the chemical analysis and isotopes provided a framework for understandings groundwater profiles and process to 100 m depth in a line parallel to the coast through the intrusion. It also provided the means to investigate the position and character of isotopically depleted groundwater plumes recharged at high altitude on the Jabal Akhdar which then crossed the coastal plain unmodified to discharge at the coast (Macumber et al., 1997).

1. NORTHERN OMAN - THE JABAL AKHDAR, AL BATINAH COASTAL PLAIN AND AL KHAWD FAN

The Al Batinah coastal plain is flanked to the south by the Jabal Akhdar forming part of the northern Oman Mountains, a high mountain range paralleling the coast, which rises to 3000 m. (Fig. 2). The Jabel Akhdar is essentially a large anticlinal structure of outcropping Permian to Cretaceous sediments - the Hajar Super Group in which the main aquifers are limestones and dolomites. The anticline is strongly incised on its northern flanks to produce large depressions in which are exposed the older pre-Permian sediments and volcanics (Late Proterozoic or Infracambrian-Autochthonous Unit A). One example of such an exposure occurs in the Saiq Plateau, which lies at an elevation of 2000 m.

The eastern limits of Jebel Akhdar are formed by the Samail Basin, eroded into ophiolitic sequences of the Samail nappe. The Samail Basin drains via Wadi Samail towards the Al Khawd Fan lying at the eastern limit of the Batinah coastal plain. The Jebel Akhdar forms an east west divide with drainage passing northwards towards the coastal plain and southwards to the interior. The coastal plain consists of a thick Tertiary to Recent alluvial aquifer, the sediment sourced from the adjacent high mountains. On the basis of geophysical evidence the aquifer may reach 1500 to 2000 m thick at the coast. Seawater intrusion into the coastal aquifer has occurred progressively with development of the coastal alluvial aquifer, and in the case of the Al

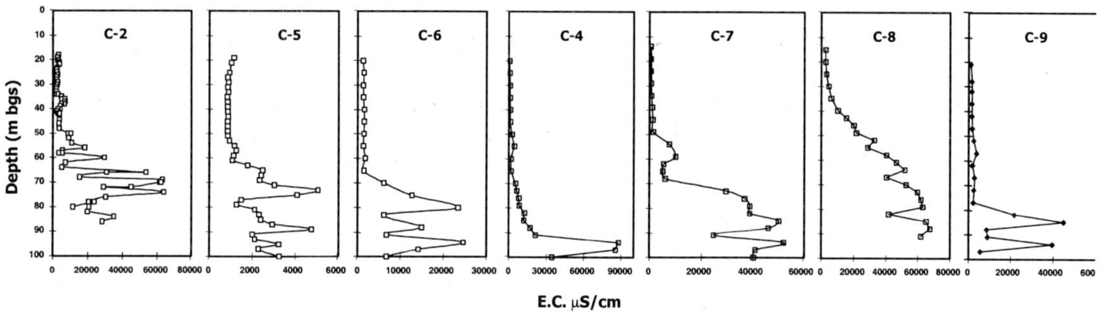

Figure 3. Electrical conductivity profiles through the C-Series bores.

Khawd Fan, reaches 6 km inland, where it is found at a depth of 150 m. The climate of northern Oman is arid with annual average rainfall being about 100 mm on the plains, but rising to over 300 mm on the higher parts of Jebel Akhdar.

Eight bores (the C-Series) drilled to a depth of 100 m were completed in a line parallel to the coast, and in two instances bores were drilled about 1 km apart at right angles to the coast (line C-5 and C-6, and line C-7 and C-8) to provide further detail of flow characteristics between the coast and the main traverse line. In all instances the bores passed at depth into saline water representing seawater intrusion into the coastal plain alluvial aquifer, but reaching only the top of the transition zone in bore C-5. Detailed electrical conductivity profiles based on samples collected during drilling are given in Fig. 3.

The cable tool approach produced a far more detailed salinity profile than that from follow-up down hole electrical conductivity logging, which by comparison gave very generalized profiles and occasionally, as was the case with C-2, showed a markedly different position for the saltwater interface in the bore compared to that actually occurring in the aquifer. A further feature of the sampling was the presence of isotopic and salinity spikes in the profiles, which represent discrete zones of 'foreign' groundwater intrusion, commonly along more permeable beds.

Stable Isotopes in the Rain and Groundwater

Stable isotopes (deuterium and oxygen-18) have been previously used to provide interpretations on the nature of rainfall, recharge and groundwater flow on the Al Batinah coastal plain and the nearby Jabal Akhdar (Gibb, 1976; PAWR, 1986). However, the relationship between deuterium and oxygen-18 in the rainfall and in the groundwater was unclear.

The most striking feature of the deuterium/oxygen-18 relationship for rainfall is a

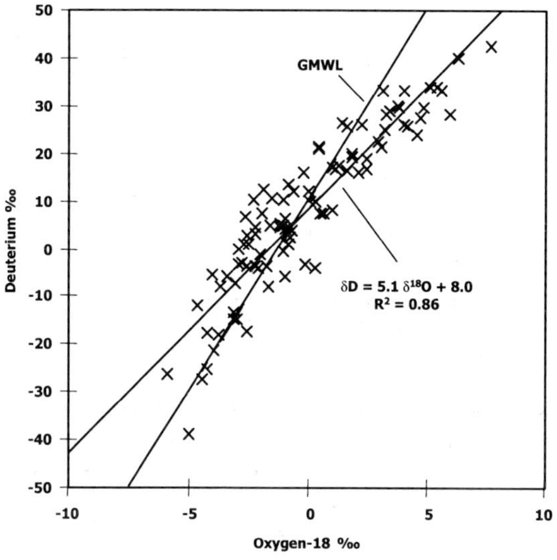

Figure 4. Stable isotopes for Jabal Akhdar-Eastern Batinah rainfall (Macumber, et. al., 1997).

non-linearity, in which the δD and δ^{18}O relationship shows a degree of duality towards the depleted end of the meteoric water line, with some data showing departures from the general trend line towards that of the GMWL (Fig. 4). This reflects a northerly intrusion of rainfall patterns normally occurring across central and southern Oman. Similar character is seen in the Rawnab freshwater lens and the O6-A cyclone in Al Wusta (Macumber 1995; Macumber et al., 1995) and in the Ibra storm of July 1995 (see details below), which swept through northern Oman causing water tables to rise significantly in the Samail Basin and across the central Al Khawd Fan (Macumber, 1997).

Since the rainfall plot does not readily fit any of the previously established meteoric water lines such as the Global Meteoric Water Line (GMWL - Craig, 1961), the Mediterranean meteoric water line (Bortolami et al., 1978) or the Northern Oman meteoric water line (PAWR, 1986), a review of the large amount of existing data was carried out and a new empirically based meteoric water line for northern Oman was established (Macumber et al., 1997; Macumber, 1998).

$\delta D = 5.1\ \delta^{18}O + 8.0$ $R^2 = 0.86$
(Macumber et al., 1997).

The new meteoric line is similar to Gibbs (1976) Bahrain MWL ($\delta D = 4.9\delta^{18}O + 9.7$) and to that obtained by Wushiki (1991) for a rainfall data set for the Saiq Plateau ($\delta D = 4.9\ \delta^{18}O + 9.0$), demonstrating the uniformity of this general relationship in precipitation from sea level at Bahrain to the 2000 m high altitude Saiq Plateau of the Jabal Akhdar.

However, the various rainfall-based meteoric water lines have proved of limited value in groundwater studies and it was considered that a water line based on the groundwater systems was a more appropriate way to study groundwater processes.

Accordingly, 50 groundwater samples were collected from aflaj and springs of Jebel Akhdar and analysis carried out at the CSIRO Laboratory in Adelaide, Australia.. Within the Jabel Akhdar/Jabel Nakhl the principal aquifers are limestones and dolomites of the Palaeozoic-Mesozoic Hajah Super Group and the pre-Permian. Hajir Limestone. From these samples, the Akhdar Water Line (Fig. 5) was erected to provide an empirical basis with which to examine the evolution and differences between groundwaters on Jebel Akhdar and on the adjacent coastal and inland plains. The relationship is:

$\delta D = 5.1\ \delta^{18}O + 3.0$ $R^2 = 0.92$
(Akhdar Water Line – Macumber et al., 1997).

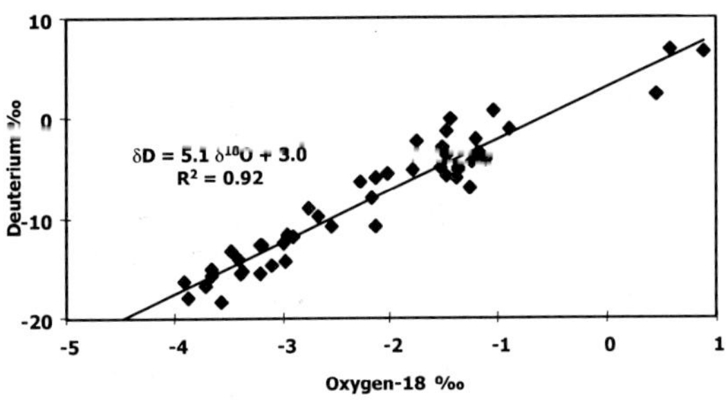

Figure 5. Stable isotope for the Akhdar Line (Macumber, et. al., 1997).

The Akhdar line is similar to the water line derived from the northern Oman (Jebel Akhdar/Batinah) rainfall data, but there is a relative 5 unit deuterium deficit. The difference is likely to be due to secondary fractionation perhaps prior to infiltration. A similar deficit is present in limestone springs recorded by Gibb (1976) from Jabal Akhdar from - $\delta D = 4.4\ \delta^{18}O + 3.6$. Using the Akhdar Line as a reference and isotopic data sets from a range of catchments and aquifers it is possible to study groundwater flow and process on and between the Jabal Akhdar and coastal plains.

Perhaps the most important factor in the identification of groundwater source comes from the altitude effect in which groundwater recharged high on the Jabal Akhdar is significantly depleted in stable isotopes relative to that occurring on the Eastern Batinah coastal plain, the Al Khawd Fan and the Samail Basin. The recognition of the high altitude signature (Fig. 6) has been central to an understanding of groundwater movement from the Jabal Akhdar to the Batinah coastal plain and to tracking its course across the plain to the coast (Macumber et al., 1997; Macumber, 1998).

Groundwater Flow in the Al Khawd Fan

The Al Khawd Fan has by far the largest catchment of any northern draining wadi, but unlike the wadis further west this is a low level ophiolitic catchment within the Samail Basin. Seawater intrusion is a significant feature of the Al Khawd Fan and water tables are at or close to sea level over much of the fan. Only in the central parts of the fan lateral to the Wadi Samail, which flows into the sea to the west of As Seeb, are they marginally above sea level (Fig. 7). Under the Ghyben-Herzberg principle, this situation would imply seawater intrusion at very shallow depths coming inland as far as the important Seeb and Al Khawd well fields in the upper and central area of the fan. However, in reality this is not the case and instead high pressures occurring deeper in the aquifer and resulting in strong vertically upwards gradients counter the seawater intrusion (Macumber, 1997). For instance, groundwater pressures at depths of 300 m in the central fan are at times 9 m or more higher than that of the immediately overlying water table (Fig. 8).

Hydrographs on the Al Khawd Fan may be subdivided into two parts, one represented by a 'spikey' response induced by local wadi recharge during wet events, best seen in shallow piezometers such as RGS-5F where the water table is mostly 0 to

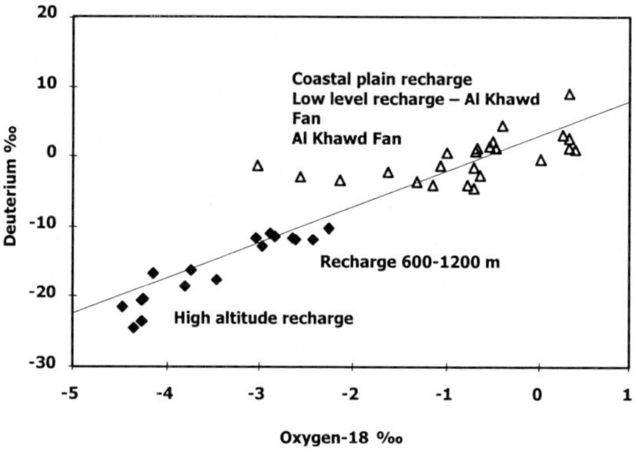

Figure 6. Elevation effect in stable isotopes from northern Oman.

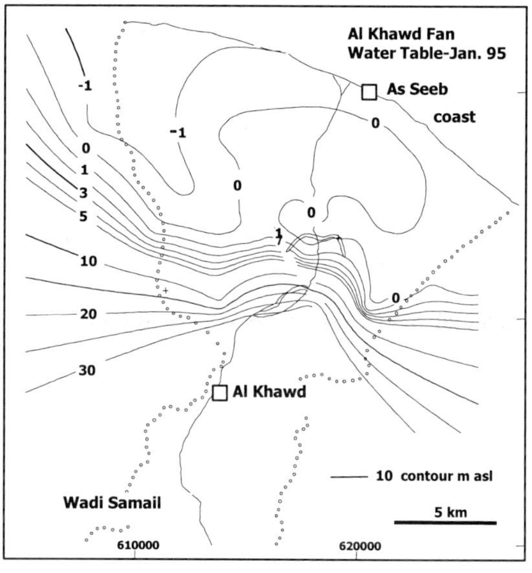

Figure 7. Water table levels on the Al Khawd Fan-January 1995.

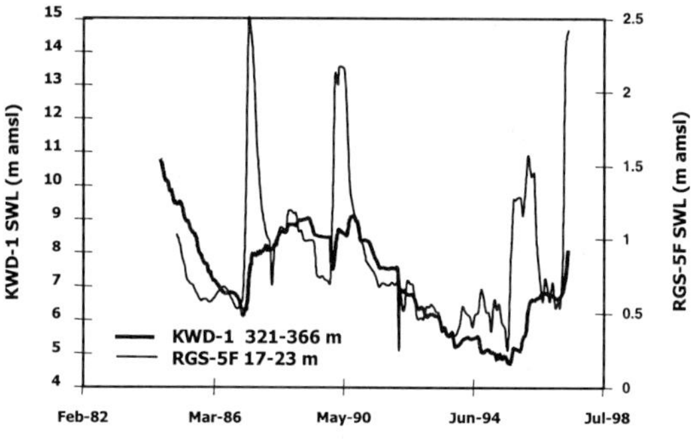

Figure 8. 'Saw-tooth' pressure responses at 321-366 m depth in the KWD-1 bore and 'spikey' responses in the shallow RGS-5F bore, superimposed on the saw-tooth pattern.

general cycles of wet and dry periods occurring across the catchment, with four cycles occurring between 1973 and 1997. They are generated within the regional and intermediate flow systems in response to recharge both at the head of the fan and further upbasin within the ophiolitic catchment of the Samail Basin.

The pattern indicates rapid recharge followed by slow decline, with an overall fall in the lowermost levels of the OW-1 piezometer of 10 m over the period 1973 to 1997; probably reflecting increased groundwater development over this time. Despite the water tables being at or close to sea level, it is the fluctuating pressure levels deeper within the alluvial aquifer of the central and lower fan (RGS-5F and KWD-1) that ultimately dictates the position and movement of the freshwater-saltwater interface, and hence the environmental and economic impacts of seawater intrusion on the Al Khawd Fan.

Gibb (1976) shows that the bulk of the groundwater in the Samail Basin is isotopically enriched, with oxygen-18 values commonly a little less than 1.5 ‰ $\delta^{18}O$, reflecting a low altitude source of recharge. These groundwaters are fed by rainfall and runoff from within the basin and its adjacent catchments. However along the western flank of the basin adjacent to the Jabal Nakhl extension of Jabal Akhdar, depleted groundwater occurs in limestone fed springs and aflaj, showing an input from a high altitude source on the nearby Jabal Nakhl. The aflaj are commonly situated on extensions of major faults and lineaments passing from the jabal into the ophiolites of the Samail Basin. These are conduits for groundwater transfer from high altitude recharge areas to low altitude discharge zones. Pumping tests (Bhatnagah, 1996) show that within the ophiolite fracture system, transmissivities range from between 1000 to 8000 m^2/d. A similar process for groundwater transfer from the Samail Basin to the Al Khawd Fan via the fractured rock ophiolitic aquifer must also occur.

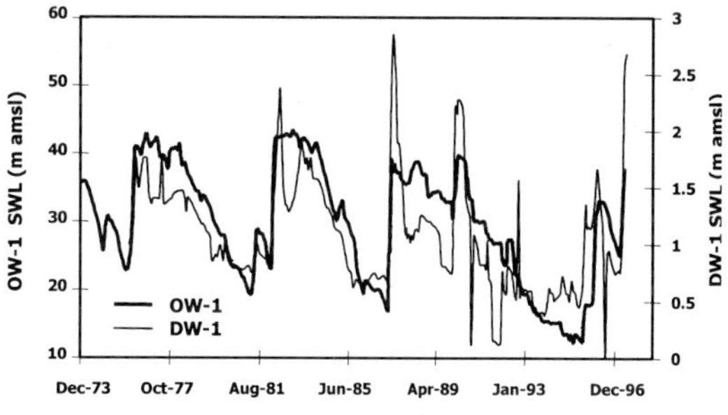

Figure 9. Cyclic saw-tooth responses in the OW-1 bore at the head of the Al Khawd Fan and also present in the DW-1 bore at the coast.

Water budgets on the Al Khawd Fan require large volumes of inflow in order to balance outflow. The Seeb and Al Khawd wellfields alone in 1996 pumped more groundwater on average (8 MCM/yr) than surface water entering the upper Al Khawd Fan from the Samail Basin (average 5.1 MCM/yr from 1983-1996, measured at the Al Khawd gauging station) Large groundwater use on the coastal plain must also be accounted for, as must the groundwater passing seawards to counter the seawater intrusion. It is difficult to see where this large additional water source comes from, if not directly through the ophiolitic basement aquifers from the Samail Basin. In having a distant source, the Al Khawd Fan shows a similar behaviour to the Maawil Trough on the Eastern Batinah coastal plain. It follows that management of the seawater intrusion on the Al Khawd Fan is closely linked to groundwater and surface water management in the Samail Basin.

Groundwater Flow in the Jabal Akhdar and Eastern Batinah Coastal Plain

Isotopically depleted groundwater in the limestone/dolomite aquifers occurs in an east west belt across Jabal Akhdar, its upper northern catchments and the Saiq Plateau. It then extends northward as a plume passing through the gap in the ophiolites near Nakhl. (Fig.10). The plume continues down the Wadi Maawil catchment, across the Batinah coastal plain to the coast, where it occurs in the coastal BM-1 bore near Barka, overlying intruded seawater (Macumber, et al., 1997; Macumber, 1998). The principle pathway from Jabal Akhdar is via the Hajir Formation in the pr-Permian Autochthonous Unit A and in the limestones/ dolomite aquifers of the Permian-Cretaceous Autochthonous Unit B, such as the Wasia and Mahil limestones. The Hajir Formation outcrops high on the Hutt Plateau in the Jabal Akhdar and along wadis in the upper catchments of the northern slopes of Jabal Akhdar.

The distribution pattern of groundwater with a high altitude signature across the Jabal Akhdar and its northern catchments and as plumes passing across the coastal plain is not repeated in the southern catchments (Macumber et al., 1997), suggesting at best only minor flow in this direction. The implications are that much of the groundwater passing northwards towards the coast is derived from recharge areas in catchments lying to the south of the surface water divide on the Jabal Akhdar including the Saiq Plateau. That is, there is a wide separation of the surface and groundwater divides on Jabal Akhdar.

Since the limestone aquifers provide preferred pathways for the northwards migration of depleted groundwater from the upper catchments of Jebel Akhdar, the plume has a characteristically lower Mg/Ca ratio than water coming from the lower altitude magnesium rich ophiolites. On this basis alone, the Maawil plume is clearly identifiable, as is a similar high altitude sourced plume which passes from the Jabal Akhdar through a further gap in the ophiolites to the west of Al Rustaq. However, while the isotopes of the western plume have a clear high altitude character, they are not as isotopically depleted as the Maawil plume.

In the upper and middle parts of the Maawil coastal plain, an indication of the vertical distribution of the plume can be had from bores occurring along the Maawil Section line of MMP (1989). These bores were sampled by PAWR (1986). In the deeper RE-3 and RE-4 bores isotopically depleted high altitude recharge water was found across a wide vertical interval extending to almost 200 m in the RE-4/4P bores (Fig. 11). While the bores were sampled over wide intervals, the depleted values for the isotopes show that there is virtually no dilution by lower altitude groundwater within the respective screened intervals. An age of ca 5,000 yr. b.p. was obtained by PAWR (1986) from the RE-4 bore, giving an approximate flow time for high altitude water to reach the central Maawil plain. Of the bores sampled, no water other than high altitude recharge water occurs, except near the coast where it merges with intruded seawater in a varyingly wide transition zone, and in bore C-6 where a lateral inflow from the west was recorded at depth, interspersed with zones of isotopically depleted groundwater.

The 5-10 km wide plume passes through the isotopically depleted C-5 and C-6 bores but is straddled to the east and west respectively by the isotopically enriched C-2 and C-4 bores (Figs. 12 and 13), indicative of low altitude recharge. Detailed chemical and isotopic samples at 5 m intervals provides data on the vertical and lateral characteristics of the plume. The chemistry was analysed at the MWR laboratory in Muscat, and the C-Series stable isotopes by Connie Weyhenmeyer of Bern University. In the C-5 bore isotopically depleted groundwater from the high Jabal Akhdar recharge source, occurred throughout its full depth to 100 m, although towards the base there is some minor mixing with isotopically enriched water at the top of the seawater intrusion transition zone. Even so, the deeper profile remains relatively depleted. In the BM-1 bore at the coast, groundwater with a high altitude source ($\delta^{18}O = -2.97$, $\delta^{2}H = -11.1$) overlies

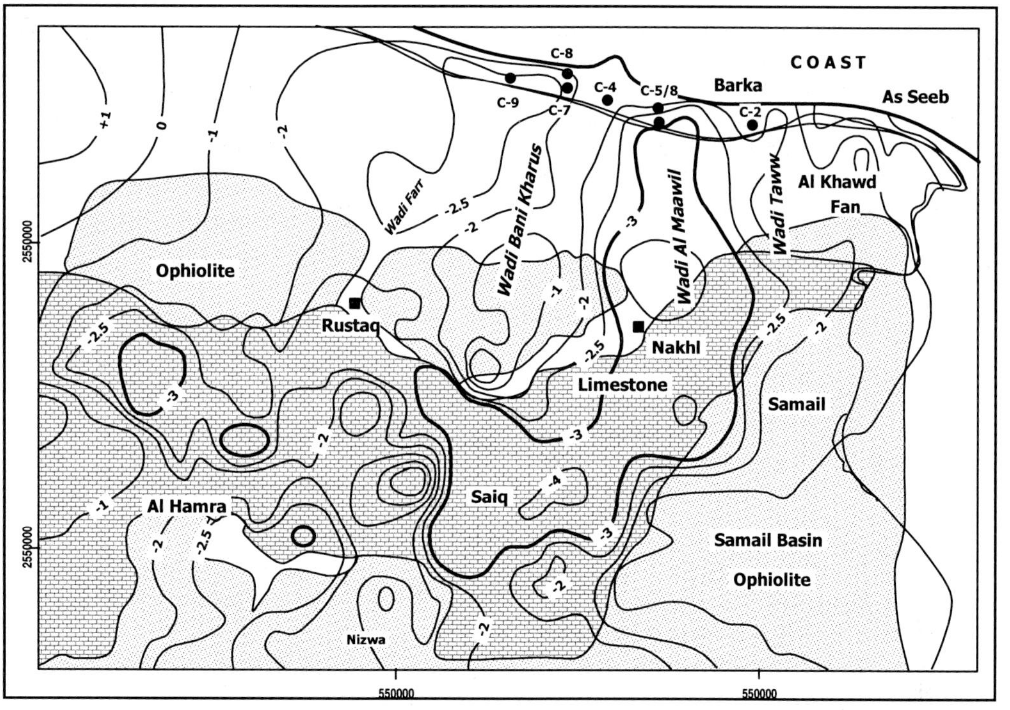

Figure 10. Stable Isotope distribution on the Jabal Akhdar and the Maawil plume.

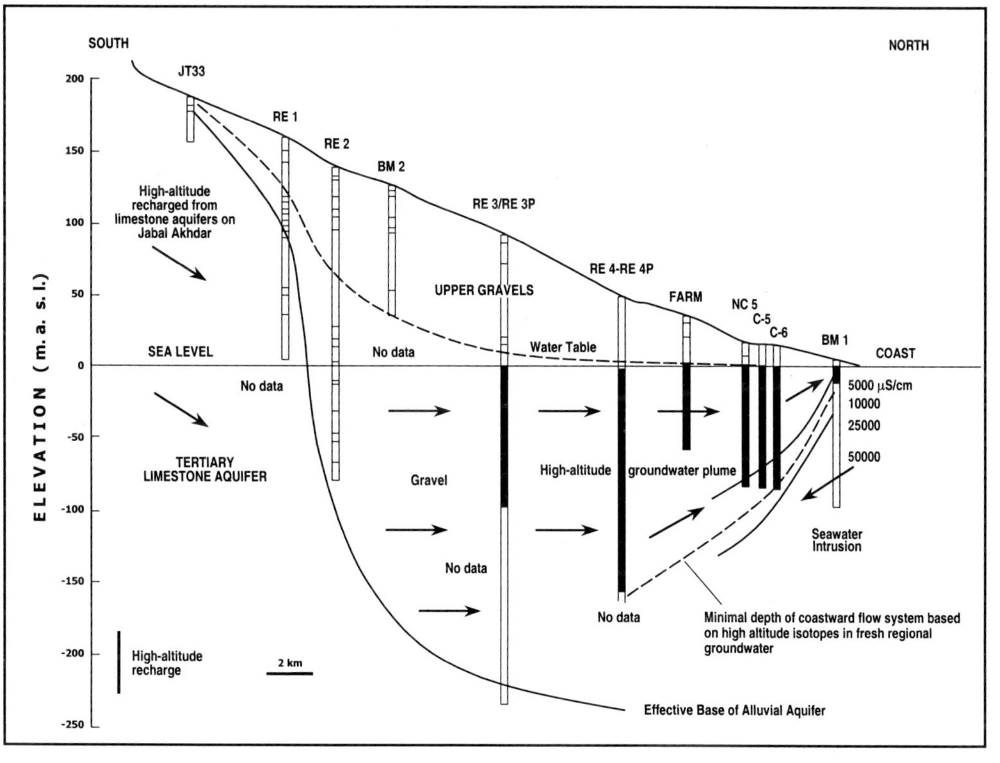

Figure 11. Occurrence and flow of isotopically depleted groundwater across the Maawil Trough.

enriched intruded seawater (δ^{18} O = 0.19, δ^{2} H = 1.8). Since virtually all water in the zone of coastal outflow lying above the transition zone at BM-1 comes from a high altitude source, this bore supports the more general conclusion that there is no significant addition of low altitude recharge to the plume on crossing the plain.

Only in the C-6 bore are there indications of inter-layering of high altitude and low altitude water, shown by the presence of low salinity isotopically enriched spikes which occur at depths of 50 m and 65 m (Fig.13). Their presence deep within the aquifer shows that it is not simply infiltration from above, but instead appears to represent the lateral migration from perhaps more westerly areas of the Maawil plain where such groundwater occurs. The enriched freshwater spikes are found in more permeable layers, which were identified by qualitative recovery tests carried out at each sampling interval. This was done by the sealing off all but the sampling interval of the aquifer with casing, then a set volume of water was removed and the extent/rate of recovery over a fixed time interval was measured. In the cases of the 50 and 65 m freshwater intervals, the recovery was rapid and virtually complete (Fig. 14).

The course of the plume is directly across the deepest part of the coastal alluvial sequence, referred to as the Maawil Trough and considered as having perhaps the highest resource potential on the Eastern Batinah. The path does not coincide with the present active wadi courses of Wadis Maawil and Taww but instead lies on the interfluve between them and this may in part explain the absence of any modification from a wadi flow contribution on the plain. The latter occurs further west along the course of Wadi Maawil at the western edge of the Maawil catchment, where groundwater occurring in the JT-10, JT-11 and JT-12 bores lying upstream of the C-4 bore, and the C-4 bore all show a similarly enriched character, indicating low altitude wadi recharge. However, this does not explain the lack of a precipitation impact, the absence of which indicates that there is in fact no meaningful modern day recharge occurring from direct precipitation on the Maawil coastal plain, sufficient to modify the character of the depleted high altitude water. Since climatologically the central Maawil plain is no different to the rest of the coastal plain, it follows that this result provides a strong basis for the more general conclusion that precipitation on the coastal plain does not contribute significantly to groundwater recharge. This is in line with the conclusions of earlier workers (Gibb, 1976; JICA, 1986 and MMP 1989), who on the basis of the near

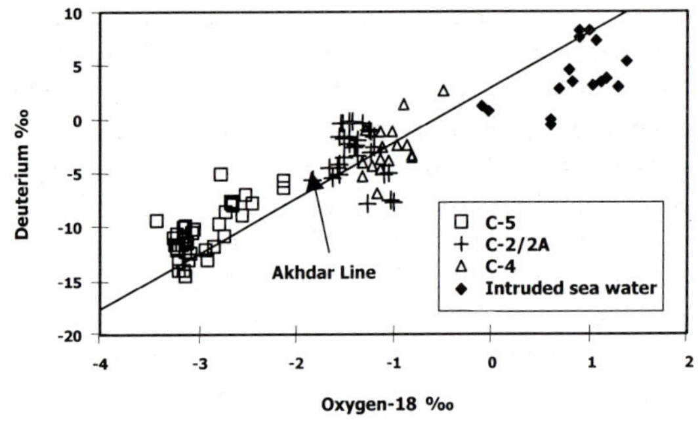

Figure 12. Stable isotope relationship for C-2, C-4 and C-5 bores and for seawater intrusion.

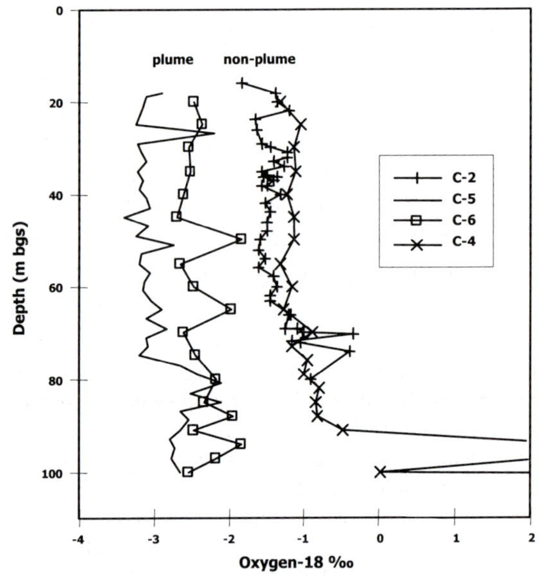

Figure 13. Oxygen-18 profiles across the Maawil plume. Bores C-2 and C-4 straddle the plume, which passes through C-5 and C-6.

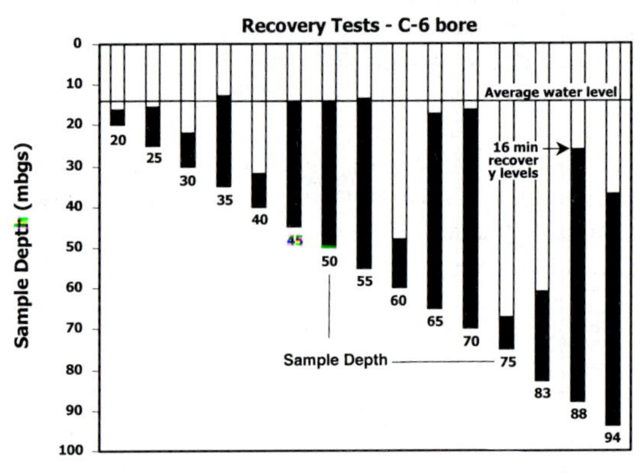

Figure 14. Qualitative recovery tests at sampling intervals in the C-6 bore.

absence of tritium in the groundwater considered that recharge on the plains was relatively minor. Gibb (1976) comments: "*that the absence of tritium from most of the Batinah alluvial plain is the most surprising and important result of the isotope study*". Gibb (1976) did not consider that a dilution of recent recharge by pre-1952 groundwater masked the recent recharge and comments on the basis of his observations of aquifer characteristics that "*the dilution hypothesis is impossible to support*". He found evidence of modern recharge water only in one bore, JT-17.

This conclusion based on stable isotopes regarding an absence of recharge on the plain from direct precipitation does not preclude recharge from wadi flow, which may become more significant in those areas away from the two established plumes passing through the lower Maawil and Farr catchments. However, wadi flow has had no impact on the plume further upbasin nearer Nakhl where the plume and wadi coincide, again supporting the tritium data suggesting limited modern recharge even from the wadi systems.

Groundwater Management

The presence of distinct zones having markedly different flow systems and recharge conditions has major implications for groundwater resource development and seawater intrusion management for the Eastern Batinah coastal plain, and perhaps the Batinah coastal plain in general. For the Eastern Batinah it is especially relevant given that the Maawil Trough is seen as perhaps having the most important groundwater resource potential of any area in the Eastern Batinah. However, the distant recharge source in the high areas of the Jabal Akhdar and its northern upper catchments, and the concomitant time lag for groundwater flow to the plains must be carefully considered should future well-field development proceed.

In the region of the central Maawil trough, there is no evidence of any meaningful modern day recharge occurring from direct precipitation on the coastal plain sufficient to modify the isotopic character of the high altitude sourced plume. Since climatologically the central Maawil plain is no different to the rest of the coastal plain, it follows that this result provides a strong basis for the more general conclusion that precipitation on the coastal plain does not contribute significantly to groundwater recharge. This is in line with the tritium data, which suggests that even recharge from wadi flow is very limited.

It follows that for both the Batinah coastal plain and the Al Khawd Fan, in any groundwater study/model carried out as part of a resource investigation or sea water intrusion investigation, the disparate flow patterns and groundwater sources must be recognized and incorporated if any meaningful analysis is to be had.

2. GROUNDWATER FLOW AND FRESHWATER LENSES IN AL WUSTA

The Al Wusta Region of Central Oman, centred on the small town of Hayma, midway between Muscat and Salalah, extends from the Saudi Arabian border to the Arabian Sea (Fig. 15). A Central Plateau formed across Tertiary marine limestone covers much of Al Wusta and extends southwards into the Najd Region of the Dhofar Province. It lies at an elevation of between 120 m and 240 m above sea level. Although essentially flat, there are a number of large depressions, several of which, at Hayma and Ma'abar, contain thin freshwater lenses. To the west of the Central Plateau, spanning the Oman-Saudi Arabian border, is the A'Rub Al Khali (Empty Quarter), with its predominantly aeolian landscape superimposed on an otherwise flat relict marine or coastal plain. Eastwards, the Central Plateau falls to the Huqf where some of the oldest rocks in Oman outcrop.

In the southeast, the Central Plateau is dissected by a number of coastward trending dry wadis, including Wadi Rawnab, which pass towards the Al Jazir coastal plain. Unlike the Batinah and Salalah plains, Al Jazir plain does not have any large fresh groundwater resource, but instead contains only a limited number of marginally potable coastal wells.

Climate

Much of Al Wusta is a hot dry desert; the climate is hyper-arid with an annual average rainfall estimated at between 20 mm and 50 mm. In summer temperatures commonly reach 50° C. The average rainfall is boosted by rare high intensity storms in response to anomalous climatic conditions, but capable of producing significant run-off and recharge. Such storms may be many years apart and at times there is no rain at all for years on end. The infrequent rain events are associated with tropical cyclones that develop over the Arabian Sea, and low-pressure frontal systems coming from either the south or from the northwest, from the direction of the Mediterranean Sea.

One major cyclonic event occurred from September 30th to October 5th 1992 when tropical cyclone O6-A passed over Al Wusta, causing rain to fall over a period of 6 days in the Wadi Rawnab catchment. During the cyclone the Wadi Rawnab flowed strongly, in places being 200-300 m wide and several metres deep. Most of the flow disappeared quickly after the rain ceased, infiltrating into the coarse sand making up the wadi floor This provided the first samples of cyclonic storm rainfall and runoff for chemical and isotopic analysis from the interior of southern Oman and provided a water line (Fig. 16) similar to that of the GMWL.

$\delta D = 8.7\,\delta^{18}O + 9.1$ $R^2 = 0.96$ 1992 Cyclonic event (Macumber et al., 1995).

$\delta D = 8.0\,\delta^{18}O + 10$ GMWL (Craig, 1961).

A feature of the cyclone was the wide isotopic range from early enriched to later depleted rain. This is deemed to be due to "rain-out" during the event. The early samples plot close to the area of monsoonal rainfall as occurs at Salalah in the far south east of Oman (Macumber et al., 1995).

Regional Groundwater Flow

Most of central and southern Oman lies within a single groundwater basin stretching from the Northern Oman Mountains southwards to the Dhofar Mountains, and westwards into Saudi Arabia and the UAE. Regional groundwater flow passes inland towards Al Wusta from recharge areas in the Northern Oman Mountains and the Dhofar Mountains, towards zones of regional groundwater discharge in central Oman situated in the Huqf and A'Rub Al Khali. The main aquifer system throughout Al Wusta is developed in Tertiary marine limestone and the regional groundwater is saline. The aquifer is unconfined with the water table generally occurring at depths of 80 to 150 m except in the A'Rub Al Khali and the Huqf, where groundwater is often at or close to the surface, resulting in large areas of active sabkha.

In the Najd the groundwater is fresh but on passing northwards gradually picks up salt, and by the time it enters Al Wusta all regional groundwater is saline. Within Al Wusta, regional groundwater salinity ranges from 5,000-6,000 mg/l TDS bordering the Najd, but quickly rising to beyond 30,000 mg/l further north. In the far north of Al Wusta salinities may reach 80,000 mg/l, reflecting additions of oil-field brines. Groundwater across Al Wusta gives a wide range of radiocarbon dates (Parker, 1985) from modern to > 30,000 years (Parker, 1985). The older dates are associated with saline regional water and indicate that this is essentially fossil, while the younger dates represent modern recharge, despite the present hyper-arid regime. Up-basin of Al Wusta, fossil to sub fossil groundwater also occurs in the Najd (Clarke et al., 1987).

Local Flow Systems

Potable or marginally potable groundwater in Al Wusta is limited to freshwater lenses which directly overly the saline regional groundwater. Many occurrences are shallow and ephemeral and rely on the rare rainstorm for their existence, or form in

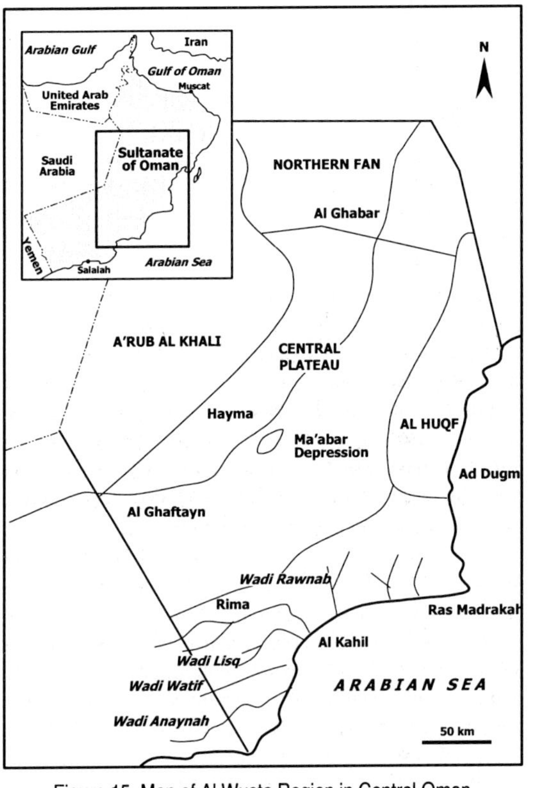

Figure 15. Map of Al Wusta Region in Central Oman.

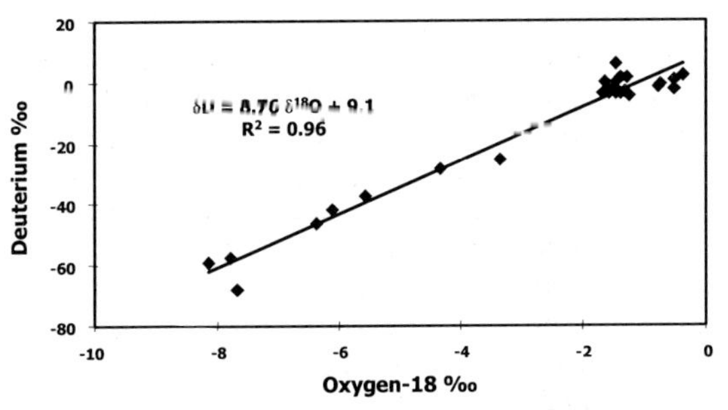

Figure 16. Meteoric water line from the 1992 tropical cyclone O6-A.

coastal situations above the seawater intrusion. Thin freshwater lenses also occur at depths of about 100 m in Tertiary marine limestone, associated with broad shallow depressions such as at Hayma, where the lens is 12 m thick, and at nearby Al Ma'abar, where several small plots covering about 4 ha are irrigated from a freshwater lens. Elsewhere, freshwater was obtained from the Tertiary limestone aquifer beneath the wadi systems draining eastwards toward the Al Jazir coastal plain.

During cyclone O6-A, the Wadi Rawnab in the upper gorge sector was 2 m deep, and observed rapid water losses into the channel gravel on passing downstream suggested that, despite the extreme aridity, fresh water lenses may form and be maintained by recharge during infrequent periods of intense rainfall associated with the occasional cyclone or frontal systems. This led to the Wadi Rawnab lens study.

Freshwater Lenses formed in broad depressions – the Ma'abar Lens

The Ma'abar Depression is a shallow 3-5 m deep basinal feature within the Central Plateau, covering an area 25 km x 4 km. It is structurally formed, consisting of two elongate asymmetric basinal synclinal folds separated by a weakly developed undulate anticlinal area where outcrop is characterized by small hillocks. In the depression, the limestone basement is overlain by a 1-2 m thick lateritic gravely infill. Within the depression are a number of smaller linked sub-basins.

In all, 19 bores including pumping and observation bores were drilled in the Ma'abar Depression covering the whole of the NE-SW oriented 25 km long depression and adjacent higher areas outside the depression. The average bore depth was 130 m (within the saline regional water), and the water table in the limestone aquifer was met at about 80-90 m below groundsurface. It showed a 12 m thick lens commencing at about 100 m below groundsurface and covering an area of 4 km^2. Only a thin freshwater band was present in the furthest northerly bore (M-4), and a similar occurrence in the far south of the depression. In the south central part of the depression there is a well developed lens associated with a sub-basin where several small irrigation farms are now established. The size of this lens is about 4 km long and 1 km wide. The source of the recharge is both precipitation falling directly in the depression and runoff from a small wadi with a 60 km^2 catchment that enters the Ma'abar depression from the south east. The restriction of the lens to that part of the depression fed by the wadi indicates the importance of catchment runoff to lens formation. In this respect, there is a minor echo of the situation occurring on the Batinah coastal plain where direct recharge from precipitation is also negligible. The lens is asymmetric being thickest on its southern side closest to the wadi.

Within the main occurrence, fresh groundwater occurs as a mainly 10-15 m thick lens of good quality water ranging from 500 mg/l to 1200 mg/l, surrounded by, and at times merging downwards into a brackish water transitional zone having a salinity of 1200 mg/l to 3,600 mg/l, and finally into saline regional groundwater with a salinity of about 11,000 mg/l to 12,000 mg/l (Fig. 17). Overall, the thickness of the Ma'abar freshwater lens was similar to that known from the Hayma Depression (9 m thick with a pumped sample being 1370 mg/l). At its thickest, the freshwater lens is about 20 m thick (bore M19), although more commonly this is the thickness of the freshwater and brackish transitional water zones combined. One unusual feature was salinity layering in the underlying saline groundwater with a slight freshening from between 115 and 120 m (Fig. 18).

Upconing of Saline Water During Pumping

In situations where there is an underlying saline aquifer there is a strong risk that the saline water will upcone and enter the pumping bore. This is the case at Ma'abar where on pumping, there is a rapid increase in salinity as water from the underlying saline parts of the aquifer flows into the bore. Upconing occurs to varying degrees in all the farm bores commencing during the early stages of the pumping. In a number of cases tested, the electrical conductivity rose rapidly from about 1000 to 2000 μ S/cm to between 5000 to 8000 μS/cm within 30 minutes of the pumps starting. This was not unexpected as the saltwater-freshwater interface lies only about 10 m below the pumping level. The extent and rate of the salinity rise was proportional to the pumping rate in some instances; at other times, it probably reflected the position of the screened interval relative to the depth of the interface between saline and fresh water. For instance, the bore FB4 pumped at a rate of 2.4.1 l/s, rose rapidly to about 5000 μS/cm then remained static (Fig.19). By contrast bore FB6 pumped at a greater rate rose more slowly to 8000 μS/cm but then continued to rise. The faster rate of rise in salinity in FB 4 is attributed to a shallower saltwater interface at 102 m, compared with an interface at 113 m in FB 6. The higher salinity level reached in FB 6 is due to the higher pumping rate.

Groundwater pumping from the Ma'abar Depression accounts for about 120,000m^3 of brackish groundwater annually with an average salinity of about 6,300 EC. Of this, about 55 % of the water (~65,000m^3) is from the freshwater lens and the remainder from the underlying saline regional groundwater. While the salinity may be considered too high for normal agricultural use, there is so far sustainability, albeit at a cost in crop (alfalfa) yield and size. However the mere presence of a green irrigated swathe in an otherwise totally desolate landscape may well justify the practice, whatever the yield. Its operation may therefore be linked as much to aesthetic values whatever the economics and could continue until the water is too saline for further agricultural use.

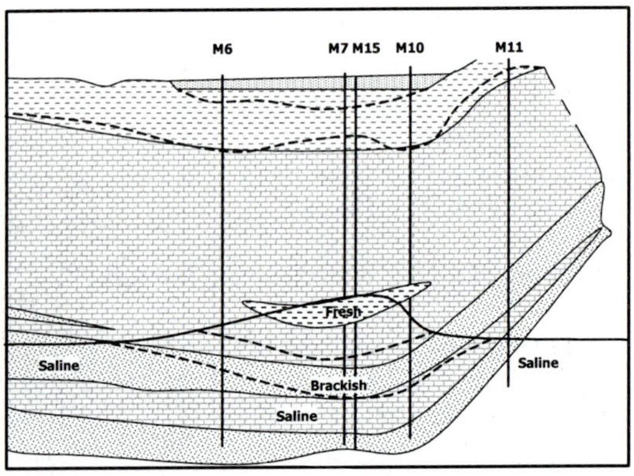

Figure 17. Freshwater lens in the Ma'abar Depression (diagrammatic).

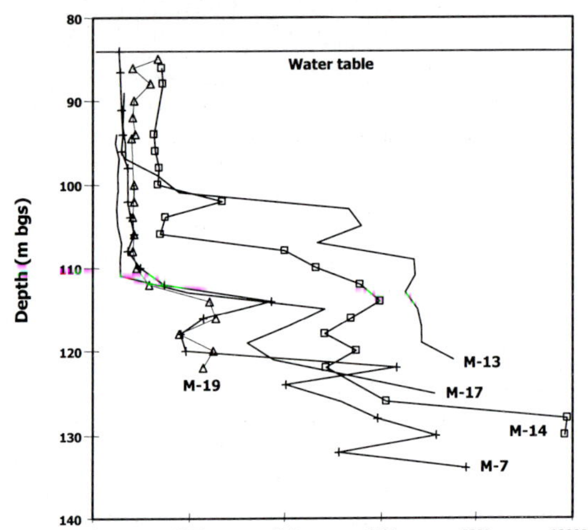

Figure 18. Salinity profiles through the Ma'abar lens.

Lenses Formed Beneath Wadi Systems-the Wadi Rawnab Lens

The Wadi Rawnab is one of a number of wadis that drain towards the Al Jazir coastal plain. The 3,400 km^2 catchment is in late Tertiary Fars Group limestone, the surface being commonly bare or overlain by a thin skeletal soil or dust mantle. The regional groundwater occurring at depths of 100 to 140 m is brackish to saline and commonly ranges from about 6,000 mg/l to 40,000 mg/l.

The catchment commences on the Central Plateau, where the tributaries are shallow broad depressions commonly cut across bare limestone. In its middle reaches, commencing about 50 km from the coast, the wadi falls abruptly from the plateau into a 30 km long and 30-40 m deep gorge (gorge tract) incised into the limestone. The Rawnab gorge is about 500 m wide on average and continues to within 15 km of the coast, terminating at a Pleistocene marine coastal terrace bordering the Al Jazir coastal plain. The wadi is incised through the terrace and then passes out onto the present day coastal plain as a broad shallow sandy depression. Within the gorge tract, the wadi has a 20 m thick unconsolidated coarse gravelly and pebbly alluvial fill. The fill is important to groundwater recharge in that it has the capacity to absorb and store large amounts of flood water during the rare occasions when the wadi flows and then slowly release it into the underlying limestone aquifer. This process is an important factor in the development of very low salinity groundwater (400 - 160 mg/l TDS) deep in the limestone aquifer beneath the Rawnab gorge. The division into plateau, gorge and coastal plain tracts is a feature of all major wadis flowing towards the Al Jazir coastal plain, Wadi Rawnab being the largest.

Groundwater in the Rawnab Catchment

Prior to the study, potable and marginally potable water were known from three localities along the Wadi Rawnab ranging from the upper catchment (Runib South bore) to the coast (WST bore) - a distance of 70 km. In all cases, the fresh water overlies saline regional groundwater. A ten bore drilling program was undertaken covering eight sites along Wadi Rawnab over a distance of 60 km (Fig. 20). Six bores were drilled in the gorge section, three in the plateau section and one coastward of the gorge on a terrace bordering the coastal plain. The results of the study are summarized in Table 1 and shown in Figures 20 and 21. Salinity profiles are given in Figures 22.

Table 1. Summary of Wadi Rawnab freshwater lens.

Bore	Position	Lens thickness (m)	Range (m)	Average salinity (mg/l)*	Yield (l/s)	D'down (m)
WCR-1	Top of gorge	100	75 - 175	362	-	-
WCR-5	Upper gorge - 500 m from WR-2 bore	78	102 - 180	327	-	-
WCR-10	Upper gorge - 32 m from WR-2 bore	78	102 - 180	370[1]	6.1	0.3
WCR-2	Middle gorge	79	81 - 160	226	4.6	0.2
WCR-9	Middle gorge - 31.5 m from WCR-2	60+	80 - 140+	213	12.5-17	1.2-1.4
WCR-3	Bottom of gorge	40	63 - 103	200	4.7	5.0
WCR-4	Coastal terrace	9	43- 52	720	-	-
WCR-6	Plateau	90+	120 - 210+	670	-	-
WCR-7	Plateau	155	120 - 275	1750	3.3	2.0
WCR-8	Plateau	125	120 - 245	1400	-	-

*For pumped bores the salinities were for samples taken during pumping, for non-pumped bores the salinities were averaged across the lens.

Pumping tests were carried out along the gorge tract at three sites WCR 2/9, WCR 3 and WCR 10, and one on the plateau tract (WCR 7) - three being for 48 hours and one (WCR 10) for six hours. The rates and final drawdowns are provided in Table 1. The best results came from the WCR2/9 borehole nest, where the freshwater lens is 80 m thick with an average salinity of 226 mg/l. The WCR-2 (observation bore) and WCR-9 (pumping bore) are 20 m apart and screened over intervals of 80 m to 140 m. Constant discharge 48 hr pumping tests were carried out at rates of 4.5, 9 and 17 l/s. In all cases steady-state was reached within less than 30 minutes after the commencement of pumping. The maximum drawdown reached within the WCR-9 pumping bore was 1.4 m after pumping for 48 hours at 17 l/s. There was no indication of increased salinity during the 48 hour pumping tests at 17 l/s on the WCR-9 bore, and instead the reverse occurred with salinity levels falling and the final water salinity being only 303 μS/cm or 213 mg/l. The salinity at the completion of pumping was 310 µS/cm or about 200 mg/l TDS. In the case of the WCR-3 bore a similar salinity decrease occurred during pumping with later pumped salinities being a remarkable 160 mg/l. Such low salinities are a clear indication of the rapidity of runoff and recharge in the gorge tract.

Within the gorge sector, marginal increases in lens salinity were only encountered at the PDO WR-2 site where continuous pumping occurs The WCR-5 and WCR-10 bores were sited 500 m apart, WCR-10 being alongside the PDO bore WR-2, and WCR-5 about 500 m away across the wadi. The WCR-5 bore had a salinity of 321 mg/l. while that of the WCR-10 was 370 mg/l, the differences between the two bores suggest a 15% increase in the lens salinity at the WR-2 site after three years daily pumping.

However, the threat of rising salinities is seen in the case of the Runib South bore, the most upstream known occurrence of the lens in the plateau section, where regular pumping was undertaken. In this bore salinities have risen from 1300 mg/l in 1980 to 1600-2000 mg/l by 1992.

A feature of the tests was the tendency for the drawdown to level out within 20 to 100 minutes of the commencement of pumping, and in the case of the WCR 9 bore when pumped at 17 l/s, the drawdown was only 1.4 m after 100 min, where it remained with minor diurnal fluctuations for the remainder of the test (Table 1). Flattening of the drawdown curves was previously noted in the case of the Ma'abar investigation and appears to be a common response to pumping of the Tertiary limestone aquifer, at least in the areas underlain by freshwater lenses. This was interpreted as indicating secondary porosity in a varyingly fissured limestone. Fissures were commonly encountered during drilling and caliper logs supported this contention, in that they showed an increased amount of fissuring in

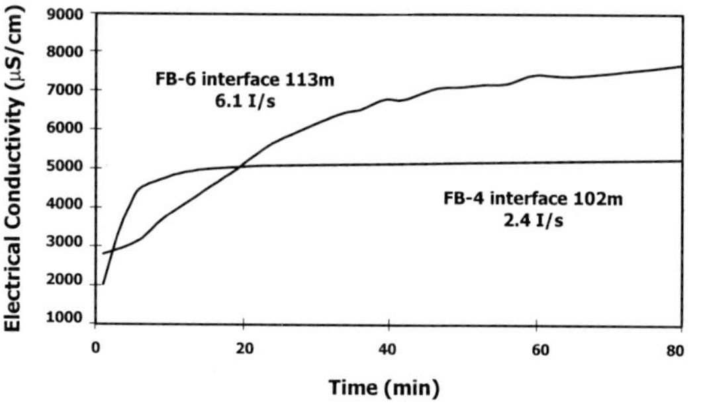

Figure 19. Salinity (as EC µS/cm) changes during pumping of the Ma'abar lens.

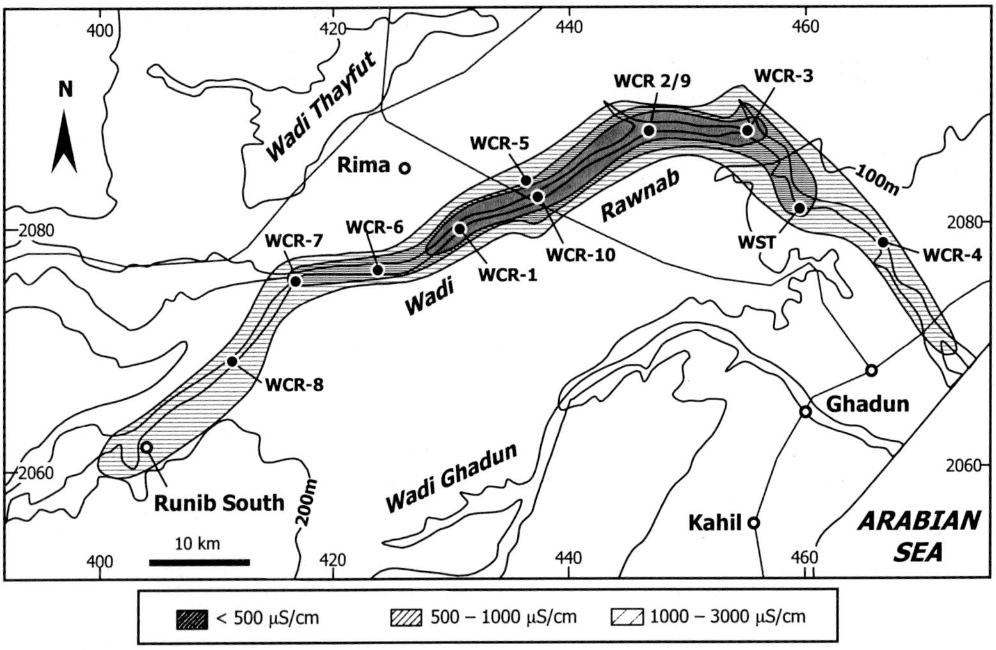

Figure 20. Borehole positions and lens salinity – Wadi Rawnab.

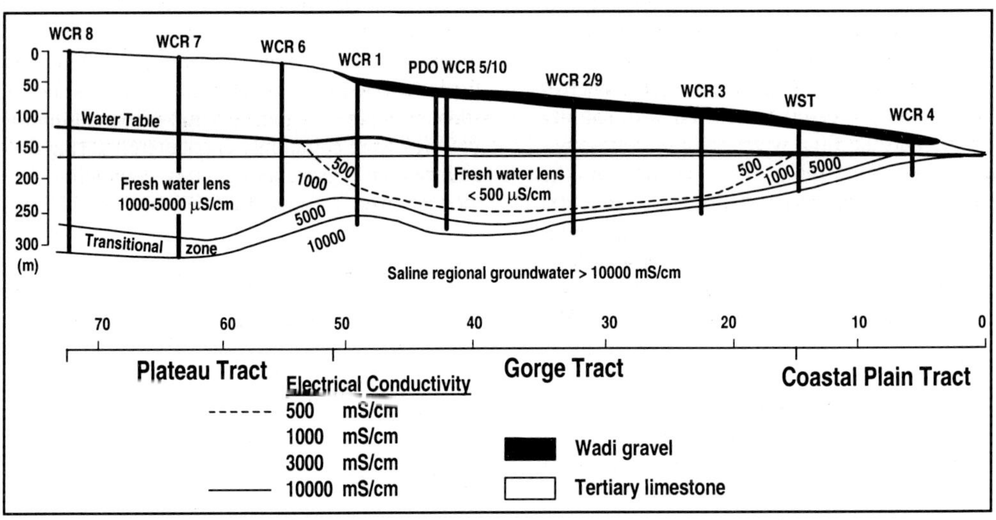

Figure 21. Salinity distribution along Wadi Rawnab.

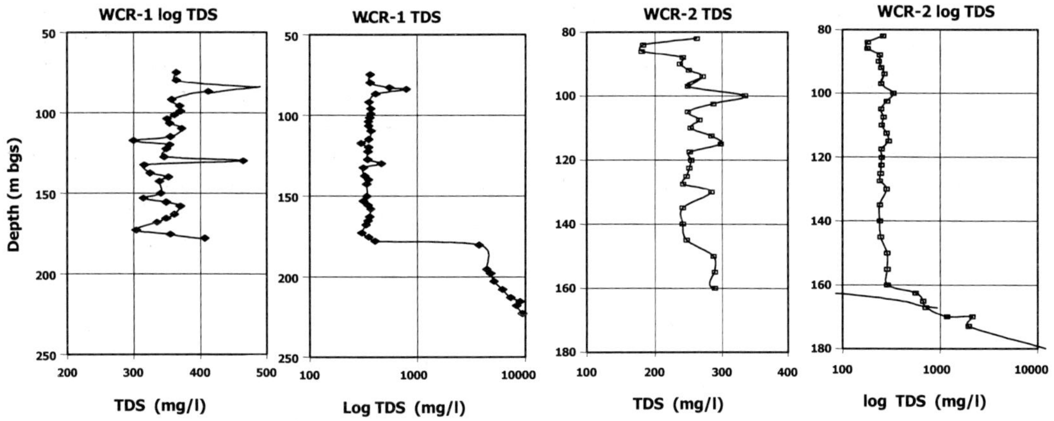

A. Detailed salinity profiles through the Rawnab freshwater lens at WCR-1 and WCR-2

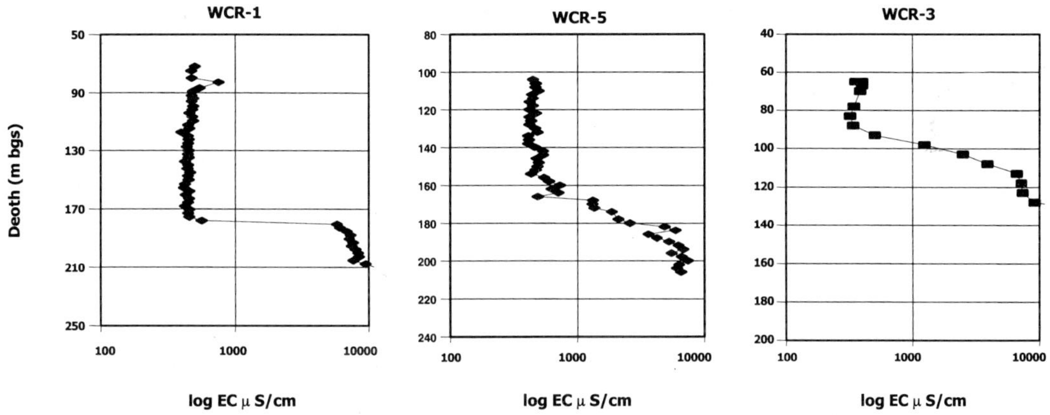

B. Electrical conductivity profile through the freshwater lens-gorge section

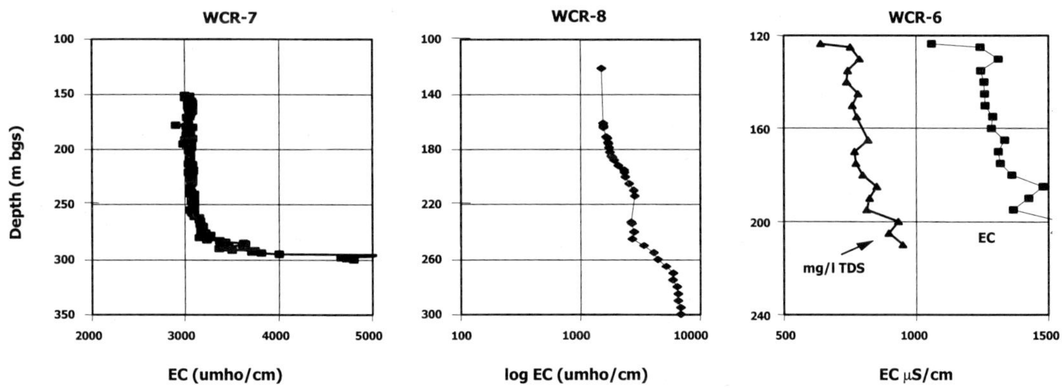

C. Electical conductivity profiles through the Wadi Rawnab lens-plateau section

Figure 22. Depth-salinity profiles in the Wadi Rawnab bores.

bores on passing down Wadi Rawnab toward the coast. Enhanced secondary porosity is attributed to the action of very low salinity aggressive groundwater on the limestone aquifer (Macumber, 1996). Since lens flow is lateral to the wadi not down it, the most recent and hence most aggressive recharge will always directly underlie the wadi with earlier recharge occurring in a zone lateral to it.

Groundwater Resource in the Rawnab Lens

The freshwater lens in Wadi Rawnab is about 70 km long (Figs. 20 and 21). On the basis of nearby PDO-WR-2 borehole data, geophysical cross-section (TDEM), and by analogy with the Ma'abar Depression (where with a minimal catchment the lens was 1 km wide), it is likely that the width of the lens is 2-3 km or more. At the PDO-WR2 site the lens is 80 m thick in each of the WCR-5 and WCR-10 bores situated 0.5 km apart on either side of the wadi.

Within the 40 km long plateau tract, the average thickness is at least 120 m, the width 2-3 km with an average salinities ranging from 670 to 1,700 mg/l. Similarly, in the gorge sector it has an average thickness of 75 m, a width of 2-3 km, a length of 30 km and a salinity of between 200 and 350 mg/l.

The volumes of water in storage can at best only be roughly estimated given the fissured nature of the unconfined aquifer, but on the basis of a 1% storage value, it would amount to more than 100 million cubic metres, with perhaps 40% being less than 350 mg/l and 60% ranging from 1700 to 670 mg/l. However, the storage value is largely speculative and requires further work for any confirmation. While the plateau water is marginally potable, shandying of the two waters from the gorge and the plateau would greatly increase the overall potability of the lens were it required.

While the volume of water in the Rawnab lens is not large in terms of aquifer systems, its true value lies in its location in an otherwise barren landscape, and its implications for likely groundwater occurrences for other wadis along the coastal plain, and in similarly very arid areas elsewhere. Its monetary value may be measured against the cost from small desalination plants in Al Wusta, where fresh water is produced at ca 3 OR/ m^3.

Groundwater Recharge to Freshwater Lenses

The relationship between the stable isotopes from the groundwater in the Wadi Rawnab lens follows closely that of the O6-A cyclonic water line and the GMWL (Fig. 23). This demonstrates that evaporation is insignificant prior to groundwater recharge, despite the very high evaporative potential in the region. This is also shown by the plots of deuterium versus chloride for both the Wadi Rawnab and the Ma'abar freshwater lenses. At Ma'abar, salinity levels show a significant range but there is no corresponding deuterium increase (Fig. 24). At Rawnab, in the gorge bores, the WCR-1 profile is significantly enriched (δD = -15 to -18 o/oo) compared to that of WCR-2 and WCR-3 (δD = -28 to-36 ‰), yet this is not matched by any comparable salinity increase (Fig. 25). In addition, the deuterium content of WCR-4 and WCR-6 lying outside the gorge tract are isotopically more depleted than WCR-1, yet are significantly more saline.

In the Wadi Rawnab gorge, recharge occurs almost instantaneously with infiltration passing rapidly into the 20 m thick wadi channel sand and thence into the limestone aquifer, presumably along preferential pathways. Recharge is aided by an initial very deep water table, being 100 m below the wadi floor.

In the hyper-arid regime where temperatures are normally high and commonly reach 50° C in summer, evaporation must account for much of the precipitation, yet there is no significant evaporative signature in the freshwater lenses. It follows that there is a clear division of process. Surface water does not linger, but is rapidly lost either by evaporation into the atmosphere or by infiltration into the groundwater systems. The hyper-arid climatic setting facilitates rapid recharge in that there is a virtual absence of soil and vegetation to impede or intercept infiltration. As a consequence, the low recharge expectation presented by small annual average rainfalls and high evaporation rates runs counter to the actual high recharge rates in a landscape experiencing only rare large storm events.

However such recharge is not general throughout the landscape and the local geomorphology is critical to this process. In both the Ma'abar and the Rawnab settings the locality of the lenses is closely linked to physiography, in that the depressions and wadi networks are large collector system for the concentration of runoff prior to its infiltration. This is best developed along the gorge tract of the wadis where the thick upper sand suite captures virtually all the wadi flow. This situation is the same in all major wadis entering the Al Jazir coastal plain, each having 20-30 km long gorge tracts. Along the Rawnab gorge, the 20 m thick coarse sand unit with an average width of 500 m and storage coefficient of 0.2 can store 2 million cubic metre of water per kilometre length of wadi, prior to releasing it into the underlying limestone aquifer.

In the case of the WCR-1 site situated closest to the head of the gorge, decreased hydraulic head with depth was measured during drilling, reflecting the presence of strong vertically downwards gradients induced by the still dissipating recharge mound from the cyclonic event three years earlier. This process could only be observed with a cable tool approach. The observation was later confirmed by the establishment of a piezometer nest at the WCR-1 site, screened at three intervals vertically across the lens. This situation was peculiar to the WCR-1 bore, and similar recharging mounds were not observed elsewhere along the gorge.

Further insight into the wadi recharge process is given by the presence of aeolian salt in the landscape and reflected by the abnormally high Cl/Br ratios (often > 1000) for much of the regional and recent groundwater including that in the Rawnab and Ma'abar lenses. Much of the remobilization of the aeolian salt occurs during the initial stages of major rainfall events, a process commonly known as "first flush". This more saline component of runoff is also the first water to be recharged in the upper-middle parts of the catchment where the flood wave forms. If the rain continues and water flows further down the catchment it gradually freshens as much of the initial high salt pick-up is lost by infiltration during the early phase of wadi flow. A similar pattern in salinity trend towards progressively fresher groundwater occurs on passing downbasin along the Wadi Rawnab lens, with the freshest water being present in the lowermost part of the gorge. In the gorge, the decreased salinity on passing down wadi is accompanied by increased isotopic depletion (Figure 25) which taken together is interpreted as indicating that only the runoff from the largest storms was capable or reaching and recharging the middle and lower parts of the gorge (Macumber, 1998). The association of more isotopically depleted water with heavier rain events has been previously noted (Dansgaard, 1964; Mazor, 1991) and is referred to as the "amount effect".

Stable Isotopes as an Indicator of Palaeo-Groundwater in Al Wusta

Stable isotope plots for regional groundwater in Al Wusta, plot roughly parallel but fall below the GMWL (Fig 26a). While radiocarbon dates for the Al Wusta groundwater give ages from modern to >30,000 (Parker, 1985), many of the older waters are likely to be minimal ages given the mixing of old and more recent water along the long flow path, especially during higher recharge periods during the episodic wet events as suggested by McClure (1976) for data from the A'Rub Al Khali. While discrete aquifer zones have been previously defined, intra-formational flow is shown by the depth of penetration of recent (hyper-arid) recharge in the Rawnab freshwater lens, which reaches 280 m to 300 m in the WRC-7 and Runib South WSW-1 bores respectively. Similarly, upwards flow from the deeper aquifers must occur within the regional discharge zones.

The saline regional groundwater in Al Wusta is fossil and plots beneath the GMWL. It represents recharge under two different climatic regimes, one producing a more isotopically enriched groundwater ($\delta^{18}O > -5.2$; $\delta D > 40$) and the other a more depleted groundwater ($\delta^{18}O < -5.2$; $\delta D < 40$). The enriched water has a high Cl/Br ratio (often > 1000), while the depleted groundwater has low Cl/Br values (Figs, 26b).

The high Cl/Br ratios (>1000) in groundwater in Al Wusta suggest recharge in an arid or hyper-arid setting as occurs today where there is re-solution prior to recharge of chloride in aeolian dust derived by deflation of sabkha (Macumber et al., 1998). This results in groundwater with salinities as low as 150 to 250 mg/l, yet with Cl/Br ratios of 1000. By contrast, the saline regional groundwater with depleted isotopes and low Cl/Br ratios is seen as representing recharge under cooler and perhaps wetter conditions than at present when there was no significant deflation of saline sabkha, probably when they were active lake systems (Macumber, 1998). That this was the case periodically in the past is seen by the enormous numbers of chert artifacts ranging upwards in age from Lower Palaeolithic times (ca 200,000 years B.P – Biagi, 1994; Macumber, 1999), showing periodic human occupation of the shorelines of former freshwater lakes in the Saiwan-Hawshi region of Al Wusta but which are now sabkha. It is also shown by the climatic cycles suggested by the work of McLure (1976).

Modern and Sub-Fossil Freshwater Lenses

Observations during the 1992 cyclone showed that significant recharge from wadis may occur under modern hyper-arid conditions. Furthermore, isotopically the lens groundwater closely matches both the 1992 cyclonic water line (Fig. 24) and the GMWL (Fig 27a), which also supports a modern age. This is also the case with groundwater from other wadis draining to the Al Jazir coastal plain (Wadi Aynaynah and Wadi Lisq-Gudam) and the coastal plain itself ('other wadis', Fig.27b).

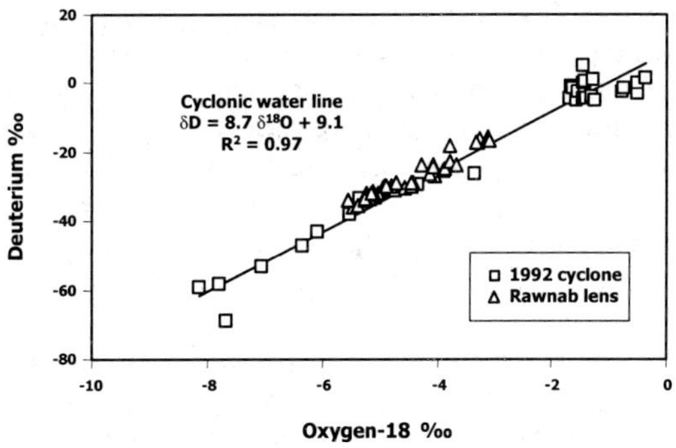

Figure 23. Deuterium vs. oxygen-18 for Wadi Rawnab lens and Cyclone O6-A.

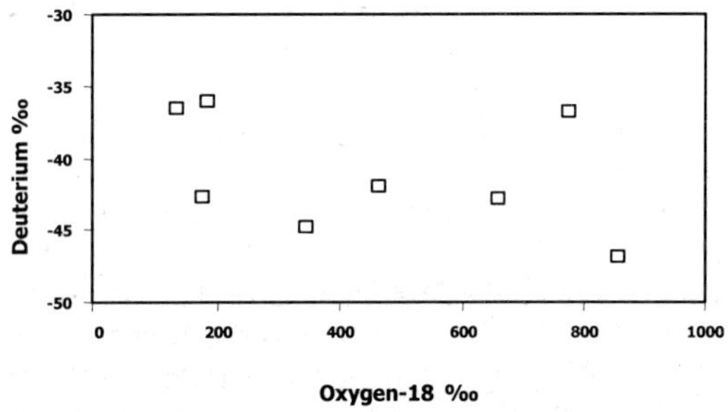

Figure 24. Deuterium vs. chloride for the Ma'abar freshwater lens.

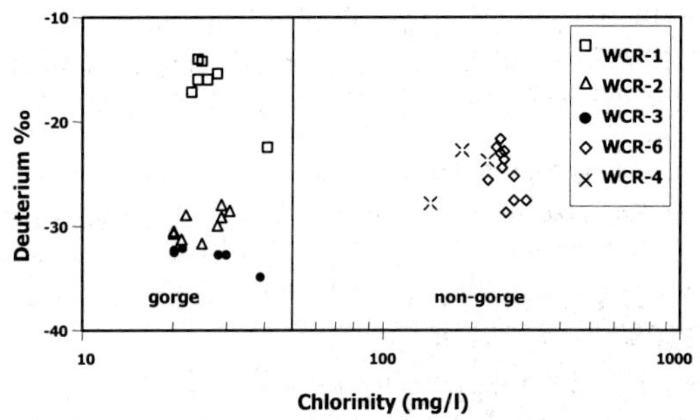

Figure 25. Deuterium vs. log chloride for the Wadi Rawnab freshwater lens.

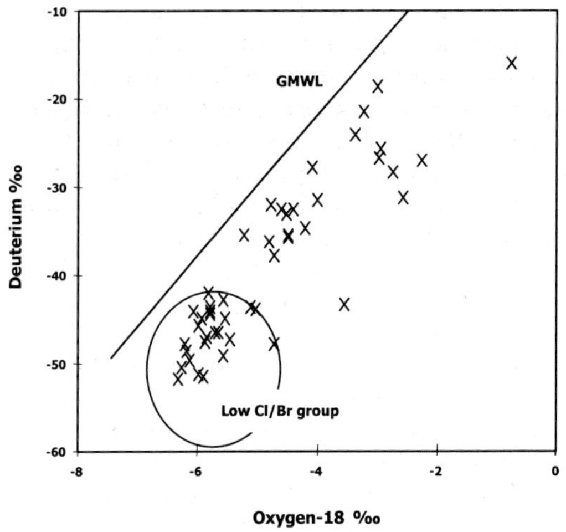

Figure 26a. Stable isotopes for regional groundwater in Al Wusta.

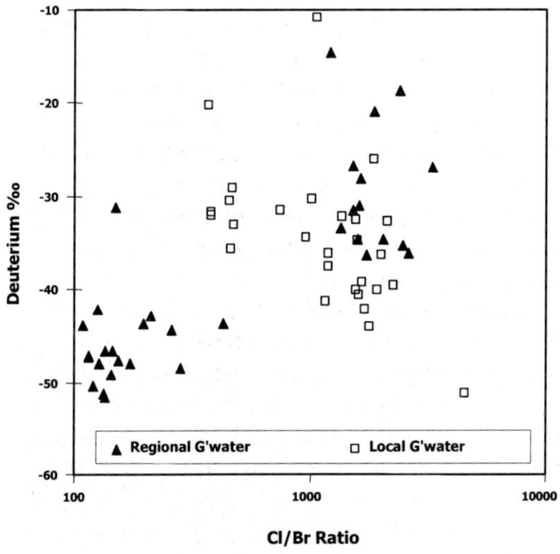

Figure 26b. Deuterium vs. Cl/Br for regional groundwater in Al Wusta.

However, while recharge also occurs to lenses beneath Ma'abar and Hayma depressions during intense rainfall events, the isotopic composition of that groundwater does not conform to the modern day cyclonic rainfall, or to storm recharge as reflected in the collective 'memory' of the Rawnab freshwater lens or to other freshwater lens types in Al Wusta. Instead it better relates to Najd and Al Wusta also fall in a line below the various modern day meteoric water lines.

While this characteristic may still arise from secondary fractionation prior to recharge, evidence both locally in the region and also from elsewhere in Oman better supports a sub-fossil origin. This distinction between modern and older groundwater is present in the case of the Al Khawd Fan, (PAWR, 1986; Macumber, 1997). In the Ibra-Wahibah Sands region to the north of Al Wusta, 'Wahibah' type groundwater with an age of 4,000- 8,000 years BP (PAWR, 1986), plots in a zone lying to the right of both the GMWL and modern day recharge associated with the Wadi Al Batha system ('Batha' type, Fig. 28). Evidence of a moist period in the Wahibah Sands has been dated to 8870 ± 120 BP (Gardner, 1988) and 8000-7000 to 5000-4000 BP Edens (1988), after which it is shown that aridity increased. The early-mid Holocene wet period in the Wahibah Sands and reflected in the ages of the Wahibah type groundwater, matches that of McLure (1976) for A'Rub Al Khali.

The Batha type groundwater is tritium-rich with an EC less than 2,000 μS/cm; it occurs along the course of Wadi al Batha and penetrates a relatively limited distance southwards beneath the Wahibah Sands. Beneath the sands the 'Wahibah type' occurs. It is more saline, has a larger deuterium deficit, and virtually no tritium (< 1 TU – PAWR, 1996). While the Batha group lies a little to the right of the GMWL, it matches the rainfall data obtained during a storm event in July 1995, where the relationship between deuterium and oxygen-18 showed a deuterium deficit of 4.2 relative to the GMWL, the relationship being:

$$\delta D = 8.1\ \delta^{18}O + 5.8$$

The slope of the July 1995 rainfall event is consistent with a southern rainfall source, but given the deuterium deficit, it appears to be a local variant of the GMWL with an evaporative shift from the GMWL seen a occurring prior to recharge; this shift is a feature of the locally recharged modern day groundwater from Wadi Al Batha (Fig. 28), however it lies well to the left of the palaeo-water line.

The plot of the Ma'abar freshwater lens and the Rawnab lens show a similar isotopic relationship to that of the Wahibah and Batha groups, with Ma'abar being more depleted than the Rawnab groundwater (Figs. 27 and 28). By comparison the isotopic composition of freshwater lenses from the gorge tracts of 'other wadis' (Lisq and Aynaynah) and from the Al Jazir coastal plain, lie on or close to the GMWL (Fig. 27b) suggestive of modern day recharge. The separation of the Ma'abar lens from the GMWL does not appear to be an evaporative diversion, since there is no increased salinity commensurate with isotopic enrichment (Fig. 24). All water samples in Fig. 24 come from within the freshwater lens. Furthermore, such a shift is not found on other demonstrably modern groundwater listed above. In addition, the Ma'abar plot $\delta D = 7.3\ \delta^{18}O - 3.7$ is very different to the characteristic evaporation line for Al Wusta with a gradient $\delta D = 5.4\ \delta^{18}O$, first noted by Parker (1985) and reproduced during the O6-A cyclone when samples were taken 11 days after the event (Macumber et al., 1995).

Finally, it seems most unlikely that Ma'abar simply represents a different storm setting to that previously recorded given that the Rawnab lens holds a collective memory of all rainfall-recharge events whether cyclonic or frontal, and that other lens types also have a similarly close relationships to the GMWL. Therefore, while a modern day evaporative shift similar to that occurring in Wadi Batha water cannot be totally discounted, the more likely explanation is that the Ma'abar (and Hayma) lenses consist predominantly of palaeo-water, albeit with modern day additions. This suggestion is supported by an age for the Hayma lens WSW-1 bore (salinity ca.1300 mg/l) of 12,900 yr BP (Parker 1985). Emphasis on the distinction - modern or fossil - is an extremely important consideration for lens sustainability and management.

The Management of Freshwater Lenses

With any freshwater lens, whatever its origin, the most important concern for management is the threat of upconing of saline water once groundwater development proceeds. This is especially the case where the lenses are thin and the extraction rate is relatively high, as occurs at Ma'abar where an initial salinity of less than 1000 μS/cm now rises to between 3,000 and 8,000 μS/cm soon after pumping commences. Risk to the lenses is exacerbated in those instances where the groundwater is sub-fossil and there is little or no opportunity for renewal.
While apparently the case of lenses at Ma'abar and perhaps Hayma, it is also the situation present throughout the regional groundwater system in the Najd.

The Al Jazir coastal plain, after the Batinah and the Salalah coastal plains, is the third largest coastal plain in Oman. However unlike the other two, the region is more arid and there is no mountainous hinterland to provide a source of fresh groundwater flow or runoff to the plain. The extent of its future development is largely determined by the availability of water. Significant supplies of fresh groundwater have so far only been proved for Wadi Rawnab.

However, the Rawnab lens is not unique, and there is strong evidence for similar lenses occurring beneath the other wadis (Lisq-Gudam, Watif and Aynaynah), which have essentially the same geomorphology. In Wadi Aynaynah for instance,

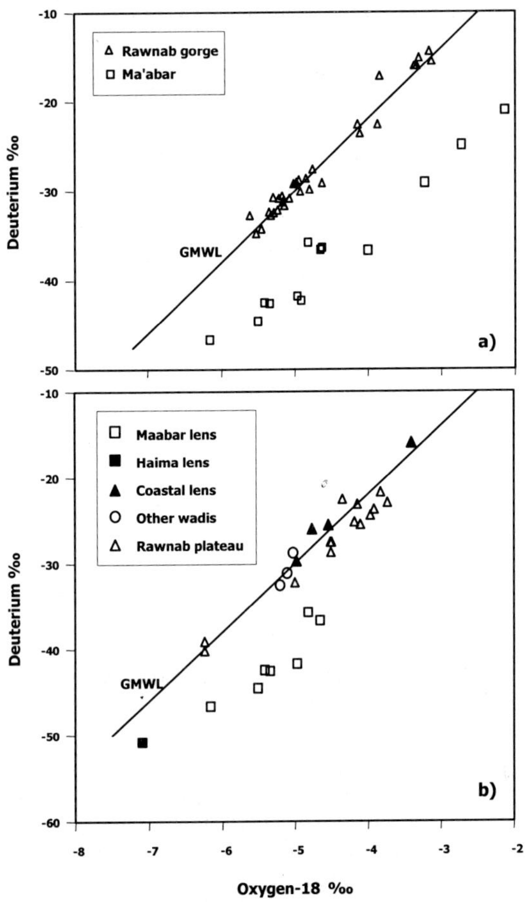

Figure 27. Ma'abar lens isotopes compared with Rawnab lens (a) and other lenses across Al Wusta (b).

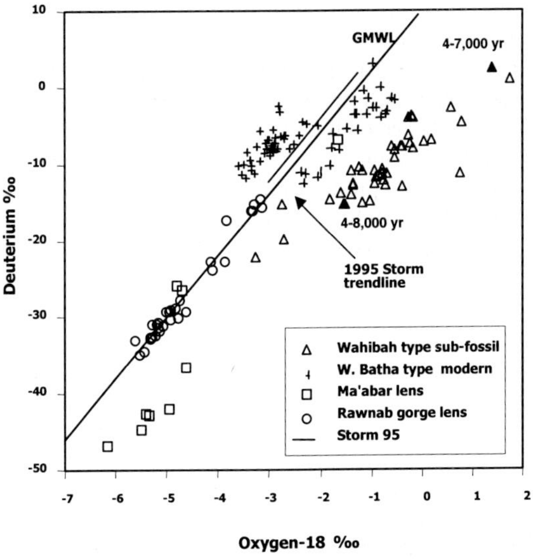

Figure 28. The isotopic identification of modern and fossil groundwater in the Al Wusta and Sharqiyah regions of Oman.

fresh water occurs in several bores near Aynaynah. The wadis, which are a renewable resource, together have a potential to supply all the coastal towns along the Al Jazir coastal plain including Kahil the local administrative centre with minimal risk to the respective lenses.

A strong potential exists for upcoming in the modern day freshwater lenses, if over pumped. To date, the salinity has risen by 25% in the Runib South bore, and there is a15 % increase to the salinity profile in the vicinity of the PDO RW-2 bore in the Rawnab gorge after daily pumping for three years. There is nevertheless an opportunity to develop a management system for keeping this threat to a minimum, or even overcoming it. For optimum results and safest practice, this requires that each wadi serve a respective part of the Al Jazir coastal plain, and in this way no single wadi will carry the burden for the whole plain. Further assurance may be had by establishing bores/pumps sites at intervals down the wadi and linked to the coastal plain by a pipeline. Individual bores can then be introduced or removed from the system as required in order to maintain the lens.

While this system of resource development and distribution will be more expensive than using any single freshwater lens by itself, it guarantees that all lenses will survive, rather than face an alternative of permanently risking any one lens on the assumption that other lenses are available should the quality in the chosen lens deteriorate beyond whatever level is desirable. Given the low periodicity of recharge events, once the lenses become saline it is difficult to see how they might be rehabilitated, should this be contemplated. To Al Wusta, the lenses are invaluable, especially given the large costs of about 3 OR per m^3 to provide large quantities of freshwater from alternative sources such as desalination.

CONCLUSION

An important feature of the groundwater systems in arid and hyper-arid Oman is the development of freshwater lenses and plumes forming discrete linear flow systems within otherwise uniform aquifers. The presence of these features requires management considerations beyond that normally required in the development of more conventional sheet aquifers in limestone or alluvial sequences.

In all systems examined, direct recharge from precipitation is not a meaningful contributor to groundwater balances, but instead recharge occurs only after concentration of runoff in wadis and depressions. Recharge is unusually high in the hyper-arid setting due to the absence of vegetation and soils to impede infiltration. The low salinity of the freshwater lenses and their isotopic composition show that, despite the extreme aridity, the only requirements to produce useful freshwater lenses is the occasional storms capable of providing runoff, and an integrated collector system that receives and holds the runoff for the short time that it takes to infiltrate, and then recharge the aquifers. The considerable depth of the fresh water lenses within the Tertiary limestone aquifer system of Al Wusta demonstrate that the absence of shallow water does not preclude fresh water at a greater depth. Before any lens investigation is considered futile, it is necessary to drill to the depth of the regional water table.

The isotopically depleted plumes provide valuable information on the groundwater source and the relative significance of various recharge processes on the Batinah coastal plain. In the case of both the Wadi Maawil and the Al Khawd Fan, groundwater flow systems generated in the limestone and ophiolitic upper catchments contribute the bulk of the groundwater appearing on the coastal plain. The distant source and age of the plumes must be considered if more intense development is undertaken in the Eastern Batinah area. In Al Khawd, pressure cycles in the regional and intermediate flow systems, recorded at the head of the fan and deep within the alluvial aquifer, are ultimately responsible for controlling the extent of seawater intrusion and protecting the A'Seeb and Al Khawd wellfields. The joint management of both upper and lower catchments is therefore a requirement for seawater intrusion control. Of the two systems – the Al Khawd Fan and the Eastern Batinah alluvial aquifers - the Al Khawd Fan has better prospects for long-term sustainability. It has a large upper catchment from which it receives run-off and recharge via the Wadi Samail, to supplement the regional flow coming from the upbasin ophiolitic catchment.

REFERENCES

Bartolomi, G.C., Ricci, B.Suzella, G.F. and Zuppi, G.M., 1978. Isotope hydrology of the Val Coraoglia, Maritime Alps, Piedmont, Italy. In: *Isotope Hydrology,* International Atomic Energy Agency, Vienna 1: 181-201.

Biagi, P., 1994. An early Palaeolithic site near Saiwan (Sultanate of Oman). *Arabian archaeology and epigraphy* 5: 81-88.

Bhatnagar, G.C.and Ravenscroft P., 1986. A hydrogeological investigation of the Al Khawd Fan with special reference to sea water intrusion. PAWR Report 20-86

Clark, I.D., P.Fritz, O.P.Quinn, P.W.Rippon, H.Nash and Sayyid Barghash Bin Ghalib and Al Said, 1987. Modern and fossil groundwater in an arid environment: a look at the hydrogeology of southern Oman. In: Isotope Techniques in Water Resources Development. International Atomic Energy Agency, Vienna, 167-187.

Craig, E., 1961. Isotopic variations in meteoric waters. *Science* 133: 1702-1703.

Dansgaard, W., 1964. Stable isotopes in precipitation. *Tellus* 16: 436-469.

Edens, C., 1988 Archaeology of the sands and the adjacent portions of the Sharqiyah. *The Journal of Oman Studies Special Report* No. 3: 113-130.

Gardner, R.A.M., 1988. Aeolianites and marine deposits of the Wahiba Sands: character and Palaeoenvironment. *The Journal of Oman Studies Special Report* No. 3: 75-95.

Gibb. A., 1976. Water resources survey of northern Oman. Final report V1 Appendix E. Water chemistry and isotope studies. Report of the Director general of Finance, Sultanate of Oman 28p.

JICA, 1986. Hydrologic observation project in the Batinah Coast of Sultanate of Oman, Final Report, Volume 4. Japan International Corporation Agency Report to the Director general of Agriculture and Fisheries, Sultanate of Oman

Macumber, P.G., 1995. Freshwater lenses in the hyper-arid region of Central Oman. International Conference on Water Management in Arid Countries, Muscat, March 1995, 480-487.

Macumber, P.G., 1996. Wadi Rawnab groundwater investigation, final report June 1996. Directorate General of Water Resources Assessment, Ministry of Water Resources, Sultanate of Oman, 65p.

Macumber, P.G., 1997. Al Khawd Fan groundwater investigation. MWR Hydrogeology Section October 1997. Report of the Directorate General of Water Resources Assessment, Ministry of Water Resources, Sultanate of Oman, 88p.

Macumber, P.G., 1998. The Cable Tool program and groundwater flow in the Eastern Batinah alluvial aquifer. Report of the Directorate of Water Resources Assessment, Ministry of Water Resources, Sultanate of Oman, 159p.

Macumber. P.G., 1999. Geo-archaeological survey of seismic lines in the Jahiz Area of Central Oman For Petroleum Development Oman LLC, May 1999.

Macumber, P.G., Barghash bin Ghalib Al-Said, Kew, G.A., and Tennakoon. T.B., 1995. Hydrogeologic implications of a cyclonic rainfall event in central Oman. In: Groundwater Quality. Editors. H. Nash and G. J. H. McCall, Chapman and Hall, London, 87-97.

Macumber, P.G., Niwas, M., Al Abadi, A.,and Seneviratne, R. 1997. A new isotopic water line for northern Oman. Proceedings of the Third Gulf Water Conference, Muscat 1997, 141-162.

Macumber, P.G., Al Abri, R and Al Akhzami, S., 1998. Hydrochemical facies in the groundwater of central and southern Oman. In: *Quaternary Deserts and Climatic Change*, A.S Alsharahan, K.W. Glennie, K.G. Whittle and C.G. St.C. Kendall editors, Balkema, Rottendam, 511-520.

Mazor, E., 1991. Applied chemical and isotopic groundwater hydrology. Open University Press.

Parker, D.H., 1985. The hydrogeology of the Cainozoic aquifers in the PDO concession areas, Sultanate of Oman.. Petroleum Development Oman LLC.

Public Authority for Water Resources (PAWR), 1986. Origin and age of groundwater in Oman. A study of environmental isotopes, Report 86-7, November 1986.

Sir M. Macdonald and Partners – MMP, 1989. Groundwater recharge scheme in Barka-Rumais area. Data analysis report – hydrogeology. Final Report vols. 2b and 3b.

Later Quaternary Paleorecharge Events in the Arabian Peninsula

STEPHEN STOKES, HELEN BRAY, ANDREW S. GOUDIE
School of Geography and the Environment, University of Oxford, Mansfield Road, Oxford OX1 3TB, England.
E-mail: stephen.stokes@geog.ox.ac.uk

WARREN W. WOOD
US Geological Survey, MS 430, Reston, VA 20192, USA. E-mail: wwwood@usgs.gov

ABSTRACT: Evidence is presented for past humid episodes in the Arabian Peninsula. Over 150 previously published radiocarbon dates on lacustrine carbonates, travertines and groundwater indicate that two widespread wet episodes occurred in the region in the past c. 40 ka. Humidity maxima occurred c. 6-10 and 25-35 ka and are in phase with insolation maxima associated with variations in the earth's precessional axis. The observed timing of the Holocene event is in agreement with a well-dated humid period in the West African Sahara. The timing of the two events implicates summer monsoon-associated precipitation as the most likely source. We confirm a monsoonal source for the Holocene event via isotopic analysis of stable isotopes in ground water from the Liwa aquifer, Abu Dhabi. The recognition of precession-associated wet periods in the preserved record of Late Quaternary Arabian environmental change provides a useful basis to consider and model the evolution of both the environmental system and past periods of ground-water recharge.

INTRODUCTION

Water scarcity remains a critical issue in many low latitude desert basins. This scarcity is often complicated further by excessive ground-water extraction and subsequent attendant water quality problems (Metha et al., 2000). Limited contemporary recharge in most such basins means that the water extracted is a finite and most likely non-renewable resource, at least in the short term. A detailed knowledge of the source and genesis of the ground-water reserves does not help in the long-term management of such resources, but it does provide a basis for understanding the evolution of such ground-water systems and allows the development of hydrological and environmental histories of such areas (Verchick, 1999).

The SE sector of the Arabian Peninsula is the central and eastern segments of the Afro-Arabian old world desert belt, which comprises the longest, virtually uninterrupted expanse of the contemporary desert (Figure 1). Climatically, this desert belt is strongly influenced by direct solar insolation; resulting in both regionally high temperatures and

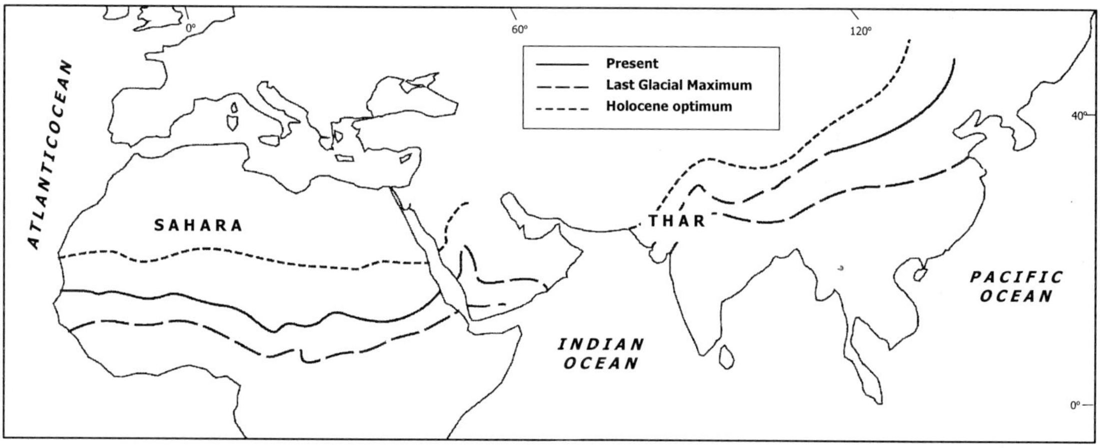

Figure 1. The Afro-Arabian desert belt. This interconnected series of sand seas stretches across the Sahara to the Thar Desert of India. Shown also are the known modern, and estimated last Glacial maximum (LGM) and mid Holocene maximum northern extents of the intertropical convergence zone (ITCZ). While the distribution of the past positions of the ITCZ is well known for the western and eastern margins of the desert belt, their estimates for the Arabian Peninsula are poorly known (modified from Yan and Petit-Maire, 1994).

Water Resources Perspectives: Evaluation, Management and Policy. Edited by A.S. Alsharhan and W.W. Wood.
Published in 2003 by Elsevier Science, Amsterdam, The Netherlands, p. 371-378.

potential evapotranspiration rates, and, over most of its extent a dominantly summer monsoonal rainfall regime related to the seasonal migration of the inter tropical convergence zone (ITCZ). Geomorphologically, the desert belt consists of a wide cross-section of contemporary aeolian (depositional and erosional) landforms that frequently interact with numerous relict landforms indicative of past intervals of increased moisture. Relict features include surface and subsurface abandoned fluvial channels, playa lake basins, and moderately developed paleosols (McClure, 1976; Wood and Imes, 1995; Glennie, 1998, Maxwell and Haynes, 2001). Over the past 20 or more years, there has been extensive research on the evolution of the western (West African) and to a lesser degree, central (East African) sectors of this desert belt (Street and Grove, 1979; Ritchie et al., 1985; Street-Perrott et al., 1990; Wendorf et al., 1994; Swézey, 2001).

A detailed understanding of the interactions between landscape, paleomonsoon rainfall, and global climate conditions for West Africa over approximately the past 20,000 ^{14}C years has emerged (Gasse et al., 1990; Street-Perrott et al., 1990; DeMenocal et al., 2000). During the latest Pleistocene (c. 30 – 10 ka) and Early Holocene, the presently hyperarid Sahara desert was characterized by numerous large and small lakes scattered across a verdant landscape which supported both large mammal and human populations. This pattern is confirmed in the Eastern Sahara (Ritchie et al., 1985). This Holocene African Humid Period' (DeMenocal et al., 2000) commenced approximately 14,500 ^{14}C years ago and terminated between 5,000 and 6,000 ^{14}C years ago resulting in a shift back to arid conditions.

The style of the climatic (arid-humid, humid-arid) transitions has recently been examined in detail by sedimentalogical and geochemical analysis of high resolution ocean cores from the adjacent tropical Atlantic (deMenocal et al., 2000). By analyzing the terrigenous fraction of these cores, which is essentially a record of aeolian dust flux, they were able to demonstrate that the transitions were rapid (c. centuries) and occurred as the summer insolation shifted across a value approximately 4.2% greater than the present day. The step-function was interpreted to be the result of changes in the intensity of the summer monsoonal rainfall regime. The intensity of the monsoonal rains during the humid period results from direct (linear) insolation forcing coupled with a strong and non-linear vegetation-albedo feedback. Similar detailed scenarios for past periods of humid conditions have not been forthcoming for the Arabian sector of the Afro-Arabian desert belt.

This paper is the first contribution of a research program that seeks a systematic paleoclimatic interpretation of past humid episodes in the Arabian sector of the Afro-Arabian desert belt based on analysis of aeolian-fluvial interactions within the Rub Al Khali. Here we collate past evidence of latest Pleistocene and Holocene humid intervals in Arabia. This evidence is then used, in combination with new isotopic analyses of paleo-ground water to infer the source, timing, and nature of past humid intervals for this region.

Direct Evidence for Episodes of Late Quaternary Humid Intervals in Arabia

In this study, we have synthesized previous radiocarbon dated records of humid intervals based on a range of independent lines of evidence. In doing so, we hope to avoid bias in our interpretations, though some must be introduced by default as we explore only those landscape components which are datable via the radiocarbon method. In this section, we systematically summarize this data. Our approach to data summarization is to collate reliable radiocarbon dated lake, spring, and speleothem records of humid conditions described by previous workers. More than 150 radiocarbon dates have been generated previously for Arabia, but no attempt has been made to interpret the complete data set.

We express dates as error-weighted combined-probability density functions, based on the sum of the individual Gaussian mean and standard deviation radiocarbon assays. The approach has the advantage of providing a visual illustration of the frequency, extent and duration of humid intervals (assuming that a direct correlation exists between episodes of lacustrine sedimentation and periods of increased humidity). Owing to incomplete reporting of radiocarbon data by some previous authors and limitations in the radiocarbon calibration curves for periods beyond c. 20,000 ^{14}C BP, we have not attempted any calibration of the radiocarbon ages into absolute calendar years.

Three major environmental archives which provide direct evidence for past humid conditions are readily identifiable: lacustrine (sabkha, playa, and interdune) carbonates; cave and spring deposits; and, (radiocarbon dates on) inorganic carbon from fossil ground water. Previous studies which exploit each of these data sources are summarized by environmental archive in the following sections.

There are other indirect sources of evidence for humid conditions based on deep marine and aeolian environmental archives. These will be described in a subsequent paper.

Radiocarbon dates on lacustrine carbonates

Interdune and lacustrine deposits are widespread within the Rub Al Khali and have been examined in some detail in both central (McClure, 1976; Sanlaville, 1992) and SE (Wood and Imes, 1995) areas of this extensive sand sea. In the present day sand sea, they are recognizable as flat-topped small mesas, benches, and mounds which represent the inverted relief of former interdune floors which are remnants of former near-surface capillary zones (Wood and Imes, 1995). Additionally they may occur as "corridors" between dunes where loose surficial sands have been removed. In places, they are observed to be inter-bedded with both alluvial (wadi) and aeolian deposits, but are typically observed as surface erosional remnants, their preferential preservation being a consequence of their relative induration (Wood and Imes, 1995).

Commonly, there is a degree of caution in evaluating radiocarbon dates on carbonates deposited within lake and capillary zone systems due to potential complications introduced by open system behavior, and the incorporation of dead carbon. Clarke and Fontes (1990) and Wood and Imes (1995) demonstrated that ages for precipitates in this region can be considered reliable due to a combination of factors including removal of carbonate via calcite precipitation during recharge; non-dissolution of old carbonates due to gypsum dissolution; and, particularly in proximity of the ultramafic rocks outcropping in the Oman mountains, the presence of alkaline and hyperalkaline ground water. As a result, when paleo-ground water reached the capillary zone it was undersaturated with respect to atmospheric carbon. In turn, CO_2 was reacquired from the atmosphere in discharge areas. It is principally this carbon (as preserved within the carbonate) which is the dated fraction.

The largest data set of radiocarbon ages for Arabian lacustrine carbonates was generated by Sanlaville (1992) who produced a set of 64 ^{14}C determinations ranging in age from 4700 to 33400 ^{14}C years BP (Figure 2a). Two clusters of ages are readily recognized: a late Pleistocene humid phase spanning c. 25-35 ^{14}C ka, and a younger latest Pleistocene-mid-Holocene phase spanning c. 5-11 ^{14}C ka. An additional minor latest Pleistocene grouping is noted between c. 13 and 15 ^{14}C ka, which when combined with the 5-11 ka cluster, exhibits a close agreement to the timing of the Holocene African Humid Period described by deMenocal et al. (2000).

The somewhat smaller (n=28) data set (also lacustrine deposits from the Rub Al Khali) produced by McClure (1976) again exhibits two main periods of humid activity; an older (21-32 ^{14}C ka) phase and a younger, exclusively Holocene (c. 6.1-8.8 ^{14}C ka) cluster (Figure 2b). In contrast, the data set of Wood and Imes (1995) (Figure 2c) is exclusively late Pleistocene in age range and includes a single main cluster spanning the period 25–33 ^{14}C ka, with some additional dates in the range 25–12 ka.

When these data are merged, the resulting combined (n = 114) probability density function exhibits a major late Pleistocene cluster centered on c. 29 ^{14}C ka, and an additional latest Pleistocene-Holocene (c. 5-11 ^{14}C ka) cluster.

Records from cave and spring-associated deposits

Precipitated carbonates (i.e., travertine) in springs and caves provide additional evidence of humid conditions in terms of both the morphological style and isotopic geochemistry of precipitation (Clark and Fontes, 1990; Burns et al., 1998; Neff et al., 2001). Clark and Fontes (1990) undertook the most detailed morphological and temporally-extended study of travertines in Arabia. They sampled and classified a range of travertines from the Nizwa region of Northern Oman. They were able to classify ^{14}C dated travertine deposits into two suites of morphologies which they believed reflected the environmental responses to changing environmental (particularly moisture) conditions. Travertines formed under arid conditions, such as those that continue to be accreted in the present time, consist of laminated and non-laminated travertines and pedogenic carbonate nodules. Isotopically, these forms of carbonate were found to be depleted in both ^{13}C and ^{18}O compared to modern–day rain water or fossil ground water).

A total of 23 carbonates with morphologies suggestive of deposition during arid conditions produce a suite of radiocarbon ages ranging from 830 to 33,270 ^{14}C yrs. BP (Figure 3a). These cluster broadly into four groups of ages from 0-4, 5-6.5, 13–16 and 24-33 ^{14}C ka. 20 radiocarbon dates were also generated for carbonates indicating humid

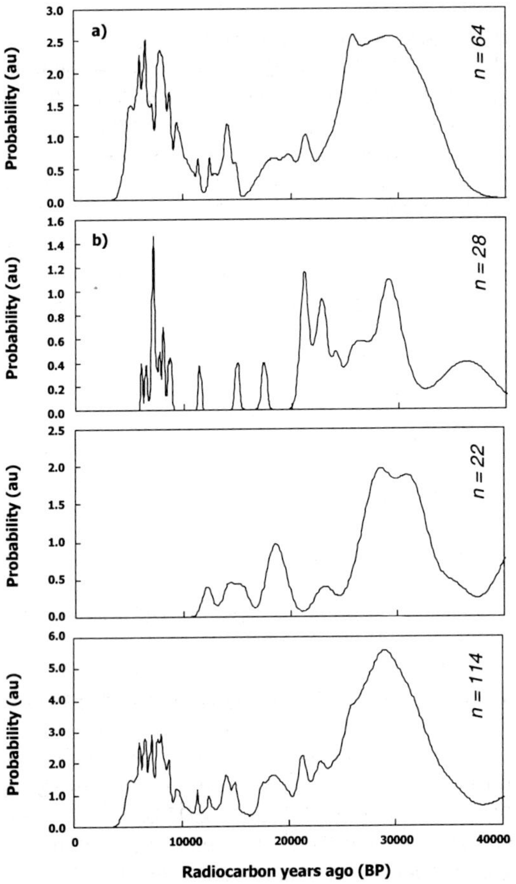

Figure 2. Radiocarbon-dated, lacustrine deposits from the Arabian Peninsula. a. After Sanlaville (1992), b. After McClure (1976), c. After Wood and Imes (1995), d. Synthesis of all data. – (au – arbitrary units).

conditions. These range in age from 6.4 to 35.8 ^{14}C ka and can be subdivided into two main groups spanning c. 6-12.5 and c. 20–32 ^{14}C ka (Figure 3b). There is general agreement between these identified humid phases and those described for the lacustrine carbonates (Figure 2d). Additionally, in their study of speleothem deposits from cave deposits in the Oman Mountains, Burns et al. (1998) found isotopic evidence for a wet period between 6,200 and 9,200 y BP, broadly corroborating the age of the younger of the two humid periods.

Direct dating of ground water

Wood and Imes (in press) dated ground water collected from depth within a ground-water mound located north of the Liwa Crescent in the Abu Dhabi Emirate. They measured the current value of soil gas ($\delta^{13}C$ −12‰ VPDB) and $\delta^{13}C$ in the water to correct for the addition of dead carbon (i.e., post-depositional). Further, they assumed that recharge was 30 °C, similar to the present to determine the temperature-dependent fraction factor between carbon dioxide gas and bicarbonate. Because of the simplicity of the system, the Pearson Method (Pearson, 1965) was used for adjusting the percent modern carbon (PMC) (Table 1). A number of factors discussed in detail in Wood and Imes (in press) may have introduced carbon to the system since the recharge event and the dates might appropriately be considered as minimum ages.

Their data set of five ages falls broadly within a time range spanning the mid-Holocene (4.2–6.8 ^{14}C ka, arithmetic mean = 5.2 ^{14}C ka) (Table 1, Figure 4). These ages agree well with evidence from lacustrine and travertine deposits of a wet phase occurring during the early to mid Holocene period. These data and ground-water modeling of the system led Wood and Imes (in press) to suggest that ground-water recharge to this aquifer terminated abruptly approximately 5000 ^{14}C yrs BP.

Table 1. Percent modern carbon, dissolved ^{13}C (VSMOW), and adjusted radiocarbon age estimates for ground water from the Liwa Aquifer, Abu Dhabi Emirates (adapted from Wood and Imes (in press)).

Field ID	Modern Carbon %	$\delta^{13}C$ (VPDB)	Uncorrected ^{14}C age
LQ-4	62.02	-4.9	4390
LQ-6	30.28	-2.9	4930
LQ-7	34.56	-4.2	6870
LQ-19	43.71	-4.4	5580
LQ-22	72.82	-4.5	4220

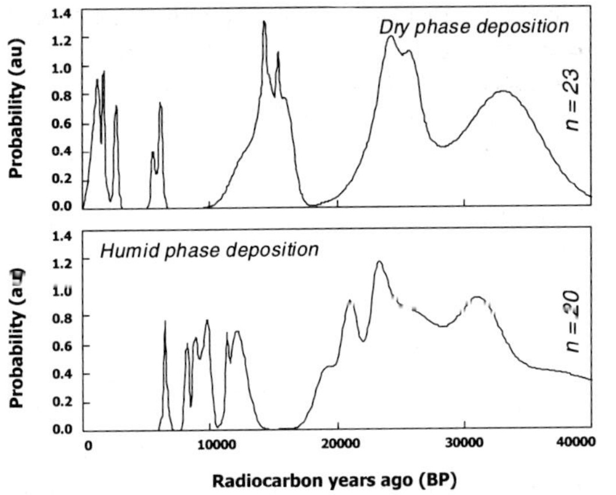

Figure 3. Periods of aridity and humidity identified via dated travertine morphologies (after Clark and Fontes, 1990). a. Periods of relative aridity, b. Periods of relative humidity.

Synthesis

The preserved sedimentary and hydrologic record collectively suggests that multiple humid (wet) periods existed in southern Arabia during the late Pleistocene. While these data have been available for some time, they have not been collated previously in a systematic fashion. The preserved record of humid episodes is clearly indicative of two major humid episodes. These episodes of increased precipitation would have resulted in ground-water recharge.

The most recent recharge event apparently ceased abruptly at approximately 5,000 ^{14}C yr BP (Figure 4). Wood and Imes (1995) estimated from precipitation/recharge relationships in contemporary desert basins that precipitation during this period in the Liwa area would have been on the order of 200±50 mm per annum, some four times greater than present-day levels.

The timing of the Holocene humid phase provides a good match to the Holocene African Humid Phase described by DeMenocal et al. (2000). Given the degree of agreement a common or related climatic forcing might be anticipated. They demonstrated via a combination of sedimentological analysis and numerical climate modeling that the Holocene African Humid Period occurred in response to changes (increases) in the monsoonal rainfall regime. This was in turn in response to summer insolation increases and an associated non-linear positive feedback relating to land surface albedo changes due to changing vegetation. The changes in insolation were a direct and in phase response to changes in the precession of earth's rotation, the third and shortest of the orbital cycles (λ=23 ka), postulated as the key climatic forcing agents within the Milankovitch or Orbital theory of global climate change. The record described by deMenocal et al. does not extent back to maxima of earlier precessional cycles. The records for Arabia, which we summarized, extend back further than the predominantly Holocene record described by deMenocal et al. (2000), though we do note that data for the Arabian Peninsula are dominated by the Holocene event, and late Pleistocene evidence is more restricted. We intend to investigate the causes of these contrasts in record preservation in subsequent research.

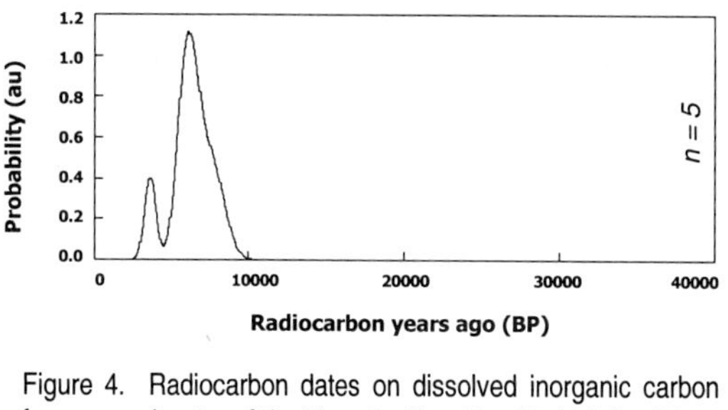

Figure 4. Radiocarbon dates on dissolved inorganic carbon from ground water of the Liwa Aquifer, Abu Dhabi (after Wood and Imes, in press).

When the various evidence for humid periods is combined into a single probability density function two humid phases are clearly recognizable, and peaks in both humid phases occur close to precessionally-influenced insolation maxima (Figure 5). Given the match in timing of the humid events and environmental inferences along the Afro-Arabian desert belt, it might be anticipated that the events share a common causation. In the next section, we explore isotopic data as a means of reconciling potential precipitation sources for humid episodes in Arabia.

Sources of paleoprecipitation

The source and seasonal timing of the precipitation has remained unresolved. The moisture could be derived either from summer monsoonal precipitation events, the likes of which are well documented for the Holocene across Africa (deMenocal et al., 2000), or from intensification of the present-day winter/spring rains (shamals), where moisture is derived from the Arabian Gulf. Here we exploit stable isotopic data from modern rainfall and ground water to consider these possible sources.

The empirical relation between the isotopes of hydrogen and oxygen in world-wide precipitation is given by the GMWL (Global Meteoric Water Line), expressed as

$$\delta^2H = 8.17\,\delta^{18}O + 10.35$$

where the δ notation is defined as: $[\{R_x/R_s\}-1] \times 1000$. R_x is the isotopic ratio of the sample, and R_s is the isotopic ratio of Vienna Standard Mean Ocean Water (VSMOW) in parts per thousand (permil, ‰). The value of 10.35 ‰ is the deuterium excess, expressed as "d." This value is controlled by the temperature-dependent equilibrium of water vapor with the water source, and the relative humidity. As a result, summer values of "d" can be as much as 15 ‰ more negative than winter values.

One of us (W.W.) has sampled and measured stable isotopes of ground water from 22 irrigation wells in the Liwa aquifer, which occurs within a Cenozoic sand body in the Abu Dhabi Emirate, approximately 100 m above sea level. The best-fit equation representing the Liwa ground-water isotopic data is $\delta^2H = 8.12\delta^{18}O - 14.49$ (Figure 6). These data exhibit a large negative value of "d" (-4.49), consistent with a warm, high humidity air mass generating the atmospheric moisture. The slope of 8.12 in addition to the lack of correlation (r^2=0.14) between chloride concentration and $\delta^{18}O$ (W. Wood, unpublished data) is consistent with the lack of significant post-precipitation evaporation.

Isotopes of rain from the current monsoons on the coast of Oman near Salalah indicate a value of "d" near zero for the 1986 data (Figure 6). These rain samples, however, represent the edge rather than the full monsoon, since only a small coastal fringe of southeastern Oman near Salalah is currently impacted by the monsoon climate. We infer that if the inter-tropical convergence zone (ITCZ) were to have significantly expanded

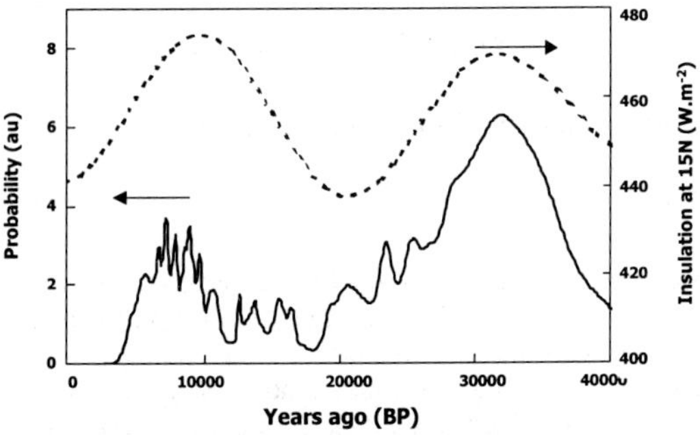

Figure 5. Evidence of humid phases in the Arabian Peninsula and its relationship to changing latest Quaternary insolation patterns. Lower curve summarizes all dated evidence of humid conditions. Upper curve plots annual insolation versus age for latitude 15ºN (after Berger and Loutre, 1991).

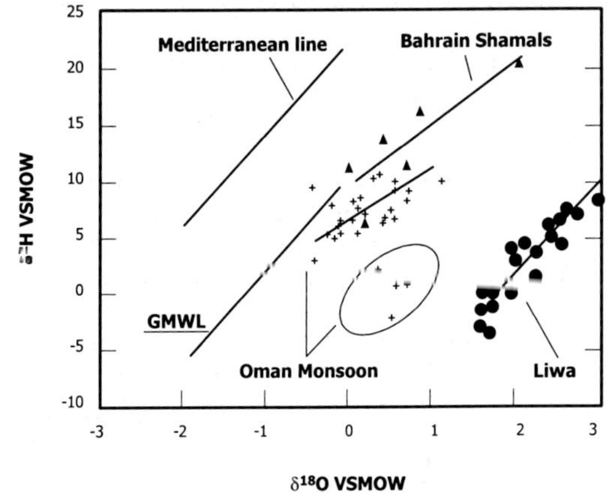

Figure 6. Oxygen and hydrogen stable isotope data from contemporary shamal, Mediterranean, and monsoonal sources and ground water from the Liwa aquifer (see text for details)

northwards (Wood and Imes, 1995), the full monsoon would have been warmer and amplified via positive feedback processes relating to land-surface albedo changes. Such air mass would have had a higher relative humidity, and under these conditions, precipitation might well have approached the "d" observed in the Liwa aquifer.

Isotopic values from the long-term International Atomic Energy Agency (IAEA) isotope station in Bahrain were used as a surrogate for values of the current shamals in the Liwa area. The best fit of the monthly average (December through April) of deuterium/oxygen isotopes from Bahrain is $\delta^2H = 5.62\ \delta^{18}O + 9.39$, a significantly heavier deuterium intercept than the isotopes observed in the Liwa samples (Figure 6). On these isotopic grounds, we dismiss the shamals as the dominant, or a significant, source of paleoprecipitation. It is also clear that precipitation derived from the Mediterranean Sea is not a source as it typically has a deuterium excess of +22 (Figure 6).

The isotopic analyses of mid-Holocene precipitation in the Liwa aquifer are consistent with expansion of the summer Oman monsoon northward rather than intensification of the winter shamals. Sedimentary evidence described in the previous section suggests that the precipitation flux in the Liwa area was similar during both the mid-Holocene and late Pleistocene wet periods and is significantly different (i.e. greater) than the present. Thus, both wet periods are inferred to have resulted from northern migration of the ITCZ and the development of a significant summer monsoonal rainfall regime.

CONCLUSIONS

There is sedimentary and hydrological evidence that collectively suggests that two paleorecharge events occurred in Arabia during the latest Pleistocene and early to mid Holocene periods. These are manifested within the landscape as widespread lacustrine deposits, precipitated cave sediments of a more restricted known distribution primarily occurring within the Oman Mountains, and inorganic carbon from ground water. The latter of these two events is comparable in timing to the Holocene African Humid Period described from analysis of deep sea core records from the tropical Atlantic Ocean.

Significant differences in temperature and seasonal relative humidity of the air masses over the summer Indian Ocean and the winter Arabian Gulf would be reflected in the isotopic signatures of the water from these sources. We identify strong

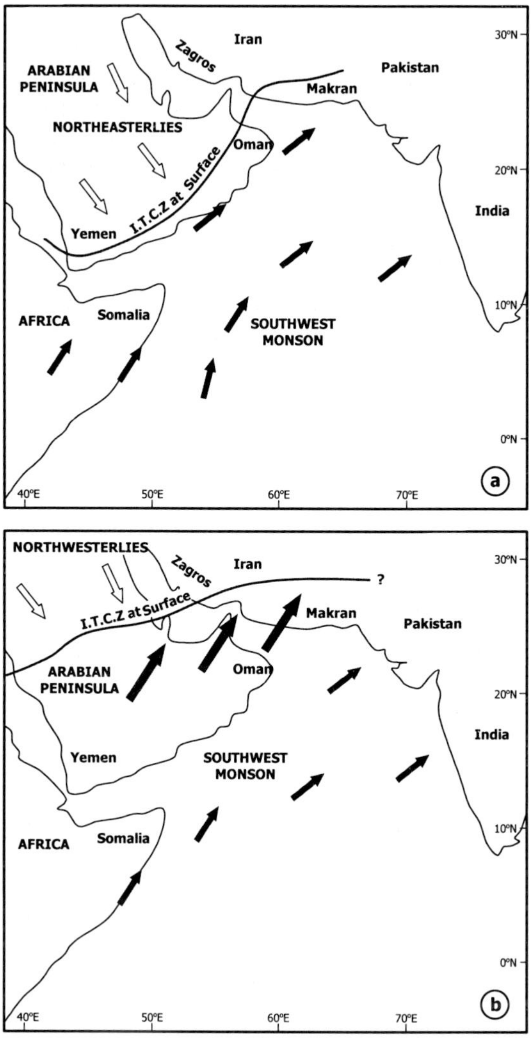

Figure 7. Modern (a) and inferred precessional maximal patterns (b) of the northernmost (summer) position if the intertropical convergence zone (ITCZ). Maximum northernmost extents of the ITCZ most recently occurred during the periods spanning c. 5-12 and 25-35 ka.

isotopic evidence to support summer monsoonal rainfall sources as the most likely cause of the increases in precipitation. For the rainfall to increase by the estimated 200 mm p.a. required by modeling studies, it is likely that the ITCZ migrated more than 500 km north of its present maximum summertime northernmost position (Figure 7), leading to resolution of the source of the Holocene precipitation.

The timing of the humid episodes is in phase with precessional maxima and implicates increases in direct summertime heating as the major forcing agent resulting in rainfall increases. It is anticipated that multiple paleorecharge events occurred in the past in phase with precessional maxima, spaced at c. 23 ka intervals.

The more extensively modeled African Holocene Humid Period has been demonstrated to exhibit a more step-like function involving rapid increase (switch on) followed by rapid decreases (shut off) in rainfall which are the consequence of a vegetation-albedo feedback amplifying the direct forcing resulting from precession. There is some evidence of a similar rapid change of climate state at the termination of the latest (early-mid Holocene) humid episode in Arabia. However, the resolution of the record of earlier climate transitions is not sufficiently resolved to confirm that a similar vegetation-albedo feedback is operative in Arabia. While it is entirely likely that such positive feedback processes were acting, further examination of long records and numerical modeling of climate change in the area are necessary to finalize and detail the observed patterns of environmental change.

REFERENCES

Berger, A., and M.F. Loutre, 1991. Insolation values for the climate of the last 10 million years. *Quaternary Science Reviews* 10: 297-317.

Burns, S.J., A. Matter, N. Frank, and A. Mangini, 1998. Speleothem-based paleoclimate record from northern Oman: *Geology* 26(6): 499-502.

Clark, I.D., J.C. Fontes, 1990. Paleoclimatic reconstruction in Northern Oman based on carbonates from hyperalkaline groundwaters. *Quaternary Research* 33: 320-336.

deMenocal, P. et al., 2000. Abrupt onset and termination of the African Humid Period – rapid climate responses to gradual insolation forcing. *Quaternary Science Reviews* 19: 347-361.

Gasse, F., R. Tehet, A. Durand, E. Gilbert, and J.C. Fontes, 1990. The arid-humid transition in the Sahara and Sahel during the last deglaciation. *Nature* 346: 141-146.

Glennie, K.W., 1998. The desert of southeast Arabia - a product of Quaternary climatic change. In *Quaternary Deserts and Climate Change,* ed. A.S. Alsharhan, K. W. Glennie, G. Whittle, and C.G. St.C. Kendall. A.A. Balkema, Rotterdam, The Netherlands, p.279-291.

Maxwell, T.A., C.V. Haynes, 2001. Sand sheet dynamics and Quaternary landscape evolution of the Selima Sand Sheet, Southern Egypt. *Quaternary Science Reviews* 20: 1623-1647.

McClure, H.A., 1976. Radiocarbon chronology of late Quaternary lakes in the Arabian Desert. *Nature* 263: 755-756.

Mehta, S., A.E. Fryar, R.M. Brady, and R.H. Morin, 2000. Modeling regional salinization of the Ogallala aquifer, Southern High Plains, TX, USA. *Journal of Hydrology* 238(1-2): 44-64.

Neff, U., S.J. Burns, A. Mangini, M. Mudelsee, D. Fleltmann, and A. Mater, 2001. Strong coherence between solar variability and the monsoon in Oman between 9 and 6 kyr ago. *Nature* 411: 290-293.

Pearson, F.J. Jr., 1965. Use of $^{13}C/^{12}C$ ratios to correct radiocarbon ages of materials initially diluted by limestone. *Proceedings of the 6th International Conference on Radiocarbon Dating, Pullman, USA,* p.357-366.

Ritchie, J.C., C.H. Eyles, and C.V. Haynes, 1985. Sediment and pollen evidence for an early to mid-Holocene humid period in the eastern Sahara. *Nature* 330: 645-647.

Sanlaville, P., 1992. Changement climatiques dans la Peninsule arabique durant le Pleistocene superieur et L'Holocene. *Paleorient* 18(1): 5-26.

Street, F.A., and A.T. Grove, 1979. Global maps of lake level fluctuations since c. 30,000 yr BP. *Quaternary Research* 12: 83-118.

Street-Perrott, F.A., J.F.B. Mitchell, D.S. Marquand, and J. Brunner, 1990. Milankovitch and albedo forcing of the tropical monsoon - a comparison of geological evidence and numerical simulations for 9000 yr BP. *Transactions of the Royal Society of Edinburgh (Earth Sciences)* 81: 407-427.

Swezey, C., 2001. Eolian sediment responses to late Quaternary climate changes: temporal and spatial patterns in the Sahara. *Palaeogeography, Palaeoclimatology, Palaeoecology* 167: 119-155.

Verchick, R.R.M., 1999. Dust bowl blues - saving and sharing the Ogallala Aquifer. *Journal of Environmental Law and Litigation* 14(1): 13-23.

Wendorf, F., R. Schild, A.E. Close, H.P. Schwarcz, G.H. Miller, R. Grun, A. Bluszcz, S. Stokes, L. Morawska, J. Huxtable, J. Lundberg, C.L. Hill, and C. McKinney, 1994. A chronology for the middle and late Pleistocene wet episodes in the Eastern Sahara. In *Late Quaternary Chronology and Paleoclimates of the Eastern Mediterranean*, ed. O. Bar-Yosef, and R.S. Kra. *Radiocarbon – Special Issue*: 147-168.

Wood, W.W., and J. L. Imes, 1995. How wet is wet? Constraints on late Quaternary climate in southern Arabian Peninsula. *Journal of Hydrology* 164: 263-268.

Wood, W.W. and J.L. Imes (2003). Dating of Holocene ground-water recharge in the Rub al Khali of Abu Dhabi - constraints on global climate-change models. In: *Desertification in the Third Millennium,* ed. A.S. Alsharhan, W.W. Wood, A.S. Goudie, A.R. Fowler, and E.M. Abdellatif. *Desertification in the Third Millennium,* A.A. Balkema, Amsterdam, The Netherlands.

Yan, Z. and Petit-Maire, N. 1994. The last 140ka in the Afro-Asian arid/semi arid transitional zone. *Palaeogeography,Palaeoclimatology, Palaeoecology* 110: 217-233

Dating of Holocene Ground-Water Recharge in Western Part of Abu Dhabi (United Arab Emirates): Constraints on Global Climate-Change Models

WARREN W. WOOD
Research Hydrologist, MS 430, National Center, U.S. Geological Survey, 12201 Sunrise Valley Drive, Reston, VA 20192, U.S.A. E-mail: wwwood@usgs.gov and School of Geography and the Environment, University of Oxford, Mansfield Road, Oxford OX1 3TB, U.K.

JEFFREY L. IMES
U.S. Geological Survey, 1400 Independence Road MS 100, Rolla 65401, USA. E-mail: jimes@srvrdmorll.er.usgs.gov

ABSTRACT: The carbon 14 (^{14}C) ages of lacustrine deposits in the southern Rub al Khali (Empty Quarter) of Saudi Arabia suggest a late Pleistocene, early Holocene paleoclimate with two wet periods, one between 6,000 and 9,000 y BP (years before present), and the other between 21,000 and 30,000 y BP. Travertine deposits in the Oman Mountains bordering the southern Rub al Khali have ^{14}C dates suggesting wet periods between 6,500 and 12,500 y BP, and between 19,000 and 33,000 y BP and dating of speleothems in northern Oman indicates a wet period between 6,200 and 9,200 y BP. These findings beg the fundamental question: "What was the rainfall flux during these wet periods?" We used a numerical ground-water flow simulation to determine that ground-water recharge must have been approximately1.4 mm/y in the Liwa Oasis area of southeastern Rub al Khali. To supply this recharge, the rainfall must have been 200 ± 50 mm/y during the period between 19,000 and 33,000 y BP.

Assuming that the water table during the younger "wet" event was at the same elevation as that during the older wet period, then the numerical model predicts that ground water in a present-day mound in the Liwa area should be between 5,500 to 9,000 years old. Radiocarbon dating of water in the core of the mound averages 5,200 y BP, consistent with the age predicted by the model. From this, it is concluded that 1) the rainfall was approximately 200 ± 50 mm/y between 6,200 and 9,000 y BP, and 2), the shift from wet to dry took place relatively rapidly between 5,500 and 6,200 y BP rather than gradual drying over thousands of years. The dates of ground water provide an independent constraint on hydrogeologic parameters used in the ground-water model and should be important in water-resource planning, global climate-change modeling, and interpretation of the age of falajes and other water related archeological features.

INTRODUCTION

Lacustrine deposits in the southern Rub al Khali (Empty Quarter) of Saudi Arabia have ^{14}C ages suggesting a paleoclimate with two wet intervals, one between 6,000 and 9,000 y BP, and the other between 21,000 and 30,000 y BP (McClure, 1976; 1984). Travertine deposits in the Oman Mountains bordering the southern Rub al Khali have ^{14}C dates suggesting wet intervals between 6,500 and 12,500 y BP, and between 19,000 and 33,000 y BP (Clark and Fontes, 1990). Burns et al., 1998, dated speleothems in northern Oman and found a wet period between 6,200 and 9,200 y BP. These studies indicate that rainfall was significantly greater during these two periods than it is at present, but how much greater? Wood and Imes (1995) addressed this question of paleo-rainfall flux by using a numerical ground-water flow model to evaluate the system. They mapped the elevation and ^{14}C dated a calcrete deposit formed by the paleo-ground-water table. From these data and the use of simulations made with a numerical ground-water model they, determined that recharge must have been approximately 1.4 mm/y to maintain the observed elevation of the paleo-ground-water table. From this recharge calculation and the use of current rainfall/recharge relations in arid and semi-arid areas (1 to 2 %) they then estimated that rainfall in the Liwa Oasis area of Abu Dhabi (Fig.1) must have been approximately 200 ± 50 mm/y in the older of the two "wet" periods. If it is assumed that the water level of the younger "wet" period was at the same elevation as that of the older "wet" period, the model suggests recharge must have stopped between 5500 and 6000 BP in order for the water levels to decline to present-day elevations. If the model is representative of the system, then the ground water in the core of the mound should be between 5,500 and 9,000 y BP. The present manuscript reports on the use of radiocarbon dating of water (not rock) as an independent constraint on the numerical model. The ^{14}C dating method was selected for this evaluation

Water Resources Perspectives: Evaluation, Management and Policy. Edited by A.S. Alsharhan and W.W. Wood.
Published in 2003 by Elsevier Science, Amsterdam, The Netherlands, p. 379-385.

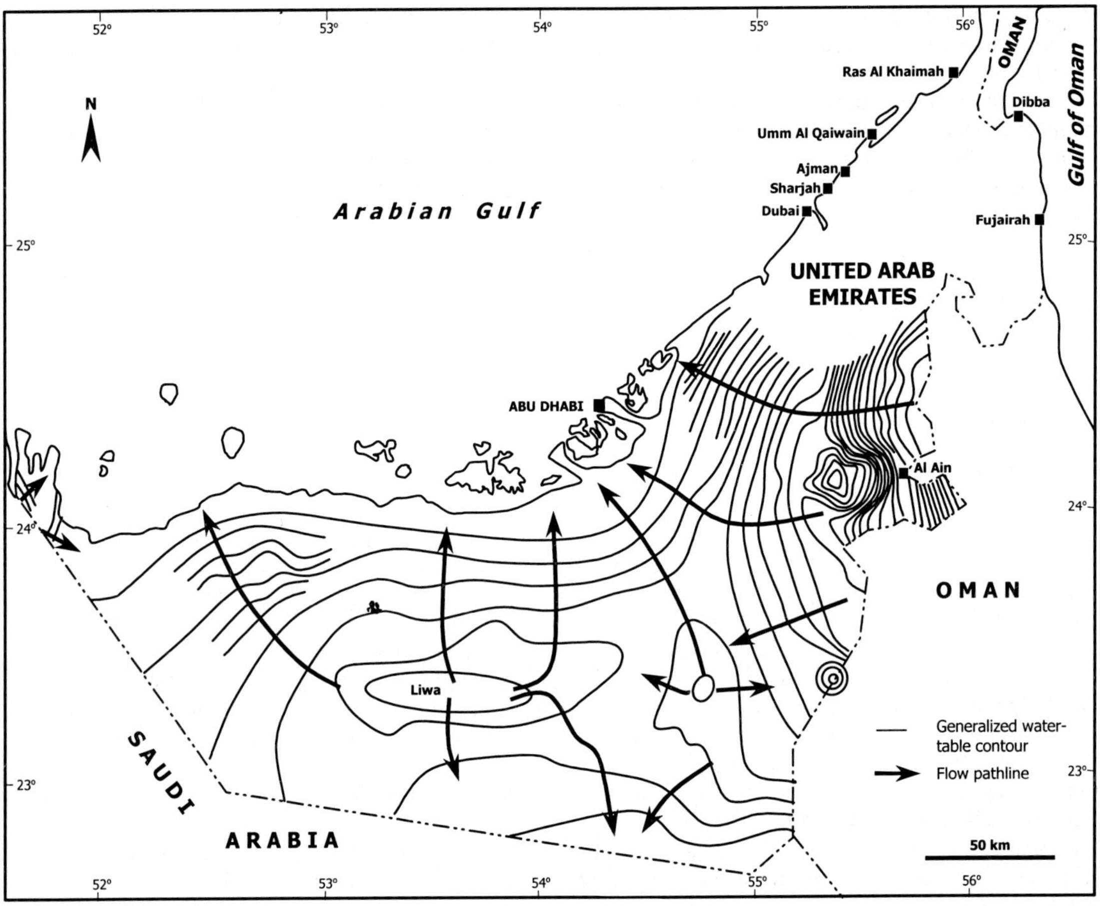

Figure 1. Generalized north/south cross section through the Liwa mound showing idealized flow lines and relative water table elevations. Evaporation and interchange with the atmosphere occurs in sabkha (salt flats) between the dunes.

because the hydrology and geochemistry of the system appeared to be suitable for its application, and the range of ages of interest fell within the ranges of the method.

Hydrogeology

The area of interest is north of the Liwa Oasis on the northeastern edge of the "Empty Quarter" in the Emirate of Abu Dhabi, approximately 150 km southwest of the city of Abu Dhabi (Fig. 1). High dunes (130 -150 meters [m] above mean sea level) with relief of 50 to 60 m dominate the topography in the area. In much of the area the dunes are separated by nearly flat (typically less than 1 m of relief) valley floors, the elevation of which is controlled by a dynamic equilibrium between the water table and eolian processes.

The dunes are composed of quartz, dolomite, calcite, and anhydrite with minor amounts of feldspar and opaque minerals (Ahmed, et al., 1998; Hadley et al., 1998). Porosity of fine eolian sand typically varies between 35 to 40 percent (Driscoll, 1986), and laboratory evaluation by the authors give an average of 38 percent, consistent with this range. Grain-size analyses of similar dunes on the road between Al Ain and the city of Abu Dhabi show a high degree of uniformity (Alsharhan et al., 1998), as would be expected with eolian sands. Aquifer hydraulic conductivity is estimated to be 1 m/d based on the arithmetic average of seven regionally distributed aquifer tests (1.75, 0.99, 1.23, 0.46, 0.96, 1.71, and 0.60 m/d) and on similar tests in eolian sediments near the coast. A value of 0.20 for specific yield (based on the average of the seven pumping tests) was used in the transient simulation. Because of the uniform grain size of the eolian deposit and the lack of cementation, the hydraulic properties were modeled as homogeneous and isotropic.

Present annual rainfall in the Crescent is extremely variable but averages 40 mm/y (UAE University, 1993). The authors observed no runoff from the dunes during a 50-mm, one-day storm in 1993. Pits dug on the edges of dunes six days after this storm showed that the wetting front was limited to a depth of less than 60 cm and appear to provide no measurable ground-water recharge. An interesting small-scale water-transport phenomenon associated with this storm was the slumping of large sheets of wet sand from the slip faces of the dunes (5 to 10 cm thick and tens of meters in area) the day after the rain.

The highest elevation of the ground-water mound is approximately 20 km north of the Liwa Oasis, corresponding with the highest surface elevation of the dunes. The aquifer is composed of uniform eolian sand with a maximum thickness of approximately 90 m that gradually thins away from the center in all directions. Miocene-age clastics form the base to the flow system that is believed to be nearly horizontal and approximately at sea level as the altitude of the sandstone varied from sea level by less than ±20 m in 12 wells drilled in the area. Ground water in the Liwa area moves toward the Arabian Gulf in the north, toward Sabkha Matti in the west, and toward the Rub al Alkali in the south (Fig. 1). Flow to the east is impeded because of another ground-water mound (Fig. 1). A north/south-generalized cross section of the ground-water mound is given in Figure 2.

Evidence for the position of the paleo-ground-water table suggests a height of 128 m above sea level at the center and 90 m at the discharge boundaries (Wood and Imes, 1995). The 1990 pre-development elevations of this mound were 105 m and 85 m, respectively. The simulated uniform steady-state-recharge rate required to maintain a water surface at the observed paleo-water-table elevation ranged from 1.1 to 1.6 mm/y, with the average estimate being 1.4 mm/y (Wood and Imes, 1995). From this recharge rate and representative rainfall/recharge rates in arid areas of the world, rainfall was estimated to be approximately 200 ± 50 mm/y during the wet interval between 20,000 and 30,000 y BP (Wood and Imes, 1995). In the transient simulation of the water-level decline in the younger wet period, initial water level was assumed to be the same as that in the older wet period. The model suggests that if recharge of 1.4 mm/y were terminated at approximately 5,500 y BP, then the water-table elevation would decay to the pre-development value observed in 1990. This model, however, while consistent with the general hydrology, is not well constrained.

Normally, models are constrained by obtaining historical water levels under various pumping stresses. This approach, however, is not applicable for this system; so we have used radiocarbon dating of the water to provide the independent confirmation of the model. If the radiocarbon dates in the core of the mound fall within the age range of 5,500 to 9,000 y BP predicted by the ground-water model and limits of the younger "wet" period, then we are willing to accept that the model is adequately constrained.

Solute Geochemistry

Dating of ground water assumes that dissolved carbon in the ground water is derived only from dissolved carbon dioxide in the rain and or soil zone at the time of recharge, and that changes in the ^{14}C activity occur only as a result of radioactive decay. Typically, however, there is some uncertainty in the ages indicated because of the addition of "old" or radioactively "dead" dissolved carbon from dissolution of calcite in the aquifer and/or the addition of modern ^{14}C by exposure of the ground water to the atmosphere (Clark and Fritz, 1997). These processes result in radiocarbon ages older or younger than the true age of recharge.

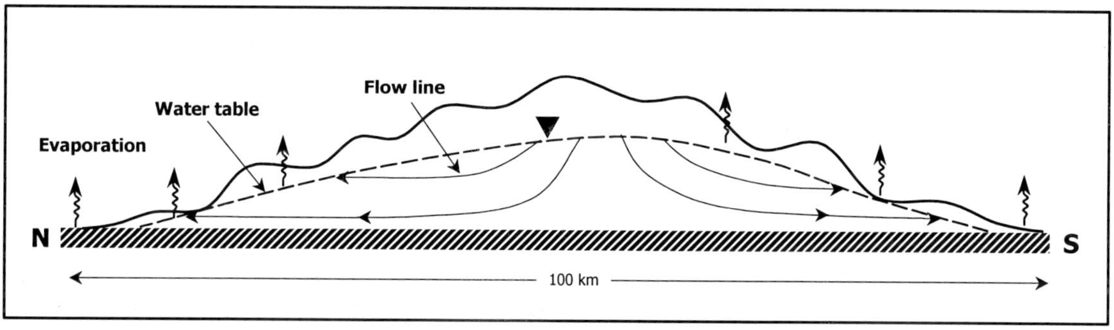

Figure 2. Location of the Liwa area in the Emirate of Abu Dhabi showing generalized configuration of the water table. The water table closed contour at Liwa is 100 m above sea level.

The conceptual model of the carbon in this system is as follows: rainfall acquires bicarbonate (HCO_3^{-1}) by interaction of atmospheric and soil carbon dioxide (CO_2) with water (H_2O):

$$^{14}CO_2 + H_2O = H^{14}CO_3^{-1} + H^{+1}. \quad (1)$$

As the water moves through the unsaturated zone and down the hydraulic gradient along the flow path, it can dissolve calcite ($CaCO_3$) that can add old or "dead" carbon:

$$H^{+1} + CaCO_3 = HCO_3^{-1} + Ca^{+2}. \quad (2)$$

However, the presence of soluble gypsum ($CaSO_4 \cdot 2H_2O$) and anhydrite ($CaSO_4$) in the unsaturated zone and the aquifer adds calcium to the solution:

$$CaSO_4 \cdot 2H_2O + H_2O = Ca^{+2} + SO_4^{-2} + 3H_2O. \quad (3)$$

The addition of calcium forces thermodynamic supersaturation of the solution with respect to calcite, and it precipitates, or prevents additional dissolution of calcite. This combined reaction is shown by the equation:

$$CaSO_4 \cdot 2H_2O + H^{14}CO_3^{-1} =$$
$$Ca^{14}CO_3\downarrow + H^{+1} + SO_4^{-2} + 2H_2O. \quad (4)$$

Support for this conceptual model can be seen as bicarbonate decreases away from the high point of the mound (Fig. 3, Table 1) and pH decreases with increasing calcium (Fig. 4, Table 1). The thermodynamic saturation indices (SI) is defined as the ratio of logarithm of K_{iap} (ion activity product) to K_{eq} (equilibrium constant) (Plummer et al., 1988). The SI of gypsum goes from a negative value (undersaturated) toward 0.0 (equilibrium) as the water moves away from the center of the ground-water mound (Table 1), consistent with this concept. Calcite is essentially at equilibrium in all the samples, which suggests that equilibrium is obtained before the water has moved far. The close correlation between increasing calcium and sulfate is shown in Table 1. These relations are consistent with the above-proposed equations. This means that in the Liwa aquifer system, the dissolution of "dead" calcite is limited because of the presence of detrital gypsum in the dunes. That is, the calcium ion derived from gypsum dissolution causes readjustment of the thermodynamic equilibrium, and bicarbonate ions present in the water tend to be removed from solution by precipitation of calcite, not added by dissolution of calcite. Fractionation of ^{14}C in the precipitation of calcite is minimal (Kalin, 2000) and neglected in this analysis.

Radiocarbon dating

PMC (percent modern carbon) and $\delta^{13}C$ values for 13 ground-water samples are given in Table 1, where $\delta^{13}C$ is defined as:

$$\delta^{13}C\ (‰VPDB) =$$
$$\{^{13}C/^{12}C\ sample/^{13}C/^{12}C\ std\} \times 1{,}000, \quad (5)$$

and VPDB is the reference carbonate. Analyses of ^{14}C were performed by AMS (accelerator mass spectroscopy) at the University of Arizona

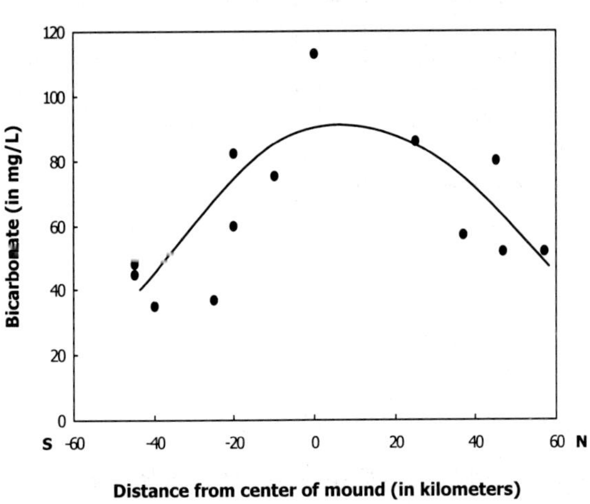

Figure 3. Graph showing the bicarbonate concentration with distance from the center of the Liwa groundwater mound.

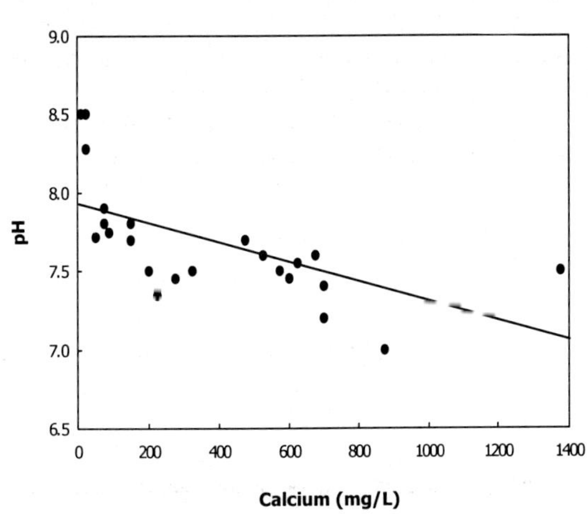

Figure 4. Graph showing decrease of pH with increasing calcium concentration in groundwater in the Liwa area.

Table 1: Chemical and isotopic analyses of groundwater from the Emirate of Abu Dhabi, United Arab Emirates.

Site Identifer	UTM east	UTM north	Date (d/m/y)	Distance from center	Ca	Mg	Na	K	Sr	HCO_3	Cl	SO_4	F	Br	NO_3	SiO_2	pH	TDS	Density	Temp
	(zone 39)		sampled	in km	mg/L	mg/L	mg/L	mg/L	mg/L	mg/L	mg/L	mg/L	mg/L	mg/L	mg/L	mg/L	mg/L	mg/L	20 °C	°C
LQ-1	767 256	2 635 100	########	57 N	975	300	2220	290	19	52	4,455	2,810	2.00	7.4	135	36	7.00	11,273	1.0082	-
LQ-2	773 785	2 621 805	########	44 N	685	160	1860	50	18	80	3,025	2,100	1.10	3.4	95	44	7.23	8,080	1.0057	29.7
LQ-47	652 289	########	20/1/93	37 N	580	200	2450	90	15	59	3,600	2,450	-	7.9	55	39		9,516	1.0001	-
LQ-4	778 773	2 600 768	########	23 N	145	63	1220	150	12	88	1,950	750	2.16	6.2	53	24	7.71	4,416	1.0023	34.6
LQ-6	784 547	2 577 688	########	0	26	9	200	60	2.6	113	230	155	2.44	0.9	33	23	8.18	795	0.9997	31.4
LQ-7	785 931	2 566 889	########	10 S	47	17	230	53	5.0	78	305	230	1.84	1.1	36	27	7.94	990	0.9997	32.5
LQ-19	764 298	2 557 633	########	20 S	65	19	240	19	7.0	103	275	325	4.60	1.5	39	27	7.82	1,068	0.9999	32.5
LQ-22	795 951	2 556 944	########	21S	205	39	375	160	12	60	530	755	2.25	1.9	43	36	7.52	2,186	1.0007	32.4
LQ-12	806 671	########	########	25 S	660	225	670	75	13	38	1,270	2,575	1.57	3.1	130	41	7.58	5,681	1.0040	-
LQ-44	693 028	########	20/1/93	39 S	650	190	2050	60	16	37	3,000	2,500	-	5	45	36		8,570	1.0030	-
LQ-40	746 026	########	20/1/93	44 S	700	185	1600	115	13	48	2,400	2,450	-	3.9	45	49		7,584	1.0028	-
LQ-43	703 395	########	20/1/93	44 S	475	160	1850	50	15	45	2,400	2,350	-	3.4	110	41		7,476	1.0055	-
LQ-41	728 491	########	20/1/93	47 S	260	90	1500	44	16	53	2,100	1,250	-	3.3	55	43		5,387	1.0025	-

Site Identifer	**$\delta\,^{2}H$ (VSMOW) o/oo**	**$\delta^{18}O$ (VSMOW) o/oo**	**Percent modern carbon**	**$\delta\,^{13}C$ (VPDB) o/oo**	**^{3}H TU**	**Adj ages 25 °C Years BP**	**Adj ages 30 °C Years BP**	**Calcite SI**	**Gypsum SI**
LQ-1	1.0	2.30	0.9567	-7.4	0.31			-0.14	0.052
LQ-2	0.0	2.10	1.0263	-6.0	0.31			-0.02	-0.096
LQ-47	-	-	1.0289	-10.8	-				
LQ-4	-3.0	1.60	0.6202	-4.9	-0.09	3,620	4,390	-0.02	-0.096
LQ-6	7.5	2.60	0.3028	-2.9	0.09	5,900	4,930	0.08	-1.922
LQ-7	4.0	2.05	0.3456	-4.2	-0.09	7,750	6,870	-0.07	-1.526
LQ-19	4.5	2.15	0.4371	-4.4	0.31	6,600	5,580	0.05	-1.279
LQ-22	6.0	2.40	0.7282	-4.5	0.31	1,590	4,220	0.08	-0.648
LQ-12	7.0	2.75	0.9851	-8.4	-	5092		0.03	0.015
LQ-44	-	-	0.7379	-18.4	-				
LQ-40	-	-	1.0104	-8.2	-				
LQ-43	-	-	0.9991	-7.6	-				
LQ-41	-	-	0.8243	-6.6	-				

Table 2. Isotope analyses of soil gas in the Emirate of Abu Dhabi, United Arab Emirates.

Site Number	UTM East	UTM North	UTM Zone	Date Sampled	CO_2 µmoles	$\delta^{13}C$ (VPDB)
LGS-1	785 983	2 572 706	39	1991	14.1	-13.09
LGS-2	782 083	2 582 873	39	1991	10.8	-13.11
WGS-1	394 323	2 675 473	40	1991	20.4	-13.99
WGS-2	386 306	2 667 925	40	1991	---	-11.43
IWGS-2	380 909	2 666 920	40	1991	---	-11.44

Accelerator Laboratory under a contract with the National Drilling Company of Abu Dhabi. It is recognized that complex methods can be used to adjust of observed PMC to obtain corrected dates (Kalin, 2000). Because of the uncertainties of the physical hydrology (local recycling of irrigation water and lack of information on the well depth and screened intervals), however, we have opted to use the simple carbon mixing method for adjustment of the PMC (Pearson, 1965). The approach assumes that changes in $\delta^{13}C$ in the water are the result of dissolution of marine carbonate that has a $\delta^{13}C$ of 0.00. Thus, the mass of any additional dead carbonate can be calculated by evaluating the change in ratio between the $\delta^{13}C$ in the initial recharge water and the $\delta^{13}C$ presently observed.

The $\delta^{13}C$ of the present soil gas was sampled at a depth 50 cm below the surface at five locations in the Emirates, including two in the Liwa Crescent (LGS-1 and 2, table 2). These five samples show little variation in $\delta^{13}C$ value and average –12.9‰ Currently, desert C-4 plants are responsible for this relatively heavy isotopic value, and it is assumed that these plants would have been present at the time of recharge - 5,500 to 9,000 y PB. To determine the $\delta^{13}C$ at the time of recharge, it is necessary to determine the temperature-dependent fraction factor between carbon dioxide gas and bicarbonate. If it is assumed that the recharge temperature was 30°C during the wet period similar to the present, a fraction factor value of 7.4 ‰ is appropriate. Using the present $\delta^{13}C$ of soil gas and this fraction factor, the $\delta^{13}C$ value of the bicarbonate in the recharge water was calculated to be approximately –5.6‰ 5,500 to 9,000 y BP. If the temperature of recharge were 5 °C cooler, then bicarbonate in the water would have had a $\delta^{13}C$ of approximately -5.1‰ The fractionation factor is also dependent on pH of the rainwater, and we have assumed that pH has not changed significantly in the past 9,000 years.

Uncorrected PMC data show a large range of values (Table 1). It is believed that much of this variation results from the difference in sampling depths, location relative to the center of the ground-water mound, and re-circulation of irrigation water. That is, the flow path leads away from the high point of the mound (Fig. 2) so that the water is exposed to evaporation and interchange with atmosphere through a series of inland sabkhas. Exposure of ground water at the surface of the interdunal sabkha increased the total dissolved solids by evaporation and introduces modern atmospheric carbon. Additionally, some of these sabkhas are used for irrigating crops of cabbages and tomatoes that are generally C-3 (Calvin) pathway plants and have $\delta^{13}C$ values close to –25 ‰ (table 1). This value is in contrast to the C-4 (Hatch-Slack) pathway plants that exhibit a $\delta^{13}C$ value of –12 ‰

Ground water moving downward (not laterally) from the mound (Fig. 2), however, does not interact with the atmosphere, is not involved with irrigation, and lacks tritium (table 1) and is believed to be suitable for dating using the ^{14}C method. The four wells nearest the crest representing the interior of the mound (LQ-4, 6, 7, 19 and 22) were selected for analyses. Correcting the observed PMC values results in ages ranging from 4,220 to 6,870 y BP (table 1), with an arithmetic average of 5,200 y BP. If the fraction factor for 25°C is used, the ages range from 1,590 to 7,750 y BP, with an arithmetic average of 5,100 y BP. The observation that the two samples furthest from the center of the mound (LQ-4 and LQ-22) yield younger age than the values calculated by the ground-water model is explained by interaction of the ground water with the atmosphere in the sabkhas distant from the center of the mound.

No formal sensitivity analysis of the ground-water model was conducted; however, the model was run a number of times with different initial elevation and recharge rates and found to be extremely sensitive to these conditions, given that the hydraulic conductivity and storage are nearly homogeneous, isotropic, and well-known. While there is some variability in the ^{14}C dates of the ground water, we believe that the greatest uncertainty in either this analysis or the earlier one (Wood and Imes, 1995) is the rainfall/recharge relationship. We have attempted to incorporate this uncertainty by giving a large error range of 50 mm/y on the rainfall. We have not however, identified a satisfactory method for further constraining this relationship because the topography, rainfall intensity, surface vegetation, porosity, depth to water, evapotranspiration and other factors controlling recharge are different in each system studied. Thus, extrapolating rainfall/recharge relationship from the present to the past is associated with a large uncertainty.

Summary and conclusions

It is clear that the ^{14}C age data from the ground water are consistent with the age of recharge calculated by the model, and the results have significant implications paleoclimate interpretations: 1) that the elevation of the water table during the 6,200 to 9,000 y BP "wet" interval was the same as 19,000 and 33,000 y BP "wet" interval, implying similar recharge and thus similar rainfall, and 2) significant recharge ceased after 5,500 y BP, although there may have been occasional recharge events. We conclude that the confluence of the two independent methods of dating justifies the assumption that elevation of the water table was the same during the two "wet" intervals. It follows that rainfall must also have been approximately 200 ± 50 mm/y in the younger wet period. The lack of significant recharge younger than 5,500 y BP suggests that the shift to drier conditions must have been fairly rapid if the climate was still in the "wet" phase at 6,200 y BP (Burns et al., 1998). The implication from the present study is that rainfall decreased from 200 mm/y to the present-day value of approximately 50 mm/y before 5,500 y BP and after 6,200 y BP rather than decreasing gradually over the last 6,000 years. Both of these conclusions have important implications for global climate models and are consistent with the model that the Pleistocene and Holocene climate change in this area resulted from a shift in the summer monsoon and intertropical convergence zone. These findings will also be useful in interpreting archeological water delivery systems and for development of the water resources.

ACKNOWLEDGMENT

We wish to thank the National Drilling Company of the Emirate of Abu Dhabi for its support of the U. S. Geological Survey (USGS) and Pierre Glynn and Ward Sanford of the USGS for their reviews. Publication of this paper was approved by the Director of the USGS and the National Drilling Company of the Emirate of Abu Dhabi.

REFERENCES

Ahmed, E. A., Soliman, M. A., Alsharhan, A. S., and Tamer, S., 1998. Mineralogical characteristics of the Quaternary sand dunes in the eastern province of Abu Dhabi, United Arab Emirates, In: *Quaternary Deserts and Climate Change*: A. S. Alsharhan, K. W. Glennie, G. L. Whittle, and C. G. St. C. Kendall, eds., A. A. Balkema, Rotterdam, p.85-90.

Burns, S. J., Matter, A, Frank, N., and Mangini, A., 1998. Speleothem-based paleoclimate record from northern Oman. *Geology* 26(6): 499-502.

Clark, I. D. and Fontes, J. C., 1990. Paleoclimactic reconstruction in Northern Oman based on carbonates from hyperalkaline ground waters. *Quaternary Research* 33: 320-336.

Clark, I. D. and Fritz P., 1997. Environmental Isotopes in Hydrogeology: Lewis Publishers, New York, 328p.

Driscoll, F. G., 1986. Groundwater and Wells, 2nd edition: Johnson Division, St. Paul, Minnesota, 1089p.

Hadley, D. G., Brouwers, E.M., and Bown T. M., 1998. Quaternary paleodunes, Arabian Gulf coast, Abu Dhabi Emirate: Age and paleoenvironmental evolution, In: *Quaternary Deserts and Climate Change*: A. S. Alsharhan, K. W. Glennie, G. L. Whittle, and C. G. St. C. Kendall, A. A. Balkema, Rotterdam, p.123-139.

Kalin, R. M., 2000. Radiocarbon dating of groundwater systems, *In:* Environmental Tracers in Subsurface Hydrology: P. Cook and A. L. Herczeg, eds., Kluwer Academic Publishers, Boston, 529p.

McClure, H. A., 1976. Radiocarbon chronology of late Quaternary lakes in the Arabian desert. *Nature* 263: 755-756.

McClure, H. A., 1984. Late Quaternary paleoenvironments of the Rub' Al Khali. Ph.D. Thesis, University of London, 245p.

Pearson, F. J., 1965. Use of C-13/C-12 ratios to correct radiocarbon ages of material initially diluted by limestone, *in* Proceeding of the 6th International Conference on Radiocarbon and Tritium Dating, Pullman, Washington, USA, 357-266.

Plummer, L. N., Parkhurst, D. L., Fleming, G. W., and Dunkle, S. A., 1988. A computer program incorporating Pitzer's equations for calculation of geochemical reactions in brines: U. S. Geological Survey Water-Resources Investigation Report 88-4153, 310p.

United Arab Emirates University, 1993. The National Atlas of the United Arab Emirates, 164p.

Wood, W. W. and Imes, J. L., 1995. How wet is wet? Constraints on late Quaternary climate in southern Arabian Peninsula. *Journal of Hydrology* 164, 263-268.